Engineering Plant-Based Food Systems

Engineering Plant-Based Food Systems

Edited by

Sangeeta Prakash

School of Agriculture and Food Sciences, The University of Queensland, Brisbane, QLD, Australia

Bhesh R. Bhandari

The University of Queensland, Brisbane, QLD, Australia

Claire Gaiani

Université de Lorraine – Laboratoire d'Ingenierie des Biomolecules, Nancy, France

ELSEVIER

ACADEMIC PRESS

An imprint of Elsevier

Academic Press is an imprint of Elsevier
125 London Wall, London EC2Y 5AS, United Kingdom
525 B Street, Suite 1650, San Diego, CA 92101, United States
50 Hampshire Street, 5th Floor, Cambridge, MA 02139, United States
The Boulevard, Langford Lane, Kidlington, Oxford OX5 1GB, United Kingdom

ISBN: 978-0-323-89842-3

For information on all Academic Press publications visit our website at https://www.elsevier.com/books-and-journals

Publisher: Nikki P. Levy
Acquisitions Editor: Nina Bandeira
Editorial Project Manager: Lena Sparks
Production Project Manager: Bharatwaj Varatharajan
Cover Designer: Miles Hitchen

Typeset by TNQ Technologies

Contents

List of contributors

Ane Aldalur, Dairy Innovation Hub, Department of Chemical Engineering, The University of Melbourne, Parkville, VIC, Australia; Bio21 Molecular Science and Biotechnology Institute, The University of Melbourne, Parkville, VIC, Australia

Mihaela Andrei, Coventry University, Coventry, United Kingdom

Nandika Bandara, Department of Food and Human Nutritional Sciences, Faculty of Agricultural and Food Sciences, University of Manitoba, Winnipeg, Manitoba, Canada; Richardson Centre for Functional Foods and Nutraceuticals, University of Manitoba, Winnipeg, Manitoba, Canada

Vasudha Bansal, Department of Food and Nutrition, Government Home Science College, Chandigarh, India

Bhesh R. Bhandari, School of Agriculture and Food Sciences, The University of Queensland, Brisbane, QLD, Australia

Muhammad Faiz Bin Muhd Faizal Abdullah Tan, School of Chemical & Life Sciences, Singapore Polytechnic, Singapore

Roman Buckow, The University of Sydney, Centre for Advanced Food Engineering, School of Chemical and Biomolecular Engineering, Darlington, NSW, Australia

Lankatillake C., STEM School of Science, RMIT University, Melbourne, VIC, Australia

Dias D., STEM School of Science, RMIT University, Melbourne, VIC, Australia

Sujit Das, Department of Rural Development and Agricultural Production, North-Eastern Hill University, Tura Campus, Shillong, Meghalaya, India

Nirali Dedhia, Centre for Technology Alternatives for Rural Areas, IIT Bombay, Mumbai, India

Bhanu Devnani, Dairy Innovation Hub, Department of Chemical Engineering, The University of Melbourne, Parkville, VIC, Australia; Bio21 Molecular Science and Biotechnology Institute, The University of Melbourne, Parkville, VIC, Australia

Pedro Elez-Martínez, Department of Food Technology, Agrotecnio Center, University of Lleida, Lleida, Spain

Gbemisola J. Fadimu, School of Science, RMIT University, Melbourne, VIC, Australia

Zhongxiang Fang, School of Agriculture and Food, University of Melbourne, Parkville, VIC, Australia

Claire Gaiani, Laboratoire d'Ingénierie des Biomolécules (LIBio), Université de Lorraine, Nancy, France

Kunal Gawai, Dairy Microbiology Department, SMC College of Dairy Science, Anand Agricultural University, Anand, Gujarat, India

Fernanda C. Godoi, Tessenderlo Innovation Center, Tessenderlo Group, Tessenderlo, Belgium

Sally L. Gras, Dairy Innovation Hub, Department of Chemical Engineering, The University of Melbourne, Parkville, VIC, Australia; Bio21 Molecular Science and Biotechnology Institute, The University of Melbourne, Parkville, VIC, Australia

Subrota Hati, Dairy Microbiology Department, SMC College of Dairy Science, Anand Agricultural University, Anand, Gujarat, India

Mital R. Kathiriya, Dairy Microbiology Department, SMC College of Dairy Science, Anand Agricultural University, Anand, Gujarat, India

Lita Katopo, Commonwealth Scientific and Industrial Research Organisation (CSIRO), Agriculture and Food, Werribee, VIC, Australia

Woojeong Kim, School of Chemical Engineering, UNSW Sydney, Sydney, NSW, Australia

William Leonard, School of Agriculture and Food, University of Melbourne, Parkville, VIC, Australia

Gloria López-Gámez, Department of Food Technology, Agrotecnio Center, University of Lleida, Lleida, Spain

Claire D. Munialo, Coventry University, Coventry, United Kingdom

Rishi Ravindra Naik, School of Chemical Engineering, UNSW Sydney, Sydney, NSW, Australia

Malik Adil Nawaz, Commonwealth Scientific and Industrial Research Organisation (CSIRO), Agriculture and Food, Werribee, VIC, Australia

Dian Widya Ningtyas, The University of Queensland, School of Agriculture and Food Sciences, Brisbane, Australia, Brawijaya University, Faculty of Agricultural Technology, Department of Food Science and Technology, Malang, Indonesia

Oladipupo Odunayo Olatunde, Department of Food and Human Nutritional Sciences, Faculty of Agricultural and Food Sciences, University of Manitoba, Winnipeg, Manitoba, Canada; Richardson Centre for Functional Foods and Nutraceuticals, University of Manitoba, Winnipeg, Manitoba, Canada

Lydia Ong, Dairy Innovation Hub, Department of Chemical Engineering, The University of Melbourne, Parkville, VIC, Australia; Bio21 Molecular Science and Biotechnology Institute, The University of Melbourne, Parkville, VIC, Australia

Sangeeta Prakash, School of Agriculture and Food Sciences, The University of Queensland, Brisbane, QLD, Australia

Jatindra K. Sahu, Food Customization Research Lab, Centre for Rural Development and Technology, Indian Institute of Technology Delhi, New Delhi, India

Cordelia Selomulya, School of Chemical Engineering, UNSW Sydney, Sydney, NSW, Australia

Narendra Shah, Centre for Technology Alternatives for Rural Areas, IIT Bombay, Mumbai, India

Nitya Sharma, Food Customization Research Lab, Centre for Rural Development and Technology, Indian Institute of Technology Delhi, New Delhi, India

Robert Soliva-Fortuny, Department of Food Technology, Agrotecnio Center, University of Lleida, Lleida, Spain

Christos Soukoulis, Environmental Research and Innovation (ERIN) Department, Luxembourg Institute of Science and Technology, Esch-sur-Alzette, Luxembourg

Ignatius Srianta, Department of Food Technology, Faculty of Agricultural Technology, Widya Mandala Catholic University Surabaya, Surabaya, Indonesia

Regine Stockmann, Commonwealth Scientific and Industrial Research Organisation (CSIRO), Agriculture and Food, Werribee, VIC, Australia

Huynh T., STEM School of Science, RMIT University, Melbourne, VIC, Australia

Jun Kiat Kovis Tay, School of Chemical & Life Sciences, Singapore Polytechnic, Singapore

Ihab Tewfik, School of Life Sciences, University of Westminster, London, United Kingdom

Tuyen Truong, School of Science, RMIT University, Melbourne, VIC, Australia

Yong Wang, School of Chemical Engineering, UNSW Sydney, Sydney, NSW, Australia

Oni Yuliarti, School of Chemical & Life Sciences, Singapore Polytechnic, Singapore

Elok Zubaidah, Department of Food Science and Technology, Faculty of Agricultural Technology, Brawijaya University, Jawa Timur, Indonesia

Preface

The last few centuries have seen an increase in food production through modern agricultural technologies and techniques with a significant effect on the quality and quantity of food produced globally but at the same time has harmed the environment. The current food system, including production and postfarm operations such as processing and distribution, accounts for over a quarter (26%) of global greenhouse gas emissions, which pose a challenge for the coming decades. People across the globe are becoming increasingly concerned about the negative impact of global greenhouse gas emissions and are realizing that it is a result of their food choices. It is clear that eating less meat and dairy products will have a much more significant impact on carbon footprint than eating plant-based foods. Therefore, an increased understanding of the environmental impacts of food production and a growing preference for healthier vegetarian foods are driving the urgency to develop sustainable food systems. Plant-based food systems offer a practical solution to this sustainable goal, resulting in an increased global demand for plant-based meat and dairy alternatives. This book, ***Engineering Plant-Based Food Systems,*** provides a broad understanding of the various plant-based foods and the technologies used to create them.

The book introduces what plant-based food means, the current trends, and its potential as a sustainable food source for the future (Chapter 1), followed by consumer expectations, sensory perception, and attitude toward plant-based foods (Chapter 2). The general health and nutrition of plant-based foods and ingredients, including the antinutritional aspect of plant-based foods and their impact on human health, are also discussed. The book also includes sustainable plant protein sources and their extraction techniques (Chapter 3), the thermal and nonthermal processing techniques used to reduce food allergies caused by plant proteins (Chapter 4), and their functionality, including color, flavor, texture, wettability, dispersibility, solubility, rheological properties, emulsifying properties, and foaming properties (Chapter 5). The subsequent chapters in the book include popular commercially available plant-based foods such as plant-based beverages (Chapter 6), plant-based gels (Chapter 7), plant-based butter-like spreads (Chapter 8), plant-based meat analog (Chapter 9), plant-based imitated fish (Chapter 10), plant-based imitated seafood (Chapter 11), fermented plant-based beverage: Kombucha (Chapter 12), and fermented plant-based food (Chapter 13), the plant materials used to manufacture them using different processing techniques, and their sensory properties. The following two chapters discuss the various functional components obtained from plants and their extraction and incorporation in plant-based food (Chapter 14) and the processing techniques to improve their bioavailability and bioaccessibility (Chapter 15). The recent developments in 3D-printed plant-based foods using plant-based ingredients are reviewed in Chapter 16, followed by a concluding Chapter 17 discussing the future of plant-based foods.

Sangeeta Prakash

Chapter 1

Plant-based food as a sustainable source of food for the future

Sangeeta Prakash[1], Claire Gaiani[2] and Bhesh R. Bhandari[1]

[1]*School of Agriculture and Food Sciences, The University of Queensland, Brisbane, QLD, Australia;* [2]*Laboratoire d'Ingénierie des Biomolécules (LIBio), Université de Lorraine, Nancy, France*

1.1 Introduction

Globally, sustainability is now a part of the everyday vocabulary, be it food, technology, workplace, or home. The **Oxford English Dictionary defines sustainability** as "the degree to which a process or enterprise can be maintained or continued while avoiding the long-term depletion of natural resources" Sustainability comprises the environment, economics, health, food, nutrition, and other related dimensions. Thus, the definition of "sustainability" can mean different things based on the context in which it is discussed. According to a United Nations report, the projected rise in the global population is estimated to reach 8.6 billion in 2030, 9.8 billion in 2050, and 11.2 billion in 2100. When considering sustainable food sources, the objective is to ensure a future when this expanded population has both enough food available to eat and access to high-quality nutritious foods.

Growth and expansion of agricultural land since the 1960s were successful in boosting food production, but they also caused many adverse environmental impacts (Foley et al., 2005). Our current food system has contributed to climate change, deforestation, soil loss, and soil pollution, alongside a considerable demand on water supply, pollution, and exploitation of certain species such as fish and seafood, to name a few. The way we choose to shop and eat has put much stress on our planet and the environment. The need of the hour is to ensure a sustainable environment for the future that is currently under threat of food insecurity. Understanding food and environmental sustainability and what we need to do will help to ensure food security for us and our future generations. Raising livestock for food usually leads to more pollution, as well as greater greenhouse gas (GHG) emissions, water use, land use, and loss of biodiversity than growing plants directly for consumption (Willett et al., 2019). In 2018, Poore and Nemecek (2018) reported that the food production accounts for 26% of global GHG emissions, which are broadly from four sources—livestock and fisheries raised for meat, dairy, eggs, fish, and seafood (31%), crop production (27%) for direct human consumption (21%) and animal feed (6%), land use (24%) for livestock (16%) and growing crops for human consumption (8%), and finally supply chains (18%) that includes retail (3%), packaging (5%), transport (6%), and food processing (4%). Fig. 1.1 summarizes the key contributors to global GHG emissions from food production. However, a recent study at the University of Illinois that collected and quantified the emission data explicitly from plant-based production and animal-based production from more than 190 countries estimated the food production to make up about 37% of global GHG emissions (Xu et al., 2021). As per their estimates, global food production contributes about 17.3 billion metric tonnes of carbon dioxide equivalent per year, of these emissions, 57% were related to the production of animal-based foods, and plant-based food production accounted for 29%. Thus, food production, especially the food type, significantly impacts GHG emissions Fig. 1.1.

According to Ritchie and Roser (2020), reducing emissions from food production while ensuring sustainable food sources for the rising global population will be one of the most significant challenges in the coming decades (Ritchie & Roser, 2020). Unlike many aspects of energy production where viable opportunities for upscaling low-carbon energy such as renewable or nuclear energy are available, ways to decarbonize agriculture are less clear. Hence, the emphasis should be on changing diets; reducing food waste; improving agricultural efficiency; and technologies that make low-carbon food alternatives scalable and affordable. For a sustainable food future, the role of diet shift has been significantly emphasized

Engineering Plant-Based Food Systems. https://doi.org/10.1016/B978-0-323-89842-3.00005-1

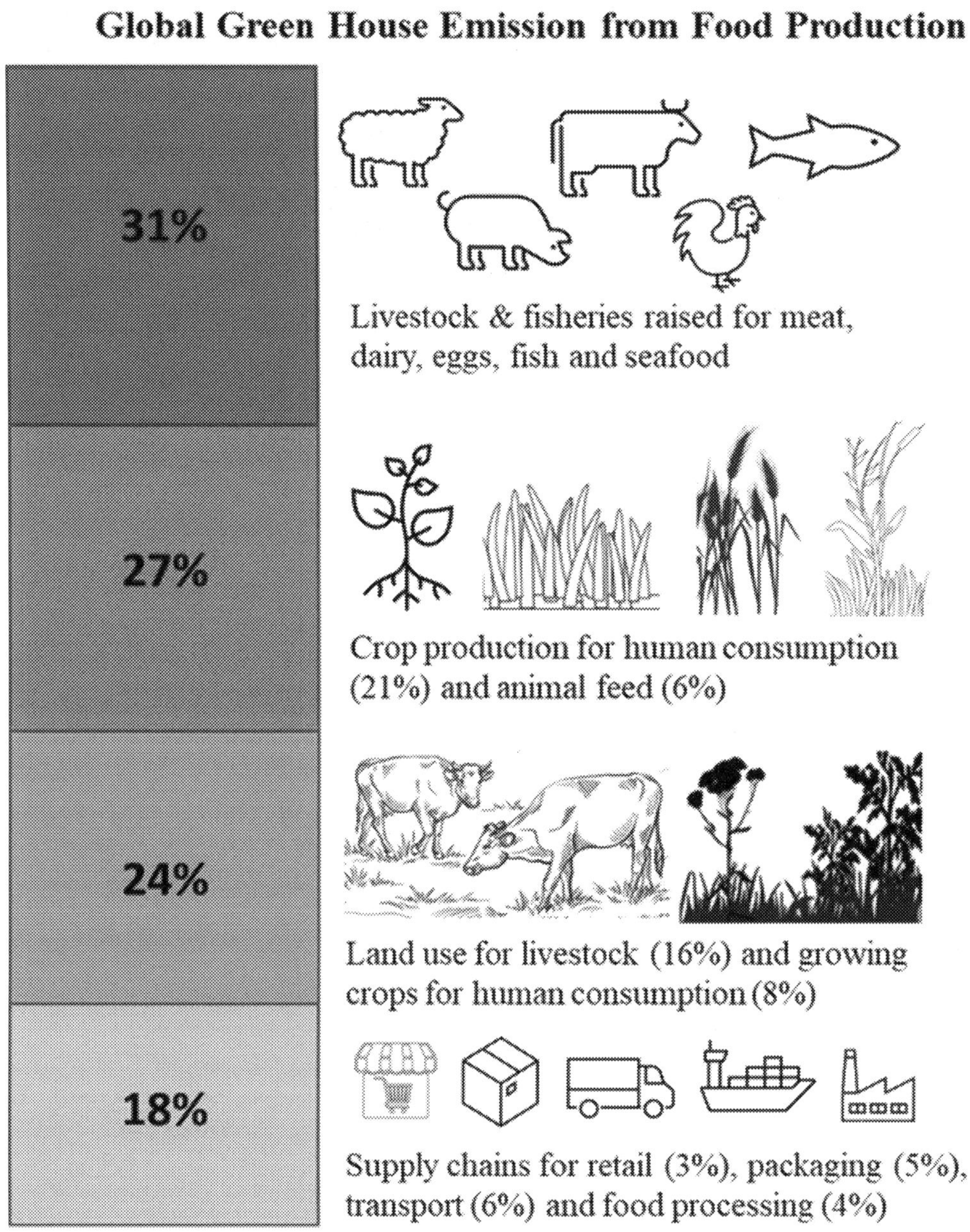

FIGURE 1.1 Main contributors to global greenhouse emission from food production.

by Ranganathan et al. (2016, p. 90), proposing shifting the diets of populations who consume high amounts of calories, protein, and animal-based foods to (1) reduce overconsumption of calories, (2) reduce overconsumption of protein by reducing consumption of animal-based foods, and (3) reduce beef consumption with an increased emphasis on plant-based foods (Ranganathan et al., 2016). Lowering meat consumption to 52 calories per person per day by 2050 would reduce the GHG mitigation gap by half (Searchinger et al., 2019). Hence, there is a growing interest in plant-based foods.

1.2 Plant-based foods and current market trends

Plant-based foods use plant-based sources, including vegetables, grains, nuts, seeds, legumes, and fruits, with no animal-source foods. "Plant-based" is a recent consumer trend of avoiding animal-based products and choosing plant-based alternatives instead, reducing the number of animal-based foods in diets overall or following dietary regimes with a sole focus on plant-based foods. Although they are very different, a plant-based diet is often referred to as a vegan diet. While a vegan diet eliminates all animal products, including dairy, meat, poultry, fish, eggs, and honey, plant-based diets do not necessarily eliminate animal products but focus on eating mostly plants, such as fruits, vegetables, nuts, seeds, and whole grains. Globally, consumers are either cutting down or altogether avoiding the consumption of animal-based food products. This trend can be related to consumers being increasingly concerned for the environment, health and wellness, ethical concerns about confining and slaughtering animals (Leiserowitz, 2020; Possidónio et al., 2021), and diversity in protein sourcing that drive the demand and growth of plant-based eating. As the global population grows, the pressure will increase on animal-based foods. More sustainable food sources will be needed, with plant-based foods expected to become an integral part of the human diet.

Consequently, the plant-based food market is booming with an estimated growth of USD 74.2 billion by 2027 (Meticulous Research, 2020) that includes dairy alternatives, meat substitutes, plant-based eggs/egg substitutes, and plant-based confectionery. Plant-based foods are not novel. Off-late has gone from being a niche industry targeting mostly vegetarians and vegans into a mainstream industry targeting everyone, causing it to grow exponentially, combined with the innovation in engineering plant-based foods. The recent coronavirus pandemic (the virus, COVID-19's origin from animal sources) has also driven several consumers to shift toward plant-based foods as they re-examined their dietary habits after the virus exposed the association between food health and immune responses. The food sector is also increasingly turning toward sustainability issues. Capitalizing on this trend, the plant-based alternative meat market proliferates and is expected to grow in the coming decades (van Vliet et al., 2020). A sustainable food system should provide sufficient, nutritious food for all within limited natural resources.

The growing plant-based alternative trend, particularly those replacing dairy products (milk, yoghurt, cheese, ice cream) and meat, is competing with the dairy and meat sector. Restaurants and food retailers now have a dedicated section for plant-based or animal-free products. As mentioned previously, plant-based foods or plant-based alternatives are not novel. They always existed—for example, texturized soy as an animal-meat substitute or the plant-based margarine as a replacement for butter; it is just that off-late they are getting enormous attention. The markets are inundated with various types of plant-based alternatives broadly classified into different categories such as plant-based meat, plant-based drinks, plant-based dairy, and plant-based frozen products. Among the various alternatives, plant-based drinks are the fastest growing segment, with China leading the plant-based milk market globally, accounting for 24.2% of the entire beverage market (Insights, 2021).

As the demand for plant-based alternatives increases worldwide, newer plant ingredients are being introduced to broaden the product range and the categories, giving the consumer a wider choice of products. For instance, the dairy alternatives market started with only soy milk. Now there has been a substantial increase in the variety of ingredients used to produce milk alternatives, such as oat, coconut, almond, rice, walnut, and hemp, to name a few. The plant-based yoghurts with live probiotics and cultures offer the same gastrointestinal benefits as dairy-based yoghurts, sometimes offering nutrients such as omega-3 fatty acids and fiber from hemp and flaxseed that even dairy-based yogurts fail to provide (Montemurro et al., 2021; Sethi et al., 2016). The global plant-based yogurt market size is projected at USD 2.89 billion by 2026 (Fior Markets, 2020). The demands for other dairy alternatives that include cheese and ice creams are also increasing. The millennials are attracted to these dairy alternatives as they are nutritious and a good option for those with lactose intolerance or milk allergy. The global dairy alternatives market is expected to be USD 47.95 billion by 2028 (Fior Markets, 2021).

The other plant-based alternative sector growing fast worldwide is plant-based meat that includes various plant-based meat types such as beef, pork, chicken, seafood, and eggs. Consumers are leaning toward plant-based meat due to increased consciousness for their health, environment, and avoiding animal cruelty. There is a growing demand for plant-based burgers, sausages, bacon, and chicken products, such as nuggets, cutlets, and patties. The worldwide plant-based meat market was USD 5.6 billion in 2020 and is expected to reach USD 14.9 billion by 2027(Research & Markets, 2021).

Thus, health that continues to take precedence in our lives and the growing environmental concerns among consumers influence the growth of the plant-based market. The role of plant-based food as a sustainable food source is discussed in the next section.

1.3 Plant-based foods are sustainable sources of foods for the future

Globally, livestock production is one of the leading causes of deforestation, with much of the land used to raise livestock for dairy and meat. Fig. 1.2 shows that for every food calorie generated, animal-based foods and ruminant meats in particular require many times more feed and land inputs and emit far more GHGs than plant-based foods (Searchinger et al., 2019). The data presented in Fig. 1.2 are from GlobAgri-WRR model (Searchinger et al., 2019) with the data sourced from (FAO, 2019).

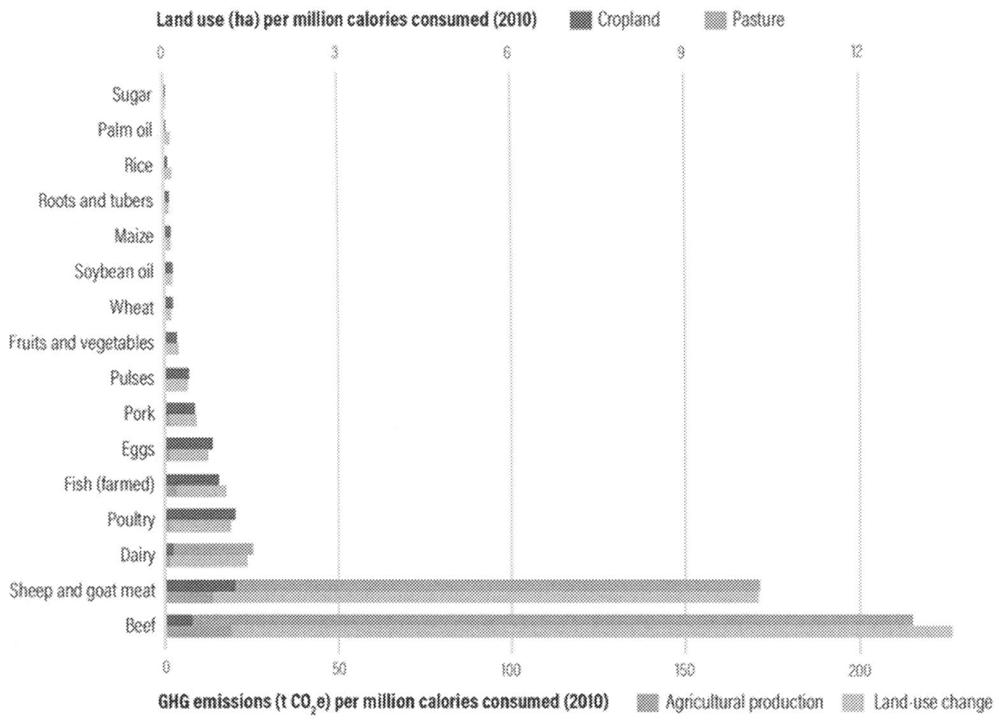

FIGURE 1.2 From Searchinger, T., Waite, R., Hanson, C., Ranganathan, J., & Matthews, E. (2019). Creating a sustainable food future: A menu of solutions to feed nearly 10 billion People by 2050. World Research Institute.https://research.wri.org/wrr-food. The land use and greenhouse gas (GHGs) emissions for every food calorie generated from some plant- and animal-based foods. Source *GlobAgri-WRR model with source data from FAO. (2019). Faostat. Available at: www.fao.org/faostat/en/. Accessed 30 November 2021. Data presented are global means, weighted by production volume. Indicators for animal-based foods include resource use to produce feed, including pasture. Tons of harvested products were converted to quantities of calories and protein using the global average edible calorie and protein contents of food types as reported in FAO. (2019). Faostat. Available at: www.fao.org/faostat/en/. Accessed 30 November 2021.*

From Fig. 1.2, meat from beef, sheep, and goat is by far the most resource-intensive food. It requires over 20 times more land and generates over 20 times more GHG emissions than pulses. Dairy's land use and GHG emissions are slightly higher than those of poultry. Pulses, fruits, vegetables, and vegetable oils are generally more resource-intensive to produce than sugars and staple crops because of their lower yields; yet, they are still favorable compared with meat, dairy, and farmed fish (Searchinger et al., 2019). It is estimated that around 100 times as much land is used to produce a kilocalorie of beef or lamb instead of plant-based substitutes (Ritchie, 2021). Overall, 26% (3.4 billion ha) and 4% (0.5 billion ha) of the Earth's ice-free surface is utilized for livestock grazing and livestock feed production, respectively (Ulian et al., 2020). Thus, eliminating beef, mutton, and dairy from the diet will significantly impact agricultural land use, freeing up the land used to feed the livestock. It is expected that if the entire world adopted a vegan diet, our total agricultural land use would shrink from 4.1 billion hectares to one billion hectares (Ritchie, 2021). Thus, compared to animal-based foods, plant-based foods consume less energy from fossil fuels, less land, and less water to grow, thereby demanding less from the environment (Sabaté & Soret, 2014). Hence, plant-based food systems are rightly the sustainable food system for the future.

The consumption of high levels of animal-based products, particularly those obtained from cows, such as beef and milk, is one of the major factors contributing to the negative impact of the modern diet on both global and individual health (Poore & Nemecek, 2018). Therefore, there is considerable interest in switching to a more plant-based diet to introduce a healthy lifestyle, increase sustainable food production, and reduce pollution, land use, and water use (Willett et al., 2019). Among the numerous cultivated plant species known to humans, the current agriculture focuses on the cultivation of large-seeded, high-yielding varieties of a few major staple crops such as rice, maize, and wheat, which provide half of the global human requirement for carbohydrates, proteins, and calories (López Noriega et al., 2012). This dependence on a few sources has decreased the dietary diversity (Webb & van Ginkel, 2009). For instance, a wide range of plant sources such as legumes largely remain underutilized as a food ingredient and mainly used as animal feed. However, currently, there is an increased emphasis on a shift from monocultures as there is a growing awareness of the importance of other more significant varieties of crops that can provide the basis for a sustainable food production system and a source of a nutritionally balanced gastronomic experience. The challenges presented by climate change (rising temperatures, intense weather occurrences, irregular rainfall and weather patterns, and increased incidents of drought) further impact agriculture productivity, which again calls for the development and promotion of underutilized crops (Massawe et al., 2015), diversifying future agriculture. Underutilized crops are an important local source of food that has not been widely adopted due to their low value, the dominance of the four major crops, i.e., rice, wheat, maize, and soybean and also because of very little and incoherent knowledge of their nutritional quality. To include the underutilized crops into the food value chain, a systematic approach is required to investigate their properties as a food ingredient through its growth, storage, and processing, underpinned by science to provide evidence-based outcomes. This provides opportunities to develop novel food products and identify their potential market through research, development, and innovation, helping promote the underutilized crops. Several plant-based ingredients are currently emerging in the market, such as sorghum and moringa, which were once only locally known.

Among the range of plant-based sources of ingredients, the most popular and much in demand is the plant-based proteins, with the global alternative protein market expected to reach USD 290 billion-plus by 2035, an 11% increase from 2020, as per a recent study by Boston Consulting Group (BCG) and Blue Horizon (Morach et al., 2021). Plant-based proteins are considered as functional ingredients with various roles in food formulations, including thickening and gelling agents, stabilizers of emulsions and foams, binding agents for fat and water, and can also be utilized to produce bioactive peptides. Research focusing on understanding protein—polysaccharide interactions and manipulating plant protein structures during protein extraction, fractionation, and modification to enhance protein functionality has recently gained momentum to create novel plant-based foods. Proteins sourced from peas, lentils, oat, hemp, and legumes are increasingly gaining popularity, more so because they are the only source of protein to the vegan/vegetarian consumers. The increasing demand for plant-based protein is mainly driven by the vital role protein plays in the human immune system, providing the essential amino acids required by humans for the immune system, building muscles and other vital functions in the body. The plant-based proteins are also popular ingredients in the food industry because of their versatile functional attributes, such as their ability to thicken, gel, emulsify, foam, and hold fluids (Loveday, 2020). However, unlike animal protein, a single source of plant-based protein, except soy protein, does not provide all the nine essential amino acids (histidine, isoleucine, leucine, lysine, methionine, phenylalanine, threonine, tryptophan, and valine) that the human body requires (van Vliet et al., 2015). Therefore, from a nutritional viewpoint, the ideal combination of proteins from diverse plant sources can supply all the essential amino acids needed to fulfill human health needs (Kumar et al., 2022).

1.4 Comparison of key nutrients from plant- and animal-based foods

Plant-based food research is still in its early stages compared to animal-based foods, including dairy products. Nutrients obtained from both animal- and plant-based foods perform a vital role in sustaining the normal physiological functions of the human body. Proteins, carbohydrates, fats, vitamins, and minerals are the primary nutrients present in food. It is quite well known that animal-based foods are a good source of various micronutrients that alone cannot be substituted by plant-based foods. The main reasons driving consumers globally toward plant-based foods are environmental and health concerns leading to a significant surge in developing a range of plant-based alternatives to replace the traditional milk, eggs, fish, and meat promoted as sustainable and healthy (Tso & Forde, 2021). However, it remains unclear if plant-based food products are healthy as claimed.

Plant-based foods are a good source of protein with the advantage of being low in fat, a source of fiber, and numerous valuable phytonutrients (high biological value), all of which are lacking in animal-based foods. Phytonutrients are a class of nutrients (polyphenols, flavonoids, isoflavonoids, terpenoids, carotenoids, anthocyanins, glucosinolates, phytosterols, resveratrol, phytoestrogens, omega-3 fatty acids, and probiotics) obtained from plants (Xiao & Bai, 2019). The phytonutrients are often called bio-actives due to their specific biological activities, e.g., antimicrobial, antioxidants, anti-inflammatory, anticancer, hepatoprotective, hypolipidemic, neuroprotective, and immuno-modulator, that support human health (Gupta & Prakash, 2014). On the other hand, animal-based foods are a good source of complete proteins, iron, calcium, vitamins A and B-12, and other micronutrients but do not contain phytonutrients and are low in fiber and high in saturated fat and cholesterol that have significant health implications. Table 1.1 summarizes in general the differences in macronutrients and micronutrients found in the plant- and animal-based foods (McMacken & Shah, 2017; Salomé et al., 2021; Tso & Forde, 2021). Table 1.1.

TABLE 1.1 Differences in macronutrients and micronutrients in plant- and animal-based foods.

		Plant-based foods	Animal-based foods
Macronutrients	Protein	Incomplete protein sources as a single plant-based protein do not contain all the nine essential amino acids	Complete protein sources contain all nine essential amino acids
	Fat	Rich in monounsaturated and polyunsaturated fats	Rich in saturated fat and trans fat
	Carbohydrates	High fiber (whole and minimally processed)	Low fiber
Micronutrients		Rich in beta carotene, vitamin K and C, folate, magnesium, and potassium	Rich source of iron, calcium, and vitamin B12
		Low in iron, calcium, zinc, riboflavin, vitamin A, and vitamin B12 (entirely missing in plant-based food)	

From McMacken, M., & Shah, S. (2017). A plant-based diet for the prevention and treatment of type 2 diabetes. *Journal of Geriatric Cardiology, 14*(5), 342–354; Salomé, M., Huneau, J. F., Le Baron, C., Kesse-Guyot, E., Fouillet, H., & Mariotti, F. (2021). Substituting meat or dairy products with plant-based substitutes has small and heterogeneous effects on diet quality and nutrient security: A simulation study in French adults (INCA3). *The Journal of Nutrition, 151*(8), 2435–2445; Tso, R., & Forde, C. G. (2021). Unintended consequences: Nutritional impact and potential pitfalls of switching from animal- to plant-based foods. *Nutrients, 13*(8), 2527.

1.5 Health benefits of plant-based foods over animal-based foods

Unhealthy lifestyles are increasingly contributing to health problems of overweight and obesity that are associated with diabetes, cardiovascular disease, hypertension, and osteoarthritis, among other health concerns. Several studies are comparing meat-based diets with plant-based diets that are discussed below.

Plant-based food consumption has many health benefits such as protection from certain types of cancer (Lanou & Svenson, 2010; Madigan & Karhu, 2018), improved neurocognitive function, and prevention and management of dementia and Alzheimer's disease (Pistollato et al., 2018). Plant-based foods have long been linked to weight management (Berkow & Barnard, 2006; Farmer et al., 2011) as vegetarians consume less total fat than their meat-eating counterparts, and their weight loss is not dependent on exercise but on increased calorie loss in contrast to nonvegetarians, wherein fewer calories are burned as food is being stored as fat. Epidemiologic studies also indicate that vegetarian diets are associated with a lower body mass index (BMI) and a lower prevalence of obesity in adults and children. Thus, it is suggested that a plant-based diet, primarily whole food, may mediate body weight (Greger, 2020; Tran et al., 2020) and associated chronic disease (diabetes, cardiovascular, hypertension), thereby reducing medical costs compared with nonvegetarian diets (Berkow & Barnard, 2006). A study conducted by Päivärinta et al., (2020) observed that replacing animal protein with plant protein sources in the diet led to an increased fiber intake and improved dietary fat quality, as well as blood lipoprotein profile, and they suggested that a flexitarian diet could provide healthy and more sustainable alternatives for the current, predominantly animal-based diets. However, research in the United States (Tuso et al., 2013) suggests healthy eating may be best achieved by encouraging a plant-based diet that consists of whole, plant-based foods discouraging meats, dairy products, eggs, and all refined and processed food. They added that plant-based diets are cost-effective, low-risk interventions that may lower body mass index, blood pressure, glycosylated hemoglobin, and cholesterol levels. Further, Pistollato et al. (2018) associated dietary patterns, characterized by high intake of plant-based foods along with other nutrients (https://www.sciencedirect.com/topics/pharmacology-toxicology-and-pharmaceutical-science/probiotic-agentprobiotics, antioxidants, soybeans, nuts, and https://www.sciencedirect.com/topics/pharmacology-toxicology-and-pharmaceutical-science/omega-3-fatty-acidomega-3 polyunsaturated fatty acids), and a low intake of https://www.sciencedirect.com/topics/pharmacology-toxicology-and-pharmaceutical-science/saturated-fatty-acidsaturated fats, animal-derived proteins, and refined sugars decreases the risk of neurocognitive impairments and eventually the onset of Alzheimer's disease (Pistollato et al., 2018). Vegetarianism as a lifestyle intervention is often linked to cancer-protective (Anand et al., 2008; Lanou & Svenson, 2010), but surprisingly very few studies have directly addressed this question. There is evidence of fruits and vegetables and some plant constituents such as fibers, antioxidants, and other phytochemicals (e.g., isoflavone) that help achieve and maintain a healthy weight, reducing the risk of cancer diagnosis and recurrence (Chang et al., 2017). An epidemiological study (Hastert et al., 2013) followed 30,797 postmenopausal women (aged between 50 and 76) with no history of breast cancer for 7 years and observed that by maintaining average body weight, limiting alcohol, and mainly eating, plant-based decreased their risk of breast cancer by 62%. In another study by Orlich et al. (2015), with approximately 80,000 subjects, an overall lower incidence of colorectal cancer was observed among vegetarians than nonvegetarians (Orlich et al., 2015). Thus, whole plant-based foods (vegetables, fruits, whole grains, beans) significantly protect against breast, prostate, colorectal, and gastrointestinal cancers (Madigan & Karhu, 2018). The plant nutrients offer protection against harmful substances found in nonvegetarian diets especially red meat and processed meat, such as saturated fats and carcinogens formed during the cooking or processing of meats (Van Hecke et al., 2017). The protective role of a vegetarian diet against some classes of cancer is increasingly becoming evident; however, large study samples are required to validate this association.

Although the advantages associated with plant-based diets are numerous, the question arises on their ability to deliver minerals and if their increased consumption leads to mineral deficiencies in the human body (Rousseau et al., 2020). A plant-based diet in individuals is often linked with lower iron stores, as iron in meat is more bioavailable than iron in plants (Tuso et al., 2013). The reduced bioaccessibility and bioavailability of iron and zinc from plant-based food are linked to the presence of mineral antinutrients such as phytic acid, polyphenols, and dietary fiber and physical barriers (surrounding macronutrients and cell wall). For example, zinc forms a strong complex with phytic acid, and iron binds to polyphenols that impact their absorption during human digestion. Among dietary fibers, mostly soluble dietary fiber (such as pectin) forms complexes with iron and zinc. This insoluble and nonabsorbable mineral chelates, formed with dietary fiber, phytic acid, and some polyphenols in the plant-based food system, are not hydrolyzed during human digestion and therefore not absorbed in the small intestine (Rousseau et al., 2020).

Additionally, mineral antinutrients can also form insoluble complexes with dietary proteins, starch, and other carbohydrates, reducing their digestibility (Rousseau et al., 2020). The other disadvantages associated with plant-based foods are the low protein intake as discussed previously, one single plant-based protein source does not provide all the essential amino acids (van Vliet et al., 2015), lower intake of calcium and vitamin D, leading to decreased bone mineralization and increased risk of fractures, vitamin B12 deficiency, and lower essential fatty acid intake (Tuso et al., 2013). A well-balanced, carefully planned, plant-based diet adequately supplemented with the essential vitamin B12, and vitamin D will overcome some of the mineral and vitamin deficiencies obtained with a plant-based diet.

1.6 Effect of plant-based food production on the environment

Eating habits have a significant alliance with human health and environmental sustainability (Willett et al., 2019). Food choices affect our health and influence the environment's state. Over the past decades, due to the growing demands, the production of animal-based foods has increased drastically. The production of enormous quantities of animal products, such as meat, fish, egg, milk, and their by-products, has been suggested to be a major reason contributing to the negative impact of the modern food supply on global environmental sustainability (McClements & Grossmann, 2021; Poore & Nemecek, 2018). As per a report by the Food and Agriculture Organization (FAO) of the United Nations (Steinfeld et al., 2006), global meat production is estimated to hit 465 million tons by 2050, almost double of the 229 million tons produced in 1999, and milk production is expected to increase from 580 to 1043 million tons. Of all the food production systems, economists estimate meat production as the most resource-intensive (Lusk & Norwood, 2009). The impact of raising livestock on agriculture production can be estimated by the fact that "for every kilogram of animal protein produced, animals consume an acreage of almost 6 kg of plant protein from grains and forage" (Pimentel & Pimentel, 2003). Scientists have estimated that the global food system is generating a third of the world's GHG emissions, of which up to 80% are associated with livestock production (Springmann, Godfray, et al., 2016), with meat production directly accounting for between 18 (Steinfeld et al., 2006) and 50% (Goodland & Anhang, 2009) of GHGs emitted to the atmosphere. The livestock industry emissions comprise mainly carbon dioxide, methane, ammonia, and nitrous oxide (Leytem et al., 2011). Carbon dioxide, naturally released from livestock, is the major contributor to global warming, followed by nitrous oxide, ammonia, and methane emissions. The nitrous oxide and ammonia emissions further cause damage to the ozone layer and acidification of the ecosystems, respectively (Goodland & Anhang, 2009; Leytem et al., 2011; Russell, 2014). Thus, the livestock industry utilizes a major share of the global natural resource for the production of animal-based foods, particularly meat, dairy products, and animal-based proteins (mostly from ruminants) for human consumption (Beverland, 2014) and are responsible for a substantial share of the environmental burden of food production (Sabaté & Soret, 2014) producing considerable amounts of untreated waste, causes pollution of ground water, land degradation, deforestation, and loss of wildlife habitat.

Over the past decades, rising household income, urbanization, growing world population, and an intent to reduce food shortage have increased global food production of heavily processed (ultra-processed), excessively high in calories, rich in sugar and saturated fat, unhealthy foods that has majorly shifted the global eating patterns (Monteiro et al., 2013). These processed and ultra-processed foods are not only environmentally unsustainable but are also increasing the risk of food-related diseases (Garnett, 2016) such as obesity, cardiovascular health (hypertension, dyslipidaemia) and disease (coronary heart diseases and cerebrovascular diseases), cancer (breast, prostate, and colorectal cancers), depression, asthma and wheezing, gastrointestinal disorders, and frailty syndrome (Monteiro et al., 2019). The negative impact of environmental degradation includes morbidity and premature mortality caused by inhaling smoke containing large amounts of fine particulate matter from biomass burning for agriculture and land clearing (Koplitz et al., 2016), reduced food security resulting from reduced quantities of food harvested due to climatic changes causing higher food prices and an increased number of malnourished people (Nelson et al., 2010; Springmann, Mason-D'Croz et al., 2016; Wheeler & Von Braun, 2013), reduced nutrient content of some crops due to rising atmospheric carbon dioxide concentrations (Myers et al., 2014), and food shortage intensified by extreme weather events such as floods and drought (Smith et al., 2014).

Climate change and increased understanding of environmental impacts of food production have resulted in an urgent need to develop sustainable food systems. Compared to animal-based food products, the production of plant-based foods is viewed as more sustainable as growing plants require less resource-intensive requiring less energy from fossil fuels, less land and less water, and environmentally impactful (Ranganathan et al., 2016). As per Poore and Nemecek (2018), the production of livestock generates the highest levels of GHG emissions, whereas the production of fruit and vegetables generates the lowest levels (Poore & Nemecek, 2018). Food production requires water, with an estimated 70% of the available freshwater stocks being used for growing food (Saxena, 2011), animal-based food products requiring as much as 10 times more water than plant crops (Stuart, 2007). Thus, the global food system needs an overhaul, which can be only achieved by consumers changing their views about food systems that can positively impact their health and environment. This change in thinking should recognize the inseparable link between human health and environmental sustainability and integrate these separate concerns into a shared global agenda to achieve healthy diets from sustainable food systems.

1.7 Summary and future direction

Globally, agriculture and food production are responsible for more than 25% of GHG emissions. The food systems can nurture human health and support environmental sustainability; however, they are threatening both as reported in EAT-Lancet Commission's report (Willett et al., 2019). Reducing the impacts of agriculture on the environment while still securing future food and nutrition is one of the most significant challenges of this century. It requires a paradigm shift from the current situation. Consumers are now interested in knowing the contents of their foods and its environmental impact (carbon labeling). Research suggests that consumers worldwide are willing to support plans and policies that promote plant-based diets and are ready to cut back on their meat, dairy, and egg intake to help protect the environment (Sanchez-Sabate & Sabaté, 2019). However, the global food system is far too diverse, and one policy will not fit all. Therefore, carefully crafted strategies considering the unique environmental and socio-economic circumstances of each geographical location worldwide should be implemented to encourage the shift to plant-based foods or reduce the consumption of animal-based foods.

Plant-based foods are gaining more popularity since they play a crucial role in sustainable, low-meat, and healthy diets. An emerging population is switching to a plant-based diet citing health and environmental benefits. However, more awareness of the health and environmental benefits of a plant-based diet instead of an animal-based diet should be backed by research. Research has increasingly proved that diets based on fruits, vegetables, legumes, nuts, and whole grains could be beneficial in reducing lifestyle-related diseases, such as coronary heart diseases and type II diabetes, can reduce the risk of developing breast, colon, and other digestive-related cancers and improving the psychological well-being (Willett et al., 2006). However, information about the adequacy of plant-based food to provide all the necessary nutrients (both macro and micronutrients) and the health consequences of following a strict plant-based food diet largely remains unclear. Hence, there is an increased need for long-term research investigating the completeness of plant-based food as a source of balanced nutrition and its impact on human health. The findings would enable developing dietary guidelines for plant-based foods, especially for the vulnerable populations such as young children, pregnant and nursing mothers, and the elderly at increased risk for nutritional deficiencies.

Over the past few years, the plant-based food industry has grown enormously. Alternatives have emerged for meat, seafood, and eggs that closely mimic their appearance, texture, and taste. The plant-based alternatives are healthier, containing lower amounts of total and saturated fat, higher amounts of fiber, and comparable amounts of protein and calories compared to their natural counterparts. However, it is essential to realize that not all plant-based foods are a rich source of nutrients. Ingredients from animal and plants are structurally and functionally different, and, in their pursuit, to mimic real meat, the animal alternatives are heavily processed with high amounts of salt and other ingredients that can cause adverse health effects. Nutritional needs vary among individuals based on their age, gender, and body metabolism. Hence, when shifting to plant-based foods, it is essential to be cautious and focus on healthy options. Further, it is crucial to conduct studies to understand consumer perceptions of plant-based alternatives that will help facilitate a mass shift toward more sustainable food consumption.

From the agricultural side, there is a range of crops rich in protein, bioactive, and other functional components that are potentially suitable for human consumption. However, they are either used as animal feed or unnoticed due to lack of awareness. There is a need to tap these underutilized crops and other locally grown high protein crops such as legumes, quinoa, seeds, and nuts. This is possible through government research and development financial incentives and supportive policies at country, regional, and continental levels for increased adoption of these underutilized and high protein crops.

However, it is crucial to understand that promoting or adopting sustainable plant-based diets alone is insufficient to cut emissions and should be practised alongside cleaner food production techniques that cause less environmental damage. One such food production approach is regenerative agriculture that practices minimum or reduced tillage, reduced use of pesticides and fertilizers, crop rotation, well-managed grazing feed, and increasing soil fertility through natural means (the use of cover-cropping) that have the potential to achieve net-zero emissions, reducing the negative impact of food production on the health of the planet. The path to a more sustainable future via the adoption of plant-based diets remains debatable, and only time will tell.

References

Anand, P., Kunnumakkara, A. B., Sundaram, C., Harikumar, K. B., Tharakan, S. T., Lai, O. S., Sung, B., & Aggarwal, B. B. (2008). Cancer is a preventable disease that requires major lifestyle changes. *Pharmaceutical Research, 25*(9), 2097–2116.

Berkow, S. E., & Barnard, N. (2006). Vegetarian diets and weight status. *Nutrition Reviews,* 64(4), 175–188.

Beverland, M. B. (2014). Sustainable eating:mainstreaming plant-based diets in developed economies. *Journal of Macromarketing, 34*(3), 369–382.

Chang, Y. J., Hou, Y. C., Chen, L. J., Wu, J. H., Wu, C. C., Chang, Y. J., & Chung, K. P. (2017). Is vegetarian diet associated with a lower risk of breast cancer in Taiwanese women? *BMC Public Health, 17*(1), 800.

FAO. (2019). *Faostat*. www.fao.org/faostat/en/ (Accessed 30 November 2021).

Farmer, B., Larson, B. T., Fulgoni, V. L., 3rd, Rainville, A. J., & Liepa, G. U. (2011). A vegetarian dietary pattern as a nutrient-dense approach to weight management: An analysis of the national health and nutrition examination survey 1999-2004. *Journal of the American Dietetic Association, 111*(6), 819–827.

Fior Markets. (2020). *Vegan yogurt Market by product (soy, almond, Coconut and others), distribution channel (Convenience store, Specialty store, E-retailer, hypermarket/supermarket), end-user, regions, global industry analysis, market size, share, growth, trends, and forecast 2019 to 2026*. Published online at: https://www.fiormarkets.com/report/vegan-yogurt-market-by-product-soy-almond-coconut-407141.html.

Fior Markets. (2021). Dairy alternatives market share, share & trends analysis report by distribution channel (convenience stores, online stores, supermarket/hypermarkets), by source (almond, rice, soy, coconut, hemp, oats), by application (cheese, Ice creams, milk, yogurt). *By Region And Forecasts*, 2021–2028. Published online at: https://www.fiormarkets.com/report/dairy-alternatives-market-share-share-trends-analysis-419730.html.

Foley, J. A., DeFries, R., Asner, G. P., Barford, C., Bonan, G., Carpenter, S. R., Chapin, F. S., Coe, M. T., Daily, G. C., Gibbs, H. K., Helkowski, J. H., Holloway, T., Howard, E. A., Kucharik, C. J., Monfreda, C., Patz, J. A., Prentice, I. C., Ramankutty, N., & Snyder, P. K. (2005). Global consequences of land use. *Science, 309*(5734), 570–574.

Garnett, T. (2016). Plating up solutions. *Science, 353*(6305), 1202–1204.

Goodland, R., & Anhang, J. (2009). *Livestock and climate change: What if the key actors in climate change are cows, pigs, and chickens? World watch*. November/December 2009.

Greger, M. (2020). A whole food plant-based diet is effective for weight loss: The evidence. *American Journal of Lifestyle Medicine, 14*(5), 500–510.

Gupta, C., & Prakash, D. (2014). Phytonutrients as therapeutic agents. *Journal of Complementary & Integrative Medicine, 11*(3), 151–169.

Hastert, T. A., Beresford, S. A., Patterson, R. E., Kristal, A. R., & White, E. (2013). Adherence to WCRF/AICR cancer prevention recommendations and risk of postmenopausal breast cancer. *Cancer Epidemiology Biomarkers and Prevention, 22*(9), 1498–1508.

Insights. (2021). Insight—COVID-19 spurs plant-based protein market in China. *The Australian Trade and Investment Commission (Austrade)*. Published online at: https://www.austrade.gov.au/news/insights/insight-covid-19-spurs-plant-based-protein-market-in-china.

Koplitz, S. N., Mickley, L. J., Marlier, M. E., Buonocore, J. J., Kim, P. S., Liu, T., Sulprizio, M. P., DeFries, R. S., Jacob, D. J., Schwartz, J., Pongsiri, M., & Myers, S. S. (2016). Public health impacts of the severe haze in equatorial asia in september—october 2015: Demonstration of a new framework for informing fire management strategies to reduce downwind smoke exposure. *Environmental Research Letters, 11*(9), 094023.

Kumar, M., Tomar, M., Punia, S., Dhakane-Lad, J., Dhumal, S., Changan, S., Senapathy, M., Berwal, M. K., Sampathrajan, V., Sayed, A. A. S., Chandran, D., Pandiselvam, R., Rais, N., Mahato, D. K., Udikeri, S. S., Satankar, V., Anitha, T., Reetu, Radha, ... Kennedy, J. F. (2022). Plant-based proteins and their multifaceted industrial applications. *LWT, 154*, 112620.

Lanou, A. J., & Svenson, B. (2010). Reduced cancer risk in vegetarians: An analysis of recent reports. *Cancer Management And Research, 3*, 1–8.

Leiserowitz. (2020). *Climate change and the American diet. Yale Program on climate change communication*. New Haven: Yale University and Earth Day Network. CT20.

Leytem, A. B., Dungan, R. S., Bjorneberg, D. L., & Koehn, A. C. (2011). Emissions of ammonia, methane, carbon dioxide, and nitrous oxide from dairy cattle housing and manure management systems. *Journal of Environmental Quality, 40*(5), 1383–1394.

López Noriega, I., Galluzzi, G., Halewood, M., Vernooy, R., Bertacchini, E., D, G., & Welch, E. (2012). *Flows under stress: Availability of plant genetic resources in times of climate and policy change. Working Paper 18*. Copenhagen, Denmark: CAFS.

Loveday, S. M. (2020). Plant protein ingredients with food functionality potential. *Nutrition Bulletin, 45*(3), 321–327.

Lusk, J. L., & Norwood, F. B. (2009). Some economic benefits and costs of vegetarianism. *Agricultural and Resource Economics Review, 38*(2), 109–124.

Madigan, M., & Karhu, E. (2018). The role of plant-based nutrition in cancer prevention. *Journal of Unexplored Medical Data, 3*, 9.

Massawe, F. J., Mayes, S., Cheng, A., Chai, H. H., Cleasby, P., Symonds, R., Ho, W. K., Siise, A., Wong, Q. N., Kendabie, P., Yanusa, Y., Jamalluddin, N., Singh, A., Azman, R., & Azam-Ali, S. N. (2015). The potential for underutilised crops to improve food security in the face of climate change. *Procedia Environmental Sciences, 29*, 140–141.

McClements, D. J., & Grossmann, L. (2021). A brief review of the science behind the design of healthy and sustainable plant-based foods. *Npj Science of Food, 5*(1), 17.

McMacken, M., & Shah, S. (2017). A plant-based diet for the prevention and treatment of type 2 diabetes. *Journal of Geriatric Cardiology, 14*(5), 342–354.

Meticulous Research®. (2020). *Plant based food market worth $74.2 billion by 2027*. Retrieved from https://www.meticulousresearch.com/pressrelease/53/plant-based-food-market-2027.

Monteiro, C. A., Cannon, G., Lawrence, M., Costa Louzada, M. L., & Pereira Machado, P. (2019). *Ultra-processed foods, diet quality, and health using the NOVA classification system*. Rome: FAO. Published online at: https://www.fao.org/fsnforum/resources/fsn-resources/ultra-processed-foods-diet-quality-and-health-using-nova-classification.

Monteiro, C. A., Moubarac, J. C., Cannon, G., Ng, S. W., & Popkin, B. (2013). Ultra-processed products are becoming dominant in the global food system. *Obesity Reviews, 14*(Suppl. 2), 21–28.

Montemurro, M., Pontonio, E., Coda, R., & Rizzello, C. G. (2021). Plant-based alternatives to yogurt: State-of-the-Art and perspectives of new biotechnological challenges. *Foods, 10*(2), 316.

Morach, B., Witte, B., Walker, D., von Koeller, E., Grosse-Holz, F., Rogg, J., Brigl, M., Dehnert, N., Obloj, P., Koktenturk, S., & Schulze, U. (2021). *Food for thought: The protein transformation*. Boston Consulting Group (BCG) and Blue Horizon Corporation (BHC). Published online at: https://www.bcg.com/publications/2021/the-benefits-of-plant-based-meats.

Myers, S. S., Zanobetti, A., Kloog, I., Huybers, P., Leakey, A. D., Bloom, A. J., Carlisle, E., Dietterich, L. H., Fitzgerald, G., & Hasegawa, T. (2014). Increasing CO_2 threatens human nutrition. *Nature, 510*(7503), 139—142.

Nelson, G. C., Rosegrant, M. W., Palazzo, A., Gray, I., Ingersoll, C., Robertson, R., Tokgoz, S., Zhu, T., Sulser, T. B., & Ringler, C. (2010). *Food security, farming, and climate change to 2050: Scenarios, results, policy options. 172*. International Food Policy Research Institute.

Orlich, M. J., Singh, P. N., Sabaté, J., Fan, J., Sveen, L., Bennett, H., Knutsen, S. F., Beeson, W. L., Jaceldo-Siegl, K., Butler, T. L., Herring, R. P., & Fraser, G. E. (2015). Vegetarian dietary patterns and the risk of colorectal cancers. *JAMA Internal Medicine, 175*(5), 767—776.

Päivärinta, E., Itkonen, S. T., Pellinen, T., Lehtovirta, M., Erkkola, M., & Pajari, A. M. (2020). Replacing animal-based proteins with plant-based proteins changes the composition of a whole nordic diet-A randomised clinical trial in healthy Finnish adults. *Nutrients, 12*(4).

Pimentel, D., & Pimentel, M. (2003). Sustainability of meat-based and plant-based diets and the environment. *The American Journal of Clinical Nutrition, 78*(3), 660S—663S.

Pistollato, F., Iglesias, R. C., Ruiz, R., Aparicio, S., Crespo, J., Lopez, L. D., Manna, P. P., Giampieri, F., & Battino, M. (2018). Nutritional patterns associated with the maintenance of neurocognitive functions and the risk of dementia and alzheimer's disease: A focus on human studies. *Pharmacological Research, 131*, 32—43.

Poore, J., & Nemecek, T. (2018). Reducing food's environmental impacts through producers and consumers. *Science, 360*(6392), 987—992.

Possidónio, C., Prada, M., Graça, J., & Piazza, J. (2021). Consumer perceptions of conventional and alternative protein sources: A mixed-methods approach with meal and product framing. *Appetite, 156*, 104860.

Ranganathan, J., Vennard, D., Waite, R., Lipinski, B., Searchinger, T., Dumas, P., Forslund, A., Guyomard, H., Manceron, S., Marajo Petitzon, E., Mouël, C., Havlík, P., Herrero, M., Zhang, X., Wirsenius, S., Ramos, F., Yan, X., Phillips, M., & Mungkung, R. (2016). *Shifting Diets for a sustainable food future*. Working Paper by World Resources Institute.

Research, & Markets. (2021). *Plant based meat market global forecast by source, product, food, regions, company Analysis*. Report ID: 5308266. Published online at https://www.researchandmarkets.com/reports/5308266/plant-based-meat-market-global-forecast-by.

Ritchie, H. (2021). *If the world adopted a plant-based diet we would reduce global agricultural land use from 4 to 1 billion hectares*. Published online at OurWorldInData.org. Retrieved from https://ourworldindata.org/land-use-diets.

Ritchie, H., & Roser, M. (2020). *Environmental impacts of food production*. Published online at OurWorldInData.org. Retrieved from https://ourworldindata.org/environmental-impacts-of-food.

Rousseau, S., Kyomugasho, C., Celus, M., Hendrickx, M. E. G., & Grauwet, T. (2020). Barriers impairing mineral bioaccessibility and bioavailability in plant-based foods and the perspectives for food processing. *Critical Reviews in Food Science and Nutrition, 60*(5), 826—843.

Russell, S. (2014). *Everything you Need to Know about agricultural emissions*. World Resources Institute. http://www.wri.org/blog/2014/05/everything-you-need-know-about-agricultural-emissions.

Sabaté, J., & Soret, S. (2014). Sustainability of plant-based diets: Back to the future. *The American Journal of Clinical Nutrition, 100*(1), 476S—482S.

Salomé, M., Huneau, J. F., Le Baron, C., Kesse-Guyot, E., Fouillet, H., & Mariotti, F. (2021). Substituting meat or dairy products with plant-based substitutes has small and heterogeneous effects on diet quality and nutrient security: A simulation study in French adults (INCA3). *The Journal of Nutrition, 151*(8), 2435—2445.

Sanchez-Sabate, R., & Sabaté, J. (2019). Consumer attitudes towards environmental concerns of meat consumption: A systematic review. *International Journal of Environmental Research And Public Health, 16*(7), 1220.

Saxena, A. M. (2011). *The vegetarian imperative*. The John Hopkins University Press.

Searchinger, T., Waite, R., Hanson, C., Ranganathan, J., & Matthews, E. (2019). *Creating a sustainable food future: A Menu of Solutions to feed nearly 10 billion People by 2050*. World Research Institute. https://research.wri.org/wrr-food.

Sethi, S., Tyagi, S. K., & Anurag, R. K. (2016). Plant-based milk alternatives an emerging segment of functional beverages: A review. *Journal of Food Science and Technology, 53*(9), 3408—3423.

Smith, K., Woodward, A., Campbell-Lendrum, D., Chadee, D., Honda, Y., Liu, Q., Olwoch, J., Revich, B., Sauerborn, R., & Aranda, C. (2014). Human health: Impacts, adaptation, and co-benefits. In *Climate change 2014: Impacts, adaptation, and vulnerability. Part A: Global and sectoral aspects. Contribution of working Group II to the fifth assessment report of the intergovernmental Panel on climate change* (pp. 709—754). Cambridge University Press.

Springmann, M., Godfray, H. C. J., Rayner, M., & Scarborough, P. (2016). Analysis and valuation of the health and climate change cobenefits of dietary change. *Proceedings of the National Academy of Sciences, 113*(15), 4146—4151.

Springmann, M., Mason-D'Croz, D., Robinson, S., Garnett, T., Godfray, H. C. J., Gollin, D., Rayner, M., Ballon, P., & Scarborough, P. (2016). Global and regional health effects of future food production under climate change: A modelling study. *The Lancet, 387*(10031), 1937—1946.

Steinfeld, H., Gerber, P., Wassenaar, T., Castel, V., Rosales, M., & De Haan, C. (2006). *Livestock's Long Shadow—environmental issues and options* (pp. 1—392). Food and Agriculture Organization (FAO), United Nations. https://www.fao.org/3/a0701e/a0701e00.htm.

Stuart, T. (2007). *The bloodless revolution: A cultural history of vegetarianism from 1600 to modern times*. New York: W.W. Norton & Company.

Tran, E., Dale, H. F., Jensen, C., & Lied, G. A. (2020). Effects of plant-based diets on weight status: A systematic review. *Diabetes, Metabolic Syndrome and Obesity: Targets and Therapy, 13*, 3433—3448.

Tso, R., & Forde, C. G. (2021). Unintended consequences: Nutritional impact and potential pitfalls of switching from animal- to plant-based foods. *Nutrients, 13*(8), 2527.

Tuso, P. J., Ismail, M. H., Ha, B. P., & Bartolotto, C. (2013). Nutritional update for physicians: Plant-based diets. *The Permanente Journal, 17*(2), 61–66.

Ulian, T., Diazgranados, M., Pironon, S., Padulosi, S., Liu, U., Davies, L., Howes, M.-J. R., Borrell, J. S., Ondo, I., Pérez-Escobar, O. A., Sharrock, S., Ryan, P., Hunter, D., Lee, M. A., Barstow, C., Łuczaj, Ł., Pieroni, A., Cámara-Leret, R., Noorani, A., ... Mattana, E. (2020). Unlocking plant resources to support food security and promote sustainable agriculture. *Plants People Planet, 2*(5), 421–445.

Van Hecke, T., Van Camp, J., & De Smet, S. (2017). Oxidation during digestion of meat: Interactions with the diet and Helicobacter pylori gastritis, and implications on human health. *Comprehensive Reviews in Food Science and Food Safety, 16*(2), 214–233.

van Vliet, S., Burd, N. A., & van Loon, L. J. (2015). The skeletal muscle anabolic response to plant- versus animal-based protein consumption. *The Journal of Nutrition, 145*(9), 1981–1991.

van Vliet, S., Kronberg, S. L., & Provenza, F. D. (2020). Plant-based meats, human health, and climate change. *Frontiers in Sustainable Food Systems, 4*(128).

Webb, P. J. R., & van Ginkel, M. (2009). In H. Eiselen (Ed.), *Fiat Panis: For a world without hunger* (pp. 410–434). Balance Publications, Hampp Media.

Wheeler, T., & Von Braun, J. (2013). Climate change impacts on global food security. *Science, 341*(6145), 508–513.

Willett, W. C., Koplan, J. P., Nugent, R., Dusenbury, C., Puska, P., & Gaziano, T. A. (2006). Prevention of chronic disease by means of diet and lifestyle changes. In D. T. Jamison, J. G. Breman, A. R. Measham, G. Alleyne, M. Claeson, D. B. Evans, P. Jha, A. Mills, & P. Musgrove (Eds.) (2nd ed.)*The international bank for reconstruction and development/the world bankDisease control priorities in developing countries*. New York and Oxford University Press.

Willett, W., Rockström, J., Loken, B., Springmann, M., Lang, T., Vermeulen, S., Garnett, T., Tilman, D., DeClerck, F., Wood, A., Jonell, M., Clark, M., Gordon, L. J., Fanzo, J., Hawkes, C., Zurayk, R., Rivera, J. A., De Vries, W., Majele Sibanda, L., ... Murray, C. J. L. (2019). Food in the anthropocene: The EAT–lancet commission on healthy diets from sustainable food systems. *The Lancet, 393*(10170), 447–492.

Xiao, J., & Bai, W. (2019). Bioactive phytochemicals. *Critical Reviews in Food Science and Nutrition, 59*(6), 827–829.

Xu, X., Sharma, P., Shu, S., Lin, T.-S., Ciais, P., Tubiello, F. N., Smith, P., Campbell, N., & Jain, A. K. (2021). Global greenhouse gas emissions from animal-based foods are twice those of plant-based foods. *Nature Food, 2*(9), 724–732.

Chapter 2

General health benefits and sensory perception of plant-based foods

Claire D. Munialo and Mihaela Andrei
Coventry University, Coventry, United Kingdom

2.1 Introduction

Food systems have been shown to be an important contributor to the global greenhouse gas emissions, as well as to the occupation and degradation of land; the disruption of the flow of nutrients; the loss of biodiversity; and the depletion of fresh water and fossil fuel (Broekema et al., 2020). Food systems determine the quality, quantity, diversity, and nutritional content of the foods and or the diet available for consumption. An ideal diet has been defined as one that is healthy, affordable, safe, containing sufficient quality and quantity, and culturally acceptable for human nutrition and health status (Broekema et al., 2020). A healthy reference diet is described to largely consist of a low quantity of red meat, processed meat, added sugar, fruits, vegetables, whole grains, nuts, legumes, unsaturated oils, a low to moderate amount of poultry and seafood, and no refined grains (Willett et al., 2019). However, the consumption of protein from animal sources has been associated with a wide range of issues, such as greenhouse gas emissions. This has resulted in a shift toward alternative sources of food proteins that include the consumption of plant-based foods. Plant-based foods constitute one of the most significant sources of nutrient in the human diet (Carvalho & Vasconcelos, 2013). According to Fardet (2017), plant-based foods can generally be classified into; (i) fruits, legumes, grains, nuts, seeds, and vegetables, (ii) their derived processed food counterparts that include breakfast cereals, pasta, bread, fruit juices, and jams; and (iii) their derived ingredients that include sugars, oleaginous seed-derived oils, as well as some spices and herbs (Fardet, 2017).

In view of sustainability, production cost, and availability, there has been a general trend in the pursuit and search for alternative food sources moving away from animal food sources (Munialo et al., 2018). Additionally, due to (i) the need to source for alternatives to conventional disease management strategies and (ii) the potential of prevention of the onset of certain diet-related diseases, which in turn would reduce health care costs, there has been a renewed interest in the health-benefiting properties of plant-based foods (Bryant, 2019). Moreover, plant-based foods are reported to play a fundamental role in promoting human health (Crosby et al., 2008).

2.2 General consumer expectations, sensory perception, and attitude toward plant-based foods

Consumers' attitudes, perceptions, expectations, and preferences of plant-based foods are influenced by several factors such as physiological, sensory, and marketing factors as illustrated in Fig. 2.1.

According to Font-i-Furnols and Guerrero, physiological factors such as attitude and beliefs about the characteristics of a certain product and the way it is produced, handled, or distributed can influence consumer perception (Font-i-Furnols & Guerrero, 2014). Consumers' perception, knowledge, and attitude toward meat production and its impact on the environment in addition to sustainability have been one of the contributors to a shift or resistance toward the consumption of plant-based foods (Saunders et al., 2011). In a recent study (unpublished results) where we evaluated the perception and knowledge of the impact of a plant-based diet on the health of 84 university students (44 females and 40 males), there were mixed results regarding the perception and attitude toward plant-based foods. The results are shown in Fig. 2.2. For instance, 94% of the participants perceived plant-based diets to be healthier than animal-based diets. Some of the

Engineering Plant-Based Food Systems. https://doi.org/10.1016/B978-0-323-89842-3.00017-8

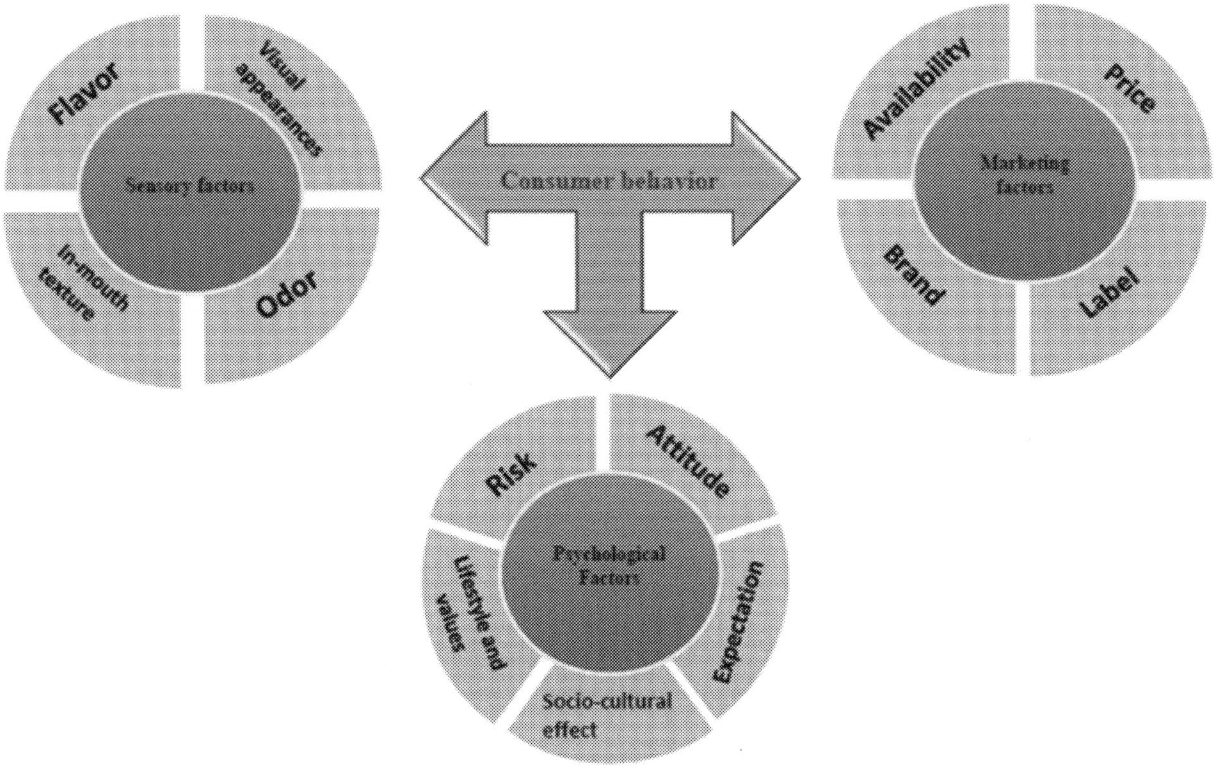

FIGURE 2.1 A simplified multidisciplinary model of the main factors affecting consumer behavior in a food domain. *Adopted from Font-i-Furnols, M., & Guerrero, L. (2014). Consumer preference, behavior and perception about meat and meat products: An overview. Meat Science, 98(3), 361–371. https://doi.org/10.1016/j.meatsci.2014.06.025.*

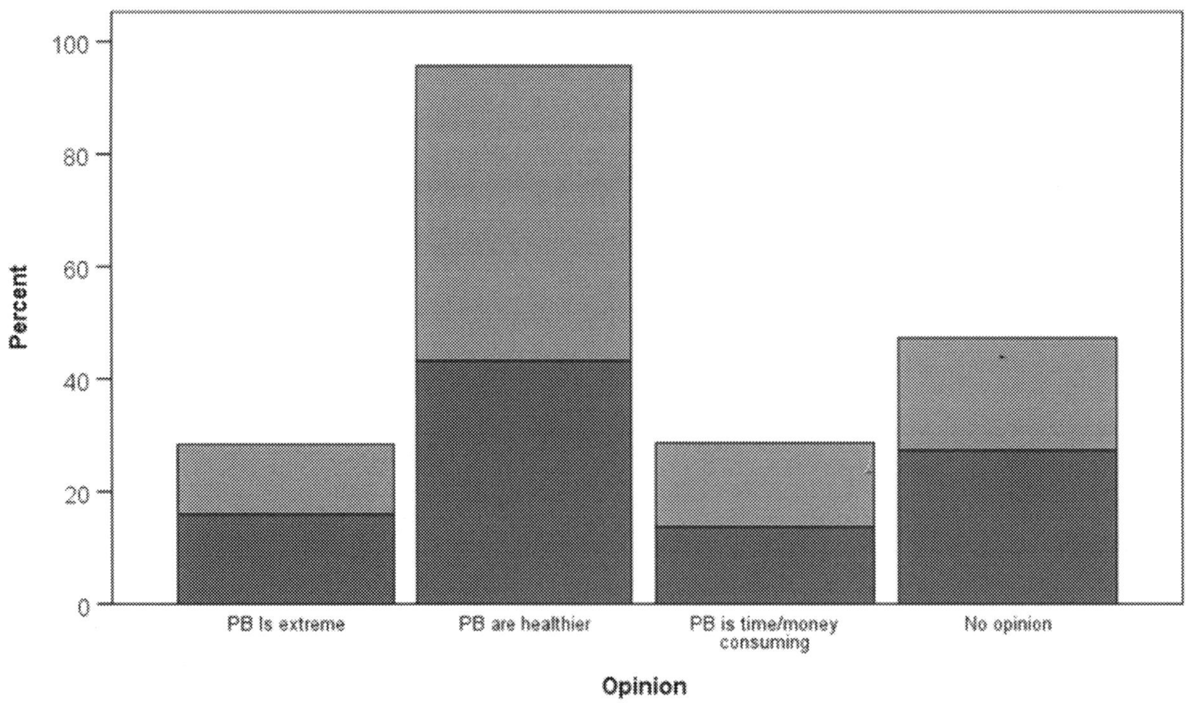

FIGURE 2.2 Participants' opinion regarding plant-based diets (unpublished results). The figure shows that 97% of the participants (both male and female) regarded a plant-based diet as being healthy.

respondents perceived plant-based foods to have the potential of lowering blood pressure, reducing the risk of cardiovascular diseases, contain less fat, and offer more nutrients compared to meats as well as are filled with micronutrients that are beneficial to human health. In our study, the majority of participants (males 44% and females 50%) believed that dietary fiber found in a plant-based diet has a beneficial impact on health. The most common statements that were associated with the perception of the health benefits of the dietary fiber included the following: "good for gut health," "aids digestion and help with bowel movement," "makes the consumer feel full and fewer calories will be consumed," "clears the internal system," and "prevents certain intestinal diseases."

The findings from our study were in line with a quantitative study that was carried out by Van Loo and co-workers on 2783 participants, which showed that consumer's perceived plant-based foods to be healthier (Van Loo et al., 2017). Contrastingly, about 6% of the participants who took part in our study believed that the fiber present in plant-based foods was not beneficial to health, which we attributed to a possible lack of knowledge regarding the benefits of fiber consumption.

A number of factors that include peer pressure may influence an individual's attitudes and perception on consuming plant-based foods. For instance, a recent study that was carried out to evaluate consumers' perceptions and acceptance of plant-based meat alternatives revealed that a number of consumers had a negative attitude toward the consumption of meat alternatives as compared to consuming meat products (Michel et al., 2021). An attempt to compare meat with corresponding meat alternatives showed that processed meat counterparts were perceived similarly to meat. Interestingly, the consumption of meat alternatives was perceived to be more appropriate in settings where an individual eats alone or with family and friends. The study concluded that in more formal settings and situations where a given level of peer pressure is likely to be present, one might not want to attract attention by eating meat alternatives or maybe afraid of being judged by their food decisions (Michel et al., 2021).

Several barriers (negative attitudes) have been identified that hinder the consumption of plant-based foods. For instance, a study by Bryant (2019) substantiated factors such as taste, convenience, not wanting to change one's current eating habits, the effect of social surroundings, and price to be the main concerns that hinder the adoption of plant-based diets (Bryant, 2019). The barriers to the consumption of plant-based foods were evaluated based on socio-demographic factors. The idea that plant-based foods are less tasty and ill nutritious (Lea et al., 2006) was shown to somewhat trouble men more than women (Michel et al., 2021). Poor nutrition was also an attribute that was attached to plant-based diets by individuals with low education, in addition to older age groups who seemed to be disinterested in changing their eating habits toward plant-based foods. Furthermore, there was some evidence that families with children also felt that meat was necessary for nutritional reasons (Kubberød et al., 2002).

The acceptance of plant-based foods is determined by several factors that pertain to the consumers and the food products. From a food product perspective, there are a number of parameters such as texture, flavor, and visual appearance that are considered to be of primary importance in the establishment of the consumer sensory perception and the hedonic response (Santagiuliana et al., 2019; Yvette et al., 2013). Whereas, from a consumer's perspective, factors such as physiological and psychological factors in addition to socially and culturally learned expectations impact the preference and/or adoption of plant-based foods. In cases where the consumers are not accustomed to the consumption of plant-based foods or where they are introduced to unfamiliar new products such as meat analogs, their expectations will mainly be based on other aspects, such as the visual appearance of the food product, properties of similar foods, descriptive information of the food product, as well as orthonasal olfactory cues (Burgess, 2016; Vidal et al., 2013). According to Santagiuliana et al. (2019), the multisensory perception of food, in general, does involve both exteroceptive (such as the visual appearance of the product) and interoceptive (such as somatosensory and gustatory perception) cues, which are stimulated prior to and during consumption of food (Santagiuliana et al., 2019). The acceptance or rejection of food products based on the confirmation or disconfirmation of expectations is mostly determined by the exteroceptive and/or interoceptive cues (Santagiuliana et al., 2019). To our knowledge, little is known about which factors contribute to the consumer's acceptance of plant-based foods. It is not yet clear how exteroceptive and interoceptive cues influence the expected and perceived sensory properties and liking of plant-based foods. We hypothesize that the expected sensory profile and liking of plant-based foods are mainly related to exteroceptive cues, such as the visual appearance and recognition of ingredients present in the product. Whereas for the acceptance of plant-based food products, both exteroceptive and interoceptive cues have to be taken into account.

One of the challenges that arise following the replacement of animal proteins with plant protein during food processing is the changes in the texture of the final product, and this has a direct impact on the sensorial properties of the food product. This, in turn, impacts the consumer perception and acceptance of the food product (Fiorentini et al., 2020). Several studies have been carried out to evaluate the sensory perception of products that were made using meat alternatives that are made mainly using plant derivatives (Elzerman et al., 2011; Fiorentini et al., 2020) or plant-derived milk (McClements et al.,

2019). Dairy milk contains a specific mix of proteins, fats, and carbohydrates that lead to its unique flavor profile, sensory attributes, and nutritional profile. Relying on plant-based ingredients to simulate the processed nuance of complex cheeses in terms of the flavor profile has been reported to be a major challenge when it comes to the application of plant-based milk in food processing (McClements et al., 2019). For instance, soybeans, which is one of the major plant-based ingredients used in plant-based cheese, have been shown in particular to exhibit flavor and mouthfeel challenges, specifically, beany flavor, and gritty mouthfeel (Short et al., 2021).

One of the primary barriers to the consumption of meat alternatives has been identified as unfamiliarity and lower sensory appeal (Hoek et al., 2011). Meat analogs have been characterized using more general attributes, such as "flavor," "taste," and "smell" as well as more specific sensory attributes such as "juiciness," "firmness/hardness," "fibrousness," "elasticity," "brittleness," "beany," "earthy," "chicken," "crumbly," "moist," and "tenderness" (Julian et al., 2021). However, how these attributes impact the preference and or the adoption of plant-based foods are not homogenous among consumers (Font-i-Furnols & Guerrero, 2014), and this makes it challenging to define these attributes and their intensities. Michel and co-workers showed that meat alternatives are often associated with "tofu," "vegan and vegetarian," as well as "disgust" (Michel et al., 2021). Soy protein—based meat analogs that were fortified with the cyanobacterium *Arthrospira platensis* (Spirulina) have also been designed, and the fortified products were analyzed by a trained sensory panel where different descriptors such as smell, color, texture, and taste attributes were used (Julian et al., 2021). The intensity of the odor, flavor, aftertaste, and color was shown to increase at a higher spirulina content. Additionally, the texture became less elastic, less fibrous, and had a softer structure with increasing spirulina concentration. Even though it has been suggested that it might be feasible to partially incorporate spirulina into a soy-based meat analog product (Julian et al., 2021), there still remain to be several technological and sensory challenges that need to be overcome in order to appease consumers. In order to effectively appease alternative meat consumers, Fiorentini and co-workers have proposed the identification and use of high-quality meat alternatives that mimic traditional meats rather than the use of meat analogs that have poor sensory attributes (Fiorentini et al., 2020). Their findings corroborate the work of Michel et al. (2021) who showed that consumers who are in favor of reducing their consumption of meat prefer meat alternatives that are similar to the meat rather than being offered meat alternatives that resemble meat.

2.3 General health and nutrition of plant-based foods and ingredients

The consumption of plant-based foods has been associated with a number of health benefits. Foods composed mainly of plants such as fruits and vegetables have been associated with a reduced risk of chronic noncommunicable diet-related diseases. However, their protective effect has been shown to be dependent on a number of factors such as the physiological state or the health status of an individual, the presence of other compounds within the food matrix owing to the complex nature of foods, and the degree of food processing, in addition to the quantity consumed (Fardet, 2017). Plant-based foods are shown to contain two critical nutrients, which are phytonutrients and fiber (Hever & Cronise, 2017). Plant-based foods are also a major source of the three main macronutrients that are present in foods, i.e., carbohydrates, proteins, and lipids. Additionally, plant-based foods are reported to possess a myriad of bioactive compounds or nutrients that have the potential protective effects toward pathophysiological processes in the human body (Fardet, 2017).

2.3.1 Carbohydrates fraction in plant-based foods

Plant-based foods are mainly composed of starch, simple sugars (glucose, fructose), disaccharides (sucrose), and oligosaccharides such as raffinose and stachyose. Whole grain cereals and legumes contain high amounts of complex carbohydrates that include both soluble and insoluble fiber, with pulses also providing oligosaccharides and slow-digesting starch (Ahnen et al., 2019). Starch, which is mainly found in grain-based foods and tubers, needs to be gelatinized before eating (Fardet, 2017). Slowly digestible starch has been reported to be more beneficial to health than rapidly digested starch, and this plays an important role in controlling glucose homeostasis in the case of diabetes (Miao et al., 2015). The digestibility of starch can be achieved in several ways such as (i) by limiting the degradation of starch at the molecular level; (ii) by entrapping starch within a fiber or protein network at a microscopic level such as in leguminous seeds (Noah et al., 1998) and pasta (Fardet et al., 1998); and (iii) by increasing the particle size of the food eaten at a macroscopic level (Fardet, 2015).

Simple sugars are generally found in fruits and, to a lesser extent, in vegetables. Simple sugars appear more slowly released from intact complex foods than from more unstructured foods such as purées. This difference is mainly attributed to the role played by the fiber network in trapping vegetable cells containing simple sugars, given that the cell wall is more disrupted in fruit purées or juices than in the natural, complex food (Fardet, 2017).

2.3.2 Protein content of plant-based foods

Protein plays an important role in the human body. Protein is regarded for its ability to assist in the management of weight, the increase in thermogenesis and satiety, enhancement of bone mineralization, and the increased high-density lipoprotein (HDL) cholesterol (Halkjær et al., 2011; Pasiakos et al., 2015). Legumes contain a relatively high protein level (Table 2.1), which makes them unique among plant-based protein providers (Ahnen et al., 2019). For instance, the boiled green lentils have been shown to contain approximately 18 g/100 g of protein (Table 2.1). Stir-fried lean component of beef, veal, lamb, and mutton on the other hand has been shown to contain about 23.2, 24.8, 21.9, and 21.5 g/100 g protein, respectively (Table 2.1). Protein from animal sources is in most cases higher than that of plant-based sources (Tables 2.1 and 2.2). Even though the protein content of the different meat is higher than the protein content of the legumes, the protein content in legumes is still considerably high, and this makes plant-based foods a good source of protein that contributes toward the improvement of human health and reduced mortality rates. An epidemiological study by Song et al. (2016) estimated that statistically, when 3% of animal proteins especially from red meat or eggs are replaced with plant protein, there would be a significant reduction in mortality rates. According to Medawar et al., the protein source itself may not be responsible for this beneficial effect but possibly the detrimental components found in meat such, e.g., heme-iron, nitrosamines, and antibiotics (Medawar et al., 2019). Contrastingly, a growing body of research also suggests that increasing the consumption of plant protein also results in an increased risk of the intake of antinutrients, such as tannins and trypsin inhibitors, that decrease pulse protein quality (Oomah et al., 2011). The antinutritional aspects of plant-based foods will be discussed in detail later on in this book chapter.

Plant-based protein sources have been shown to generally have a lower leucine content compared to animal-based protein sources (Berrazaga et al., 2019). Additionally, plant-based protein sources are reported to be deficient in some essential amino acids, such as lysine, which is deficient in cereals (Boye et al., 2010). Limiting an essential amino acid in the human body has an impact on the other amino acids being properly used for protein synthesis. As a result, the amino acids become deaminated and oxidized and consequently, irreversibly eliminated (Joint & Organization, 2007). There is

TABLE 2.1 Some of the nutritional composition of half a cup (approximately 100 g) common legumes (boiled without salt).

Legume	Protein (g)	Folate (μg)	Magnesium (mg)	Potassium (mg)	Phosphorus(mg)	Selenium (μg)
Black bean	15.2	256	120	611	241	3.8
Navy bean	15	255	96	708	262	5.3
Kidney bean	15.4	230	74	717	244	1.9
Pinto bean	15.4	294	86	746	35.9	10.6
Garbanzo bean	14.5	282	79	477	276	6.1
Green pea	8.6	101	62	434	187	3
Green lentil	17.9	358	71	356	356	5.5

Adopted from the work of Oomah, B. D., Caspar, F., Malcolmson, L. J., & Bellido, A. S. (2011). Phenolics and antioxidant activity of lentil and pea hulls. Food Research International, *44*(1), 436–441. https://doi.org/10.1016/j.foodres.2010.09.027.

TABLE 2.2 Average nutrient composition (per 100 g) of the stir-fried lean component of red meat.

Meat	Protein (g)	Folate (μg)	Magnesium (mg)	Potassium (mg)	Phosphorus (mg)	Selenium (μg)
Beef	23.2	ND	25	363	215	17
Veal	24.8	ND	26	362	260	<10
Lamb	21.9	ND	28	344	194	14
Mutton	21.5	ND	28	365	290	<10

Adopted from the work of Wisessing, A., & Choowongkomon, K. (2012). Amylase inhibitors of plants: Structures, functions and applications. *Functional Plant Science and Biotechnology*, *6*, 31–41. The study focused on both raw and cooked Australian red meat and the folate composition in cooked meat was not included thus, ND is used in this context to denote not determined.

also a possibility that limiting amino acids could influence the accretion of protein within the body (Berrazaga et al., 2019). The other issue that has been highlighted when it comes to proteins from plant sources is the digestibility of the proteins, which tends to be lower especially when eaten raw (Berrazaga et al., 2019). The lower digestibility of proteins from plant sources has been attributed to (i) the structural differences between animal and plant proteins. Plant proteins are reported to contain relatively more β-sheet structures and lower α-helixes compared to animal proteins, which makes proteins from plant sources more resistant to digestion in the digestive system (Carbonaro et al., 2012; Nguyen et al., 2015), and (ii) more fibers have been shown to be present in plant-based protein, which has a lower proteolytic digestibility (Pam Ismail et al., 2020). Oral processing has been used as a strategy to mitigate against the low digestibility of plant-based proteins especially in cases where plant-based foods are consumed raw. In vitro, oral processing was shown to result in increased protein digestibility, an indication that physical breakdown of the plant cell walls and starch is significant in increasing the digestibility and absorption of proteins from plant sources in the small intestine (Lü et al., 2004).

Proteins also play an important role in food processing due to their functional properties. Gel formation by proteins is of great significance in the generation of texture in food (Munialo et al., 2015). Proteins from animal sources are often used as structuring and texture agents in foods. However, in the recent past, proteins from plant-based sources have been used in food processing (Munialo, Martin, et al., 2014; Munialo et al., 2015; Munialo, van der Linden, & de Jongh, 2014). Food processing in most cases does involve the heating of the food products. Heating of globular proteins above their denaturation temperature results in partial unfolding of the proteins, which results in the exposure of interaction sites and resulting in intermolecular interactions that culminate in the clustering of protein aggregates to form a spatial gel network (Clark et al., 2001). The heating of proteins at a pH away from the isoelectric point (IEP) may result in aggregation of the proteins, and this results in the formation of fibrillar aggregates (fibrils) (Tang et al., 2012).

Fibrils have the potential of being used in low-caloric products due to their functionality, which ranges from gelling agents to stabilizers of foams and emulsions. Additionally, fibrils have been reported to have the ability to form entanglement networks that could increase bulk viscosity with relatively little protein, making them potentially economical ingredients for modifying texture in food products (Loveday et al., 2012). Munialo and co-workers showed the potential of pea proteins being assembled into fibrillar aggregates (Munialo, Martin, et al., 2014). In this study, fibrils were made at pH 2.0 followed by the characterization of the gelation properties of these fibrillar aggregates, and the results compared with fibrils made from soy proteins. Fibrils made using pea protein had a strength comparable (measured in terms of storage (elastic) modulus being higher than the loss (viscous) modulus) to that of soy protein-based fihad a s gels but were weaker than fibrillar gels made from whey protein (animal protein). The results from this study suggested that the fibril formation mechanism in plant storage proteins is generic as pea protein fibrils were comparable to soy protein fibrils. However, the gel strength of fibrils made from plant-based proteins was lower than that of animal-based proteins, and this shows some of the technological challenges that the food industry is faced with when it comes to the exchangeability of proteins and replacing animal-based proteins with plant-based proteins.

2.3.3 Lipids

Plant-based foods are a good source of lipids. The food structure features influence the rate at which the lipid fraction of a plant-based diet is digested. For instance, the release of lipids in almonds has been shown to be increased by chewing (Ellis et al., 2004). The proportion of lipid lost in the stool was reported to be higher after 10 chews compared to after 40 chews in humans (Cassady et al., 2009). Higher fat losses have also been found after eating whole peanut seeds compared with the consumption of peanut oil, butter, or flour, which proposes the possibility of reduced lipid availability (Traoret et al., 2008). These findings suggest that most oilseeds could exhibit a similar fate in terms of lipid digestion depending on food structure integrity even though little is known about the effect of plant-based food structure on lipid bioavailability in humans (Fardet, 2017).

Plant-based foods are also a good source of plant sterols. Plant sterols have been demonstrated to have the ability to decrease total serum and low-density lipoprotein cholesterol when sterol-rich foods providing 1.5–2 g of plant sterols per day are consumed (Miettinen et al., 1995). Additionally, experimental and epidemiological studies have suggested that dietary plant sterols may offer protection against several types of cancer (Piironen et al., 2003).

2.3.4 Phytonutrients and phytochemicals present in plant-based foods

Plant-based foods are associated with potentially positive health effects on chronic diseases. Several epidemiological surveys have shown that frequent ingestion of plant-based foods, which contain abundant phytochemicals has the potential of lowering the risk of the onset of a number of diseases (Murakami, 2013). Phytonutrients are reported to work synergistically in the reduction of inflammation and oxidation, and this provides protection from the initiation and progression of diseases (Teresa et al., 2020, pp. 1–22). The unique chemical modifications present in the structure of phytochemicals or phytonutrients make them: (i) excellent protein modulators, (ii) activators of intracellular signal cascade systems, and (iii) intercalating agents (Murakami, 2013).

The positive health effect of plant-based foods is mainly attributed to the presence of a variety of phytochemicals (such as minerals, vitamins, trace elements, carotenoids, polyphenols, and phytosterols) especially in unrefined and/or minimally processed plant-based foods (Fardet & Chardigny, 2013). There are four main classes of polyphenols present in plant-based foods. These are flavonoids, lignans, phenolic acids, and stilbenes. However, other polyphenolic compounds, such as curcumin, lignins, saponins, and phytosterols, among others, are also present in plant-based foods (Fardet, 2017). A few of these phytochemicals and their health benefits will be discussed in this book chapter.

Curcumin, which is the major yellow pigment in turmeric (*Curcuma longa,* Zingiberaceae), has been reported to possess versatile biological and physiological activities such as antineurodegenerative, anti-Alzheimer's disease, anti-inflammatory, antiobesity, and antioxidative activities (Murakami, 2013). Curcumin and curcuminoids are also reported to have the potential of lowering cholesterol levels in the blood, reducing the aggregation of platelets, inhibiting the proliferation of cancer cells, and increasing the flow of bile from the gallbladder, which improves digestion (Sundaram et al., 2007). The desirable preventive or putative therapeutic properties of curcumin have been attributed to its antioxidant and anti-inflammatory properties (Murakami, 2013).

Resveratrol, which is a phytoalexin that is commonly found in grapes, is classified as a stilbene-type polyphenol. Resveratrol has been shown to result in marked chemopreventive effects that include its ability to inhibit the three main stages of tumor, i.e., initiation, promotion, and progression (Jang et al., 1997). Resveratrol has also been hypothesized to be involved in the French paradox, where French individuals have been shown to have an extremely low incidence of cardiovascular diseases, despite their diet being high in saturated fat (Ferrières, 2004). Pietrocola et al. carried out research on the possible active components in red wine apart from resveratrol given that its concentration in red wine is rather too low for it to account alone for the French paradox. A selection of phenolic compounds that are found in red wine was explored, which included anthocyanins, monophenols, stilbenoids, glucosides, and flavonoids (mainly the catechin, epicatechin, myricetin, and quercetin). The research findings showed that all of those components were capable of stimulating autophagy, though with varying potencies (Pietrocola et al., 2012). Due to the complex and occasionally contradictory mechanisms of action, modulation of autophagy by resveratrol is now attracting the attention of many researchers.

2.3.5 Fiber present in plant-based foods

Fiber, which is generally composed of cellulose in combination with other components such as lignin, (Chand & Fahim, 2008) can be found in multiple varieties in all intact plant-based foods. Dietary fiber, which includes oligosaccharides and carbohydrate polymers that lack the ability to be hydrolyzed by endogenous enzymes that are present in the small intestines of humans, provides powerful protection of the gastrointestinal tract (McRorie & McKeown, 2017). Persons who consume a high amount of dietary fiber were reported to have a reduced risk of coronary heart disease (CHD), stroke, peripheral vascular disease, and overall mortality (Willett et al., 2019). Major risk factors for CHD, such as diabetes, hypertension, dyslipidaemia, and obesity, have also been reported to be less common in individuals that consume high levels of fiber (Anderson et al., 2009). Increasing the intake of dietary fiber has also been shown to result in the improvement of serum lipoprotein values, a significant reduction in blood pressure, improvement of blood glucose control for people with diabetes, as well as weight loss (Fardet, 2017). The soluble fiber was hypothesized to exert primary effects on serum cholesterol and low-density lipoprotein cholesterol values due to binding of bile acids in the small intestine and increasing their fecal excretion (McRorie & McKeown, 2017). Hypocholesterolemia due to the attenuation of cholesterol synthesis has been linked to fiber fermentation in the colon, which results in the production of the short-chain fatty acid propionate (Anderson et al., 2009). The intake of dietary fiber has been linked to up to a 40% reduction in the risk of colorectal cancer (Aune et al., 2011). Additionally, high fiber intake from legumes, grains, vegetables, and fruits is a prominent feature of plant-based foods and has been hypothesized to induce beneficial metabolic processes. These include: (i) the upregulation of carbohydrate fermentation and downregulation of protein fermentation; (ii) an improved gut hormonal-driven appetite regulation; and (iii) the prevention of chronic diseases such as obesity and type 2 diabetes mellitus, which occurs as a result of the digestion being slowed down and improved lipid control (Medawar et al., 2019).

2.3.6 Vitamins, minerals, and trace elements

Table 2.1 shows an example of minerals and trace elements that are found in legumes. The listed legumes were shown to contain higher levels of magnesium when cooked compared to cooked meat. The levels of potassium in some legumes are more than double the amount of potassium in lean red meat. Additionally, reasonably sufficient amounts of selenium and phosphorous are composed in legumes compared to lean red meat. These findings suggest that plant-based foods can be a good source of trace elements and vitamins that are vital for the optimal functioning of the human body. Plant-based foods are also reported to contain high levels of vitamin C and K and folate. However, a major criticism of plant-based foods is the increased risk of deficiency of specific micronutrients, especially vitamin B12, which is a mainly animal-derived nutrient. Vitamin B12 deficiency has been shown to be detrimental to neurological and cognitive health (Medawar et al., 2019). The use of supplements or fortified foods has been shown to be effective strategies in mitigating against vitamin B12 deficiencies in individuals who solely consume plant-based foods (Madry et al., 2012; Rizzo et al., 2016). Iron deficiency is another commonly assumed risk for plant-based dieters. A meta-analysis of 24 studies was shown to propose that serum ferritin levels were lower in vegetarians on average (Medawar et al., 2019). Thus, it is recommended that in order to sustain optimal serum ferritin levels, well-monitored supplementation strategies are needed (Medawar et al., 2019).

2.4 Antinutritional aspect of plant-based foods

Several antinutrients are present in plant-based foods. Table 2.3 shows a summary of some of the antinutrients present in plant-based foods. The highest concentrations of antinutrients are reported to be in grains, legumes, and nuts (Table 2.3), although a certain concentration of antinutrients can also be found in fruits, leaves, and roots of certain varieties of plants.

The major antinutrients in plant-based foods include tannins, vicine, convicine, lectins, phytates, and oxalates among others (Popova & Mihaylova, 2019). In most cases, antinutrients are a major concern in cases where an individual's diet is composed exclusively of uncooked plant foods. Popova and Mihaylova (2019) have shown that low levels of phytic acid, phenolic compounds, lectins, enzyme inhibitors, and saponins have the ability to reduce blood glucose and/or plasma cholesterol and triacylglycerol (Popova & Mihaylova, 2019).

2.4.1 Tannins

Tannins, which are oligomers of flavan-3-ols and flavan-3, 4-diols, are mainly concentrated in the bran fraction of legumes even though green tea, grapes, and chocolate are reported to also be rich in tannins (Lü et al., 2004). The antinutritional properties of tannins are exhibited when the digestion of various nutrients are impaired, and this prevents the body from absorbing beneficial bioavailable substances (Popova & Mihaylova, 2019). Tannins have been documented to have the potential ability to hinder iron absorption in the digestive tract given that tannins can bind with iron present in plant-based foods and this may render it unavailable for absorption (Delimont et al., 2017). Tannins have been reported to also have the

TABLE 2.3 Antinutrients in different foods.

Source	Amount	Type
Grains (barley, wheat, oat, rye, corn, millet, kamut, spelt, and sorghum)	35–270 mg/100 g 50–74 mg/g	Oxalates, phytic acid
Pseudo-grains (quinoa, amaranth, buckwheat, Teff, and wheat)	0.5–7.3 g/100 g 0.04–2.14 ppm	Phytic acid, lectins, saponins, goitrogens
Legumes (lentils, chick peas, peanuts, soya, beans)	8 mg/kg 386–714 mg/100 g 106–170 mg/100 g 1.8–18 mg/g 6.7 mg/100 g 2–200 mg/100 g	Oxalates, phytic acid, saponins, tannins, trypsin inhibitor, cyanide
Seeds (flaxseed, sesame, sunflower, poppy seed, pumpkin) nuts (such as almonds, Brazil nuts, cashew, hazelnut, macadamia, pignola, pistachio, walnuts, among others)	1–10.7 g/100 g 0.251 mg/mL 140–370 ppm 150–9400 mg/100 g 37–144 μg/g 40–490 mg/100g	Phytic acid, alpha-amylase inhibitor, cyanide, phytic acid, lectins, oxalates
Nightshades (eggplant, pepper, potato, tomato)	0.82–4.48 mg/100 g 0.19 mg/100 g 0.16–0.25 mg/100 g 1.6–10.5 mg/100 g	Phytic acid, tannins, saponins, cyanide
Tubers (carrot, manioc (or tapioca), Jerusalem artichoke, sweet potato, yam	0.4–2.3 mg/100 g 4.18–6.72 mg/100 g 0.06–0.08 mg/100 g	Oxalates, tannins, phytates

This data has been adopted from the work of Popova, A., & Mihaylova, D. (2019). Antinutrients in plant-based foods: A review. Open Biotechnology Journal, 13(1), 68–79. https://doi.org/10.2174/1874070701913010068.

ability to bind and shrink proteins. When tannins bind to proteins, this results in the formation of tannin–protein complexes. These complexes have been hypothesized to cause the inactivation of digestive enzymes in addition to the reduction in protein digestibility as a result of protein substrate and ionisable iron interaction Herbivores (1991). Excessive consumption of tannins rich foods can also result in the inactivation of the enzymes that are responsible for protein absorption (Popova & Mihaylova, 2019).

2.4.2 Lectins

Lectins are either proteins or glycoproteins of nonimmune origin, which are abundant in plants. Lectins are reported to have the ability to bind to either carbohydrates or glycoconjugates, without modifying them (Ambrosi et al., 2005; Sharon & Lis, 2003; Singh & Sai, 2012). Lectins have the ability to bypass the human defense system and traveling all over the body and may result in the causation of diseases such as Crohn's disease and colitis, which are the result of the surface of the small intestine breaking down (Yasuoka et al., 2003). Acne, inflammation, migraines, or joint pains can also be caused by the consumption of lectins (Singh & Sai, 2012). Lectins are usually found in the hull of grains, and as such, choosing white rice can lower the lectin intake. Heating plant sources in the process of cooking can also significantly lower the level of lectins in them (Zubcevic et al., 2016).

2.4.3 Oxalates

Oxalic acid can form soluble sodium and potassium soluble salts or insoluble iron, calcium, or magnesium salts or esters that are referred to as oxalates (Popova & Mihaylova, 2019). Oxalates are commonly found in leafy vegetable even though they can as well be synthesized in the body (Francesco, 2015). Examples of leafy vegetables that contain relatively high amounts of oxalates include cruciferous vegetables such as kale, broccoli, cauliflower, and radishes, as well as spinach, chard, parsley rhubarb, beets, and beans (Mamboleo, 2015). In people who are sensitive, minute quantities of oxalates can result in burning in the ears, eyes, and throat. Whereas, large amounts can result in nausea, muscle weakness, and abdominal pain. Additionally, calcium oxalate can have a damaging effect on human nutrition and health by accumulating kidney stones due to it preventing calcium from being absorbed in the body by binding with it (Popova & Mihaylova, 2019).

2.4.4 Saponins

Saponins are naturally occurring substances that have various biological effects. When cholesterol is present, saponins have the potential of exhibiting a strong hypocholesterolemic effect. Additionally, saponins can also lead to hypoglycemia (Barky Hussein et al., 2017), the development of a leaky gut, impairment of the uptake of minerals and vitamins in the gut, or protein digestion (Johnson et al., 1986). Saponins can also interact with the cholesterol group of erythrocyte membranes, which may lead to hemolysis (Popova & Mihaylova, 2019).

2.4.5 Phytates

Dietary phytate has been associated with certain health benefits, including blood glucose and lipid-lowering effects, anticancer activity, antioxidant properties, and anticalcification. The presence of phytates in a diet may have an impact on the solubility, digestibility, and functionality of proteins and carbohydrates in addition to impairing the bioavailability of minerals (Popova & Mihaylova, 2019). Phytase, which is a digestive enzyme, has the ability to unlock any phosphorus that is stored as phytic acid. However, in cases where phytase is absent, the presence of phytic acid can bind to minerals such as calcium, iron, magnesium, and zinc, which impedes their absorption (Golam Masum Akond et al., 2011). This may result in the formation of insoluble salts, which have a reduced ability to be absorbed by the gastrointestinal tract. Consequently, the bioavailability of minerals is lowered (Popova & Mihaylova, 2019). Additionally, phytates have been reported to inhibit digestive enzymes such as amylase, pepsin, and trypsin (Kumar et al., 2010).

2.4.6 Exorphins

Gliadins, which are the alcohol-soluble prolamins (proteins) of cereal grains, can be further degraded in the gastrointestinal tract to result in the collection of opioid-like polypeptides. The polypeptides are referred to as exorphins (Tatham & Shewry, 2008). A dietary overload of food-derived exorphins has been shown to affect behavioral traits such as memory and pain perception in addition to producing schizophrenia-like symptoms through their actions on central nervous system (Lister et al., 2015). Exorphins have also been reported to have the ability to increase the intestinal transit and gastric emptying (Lister et al., 2015).

2.4.7 Trypsin inhibitors

Trypsin inhibitors contribute to the substantial loss of trypsin and chymotrypsin in the gut, thus preventing the digestion of protein (Popova & Mihaylova, 2019). Excess synthesis of trypsin and an increased burden on sulfur-containing amino acids requirement of the human body are mainly due to the release of cholecystokinin triggered by trypsin inhibitors (Tibe & Amarteifio, 2007).

2.4.8 Alpha-amylase inhibitors

Several acquired disorders such as diabetic ketoacidosis, hyperosmolar coma, and hypoglycemia are associated with the uptake and metabolism of carbohydrates (Francesco, 2015). One of the commonly used strategies in treatment of disorders in carbohydrate uptake involve the inhibition of α-amylase (de Sales et al., 2012). Amylase inhibitors bind to α-amylases, and this makes them inactive (Wisessing & Choowongkomon, 2012). However, the instability of amylase inhibitor under gastrointestinal tract conditions in addition to being a very heat-liable constituent has been shown to result in its failure to be used as a starch blocker (Tysoe et al., 2016).

2.4.9 Protease inhibitors

Proteases are enzymes that are found in all cells and tissues and play a vital role in cell process regulation. Protease inhibitors that bind to their target proteins reversibly or irreversibly are commonly present in raw cereals and legumes (Popova & Mihaylova, 2019). Protease inhibitors' antinutrient activity has been associated with the inhibition of growth, poor food utilization, and pancreatic hypertrophy (Adeyemo & Onilude, 2013; Salunkhe & Chavan, 1989). High levels of protease inhibitors have been reported to lead to increased secretion of digestive enzymes by the pancreas (Logsdon & Ji, 2013).

2.5 The impact of antinutrients on human health

When it comes to antinutrients present in plant-based foods, it is important for consumers to focus on a number of factors such as: (i) the potential health benefits of these antinutrients; (ii) the dosage intake in order for one to find the balance between beneficial and hazardous effects of plant bioactives and antinutrients; and (iii) the chemical structure, time of exposure, and the interaction with other dietary components (Popova & Mihaylova, 2019). Additionally, one needs to be aware of the antinutritional factors that have negative effects or nonnutritive compounds versus those that have positive health effects. Antinutrients are valuable active ingredients that are used in the production of food and drinks. Low levels of lectins, phytic acid, phenolic compounds, enzyme inhibitors, and saponins have been reported to contribute to the reduction of blood glucose and/or plasma cholesterol and triacylglycerols. Saponins have been shown to effectively act in the maintenance of liver function and prevention of osteoporosis, in addition to the agglutination of platelets (Kao et al., 2008). A reduction in the risk of cancer has been associated with phenolic compounds from plant sources, such as phytic acid, protease inhibitors, lignans, saponins, and phytoestrogens (Popova & Mihaylova, 2019). Tannins, on the other hand, have been demonstrated to possess possible antiviral, antibacterial, and antiparasitic effects (Akiyama et al., 2001; Lü et al., 2004). These findings show the potentially safe health benefits of the antinutrients even though they lack nutritional value. On the contrary, several antinutrients, such as lignans and phytoestrogens, have been linked to the induction of infertility in humans (de Sales et al., 2012).

2.6 Summary

In view of sustainability, cost, and availability, plant-based foods are increasingly becoming a common part of the human diet. Plant-based foods are associated with a number of health and nutritional benefits such as reduction in obesity, cancers, as well as the risk for cardiovascular diseases. Plant-based foods also have considerable levels of macro and micronutrients in addition to the presence of several bioactive compounds such as phytochemicals, which are beneficial to human health if the processing applied to the foods is not too drastic. Bioavailability can impair the nutritional quality of plant-based foods.

Plant-based foods are characterized by the presence of a solid food structure due to the fiber network surrounding cells, and this can impair the digestibility and bioavailability of these foods. Additionally, the presence of antinutritional factors can alter the nutritional properties of plant-based foods. Plant-based foods also contain a number of antinutritive factors such as oxalates, which can interfere with the absorption of a number of nutrients within the human body especially when these foods are consumed raw. Thus, in order to develop more healthful plant-based foods, there is a need for individuals to plan to diversify the foods in their diet. Additionally, one needs to ensure that the food is correctly processed in a way that favors the maximum diversity and availability of protective components that are beneficial to human health.

References

Adeyemo, S. M., & Onilude, A. A. (2013). Enzymatic reduction of anti-nutritional factors in fermenting soybeans by *Lactobacillus plantarum* isolates from fermenting cereals. *Nigerian Food Journal*, 84–90. https://doi.org/10.1016/s0189-7241(15)30080-1

Ahnen, R. T., Jonnalagadda, S. S., & Slavin, J. L. (2019). Role of plant protein in nutrition, wellness, and health. *Nutrition Reviews, 77*(11), 735–747. https://doi.org/10.1093/nutrit/nuz028

Akiyama, H., Fujii, K., Yamasaki, O., Oono, T., & Iwatsuki, K. (2001). Antibacterial action of several tannins against *Staphylococcus aureus*. *Journal of Antimicrobial Chemotherapy, 48*(4), 487–491. https://doi.org/10.1093/jac/48.4.487

Ambrosi, M., Cameron, N. R., & Davis, B. G. (2005). Lectins: Tools for the molecular understanding of the glycocode. *Organic and Biomolecular Chemistry, 3*(9), 1593–1608. https://doi.org/10.1039/b414350g

Anderson, J. W., Baird, P., Davis, R. H., Ferreri, S., Knudtson, M., Koraym, A., Waters, V., & Williams, C. L. (2009). Health benefits of dietary fiber. *Nutrition Reviews, 67*(4), 188–205. https://doi.org/10.1111/j.1753-4887.2009.00189.x

Aune, D., Chan, D. S. M., Lau, R., Vieira, R., Greenwood, D. C., Kampman, E., & Norat, T. (2011). Dietary fibre, whole grains, and risk of colorectal cancer: Systematic review and dose-response meta-analysis of prospective studies. *BMJ (Online), 343*(7833), 1082. https://doi.org/10.1136/bmj.d6617

Barky Hussein, S. A., Alm-Eldeen, A. A., Hafez, Y. A., & Mohamed, T. M. (2017). Saponins and their potential role in diabetes mellitus. *Diabetes Management, 7*(1), 148–158.

Berrazaga, I., Micard, V., Gueugneau, M., & Walrand, S. (2019). The role of the anabolic properties of plant-versus animal-based protein sources in supporting muscle mass maintenance: A critical review. *Nutrients, 11*(8). https://doi.org/10.3390/nu11081825

Boye, J., Zare, F., & Pletch, A. (2010). Pulse proteins: Processing, characterization, functional properties and applications in food and feed. *Food Research International, 43*(2), 414–431. https://doi.org/10.1016/j.foodres.2009.09.003

Broekema, R., Tyszler, M., Van 'T Veer, P., Kok, F. J., Martin, A., Lluch, A., & Blonk, H. T. J. (2020). Future-proof and sustainable healthy diets based on current eating patterns in The Netherlands. *American Journal of Clinical Nutrition, 112*(5), 1338–1347. https://doi.org/10.1093/ajcn/nqaa217

Bryant, C. J. (2019). We can't keep meating like this: Attitudes towards vegetarian and vegan diets in the United Kingdom. *Sustainability, 11*(23).

Burgess, P. (2016). Consumers' mindset: Expectations, experience, and satisfaction. In *Integrating the packaging and product experience in food and beverages: a road-map to consumer satisfaction* (pp. 161–181). Elsevier Inc. https://doi.org/10.1016/B978-0-08-100356-5.00008-5

Carbonaro, M., Maselli, P., & Nucara, A. (2012). Relationship between digestibility and secondary structure of raw and thermally treated legume proteins: A Fourier transform infrared (FT-IR) spectroscopic study. *Amino Acids, 43*(2), 911–921. https://doi.org/10.1007/s00726-011-1151-4

Carvalho, S. M. P., & Vasconcelos, M. W. (2013). Producing more with less: Strategies and novel technologies for plant-based food biofortification. *Food Research International, 54*(1), 961–971. https://doi.org/10.1016/j.foodres.2012.12.021

Cassady, B. A., Hollis, J. H., Fulford, A. D., Considine, R. V., & Mattes, R. D. (2009). Mastication of almonds: Effects of lipid bioaccessibility, appetite, and hormone response. *American Journal of Clinical Nutrition, 89*(3), 794–800. https://doi.org/10.3945/ajcn.2008.26669

Chand, N., & Fahim, M. (Eds.). (2008). *Woodhead publishing series in composites science and engineering, tribology of natural fiber polymer composites*. https://doi.org/10.1533/9781845695057.1

Clark, A. H., Kavanagh, G. M., & Ross-Murphy, S. B. (2001). Globular protein gelation - Theory and experiment. *Food Hydrocolloids, 15*(4–6), 383–400. https://doi.org/10.1016/S0268-005X(01)00042-X

Crosby, K. M., Jifon, J., & Leskovar, D. (2008). Agronomy and the nutritional quality of vegetables. In *Improving the health-promoting properties of fruit and vegetable products* (pp. 392–411). Elsevier Ltd. https://doi.org/10.1533/9781845694289.4.392

De Sales, P. M., de Souza, P. M., Simeoni, L. A., Magalhães, P.d. O., & Silveira, D. (2012). α-Amylase inhibitors: A review of raw material and isolated compounds from plant source. *Journal of Pharmacy and Pharmaceutical Sciences, 15*(1), 141–183. https://doi.org/10.18433/j35s3k

Delimont, N. M., Haub, M. D., & Lindshield, B. L. (2017). The impact of tannin consumption on iron bioavailability and status: A narrative review. *Current Developments in Nutrition, 1*(2), 1–12. https://doi.org/10.3945/cdn.116.000042

Ellis, P. R., Kendall, C. W. C., Ren, Y., Parker, C., Pacy, J. F., Waldron, K. W., & Jenkins, D. J. A. (2004). Role of cell walls in the bioaccessibility of lipids in almond seeds. *American Journal of Clinical Nutrition, 80*(3), 604–613. https://doi.org/10.1093/ajcn/80.3.604

Elzerman, J. E., Hoek, A. C., van Boekel, M. A. J. S., & Luning, P. A. (2011). Consumer acceptance and appropriateness of meat substitutes in a meal context. *Food Quality and Preference, 22*(3), 233–240. https://doi.org/10.1016/j.foodqual.2010.10.006

Fardet, A. (2015). A shift toward a new holistic paradigm will help to preserve and better process grain products' food structure for improving their health effects. *Food and Function, 6*(2), 363–382. https://doi.org/10.1039/c4fo00477a

Fardet, A. (2017). New concepts and paradigms for the protective effects of plant-based food components in relation to food complexity. In *Vegetarian and plant-based diets in health and disease prevention* (pp. 293–312). Elsevier Inc. https://doi.org/10.1016/B978-0-12-803968-7.00016-2

Fardet, A., & Chardigny, J. M. (2013). Plant-based foods as a source of lipotropes for human nutrition: A survey of in vivo studies. *Critical Reviews in Food Science and Nutrition, 53*(6), 535–590. https://doi.org/10.1080/10408398.2010.549596

Fardet, A., Hoebler, C., Baldwin, P. M., Bouchet, B., Gallant, D. J., & Barry, J. L. (1998). Involvement of the protein network in the in vitro degradation of starch from spaghetti and lasagne: A microscopic and enzymic study. *Journal of Cereal Science, 27*(2), 133–145. https://doi.org/10.1006/jcrs.1997.0157

Ferrières, J. (2004). The French paradox: Lessons for other countries. *Heart, 90*(1), 107–111. https://doi.org/10.1136/heart.90.1.107

Fiorentini, M., Kinchla, A. J., & Nolden, A. A. (2020). Role of sensory evaluation in consumer acceptance of plant-based meat analogs and meat extenders: A scoping review. *Foods, 9*(9). https://doi.org/10.3390/foods9091334

Font-i-Furnols, M., & Guerrero, L. (2014). Consumer preference, behavior and perception about meat and meat products: An overview. *Meat Science, 98*(3), 361–371. https://doi.org/10.1016/j.meatsci.2014.06.025

Francesco, T. (2015). Clinical and diagnostic aspects of gluten related disorders. *World Journal of Clinical Cases*, 275. https://doi.org/10.12998/wjcc.v3.i3.275

Golam Masum Akond, A. S. M., Crawford, H., Berthold, J., Talukder, Z. I., & Hossain, K. (2011). Minerals (Zn, Fe, Ca and Mg) and antinutrient (Phytic Acid) constituents in common bean. *American Journal of Food Technology, 6*(3), 235–243. https://doi.org/10.3923/ajft.2011.235.243

Halkjær, J., Olsen, A., Overvad, K., Jakobsen, M. U., Boeing, H., Buijsse, B., Palli, D., Tognon, G., Du, H., Van Der A, D. L., Forouhi, N. G., Wareham, N. J., Feskens, E. J. M., Sørensen, T. I. A., & Tjønneland, A. (2011). Intake of total, animal and plant protein and subsequent changes in weight or waist circumference in European men and women: The Diogenes project. *International Journal of Obesity, 35*(8), 1104–1113. https://doi.org/10.1038/ijo.2010.254

Herbivores. (1991). In *Herbivores: Their interactions with secondary plant metabolites, 1* pp. 355–388).

Hever, J., & Cronise, R. J. (2017). Plant-based nutrition for healthcare professionals: Implementing diet as a primary modality in the prevention and treatment of chronic disease. *Journal of Geriatric Cardiology, 14*(5), 355–368. https://doi.org/10.11909/j.issn.1671-5411.2017.05.012

Hoek, A. C., Luning, P. A., Weijzen, P., Engels, W., Kok, F. J., & de Graaf, C. (2011). Replacement of meat by meat substitutes. A survey on person- and product-related factors in consumer acceptance. *Appetite, 56*(3), 662–673. https://doi.org/10.1016/j.appet.2011.02.001

Jang, M., Cai, L., Udeani, G. O., Slowing, K. V., Thomas, C. F., Beecher, C. W. W., Fong, H. H. S., Farnsworth, N. R., Kinghorn, A. D., Mehta, R. G., Moon, R. C., & Pezzuto, J. M. (1997). Cancer chemopreventive activity of resveratrol, a natural product derived from grapes. *Science, 275*(5297), 218–220. https://doi.org/10.1126/science.275.5297.218

Johnson, I. T., Gee, J. M., Price, K., Curl, C., & Fenwick, G. R. (1986). Influence of saponins on gut permeability and active nutrient transport in vitro. *Journal of Nutrition, 116*(11), 2270–2277. https://doi.org/10.1093/jn/116.11.2270

Joint, F., & Organization, W. H.. (2007). *Protein and amino acid requirements in human nutrition: Report of a joint FAO/WHO/UNU expert consultation*. World Health Organization.

Julian, M. D., Jochen, W., Kinchla, Nolden, & Lutz, G. (2021). Methods for testing the quality attributes of plant-based foods: Meat- and processed-meat analogs. *Foods, 10*(2), 260. https://doi.org/10.3390/foods10020260

Kao, T. H., Huang, S. C., Inbaraj, B. S., & Chen, B. H. (2008). Determination of flavonoids and saponins in *Gynostemma pentaphyllum* (Thunb.) Makino by liquid chromatography-mass spectrometry. *Analytica Chimica Acta, 626*(2), 200–211. https://doi.org/10.1016/j.aca.2008.07.049

Kubberød, E., Ueland, Ø., Tronstad, A., & Risvik, E. (2002). Attitudes towards meat and meat-eating among adolescents in Norway: A qualitative study. *Appetite, 38*(1), 53–62. https://doi.org/10.1006/appe.2002.0458

Kumar, V., Sinha, A. K., Makkar, H. P. S., & Becker, K. (2010). Dietary roles of phytate and phytase in human nutrition: A review. *Food Chemistry, 120*(4), 945–959. https://doi.org/10.1016/j.foodchem.2009.11.052

Lea, E. J., Crawford, D., & Worsley, A. (2006). Public views of the benefits and barriers to the consumption of a plant-based diet. *European Journal of Clinical Nutrition, 60*(7), 828–837. https://doi.org/10.1038/sj.ejcn.1602387

Lister, J., Fletcher, P. J., Nobrega, J. N., & Remington, G. (2015). Behavioral effects of food-derived opioid-like peptides in rodents: Implications for schizophrenia? *Pharmacology Biochemistry and Behavior, 134*, 70–78. https://doi.org/10.1016/j.pbb.2015.01.020

Logsdon, C. D., & Ji, B. (2013). The role of protein synthesis and digestive enzymes in acinar cell injury. *Nature Reviews Gastroenterology and Hepatology, 10*(6), 362–370. https://doi.org/10.1038/nrgastro.2013.36

Loveday, S. M., Su, J., Rao, M. A., Anema, S. G., & Singh, H. (2012). Whey protein nanofibrils: Kinetic, rheological and morphological effects of group IA and IIA cations. *International Dairy Journal, 26*(2), 133–140. https://doi.org/10.1016/j.idairyj.2012.03.001

Lü, L., Liu, S. W., Jiang, S. B., & Wu, S. G. (2004). Tannin inhibits HIV-1 entry by targeting gp41. *Acta Pharmacologica Sinica, 25*(2), 213–218.

Madry, E., Lisowska, A., Grebowiec, P., & Walkowiak, J. (2012). The impact of vegan diet on B-12 status in healthy omnivores: Five-year prospective study. *Acta Scientiarum Polonorum, Technologia Alimentaria, 11*(2), 209–212. http://www.food.actapol.net/pub/11_2_2012.pdf.

Mamboleo, T. F. (2015). *Nutrients and antinutritional factors at different maturity stages of selected indigenous African green leafy vegetables*.

McClements, D. J., Newman, E., & McClements, I. F. (2019). Plant-based milks: A review of the science underpinning their design, fabrication, and performance. *Comprehensive Reviews in Food Science and Food Safety, 18*(6), 2047–2067. https://doi.org/10.1111/1541-4337.12505

McRorie, J. W., & McKeown, N. M. (2017). Understanding the physics of functional fibers in the gastrointestinal tract: An evidence-based approach to resolving enduring misconceptions about insoluble and soluble fiber. *Journal of the Academy of Nutrition and Dietetics, 117*(2), 251–264. https://doi.org/10.1016/j.jand.2016.09.021

Medawar, E., Huhn, S., Villringer, A., & Veronica Witte, A. (2019). The effects of plant-based diets on the body and the brain: A systematic review. *Translational Psychiatry, 9*(1). https://doi.org/10.1038/s41398-019-0552-0

Miao, M., Jiang, B., Cui, S. W., Zhang, T., & Jin, Z. (2015). Slowly digestible starch—a review. *Critical Reviews in Food Science and Nutrition, 55*(12), 1642–1657. https://doi.org/10.1080/10408398.2012.704434

Michel, F., Hartmann, C., & Siegrist, M. (2021). Consumers' associations, perceptions and acceptance of meat and plant-based meat alternatives. *Food Quality and Preference*, 87. https://doi.org/10.1016/j.foodqual.2020.104063

Miettinen, T. A., Puska, P., Gylling, H., Vanhanen, H., & Vartiainen, E. (1995). Reduction of serum cholesterol with sitostanol-ester margarine in a mildly hypercholesterolemic population. *New England Journal of Medicine, 333*(20), 1308–1312. https://doi.org/10.1056/NEJM199511163332002

Munialo, C. D., Euston, S. R., & de Jongh, H. H. J. (2018). Protein gels. In *Proteins in food processing* (2nd ed., pp. 501–521). Elsevier Inc. https://doi.org/10.1016/B978-0-08-100722-8.00020-6

Munialo, C. D., Martin, A. H., Van Der Linden, E., & De Jongh, H. H. J. (2014). Fibril formation from pea protein and subsequent gel formation. *Journal of Agricultural and Food Chemistry, 62*(11), 2418–2427. https://doi.org/10.1021/jf4055215

Munialo, C. D., van der Linden, E., Ako, K., & de Jongh, H. H. J. (2015). Quantitative analysis of the network structure that underlines the transitioning in mechanical responses of pea protein gels. *Food Hydrocolloids, 49*, 104–117. https://doi.org/10.1016/j.foodhyd.2015.03.018

Munialo, C. D., van der Linden, E., & de Jongh, H. H. J. (2014). The ability to store energy in pea protein gels is set by network dimensions smaller than 50nm. *Food Research International, 64*, 482–491. https://doi.org/10.1016/j.foodres.2014.07.038

Murakami, A. (2013). Modulation of protein quality control systems by food phytochemicals. *Journal of Clinical Biochemistry and Nutrition, 52*(3), 215–227. https://doi.org/10.3164/jcbn.12-126

Nguyen, T. T. P., Bhandari, B., Cichero, J., & Prakash, S. (2015). Gastrointestinal digestion of dairy and soy proteins in infant formulas: An in vitro study. *Food Research International, 76*, 348–358. https://doi.org/10.1016/j.foodres.2015.07.030

Noah, L., Guillon, F., Bouchet, B., Buléon, A., Molis, C., Gratas, M., & Champ, M. (1998). Digestion of carbohydrate from white beans (*Phaseolus vulgaris* L.) in healthy humans. *Journal of Nutrition, 128*(6), 977–985. https://doi.org/10.1093/jn/128.6.977

Oomah, B. D., Caspar, F., Malcolmson, L. J., & Bellido, A. S. (2011). Phenolics and antioxidant activity of lentil and pea hulls. *Food Research International, 44*(1), 436–441. https://doi.org/10.1016/j.foodres.2010.09.027

Pam Ismail, B., Senaratne-Lenagala, L., Stube, A., & Brackenridge, A. (2020). Protein demand: Review of plant and animal proteins used in alternative protein product development and production. *Animal Frontiers, 10*(4), 53–63. https://doi.org/10.1093/af/vfaa040

Pasiakos, S. M., Lieberman, H. R., & Fulgoni, V. L. (2015). Higher-protein diets are associated with higher HDL cholesterol and lower BMI and waist circumference in US adults. *Journal of Nutrition, 145*(3), 605–614. https://doi.org/10.3945/jn.114.205203

Pietrocola, F., Mariño, G., Lissa, D., Vacchelli, E., Malik, S. A., Niso-Santano, M., Zamzami, N., Galluzzi, L., Maiuri, M. C., & Kroemer, G. (2012). Pro-autophagic polyphenols reduce the acetylation of cytoplasmic proteins. *Cell Cycle, 11*(20), 3851–3860. https://doi.org/10.4161/cc.22027

Piironen, V., Toivo, J., Puupponen-Pimiä, R., & Lampi, A. M. (2003). Plant sterols in vegetables, fruits and berries. *Journal of the Science of Food and Agriculture, 83*(4), 330–337. https://doi.org/10.1002/jsfa.1316

Popova, A., & Mihaylova, D. (2019). Antinutrients in plant-based foods: A review. *Open Biotechnology Journal, 13*(1), 68–79. https://doi.org/10.2174/1874070701913010068

Rizzo, G., Laganà, A. S., Rapisarda, A. M. C., La Ferrera, G. M. G., Buscema, M., Rossetti, P., Nigro, A., Muscia, V., Valenti, G., Sapia, F., Sarpietro, G., Zigarelli, M., & Vitale, S. G. (2016). Vitamin B12 among vegetarians: Status, assessment and supplementation. *Nutrients, 8*(12). https://doi.org/10.3390/nu8120767

Salunkhe, D. K., & Chavan, J. (1989). *Dietary tannins: Consequences and remedies.*

Santagiuliana, M., Bhaskaran, V., Scholten, E., Piqueras-Fiszman, B., & Stieger, M. (2019). Don't judge new foods by their appearance! How visual and oral sensory cues affect sensory perception and liking of novel, heterogeneous foods. *Food Quality and Preference, 77*, 64–77. https://doi.org/10.1016/j.foodqual.2019.05.005

Saunders, C. M., Guenther, Meike, Tait, P. R., Kaye-Blake, William, Saunders, Miller, Sini, Abell, & Walter, L. (2011). Consumer attitudes towards sustainability attributes on food labels in the UK and Japan. In *85th Annual Conference.*

Sharon, N., & Lis, H. (2003). *Lectins.*

Short, E. C., Kinchla, A. J., & Nolden, A. A. (2021). *Plant-based cheeses: A systematic review of sensory evaluation studies and strategies to increase consumer acceptance, 10.* https://doi.org/10.3390/foods10040725

Singh, H., & Sai, P. (2012). Insight of lectins-A review. *International Journal of Scientific and Engineering Research.*

Song, M., Fung, T. T., Hu, F. B., Willett, W. C., Longo, V. D., Chan, A. T., & Giovannucci, E. L. (2016). Association of animal and plant protein intake with all-cause and cause-specific mortality. *JAMA Internal Medicine, 176*(10), 1453–1463. https://doi.org/10.1001/jamainternmed.2016.4182

Sundaram, C., Malani, N., & Ichikawa, H. (2007). Curcumin: The Indian solid Gold. In *The molecular targets and therapeutic uses of curcumin in health and disease* (pp. 1–75).

Tang, C. H., Wang, S. S., & Huang, Q. (2012). Improvement of heat-induced fibril assembly of soy β-conglycinin (7S Globulins) at pH 2.0 through electrostatic screening. *Food Research International, 46*(1), 229–236. https://doi.org/10.1016/j.foodres.2011.11.030

Tatham, A. S., & Shewry, P. R. (2008). Allergens to wheat and related cereals. *Clinical and Experimental Allergy, 38*(11), 1712–1726. https://doi.org/10.1111/j.1365-2222.2008.03101.x

Teresa, P., Davide, B., Antonella, C., & Rosaria, P. M. (2020). *Phytonutrients: Structure-function relationship, health benefits, stability, and fate during processing.* Elsevier BV. https://doi.org/10.1016/b978-0-12-815354-3.00002-2

Tibe, O., & Amarteifio, J. (2007). Trypsin inhibitor activity and condensed tannin content in Bambara groundnut (*Vigna subterranea* (L.) Verdc) grown in Southern Africa. *Journal of Applied Sciences and Environmental Management*, 2.

Traoret, C. J., Lokko, P., Cruz, A. C. R. F., Oliveira, C. G., Costa, N. M. B., Bressan, J., Alfenas, R. C. G., & Mattes, R. D. (2008). Peanut digestion and energy balance. *International Journal of Obesity, 32*(2), 322–328. https://doi.org/10.1038/sj.ijo.0803735

Tysoe, C., Williams, L. K., Keyzers, R., Nguyen, N. T., Tarling, C., Wicki, J., Goddard-Borger, E. D., Aguda, A. H., Perry, S., Foster, L. J., Andersen, R. J., Brayer, G. D., & Withers, S. G. (2016). Potent human alpha;-amylase inhibition by the β-defensin-like protein helianthamide. *ACS Central Science, 2*(3), 154–161. https://doi.org/10.1021/acscentsci.5b00399

Van Loo, E. J., Hoefkens, C., & Verbeke, W. (2017). Healthy, sustainable and plant-based eating: Perceived (mis)match and involvement-based consumer segments as targets for future policy. *Food Policy, 69*, 46–57. https://doi.org/10.1016/j.foodpol.2017.03.001

Vidal, L., Barreiro, C., Gómez, B., Ares, G., & Giménez, A. (2013). Influence of information on consumers' evaluations using check-all-that-apply questions and sorting: A case study with milk desserts. *Journal of Sensory Studies, 28*(2), 125–137. https://doi.org/10.1111/joss.12030

Willett, W., Rockström, J., Loken, B., Springmann, M., Lang, T., Vermeulen, S., Garnett, T., Tilman, D., DeClerck, F., Wood, A., Jonell, M., Clark, M., Gordon, L. J., Fanzo, J., Hawkes, C., Zurayk, R., Rivera, J. A., De Vries, W., Majele Sibanda, L., … Murray, C. J. L. (2019). Food in the anthropocene: The EAT–Lancet Commission on healthy diets from sustainable food systems. *The Lancet, 393*(10170), 447–492. https://doi.org/10.1016/S0140-6736(18)31788-4

Wisessing, A., & Choowongkomon, K. (2012). Amylase inhibitors of plants: Structures, functions and applications. *Functional Plant Science and Biotechnology, 6*, 31–41.

Yasuoka, T., Sasaki, M., Fukunaga, T., Tsujikawa, T., Fujiyama, Y., Kushima, R., & Goodlad, R. A. (2003). The effects of lectins on indomethacin-induced small intestinal ulceration. *International Journal of Experimental Pathology, 84*(5), 231–237. https://doi.org/10.1111/j.1365-2613.2003.00360.x

Yvette, P., Hicran, K., & Allen, F. E. (2013). Food structure: Roles of mechanical properties and oral processing in determining sensory texture of soft materials. *Current Opinion in Colloid & Interface Science*, 324–333. https://doi.org/10.1016/j.cocis.2013.03.009

Zubcevic, N., Damir, S., Focak, M., & Rukavina, D. (2016). Efekti biljnih lektina na aglutinaciju eritrocita kod ljudi. *Serbian Journal of Experimental and Clinical Research, 17*(3), 207–213. https://doi.org/10.1515/SJECR-2016-0031

Section I

Plant-based proteins

Chapter 3

Sustainable plant-based protein sources and their extraction

Ane Aldalur[1,2], Bhanu Devnani[1,2], Lydia Ong[1,2] and Sally L. Gras[1,2]

[1]*Dairy Innovation Hub, Department of Chemical Engineering, The University of Melbourne, Parkville, VIC, Australia;* [2]*Bio21 Molecular Science and Biotechnology Institute, The University of Melbourne, Parkville, VIC, Australia*

3.1 Introduction

Protein consumption has been growing worldwide, with an increase from 26 to 29 kg/year per capita occurring between 2000 and 2018 (FAOSTAT, 2021), and demand is expected to further increase (FIAL, 2019). Plant proteins account for ~60% of total protein intake, but the proportion of plant-based foods differs significantly between developed and developing countries. Developing counties sourced most of their protein from plants in 2018 (63%), while developed countries sourced only 39% of protein from plant sources (FAOSTAT, 2021). Interest in plant proteins is increasing in developed countries, however, with up to 29% of consumers seeking to raise their intake of plant-based protein (GlobalData, 2020). Increased demand for plant proteins is being driven by a greater awareness of food sustainability, an interest in alternative diets, animal welfare, and the health benefits of increased plant intake (Jeske et al., 2018).

Currently the most popular plant protein source is soy, but other plant materials such as pea, wheat, oat, almond, and rapeseed are increasing in popularity (Mäkinen, Wanhalinna, et al., 2016; Sha & Xiong, 2020). Interest is also growing in protein extraction from by-products or residues produced during the harvest or primary processing of plant materials, which have the potential to generate a more environmentally and economically sustainable agroindustry (Contreras et al., 2019).

The processes available to extract proteins from plant materials and produce protein concentrates (≥60% protein purity) or isolates (≥90% protein purity) are illustrated in Fig. 3.1. Plant materials are first preprocessed to remove specific components or to weaken the cellular structure; they are then milled to reduce particle size. The ground material is then further processed by dry or wet extraction techniques to obtain protein products. The most popular commercial method for protein extraction from plant materials is alkaline extraction, followed by isoelectric precipitation (GEA, 2020). This process can produce high-protein-purity isolates but is energy and water intensive, and the functionality of proteins extracted is also generally poor (Berghout et al., 2015; Nasrabadi et al., 2021). In recent years, many authors have explored novel protein extraction techniques and examined the environmental impact of these processes (Du et al., 2020; Yue, Zhu, et al., 2021; Zhang et al., 2018).

The functional properties of plant protein concentrates and isolates produced by the methods in Fig. 3.1 can be utilized in several ways. Plant proteins are particularly popular as alternatives to animal proteins in dairy- and meat-like products (Sha & Xiong, 2020; Vogelsang-O'Dwyer et al., 2021). For instance, pea protein isolate has been utilized as the main dry ingredient to formulate plant-based hamburgers (Beyond Meat) and milk (Australia's Own). Plant protein isolates and concentrates from soy, wheat, and lupin have also been used to increase the protein content of breads (Herman Brot), muesli bars (Carman's), and breakfast cereals (Kellogg's). Sports nutrition is another niche market where high-protein-plant-based products, like protein powders (Evolve Sports Nutrition), have been commercialized. Producing plant-based products, especially dairy and meat analogs, that meet consumer expectations, however, remains a major technical challenge and further process knowledge and process optimization is still required.

In this chapter, we provide an overview of the protein extraction processes studied to date, with particular emphasis on sustainability. Preprocessing steps are considered, together with dry and wet extraction. Methods that can improve protein functionality and novel hybrid dry and wet extraction methods are also examined.

Engineering Plant-Based Food Systems. https://doi.org/10.1016/B978-0-323-89842-3.00004-X

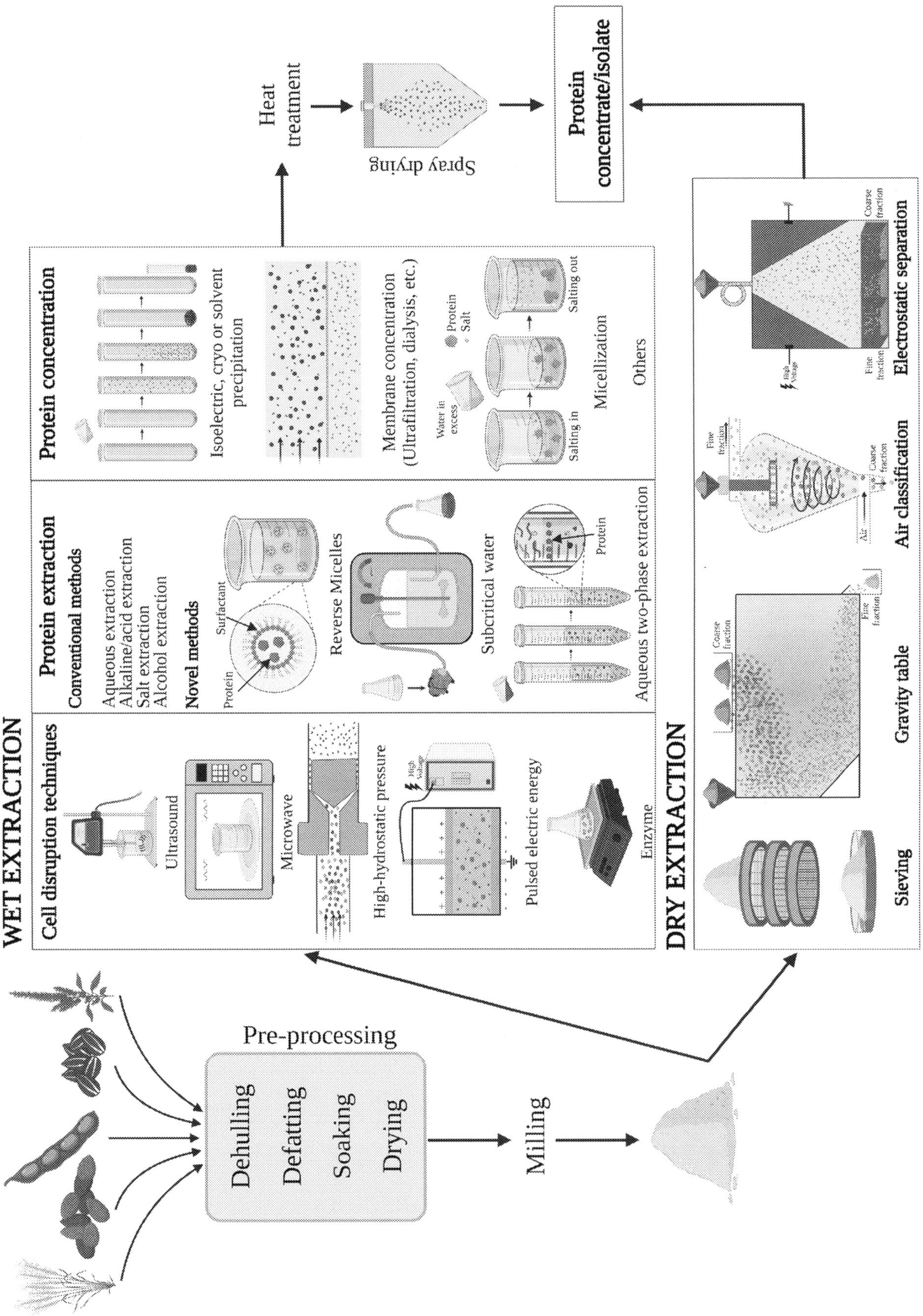

FIGURE 3.1 Schematic dry and wet protein extraction processes. Illustrated using the drawing software from Biorender.com.

3.2 Preprocessing steps and milling

The preprocessing steps involved in protein extraction include dehulling, defatting, soaking, and drying (Fig. 3.1). These steps are employed to initially remove specific components from plant materials, such as oils or seed coats. They can also be used to weaken plant cell structures by the addition of water or temperature treatment prior to disruption by mechanical methods, such as milling. The main aim is to facilitate protein extraction and achieve a higher protein yield, purity, and/or functionality.

Knowledge of plant composition and protein location is essential for the selection of individual preprocessing steps, together with milling and protein extraction steps and helps to inform the design of a purification strategy. Five plant materials are examined in the following sections: legumes, cereals, nuts, oilseeds, and pseudocereals, and these are illustrated in Fig. 3.2. These are collectively referred to here as seeds, while the term oilseed is utilized for sunflower, rapeseed, and similar plant sources.

3.2.1 The structure, microstructure, and composition of plant materials used for protein extraction

The schematic structures, microstructures, and compositions of a variety of plant protein sources including legumes, cereals, nuts, oilseeds, and pseudocereals are illustrated in Fig. 3.2. The structure of these plant materials varies greatly

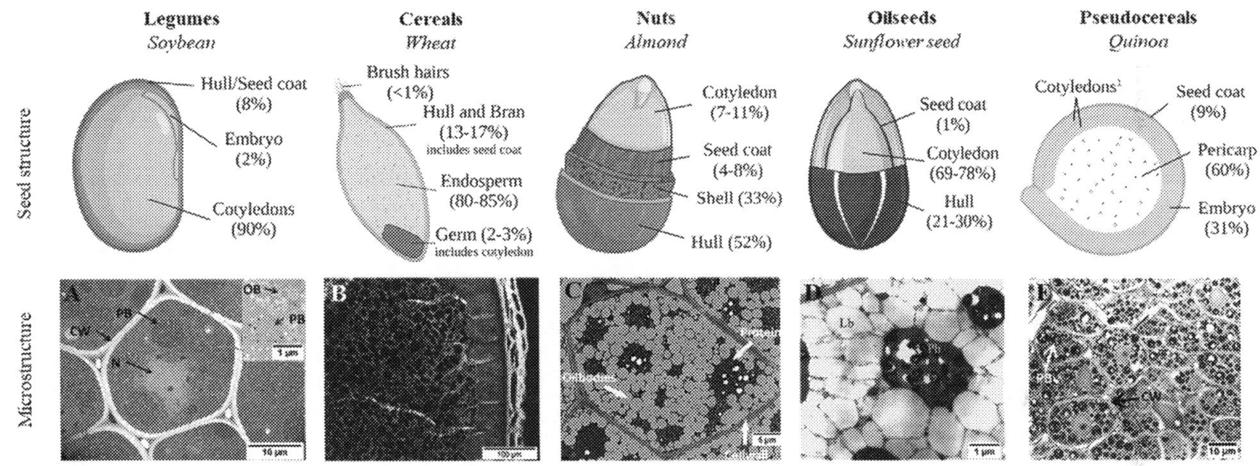

Composition (%, w/w db)	Soybean	Wheat	Almond	Sunflower seed	Quinoa
Protein	52	13	21	24	15
Fat	9	3	53	53	7
Carbohydrates	15	66	6	2	67
Starch	*6*	*64*	*0*	*0*	*61*
Fibre	17	13	11	11	14
Ash	6	2	3	4	2

FIGURE 3.2 Schematic structure of seeds, microstructure, and estimate composition of different food group examples used for protein extraction. Seed structures were illustrated using the drawing software from Biorender.com. Seed structure illustrations include the weight percentage of each part of the seed. Micrographs in second row: (A) Transmission electron microscopy image of soybean cotyledon cell cross-section (https://doi.org/10.1021/jf902298aCampbell & Glatz, 2009); (B) Optical microscopy image of wheat grain stained with Calcofluor and Acid Fuchsin showing protein in red, β-glucan in blue, starch granules unstained in black and outer layers in green due to autofluorescence (Jääskeläinen et al., 2013); (C) Transmission electron microscopy image of almond cotyledon cell (https://doi.org/10.1111/ijfs.13192Grundy et al., 2016); (D) Transmission electron microscopy image of sunflower seed cotyledon cells (https://doi.org/10.1007/s00425-005-1541-5Walters et al., 2005); (E) Light microscopy image of quinoa cotyledon cells stained with Coomassie brilliant blue showing protein bodies in black (https://doi.org/10.1006/anbo.1998.0704Prego et al., 1998). Scale bars are as indicated in figures: (A) 10 μm (1 μm for enlarged image); (B) 100 μm; (C) 5 μm; (D) 1 μm; and (E) 10 μm. CW, cell wall; PB/Pb, protein body; N, cell nucleus; OB/Lb, oil/lipid body; db, dry basis. [1] Cotyledon weight percentage of the total seed is included in the embryo. *With permission from Campbell, K. A., & Glatz, C. E. (2009). Mechanisms of aqueous extraction of soybean oil.* Journal of Agricultural and Food Chemistry, *57(22), 10904–10912. https://doi.org/10.1021/jf902298a; Jääskeläinen, A. S., Holopainen-Mantila, U., Tamminen, T., & Vuorinen, T. (2013). Endosperm and aleurone cell structure in barley and wheat as studied by optical and Raman microscopy.* Journal of Cereal Science, *57(3), 543–550. https://doi.org/10.1016/j.jcs.2013.02.007; Grudniewska, A., De Melo, E. M., Chan, A., Gniłka, R., Boratyński, F., & Matharu, A. S. (2018). Enhanced protein extraction from oilseed cakes using glycerol-choline chloride deep eutectic solvents: A biorefinery approach.* ACS Sustainable Chemistry and Engineering, *6(11), 15791–15800. https://doi.org/10.1021/acssuschemeng.8b04359; Walters, C., Landré, P., Hill, L., Corbineau, F., & Bailly, C. (2005). Organization of lipid reserves in cotyledons of primed and aged sunflower seeds.* Planta, *222(3), 397–407. https://doi.org/10.1007/s00425-005-1541-5 and modified from Prego, I., Maldonado, S., & Otegui, M. (1998). Seed structure and localization of reserves in Chenopodium quinoa.* Annals of Botany, *82(4), 481–488. https://doi.org/10.1006/anbo.1998.0704.* For interpretation of the references to color in this figure legend, please refer online version of this title.

depending on the type of plant material, species, and variety. In general, seeds are protected by outer layers, such as the hull, shell, and/or seed coat, which represent between 5% and 50% of their total weight (Fig. 3.2) and are composed mainly of insoluble fibers (Prgomet et al., 2017; Zhong et al., 2018). Within the seed coat lies the cotyledon (in legumes, nuts, pseudocereals, and oilseeds) or endosperm (in cereals) (Zhu et al., 2021b).

Plant protein is often located in dense protein bodies alongside starch granules and oil bodies packed within a fiber-rich cell wall (Zhu et al., 2021b). Different parts of the same seed contain cells with a very different microstructure and composition (i.e., bran, endosperm, and cotyledon). For soybeans, almonds, and sunflower seeds, the highest amount of protein occurs in the cotyledon (Fig. 3.2A,C, and D, respectively (Campbell & Glatz, 2009; Grundy et al., 2016; Walters et al., 2005)). For wheat, however, protein bodies are mainly located in the aleurone layer, which is part of the bran (Fig. 3.2B, in red (Jääskeläinen et al., 2013)), while the embryo and cotyledon cells of quinoa store the majority of the protein (Fig. 3.2E, in black (Prego et al., 1998)).

The composition of the main macronutrients of different plant materials varies widely: protein ranges from 4% to 50%, fat from 1% to 61%, and carbohydrates (nonfiber) from 2% to 65% (Day, 2013; Grundy et al., 2016). Oil-rich legumes (i.e., soybean, lupin, chickpeas), nuts, and oilseeds are especially rich in fat, while starch-rich legumes (i.e., pea, lentil, faba bean), cereals, and pseudocereals have high amounts of carbohydrates (Fig. 3.2). Knowledge of the structure and composition of the target plant material is essential to the development of a successful series of preprocessing, milling, and protein extraction process.

3.2.2 Preprocessing steps

The preprocessing steps selected to process plant materials prior to milling will depend on:

- the nature of the plant material;
- the protein extraction method selected; and
- the desired product properties.

The most common preprocessing steps carried out before dry extraction (i.e., air classification or electrostatic separation, Section 3.3) include seed dehulling and defatting, which requires no water. In contrast, wet extractions typically require soaking and thermal treatment.

Dehulling is the process used to separate the fiber-rich seed coat from the kernel and is commonly carried out for all kinds of plant seeds prior to further processing (Rommi et al., 2014; Saldanha do Carmo et al., 2020). Abrasion and attrition mills are usually employed to detach the hull, which is discarded by air separation or sieving (Fernando, 2021). Hull removal results in a protein enriched and fiber depleted starting material, and this process has been found to improve the protein content and yield of concentrates and isolates in pea, lupin, and several oilseeds during subsequent protein extraction (Murru & Calvo, 2020; Pelgrom, Wang, et al., 2015; Rommi et al., 2014).

Dehulling is not always necessary, as hulls can be well separated from the protein fraction during air classification in the case of faba bean and barley concentrates (Saldanha do Carmo et al., 2020; Srinivasan et al., 2010). Nevertheless, the removal of the hulls can result in a more neutral color for protein-enriched fractions of faba beans, which can be advantageous for use in protein powders and high protein beverages (Saldanha do Carmo et al., 2020). More importantly, dehulling has been reported to dramatically decrease the levels of antinutrient factors like trypsin inhibitor activity and phenolic compounds, since these are mainly located in the seed coat (Reinkensmeier et al., 2015; Singh, 1993). The reduction in such antinutrient factors is linked to a better in vitro protein digestibility (Embaby, 2010) and a reduction in the bitterness of protein concentrates (Singh & Singh, 1992).

The effect of dehulling on the functionality of the proteins obtained by dry fractionation is not yet clear. According to (Fernando, 2021), dehulled pulses had better wettability, dispersibility, foaming, and emulsifying properties. Yet other authors did not observe any differences in the solubility, water and fat holding capacity, gelation behavior, and foaming properties, reporting only slightly improved emulsion capacity for dehulled peas and faba beans (Saldanha do Carmo et al., 2020).

It is worth noting that protein concentrates typically contain ~60% protein, with the remaining mass consisting of a mixture of starch, fiber, fat, and ash; these other components have a significant impact on the functionality of the protein concentrates. Therefore, the lack of consensus on the effect of dehulling in protein functionality may be linked to differences in composition of the concentrates.

Several novel technologies, such as ultrasound, microwave, and enzymatic processing, have also been shown to aid the dehulling process but have not yet been commercialized due to a range of reasons including their energy-intensive nature (Fernando, 2021).

Defatting is the removal of oil from the seeds prior to protein extraction. Due to the high oil content of some plant materials, such as oilseeds, nuts, and oil-rich legumes (Fig. 3.2); these are first used by oil extracting industries to manufacture food grade oils. At an industrial scale, extraction is usually performed either through:

1. mechanical pressing (Geow et al., 2021),
2. the use of food grade organic solvents (i.e., hexane, isopropyl alcohol, or acetone) (Mwaurah et al., 2020), or
3. novel techniques such as supercritical CO_2 (Sibakov et al., 2011).

In oil-rich legumes, such as soybean and lupin (~7%–20% oil), oil extraction facilitates the subsequent protein fractionation stages, since it prevents particle agglomeration caused by exposed lipids (Xing et al., 2018). It also increases the extractability of hydrophobic proteins that otherwise naturally associate with oil bodies. Removing fat from lupin flour (original lipid content, 7.3% dry basis (db)) increased the protein content of the concentrates obtained by air classification from 45% db to 59% db. In flour with lower lipid content, such as pea flour (original lipid content 1.2% db), the protein content was not increased by fat removal (Pelgrom, Wang, et al., 2015).

Fat removal is not only important for oil-rich legumes and seeds but also for other plant materials where the fat concentration is higher than ~5% db. It has been demonstrated that defatting of oats (original fat content, 6%–10%) enhanced milling performance, which further reduced the particle size, leading to improved protein separation during air classification (Sibakov et al., 2011). Fat removal also increased the protein concentration obtained by wet extraction from 68% to 89% and altered the secondary structure and functionality of oat proteins (Yue, Gu, et al., 2021).

Defatting prior to protein extraction, either through dry or wet methods, has therefore been observed to aid protein recovery from a range of raw materials.

The **soaking** of seeds in water increases seed size, due to water absorption and loosens the cellular structure. Soaking can be carried prior to wet or dry protein extraction to either aid the dehulling and milling processes or to increase protein digestibility.

Soaking is usually performed in water for 1–16 h at temperatures between 10 and 40°C (Pan & Tangratanavalee, 2003). For dry extraction methods, soaking is followed by a drying process, which significantly changes the structure of the seed but increases energy consumption. The seed coat does not shrink in proportion to the remaining kernel leading to a gap between these two structural features, which can facilitate dehulling for certain pulses (Fernando, 2021). A higher absorption of moisture by navy and pinto beans also leads to greater seed coat removal (Anton et al., 2008).

Soaking is known to impact on the energy requirements and efficiency of milling, with the impact varying with the type of processing (dry/wet). The soaking of soybeans prior to wet milling (grinding with water), for example, resulted in greater protein recovery, lower grinding time, reduced energy consumption, and smaller particle sizes following grinding (Pan & Tangratanavalee, 2003; Vishwanathan et al., 2011).

Soaking is not always beneficial. In case of oil rich legumes, soaking significantly impaired milling and dry protein fractionation, resulting in a lower final protein concentration and yield (Pelgrom, Wang, et al., 2015). The protein bodies in these legumes may have disintegrated during soaking, affecting the integrity of the structure and altering the native particle density. A higher milling energy was also required to mill presoaked materials, due to their higher moisture content, which enhanced ductility (Dijkink & Langelaan, 2002). It is therefore essential to consider the impact of soaking across all processing steps.

Soaking can have a beneficial impact on protein bioavailability, however, as it can reduce antinutrients (phytates, tannins) present in pulses, increasing their in-vitro digestibility (Khandelwal et al., 2010; Luo & Xie, 2013).

3.2.3 Milling

Milling involves the mechanical disruption of plant materials, resulting in fine particles and is a crucial step for protein extraction, due to its direct impact on particle size. The behavior of plant materials during milling is influenced by several properties including chemical composition, structure, and the degree of adhesion between cellular components (Schutyser & van der Goot, 2011; Tyler, 1984). Grinding exposes and disentangles cellular components, such as starch granules, protein bodies, and cell walls (Fig. 3.2), which allows fractionation on the basis of size, density, and/or charge differences (Assatory et al., 2019). Milling can be performed in the presence or absence of a liquid (i.e., a wet or dry grinding process, respectively) allowing flexibility.

Wet milling has been used commercially to generate macronutrient (starch, protein, fiber, and fat) rich fractions from corn and wheat (Wronkowska, 2016). On a laboratory scale, this method has also been utilized to generate protein isolates from legumes (chickpea, cowpea etc.) and pseudocereals (quinoa) (Espinosa-Ramírez & Serna-Saldívar, 2019; Ningsanond & Ooraikul, 1989). For example, chickpea protein has been isolated from the by-product stream obtained following starch removal from wet milled chickpea (Espinosa-Ramírez & Serna-Saldívar, 2019). This protein isolate had properties comparable to conventional alkali extraction-isoelectric precipitation, including solubility, emulsification and foaming capacity. Saponins, a starch-rich fraction and the protein-rich embryo have also been successfully separated from quinoa using an optimized wet milling and sieving process, recovering ~96.5% of the protein originally contained in the seeds (Mufari et al., 2018).

The ready availability of wet milling techniques means these methods can easily be applied to wide range of starch and protein rich plant materials. Wet milling is less utilized compared to dry milling, however, as it is capital, resource and energy intensive, owing to the consumption of large volumes of water typically between 1 and 15 kg of water per kg of plant material (Espinosa-Ramírez & Serna-Saldívar, 2019; Wronkowska, 2016). The following discussion is therefore focused on dry milling.

There are different type of mills, such as stone, attrition, roller, colloidal, hammer, knife, ball, pin, roller, centrifugal, vibratory, disk, and jet mills (Table 3.1), which use frictional, compression, and impact forces to grind plant materials, are frequently used in combination to achieve optimal particle size and distribution (Fernando & Manthey, 2021; Zhu et al., 2021a). The milling type directly affects the physicochemical properties of particles and consequently protein separation. Fig. 3.3A shows the particle size distribution achieved using three different mills to process rapeseed oil cake (pregrinded in a knife mill with a 2 mm screen size (Kdidi et al., 2019). Ball milling produced a much more heterogeneous particle size distribution compared to impact or jet milling, leading to a lower protein concentration following fractionation. Jet milling effectively reduces particle size but can partially disrupt the native structure and functionality of proteins, due to the heat generated by frictional forces (Silventoinen et al., 2021). Other milling methods, including hammer milling and pin milling, have been used to produce a particle size (D_{50}) of around 68 and 48 μm for navy beans, respectively. The smaller particle size resulted in a more effective protein separation due to the better disentanglement of navy bean proteins from starch granules (Vitelli et al., 2020).

Milling operational variables (i.e., screen size grid, classifier wheel speed, temperature, time, air flow, and feed rate) also affect the particle size and size distribution of the ground material during milling (Schutyser et al., 2015). It has been suggested that classifier wheel speed and air flow are the two parameters that impact most significantly on final particle size (Pelgrom et al., 2013). Increasing the classifier wheel speed at constant air flow lengthens the residence time of particles in the mill, which results in a more extensive milling process, generating a smaller particle size and a more homogeneous particle distribution (Fig. 3.3B).

The impact of other parameters, such as milling time and temperature on the milling of wheat bran, have also been examined and are plotted in Fig. 3.3C–E. Increasing the milling time from 1 to 8 min (where the milling time was equal to the residence time) resulted in a reduction of the median particle size from 287 to 37 μm (Hemery et al., 2011) (Fig. 3.3C). Interestingly, milling at cryogenic temperatures (with liquid nitrogen) did not follow the same trend observed at ambient temperature; particle size decreased faster during the first 2 min (the D_{50} after 2 min at ambient and cryogenic temperatures was 121 vs. 83 μm, respectively) but then reduced much less with longer milling time (the D_{50} after 8 min at ambient and cryogenic temperatures 37 vs. 59 μm, respectively) (Fig. 3.3D). Temperatures below zero also affected milled particle size and distribution (Fig. 3.3E). Reducing the temperature from −10 to −60°C had a major effect on the behavior of different layers of the wheat bran, making these more brittle and easier to reduce in size (Hemery et al., 2010). The temperature drop from −60 to −140°C, however, did not further affect the size or distribution of wheat bran particles. The authors concluded that cryogenic milling was a convenient method to reduce particle size in a relatively short time, decreasing energy usage during milling, while ambient temperature grinding can be better for disentangling different cellular structures (Hemery et al., 2011).

Although a finer particle size distribution has been linked to increased protein purity, milling needs to be carefully controlled to minimize:

1. the reduction in yield due to particle aggregation (owing to poor cohesive flow behavior with smaller particle sizes) (Schutyser & van der Goot, 2011) and wall fouling effects (Wang et al., 2016; Xing et al., 2018)
2. the damage to bigger particles, such as the starch granules, which limits their functionality and further usage (Boye et al., 2010; Pelgrom et al., 2013).

The reduction of plant materials to powders is one of the most energy-consuming steps in the protein extraction process. In general, the smaller the particle size in milled flours, the higher the energy consumption of milling (Protonotariou et al., 2021). This increase is not proportional though, and the energy required to generate particles below ~50 μm increases exponentially (Schutyser & van der Goot, 2011). For example, the energy needed to grind wheat flour to 20 μm was 10 times higher than generating ~70 μm particles from sunflower oil cake (Table 3.1). Energy consumption

TABLE 3.1 Energy consumption, water usage, and final particle size of a variety of plant materials produced using different milling methods.

Milling type	Plant material	Energy consumption (kWh/t)	Water usage (kg/kg plant material)	Feed particle size (mm)	Particle size, D50 (μm)	References
Wet milling						
Impact mill	Soybean	276	1.6	Bean[a]	167	Vishwanathan et al. (2011)
Stone mill		332	1.7	Bean[a]	218	
Colloid mill		29	1.3	Bean[a]	206	
Dry milling						
Hammer mill	Sunflower meal	0.4	–	Kernel	573	Banjac et al. (2017)
		1.7	–	Kernel	453	
		6.6	–	Kernel	343	
Roller mill	Soft wheat grain	3.7	–	2.88	720	Pujol et al. (2000)
	Hard wheat grain	5	–	2.85	860	
	Durum wheat	5.4	–	2.65	1045	
Impact mill	Rapeseed oil cake	189	–	<2[b]	89.7	Basset et al. (2016)
Impact mill	Sunflower oil cake	298	–	<2[b]	69.5	Barakat et al. (2015)
Jet mill	Wheat flour	2950	–	0.127	20.4	Protonotariou et al. (2021)

[a] *Soybeans were pre-soaked for 3 h before milling.*
[b] *<2, samples were first pre-milled using a grid with screen size <2 mm, but particle size was not measured.*

depends on final desired particle size, grinding method, milling conditions, and the particle size and characteristics of the starting material.

Jet milling is usually the most energy intensive process, due to the need for air compression (Protonotariou et al., 2021). Material properties influencing energy consumption include protein content, vitreousness (an optical property attributed to differences in endosperm porosity), hardness (Pujol et al., 2000), and particle size (Schutyser & van der Goot, 2011). It is therefore important to consider the different aspects of the milling process in order to minimize energy consumption both during the milling and later throughout the protein extraction process.

3.3 Dry protein extraction

Dry protein extraction techniques are solvent-free methods used to produce protein-rich fractions from raw materials such as pulses, cereals, defatted oilseed meals, and pseudocereals (Laguna et al., 2018; Pelgrom et al., 2014; Ruiz et al., 2016; Silventoinen et al., 2018). The protein concentrates obtained are typically close to 60% protein (GEA, 2020) but can reach up to 75% protein (Table 3.2). Concentrates obtained by dry fractionation could play an important role, as the need to obtain high-quality plant-based proteins with high functionality increases. These methods offer affordability, improved sustainability compared to wet extraction techniques, and produce proteins with good functional properties. Table 3.2 provides an overview of dry protein extraction methods used to extract protein from different plant materials.

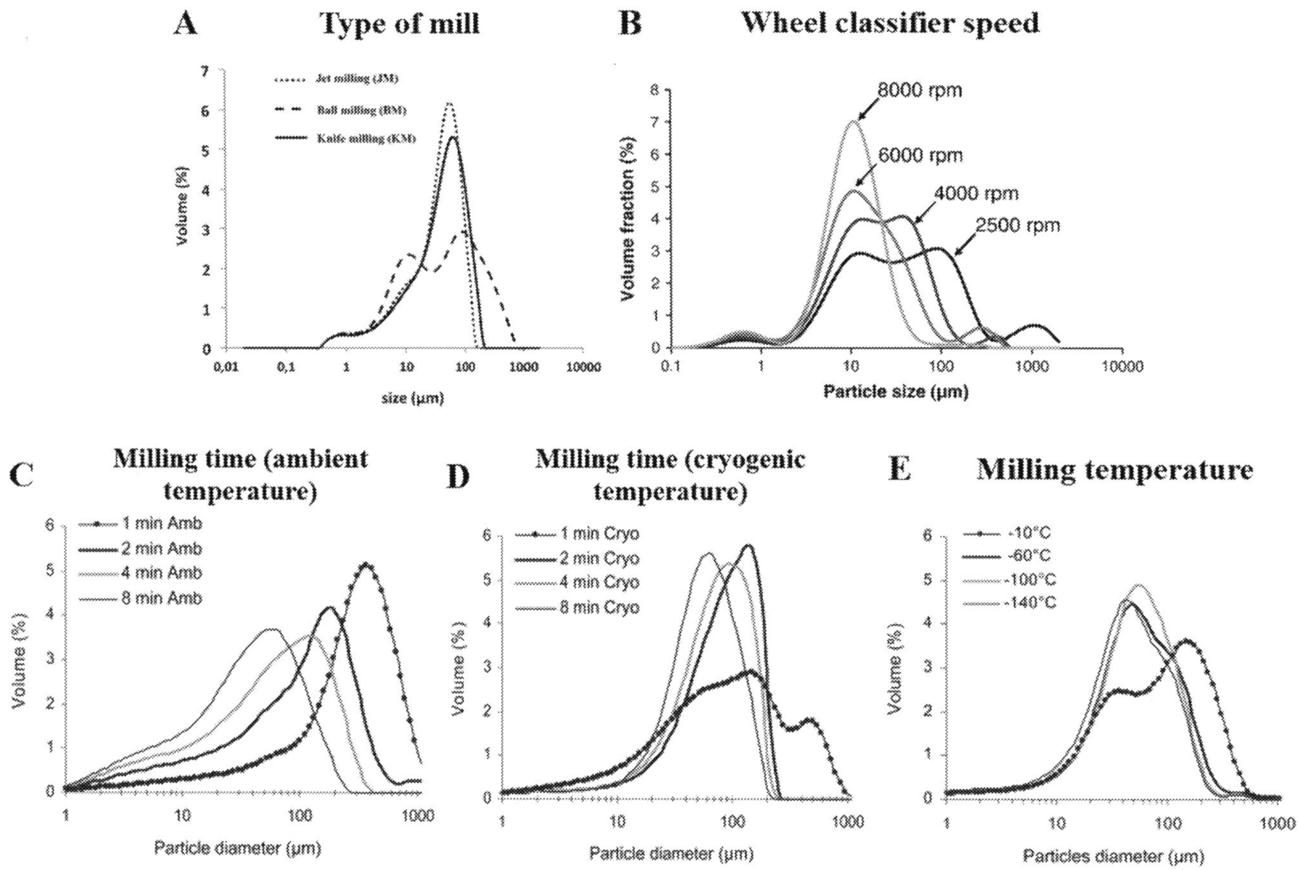

FIGURE 3.3 The effect of different operational conditions on the particle size distribution produced by milling: (A) type of mill (jet, ball and knife) for rapeseed oil cake; (B) wheel classifier speed (2500, 4000, 6000 and 8000 rpm) in an impact mill for lupine seeds; (C) milling time (1, 2, 4 and 8 min) in a vibratory micro-mill at ambient temperature for wheat bran; (D) milling time (1, 2, 4, and 8 min) in an impact mill at cryogenic temperatures for wheat bran; (E) milling temperature (−140, −100, −60 and −10°C) in a pilot scale impact mill for wheat bran. Amb, ambient temperature; Cryo, cryogenic temperature; rpm, revolutions per minute. *From (A) Kdidi et al. (2019); (B) Wang, J., Zhao, J., De Wit, M., Boom, R. M., & Schutyser, M. A. I. (2016). Lupine protein enrichment by milling and electrostatic separation. Innovative Food Science and Emerging Technologies, 33, 596–602. https://doi.org/10.1016/j.ifset.2015.12.020; (C–E) Hemery, Y., Chaurand, M., Holopainen, U., Lampi, A. M., Lehtinen, P., Piironen, V., Sadoudi, A., & Rouau, X. (2011). Potential of dry fractionation of wheat bran for the development of food ingredients, part I: Influence of ultra-fine grinding. Journal of Cereal Science, 53(1), 1–8. https://doi.org/10.1016/j.jcs.2010.09.005.*

Four different measures are usually used to assess the adequacy of dry extraction processes: the final protein concentration (P_P), protein yield (ϕ_p), protein separation efficiency (PSE), and protein shift percentage (δ) (Table 3.2). It is important these measures are considered together to optimize the extraction process, since the final protein concentration and yield are often negatively correlated, with high-purity products achieved at the expense of protein yield. The protein yield is defined as the mass percentage of protein that is recovered from the original flour. PSE, which is also called protein recovery, is defined as the protein that is recovered from the total protein in the original flour (Tyler et al., 1981), calculated as follows.

$$\mathrm{PSE} = \frac{P_P \times \Phi_P}{P_{\mathrm{flour}} \times \Phi_{\mathrm{flour}}} \tag{3.1}$$

The protein shift percentage is a measure of the protein that shifts form the original flour to the protein-rich fraction (Schutyser & van der Goot, 2011), calculated as:

$$\delta = \frac{\left(P_p - P_{\mathrm{flour}}\right) \times \phi_P}{P_{\mathrm{flour}}} \tag{3.2}$$

TABLE 3.2 Overview of dry protein extraction methods used in different plant materials.

Extraction method	Plant protein source	Extraction efficiency				Advantages	Limitations	Energy consumption	Current scale	References
		P_P	Φ_P	PSE[a]	δ[a]					
Sieving	Cereals Legumes Pseudocereals Oil-bearing fruits Oilseed meals	10% −15% 26% 23% −33% 8% −11% 39% −49%	24% −51% ND 23% −30% 18% −29% 23% −51%	28% −52% ND 47% −67% 31% −70% 27% −66%	2% −12% ND 24% −38% 13% −41% 5% −15%	Very simple process can be used as a pretreatment process combined with other dry extraction methods to increase extraction efficiency	Low protein concentration and protein yield Clogging of the sieve	ND	Industrial scale	Sundberg and Aman (1994), Maaroufi et al. (2000), Srinivasan et al. (2010); Lammi et al. (2018), Hansen et al. (2017), Opazo-Navarrete et al. (2018a), Opazo-Navarrete et al. (2018b), Vidosavljević et al. (2019)
Gravity separation	Oilseed meals	42% −45%	60% −75%	72% −89%	11% −16%	Currently used in grain industry for cleaning steps Easy process to adapt to protein fractionation	Low protein shift Limited data available	ND	Pilot scale	Hahn et al. (2014), Murru and Calvo (2020)
Air classification	Cereals Legumes Oilseed meals	22% −73% 28% −73% 40% −52%	5% −22% 18% −82% 12% −62%	18% −59% 18% −90% 17% −93%	9% −37% 12% −52% 5% −30%	Good extraction method to separate starch-rich legumes due to size differences between starch granules and protein bodies	May not separate particles with similar physical characteristics The presence of oil can significantly reduce the efficiency of protein separation. Antinutrients are not necessarily separated by this method and must be considered for certain applications (higher trypsin inhibitor activity in concentrates vs. isolates)	0.7−12 kWh/t	Industrial scale	Tyler et al. (1981), Wu and Abbott (2003), Sessa (2004), Sibakov et al. (2011), Banjac et al. (2013), Banjac et al. (2017), Pelgrom et al. (2014), Pelgrom et al. (2013), Reinkensmeier et al. (2015), Hansen et al. (2017), Rempel et al. (2019), Saldanha do Carmo et al. (2020), Silventoinen et al. (2021), Silventoinen et al. (2019),

TABLE 3.2 Overview of dry protein extraction methods used in different plant materials.—cont'd

Extraction method	Plant protein source	Extraction efficiency				Advantages	Limitations	Energy consumption	Current scale	References
		P_P	Φ_P	PSE[a]	δ[a]					
										Silventoinen et al. (2018), Vidosavljević et al. (2019)
Electrostatic separation	Cereals Legumes Oilseed meals	19%–20% 37%–52% 44%–51%	41%–57% 17%–62% 30%–59%	58%–69% 27%–67% 49%–93%	9%–12% 4%–20% 10%–34%	Unlike air classification or sieving, an electrostatic method could effectively separate particles that are similar in size and density but different in charge	Protein and starch particles can be similarly charged and difficult to separate in some raw materials	6–11 kWh/t	Pilot scale	Barakat et al. (2015), Hemery et al. (2011), Kdidi et al. (2019), Laguna et al. (2018), Tabtabaei et al. (2016a), Tabtabaei et al. (2016b), Tabtabaei et al. (2019), Wang et al. (2015), Wang et al. (2016), Xing et al. (2018)
Air classification and electrostatic separation	Legumes	49%–72%	9%–13%	21%–28%	12%–15%	Improves the purity of the protein concentrates obtained from starch-rich legumes	Low yields and efficiency	7–23 kWh/t	Lab scale	Pelgrom, Wang, et al. (2015), Xing et al. (2020), Zhu et al. (2021)
Sieving and air classification	Cereals	10%–41%	33%–88%	39%–93%	5%–13%	Useful to separate hulls from seeds	Similar to air classification limitations	ND	Pilot scale	Srinivasan et al. (2010), (2005), (2009)

P_P, protein concentration of the protein-rich fraction; PSE, protein separation efficiency; δ, protein shift percentage; Φ_P, mass yield of the protein-rich fraction; *ND*, not determined.
[a]*Most of the values for PSE and δ have been calculated from the protein concentrations and yield reported by other authors, using the equations given in Section 3. Dry protein extraction.*

where the P_{flour} and ϕ_{flour} are the protein concentration and yield of the original flour, respectively.

Prior to dry protein extraction, the plant materials are preprocessed and milled (as described in Section 3.2). This results in small particles, which can be separated based on differences in physicochemical properties such as size, shape, density, or dielectric constant. Some commonly used fractionation techniques include sieving, gravity separation, air classification, and electrostatic separation, performed either alone or in combination (Table 3.2).

The process of *sieving* separates particles primarily based on their size and shape, utilizing a set of sieves with various mesh sizes (0.15−1 mm aperture) (Fig. 3.1). Size classification effectively concentrates protein in fine fractions because protein bodies are often smaller than other plant structures (Fig. 3.2). A drawback is the limited capacity to obtain products with a final protein concentration higher than 50% db (P_P, Table 3.2); the protein shift is also especially low ($\leq$15% db) for cereals and oilseed meals (δ, Table 3.2). Sieving is commonly used prior to other dry extraction techniques (Murru & Calvo, 2020; Srinivasan et al., 2010), due to the simplicity of this process and low energy consumption.

Gravity separation has been historically used in the mineral processing and grain industries as a cleaning step (Falconer, 2003), and the application of this method to extract plant protein is relatively new. Separation is achieved based on the difference in density of the ground plant material (Krueger et al., 2007), and it is carried out on a gravity table, which consists of a bed of fluidized and vibrating particles on an inclined oscillating deck. When the operational variables are optimized, denser particles are collected at the top of the deck, while lighter particles concentrate at the bottom (Fig. 3.1) (Kannan et al., 2017; Murru & Calvo, 2020).

Gravity separation has been effectively used to fractionate protein from defatted sunflower meal using a combination of sieving and gravity separation, achieving a product with 42%−45% protein and 72%−89% PSE (Table 3.2) (Murru & Calvo, 2020). Protein-enriched products with up to 48% P_P have also been obtained from rapeseed meal for animal nutrition on an industrial scale using this method (Hanh et al., n.d.). Gravity separation can easily be adapted to fractionate protein from grain, as the grain industry already uses such machinery, but further research is needed to understand the benefits of using gravity separation instead of other dry extraction methods with a range of dry plant materials.

Air classification is the most used dry extraction technique for the manufacture protein concentrates. This method has been widely applied to cereals (Sibakov et al., 2011; Silventoinen et al., 2018, 2021), pulses (Pelgrom et al., 2014; Rempel et al., 2019; Tyler et al., 1981), and defatted oilseed meals (Banjac et al., 2013; Wu & Abbott, 2003; Vidosavljević et al., 2019) and typically results in a final protein concentration ranging from 22% to 73% (P_P, Table 3.2).

During air classification, air is pumped into a chamber, allowing segregation based on particle size and density (Fig. 3.1). The process results in two fractions:

1. a fine fraction that is enriched in protein, and
2. a coarse fraction that is depleted of protein.

The cut-size is a major determinant of separation efficiency and is defined as the particle size that has equal probability to end in either the coarse or fine fraction (Tyler et al., 1984). Bigger and denser particles (like starch granules; ranging 2−40 μm in size) tend to concentrate in the coarse fraction above the cut-size, whereas smaller and lighter particles (such as protein bodies; ranging 1−20 μm in size) tend to concentrate in the fine fraction below the cut size (Schutyser & van der Goot, 2011).

Other uncontrolled factors also influence separation by air classification. These include particle collisions and air turbulence, which results in coarse particles appearing in the fine fraction and vice versa (Shapiro & Galperin, 2005). The degree of powder dispersion in the classifier also affects separation. Smaller particles, in particular, are likely to adhere to larger particles reducing the quality of separation (Dijkink et al., 2007). Despite these factors, air classification has a very good PSE (up to 90%) and protein shift for legumes (δ up to 52%; Table 3.2), especially for starch-rich pulses, such as peas and faba beans, where the difference in particle size of cellular components is well defined (Fig. 3.2). As a result of high protein extraction efficiencies, air classification has long been used in industry to obtain protein concentrates from legumes, cereals, and oilseeds (Swain, 1975).

Another technique that has attracted a lot of attention in the past 5 years is *electrostatic separation*, where differences in charge between different food components act as the main driving force for separation. Proteins can be much more highly charged than dietary fiber, owing to the presence of ionizable C and N terminal groups in their chemical structure (Assatory et al., 2019). Prior to fractionation, raw material particles are charged through the process of tribo-electrification (Fig. 3.1), which results in frictional forces between particles and between the particles and external materials.

The main factors affecting the tribo-electrification process are the residence time and material of the pipe, together with the particle size of the milled material and the concentration of the plant flour. The residence time on the tribopipe in turn depends mainly on the design (length, diameter, and shape) and gas flow used. When the residence time is prolonged, the probability of the particles colliding with each other or the pipe wall increases, which improves the efficiency of particle charging (Tabtabaei et al., 2016b). The residence time must be optimized for each process though, to avoid overcharging and particle aggregation, which can greatly impair subsequent separation processes.

Different materials can be used in the inner wall of the tribopipe, influencing the particle charge (i.e., whether the particle gets more positively or negatively charged). Materials that can be used include synthetic polymers (polytetrafluoroethylene (PTFE), polyvinyl chloride (PVC), nylon, or polystyrene), metals, rubber, wood or silk. PTFE has been found to be the most suitable material for the extraction of navy bean proteins (Tabtabaei et al., 2016b), while pipe material was found to have little effect on the separation of lupin flour (Xing et al., 2021). The particle size impacts on separation by electrostatic separation, as smaller particles get more charged than larger particles, due to an increased surface to volume ratio and a higher number of collisions between the wall and small particles during turbulent gas flow (Wang et al., 2014).

The concentration of the ground plant material also affects particle diffusion in the tube, impacting electrostatic separation (Chen et al., 2014), with higher concentration of particles increasing the probability of particle—particle collisions and lower concentrations favoring particle—wall collisions. Once the particles have been charged, they are separated under the influence of an electric field according to the charge magnitude and type (Tabtabaei et al., 2016a). A more detailed description of the mechanism and technical operating conditions of electrostatic separation is available in a recently published review from (Zhu et al., 2021a).

Electrostatic separation technology has been successfully used to produce protein-enriched concentrates from cereal brans, legumes, and oilseed cakes. The extraction efficiency for legumes is much lower for this method than air classification (Table 3.2) and varies with the type of pulse used. For instance, the PSE for navy bean concentrates was ~43% (Tabtabaei et al., 2016b, 2019) but 62% for soybean concentrates (Xing et al., 2018). The original composition of these two pulses is very different: navy beans are rich in starch (~32% w/w), while soybeans have less starch (~5% w/w) and are rich in oil (~9—19% w/w), explaining these differences.

Electrostatic separation in starch-rich legumes is typically not efficient because both starch and protein bodies are charged in similar manner during the tribo-electric charging process (Pelgrom, Wang, et al., 2015). This problem can be easily solved by adding a preprocessing step that segregates by particle size or density, either by air classification or sieving. For oilseed meals and cereals, the separation efficiency obtained using electrostatic separation or air classification is comparable (Table 3.2). For rapeseed and sunflower meals, a final protein concentration of between 44% and 51% (w/w, db) has been reported using electrostatic separation, with PSE values ranging between 49% and 93% (Barakat et al., 2015; Kdidi et al., 2019; Laguna et al., 2018). High PSE values have also been obtained for wheat bran using electrostatic separation (58% (Wang et al., 2015)), although the final protein concentration was lower than achieved by using air classification (P_P 20% vs. 31% w/w, db (Silventoinen et al., 2021)); nevertheless, the protein shift in both techniques was similar (~8.5%).

The dry extraction methods of sieving, gravity separation, air classification, and electrostatic separation are frequently combined to maximize separation efficiency. Air classification, for example, can improve PSE when performed prior to electrostatic separation for starch-rich legumes (Pelgrom, Wang, et al., 2015; Xing et al., 2020; Zhu et al., 2021b), overcoming the problem of common charging during the tribo-electric charging process of electrostatic separation (Pelgrom, Wang, et al., 2015). Protein has been concentrated to up to 72% protein for pea flours (Zhu et al., 2021b), with other studies reporting purity of either 49% or 62% ((Pelgrom, Wang, et al., 2015; Xing et al., 2020), respectively; P_P, Table 3.2).

The combination of sieving and air classification, also known as the elusive process, is also common. This approach has been successfully use to separate the protein-rich fraction from fiber in distillers dried grains with solubles, a dried by-product of ethanol production originating from corn and other cereals (Srinivasan et al., 2005). The combination of these two processes resulted in a protein content between 10% and 41% (P_P, Table 3.2), reducing the fiber content considerably.

Another strategy is the repeated application of the same extraction method; this approach has been used by several authors to improve PSE (Laguna et al., 2018; Silventoinen et al., 2019; Tabtabaei et al., 2017; Tyler et al., 1981; Wang et al., 2016). Repeated extraction increases yield, but protein concentration can be slightly reduced. For instance, a two-step electrostatic separation process applied to navy beans improved the PSE by 15.1% (from 44.8% to 59.9%) with a reduction in protein from ~41% to ~38% (Tabtabaei et al., 2017). Similarly, a second electrostatic separation step used to process the fine fractions of pea protein enhanced the yield from 6.7% to 15.8% (w/w, db), reducing the protein concentration from 66.9% to 63.4% (Xing et al., 2020). Multiple air classification steps also increased the PSE in rice bran from 38% to 40% (Silventoinen et al., 2019). These studies indicate that the most important measures of extraction efficiency should be used to guide dry extraction methods and the way that these operations are applied.

Regarding the structure and functionality of protein concentrates, it is generally agreed that the proteins extracted by dry fractionation suffer little or no damage to their structure, keeping their native functionality (Assatory et al., 2019).

Nevertheless, there are a limited number of studies confirming the functional properties of proteins extracted by dry methods, with most studies performed following air classification. Functional attributes that are important in food technology include solubility, emulsifying properties, gelling and foaming properties, water- and fat-binding capacities, and viscosity (Foegeding & Davis, 2011).

Example studies of protein functionality include protein concentrates obtained from barley endosperm, rye, wheat, and rice bran by air classification. These concentrates had a water solubility of 17%, 35%, 45%, and 67% at their native pH (6.7−7) (Silventoinen et al., 2019, 2021). Proteins from rice contain a high proportion of albumins (water-soluble proteins) potentially explaining their higher solubility, while proteins including prolamins, glutelins, and globulins are higher in other cereal concentrates. For legumes, the protein solubility of concentrates obtained by dry extraction ranges between 40% and 85% depending on the type of pulse (Keivaninahr et al., 2021; Pelgrom, Boom, & Schutyser, 2015; Vogelsang et al., 2020). A high solubility also occurs concurrently with good foam stability and a low viscosity in lupin concentrates with 59% of protein (Pelgrom et al., 2014).

It is important to note that the functionality of concentrates is determined not only by the proteins, which usually account for ~60% of the product, but also by other compounds such as starch, fibers, and fat that remain in the concentrate. Starch and fibers have been thought to contribute to the high water- and oil-binding capacities of protein-enriched fractions from legumes, cereals, and pseudocereals (Opazo-Navarrete et al., 2018b; Pelgrom et al., 2013; Silventoinen et al., 2021), illustrating the need to characterize all components within the concentrate in order to understand these complex preparations. Overall, protein concentrates obtained by dry methods seem to have suitable functional properties to apply in different foods such as, baked goods, as dairy, egg or meat substitutes, as texturized products and as high protein beverages.

The dry fractionation methods detailed above have been used in a variety of industries that sort diverse materials from minerals to grains (Falconer, 2003). Yet wet extraction techniques (Section 3.4) are more commonly used by the food industry to obtain protein concentrates and isolates. For certain plant materials and food applications, however, dry protein extraction methods should be considered, as they are potentially more sustainable and cost-effective. For example, protein fractionation by air classification or electrostatic separation has an energy consumption below 12 kWh per tonne of processed material (Table 3.2), and no water is used during processing. The potential application of dry methods for plant protein extraction, particularly at the pilot and industrial scale, is therefore a promising area of future research.

3.4 Wet protein extraction

Wet protein extraction techniques use water to extract plant proteins. Unlike dry fractionation, these techniques can produce high-purity protein products but involve a high amount of water and energy. At the industrial scale, plant products are subject to wet protein extraction and concentration, followed by heat treatment, such as pasteurization, to eliminate pathogens and then spray dried to obtain a final powdered product (Fig. 3.1) (GEA, 2020). The latter spray drying steps are the most energy-intensive part of the extraction process, requiring 1333 kWh per ton of evaporated water (Schutyser et al., 2015). In this section, we discuss conventional methods used in industry and the latest extraction techniques that are currently being studied at a laboratory scale (Table 3.3).

3.4.1 Conventional wet protein extraction techniques

Conventional wet extraction of plant proteins is primarily based on aqueous solubilization, followed by concentration and recovery using different precipitation techniques, such as isoelectric precipitation, cryoprecipitation, and/or membrane filtration techniques, including ultrafiltration or dialysis (Fig. 3.1). These methods have been successfully used to purify proteins from almost all plant groups including cereals, legumes, nuts, and oilseeds (Table 3.3). Based on their solubility, plant proteins have been broadly classified into albumins (water soluble), globulins (salt soluble), glutelins (dilute alkali/acid soluble), and prolamins (aqueous-alcohol soluble) (Osborne, 1924). Organic or inorganic food grade alkalis (NaOH, Na_2CO_3, and $NaHCO_3$), acids (HCl, citric acid), salts (NaCl, $Na_6[(PO_3)_6]$), and alcohols (ethanol, 2-propanol) are often used in conjunction with each other, along with water, to enhance protein solubility, creating the methods of alkaline or acid extraction, salt extraction, as well as alcohol extraction (Sathe et al., 2018; Zayas, 1997).

Alkaline and acidic extractions make use of changes in protein charge and solubility with pH. At their isoelectric point (IEP) (mostly pH 4−5), proteins do not possess a net charge, and hydrophobic interactions dominate protein−protein interactions, leading to protein aggregation and precipitation. As the pH is increased or decreased beyond the IEP, proteins acquire a negative (pH > IEP) or positive (pH < IEP) charge. Under both these conditions, electrostatic repulsions between proteins increase, facilitating protein−solvent interactions. In general, the further the solution pH from the IEP, the

TABLE 3.3 Overview of conventional and novel wet protein extraction techniques applied to different plant materials.

Extraction method	Plant protein source	Extraction efficiency[a]			Protein structure and functionality[b]	Advantages	Limitations	Sustainability	Current scale	References
Alkaline extraction	Lupin Defatted rapeseed meal Defatted hemp meal Oat Soy Rye Pea	60% 37% 51% 62%–68% 52% 47%–74% 46%–70%			Structural changes occur due to irreversible unfolding during the extraction process. The degree of denaturation and change in functionality depend on the plant protein source and extraction parameters	High-purity protein isolates (>90%) Simple and rapid Low cost Reduction of antinutritional compounds Digestibility	Selective precipitation of globulins leads to wastage of other water-soluble proteins (albumins) Loss of native protein structure and functionality	High water and energy requirement Generation of waste streams containing a mixture of alkalis and acids	Industrial scale	Russin et al. (2007), Berghout et al. (2014), Stone et al. (2015), Drakos et al. (2017), Hadnadev et al. (2017), Kalaydzhiev et al. (2020), Gao et al. (2020), Tanger et al. (2020), Hadidi et al. (2020), Momen et al. (2021), Yue et al. (2021a)
Saline extraction	Lupin Pea Almond Oat Hemp Pigeon pea Cowpea	30% 25%–70% 50% 25% 40% 40% 37%			Native protein structure is better preserved than in alkaline extraction	Milder than alkaline extraction Low cost	Long extraction time	High water and energy requirement Generation of waste streams containing salts and alkalis	Industrial scale	Mwasaru et al. (1999), Rodriguez-Ambriz et al. (2005), Sharma et al. (2010), Stone et al. (2015), Tanger et al. (2020), Hadnadev et al. (2017)
Reverse micelles		FEE[c]	BEE[c]	TEE[c]	Molecular unfolding and changes in the secondary structures of proteins (higher α-helical and lower β-sheet and turn structures) Improved[d]: H_2O and N solubility—fat absorption capacity—foaming and emulsifying capacity and stability—surface hydrophobicity—hardness, gumminess, adhesiveness and chewiness of gels Reduced[d]: water absorption capacity—cohesiveness of gels	Extraction of specific protein types Reduced processing times (1 vs. 3 –4 days) The modification of functional properties of proteins Simultaneous extraction of oils and proteins	Most surfactants and solvents not authorized for food use Difficult to separate the surfactant and protein Currently more expensive and highly variable extraction efficiencies	High energy and water use, particularly for backward extraction Recovery and reusability of surfactants and solvents is limited	Lab scale	Sun et al. (2008, 2009), Zhu et al. (2010), (2009), Chen et al. (2014), (2013), Guo et al. (2015), Zhao et al. (2019a), Zhao et al. (2019b), (2018), (2015), He et al. (2015), Zhang et al. (2017), Wang, Zhang, et al. (2018), Wang et al. (2020b)
	Defatted wheat germ Soy Peanut Walnut Tea residues Black turtle bean Grape seeds	37% –57% 70% 99% –82%	80% 92% 80%	30% –46% 65% –72% 42% 53% 82%						

Subcritical water extraction	(Black) Rice bran Wheat bran Soybean Defatted flax-seed meal Rapeseed meal Zein protein	75%–100%–60% 45%–77% 68% 71%–90%[f]	Very dependent on plant protein source, temperature, pressure and duration of extraction Improved[e]: H_2O solubility (higher degree of hydrolysis)—Emulsifying capacity and stability (extractions <200°C)—Foaming capacity and stability (extraction at 120°C)	Extraction solvent is water, so food safe Shorter treatment (from minutes to a few hours) Low cost Modification of functional properties of proteins	Nonspecific hydrolysis of proteins	High energy and water use Does not require the use of organic solvents	Lab scale	Ho et al. (2007), Kataoka et al. (2008), Lu et al. (2016), Pińkowska (2014), Pińkowska et al. (2013), Ramachandraiah et al. (2017), Sereewatthanawut et al. (2008), Watchararuji et al. (2008), Wiboonsirikul et al. (2007a), Wiboonsirikul et al. (2007b), Wiboonsirikul et al. (2008), Wiboonsirikul et al. (2013), Zhang et al. (2018)
Aqueous two-phase extraction	Oats Rapeseed Primrose Algae Pumpkin Pomogranate peel	14%–42% 13%–30% 11%–62% ≤ 92% 0.91 g isolate/g protein in raw material 0.2 g protein/g raw material	May affect the tertiary and secondary structure of protein Similar[d]: Thermal stability— H_2O solubility Reduced foaming capacity[d] Improved foaming stability[d]	Simple and rapid Easy scale up with commercially available equipment Availability of wide variety of low-cost, food-grade solvents (NADES) Allows simultaneous separation of multiple components	Lower protein content of concentrates (<60%)	High energy and water use nontoxic and biodegradable solvents (NADES) prevent the generation of hazardous wastewater streams	Lab scale	Liu et al. (2017); Grudniewska et al. (2018); Suarez Ruiz et al. (2018), Hernández-Corroto et al. (2020), Yue et al. (2021b)

[a]Extraction efficiency is referred to the equivalent of Protein Separation Efficiency (PSE) in Table 3.2, also called protein recovery.
[b]The protein function ality will be highly determined by the type of plan material and extraction conditions but the column aims to highlight the potential changes that wet extraction techniques cause.
[c]FEE, Forward extraction efficiency; BEE, Backward extraction efficiency; TEE, Total extraction efficiency.
[d]Compared to conventional methods.
[e]Compared to the native protein.
[f]Amino acid recovery.

better the solubility and higher the protein extraction yield (Deleu et al., 2019; Zayas, 2012). Alkalis have two advantages, firstly they can degrade the cell matrix with high efficiency; secondly, most proteins have an IEP in acidic range, resulting in high protein solubility under alkaline conditions (Pojić et al., 2018). Salt-based extractions are based on the process of "salting in," where the enhanced electrostatic interactions between the charged groups of proteins and salt counterions, are responsible for increased protein solvation (Fig. 3.1). Whereas, in alcohol-based extractions, the hydrophobic interactions between the protein and alkyl chain of alcohol help to solubilize the protein (Zayas, 1997).

Conventional wet extraction involves the suspension of ground plant flour in an aqueous phase (also containing a salt in the case of salt extraction) with varying solid:solvent ratios (usually ranging between 1:1 and 1:50). The pH can then be adjusted to pH 8−12 using an alkali, enabling so called alkaline extraction or to pH two to six using acid, in the case of acid extraction. Protein solubilization is achieved with constant stirring, the extent of which can be modulated by regulating the temperature (25−95°C) and duration of extraction (30−180 min) (Boye et al., 2010; Hadnadjev et al., 2017; Sari et al., 2015; Sathe et al., 2018). The suspension is then centrifuged (typically 1000−20,000 g for 10−30 min) to obtain the supernatant containing soluble proteins, which can be recovered through various techniques. In case of alkali and acid extracts, techniques such as isoelectric precipitation and cryoprecipitation (refrigeration) are often used, with the most common being alkali extraction-isoelectric precipitation (AE-IP). Alternatively, ultrafiltration, where the extract is passed through a membrane to retain the specific proteins of interest, can also be used to concentrate proteins (Boye et al., 2010). This latter method has the potential advantage of being a less energy-intensive method than spray drying (Schutyser et al., 2015).

In the presence of salt, as in the case with salt extraction, dialysis and micellar precipitation are often employed for protein concentration, resulting in the methods of salt-extraction dialysis (SE-D) and micellization, respectively (Fig. 3.1). Dialysis is a membrane-based process, where the protein solution is dialyzed against deionized water through a membrane with a molecular weight cut-off lower than the size of the protein. Micellar precipitation involves dilution of the protein extract with cold water (4°C), resulting in precipitation of highly stable protein micelles that form as a result of hydrophobic interactions between proteins (Arntfield et al., 1985). The protein recovered from micellar precipitation is then washed with distilled water, adjusted to neutral pH, and dried. While all wet extraction methods affect protein structure to some extent (Table 3.3), several studies have reported micellization to be the mildest extraction procedure, resulting in the least protein denaturation, as compared to SE-D and AE-IP, based on studies of yellow pea, cowpea, pigeon pea, and chickpea proteins (Mwasaru et al., 1999; Paredes-Lopez et al., 1991; Stone et al., 2015).

Several process parameters greatly influence the protein yield, purity, structure, and functionality obtained by conventional wet extraction. Initially, the particle size of the raw materials influences extraction (Section 3.2); other factors have also been determined through the study of food crops including cereals, legumes, and oilseeds (Hadnadjev et al., 2017; Momen et al., 2021; Rodrigues et al., 2012; Sari et al., 2015). In general, a lower solid:solvent ratio (i.e., greater volumes of solvent), along with a higher pH, has been linked to higher protein yields. For instance, (Drakos et al., 2017) recovered a higher amount (74% vs. 46%) of rye proteins under alkaline conditions when the solid to solvent ratio was decreased from 1:6 to 1:12. Similarly, (Ruiz et al., 2016) reported a greater extractability of quinoa proteins at higher pH (52% at pH 11 vs. 36% at pH 8). This increase in pH, however, also lead to increased denaturation. Similar results have been reported for many other proteins, including those from wheat, yellow pea, and lentils, though the impact of extraction pH on the protein structure varies (Gao et al., 2020; Jarpa-Parra et al., 2014; Liu et al., 2013). For instance, an extraction pH of ~9.5−10 did not seem to affect the secondary structure for yellow pea proteins but hydrolyzed lentil proteins.

Other factors known to aid protein recovery during wet extraction include, a higher temperature, and longer extraction, as demonstrated for sunflower seeds (Ordóñez et al., 2001) and rice bran (Gupta et al., 2008). These parameters need to be carefully regulated though, as strong alkaline conditions (pH > 10) and higher temperatures (> 40°C) have been shown to:

1. cause extensive protein denaturation through hydrolysis; and
2. trigger certain undesirable chemical reactions (i.e., sulfhydryl oxidation, isomerization cross-linking, protein−polyphenol reactions, Maillard reaction) leading to degradation of protein nutritional and organoleptic quality, as observed for soy, wheat, and hemp proteins (Deleu et al., 2019; Friedman et al., 1984; Liu et al., 2013; Mäkinen, Zannini, et al., 2016; Potin et al., 2019).

The advantages associated with conventional methods of wet extraction (Table 3.3) include the preparation of highly pure proteins containing > 90% protein (Schutyser & van der Goot, 2011) and the low cost and simplicity of operations (Zhang et al., 2015). Further, some isolates produced by wet extraction, such as those from cowpea, canola, and faba beans, have been reported to have decreased antinutritional factors, which may lead to improved digestibility (Hadnadjev et al., 2017; Momen et al., 2021). Owing to all these advantages, these methods have been widely used at the industrial scale (Table 3.3), to develop highly pure plant protein powders from all food groups, including from legumes (i.e., soy, pea), cereals (i.e., rice, wheat), and oilseeds (i.e., canola, sunflower) (Sozer & Poutanen, 2015).

The main disadvantages of wet extraction methods include the resource- and energy-intensive nature of these processes, i.e., 22.4 kg of hexane, 80 kg of water, and 40 g each of NaOH and HCl are required to produce 1 kg of lupin protein isolate from lupin seeds (Berghout et al., 2015). The use of chemicals, along with large volumes of water and generation of salt-containing effluents, raise environmental, and sustainability concerns (Sari et al., 2013; Schutyser & van der Goot, 2011). Additionally, as discussed previously, harsh processing conditions might adversely affect the nutritional and functional attributes, generating lower quality isolates. Process optimization, together with novel and more sustainable protein extraction methods, is therefore required to overcome these disadvantages.

3.4.2 Novel wet protein extraction techniques

3.4.2.1 Reverse micelles

Reverse micelles (RMs), also known as water-in-oil microemulsions, are nanometer-sized aggregates of surfactant molecules that have an aqueous inner core within a nonpolar solvent system (Fig. 3.1). RMs were initially identified as a promising method for protein and enzyme extraction without loss of activity in the 1980s (Barbaric & Luisi, 1981; Leser et al., 1986; Leser & Luisi, 1989). Yet despite much recent research, there are still several challenges to overcome for the application of this technique to protein extraction.

The application of RM to wet extraction of plant proteins from varied sources and associated extraction efficiencies is described in Table 3.3. The protein extraction process for RM consists of two steps:

1. forward extraction, where plant proteins are solubilized in the aqueous inner core of the RM; and
2. backward extraction, where the proteins are recovered from the RM into an aqueous phase.

Several factors determine the efficiency of protein extraction, starting with the choice of initial nonpolar solvent. Electrostatic interactions between plant proteins and RM surfactant are also key to the initial forward extraction. Parameters affecting these interactions therefore need consideration and optimization, including: the type and concentration of the surfactant, pH, ionic strength, water to surfactant ratio (W_0), nature and concentration of the plant protein, properties of the nonpolar solvent, and the extraction time and temperature (Sankaran et al., 2019; Sun & Bandara, 2019).

Several types of surfactants, such as cationic, anionic, zwitterionic, nonionic, and mixed surfactants, can be used to produce RM. The most popular choice for protein extraction from legumes, cereals, and nuts has been the anionic surfactant sodium di-2-ethylhexylsulfosuccinate (AOT), as aggregates of this surfactant can include large quantities of water within the inner core (Du et al., 2020; Liu et al., 2014; Zhu et al., 2009). This property is measured by the ratio of water to surfactant (W_0) and determines the size of the RM. Larger RMs have the capacity to solubilize larger and more concentrated proteins within their core, which ultimately increases the forward extraction efficiency (Sun et al., 2008).

Ionic strength has also been reported to affect RM dimensions, reducing RM size at higher salt concentrations (Wang, Liu, et al., 2018). Both ionic strength and pH play an important role, due to their impact on protein surface charge. With AOT, for example, when the pH is higher than the IEP of the proteins, both surfactant and protein will be negatively charged, leading to repulsion and reduced extraction efficiency.

During back extraction, the proteins are recovered using a fresh aqueous solution with a specific pH and ionic strength. The extraction efficiency of this process is typically low and requires both high amounts of water and energy (Table 3.3). Therefore, several authors have developed alternative backward extraction strategies, such as the addition of silica to trap both water and proteins (Leser et al., 1993) or the vaporization of the nonpolar solvent followed by protein precipitation by the sequential addition of acetone and then a water and isooctane mixture (Sun et al., 2009).

The structure and functionality of the proteins extracted by RM have been reported to have desirable characteristics for potential food applications. Several authors have demonstrated that RM extraction altered plant protein tertiary and secondary structure, increasing α-helical content, and reducing β-sheet and turn structures (Chen et al., 2013; Zhao et al., 2019b; Zhu et al., 2010). These conformational changes affected the functional properties of plant proteins. Compared to proteins extracted by conventional techniques, soybean and wheat germ proteins exhibited improved protein solubility, oil absorption capacity, foaming capacity and stability, and emulsifying properties but demonstrated reduced water absorption capacity (Zhao et al., 2018; Zhu et al., 2010). The textural properties of gels made from soybean proteins extracted by RM were also reported to have increased hardness, gumminess, adhesiveness, and chewiness (Zhao et al., 2015).

Extraction by RM has some advantages compared to conventional methods. In addition to improved protein functionality, the protein extraction process can be carried out together with fat extraction, reducing overall extraction times. Extraction by RM is also been reported to be more efficient and has fewer steps than some wet techniques (He et al., 2015), although many studies have only examined non-food-approved nonpolar solvents and surfactants. Research is therefore needed to investigate the potential of this extraction method using edible surfactants and solvents. RM experiments have also been restricted to laboratory scale, and the technique is currently expensive compared to other commercial methods. Other considerations include the potential reuse of surfactant and solvent streams and the large requirement for water and energy. Overall, the extraction of plant proteins using RM seems a promising method to improve the current functionality of protein isolates, but challenges still need to be overcome to enable potential uptake at larger scale.

3.4.2.2 Subcritical water extraction

Subcritical water extraction has been widely studied for the potential extraction of high-value bioactive compounds and essential oils for use in the pharmaceutical and food industries (Zhang et al., 2020). This process uses pressurized water in a liquid form above atmospheric boiling temperature (100°C) and below its critical temperature (374°C) (Fig. 3.1). Under these conditions, the physicochemical properties of water are radically modified:

1. the dielectric constant of water becomes similar to some organic solvents, such as acetone, methanol, or ethanol, allowing the interaction and dissolution of nonpolar compounds (Carr et al., 2011)
2. the ionic product of water increases due to the high temperature, which leads to a higher concentration of hydrogen ions and hydroxide ions, increasing the capability of water to act as an acid or basic catalyst in hydrolysis reactions (Marcet et al., 2016).

Several authors have applied this technology to extract proteins, peptides, and amino acids from cereals, pulses, and oilseeds (Pińkowska et al., 2013; Ramachandraiah et al., 2017; Wiboonsirikul et al., 2008). It has also been used to modify the functional properties of plant protein isolates (Zhang et al., 2018; Zhang, Wu, et al., 2014) (Table 3.3).

The severe pressure and temperature applied in subcritical conditions can hydrolyze the cell walls of plant materials, liberating proteins and other substances. The solubility of proteins is then favored by the enhanced properties of the subcritical water (Du & Kruse, 2021). Several chemical reactions occur in the high-pressure and -temperature environment including:

- dissolution;
- aggregation;
- disaggregation; and
- degradation of proteins (Sunphorka et al., 2012).

The extent of aggregation, hydrolysis, and extraction efficiency of proteins will depend on the processing conditions used. In addition to the pressure and temperature, the extraction time, pH, and ratio of plant material to water must be considered. In general, higher protein extraction yields are achieved with increasing temperatures between 100 and 200°C, as observed for wheat and rice brans (Table 3.3) (Kataoka et al., 2008; Wiboonsirikul et al., 2008). Above ~200°C, however, the degradation of proteins in plant sources, such as rice bran and soybean, leads to an increase in the concentration of low molecular peptides (lower than 10 kDa) and amino acids (Ramachandraiah et al., 2017; Sereewatthanawut et al., 2008).

Extraction times are closely linked to the temperature selected: at higher temperatures, extraction times should be shorter to avoid protein degradation, whereas longer times are more suitable for milder temperatures (Wiboonsirikul et al., 2007). A more alkaline environment (pH 9) and less concentrated material to solvent ratio (1:10) can also improve extraction efficiency for some proteins, such as those in flaxseed (Ho et al., 2007). Overall, given the variation in plant material and target protein properties, it is necessary to optimize the subcritical water extraction conditions for the specific plant materials to maximize the extraction efficiency and minimize protein degradation.

Subcritical water extraction greatly affects the structure and functional properties of the extracted proteins. Protein structure is easily lost when either high pressures or temperatures are applied. The combined effect of both conditions also hydrolyses proteins into peptides, amino acids, or further into organic acids (Di Domenico Ziero et al., 2020). Protein solubility improves during subcritical water extraction, due to protein hydrolysis, although partial aggregation can also occur during processing, reducing solubility (Wiboonsirikul et al., 2007; Zhang, Wu, et al., 2014). The emulsifying properties of wheat bran, rice bran, and soybean proteins improved with increasing extraction temperature, due to the unfolding of proteins and exposure of hydrophobic groups (Hata et al., 2008; Kataoka et al., 2008; Lu et al., 2016). Extracts obtained at higher temperatures, however, did not have any emulsifying activity, potentially due to the degradation of saccharides and protein conjugates at temperatures higher than 200°C. Similarly, foaming capacity and the stability of soy and zein protein isolates improved under subcritical water conditions, likely as a result of changes in protein aggregation, disaggregation, and degradation (Zhang et al., 2015; Zhang, Wu, et al., 2014).

The main advantage of the subcritical water extraction is the use of water as the extraction solvent, which makes this technology more sustainable than other conventional extraction processes (Section 3.4.1). Additionally, the extraction processes are shorter (from minutes to few hours) compared to other technologies. A disadvantage of the subcritical water extraction process is that it requires large amounts of water and energy, even though the energy consumption has been optimized with the introduction of novel designs of reactors (Di Domenico Ziero et al., 2020). Extraction also results in the loss of native protein structure, and the process must be optimized to avoid the degradation of proteins into amino or organic acids.

Subcritical water extraction has low technical readiness for commercial application and has only been trailed at a lab scale. Most studies have analyzed liquid extracts obtained directly from the subcritical water extraction process, without considering necessary concentration and drying steps. Overall, subcritical water extraction is a promising technology, but further development is needed.

3.4.2.3 Aqueous two-phase extraction systems

Proteins can be extracted using aqueous two-phase systems (ATPS), using two water-based solutes, which when mixed in appropriate ratios spontaneously separate into two immiscible phases (Fig. 3.1) (Raghavarao et al., 2003). During the extraction process, the proteins preferentially partition into one of the two phases; they are then precipitated from the protein-rich phase by addition of water, alcohol (i.e., ethanol), inorganic salt (i.e., K_3PO_4) or by adjusting the solution to the IEP of the protein. Though this process has been used extensively to extract and purify biomolecules, such as pigments, polyphenols, enzymes (i.e., papain and bromelain (Coelho et al., 2013; Pang et al., 2017)), and recombinant/allergenic proteins (from corn or wheat) (Gu, 2006; Lores et al., 2017); the utilization of ATPS for plant proteins in food applications is limited to certain cereals (oats), oilseeds (rapeseed, primrose, pumpkin), algae, and fruit peels (pomegranate) (Grudniewska et al., 2018; Hernández-Corroto et al., 2020; Liu et al., 2017; Suarez Ruiz et al., 2018; Yue, Zhu, et al., 2021).

Solutions used for ATPS include polymers (i.e., polyethylene glycol, dextran), salts (i.e., Na salts of citrate, phosphate), or ionic liquids ions that are liquid at temperatures $\leq$100°C such as alkylammonium salts (Ghandi, 2014; Iqbal et al., 2016). Environmentally friendly solvents are desired to reduce salty by-product streams and the impact of nonbiodegradable polymers and ionic liquids. Deep eutectic solvents (DESs) are one more sustainable example, which are defined by a low melting point than their two component solvents and require a hydrogen bond donor (i.e., choline chloride) and hydrogen bond acceptor (i.e., alcohols, acids) (Paiva et al., 2014). Naturally occurring DESs (NADES) made from plant metabolites, such as sugars (fructose, sucrose), alcohols (glycerol), acids (citric acid, malic acid), and choline chloride in particular have gained recent attention because of their nontoxic, nonflammable, recyclable, and biodegradable nature, as well as their generally recognized as safe (GRAS) status (Cunha & Fernandes, 2018; Grudniewska et al., 2018).

To extract proteins, the ground plant material is first suspended with ATPS and the suspension vigorously shaken (either manually or mechanically) for 30−180 min at a particular temperature (25−140°C). In general, higher temperatures and prolonged extraction times enhance protein recovery. For instance, increasing the extraction temperature from 60 to 140°C enhanced the protein content from 36% to 48% and doubled the yield for rapeseed proteins (Grudniewska et al., 2018). Similarly, greater protein recovery was observed with higher extraction times of 90 min compared to 60 min for oat proteins extracted at 80°C with several deep eutectic solvents (Yue, Zhu, et al., 2021). A further increase in the extraction time of up to 120 min, however, adversely affected protein yield and purity, potentially due to protein denaturation and simultaneous extraction of other components, such as β-glucan. These experiments demonstrate how extraction parameters need to be carefully optimized and controlled. The fraction of proteins that enters the two phases depends on protein composition, molecular weight, and charge, which determine solvent interactions, including electrostatic and hydrophobic interactions.

The extracted protein composition, structure, and functionality have been shown to vary with different solvents, owing to different protein−solvent interactions and solubility (Yue, Zhu, et al., 2021). Extraction using ATPS can also alter the secondary structure of proteins, similar to conventional extraction techniques (Section 3.4.1), with the degree of denaturation dependent on extraction conditions and solvents. ATPS has been shown to provide similar thermal stability (~105−111°C), protein solubility (10%−25%), and foaming stability (~80%) but lower foaming capacity (up to 150% with ATPS vs. up to 200% with AE-IP) than AE-IP for oat proteins (Yue, Gu, et al., 2021).

Similar to RMs, this method provides the advantage of simultaneous extraction of different biomolecules. For example, several different algal pigments have been successfully extracted in a polymer-rich phase, with proteins isolates occurring at the solvent interface using ATPS (Suarez Ruiz et al., 2018). Other advantages include short extraction times, easy-scale up, low costs, and the potential high availability of wide range of food-grade, biodegradable solvents (particularly NADES). Further, the extraction process using NADES, followed by precipitation of proteins through water addition, prevents the generation of acid/alkaline and salt-containing by-product streams, making this an environmentally friendly option compared to other to conventional extraction processes. Protein content and recovery are much lower than conventional methods (<60% vs. 75%–95% and 11%–60% vs. 25%–75%, respectively; Table 3.3), and the process still requires large amounts of water (~1:50) (Grudniewska et al., 2018; Yue, Zhu, et al., 2021). Nevertheless, APTS has potential as a more sustainable method. The limited studies performed to date highlight the need for further research and development to evaluate the suitability of this technique for larger scale commercial application.

3.4.3 Cell disruption techniques

Several cell disruption techniques can be applied to enhance protein yield and functionality. These can be employed either alone or synergistically during the wet extraction process. Such methods include ultrasound (US), microwave (MW), high hydrostatic pressure (HHP), pulsed electric energy (PEE), or the use of enzymes (Fig. 3.1 and Table 3.4). These methods work on the principle of increasing cell porosity, facilitating solvent diffusion into cells and increase protein release into the surrounding medium. Several studies have recently reviewed the effect of each of these techniques on the recovery of plant proteins from different sources (Contreras et al., 2019; Gençdağ et al., 2021; Kumar et al., 2021; Pojić et al., 2018; Rahman & Lamsal, 2021; Sari et al., 2015).

Each of these cell disruption techniques has their own advantages and disadvantages (Table 3.4). While the mechanical techniques (US, MW, PEE, and HHP) offer the potential to improve the overall process efficiency by reducing the processing time and solvent requirements, they could lead to undesirable outcomes, such as the production of free radicals and protein oxidation (Gençdağ et al., 2021; Rahman & Lamsal, 2021). US and MW, additionally, could cause heat induced denaturation of proteins (Kumar et al., 2021; Rahman & Lamsal, 2021). Enzymatic treatment can be used to better preserve the native structure of proteins, particularly when carbohydrases are used to disrupt matrices involving both carbohydrate and protein (Pojić et al., 2018; Sari et al., 2015).

Owing to their energy and capital-intensive nature, these techniques have largely been restricted to laboratory scale to date (Table 3.4). Further, while several studies have investigated the influence of ultrasound and microwave on the extractability of proteins, the data relating to application of HHP and PEE are scarce. More studies are therefore required to elucidate the impact of these methods on protein structure and function to assist the development of these processes.

3.5 Methods to improve the functionality of protein isolates

The application of plant proteins in food products is currently limited beyond the full potential of these ingredients due to their reduced functional properties compared to proteins of animal origin. These limitations typically include a lower solubility and higher susceptibility to pH and temperature changes (Nasrabadi et al., 2021) or result from a loss of structure and functionality during the extraction process. The latter is common in proteins extracted by a combination of alkaline extraction and isoelectric precipitation (Momen et al., 2021), which is currently the most common method for plant protein extraction at an industrial scale (GEA, 2020).

Many authors have investigated novel chemical, physical, and biological methods to improve the functionality and nutritional properties of plant proteins. Most studies have been conducted on extracted plant protein concentrates and isolates (García Arteaga et al., 2021; Mir et al., 2019; Wang, Sun, et al., 2018). However, these processes could also be applied to plant materials during the extraction process (Fig. 3.1 and Section 3.4.3), which may simultaneously assist protein extraction by disrupting cell walls and alter the structure and functionality of proteins (Wang et al., 2020). Chemical, physical, and biological modifications are considered here in turn.

Chemical modification consists of the incorporation, or elimination, of functional groups on the protein structure through the addition of chemical agents and is a highly efficient method for modifying plant proteins. Some of the most popular chemical reactions used to modify plant proteins include glycosylation, acylation, succinylation, phosphorylation, cationization, deamidation, and pH shifting treatments (Nasrabadi et al., 2021). The structure and functionality of cereals, oilseeds, legumes, and other plant protein isolates have been successfully modified by such chemical modulation (Wang et al., 2019; Wang, Sun, et al., 2018). For instance, phosphorylation altered the secondary and primary structure of soy and pea proteins, improving solubility, emulsifying, and foaming properties and oil absorption capacity (Liu et al., 2020; Sánchez-Reséndiz et al., 2018). Glycosylation of soy protein isolate with several saccharides, such as maltodextrin, xylose, fructose, or lactose, led to enhanced solubility and water holding capacity but varying emulsification properties (Wang, Sun, et al., 2018; Wang et al., 2013; Zhang, Wu, et al., 2014). Interestingly, glycosylation has also been used to reduce the beany flavor of pulse protein isolates (Zhang et al., 2019), which could be especially helpful in the application of pulse protein isolates. Both regulatory and clean label concerns need to be addressed, however, for this method.

TABLE 3.4 Overview of novel cell disruption techniques applied during wet protein extraction to enhance protein yield and/or functionality.

Disruption technique	Principle of disruption	Advantages	Limitations	Scale	References
Ultrasound	Shear induced disruption through cavitation, achieved by application of sound waves of a frequency >20 kHz	Easy installation Reduced processing time Low maintenance costs Potential to reduce allergenicity	Protein denaturation/oxidation may occur Different yields obtained at lab and pilot scale Possible metal contamination Need for safety considerations to prevent health hazards associated with high sound levels during operation Energy intensive	Lab scale	Kumar et al. (2021), Rahman and Lamsal (2021), Ochoa-Rivas et al. (2016)
Microwave	Increased cell porosity due to pressure generation by heated water molecules inside the cells, achieved via electromagnetic waves in frequency range of 300 MHz to 300 GHz	Potential to reduce allergenicity Reduced processing time Low energy and cost	High amount of heat generated may denature proteins	Lab scale	Kumar et al. (2021), Gençdağ et al. (2021), Pojić et al. (2018)
High-hydrostatic pressure	Shear induced disruption through hydrodynamic cavitation achieved at pressures >35 MPa	Potential to reduce allergenicity Thermal denaturation of proteins is prevented, as extraction can be performed at ambient temperatures	High energy and investment cost	Lab scale	Kumar et al. (2021), Pojić et al. (2018), Preece et al. (2017)
Pulsed electric energy (including pulsed electric field, high voltage electrical discharge and ohmic heating)	Tissue damage through electroporation by application of high amplitude (100–300 V/cm to 10–50 kV/cm) pulses for short duration (ns to ms)	Low operating cost Thermal denaturation of proteins is prevented, as extraction can be performed at ambient temperatures	Energy intensive Can generate undesirable reactive species, such as free radicals	Lab scale	Kumar et al. (2021), Gençdağ et al. (2021), Pojić et al. (2018), Vorobiev and Lebovka (2016)
Enzyme	Cell wall and protein degradation, achieved by carbohydrases and proteases	Use of cell wall degrading enzymes at near neutral pH (6–7) preserves the nativity of proteins Simultaneous extraction of other components Thermal denaturation of proteins is prevented, as extraction is performed near ambient temperatures (25–60°C)	Cost, time, and energy intensive	Lab scale	Kumar et al. (2021), Pojić et al. (2018), Sari et al. (2015)

Physical modification involves the use of energy in various forms (i.e., thermal, pressure, radiation) to alter the structure of proteins. This method includes heating (conventional, microwave, ohmic, or infrared), sonication, extrusion, radiation (gamma, ultraviolet, or electron beam), pulsed-electric field, or high-pressure treatments. Many of these techniques are also used as extraction-assisting techniques (Fig. 3.1, Table 3.4 and Section 3.4.3) and have been reported to improve several functional, organoleptic, and nutritional properties of plant proteins (Nasrabadi et al., 2021). Thermal processing is one of the most popular physical approaches, due to its potential to unfold proteins. Thermal processing has been reported to modify protein solubility, thermal stability, water holding capacity, gelling, emulsifying, and foaming properties for several plant proteins (Akharume et al., 2021).

Sonication is a novel and sustainable technology the uses high-frequency waves (>16 kHz) to induce changes in the secondary, tertiary, and quaternary structure of proteins, which can be used to modify plant proteins. Sonication has been used to improve the solubility, water, and oil holding capacity and emulsifying and foaming properties of pulses, cereals, nuts, pseudocereals, and oilseeds (Mir et al., 2019; Nazari et al., 2018; Zhu et al., 2018). Sonication also has the potential to reduce allergenicity and antinutrient activity, as well as enhancing protein digestibility (Rahman & Lamsal, 2021; Vanga et al., 2020). The use of physical methods to modify proteins is growing in popularity, due to the versatility of the modified proteins, clean labeling, efficiency, and sustainability of many of these techniques.

Biological modification is another potentially environmentally and consumer friendly approach to protein modification that uses enzymatic and fermentation reactions to modulate protein structure and properties. Enzymes can be used to hydrolyze specific peptide bonds (i.e., pepsin) (Wouters et al., 2016) or to catalyze the cross-linking of several proteins (i.e., transglutaminase) (Sun & Arntfield, 2011). The changes in the functional properties obtained by enzymatic modification depend on the type of protein, enzyme, and reaction conditions (Arteaga et al., 2020; Wouters et al., 2016). A variety of enzymes such as alcalase, papain, esperase, bromelain, trypsin, and chymotrypsin have shown potential as aids to improve protein solubility, as well as emulsion and foaming capacity for pea protein isolates (Arteaga et al., 2020). Bitterness is a common problem of protein hydrolysates, which arise due to low-molecular-weight peptides containing high amounts of hydrophobic amino acids (Panyam & Kilara, 1996). Enzymatic cross-linking has also been shown to improve texture, emulsifying, gelation, and foaming properties, together with improved viscosity for several legumes and cereals (Akharume et al., 2021; Sun & Arntfield, 2011).

Another biological approach is fermentation, which can be applied to protein isolates using bacteria, yeast, and molds (Meinlschmidt et al., 2016). This method has been reported to not only to improve functional properties but also to improve digestibility and decrease antinutritional factors and allergenicity, in addition to bitter and beany off-flavors (García Arteaga et al., 2021; Meinlschmidt et al., 2016; Shi et al., 2021).

Plant protein modification is an important step within the plant protein production process that allows high-quality protein products with superior functionality. This interesting topic has been thoroughly reviewed by (Nasrabadi et al., 2021) and (Akharume et al., 2021). These recent reviews also detail other approaches to protein modification, including complexation and amyloid fibril formation, which are further ways to modify protein structure and function, although the latter requires careful consideration (Raynes et al., 2014). Most studies conducted to date have been carried out using protein isolates as raw materials. Integrating the protein modulation stage into the protein extraction process (Fig. 3.1) would potentially improve the overall efficiency of protein extraction and may produce protein with improved functional properties for use in the food industry.

3.6 Novel hybrid dry and wet extraction methods

Both dry and wet protein extraction methods have several advantages, as discussed in Sections 3.3 and 3.4. Protein concentrates obtained by dry extraction typically maintain their native structure and functionality; the process is also energy efficient and uses no water. The maximum protein concentration that can be achieved by dry extraction, however, is ~70% and more commonly ~60% (Table 3.2). Using conventional wet extraction methods, protein purity can be > 90%, but the protein structure and functionality are often affected by extraction, which consumes higher amounts of energy, water, and other solvents. Considering the positive aspects of both methods, a few authors have recently explored a two-step hybrid dry and wet protein extraction method for different legumes and quinoa seeds.

A combined dry and wet extraction process has been used to:

1. increase the purity of protein concentrates (Föste et al., 2015; Ruiz et al., 2016),
2. improve the overall protein recovery (Dumoulin et al., 2021),
3. obtain protein fractions with specific functional properties (Pelgrom, Boom, & Schutyser, 2015), and
4. improve overall sustainability.

This latter point is most important and has been covered well in a sustainability assessment of wet, dry, and hybrid methods (Berghout et al., 2015).

A hybrid extraction process has successfully been applied to quinoa, involving either air classification followed by saline extraction and ultrafiltration (Ruiz et al., 2016) or sieving followed by alkaline extraction and isoelectric precipitation (Föste et al., 2015). Both studies aimed to initially separate the embryo, which contains the most concentrated source of protein in the quinoa seed (~57%) (Ando et al., 2002), from the perisperm (Fig. 3.2). A better separation of these two structures was obtained during milling when the moisture content of the seeds was adjusted to 15%. After milling and dry fractionation, the protein purity of the enriched fraction was 24% and 28% db, respectively (for (Ruiz et al., 2016) and (Föste et al., 2015), where the initial protein content in the flour was 14.5% and 11.8% db, respectively).

In the study involving saline extraction, the milled quinoa particles were reduced from 559 to 30 μm and then suspended in saline (1:5 w/w, 0.5M NaCl), resulting in three phases, where the proteins were mainly located in the top liquid layer, which could then be ultrafiltered. The hybrid process resulted in lower protein purities than conventional wet extraction (59% vs. 88%−91% db) but higher protein recovery (62% vs. 36%) and used 98% less water (Ruiz et al., 2016). It should be noted, however, that water reduction was calculated on the basis of the starting material instead of the concentrate obtained, which would be a more accurate measure.

In the study involving alkaline extraction, the protein-rich fraction was suspended in an alkaline solution (1:10 w/v, pH 10) and concentrated by isoelectric precipitation at pH 4. The final protein content obtained by the hybrid extraction process was 68% db, compared to 52% db when the initial dry fractionation was omitted (Föste et al., 2015).

The proteins from yellow peas, lupins, and faba beans have also been extracted by a combination of dry and wet extraction. Dry extraction was carried out using air classification, but the following wet extraction differed, depending on the aim of the study. (Pelgrom, Boom, & Schutyser, 2015) dispersed the protein-rich fine fractions in water (1:5 w/w) and obtained four different phases. The protein concentration of the two top layers was 59% db, and these were further concentrated to 67% db by ultrafiltration, resulting in a protein recovery of 63% from the original pea flour. (Berghout et al., 2015) further processed lupin fine fractions with alkaline extraction (1:15 w/w, pH 9) and isoelectric precipitation, obtaining a protein purity of ~83% and protein recovery of 66%. In order to increase protein extraction recovery from faba beans, (Dumoulin et al., 2021) continued the air classification process by extracting the proteins from the protein-deprived coarse fraction (protein purity 22% db) using alkaline extraction (1:10 w/v, pH 9.5) and cryo-precipitation (4°C). This hybrid approach produced two faba protein-enriched fractions: (1) the fine fraction from air classification (54% db) with native functionality and (2) the wet extracted protein concentrate (61% db), with a combined recovery of 88% of the proteins initially contained in faba beans.

There is little scientific literature regarding the functionality of plant proteins extracted by hybrid methods. It could be hypothesized that the wet extraction process will mainly determine the structure and functionality of the proteins, due to the more severe nature of this process. Differences in functionality have been observed between fractionation processes. The heat-induced gelatinization and microstructure of gels formed from yellow pea flour formed from two different extractions methods differed; these were dry extracted fine and coarse fractions or a fraction obtained by a dry−wet hybrid method (Pelgrom, Boom, & Schutyser, 2015). The gel obtained from the hybrid method formed a continuous protein network and was harder than the gel obtained from the fine fraction, due to a higher protein content. Such comparisons are complex, however, due to differences in composition. For example, the same gel was weaker than the original flour and coarse flour due to their higher starch content. Further research is needed to understand the changes in structure and functionality of proteins obtained from dry and wet hybrid extractions independently of other compositional changes in these mixtures.

Hybrid extraction methods have both advantages and drawbacks. A key feature is the high protein recovery, although the protein purity obtained is lower than for conventional wet methods. Initial dry fractionation steps reduce the amount of starch and fiber in the plant materials, which results in far less water and energy use. Estimates suggest that the amount of water, NaOH, and HCl can be halved during the production of a similar protein concentrate (Berghout et al., 2015). The main drawback of combined methods is the greater concentration of starting material required (approximately double) since the mass flow into the wet extraction is approximately halved. The side streams are rich in starch and fiber and still contain a high amount of protein, and these ingredients could be isolated and valorized to increase overall process efficiency and profitability. In conclusion, hybrid methods involving both dry and wet methods of extraction are a promising approach that may allow more sustainable extraction processes, producing concentrates with an adequate protein purity for food applications.

3.7 Summary and future outlook

Consumer demand for novel and sustainably developed plant proteins is increasing, driven by a growing awareness of food sustainability, alternative diets, animal welfare, and the health benefits associated with consumption of plant-based products. Increasing the production of plant proteins to meet this demand requires:

- a comprehensive understanding of raw plant materials, particularly from emerging plant sources, such as algae, leafy greens, and agri-food residues (i.e., seeds and peels from fruit and vegetables) and
- increased know-how on how these materials may be processed and utilized.

The use of agri-food residues as feedstocks for such protein production has the potential to transform the agri-food industry into a circular economy. It is expected that most of this change will occur within specific food industries or within specific geographical areas, where mass agricultural residues are produced.

An ideal protein extraction process will generate high yields of protein and ingredients that are functional for use in a variety of food products, such as baked goods or egg, dairy, and meat analogs or tailored for a specific final application. The selection or development of extraction processes starts with a consideration of the characteristics (e.g., structural features) and composition of the raw source materials. The sustainability and economically feasibility of the process are also critically important. To achieve these goals the process should involve:

- the minimal processing of raw materials, i.e., as few processing steps as possible and
- the efficient utilization of resources, i.e., the efficient use of energy and water.

Extracted protein ingredients include both highly purified proteins, or isolates, as well as protein concentrates, which have a lower protein purity.

Highly purified protein can be produced using current wet extraction techniques, but these are not necessarily sustainable. In contrast, protein concentrates may be obtained by dry or hybrid extraction techniques, which are potentially more environmentally sustainable. The optimization and upscaling of dry fractionation, however, require a greater understanding of the impact of this technique on protein structure and functionality.

Novel wet extraction techniques offer further useful routes to protein ingredients and have the advantage of producing an additional product streams, such as oils or other bioactive compounds that may be simultaneously extracted with protein. Aqueous two-phase extraction, for example, has been successfully used to separate several algal pigments, together with protein isolates; RMs have also been used to extract both fat and protein from oilseeds (Leser & Luisi, 1989; Suarez Ruiz et al., 2018). Simultaneous extraction can also improve process efficiency and process economics.

The reduction of solvent use, the adoption solvent recycling, or the transition to processes that use biodegradable solvents could also reduce the environmental impact of both hybrid and novel wet extraction techniques.

A key limitation of many novel extraction processes is that most studies have been conducted at laboratory scale. The optimization and upscaling of these methods is required to fully evaluate both environmental impact and commercial viability. Research is also required to further assess the potential for side streams valorization.

A quicker route to plant protein production could also involve the repurposing of existing processes from within the food industry, such as membrane filtration. This equipment is readily available, and while studies are required to apply these techniques to plant proteins, there is already considerable existing technical knowledge concerning the operation and economics of these techniques. This approach may also make the transition to more sustainable and commercially viable protein extraction processes possible in the immediate future.

References

Akharume, F. U., Aluko, R. E., & Adedeji, A. A. (2021). Modification of plant proteins for improved functionality: A review. *Comprehensive Reviews in Food Science and Food Safety, 20*(1), 198–224. https://doi.org/10.1111/1541-4337.12688

Ando, H., Chen, Y. C., Tang, H., Shimizu, M., Watanabe, K., & Mitsunaga, T. (2002). Food components in fractions of quinoa seed. *Food Science and Technology Research, 8*(1), 80–84. https://doi.org/10.3136/fstr.8.80

Anton, A. A., Ross, K. A., Beta, T., Gary Fulcher, R., & Arntfield, S. D. (2008). Effect of pre-dehulling treatments on some nutritional and physical properties of navy and pinto beans (*Phaseolus vulgaris* L.). *LWT - Food Science and Technology, 41*(5), 771–778. https://doi.org/10.1016/j.lwt.2007.05.014

Arntfield, S. D., Ismond, M. A. H., & Murray, E. D. (1985). The fate of antinutritional factors during the preparation of a fababean protein isolate using a micellization technique. *Canadian Institute of Food Science and Technology Journal*, 137–143. https://doi.org/10.1016/s0315-5463(85)71771-3

Arteaga, G., Guardia, M., Muranyi, I., Eisner, P., & Schweiggert-Weisz, U. (2020). Effect of enzymatic hydrolysis on molecular weight distribution, techno-functional properties and sensory perception of pea protein isolates. *Innovative Food Science & Emerging Technologies, 65*.

Assatory, A., Vitelli, M., Rajabzadeh, A. R., & Legge, R. L. (2019). Dry fractionation methods for plant protein, starch and fiber enrichment: A review. *Trends in Food Science and Technology, 86*, 340–351. https://doi.org/10.1016/j.tifs.2019.02.006

Avila Ruiz, G., Arts, A., Minor, M., & Schutyser, M. (2016). A hybrid dry and aqueous fractionation method to obtain protein-rich fractions from quinoa (*Chenopodium quinoa* Willd). *Food and Bioprocess Technology, 9*(9), 1502–1510. https://doi.org/10.1007/s11947-016-1731-0

Banjac, V., Colovic, R. R., Vukmirovic, D. M., Sredanovic, S. A., Colovic, D. S., Levic, J. D., & Teodosin, S. J. (2013). Protein enrichment of sunflower meal by air classification. *Food and Feed Research, 40*, 77–83.

Barakat, A., Jérôme, F., & Rouau, X. (2015). A dry platform for separation of proteins from biomass-containing polysaccharides, lignin, and polyphenols. *ChemSusChem, 8*(7), 1161–1166. https://doi.org/10.1002/cssc.201403473

Barbaric, S., & Luisi, P. L. (1981). Micellar solubilization of biopolymers in organic solvents. 5. Activity and conformation of α-chymotrypsin in isooctane-AOT reverse micelles. *Journal of the American Chemical Society, 103*(14), 4239–4244. https://doi.org/10.1021/ja00404a044

Berghout, J. A. M., Pelgrom, P. J. M., Schutyser, M. A. I., Boom, R. M., & Van Der Goot, A. J. (2015). Sustainability assessment of oilseed fractionation processes: A case study on lupin seeds. *Journal of Food Engineering, 150*, 117–124. https://doi.org/10.1016/j.jfoodeng.2014.11.005

Boye, J., Zare, F., & Pletch, A. (2010). Pulse proteins: Processing, characterization, functional properties and applications in food and feed. *Food Research International, 43*(2), 414–431. https://doi.org/10.1016/j.foodres.2009.09.003

Campbell, K. A., & Glatz, C. E. (2009). Mechanisms of aqueous extraction of soybean oil. *Journal of Agricultural and Food Chemistry, 57*(22), 10904–10912. https://doi.org/10.1021/jf902298a

Carr, A. G., Mammucari, R., & Foster, N. R. (2011). A review of subcritical water as a solvent and its utilisation for the processing of hydrophobic organic compounds. *Chemical Engineering Journal, 172*(1), 1–17. https://doi.org/10.1016/j.cej.2011.06.007

Chen, X., Ru, Y., Chen, F., Wang, X., Zhao, X., & Ao, Q. (2013). FTIR spectroscopic characterization of soy proteins obtained through AOT reverse micelles. *Food Hydrocolloids, 31*(2), 435–437. https://doi.org/10.1016/j.foodhyd.2012.11.017

Chen, Z., Wang, L., Wang, R., Li, Y., & Chen, Z. (2014). Triboelectric separation of aleurone cell-cluster from wheat bran fragments in nonuniform electric field. *Food Research International, 62*, 111–120. https://doi.org/10.1016/j.foodres.2014.02.043

Coelho, D. F., Silveira, E., Pessoa Junior, A., & Tambourgi, E. B. (2013). Bromelain purification through unconventional aqueous two-phase system (PEG/ammonium sulphate). *Bioprocess and Biosystems Engineering, 36*(2), 185–192. https://doi.org/10.1007/s00449-012-0774-5

Contreras, M.d. M., Lama-Muñoz, A., Manuel Gutiérrez-Pérez, J., Espínola, F., Moya, M., & Castro, E. (2019). Protein extraction from agri-food residues for integration in biorefinery: Potential techniques and current status. *Bioresource Technology, 280*, 459–477. https://doi.org/10.1016/j.biortech.2019.02.040

Cunha, S. C., & Fernandes, J. O. (2018). Extraction techniques with deep eutectic solvents. *TrAC - Trends in Analytical Chemistry, 105*, 225–239. https://doi.org/10.1016/j.trac.2018.05.001

Day, L. (2013). Proteins from land plants - potential resources for human nutrition and food security. *Trends in Food Science and Technology, 32*(1), 25–42. https://doi.org/10.1016/j.tifs.2013.05.005

Deleu, L. J., Lambrecht, M. A., Van de Vondel, J., & Delcour, J. A. (2019). The impact of alkaline conditions on storage proteins of cereals and pseudo-cereals. *Current Opinion in Food Science, 25*, 98–103. https://doi.org/10.1016/j.cofs.2019.02.017

Di Domenico Ziero, H., Buller, L. S., Mudhoo, A., Ampese, L. C., Mussatto, S. I., & Carneiro, T. F. (2020). An overview of subcritical and supercritical water treatment of different biomasses for protein and amino acids production and recovery. *Journal of Environmental Chemical Engineering, 8*(5). https://doi.org/10.1016/j.jece.2020.104406

Dijkink, B. H., & Langelaan, H. C. (2002). Milling properties of peas in relation to texture analysis. Part I. Effect of moisture content. *Journal of Food Engineering, 51*(2), 99–104. https://doi.org/10.1016/S0260-8774(01)00043-7

Dijkink, B. H., Speranza, L., Paltsidis, D., & Vereijken, J. M. (2007). Air dispersion of starch-protein mixtures: A predictive tool for air classification performance. *Powder Technology, 172*(2), 113–119. https://doi.org/10.1016/j.powtec.2006.10.039

Drakos, A., Malindretou, K., Mandala, I., & Evageliou, V. (2017). Protein isolation from jet milled rye flours differing in particle size. *Food and Bioproducts Processing, 104*, 13–18. https://doi.org/10.1016/j.fbp.2017.04.004

Du, L., & Kruse, A. (2021). Cell disruption and value-added substances extraction from Arthrospira platensis using subcritical water. *The Journal of Supercritical Fluids, 171*, 105193. https://doi.org/10.1016/j.supflu.2021.105193

Dumoulin, L., Jacquet, N., Malumba, P., Richel, A., & Blecker, C. (2021). Dry and wet fractionation of plant proteins: How a hybrid process increases yield and impacts nutritional value of faba beans proteins. *Innovative Food Science and Emerging Technologies*, 72. https://doi.org/10.1016/j.ifset.2021.102747

Du, Y., Zhang, Q., Zhao, X., & Chen, F. (2020). Effect of reverse micelle on physicochemical properties of soybean 7S globulins. *Journal of Food Engineering*, 282. https://doi.org/10.1016/j.jfoodeng.2020.110026

Embaby, H. E. S. (2010). Effect of soaking, dehulling, and cooking methods on certain antinutrients and in vitro protein digestibility of bitter and sweet lupin seeds. *Food Science and Biotechnology, 19*(4), 1055–1062. https://doi.org/10.1007/s10068-010-0148-1

Espinosa-Ramírez, J., & Serna-Saldívar, S. O. (2019). Wet-milled chickpea coproduct as an alternative to obtain protein isolates. *LWT*, 115. https://doi.org/10.1016/j.lwt.2019.108468

Falconer, A. (2003). Gravity separation: Old technique/new methods. *Physical Separation in Science and Engineering, 12*(1), 31–48. https://doi.org/10.1080/1478647031000104293

FAOSTAT. (2021). *Food security indicators.*

Fernando, S. (2021). Production of protein-rich pulse ingredients through dry fractionation: A review. *LWT*, 141. https://doi.org/10.1016/j.lwt.2021.110961

Fernando, S., & Manthey, F. A. (2021). Milling method affects the physical properties of black bean flour. *Cereal Chemistry, 98*(3), 749–758. https://doi.org/10.1002/cche.10418

FIAL. (2019). *Protein market: Size of the prize analysis for Australia.*

Foegeding, E. A., & Davis, J. P. (2011). Food protein functionality: A comprehensive approach. *Food Hydrocolloids, 25*(8), 1853–1864. https://doi.org/10.1016/j.foodhyd.2011.05.008

Föste, M., Elgeti, D., Brunner, A. K., Jekle, M., & Becker, T. (2015). Isolation of quinoa protein by milling fractionation and solvent extraction. *Food and Bioproducts Processing, 96*, 20–26. https://doi.org/10.1016/j.fbp.2015.06.003

Friedman, M., Gumbmann, M. R., & Masters, P. M. (1984). *Protein-alkali reactions: Chemistry, toxicology, and nutritional consequences* (pp. 367–412). Springer Science and Business Media LLC. https://doi.org/10.1007/978-1-4684-4790-3_18

Gao, Z., Shen, P., Lan, Y., Cui, L., Ohm, J. B., Chen, B., & Rao, J. (2020). Effect of alkaline extraction pH on structure properties, solubility, and beany flavor of yellow pea protein isolate. *Food Research International*, 131. https://doi.org/10.1016/j.foodres.2020.109045

García Arteaga, V., Leffler, S., Muranyi, I., Eisner, P., & Schweiggert-Weisz, U. (2021). Sensory profile, functional properties and molecular weight distribution of fermented pea protein isolate. *Current Research in Food Science, 4*, 1–10. https://doi.org/10.1016/j.crfs.2020.12.001

GEA. (2020). *Getting Started in plant protein separation – the basics.*

Gençdağ, E., Görgüç, A., & Yõlmaz, F. M. (2021). Recent advances in the recovery techniques of plant-based proteins from agro-industrial by-products. *Food Reviews International, 37*(4), 447–468. https://doi.org/10.1080/87559129.2019.1709203

Geow, C. H., Tan, M. C., Yeap, S. P., & Chin, N. L. (2021). A review on extraction techniques and its future applications in industry. *European Journal of Lipid Science and Technology, 123*(4). https://doi.org/10.1002/ejlt.202000302

Ghandi, K. (2014). A review of ionic liquids, their limits and applications. *Green and Sustainable Chemistry, 4*, 44–53. https://doi.org/10.4236/gsc.2014.41008

GlobalData. (2020). *Top Trends in Dairy 2020. Exploring the latest consumer and innovation trends, and future opportunities.*

Grudniewska, A., De Melo, E. M., Chan, A., Gniłka, R., Boratyński, F., & Matharu, A. S. (2018). Enhanced protein extraction from oilseed cakes using glycerol-choline chloride deep eutectic solvents: A biorefinery approach. *ACS Sustainable Chemistry and Engineering, 6*(11), 15791–15800. https://doi.org/10.1021/acssuschemeng.8b04359

Grundy, M. M. L., Lapsley, K., & Ellis, P. R. (2016). A review of the impact of processing on nutrient bioaccessibility and digestion of almonds. *International Journal of Food Science and Technology, 51*(9), 1937–1946. https://doi.org/10.1111/ijfs.13192

Gu, Z. (2006). *Recovery and purification of recombinant protein from transgenic corn by aqueous two-phase partitioning.*

Gupta, S., Chandi, G. K., & Sogi, D. S. (2008). Effect of extraction temperature on functional properties of rice bran protein concentrates. *International Journal of Food Engineering, 4*(2). https://doi.org/10.2202/1556-3758.1165

Hadnadjev, M., Dapcevic-Hadnadjev, T., Pojic, M., Saric, B., Misan, A., Jovanov, P., & Sakac, M. (2017). Progress in vegetable proteins isolation techniques: A review. *Food and Feed Research*, 11–21. https://doi.org/10.5937/FFR1701011H

Hahn, J.J., C. Lera Calvo, M. Murru, inventors. Protein enrichment. (2014) US Pat. No. WO2014179572.

Hata, S., Wiboonsirikul, J., Maeda, A., Kimura, Y., & Adachi, S. (2008). Extraction of defatted rice bran by subcritical water treatment. *Biochemical Engineering Journal, 40*(1), 44–53. https://doi.org/10.1016/j.bej.2007.11.016

Hemery, Y., Chaurand, M., Holopainen, U., Lampi, A. M., Lehtinen, P., Piironen, V., Sadoudi, A., & Rouau, X. (2011). Potential of dry fractionation of wheat bran for the development of food ingredients, part I: Influence of ultra-fine grinding. *Journal of Cereal Science, 53*(1), 1–8. https://doi.org/10.1016/j.jcs.2010.09.005

Hemery, Y. M., Mabille, F., Martelli, M. R., & Rouau, X. (2010). Influence of water content and negative temperatures on the mechanical properties of wheat bran and its constitutive layers. *Journal of Food Engineering, 98*(3), 360–369. https://doi.org/10.1016/j.jfoodeng.2010.01.012

Hernández-Corroto, E., Plaza, M., Marina, M. L., & García, M. C. (2020). Sustainable extraction of proteins and bioactive substances from pomegranate peel (*Punica granatum* L.) using pressurized liquids and deep eutectic solvents. *Innovative Food Science and Emerging Technologies*, 60. https://doi.org/10.1016/j.ifset.2020.102314

He, S., Shi, J., Walid, E., Zhang, H., Ma, Y., & Xue, S. J. (2015). Reverse micellar extraction of lectin from black turtle bean (*Phaseolus vulgaris*): Optimisation of extraction conditions by response surface methodology. *Food Chemistry, 166*, 93–100. https://doi.org/10.1016/j.foodchem.2014.05.156

Ho, C. H. L., Cacace, J. E., & Mazza, G. (2007). Extraction of lignans, proteins and carbohydrates from flaxseed meal with pressurized low polarity water. *LWT - Food Science and Technology, 40*(9), 1637–1647. https://doi.org/10.1016/j.lwt.2006.12.003

Iqbal, M., Tao, Y., Xie, S., Zhu, Y., Chen, D., Wang, X., Huang, L., Peng, D., Sattar, A., Shabbir, M. A. B., Hussain, H. I., Ahmed, S., & Yuan, Z. (2016). Aqueous two-phase system (ATPS): An overview and advances in its applications. *Biological Procedures Online, 18*(1), 1–18. https://doi.org/10.1186/s12575-016-0048-8

Jääskeläinen, A. S., Holopainen-Mantila, U., Tamminen, T., & Vuorinen, T. (2013). Endosperm and aleurone cell structure in barley and wheat as studied by optical and Raman microscopy. *Journal of Cereal Science, 57*(3), 543–550. https://doi.org/10.1016/j.jcs.2013.02.007

Jarpa-Parra, M., Bamdad, F., Wang, Y., Tian, Z., Temelli, F., Han, J., & Chen, L. (2014). Optimization of lentil protein extraction and the influence of process pH on protein structure and functionality. *LWT - Food Science and Technology, 57*(2), 461–469. https://doi.org/10.1016/j.lwt.2014.02.035

Jeske, S., Zannini, E., & Arendt, E. K. (2018). Past, present and future: The strength of plant-based dairy substitutes based on gluten-free raw materials. *Food Research International, 110*, 42–51. https://doi.org/10.1016/j.foodres.2017.03.045

Kannan, A. S., Jareteg, K., Lassen, N. C. K., Carstensen, J. M., Hansen, M. A. E., Dam, F., & Sasic, S. (2017). Design and performance optimization of gravity tables using a combined CFD-DEM framework. *Powder Technology, 318*, 423–440. https://doi.org/10.1016/j.powtec.2017.05.046

Kataoka, M., Wiboonsirikul, J., Kimura, Y., & Adachi, S. (2008). Properties of extracts from wheat bran by subcritical water treatment. *Food Science and Technology Research, 14*(6), 553–556. https://doi.org/10.3136/fstr.14.553

Kdidi, S., Vaca-Medina, G., Peydecastaing, J., Oukarroum, A., Fayoud, N., & Barakat, A. (2019). Electrostatic separation for sustainable production of rapeseed oil cake protein concentrate: Effect of mechanical disruption on protein and lignocellulosic fiber separation. *Powder Technology, 344*, 10–16. https://doi.org/10.1016/j.powtec.2018.11.107

Keivaninahr, F., Gadkari, P., Zoroufchi Benis, K., Tulbek, M., & Ghosh, S. (2021). Prediction of emulsification behaviour of pea and faba bean protein concentrates and isolates from structure-functionality analysis. *RSC Advances, 11*(20), 12117–12135. https://doi.org/10.1039/d0ra09302e

Khandelwal, S., Udipi, S. A., & Ghugre, P. (2010). Polyphenols and tannins in Indian pulses: Effect of soaking, germination and pressure cooking. *Food Research International, 43*(2), 526–530. https://doi.org/10.1016/j.foodres.2009.09.036

Krueger, N. A., Bern, C. J., Misra, M. K., & Adam, K. M. (2007). Gravity table sorting of commodity corn. *Applied Engineering in Agriculture, 23*(3), 319–325.

Kumar, M., Tomar, M., Potkule, J., Verma, R., Punia, S., Mahapatra, A., Belwal, T., Dahuja, A., Joshi, S., Berwal, M. K., Satankar, V., Bhoite, A. G., Amarowicz, R., Kaur, C., & Kennedy, J. F. (2021). Advances in the plant protein extraction: Mechanism and recommendations. *Food Hydrocolloids*, 115. https://doi.org/10.1016/j.foodhyd.2021.106595

Laguna, O., Barakat, A., Alhamada, H., Durand, E., Baréa, B., Fine, F., Villeneuve, P., Citeau, M., Dauguet, S., & Lecomte, J. (2018). Production of proteins and phenolic compounds enriched fractions from rapeseed and sunflower meals by dry fractionation processes. *Industrial Crops and Products, 118*, 160–172. https://doi.org/10.1016/j.indcrop.2018.03.045

Leser, M. E., koci, K., & Luisi, P. L. (1993). Reverse micelles in protein separation: The use of silica for the back-transfer process. *Biotechnology and Bioengineering, 41*(4), 489–492. https://doi.org/10.1002/bit.260410413

Leser, M. E., & Luisi, P. L. (1989). The use of reverse micelles for the simultaneous extraction of oil and proteins from vegetable meal. *Biotechnology and Bioengineering, 34*, 1140–1146.

Leser, M. E., Wei, G., Luisi, P. L., & Maestro, M. (1986). Application of reverse micelles for the extraction of proteins. *Biochemical and Biophysical Research Communications, 135*(2), 629–635. https://doi.org/10.1016/0006-291X(86)90039-2

Liu, F., Chen, Z., Wang, L., & Wang, R. (2013). Effects of protein solubilisation and precipitation pH values on the functional properties of defatted wheat germ protein isolates. *International Journal of Food Science and Technology, 48*(7), 1490–1497. https://doi.org/10.1111/ijfs.12117

Liu, Y., Wang, D., Wang, J., Yang, Y., Zhang, L., Li, J., & Wang, S. (2020). Functional properties and structural characteristics of phosphorylated pea protein isolate. *International Journal of Food Science and Technology, 55*(5), 2002–2010. https://doi.org/10.1111/ijfs.14391

Liu, F., Wang, X., Zhao, X., Hu, H., Chen, F., & Sun, Y. (2014). Surface properties of walnut protein from AOT reverse micelles. *International Journal of Food Science and Technology, 49*(2), 626–633. https://doi.org/10.1111/ijfs.12345

Liu, R. L., Yu, P., Ge, X. L., Bai, X. F., Li, X. Q., & Fu, Q. (2017). Establishment of an aqueous PEG 200-based deep eutectic solvent extraction and enrichment method for pumpkin (*Cucurbita moschata*) seed protein. *Food Analytical Methods, 10*(6), 1669–1680. https://doi.org/10.1007/s12161-016-0732-y

Lores, H., Romero, V., Costas, I., Bendicho, C., & Lavilla, I. (2017). Natural deep eutectic solvents in combination with ultrasonic energy as a green approach for solubilisation of proteins: Application to gluten determination by immunoassay. *Talanta, 162*, 453–459. https://doi.org/10.1016/j.talanta.2016.10.078

Lu, W., Chen, X. W., Wang, J. M., Yang, X. Q., & Qi, J. R. (2016). Enzyme-assisted subcritical water extraction and characterization of soy protein from heat-denatured meal. *Journal of Food Engineering, 169*, 250–258. https://doi.org/10.1016/j.jfoodeng.2015.09.006

Luo, Y. W., & Xie, W. H. (2013). Effect of different processing methods on certain antinutritional factors and protein digestibility in green and white faba bean (*Vicia faba* L.). *CYTA - Journal of Food, 11*(1), 43–49. https://doi.org/10.1080/19476337.2012.681705

Mäkinen, O. E., Wanhalinna, V., Zannini, E., & Arendt, E. K. (2016). Foods for special dietary needs: Non-dairy plant-based milk substitutes and fermented dairy-type products. *Critical Reviews in Food Science and Nutrition, 56*(3), 339–349. https://doi.org/10.1080/10408398.2012.761950

Mäkinen, O. E., Zannini, E., Koehler, P., & Arendt, E. K. (2016). Heat-denaturation and aggregation of quinoa (*Chenopodium quinoa*) globulins as affected by the pH value. *Food Chemistry, 196*, 17–24. https://doi.org/10.1016/j.foodchem.2015.08.069

Marcet, I., Álvarez, C., Paredes, B., & Díaz, M. (2016). The use of sub-critical water hydrolysis for the recovery of peptides and free amino acids from food processing wastes. Review of sources and main parameters. *Waste Management, 49*, 364–371. https://doi.org/10.1016/j.wasman.2016.01.009

Meinlschmidt, P., Ueberham, E., Lehmann, J., Schweiggert-Weisz, U., & Eisner, P. (2016). Immunoreactivity, sensory and physicochemical properties of fermented soy protein isolate. *Food Chemistry, 205*, 229–238. https://doi.org/10.1016/j.foodchem.2016.03.016

Mir, N. A., Riar, C. S., & Singh, S. (2019). Structural modification of quinoa seed protein isolates (QPIs) by variable time sonification for improving its physicochemical and functional characteristics. *Ultrasonics Sonochemistry*, 58. https://doi.org/10.1016/j.ultsonch.2019.104700

Momen, S., Alavi, F., & Aider, M. (2021). Alkali-mediated treatments for extraction and functional modification of proteins: Critical and application review. *Trends in Food Science and Technology, 110*, 778–797. https://doi.org/10.1016/j.tifs.2021.02.052

Mufari, J. R., Miranda-Villa, P. P., & Calandri, E. L. (2018). Quinoa germ and starch separation by wet milling, performance and characterization of the fractions. *LWT, 96*, 527–534. https://doi.org/10.1016/j.lwt.2018.06.010

Murru, M., & Calvo, C. L. (2020). Sunflower protein enrichment. Methods and potential applications. *OCL - Oilseeds and Fats, Crops and Lipids*, 27. https://doi.org/10.1051/ocl/2020007

Mwasaru, M. A., Muhammad, K., Bakar, J., & Man, Y. B. C. (1999). Effects of isolation technique and conditions on the extractability, physicochemical and functional properties of pigeonpea (*Cajanus cajan*) and cowpea (*Vigna unguiculata*) protein isolates. I. Physicochemical properties. *Food Chemistry, 67*(4), 435–443. https://doi.org/10.1016/S0308-8146(99)00150-8

Mwaurah, P. W., Kumar, S., Kumar, N., Attkan, A. K., Panghal, A., Singh, V. K., & Garg, M. K. (2020). Novel oil extraction technologies: Process conditions, quality parameters, and optimization. *Comprehensive Reviews in Food Science and Food Safety, 19*(1), 3–20. https://doi.org/10.1111/1541-4337.12507

Nasrabadi, N., Sedaghat, A., Doost, R., & Mezzenga, R. (2021). Modification approaches of plant-based proteins to improve their techno-functionality and use in food products. *Food Hydrocolloids*, 118.

Nazari, B., Mohammadifar, M. A., Shojaee-Aliabadi, S., Feizollahi, E., & Mirmoghtadaie, L. (2018). Effect of ultrasound treatments on functional properties and structure of millet protein concentrate. *Ultrasonics Sonochemistry, 41*, 382–388. https://doi.org/10.1016/j.ultsonch.2017.10.002

Ningsanond, S., & Ooraikul, B. (1989). Physicochemical and functional properties of dry and wet milling products of red cowpeas. *Starch - Stärke, 41*(7), 255–261. https://doi.org/10.1002/star.19890410705

Opazo-Navarrete, M., Tagle Freire, D., Boom, R. M., Janssen, A. E. M., & Schutyser, M. A. I. (2018). Dry fractionation of quinoa sweet varieties Atlas and Riobamba for sustainable production of protein and starch fractions. *Journal of Food Composition and Analysis, 74*, 95–101. https://doi.org/10.1016/j.jfca.2018.09.009

Ordóñez, C., Asenjo, M. G., Benitez, C., & González, J. L. (2001). Obtaining a protein concentrate from integral defatted sunflower flour. *Bioresource Technology, 78*(2), 187–190. https://doi.org/10.1016/s0960-8524(00)00184-x

Osborne, T. B. (1924). *The vegetable proteins*. Longmans, Green and Company.

Paiva, A., Craveiro, R., Aroso, I., Martins, M., Reis, R. L., & Duarte, A. R. C. (2014). Natural deep eutectic solvents - solvents for the 21st century. *ACS Sustainable Chemistry and Engineering, 2*(5), 1063–1071. https://doi.org/10.1021/sc500096j

Pang, J., Sha, X., Chao, Y., Chen, G., Han, C., Zhu, W., Li, H., & Zhang, Q. (2017). Green aqueous biphasic systems containing deep eutectic solvents and sodium salts for the extraction of protein. *RSC Advances, 7*(78), 49361–49367. https://doi.org/10.1039/c7ra07315a

Pan, Z., & Tangratanavalee, W. (2003). Characteristics of soybeans as affected by soaking conditions. *LWT - Food Science and Technology, 36*(1), 143–151. https://doi.org/10.1016/S0023-6438(02)00202-5

Panyam, D., & Kilara, A. (1996). Enhancing the functionality of food proteins by enzymatic modification. *Trends in Food Science and Technology, 7*(4), 120–125. https://doi.org/10.1016/0924-2244(96)10012-1

Paredes-Lopez, O., Ordorica-Falomir, C., & Olivares-Vazquez, M. R. (1991). Chickpea protein isolates: Physicochemical, functional and nutritional characterization. *Journal of Food Science, 56*(3), 726–729. https://doi.org/10.1111/j.1365-2621.1991.tb05367.x

Pelgrom, P. J. M., Berghout, J. A. M., van der Goot, A. J., Boom, R. M., & Schutyser, M. A. I. (2014). Preparation of functional lupine protein fractions by dry separation. *LWT - Food Science and Technology, 59*(2), 680–688. https://doi.org/10.1016/j.lwt.2014.06.007

Pelgrom, P. J. M., Boom, R. M., & Schutyser, M. A. I. (2015). Functional analysis of mildly refined fractions from yellow pea. *Food Hydrocolloids, 44*, 12–22. https://doi.org/10.1016/j.foodhyd.2014.09.001

Pelgrom, P. J. M., Vissers, A. M., Boom, R. M., & Schutyser, M. A. I. (2013). Dry fractionation for production of functional pea protein concentrates. *Food Research International, 53*(1), 232–239. https://doi.org/10.1016/j.foodres.2013.05.004

Pelgrom, P. J. M., Wang, J., Boom, R. M., & Schutyser, M. A. I. (2015). Pre- and post-treatment enhance the protein enrichment from milling and air classification of legumes. *Journal of Food Engineering, 155*, 53–61. https://doi.org/10.1016/j.jfoodeng.2015.01.005

Pińkowska, H., Wolak, P., & Oliveros, E. (2013). Application of Doehlert matrix for determination of the optimal conditions of hydrothermolysis of rapeseed meal in subcritical water. *Fuel, 106*, 258–264. https://doi.org/10.1016/j.fuel.2012.12.023

Pojić, M., Mišan, A., & Tiwari, B. (2018). Eco-innovative technologies for extraction of proteins for human consumption from renewable protein sources of plant origin. *Trends in Food Science and Technology, 75*, 93–104. https://doi.org/10.1016/j.tifs.2018.03.010

Potin, F., Lubbers, S., Husson, F., & Saurel, R. (2019). Hemp (*Cannabis sativa* L.) protein extraction conditions affect extraction yield and protein quality. *Journal of Food Science, 84*(12), 3682–3690. https://doi.org/10.1111/1750-3841.14850

Prego, I., Maldonado, S., & Otegui, M. (1998). Seed structure and localization of reserves in *Chenopodium quinoa*. *Annals of Botany, 82*(4), 481–488. https://doi.org/10.1006/anbo.1998.0704

Prgomet, I., Gonçalves, B., Domínguez-Perles, R., Pascual-Seva, N., & Barros, A. (2017). Valorization challenges to almond residues: Phytochemical composition and functional application. *Molecules, 22*(10), 1774. https://doi.org/10.3390/molecules22101774

Protonotariou, S., Ritzoulis, C., & Mandala, I. (2021). Jet milling conditions impact on wheat flour particle size. *Journal of Food Engineering*, 294. https://doi.org/10.1016/j.jfoodeng.2020.110418

Pujol, R., Létang, C., Lempereur, I., Chaurand, M., Mabille, F., & Abecassis, J. (2000). Description of a micromill with instrumentation for measuring grinding characteristics of wheat grain. *Cereal Chemistry, 77*(4), 421–427. https://doi.org/10.1094/CCHEM.2000.77.4.421

Raghavarao, K. S. M. S., Ranganathan, T. V., Srinivas, N. D., & Barhate, R. S. (2003). Aqueous two phase extraction? An environmentally benign technique. *Clean Technologies and Environmental Policy*, 136–141. https://doi.org/10.1007/s10098-003-0193-z

Rahman, M. M., & Lamsal, B. P. (2021). Ultrasound-assisted extraction and modification of plant-based proteins: Impact on physicochemical, functional, and nutritional properties. *Comprehensive Reviews in Food Science and Food Safety, 20*(2), 1457–1480. https://doi.org/10.1111/1541-4337.12709

Ramachandraiah, K., Koh, B. B., Davaatseren, M., & Hong, G. P. (2017). Characterization of soy protein hydrolysates produced by varying subcritical water processing temperature. *Innovative Food Science and Emerging Technologies, 43*, 201–206. https://doi.org/10.1016/j.ifset.2017.08.011

Raynes, J. K., Carver, J. A., Gras, S. L., & Gerrard, J. A. (2014). Protein nanostructures in food - should we be worried? *Trends in Food Science and Technology, 37*(1), 42–50. https://doi.org/10.1016/j.tifs.2014.02.003

Reinkensmeier, A., Bußler, S., Schlüter, O., Rohn, S., & Rawel, H. M. (2015). Characterization of individual proteins in pea protein isolates and air classified samples. *Food Research International, 76*(1), 160–167. https://doi.org/10.1016/j.foodres.2015.05.009

Rempel, C., Geng, X., & Zhang, Y. (2019). Industrial scale preparation of pea flour fractions with enhanced nutritive composition by dry fractionation. *Food Chemistry, 276*, 119–128. https://doi.org/10.1016/j.foodchem.2018.10.003

Rodrigues, I. M., Coelho, J. F. J., & Carvalho, M. G. V. S. (2012). Isolation and valorisation of vegetable proteins from oilseed plants: Methods, limitations and potential. *Journal of Food Engineering, 109*(3), 337–346. https://doi.org/10.1016/j.jfoodeng.2011.10.027

Rommi, K., Hakala, T. K., Holopainen, U., Nordlund, E., Poutanen, K., & Lantto, R. (2014). Effect of enzyme-aided cell wall disintegration on protein extractability from intact and dehulled rapeseed (*Brassica rapa* L. and *Brassica napus* L.) press cakes. *Journal of Agricultural and Food Chemistry, 62*(32), 7989–7997. https://doi.org/10.1021/jf501802e

Ruiz, G. A., Xiao, W., Van Boekel, M., Minor, M., & Stieger, M. (2016). Effect of extraction pH on heat-induced aggregation, gelation and microstructure of protein isolate from quinoa (*Chenopodium quinoa* Willd). *Food Chemistry, 209*, 203–210. https://doi.org/10.1016/j.foodchem.2016.04.052

Saldanha do Carmo, C., Silventoinen, P., Nordgård, C. T., Poudroux, C., Dessev, T., Zobel, H., Holtekjølen, A. K., Draget, K. I., Holopainen-Mantila, U., Knutsen, S. H., & Sahlstrøm, S. (2020). Is dehulling of peas and faba beans necessary prior to dry fractionation for the production of protein- and starch-rich fractions? Impact on physical properties, chemical composition and techno-functional properties. *Journal of Food Engineering, 278*. https://doi.org/10.1016/j.jfoodeng.2020.109937

Sánchez-Reséndiz, A., Rodríguez-Barrientos, S., Rodríguez-Rodríguez, J., Barba-Dávila, B., Serna-Saldívar, S. O., & Chuck-Hernández, C. (2018). Phosphoesterification of soybean and peanut proteins with sodium trimetaphosphate (STMP): Changes in structure to improve functionality for food applications. *Food Chemistry, 260*, 299–305. https://doi.org/10.1016/j.foodchem.2018.04.009

Sankaran, R., Bong, J. H., Chow, Y. H., Wong, F. W. F., Ling, T. C., & Show, P. L. (2019). Reverse micellar system in protein recovery - a review of the latest developments. *Current Protein and Peptide Science, 20*(10), 1012–1026. https://doi.org/10.2174/1389203720666190628142203

Sari, Y. W., Bruins, M. E., & Sanders, J. P. M. (2013). Enzyme assisted protein extraction from rapeseed, soybean, and microalgae meals. *Industrial Crops and Products, 43*(1), 78–83. https://doi.org/10.1016/j.indcrop.2012.07.014

Sari, Y. W., Mulder, W. J., Sanders, J. P. M., & Bruins, M. E. (2015). Towards plant protein refinery: Review on protein extraction using alkali and potential enzymatic assistance. *Biotechnology Journal, 10*(8), 1138–1157. https://doi.org/10.1002/biot.201400569

Sathe, S. K., Zaffran, V. D., Gupta, S., & Li, T. (2018). Protein solubilization. *Journal of the American Oil Chemists' Society, 95*(8), 883–901. https://doi.org/10.1002/aocs.12058

Schutyser, M. A. I., Pelgrom, P. J. M., van der Goot, A. J., & Boom, R. M. (2015). Dry fractionation for sustainable production of functional legume protein concentrates. *Trends in Food Science and Technology, 45*(2), 327–335. https://doi.org/10.1016/j.tifs.2015.04.013

Schutyser, M. A. I., & van der Goot, A. J. (2011). The potential of dry fractionation processes for sustainable plant protein production. *Trends in Food Science and Technology, 22*(4), 154–164. https://doi.org/10.1016/j.tifs.2010.11.006

Sereewatthanawut, I., Prapintip, S., Watchiraruji, K., Goto, M., Sasaki, M., & Shotipruk, A. (2008). Extraction of protein and amino acids from deoiled rice bran by subcritical water hydrolysis. *Bioresource Technology, 99*(3), 555–561. https://doi.org/10.1016/j.biortech.2006.12.030

Shapiro, M., & Galperin, V. (2005). Air classification of solid particles: A review. *Chemical Engineering and Processing: Process Intensification, 44*(2), 279–285. https://doi.org/10.1016/j.cep.2004.02.022

Sha, L., & Xiong, Y. L. (2020). Plant protein-based alternatives of reconstructed meat: Science, technology, and challenges. *Trends in Food Science and Technology, 102*, 51–61. https://doi.org/10.1016/j.tifs.2020.05.022

Shi, Y., Singh, A., Kitts, D. D., & Pratap-Singh, A. (2021). Lactic acid fermentation: A novel approach to eliminate unpleasant aroma in pea protein isolates. *LWT*, 150. https://doi.org/10.1016/j.lwt.2021.111927

Sibakov, J., Myllymäki, O., Holopainen, U., Kaukovirta-Norja, A., Hietaniemi, V., Pihlava, J. M., Poutanen, K., & Lehtinen, P. (2011). Lipid removal enhances separation of oat grain cell wall material from starch and protein. *Journal of Cereal Science, 54*(1), 104–109. https://doi.org/10.1016/j.jcs.2011.04.003

Silventoinen, P., Kortekangas, A., Ercili-Cura, D., & Nordlund, E. (2021). Impact of ultra-fine milling and air classification on biochemical and techno-functional characteristics of wheat and rye bran. *Food Research International*, 139. https://doi.org/10.1016/j.foodres.2020.109971

Silventoinen, P., Rommi, K., Holopainen-Mantila, U., Poutanen, K., & Nordlund, E. (2019). Biochemical and techno-functional properties of protein- and fibre-rich hybrid ingredients produced by dry fractionation from rice bran. *Food and Bioprocess Technology, 12*(9), 1487–1499. https://doi.org/10.1007/s11947-019-02307-w

Silventoinen, P., Sipponen, M. H., Holopainen-Mantila, U., Poutanen, K., & Sozer, N. (2018). Use of air classification technology to produce protein-enriched barley ingredients. *Journal of Food Engineering, 222*, 169–177. https://doi.org/10.1016/j.jfoodeng.2017.11.016

Singh, U. (1993). Protein quality of pigeonpea (*Cajanus cajan* (L.) Millsp.) as influenced by seed polyphenols and cooking process. *Plant Foods for Human Nutrition, 43*(2), 171–179. https://doi.org/10.1007/BF01087921

Singh, U., & Singh, B. (1992). Tropical grain legumes as important human foods. *Economic Botany, 46*(3), 310–321. https://doi.org/10.1007/BF02866630

Sozer, N., & Poutanen, K. (2015). Plant protein ingredients for future foods. *Agro Food Industry Hi-Tech, 26*(5), 56–59. http://www.teknoscienze.com/Contents/Riviste/Sfogliatore/AF5_2015/files/assets/common/downloads/AF5_2015.pdf.

Srinivasan, R., Hicks, K. B., Challa, R. K., Wilson, J., Kurantz, M., & Moreau, R. A. (2010). Fractionation of barley flour using elusieve processing: A combination of sieving and air classification. *Transactions of the ASABE, 53*(2), 503–508. http://asae.frymulti.com/azdez.asp?JID=3&AID=29560&CID=t2010&v=53&i=2&T=2.

Srinivasan, R., Moreau, R. A., Rausch, K. D., Belyea, R. L., Tumbleson, M. E., & Singh, V. (2005). Separation of fiber from distillers dried grains with solubles (DDGS) using sieving and elutriation. *Cereal Chemistry, 82*(5), 528–533. https://doi.org/10.1094/CC-82-0528

Stone, A. K., Karalash, A., Tyler, R. T., Warkentin, T. D., & Nickerson, M. T. (2015). Functional attributes of pea protein isolates prepared using different extraction methods and cultivars. *Food Research International, 76*(1), 31–38. https://doi.org/10.1016/j.foodres.2014.11.017

Suarez Ruiz, C. A., Emmery, D. P., Wijffels, R. H., Eppink, M. H. M., & van den Berg, C. (2018). Selective and mild fractionation of microalgal proteins and pigments using aqueous two-phase systems. *Journal of Chemical Technology and Biotechnology, 93*(9), 2774–2783. https://doi.org/10.1002/jctb.5711

Sun, X. D., & Arntfield, S. D. (2011). Gelation properties of salt-extracted pea protein isolate catalyzed by microbial transglutaminase cross-linking. *Food Hydrocolloids, 25*(1), 25–31. https://doi.org/10.1016/j.foodhyd.2010.05.002

Sun, X., & Bandara, N. (2019). Applications of reverse micelles technique in food science: A comprehensive review. *Trends in Food Science and Technology, 91*, 106–115. https://doi.org/10.1016/j.tifs.2019.07.001

Sunphorka, S., Chavasiri, W., Oshima, Y., & Ngamprasertsith, S. (2012). Kinetic studies on rice bran protein hydrolysis in subcritical water. *Journal of Supercritical Fluids, 65*, 54–60. https://doi.org/10.1016/j.supflu.2012.02.017

Sun, X. H., Zhu, K. X., & Zhou, H. M. (2008). Protein extraction from defatted wheat germ by reverse micelles: Optimization of the forward extraction. *Journal of Cereal Science, 48*(3), 829–835. https://doi.org/10.1016/j.jcs.2008.06.006

Sun, X. H., Zhu, K. X., & Zhou, H. M. (2009). Optimization of a novel backward extraction of defatted wheat germ protein from reverse micelles. *Innovative Food Science and Emerging Technologies, 10*(3), 328–333. https://doi.org/10.1016/j.ifset.2009.01.006

Swain, R. B. (1975). *Process for producing protein concentrate using air classification*. United States Pat: The Procter & Gamble Company, assignee.

Tabtabaei, S., Jafari, M., Rajabzadeh, A. R., & Legge, R. L. (2016a). Development and optimization of a triboelectrification bioseparation process for dry fractionation of legume flours. *Separation and Purification Technology, 163*, 48–58. https://doi.org/10.1016/j.seppur.2016.02.035

Tabtabaei, S., Jafari, M., Rajabzadeh, A. R., & Legge, R. L. (2016b). Solvent-free production of protein-enriched fractions from navy bean flour using a triboelectrification-based approach. *Journal of Food Engineering, 174*, 21–28. https://doi.org/10.1016/j.jfoodeng.2015.11.010

Tabtabaei, S., Konakbayeva, D., Rajabzadeh, A. R., & Legge, R. L. (2019). Functional properties of navy bean (*Phaseolus vulgaris*) protein concentrates obtained by pneumatic tribo-electrostatic separation. *Food Chemistry, 283*, 101–110. https://doi.org/10.1016/j.foodchem.2019.01.031

Tabtabaei, S., Vitelli, M., Rajabzadeh, A. R., & Legge, R. L. (2017). Analysis of protein enrichment during single- and multi-stage tribo-electrostatic bioseparation processes for dry fractionation of legume flour. *Separation and Purification Technology, 176*, 48–58. https://doi.org/10.1016/j.seppur.2016.11.050

Tyler, R. T. (1984). Impact milling quality of grain legumes. *Journal of Food Science, 49*(3), 925–930. https://doi.org/10.1111/j.1365-2621.1984.tb13243.x

Tyler, R. T., Youngs, C. G., & Sosulski, F. W. (1981). Air Classification of Legumes. I. Separation efficiency, yield, and composition of the starch and protein fractions. *Cereal Chemistry, 58*, 144–148.

Tyler, R. T., Youngs, C. G., & Sosulski, F. W. (1984). Air classification of legumes: Cut-size effects. *Canadian Institute of Food Science and Technology Journal*, 71–78. https://doi.org/10.1016/s0315-5463(84)72359-5

Vanga, S. K., Wang, J., & Raghavan, V. (2020). Effect of ultrasound and microwave processing on the structure, in-vitro digestibility and trypsin inhibitor activity of soymilk proteins. *LWT*, 131. https://doi.org/10.1016/j.lwt.2020.109708

Wu, Y. V., & Abbott, T. P. (2003). Protein enrichment of defatted salicornia meal by air classification. *Journal of the American Oil Chemists' Society, 80*(2), 167–169. https://doi.org/10.1007/s11746-003-0671-7

Vidosavljević, S., Bojanić, N., Stojkov, V., Čolović, R., Đuragić, O., Fišteš, A., & Banjac, V. (2019). Comparison of two dry fractionation processes for protein enrichment of sunflower meal. *Food and Feed Research*, 209–217. https://doi.org/10.5937/ffr1902209v

Vishwanathan, K. H., Singh, V., & Subramanian, R. (2011). Wet grinding characteristics of soybean for soymilk extraction. *Journal of Food Engineering, 106*(1), 28–34. https://doi.org/10.1016/j.jfoodeng.2011.04.002

Vitelli, M., Rajabzadeh, A. R., Tabtabaei, S., Assatory, A., Shahnam, E., & Legge, R. L. (2020). Effect of hammer and pin milling on triboelectrostatic separation of legume flour. *Powder Technology, 372*, 317–324. https://doi.org/10.1016/j.powtec.2020.06.007

Vogelsang-O'Dwyer, M., Zannini, E., & Arendt, E. K. (2021). Production of pulse protein ingredients and their application in plant-based milk alternatives. *Trends in Food Science and Technology, 110*, 364–374. https://doi.org/10.1016/j.tifs.2021.01.090

Vogelsang-Dwyer, M., Petersen, I. L., Joehnke, M. S., Sorensen, J. C., Bez, J., Detzel, A., Busch, M., Krueger, M., Arendt, E. K., & Zannini, E. (2020). Comparison of faba bean protein ingredients produced using dry fractionation and isoelectric precipitation: Techno-functional. Nutritional and environmental performance. *Foods, 9*(3).

Walters, C., Landré, P., Hill, L., Corbineau, F., & Bailly, C. (2005). Organization of lipid reserves in cotyledons of primed and aged sunflower seeds. *Planta, 222*(3), 397–407. https://doi.org/10.1007/s00425-005-1541-5

Wang, J., De Wit, M., Schutyser, M. A. I., & Boom, R. M. (2014). Analysis of electrostatic powder charging for fractionation of foods. *Innovative Food Science and Emerging Technologies, 26*, 360–365. https://doi.org/10.1016/j.ifset.2014.06.011

Wang, L., Liu, K., Chen, F., & Yin, L. (2018). Size changes of reverse micelles after extraction of peanut protein and their forward extraction rates. *Grain & Oil Science and Technology*, 32–39. https://doi.org/10.3724/SP.J.1447.GOST.2018.18013

Wang, J., Smits, E., Boom, R. M., & Schutyser, M. A. I. (2015). Arabinoxylans concentrates from wheat bran by electrostatic separation. *Journal of Food Engineering, 155*, 29–36. https://doi.org/10.1016/j.jfoodeng.2015.01.008

Wang, L. H., Sun, X., Huang, G. Q., & Xiao, J. X. (2018). Conjugation of soybean protein isolate with xylose/fructose through wet-heating Maillard reaction. *Journal of Food Measurement and Characterization, 12*(4), 2718–2724. https://doi.org/10.1007/s11694-018-9889-y

Wang, Y. R., Yang, Q., Fan, J. L., Zhang, B., & Chen, H. Q. (2019). The effects of phosphorylation modification on the structure, interactions and rheological properties of rice glutelin during heat treatment. *Food Chemistry*, 297. https://doi.org/10.1016/j.foodchem.2019.124978

Wang, X. B., Zhang, Y. H., & Jiang, L. Z. (2013). Improvement of emulsifying properties of soybean protein isolate through glycosylation modification. In , *Vols. 781–784. Advanced materials research* (pp. 1495–1499). https://doi.org/10.4028/www.scientific.net/AMR.781-784.1495

Wang, F., Zhang, Y., Xu, L., & Ma, H. (2020). An efficient ultrasound-assisted extraction method of pea protein and its effect on protein functional properties and biological activities. *LWT*, 127. https://doi.org/10.1016/j.lwt.2020.109348

Wang, Z., Zhang, C., Zhang, T., Ju, X., & He, R. (2018). Effects of acylation and glycation treatments on physicochemical and gelation properties of rapeseed protein isolate. *RSC Advances, 8*(70), 40395–40406. https://doi.org/10.1039/C8RA07912A

Wang, J., Zhao, J., De Wit, M., Boom, R. M., & Schutyser, M. A. I. (2016). Lupine protein enrichment by milling and electrostatic separation. *Innovative Food Science and Emerging Technologies, 33*, 596–602. https://doi.org/10.1016/j.ifset.2015.12.020

Wiboonsirikul, J., Hata, S., Tsuno, T., Kimura, Y., & Adachi, S. (2007). Production of functional substances from black rice bran by its treatment in subcritical water. *LWT - Food Science and Technology, 40*(10), 1732–1740. https://doi.org/10.1016/j.lwt.2007.01.003

Wiboonsirikul, J., Kimura, Y., Kanaya, Y., Tsuno, T., & Adachi, S. (2008). Production and characterization of functional substances from a by-product of rice bran oil and protein production by a compressed hot water treatment. *Bioscience, Biotechnology and Biochemistry, 72*(2), 384–392. https://doi.org/10.1271/bbb.70464

Wouters, A. G. B., Rombouts, I., Fierens, E., Brijs, K., & Delcour, J. A. (2016). Relevance of the functional properties of enzymatic plant protein hydrolysates in food systems. *Comprehensive Reviews in Food Science and Food Safety, 15*(4), 786–800. https://doi.org/10.1111/1541-4337.12209

Wronkowska, M. (2016). Wet-milling of cereals. *Journal of Food Processing and Preservation, 40*(3), 572–580. https://doi.org/10.1111/jfpp.12626

Xing, Q., de Wit, M., Kyriakopoulou, K., Boom, R. M., & Schutyser, M. A. I. (2018). Protein enrichment of defatted soybean flour by fine milling and electrostatic separation. *Innovative Food Science and Emerging Technologies, 50*, 42–49. https://doi.org/10.1016/j.ifset.2018.08.014

Xing, Q., Kyriakopoulou, K., de Wit, M., Boom, R. M., & Schutyser, M. A. I. (2021). Effect of tube wall material on electrostatic separation of plant raw-materials. *Journal of Food Process Engineering, 44*(1). https://doi.org/10.1111/jfpe.13575

Xing, Q., Utami, D. P., Demattey, M. B., Kyriakopoulou, K., de Wit, M., Boom, R. M., & Schutyser, M. A. I. (2020). A two-step air classification and electrostatic separation process for protein enrichment of starch-containing legumes. *Innovative Food Science and Emerging Technologies, 66*. https://doi.org/10.1016/j.ifset.2020.102480

Yue, J., Gu, Z., Zhu, Z., Yi, J., Ohm, J. B., Chen, B., & Rao, J. (2021). Impact of defatting treatment and oat varieties on structural, functional properties, and aromatic profile of oat protein. *Food Hydrocolloids, 112*. https://doi.org/10.1016/j.foodhyd.2020.106368

Yue, J., Zhu, Z., Yi, J., Lan, Y., Chen, B., & Rao, J. (2021). Structure and functionality of oat protein extracted by choline chloride–dihydric alcohol deep eutectic solvent and its water binary mixtures. *Food Hydrocolloids, 112*. https://doi.org/10.1016/j.foodhyd.2020.106330

Zayas, J. F. (1997). *Solubility of Proteins*, 6–75.

Zayas, J. F. (2012). *Functionality of proteins in food*. Springer Science & Business Media.

Zhang, Q., Li, L., Lan, Q., Li, M., Wu, D., Chen, H., Liu, Y., Lin, D., Qin, W., Zhang, Z., Liu, J., & Yang, W. (2019). Protein glycosylation: A promising way to modify the functional properties and extend the application in food system. *Critical Reviews in Food Science and Nutrition, 59*(15), 2506–2533. https://doi.org/10.1080/10408398.2018.1507995

Zhang, C., Sanders, J. P. M., Xiao, T. T., & Bruins, M. E. (2015). How does alkali aid protein extraction in green tea leaf residue: A basis for integrated biorefinery of leaves. *PLoS One, 10*(7). https://doi.org/10.1371/journal.pone.0133046

Zhang, Q.-T., Tu, Z.-C., Wang, H., Huang, X.-Q., Fan, L.-L., Bao, Z.-Y., & Xiao, H. (2014). Functional properties and structure changes of soybean protein isolate after subcritical water treatment. *Journal of Food Science and Technology*. https://doi.org/10.1007/s13197-014-1392-9

Zhang, J., Wen, C., Zhang, H., Duan, Y., & Ma, H. (2020). Recent advances in the extraction of bioactive compounds with subcritical water: A review. *Trends in Food Science and Technology, 95*, 183–195. https://doi.org/10.1016/j.tifs.2019.11.018

Zhang, J., Wen, C., Zhang, H., Zandile, M., Luo, X., Duan, Y., & Ma, H. (2018). Structure of the zein protein as treated with subcritical water. *International Journal of Food Properties, 21*(1), 128–138. https://doi.org/10.1080/10942912.2017.1414839

Zhang, J., Wu, N., Lan, T., & Yang, X. (2014). Improvement in emulsifying properties of soy protein isolate by conjugation with maltodextrin using high-temperature, short-time dry-heating Maillard reaction. *International Journal of Food Science and Technology, 49*(2), 460–467. https://doi.org/10.1111/ijfs.12323

Zhao, X., Liu, H., Zhang, X., Zhu, H., & Ao, Q. (2019). Surface structure and volatile characteristic of peanut proteins obtained through AOT reverse micelles. *Colloids and Surfaces B: Biointerfaces, 173*, 860–868. https://doi.org/10.1016/j.colsurfb.2018.10.070

Zhao, X., Zhang, X., Liu, H., Zhang, G., & Ao, Q. (2018). Functional, nutritional and flavor characteristic of soybean proteins obtained through reverse micelles. *Food Hydrocolloids, 74*, 358–366. https://doi.org/10.1016/j.foodhyd.2017.08.024

Zhao, X., Zhu, H., & Chen, J. (2015). Effects of sodium bis(2-ethylhexyl) sulfo succinate (AOT) reverse micelles on physicochemical properties of soy protein. *Food and Bioproducts Processing, 94*, 500–506. https://doi.org/10.1016/j.fbp.2014.07.009

Zhong, L., Fang, Z., Wahlqvist, M. L., Wu, G., Hodgson, J. M., & Johnson, S. K. (2018). Seed coats of pulses as a food ingredient: Characterization, processing, and applications. *Trends in Food Science and Technology, 80*, 35–42. https://doi.org/10.1016/j.tifs.2018.07.021

Zhu, K. X., Sun, X. H., Chen, Z. C., Peng, W., Qian, H. F., & Zhou, H. M. (2010). Comparison of functional properties and secondary structures of defatted wheat germ proteins separated by reverse micelles and alkaline extraction and isoelectric precipitation. *Food Chemistry, 123*(4), 1163–1169. https://doi.org/10.1016/j.foodchem.2010.05.081

Zhu, K. X., Sun, X. H., & Zhou, H. M. (2009). Optimization of ultrasound-assisted extraction of defatted wheat germ proteins by reverse micelles. *Journal of Cereal Science, 50*(2), 266–271. https://doi.org/10.1016/j.jcs.2009.06.006

Zhu, H. G., Tang, H. Q., Cheng, Y. Q., Li, Z. G., & Tong, L. T. (2021a). Electrostatic separation technology for obtaining plant protein concentrates: A review. *Trends in Food Science and Technology, 113*, 66–76. https://doi.org/10.1016/j.tifs.2021.04.044

Zhu, H. G., Tang, H. Q., Cheng, Y. Q., Li, Z. G., & Tong, L. T. (2021b). Novel electromagnetic separation technology for the production of pea protein concentrate. *Innovative Food Science and Emerging Technologies, 70*. https://doi.org/10.1016/j.ifset.2021.102668

Zhu, Z., Zhu, W., Yi, J., Liu, N., Cao, Y., Lu, J., Decker, E. A., & McClements, D. J. (2018). Effects of sonication on the physicochemical and functional properties of walnut protein isolate. *Food Research International, 106*, 853–861. https://doi.org/10.1016/j.foodres.2018.01.060

Banjac, V., Pezo, L., Pezo, M., Vukmirovic, D., Colovic, D., Fistes, A., & Colovic, R. (2017). Optimization of the classification process in the zigzag air classifier for obtaining a high protein sunflower meal – Chemometric and CFD approach. *Advanced Powder Technology, 28*(3), 1069–1078. https://doi.org/10.1016/j.apt.2017.01.013

Basset, C., Kedidi, S., & Barakat, A. (2016). Chemical- and solvent-free mechanophysical fractionation of biomass induced by tribo-electrostatic charging: separation of proteins and lignin. *ACS Sustainable Chemistry & Engineering, 4*, 4166–4173. https://doi.org/10.1021/acssuschemeng.6b00667

Berghout, J. A. M., Boom, R. M., & Van Der Goot, A. J. (2014). The potential of aqueous fractionation of lupin seeds for high-protein foods. *Food Chemistry, 159*, 64–70.

Guo, Z., Chen, F., Yang, H., & Liu, K. (2015). Kinetics of protein extraction in reverse micelle. *International Journal of Food Properties, 18*(8), 1707–1718.

Hadidi, M., Khaksar, F. B., & Pagan, J. (2020). Application of Ultrasound-Ultrafiltration-Assisted alkaline isoelectric precipitation (UUAAIP) technique for producing alfalfa protein isolate for human consumption: Optimization, comparison, physicochemical, and functional properties. *Food Research International, 130*, 108907.

Hansen, J.Ø., Skrede, A., & Mydland, L. T. (2017). Fractionation of rapeseed meal by milling, sieving and air classification—Effect on crude protein, amino acids and fiber content and digestibility. *Animal Feed Science and Technology, 230*, 143–153.

Kalaydzhiev, H., Georgiev, R., Ivanova, P., Stoyanova, M., & Silva, C. L. M. (2020). Enhanced solubility of rapeseed meal protein isolates prepared by sequential isoelectric precipitation. *Foods, 9*(6).

Lammi, S., Barakat, A., Mayer-Laigle, C., Djenane, D., & Gontard, N. (2018). Dry fractionation of olive pomace as a sustainable process to produce fillers for biocomposites. *Powder Technology, 326*, 44–53.

Maaroufi, C., Melcion, J. P., Monredon, F. D., Giboulot, B., Guibert, D., & Le Guen, M. P. (2000). Fractionation of pea flour with pilot scale sieving. I. Physical and chemical characteristics of pea seed fractions. *Animal Feed Science and Technology, 85*, 61–78.

Ochoa-Rivas, A., Nava-Valdez, Y., & Serna-Saldívar, S. O. (2016). Microwave and ultrasound to enhance protein extraction from peanut flour under alkaline conditions: Effects in yield and functional properties of protein isolates. *Food and Bioprocess Technology, 10*(3), 543–555.

Opazo-Navarrete, M., Schutyser, M. A. I., & Boom, R. M. (2018a). Effect of pre-treatment on in vitro gastric digestion of quinoa protein (Chenopodium quinoa Willd.) obtained by wet and dry fractionation. *International Journal of Food Sciences and Nutrition, 69*(1), 1–11.

Pińkowska, H. (2014). Application of the Doehlert matrix for the determination of the optimal conditions of hydrolysis of soybean protein in subcritical water. *Industrial & Engineering Chemistry Research, 53*(4), 1320–1326.

Preece, K. E., & Hooshyar, N. (2017). Whole soybean protein extraction processes: A review. *Innovative Food Science & Emerging Technologies, 43*, 163–172.

Rodríguez-Ambriz, S. L., Martínez-Ayala, A. L., & Millán, F. (2005). Composition and functional properties of lupinus campestris protein isolates. *Plant Foods for Human Nutrition, 60*(3), 99–107.

Russin, T. A., & Arcand, Y. (2007). Particle size effect on soy protein isolate extraction. *Journal of Food Processing and Preservation, 31*(3), 308–319.

Sessa, D. J. (2004). Processing of soybean hulls to enhance the distribution and extraction of value-added proteins. *Journal of the Science of Food and Agriculture, 84*(1), 75–82.

Sharma, G. M., Su, M., Joshi, A. U., & Roux, K. H. (2010). Functional properties of select edible oilseed proteins. *Journal of Agricultural and Food Chemistry, 58*(9), 5457–5464.

Srinivasan, R., To, F., & Columbus, E. (2009). Pilot scale fiber separation from distillers dried grains with solubles (DDGS) using sieving and air classification. *Bioresource Technology, 100*(14), 3548–3555.

Sundberg, B., & Åman, P. (1994). Fractionation of different types of barley by roller milling and sieving. *Journal of Cereal Science, 19*, 179–184.

Tanger, C., Engel, J., & Kulozik, U. (2020). Influence of extraction conditions on the conformational alteration of pea protein extracted from pea flour. *Food Hydrocolloids, 107*.

Vorobiev, E., & Lebovka, N. (2016). *Application of Pulsed Electric Energy for Lignocellulosic Biorefinery* (pp. 1–19). Springer International Publishing.

Wang, N., Wang, J., Li, Y., & Li, L. (2020b). Reverse microemulsion prepared by AOT/CTAB/SDS/Tween80 for extraction of tea residues protein. *Journal of Molecular Liquids, 320*.

Watchararuji, K., Goto, M., Sasaki, M., & Shotipruk, A. (2008). Value-added subcritical water hydrolysate from rice bran and soybean meal. *Bioresource Technology, 99*(14), 6207–6213.

Wiboonsirikul, J., Kimura, Y., Kadota, M., Morita, H., & Tsuno, T. (2007). Properties of Extracts from Defatted Rice Bran by Its Subcritical Water Treatment. *Journal of Agricultural and Food Chemistry, 55*, 8759–8765.

Wiboonsirikul, J., Mori, M., & Khuwijitjaru, P. (2013). Properties of extract from Okara by Its subcritical water treatment. *International Journal of Food Properties, 16*(5), 974–982.

Zhang, X., Hou, Y., & Zhang, F. (2017). Protein extraction from grape seeds by reverse micelles: Optimization of the forward extraction. *OALib, 04*(06), 1–12.

Zhao, X., Liu, H., & Zhang, X. (2019). Comparison of structures of walnut protein fractions obtained through reverse micelles and alkaline extraction with isoelectric precipitation. *Int J Biol Macromol, 125*, 1214–1220.

Chapter 4

Reducing allergenicity in plant-based proteins

Gbemisola J. Fadimu[1], Oladipupo Odunayo Olatunde[2,3], Nandika Bandara[2,3] and Tuyen Truong[1]
[1]School of Science, RMIT University, Melbourne, VIC, Australia; [2]Department of Food and Human Nutritional Sciences, Faculty of Agricultural and Food Sciences, University of Manitoba, Winnipeg, Manitoba, Canada; [3]Richardson Centre for Functional Foods and Nutraceuticals, University of Manitoba, Winnipeg, Manitoba, Canada

4.1 Introduction

Food allergy is a hypersensitive reaction (allergic reactions) of the immune system when exposed to proteins, leading to adverse health effects (Verhoeckx et al., 2015). A range of symptoms from oral allergy syndrome to skin reaction, swelling of internal organs, confusion, and life-threatening anaphylaxis have been documented from the reported cases of people reacting to food allergy (Lucas et al., 2004; Mancuso & Berdondini, 2001; Novembre et al., 1995; Worm et al., 2014). Food allergies occur via the production of allergen-specific immunoglobulin E (IgE) and subsequent interaction with allergenic proteins (Fig. 4.1), otherwise known as "IgE-mediated type 1 allergy." However, a limited non-IgE-mediated food allergy has been reported (Shahali & Dadar, 2018). There are two stages involved in food allergy development: (i) the allergic sensitization arising from the immunological priming of susceptible subjects to a specific food protein and (ii) the allergic reaction elicited from the sensitized subjects exposed to sufficient levels of the inducing allergens (Verhoeckx et al., 2015). These reactions can vary depending on the individual, from mild to fatal (Perry & Pesek, 2013; Sicherer & Wood, 2013).

Food allergy has been recognized as an important health problem, which its prevalence in human varies depending on the age group, mostly affecting about 6% of infants less than 3 years of age and decreasing over the first decade (Sampson, 2004). Different regulatory bodies worldwide have identified different food allergen groups, and the common thing among them is that they are proteins (with only a few exceptions). The most common allergenic foods identified around the world are hen eggs, cow milk, tree nuts, peanuts, wheat, soy, fish, and shellfish (Hefle et al., 1996; Sampson, 2004). However, some other foods, including gluten-containing cereals, crustaceans, mollusks, soybeans, milk, celery, mustard, sesame, and lupin, have been reported to mediate allergic reactions in human (Commission-Directive-2006/142/EC, 2006). Several factors such as the age group, the method of food preparation, dietary habit, time, change in geographical location influence the extent to which allergy is associated with any particular foods (Lucas et al., 2004; Verhoeckx et al., 2015). Nevertheless, most food allergens are proteinaceous in nature and are very potent. Other food allergens, including polymers, such as polysaccharides, usually generate poor antibody responses (Mills et al., 2004).

Globally, the demand for protein is continuously increasing. The demand rose to approximately 202 million tonnes in 2017 (Henchion et al., 2017), increasing with the increasing human population. Current animal production systems cannot meet the increasing demand for proteins. Therefore, plant-based alternatives are necessary to provide sufficient protein to feed the world (Bizzozero, 2016). The demand for plant proteins is expected to increase from $4.6 billion in 2018 to a whopping $85 billion by 2030 (Byington, 2020). Nevertheless, there are many challenges facing plant proteins, including the method of extraction, techno-functional properties, and their inherent allergenicity, limiting their application and utilization. Hence, it is essential to circumvent these challenges to meet the increasing demand for plant protein by food manufacturers and food service operators. From a Food Technologist's standpoint, understanding the critical techno-functional properties, allergens, and nutritional properties of plant proteins is imperative to develop novel approaches for producing protein ingredients with enhanced functionality, digestibility, and nutrient bioavailability (Bandara et al., 2018).

Engineering Plant-Based Food Systems. https://doi.org/10.1016/B978-0-323-89842-3.00012-9

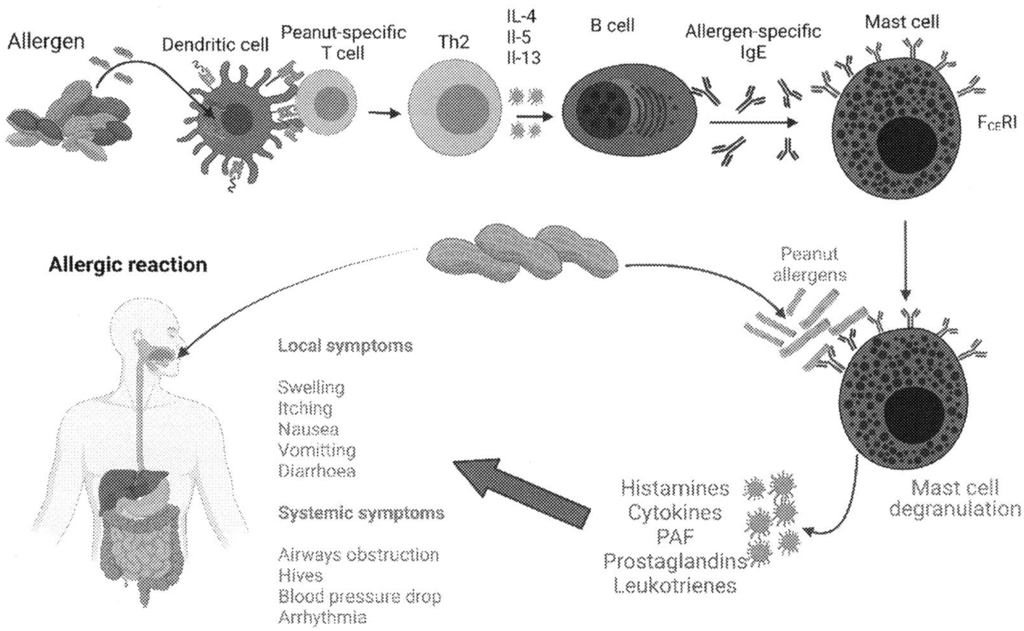

FIGURE 4.1 Mechanism of an allergic reaction. *Adapted from Burks, A. W. (2008). Peanut allergy.* The Lancet, 371*(9623), 1538–1546. https://doi.org/10.1016/S0140-6736(08)60659-5.*

Allergens in plant proteins are more of the major factors inhibiting their wide applications. Mills and Shewry (2004) document that plant protein allergens are more life-threatening when compared to other food allergens, affecting at least one in 10 children and one in 100 adults. Some allergic reaction cases have been documented after the ingestion of tree nuts, peanuts, seeds, fresh fruits, cereal, celery, and legumes by allergic subjects (Lucas et al., 2004; Shahali et al., 2018). Undoubtedly, reducing allergic substances in plant proteins has become an integral part of its preparation and utilization.

Several interventions (thermal and nonthermal) have been adopted to circumvent the allergenicity of plant protein (Fig. 4.2). These processing techniques involve the modifications of the structure of allergens at the molecular level (via deamidation, aggregation, and hydrolysis) or the interactions of allergens with other food components (formation of complexes, Maillard reaction) (El Mecherfi et al., 2020; Mills et al., 2004; Taheri-Kafrani et al., 2009; Villemin et al., 2019). Moist heat (boiling, autoclaving extrusion, frying, and retorting) and dry heat (including baking, microwaving, and roasting) are the two types of thermal treatment used for reducing allergenicity in plant proteins (Li et al., 2016). The heat generated in the thermal processes can induce conformational changes in the plant protein molecular structure, altering its allergenicity (Mondoulet et al., 2005). However, these processes have been reported to cause significant changes in the plant proteins' techno-functional and organoleptic properties. Therefore, nonthermal technologies have gained researchers' attention due to minimization of adverse impact on plant proteins. Pulsed electric fields (PEFs), pulsed ultraviolet (PUV)

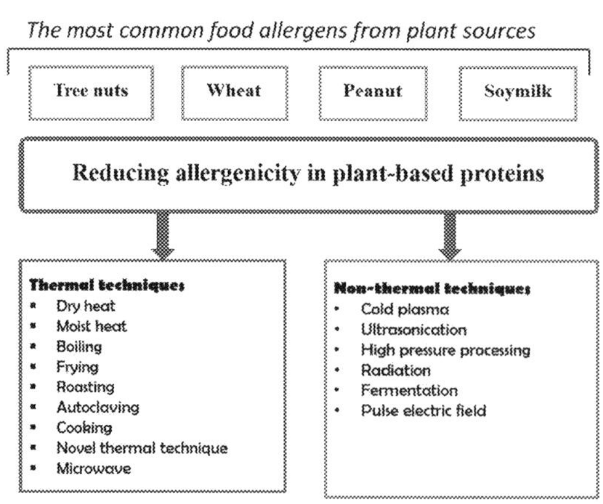

FIGURE 4.2 An overview of thermal and nonthermal food processing technologies used to eliminate allergenicity in plant-based proteins.

light, gamma irradiation, high-pressure processing, high-intensity ultrasound (HIU), genetic modification, cold plasma, enzymatic hydrolysis, and fermentation are promising nonthermal technologies have shown potentials in reducing allergens in plant protein (Li et al., 2016; Shriver & Yang, 2011). The various sources of plant protein allergens and their effects are discussed in this chapter. This chapter presents an overview of thermal and nonthermal processing techniques used to reduce food allergies caused by plant proteins and their limitations in food industry applications.

4.2 Sources of allergen and their effect

Food allergens occur naturally in foods. According to the World Health Organization (WHO) and Food and Agriculture Organization (FAO) of the United Nations, eight foods sources, including eggs, fish, milk, tree nuts, wheat, peanuts, crustacean shellfish, and soybeans, are the major sources of food allergens that frequently produce adverse immunological responses among the vast majority of sensitive individuals (FAO/WHO, 2007). Among these sources, plant sources are the most popular, owing to their popularity in a wide range of cuisines across the world, availability, and low cost. This section highlights various allergens occurring in plant foods.

4.2.1 Tree nuts

Tree nuts are dried fruits with hard, inedible shells and seeds, and their consumption has several positive effects on cardiovascular systems, including cholesterol, fasting blood sugar, and triglycerides lowering potentials (Del Gobbo et al., 2015). They are among the eight most common sources of food allergens, and their allergic reactions can be very severe (Sicherer & Sampson, 2018). Among all tree nuts, cashew, macadamia, walnut, pine out, almond, hazelnut, pistachio, Brazil nut, and pecan are responsible for the vast majority of allergic reaction cases reported (Alasalvar & Shahidi, 2008). Generally, tree nuts allergy is mainly caused by storage proteins in the nut, such as albumins, vicillins, legumins, and profilins, and other plant defense proteins (Sicherer et al., 2001).

4.2.2 Peanuts

Allergic reactions induced by peanut allergens are known to have the highest mortality rate among sensitive individuals (Bock et al., 2007). This may be attributed to the sensitive nature of the allergens, as ingestion of even a trace amount could trigger life-threatening symptoms. Peanut allergy is caused by specific proteins in the nuts and is classified into four categories: profilin (Ara h 5), cupin (Ara h 1), Bet v-1 (Ara h 8), and prolamin (Ara h 2, 6, 7, 9) (Mueller et al., 2014). Presently, 13 different proteins with a high affinity for peanut-specific immunoglobulin (IgE) have been recognized (Allergen Nomenclature, 2021). Of all these, Ara h 1, Ara h 2, Ara h 3, and Ara h six are the predominant peanut allergens that trigger an immunological response in over 50% of allergic individuals (Mueller et al., 2014).

4.2.3 Wheat

Wheat is by far the most popular cereal in the world and is an important source of carbohydrates. The major protein in wheat is gluten, and it exhibits viscoelastic and adhesive properties after hydration, making it a suitable raw material for the production of a wide range of products consumed worldwide (Shewry & Hey, 2015). However, one of the problems associated with its consumption, especially by food allergic individuals, is allergenic substances. According to Šotkovský et al. (2011), about 27 allergenic substances have been identified in wheat. Of all these wheat allergens, ώ-3 gliadin has been reported to induce the most severe food dependent, exercise-induced anaphylaxis (FDEIA) (Akagawa et al., 2007). Other proteins in wheat, such as prolamin, glutelin, albumin and globulin, may also induce allergic reactions.

4.2.4 Soybeans

Soybean is an important source of cheap proteins in different parts of the world. In addition to its good-quality protein, it is also a rich source of minerals, carbohydrates, and relatively low-fat content (Huang, Hsu, et al., 2014). Soybean contains some allergenic substances, which may trigger an overreaction of the immune system in sensitive individuals. The dominant allergen in soybean is GlymBd 30 k. Others include subunits of glycinin (G1 and G2), as well β-conglycin (Peñas et al., 2011).

4.3 Techniques employed to reduce/remove allergenicity of plant proteins

Fig. 4.2 depicts the most common food allergens from plant sources and various techniques, which can be categorized as thermal and nonthermal food processing technologies, to eliminate the food allergens in plant-based proteins.

4.3.1 Thermal techniques

Thermal treatments are processing techniques applied to foods to alter their physicochemical and sensory characteristics and enhance the texture, microbial integrity, and digestibility of food. It is the oldest form of food processing and usually involves the direct or indirect application of heat to foods. Application of heat treatment to proteins may alter the native structure of protein, breakdown of peptide bonds, formation of aggregates by noncovalent bonds and disulfide bonds, protein denaturation, and reaction with other food components such as carbohydrates and lipids. Accordingly, these heat-induced protein modifications indicate that heat treatment could either reduce (loss of epitopes) or increase (generation or exposure of epitopes) allergenicity in foods (Wal, 2003), since the majority of known food allergens are proteins (Bannon, 2004). This section will highlight various thermal processing techniques employed in food systems and their impact on various food allergens in plant proteins.

4.3.1.1 Moist heat

Moist heat involves the transfer of heat to food in the form of steam or boiling water. It is carried out solely to destroy pathogenic microbes in foods. In the process of enabling microbiological safety of foods, it may also cause structural modification of the protein, which is assumed to reduce the allergenicity of plant proteins. Summary of impacts of thermal and nonthermal processing on reduction of plant protein allergens is presented in Table 4.1. Generally, allergenicity-reduction ability at elevated temperature may be attributed to protein denaturation, which modifies the spatial configuration of the protein and causes the destruction of IgE-reactive epitopes (Paschke, 2009). Studies involving the use of moist heat to reduce allergenicity in plant proteins have been well documented, and it has been established to be a promising approach for making plant proteins safe for allergic individuals (Vanga et al., 2017). Boiling and autoclaving are the two most important moist heat treatments, and their impact on reducing plant protein allergenicity are highlighted in the next section.

TABLE 4.1 Summary of various impacts of thermal and nonthermal treatments on plant protein allergens.

Plant protein	Treatment	Target allergens	Effect of treatment on allergens	References
Peanut	Boiling (100°C, 30 min)	Ara h 1, Ara h 2	Decreased	(Mondoulet et al., 2005)
Peanut	Autoclaving (121°C, 20 min, 2 bar)	Ara h 6	Deceased	(De Jongh et al., 2020)
Peanut	Frying	Ara h 1, Ara h 2	Decreased	(Comstock et al., 2016)
Peanut	Roasting	Ara h 1, Ara h 2	Increased	(Maleki et al., 2000)
Peanut	Roasting		Increased	(Zhang et al., 2019)
Peanut	HPP (180 MPa)	Ara h 2	Decreased	(Hu et al., 2011)
Peanut kernels	Ultrasound (50 Hz, 1 h)	Ara h 1, Ara h 2	Decreased	(Hao et al., 2013)
Peanut	Cold plasma (80 kV, 60 min)	Ara h 1	Decreased	(Venkataratnam et al., 2019)
Defatted peanut	Cold plasma (32 kV, 118 s)	Ara h 1, Ara h 2	Decreased	(Venkataratnam et al., 2019)
Peanut protein	Irradiation (10 kGy, 10°C)	Ara h 6	Decreased	(Li et al., 2013)
Crushed peanut	HPP (800 MPa, 10 min)		No effect	(Huang et al., 2014)
Lentil	Boiling (100°C, 30 min)	Len c 1	Decreased	(Cuadrado et al., 2009)
Chickpea	Boiling (100°C, 30 min)		No effect	(Cuadrado et al., 2011)

TABLE 4.1 Summary of various impacts of thermal and nonthermal treatments on plant protein allergens.—cont'd

Plant protein	Treatment	Target allergens	Effect of treatment on allergens	References
Chickpea	Autoclaving (121°C, 15 min, 1.2 atm)		Decreased	(Cuadrado et al., 2011)
Lupine flour	Autoclaving (138°C, 30 min)		Eliminated	(Álvarez-Álvarez et al., 2005)
Soybean	Cooking (30 min)		No effect	(Cabanillas et al., 2018)
Soybean protein	Microwave		Decreased	(Li, Zhu, et al., 2016)
Soymilk	Fermentation (using *Enterococcus faecalis VB43*)		Decreased	(Biscola et al., 2017)
Soybean protein	Ultrasound (20 kHz, 600 W, 50%, 15 min)		Decreased	(Li, Zhu, et al., 2016)
Soybean whey	HPP and enzyme hydrolysis	Gly m 1	Decreased	Peñas et al. (2006b)
Soybean protein isolate	Cold plasma (11 kV, 10 min)	Gly m 5, Gly m 6	Decreased	(Meinlschmidt et al., 2016)
Cashew nut	Roasting (200°C, 15 min)		No effect	(Venkatachalam et al., 2008)
Almond	Roasting (137.7–168.3°C		No effect	(Venkatachalam et al., 2008)
Hazelnut	Microwave (630 W, 10 min)		No effect	(Wigotzki et al., 2000)
Walnut	Microwave (500 W, 3 min)		No effect	(Su et al., 2004)
Walnut	HPP (650 MPa, 15 min, 100°C)		Decreased	(Yang et al., 2017)
Pea flour	Fermentation (using lactic acid bacteria and fungi)		Decreased	(Barkholt et al., 1998)
Wheat protein	Cold plasma (10 min)		Decreased	(Meinlschmidt et, al., 2016)
Black bean, kidney bean, peanut	Irradiation (25 kGy) and cooking (121°C, 15 min)		Decreased	(Kasera et al., 2012)

4.3.1.2 Boiling and autoclaving

Boiling and autoclaving are moist heat techniques commonly employed in food systems. While boiling is essentially applied to food to alter their organoleptic and digestibility properties, autoclaving is mostly used for sterilization purposes. Studies on their effects in reducing allergenicity in plant-based materials have been conducted. Boiling treatment has been observed to have minor impacts on allergens from peas (Bogdanov et al., 2016), chickpea and lentil (Cuadrado et al., 2009), lupin (Álvarez-Álvarez et al., 2005), and soybean (Burks et al., 1992). Only a few studies demonstrated boiling treatment to be a good approach for reducing allergenicity. According to Mondoulet et al. (2005), boiling of peanut extract for 30 min at 100°C decreased its major allergens (Ara h 1and Ara h 2) composition.

In most of the studies conducted to date, autoclaving has been the most effective and consistent moist heat treatment for reducing or eliminating allergenic potentials in plant proteins. In a study conducted on lentils using different heat treatments, boiling of lentils for 30 min eliminated the major allergen (Len c 1) but did not affect the allergenic potential of lentils on the allergenic protein. However, autoclaving (30 min at 2.6 atm) caused a significant reduction in the allergenic potential of lentils (Cuadrado et al., 2009). In another study, Cuadrado et al. (2011) showed that boiling of chickpea for 30 min had no significant impact on immunoreactive proteins, while a significant reduction was observed after 60 min. When the same chickpea was subjected to autoclaving for 15 min at 1.2 atm, a significant reduction in immunoreactive proteins was found, and a further reduction was observed as autoclaving time and pressure increased. De Jongh et al.

(2020) investigated the effect of different thermal treatments on the conformational stability of peanut allergens (Ara h 6). They found that only autoclaving at 121°C for 20 min at 2 bars caused irreversible denaturation of the allergens, resulting in a significant reduction of IgE-binding potential of the allergen. Complementarily, Cabanillas et al. (2012) found that autoclaving of peanut allergens for 30 min at 2.56 atm reduced its allergenic content. According to Álvarez-Álvarez et al. (2005), autoclaving lupine flour for 30 min at 138°C eliminated potent allergens in the flour. The potentials of pressure treatment, like autoclaving in causing a reduction in allergenic potentials, may be attributed to its effect on protein secondary structure (unfolding) and enhanced digestibility (Cabanillas et al., 2012).

4.3.1.3 Cooking

Cooking is one of the oldest processing methods and has been applied extensively to various foods to reduce their allergenicity composition. In most of the results presented to date, pressure cooking has been reported to be the most effective in reducing allergenicity. For example, in a study by Cabanillas et al. (2018), cooking soybean for 30 min had no impact on IgE reactivity. However, a combination of pressure and heat caused fragmentation of proteins, which subsequently caused a reduction in IgE-binding capacity. Still, well-known soybean allergen like 7S globulins survived the heat treatment.

4.3.1.4 Frying

Frying is a processing method that utilizes oil for cooking foods. The majority of studies on the effect of frying on allergenicity are reported for only peanuts and tree nuts. An earlier study performed on the frying effect on the allergenicity component of peanut reported that IgE reactivity of proteins in peanut decreased following frying Beyer et al. (2001). A recent study also Comstock et al. (2016) reported that frying reduced two major allergens (Ara h one and Ara h 2) in peanut Comstock et al. (2016). A later study conducted did not show any major difference in allergenicity of raw and fried peanut (Cabanillas et al., 2018). On the other hand, the antigenicity of proteins in tree nuts such as walnut, cashew, brazil nut, and almond have been reported to be resistant to even extreme frying temperatures up to 191°C (Sharma et al., 2009; Su et al., 2004).

4.3.2 Dry heat

Dry heat technique can alter the protein configurations, which, in turn, may lower the allergenicity of the protein allergens. Roasting is the most common dry heat treatment applied to different plant proteins, and contradictory findings have been reported about its effect on allergenicity. In most studies conducted on plant protein, roasting has been reported to elevate the allergenicity of plant proteins rather than decreasing it (Beyer et al., 2001; Schwager et al., 2017). This effect has been attributed to the Maillard reaction, which is common in roasted products and involves interactions between free amino acids in proteins and reducing sugars in foods. These interactions often lead to the development of aggregates, which may influence allergenicity and the rate of digestion of allergenic compounds in the stomach (Thomas et al., 2007). In a study undertaken to evaluate the impact of roasting on the allergenicity of peanut allergen, reaction between proteins and carbohydrates (Maillard reaction) was reported to increase the IgE binding capability of Ara h one and Ara h two by over 90 times in comparison to unroasted (raw) peanut (Maleki et al., 2000). A more recent study by Zhang et al. (2019) on the allergenicity of peanut protein indicated that roasted peanuts showed higher allergenicity compared to raw and those processed using other thermal treatments.

Regarding other plant proteins, it appears that roasting did not cause any undesirable increase in allergenicity. Venkatachalam et al. (2008) reported that allergic compounds in cashew nut flour remain unchanged after roasting for 15 min at 200°C. Similarly, the roasting of almonds within 137.7−168.3°C also did not cause a considerable increase in allergenicity of the nuts (Venkatachalam et al., 2006; Sharma et al., 2009).

Taken together, the impact of roasting on protein structure and consequently on allergenicity varies among different plant proteins. While roasting caused the undesirable increase in the reactivity of allergen in peanut proteins, it did not affect proteins in almond and cashew nuts. This could mean that the effect of roasting depends on the type of epitope present in different proteins, and this should be considered when selecting thermal processing for reducing allergenicity in plant proteins in other to achieve the most desirable results.

4.3.3 Novel thermal techniques

There are innovative thermal techniques developed to minimize the environmental impacts and emission of greenhouse gases (GHG) of conventional thermal processing. Microwave, dielectric heating, radio frequency, and ohmic heating are examples of novel thermal processing. This innovation resulted in not only the development of energy-efficient and environmentally friendly technologies for use in food industries around the world but also reducing the cost of energy, improving quality of food products, and reducing wastage of natural resources in comparison to conventional thermal techniques (Pereira & Vicente, 2010). However, the impacts of these novel thermal techniques on food components, especially proteins and foods allergens, have not been fully elucidated. Studies conducted using these techniques to reduce allergenicity are only available for microwave technique, as the following discussion.

Microwave is a type of electromagnetic radiation with wavelength within 300 MHz and 3000 GHz bands. They have a wide range of applications in the medical, food, and telecommunication industries. In food industries, they are majorly used for precooking, pasteurizing, drying, and thawing of frozen foods (Vanga et al., 2017). This is due to the fact that the microwave is easy to use, requires less time to start, instantly generates heat, heat foods more efficiently, and has negligible impacts on sensory and nutritional qualities of food (Sun, 2014). Interactions between microwaves and proteins have been reported to have some effects on the surface properties of proteins due to differences in bonds formations between molecules of water and other ions (Meda et al., 2017). These impact on protein surface properties suggest that microwaves could potentially be useful in reducing the allergenicity of protein allergens since most processes are known to reduce allergenicity affect the structural or chemical integrity of the proteins. Modification of IgE binding activity of proteins microwave may be linked to modifying the protein epitopes via changes in the conformational configuration of the proteins (Vanga et al., 2017).

Several studies have been conducted to evaluate the effects of microwave heating on allergenic proteins in foods. In a study conducted by Wigotzki et al. (2000) to evaluate the effect of microwave on the allergenicity of protein allergens from four hazelnut varieties, microwave treatment (630 W for 10 min) had no significant impact on the allergenicity of their proteins. This may be due to the presence of heat-resistant allergens in hazelnut. Venkatachalam et al. (2002) investigated the impact of three different heat treatment techniques, including blanching, microwave, and roasting, on the allergenicity of whole almond seeds and defatted almond flour. The authors demonstrated that all the heat treatments applied, including microwave (50% power for 1, 2, and 3 min), had minimal impact on epitopes of almond proteins. This revealed that epitopes in almond proteins are resistant to heat. In a similar study, Su et al. (2004) demonstrated that allergenicity of walnut, almonds, and cashew nuts remained stable after microwave treatment for 1 and 3 min at 500 W. In a more recent study, Li et al. (2016) evaluated the effect of four different physical techniques such as high hydrostatic pressure, high-intensity ultrasound, microwaving, and high-pressure homogenization for reducing allergenicity of soybean protein isolate. It was reported that microwaving caused about a 24.7% reduction in allergenicity of the isolate. It was assumed that this reduction was caused by the heat lability of soybean protein allergens. This finding is in agreement with earlier findings of Vieths et al. (1995), who reported a decrease in allergenic potentials of soybeans subjected to microwave treatment. Thus, the susceptibility of plant proteins allergens to microwave treatment seems to depend on the epitopes of the protein allergens and structural pattern of the proteins.

4.3.4 Nonthermal technologies

Nonthermal technologies have shown promising potential in reducing the allergies in plant protein via conformation changes or induced aggregation, aiming to change the allergen structures and reduce the allergen levels and allergenic potency. The modes of action of nonthermal processes commonly used in reducing allergenicity in plant proteins are discussed below.

4.3.4.1 Fermentation

Fermentation is the old nonthermal process employed for the reduction of protein allergenicity, mainly plant protein. Fermentation is a process that led to desirable biochemical changes in food via the activities of enzymes or microorganisms (Campbell-Platt, 1987). Depending on the food material, fermentation may involve the breakdown of polysaccharides into smaller molecules such as di- or mono-saccharides and also proteins into smaller peptides, which are easily digested or absorbed in the body. However, food fermentation could either be spontaneous (uncontrolled) or nonspontaneous (controlled) (Martínez-Álvarez et al., 2017). During food fermentation, endogenous and exogenous proteases from microorganisms (bacteria, yeasts, or molds) act on the hydrolysis of food components. The mechanisms of fermentation in degrading or reducing allergenicity in plant protein involve the breakdown of plant protein allergens via the proteolytic activities of the enzymes or fermenting organisms into amino acids and peptides, losing the IgE-binding ability. Additionally, fermentation lowers allergenicity and contributes significantly to improving the organoleptic and nutritional values of resulting products (Ko et al., 2010). The most important microorganisms responsible for the transformation and maturation of the product are lactic acid bacteria (LAB). Those belong to genera Lactobacillus, Staphylococcus, Pediococcus, Enterococcus, Staphylococcus, Micrococcus, and Leuconostoc (De Vuyst et al., 2008). The main components of LAB proteolytic systems are proteases associated with the bacterial envelope, an amino acid and peptide transport system, and a range of intracellular peptidases (Bachmann et al., 2012). LAB proteolytic activity can affect allergenic epitopes, reducing the allergenicity of many food components (El Mecherfi et al., 2020).

Spontaneous fermentation, otherwise known as natural fermentation, is the oldest method of fermentation. It is usually carried out by the resident microorganisms or microflora naturally present in food (Rajauria et al., 2016). Due to the traditional process employed or properties of the raw material, the sensory, nutritional, or functional qualities of spontaneously fermented food are not distinct from the controlled process (Ojha et al., 2015; Rajauria et al., 2016). Spontaneously fermented plant protein or plant protein products are made from a variety of fruits, citrus, root, crops, grains, pulses, marine algae, and raw cane sugar fermented for a long period of time (>1 month) at room temperature (Fujimura et al., 2018). The efficacy of spontaneous fermentation in degrading allergenicity has been documented. Kobayashi et al. (2004) confirmed that the degradation of allergenicity in soy sauce affects spontaneous fermentation of microbial proteolytic enzymes for 3 months.

Nonspontaneous fermentation, also known as controlled fermentation, is implemented to assure the uniformity of the final fermented product since the microflora of raw materials may vary depending on their nature or geographical location (Martínez-Álvarez et al., 2017). Nonspontaneous fermentation involves using a starter culture or enzymes to jump-start the fermentation of food to produce stable products with superior nutritional and sensory properties (Ojha et al., 2015). Nonspontaneous fermentation has been implemented to reduce allergenicity in plant-based milk. Biscola et al. (2017) documented a significant reduction in the immunoreactivity of allergenic proteins β-conglycinin (7S) and glycinin (11S) in soymilk fermented with *Enterococcus faecalis* VB43.

The influence of nonspontaneous fermentation on the plant-based meal or protein isolate or hydrolysate has been examined. In the study of Meinlschmidt, Schweiggert-Weisz et al. (2016), *Actinomucor elegans* DSM 1174, *Rhizopus oryzae* DSM 2200, and *Lactobacillus perolens* DSM 12744 were used in the liquid-state fermentation of different soy protein hydrolysates prepared using Alcalase, papain, and the enzyme combination (papain and flavorzyme). The reduction in the major soy allergens (Gly m5 and Gly m6) in the hydrolysate with superior techno-functional properties was reported regardless of the enzymes used for producing the soybean hydrolysate (Meinlschmidt, Schweiggert-Weisz et al., 2016). Additionally, solid-state fermentation of soybean meal (SBM) with a mixture of *Bacillus subtilis* CICC No. 20641, *Lactobacillus casei* CGMCC No. 1. 539, and yeast at a ratio of 2:1:1 (v/v/v) at 30°C for 3 days reduced the potential allergenicity of SBM (Yang et al., 2018). The authors reported that the reduction in the allergenicity of SBM after fermentation was attributed to the degradation of allergen proteins into low-molecular-weight polypeptides, in which allergenic epitopic sequences of β-conglycinin and glycinin were destroyed. Barkholt et al. (1998) documented the efficacy of three LABs species (*Pediococcus pentosaceus, Lactococcus raffinolactis, and Lactobacillus plantarum*), and two fungi (*Rhizopus microsporus*, var. oligosporus and *Geotrichum candidum*) in reducing allergenicity in pea flour, in which 10% reductions in allergens was recorded for the fermented pea floor when all the microorganisms used except *Geotrichum candidum*.

The application of nonspontaneous fermentation in reducing allergy in cereals has also been investigated. *Lactobacillus alimentarius* 15M, Lacto*bacillus brevis* 14G, *Lactobacillus sanfranciscensis* 7A, and *Lactobacillus hilgardii* 51B used in fermenting wheat and rye dough degraded some IgE-binding proteins (albumins/globulins and gliadins mainly) with respect to that chemically acidified (Rizzello et al., 2006).

The degradation of allergens in the plant-based product by the enzymatic hydrolysis using peptidase has also been investigated. However, the efficacy of enzymes in reducing allergenicity depends on several factors such as enzyme/protein ratio, type of enzymes, time, concentration. In the study of Yu et al. (2011), a-chymotrypsin and trypsin were used for reducing allergenicity in both raw and roasted peanut kernels via enzymatic hydrolysis. The authors reported that chymotrypsin and trypsin effectively degraded the extractable fraction of the two major peanut allergens Ara h one and Ara h 2. Furthermore, Ara h one and Ara h two in roasted peanut kernel were significantly degraded by the combined treatment of ultrasound (sonication for 1–5 h) and protease hydrolysis using trypsin or a-chymotrypsin at different enzyme/peanut ratios (0%–0.30% w/w) (Li et al., 2013). The ultrasound–enzyme combination also significantly lowered IgE binding of peanut extract (Li et al., 2013). Alcalase from B*acillus licheniformis,* and papain from the latex of papaya fruit showed higher efficacy in reducing allergenic gliadins in wheat flour when compared to flavorzyme, pepsin, trypsin or α-chymotrypsin (Li, Yu, et al., 2016). Hydrolysates from peanut flour produced using alcalase, pepsin or flavorzyme for 60 min had reduced IgE binding capacity compared with nonhydrolyzed counterparts (Shi et al., 2013). Yu et al. (2015) reported that enzymatic hydrolysis using Alcalase effectively reduced the allergenic protein (Ara h one and Ara h 2) in roasted peanut kernels in comparison with the nonhydrolyzed counterpart. Similarly, IgE-binding activities were lowered in alcalase treated roasted peanuts (Yu et al., 2015). When alpha-chymotrypsin, trypsin or the combination were used in hydrolyzing home-made and commercial peanut butter, Ara h one and Ara h two were significantly reduced, in which the hydrolysis with the combined enzymes showed the highest efficacy (Yu et al., 2013).

4.3.4.2 Ultrasonication

Ultrasonication is a technique that utilizes the frequency of sound waves above 20 kHz, a frequency higher than the detection level of human audition (Kadam et al., 2015). Ultrasound waves carry high energy while traveling through a medium (Mason et al., 2005). Ultrasound waves compress and stretch during its movement in a medium, thereby forming localized negative pressure. Working principles and food applications of ultrasonication were extensively described elsewhere (Mason et al., 2005; Torley et al., 2020). Ultrasound can alter the secondary, tertiary, and quaternary structures of proteins (Jiang et al., 2014; Saleem & Ahmad, 2016; Vidal et al., 2018; Yuanqing et al., 2020). Generally, changes in the structural configuration of proteins following ultrasound treatment may be due to the ability of ultrasound energy to disrupt the intermolecular interactions between protein chains as well as disulfide bonds in proteins (Jin et al., 2015). The impact of ultrasound on the protein structure can be attributed to extremely (localized) high temperature (above 1000°C) and pressures (ranging from 50 to 500 MPa) generated by cavitation, which occurs when food samples are subjected to ultrasound treatment (Islam et al., 2014; Tao & Sun, 2015). This extremely high acoustic energy generated during ultrasound treatment can disrupt the noncovalent bonds (e.g., hydrogen bonds) and disulfide bonds in proteins. These disruptions play a vital role in the conformational changes in the secondary and tertiary structures of proteins (O'Sullivan et al., 2017). Additionally, cavitation generated during ultrasound treatment can break down water molecules to generate reactive species such as H and OH(Rahman et al., 2020). These reactive species could induce protein oxidation (Olatunde, Singh, et al., 2021). Oxidization of protein can cause secondary and tertiary structure changes either via protein aggregation or crosslinking (Olatunde, Singh, et al., 2021; Rahman et al., 2020). This phenomenon might also be associated with the efficacy of ultrasound in decreasing potential protein allergenicity (Xu et al., 2017).

Despite the potential of ultrasound in decreasing protein allergenicity, its application, particularly in plant-based protein, has not been fully exploited. Li, Zhu, et al. (2016) reported a 29.8% reduction in the allergenicity of soy protein isolate after ultrasound using a 20 kHz and 600 W at an amplitude of 50% for 15 min. Ultrasound treatment coupled with enzymatic hydrolysis has been examined to further increase the efficacy of ultrasound in reducing allergenicity. Li et al. (2013) documented that ultrasound treatment alone for 1 h at 50 Hz decreased IgE-binding affinity of roasted peanut kernels allergens (Ara h1 and Ara h2) by 10%. When it was combined with protease digestion using trypsin or α-chymotrypsin at different enzyme/peanut ratios (0%, 0.1%, 0.15%, 0.20%, 0.25%, or 0.30% w/w), the solubility of proteins was significantly increased, and IgE reactivity was further decreased by almost 100%. The authors proposed that the ultrasonication might reduce the original allergens by molecular destruction due to shear force (Li et al., 2013). Therefore, it is imperative for direct research toward the application of ultrasound either as a single nonthermal treatment or in combination with other technologies for reducing allergenicity in plant-based protein. Additionally, the mechanism behind this phenomenon should be properly investigated.

4.3.4.3 High-pressure processing

High hydrostatic pressure processing (HHP) ,or ultra-high pressure (UHP), involves the exposure of food materials to pressure (100–1000 MPa), and the use of HPP in food processing has improved dramatically in the last 20 years (Norton & Sun, 2008). The technique has been exploited to diverse consumer demands with the generation of novel foods, tastes, and textures (Garcia-Mora et al., 2015). Lou et al. (2015) and Abera (2019) provided an updated review of working principles and applications of HPP in food systems.

Proteins are stabilized in their native state by covalent bonds (including disulfide bridges), electrostatic interactions (ion pairs, polar groups), hydrogen bridges, and hydrophobic interactions. The covalent bonds in proteins are usually not affected by HPP, especially at low temperatures (0–40°C); hence, the primary structure of proteins is intact during HPP treatment (Goyal et al., 2018). However, HPP treatments affect the secondary, tertiary and quaternary structures via hydrogen ion and electrostatic interactions, hydrophobic and hydrogen bonding, and hydrophobic interactions, respectively (Goyal et al., 2018). At a pressure <300 MPa, these changes in protein structures are reversible, but at pressures >300 MPa, an irreversible denaturation of proteins can occur. At pressure above 700 MPa, HPP may give rise to irreversible protein denaturation by disrupting the secondary structure of the proteins (Balny & Masson, 1993).

The conformational changes induced by HPP on plant-based proteins can modulate their digestibility and allergenicity. It was shown that HPP treatment (350 MPa for 16 min) showed the highest efficacy in reducing soy protein isolate (SPI) allergens when compared to other food processing technologies (e.g., microwave, high-intensity ultrasound, and high-pressure homogenization) (Li, Zhu, et al., 2016).

The observed effect was attributed to the steric epitope destruction, molecular interactions including free sulfuryl group, hydrophobicity, and zeta potential and also altered distributions of secondary structures induced by HPP. Huang, Yang et al. (2014) investigated the influence of HPP at different conditions (200, 400, 600, and 800 MPa for 5 and 10 min) on the allergenicity of crushed peanut. No significant changes in the peanut allergenicity were observed when 200 MPa was used regardless of the treatment time. At the same time, the immune reactivity of crushed peanut was found to decrease in pressure- and time-dependent manner when the pressurization level was achieved at 400, 600, and 800 MPa. The reduced allergen in crushed peanut after HPP treatment was attributed to the induced protein conformational modifications, protein denaturation, allergen removal by extraction of allergens, and the promotion of enzymatic hydrolysis to alter the sensitization of the allergens (Huang, Yang et al., 2014). A similar observation was documented by Hu et al. (2011) when peanut was subjected to various HPP treatments (60, 90, 120, 150, and 180 MPa), in which the decrease in Ara h two binding to IgE with increasing pressure was reported. It was proposed that the changes in binding characteristics of Ara h two were linked to the altered structure and surface hydrophobicity of the allergen protein mediated by the high-pressure treatments. Furthermore, HPP treatments (200−500 MPa, 5−20 min) caused the reduction in allergenicity of soy protein isolated from infant formula in which 300 MPa for 15 min could decrease 48.6% allergen (Li et al., 2012). The immunoreactivity of peanuts allergens (Ara h two and 6) was decreased by HPP (150−800 MPa for 10 min), which was attributed to adiabatic heating and protein aggregation during HPP (Johnson et al., 2010). However, HPP shows great potential in reducing allergenicity in protein-based protein; its efficacy largely depends on the variation of allergen microstructure in different plant proteins.

HPP combined with enzymatic process. To further increase the efficacy of HPP in reducing the allergenicity of plant-based protein, its combined application with the enzymatic process has been implemented. The conformation changes induced by HPP on food proteins can expose the active site for the proteolytic enzymes to effectively cleave the proteins. The allergenicity of soybean whey hydrolyzed with either alcalase, neutrase, corolase 7089, and corolase PNL in combination with HPP (100, 200 and 300 MPa) for 15 min was evaluated by Peñas et al. (2006a). The immunochemical response of soybean whey to anti-Gly m one monoclonal antibodies decreased with the combined treatment of HPP and enzymatic hydrolysis, particularly when high pressure was used. A study performed on rice grains reported that the content of allergenic proteins decreased when subjected to HPP treatment (300 MPa at 20°C for 30 min). Interestingly, an almost complete disappearance in allergens from rice grains was recorded when HPP was used in the presence of protease-N (Kato et al., 2000).

HPP combined with thermal process. HPP in combination with thermal process in reducing allergenicity in plant-based protein has also been investigated. Husband et al. (2011) documented that the combined treatment of HPP (700 MPa for 10 min) and heat (115 and 118°C) shown high efficiency in reducing allergenicity in both apple and celeriac (Mal d 1, Mal d 3, and Api g 1) as compared to that treated with only heat. Ara h two in peanut after exposure to the combined treatment of pressure (100−600 MPa) and heat (25−75°C for 2.5−30 min) was significantly reduced, in which the highest reduction in IgE binding was observed when the samples were subjected to 55°C, 600 MPa for more than 10 min (Long et al., 2016). The efficacy of the combined treatment of HPP and thermal treatment (650 MPa, 100°C for 15 min) in reducing allergenicity in walnut compared with boiling treatment was investigated by Yang et al. (2017). The authors reported 86.37% reduction in the IgE binding capacity of walnut subjected to HPP and thermal treatment as compared to $<50\%$ reduction reported for walnut subjected to boiling. When mice were fed with processed, treated walnut (WAH), the allergenic reaction was characterized using IgE binding capacity, histamine levels in serum, and cytokine mRNA expressions. A significant reduction in serum specific IgE binding capacity (41.04%) and histamine levels (19.76%) were observed in mice fed with WAH, accompanied by a 36.79% and 34.39% repression in both IL-4 and IFN-γ mRNA expressions in ileums, respectively (Yang et al., 2017).

4.3.4.4 Pulsed electric field

Pulsed electric field (PEF) is one of the novel nonthermal technologies that can effectively inactivate microorganisms and enzymes inactivation in food (Olatunde & Benjakul, 2018). PEF is based on electroporation and cell disintegration by applying repeated pulses when the food is placed between the two parallel electrodes (Shiekh & Benjakul, 2020b). In PEF processing, microsecond high-voltage pulses in the range of 10−60 kV are employed (Kumar et al., 2016). The application of high-voltage pulses induces pores named "electroporation" in cell membranes, initiating a loss of barrier function, intracellular content leakage, and loss of vitality (Olatunde et al., 2018). Studies have indicated that nonthermal treatment in the form of electromagnetic waves and electric fields could modify the structure of proteins (Gomaa et al., 2013; Li, 2012), thereby modifying the native structure of the protein. Only secondary and tertiary protein structures are affected by the electromagnetic wave and electric fields (Han et al., 2018). The mechanism of alteration of protein structure may be linked to energy absorption potential by polar groups of proteins and generation of free radicals, which causes unfolding of proteins (Han et al., 2018) as well as protein oxidation (Olatunde, Singh, et al., 2021). Recently, Tobajas et al. (2020) documented the efficacy of PEF (25 kV/cm) without and with heat treatment (50°C) in denaturing Pru p 3, which is the major allergenic protein of peach. Nevertheless, the efficacy of PEF in reducing allergenicity in food protein is still lacking or in its early stage. Therefore, it is recommended that the mechanism and application of PEF in allergen reduction in plant-based protein should be investigated.

4.3.4.5 Cold plasma processing

The excitement of any gas (combined or single) with high electric field strength or any form of energy will change its state to the ionized form, known as plasma. During this process, diverse species such as negative and positive ions, radicals, neutral and excited molecules, electrons, and quanta of electromagnetic radiation (i.e., visible light and ultraviolet) are produced (Fukuda et al., 2019). CPP technology has evolved as a novel nonthermal technology for microbial decontamination, cancer treatment, and wound healing (Graves, 2014; Shaw et al., 2015; von Woedtke et al., 2013). The application of CPP technology has been extended to the preservation of foods including meat (Moutiq et al., 2019), fish (Olatunde et al., 2019a, b), shrimps (Shiekh & Benjakul, 2020a), and fresh agricultural produce (Hojnik et al., 2017), which is attributed to their excellent antimicrobial properties of the reactive species generated. Nevertheless, these reactive species could induce some conformation changes in food protein or accelerate protein oxidation (Olatunde, Shiekh et al., 2021).

Protein oxidation is the covalent modification of a protein-mediated by reactive species produced by CPP, which could be direct (interaction of the protein with reactive species) or indirect (interaction of the protein via the secondary oxidation products) (Zhang et al., 2013). The backbone of protein and side chain of amino acids undergoes oxidation in the presence of reactive species (Estévez, 2011). These could induce protein−protein cross-linkages and/or protein fragmentation (Zhang et al., 2013). Single backbone cleavage mediated by the oxidation of prolyl, glutamyl, and aspartyl side chains could induce protein fragmentation (Essex et al., 2001). The modifications of sulfur containing amino acids (methionine and cysteine) could also be initiated during protein oxidation, and these modifications can change the protein properties (chemical and physical), including structure, enzyme activities, susceptibility to proteolysis, conformation, and solubility (Zhang et al., 2013).

The modifications of protein, either by crosslinking or fragmentation induced by protein oxidation, influence the allergenicity of plant-based proteins (Gavahian & Khaneghah, 2020). In the study of Meinlschmidt, Ueberham, et al. (2016), SPI was treated with CPP generated with different voltages (9, 10, and 11 kV) and times (1, 2.5, 5, 7.5, and 10 min). A degradation of SPI principal allergens (b-conglycinin (Gly m5) and glycinin (Gly m6)) was observed with generation of new protein bands at 50 kDa, indicating the formation of insoluble aggregates after the plasma treatment. A complete reduction (100%) in immunoreactivity of SPI was recorded after 10 min CPP treatment (Meinlschmidt, Ueberham, et al., 2016). CPP treatment also caused a decrease by 37% in wheat allergens (Nooji, 2011). Li et al. (2017) demonstrated the efficacy of CPP treatment (51 W for 21 min) in eliminating soybean antinutritional compounds. Soybean trypsin inhibitor activities of soymilk were reduced by 86.1% after CPP treatment, which was attributed to CPP-induced conformational changes and oxidative reactions. Ara h one in dry, defatted peanut flour, whole peanut reduced after CPP treatment at a voltage of 80 kV for different treatment durations (0, 15, 30, 45, and 60 min) in a treatment time-dependent manner (Venkataratnam et al., 2019). The reduced allergenicity in the aforementioned samples was attributed to the conformational changes induced by CPP reactive species (Venkataratnam et al., 2019). Similarly, soybean agglutinin (SBA) activity was decreased after CPP treatment (40 kV, 12 kHz at 1, 2, 3, and 4 min) in both SBA model system and soymilk. This resulted from conformational changes as well as oxidation of aromatic amino acid, cleavage of peptide bonds and the breakage of polypeptide leading to fragmentation and/or denaturation of SBA (Liu et al., 2021). CPP (32 kV for 118 s) also showed high potential in reducing the protein allergy in peanut, in which the highest reduction in antigenicity was 65% for Ara h 1% and 66% Ara h 2. The reduced peanut protein solubility induced by the aggregation of protein-mediated by CPP causes changes in allergen structure, leading to reduced antigenicity (Venkataratnam et al., 2020). The IgE-binding level of SPI was decreased by 75% when SPI was subjected to CPP treatment at varying frequencies (80, 100, and 120 Hz) and treatment time (1, 2,5, and 10 min), compared to the control (Zhang et al., 2021). In summary, the research findings demonstrate that the changes in the protein structure as well as oxidation induced by CPP play a significant role in reducing allergen in plant-based protein.

4.3.4.6 Radiation

Radiation has gained researchers interest as a promising nonthermal food preservation technique owing to its benefits in minimal changes in the nutritional and sensory properties of treated foods (Rahaman et al., 2016). However, the use of radiation to effect conformational changes in proteins is not recognized as a nonthermal method for modulating immune protein reactivity (Luo et al., 2013).

Radiation has been documented to be an effective nonthermal technology for reducing allergy in plant-based protein. When irradiation was applied at different doses (1, 10, and 25 kGy) in reducing agglutinin from wheat germ, it appeared that radiation caused initial fragmentation of polypeptide chain followed by formation of large insoluble amorphous aggregates, which resulted in lowered allergenicity Vaz et al. (2012). In addition, radiation inhibited cell agglutination, affecting its binding to the gut and reducing the allergic effect (Vaz et al., 2012). A significant decrease in immunoglobulin G (IgG) binding activity with increasing radiation dose (1, 3, 5, and 10 kGy at 10°C) was reported in reducing isolated peanut allergen Ara h six and allergen in whole peanut extract Luo et al. (2013). The efficacy of allergen reduction was higher when radiation was applied on isolated allergens as compared to the whole protein extract. This was due to the complex nature of the whole peanut extract, which helps in protecting the epitopes on Ara h6 (Luo et al., 2013). The modification induced by radiation (10 and 50 kGy) on the conformation of epitopes found in peanuts results in the profound reduction of IgE-binding capacity to several antigens and the lowered allergenic cytokine production (IL-4) by splenocytes of sensitized mice allergy (Oh et al., 2009).

As a panacea to further increase the efficacy of radiation in reducing allergy in plant-based protein, combined treatment of radiation with several other processing technologies has been implemented. The allergenicity of different legume (kidney bean, black gram, and peanut) proteins subjected to the combined treatment of irradiation (5, 10, 15, 20, and 25 kGy) and cooking (121°C for 15 min) (Kasera et al., 2012). The combined treatment effectively reduced IgE binding of both soluble and insoluble proteins, which resulted in attenuating allergenicity of the aforementioned legume proteins (Kasera et al., 2012). The conformational changes induced on lectin protein treated with gamma irradiations (0.02−35 kGy) lead to misfolding and aggregation of protein via the modification of the hydrophobic amino acids, reducing the allergenicity of proteins in lectins (Vaz et al., 2011).

4.4 Concluding remarks

Modification of food allergens in plant-based proteins through protein denaturation, oxidation, structural modification using thermal and nonthermal technologies holds potential in reducing the allergenicity of plant-based proteins. It appears that the structural patterns of plant proteins differ from one another; therefore, they can react differently to both thermal and nonthermal induced conformational changes. This can partly explain why empirical results to lower the food allergens in tree nuts, peanut, soybean, and wheat have been contradictory, primarily due to the discrepancies in epitope and structural pattern of the plant proteins. In addition, the understanding of the mechanism involves in each technique in reducing or lowering allergens in plant protein is still in infancy. As such, there is a growing interest in evaluating the peculiarity of proteins from different plants and optimizing the parameters of both thermal and nonthermal technologies to explore them in food industries entirely. Furthermore, designing combined processes based on the mode of action of different plant proteins could be the future direction for researchers to increase the efficacy of these technologies, eradicating allergenicity in plant protein.

References

Abera, G. (2019). Review on high-pressure processing of foods. *Cogent Food Agriculture, 5*(1), 1568725. https://doi.org/10.1080/23311932.2019.1568725

Akagawa, M., Handoyo, T., Ishii, T., Kumazawa, S., Morita, N., & Suyama, K. (2007). Proteomic analysis of wheat flour allergens. *Journal of Agricultural Food Chemistry, 55*(17), 6863−6870.

Alasalvar, C., & Shahidi, F. (2008). *Tree nuts: Composition, phytochemicals, and health effects*. CRC press.

Allergen Nomenclature, S.-C. (2021). *Arachis hypogaea* (peanut). Available from: http://www.allergen.org/search.php?allergensource=peanut . May 2021 2021.

Álvarez-Álvarez, J., Guillamón, E., Crespo, J. F., Cuadrado, C., Burbano, C., Rodríguez, J., … Muzquiz, M. (2005). Effects of extrusion, boiling, autoclaving, and microwave heating on lupine allergenicity. *Journal of Agricultural Food Chemistry, 53*(4), 1294−1298.

Bachmann, H., Starrenburg, M. J., Molenaar, D., Kleerebezem, M., & van Hylckama Vlieg, J. E. (2012). Microbial domestication signatures of Lactococcus lactis can be reproduced by experimental evolution. *Genome Research, 22*(1), 115−124.

Balny, C., & Masson, P. (1993). Effects of high pressure on proteins. *Food Reviews International, 9*(4), 611−628. https://doi.org/10.1080/87559129309540980

Bandara, N., Akbari, A., Esparza, Y., & Wu, J. (2018). Canola protein: A promising protein source for delivery, adhesive, and material applications. *Journal of the American Oil Chemists' Society, 95*(8), 1075−1090.

Bannon, G. A. (2004). What makes a food protein an allergen? *Current Allergy Asthma Reports, 4*(1), 43−46.

Barkholt, V., Jergensen, P., Sørensen, D., Bahrenscheer, J., Haikara, A., Lemola, E., … Frøkiær, H. (1998). Protein modification bv fermentation: Effect of fermentation on the potential allergenicity of pea. *Allergy, 53*, 106−108.

Beyer, K., Morrowa, E., Li, X.-M., Bardina, L., Bannon, G. A., Burks, A. W., & Sampson, H. A. (2001). Effects of cooking methods on peanut allergenicity. *Journal of Allergy Clinical Immunology, 107*(6), 1077–1081.

Biscola, V., de Olmos, A. R., Choiset, Y., Rabesona, H., Garro, M. S., Mozzi, F., … Franco, B. (2017). Soymilk fermentation by *Enterococcus faecalis* VB43 leads to reduction in the immunoreactivity of allergenic proteins β-conglycinin (7S) and glycinin (11S). *Beneficial Microbes, 8*(4), 635–643.

Bizzozero, J. (2016). *Demand booming for plant-based proteins in food, beverages*. Retrieved from: https://www.naturalproductsinsider.com/regulatory/demand-booming-plant-based-proteins-food-beverages.

Bock, S. A., Muñoz-Furlong, A., & Sampson, H. A. (2007). Further fatalities caused by anaphylactic reactions to food, 2001-2006. *Journal of Allergy Clinical Immunology, 119*(4), 1016.

Bogdanov, I. V., Shenkarev, Z. O., Finkina, E. I., Melnikova, D. N., Rumynskiy, E. I., Arseniev, A. S., & Ovchinnikova, T. V. (2016). A novel lipid transfer protein from the pea Pisum sativum: Isolation, recombinant expression, solution structure, antifungal activity, lipid binding, and allergenic properties. *BMC Plant Biology, 16*(1), 1–17.

Burks, A. W., Williams, L. W., Thresher, W., Connaughton, C., Cockrell, G., & Helm, R. M. (1992). Allergenicity of peanut and soybean extracts altered by chemical or thermal denaturation in patients with atopic dermatitis and positive food challenges. *Journal of Allergy Clinical Immunology, 90*(6), 889–897.

Byington, L. (2020). *Tracking the plant-based protein movement*. Retrieved from: https://www.fooddive.com/news/plant-based-protein-tracker/564886/#:~:text=Growth%20of%20plant%2Dbased%20protein,is%20primed%20for%20continued%20growth (Accessed).

Cabanillas, B., Cuadrado, C., Rodriguez, J., Dieguez, M. C., Crespo, J. F., & Novak, N. (2018). Boiling and pressure cooking impact on IgE reactivity of soybean allergens. *International Archives of Allergy Immunology, 175*(1–2), 36–43.

Cabanillas, B., Maleki, S. J., Rodríguez, J., Burbano, C., Muzquiz, M., Jiménez, M. A., … Crespo, J. F. (2012). Heat and pressure treatments effects on peanut allergenicity. *Food Chemistry, 132*(1), 360–366.

Campbell-Platt, G. (1987). *Fermented foods of the world. A dictionary and guide*. Butterworths.

Commission-Directive-2006/142/EC. (2006). Commission Directive 2006/142/EC of 22 December 2006 amending Annex IIIa of Directive 2000/13/EC of the European Parliament and of the Council listing the ingredients which must under all circumstances appear on the labelling of foodstuffs. *Journal of Official and Journal of European Union, 50*, 110–111.

Comstock, S. S., Maleki, S. J., & Teuber, S. S. (2016). Boiling and frying peanuts decreases soluble peanut (*Arachis hypogaea*) allergens Ara h 1 and Ara h 2 but does not generate hypoallergenic peanuts. *PLoS One, 11*(6), e0157849.

Cuadrado, C., Cabanillas, B., Pedrosa, M. M., Muzquiz, M., Haddad, J., Allaf, K., … Burbano, C. (2011). Effect of instant controlled pressure drop on IgE antibody reactivity to peanut, lentil, chickpea and soybean proteins. *International Archives of Allergy Immunology, 156*(4), 397–404.

Cuadrado, C., Cabanillas, B., Pedrosa, M. M., Varela, A., Guillamón, E., Muzquiz, M., … Burbano, C. (2009). Influence of thermal processing on IgE reactivity to lentil and chickpea proteins. *Molecular Nutrition Food Research, 53*(11), 1462–1468.

De Jongh, H. H., De Jong, G. A., Apostolovic, D., Taylor, S. L., Baumert, J. L., & Koppelman, S. J. (2020). Effect of heat treatment on the conformational stability of intact and cleaved forms of the peanut allergen Ara h 6 in relation to its IgE-binding potency. *Food Chemistry, 326*, 127027.

De Vuyst, L., Falony, G., & Leroy, F. (2008). Probiotics in fermented sausages. *Meat Science, 80*(1), 75–78.

Del Gobbo, L. C., Falk, M. C., Feldman, R., Lewis, K., & Mozaffarian, D. (2015). Effects of tree nuts on blood lipids, apolipoproteins, and blood pressure: Systematic review, meta-analysis, and dose-response of 61 controlled intervention trials. *The American Journal of Clinical Nutrition, 102*(6), 1347–1356.

El Mecherfi, K.-E., Todorov, S. D., de Albuquerque, C., Albuquerque, M., Denery-Papini, S., Lupi, R., … Larré, C. (2020). Allergenicity of fermented foods: Emphasis on seeds protein-based products. *Foods, 9*(6), 792.

Essex, D. W., Li, M., Miller, A., & Feinman, R. D. (2001). Protein disulfide isomerase and sulfhydryl-dependent pathways in platelet activation. *Biochemistry, 40*(20), 6070–6075.

Estévez, M. (2011). Protein carbonyls in meat systems: A review. *Meat Science, 89*(3), 259–279.

FAO/WHO. (2007). *FAO/WHO Expert Consultation on the safety assessment of foods derived from recombinant-DNA animals: World Health Organization*. Headquarters Geneva, Switzerland: World Health Organization, 26 February–2 March 2007: report.

Fujimura, T., Hori, A., Torii, H., Kishida, S., Matsuura, Y., & Kawamoto, S. (2018). Intake of a fermented plant product attenuates allergic symptoms without changing systemic immune responses in a mouse model of Japanese cedar pollinosis. *World Allergy Organization Journal, 11*(1), 1–9.

Fukuda, S., Kawasaki, Y., & Izawa, S. (2019). Ferrous chloride and ferrous sulfate improve the fungicidal efficacy of cold atmospheric argon plasma on melanized Aureobasidium pullulans. *Journal of Bioscience and Bioengineering, 128*(1), 28–32.

Garcia-Mora, P., Peñas, E., Frias, J., Gomez, R., & Martinez-Villaluenga, C. (2015). High-pressure improves enzymatic proteolysis and the release of peptides with angiotensin I converting enzyme inhibitory and antioxidant activities from lentil proteins. *Food Chemistry, 171*, 224–232. https://doi.org/10.1016/j.foodchem.2014.08.116

Gavahian, M., & Khaneghah, A. M. (2020). Cold plasma as a tool for the elimination of food contaminants: Recent advances and future trends. *Critical Reviews in Food Science and Nutrition, 60*(9), 1581–1592.

Gomaa, A., Sedman, J., & Ismail, A. (2013). An investigation of the effect of microwave treatment on the structure and unfolding pathways of β-lactoglobulin using FTIR spectroscopy with the application of two-dimensional correlation spectroscopy (2D-COS). *Vibrational Spectroscopy, 65*, 101–109.

Goyal, A., Sharma, V., Upadhyay, N., Sihag, M., & Kaushik, R. (2018). High pressure processing and its impact on milk proteins: A review. *Journal of Dairy Science and Technology, 2*(1), 12–20.

Graves, D. B. (2014). Low temperature plasma biomedicine: A tutorial review. *Physics of Plasmas, 21*(8), 080901. https://doi.org/10.1063/1.4892534

Han, Z., Cai, M.-j., Cheng, J.-H., & Sun, D.-W. (2018). Effects of electric fields and electromagnetic wave on food protein structure and functionality: A review. *Trends in Food Science & Technology, 75*, 1–9.

Hao, L., Jianmei, Y., Mohamed, A., & Ipek, G. (2013). Reduction of major peanut allergens Ara h 1 and Ara h 2, in roasted peanuts by ultrasound assisted enzymatic treatment. *Food Chemistry, 15*(141 (2)), 762–768.

Hefle, S. L., Nordlee, J. A., & Taylor, S. L. (1996). Allergenic foods. *Critical Reviews in Food Science and Nutrition, 36*(S1), 69–89.

Henchion, M., Hayes, M., Mullen, A. M., Fenelon, M., & Tiwari, B. (2017). Future protein supply and demand: Strategies and factors influencing a sustainable equilibrium. *Foods, 6*(7), 53.

Hojnik, N., Cvelbar, U., Tavčar-Kalcher, G., Walsh, J. L., & Križaj, I. (2017). Mycotoxin decontamination of food: Cold atmospheric pressure plasma versus "classic" decontamination. *Toxins, 9*(5), 151. https://doi.org/10.3390/toxins9050151

Huang, H., Hsu, C., Yang, B. B., & Wang, C. (2014b). Potential utility of high-pressure processing to address the risk of food allergen concerns. *Comprehensive Reviews in Food Science & Food Safety, 13*(1), 78–90.

Huang, H.-W., Yang, B. B., & Wang, C.-Y. (2014a). Effects of high pressure processing on immunoreactivity and microbiological safety of crushed peanuts. *Food Control, 42*, 290–295.

Hu, C.q., Chen, H.b., Gao, J.y., Luo, C.p., Ma, X.j., & Tong, P. (2011). High-pressure microfluidisation-induced changes in the antigenicity and conformation of allergen Ara h 2 purified from Chinese peanut. *Journal of the Science of Food and Agriculture, 91*(7), 1304–1309.

Husband, F. A., Aldick, T., Van der Plancken, I., Grauwet, T., Hendrickx, M., Skypala, I., & Mackie, A. R. (2011). High-pressure treatment reduces the immunoreactivity of the major allergens in apple and celeriac. *Molecular Nutrition and Food Research International, 55*(7), 1087–1095.

Islam, M. N., Zhang, M., & Adhikari, B. (2014). The inactivation of enzymes by ultrasound—a review of potential mechanisms. *Food Reviews International, 30*(1), 1–21.

Jiang, L., Wang, J., Li, Y., Wang, Z., Liang, J., Wang, R., … Zhang, M. (2014). Effects of ultrasound on the structure and physical properties of black bean protein isolates. *Food Research International, 62*, 595–601.

Jin, J., Ma, H., Wang, K., Yagoub, A. E.-G. A., Owusu, J., Qu, W., … Ye, X. (2015). Effects of multi-frequency power ultrasound on the enzymolysis and structural characteristics of corn gluten meal. *Ultrasonics Sonochemistry, 24*, 55–64.

Johnson, P. E., Van der Plancken, I., Balasa, A., Husband, F. A., Grauwet, T., Hendrickx, M., … Mackie, A. R. (2010). High pressure, thermal and pulsed electric-field-induced structural changes in selected food allergens. *Molecular Nutrition and Food Research International, 54*(12), 1701–1710.

Kadam, S. U., Tiwari, B. K., Álvarez, C., & O'Donnell, C. P. (2015). Ultrasound applications for the extraction, identification and delivery of food proteins and bioactive peptides. *Trends in Food Science and Technology, 46*(1), 60–67.

Kasera, R., Singh, A. B., Kumar, R., Lavasa, S., Prasad, K. N., & Arora, N. (2012). Effect of thermal processing and γ-irradiation on allergenicity of legume proteins. *Food and Chemical Toxicology, 50*(10), 3456–3461.

Kato, T., Katayama, E., Matsubara, S., Omi, Y., & Matsuda, T. (2000). Release of allergenic proteins from rice grains induced by high hydrostatic pressure. *Journal of Agricultural and Food Chemistry, 48*(8), 3124–3129.

Kobayashi, M., Hashimoto, Y., Taniuchi, S., & Tanabe, S. (2004). Degradation of wheat allergen in Japanese soy sauce. *International Journal of Molecular Medicine, 13*(6), 821–827.

Ko, Y., Kim, E., Bahk, J., Lee, S., Oh, C., & Ryu, C. (2010). Decrease of wheat allergen by treatment with fermented soybean extract. *Journal of Allergy and Clinical Immunology, 125*(2), AB221.

Kumar, D., Chatli, M. K., Singh, R., Mehta, N., & Kumar, P. (2016). Antioxidant and antimicrobial activity of camel milk casein hydrolysates and its fractions. *Small Ruminant Research, 139*, 20–25.

Li, Y.-Q. (2012). Structure changes of soybean protein isolates by pulsed electric fields. *Physics Procedia, 33*, 132–137.

Liu, Z.-W., Niu, D., Zhou, Y.-X., Cheng, J.-H., Bekhit, A. E.-D., & Aadil, R. M. (2021). Oxidation induced by dielectric-barrier discharge (DBD) plasma treatment reduces soybean agglutinin activity. *Food Chemistry, 340*, 128198.

Li, J., Xiang, Q., Liu, X., Ding, T., Zhang, X., Zhai, Y., & Bai, Y. (2017). Inactivation of soybean trypsin inhibitor by dielectric-barrier discharge (DBD) plasma. *Food Chemistry, 232*, 515–522.

Li, H., Yu, J., Ahmedna, M., & Goktepe, I. (2013). Reduction of major peanut allergens Ara h 1 and Ara h 2, in roasted peanuts by ultrasound assisted enzymatic treatment. *Food Chemistry, 141*(2), 762–768.

Li, Y., Yu, J., Goktepe, I., & Ahmedna, M. (2016). The potential of papain and alcalase enzymes and process optimizations to reduce allergenic gliadins in wheat flour. *Food Chemistry, 196*, 1338–1345.

Li, H., Zhu, K., Zhou, H., & Peng, W. J. (2012). Effects of high hydrostatic pressure treatment on allergenicity and structural properties of soybean protein isolate for infant formula. *Food Chemistry, 132*(2), 808–814.

Li, H., Zhu, K., Zhou, H., Peng, W., & Guo, X. (2016). Comparative study of four physical approaches about allergenicity of soybean protein isolate for infant formula. *Food Agricultural Immunology, 27*(5), 604–623.

Long, F., Yang, X., Sun, J., Zhong, Q., Wei, J., Qu, P., & Yue, T. J. (2016). Effects of combined high pressure and thermal treatment on the allergenic potential of peanut in a mouse model of allergy. *Innovative Food Science and Emerging Technologies, 35*, 133–138.

Lou, F., Neetoo, H., Chen, H., & Li, J. (2015). High hydrostatic pressure processing: A promising nonthermal technology to inactivate viruses in high-risk foods. *Annual Review of Food Science Technology, 6*, 389–409.

Lucas, J., Grimshaw, K. E., Collins, K., Warner, J. O., & Hourihane, J. O. B. (2004). Kiwi fruit is a significant allergen and is associated with differing patterns of reactivity in children and adults. *Clinical and Experimental Allergy, 34*(7), 1115–1121.

Luo, C., Hu, C., Gao, J., Li, X., Wu, Z., Yang, A., & Chen, H. (2013). A potential practical approach to reduce Ara h 6 allergenicity by gamma irradiation. *Food Chemistry, 136*(3–4), 1141–1147.

Maleki, S. J., Chung, S.-Y., Champagne, E. T., & Raufman, J.-P. (2000). The effects of roasting on the allergenic properties of peanut proteins. *Journal of Allergy Clinical Immunology, 106*(4), 763–768.

Mancuso, G., & Berdondini, R. (2001). Oral allergy syndrome from kiwi fruit after a lover's kiss. *Contact Dermatitis, 45*(1).

Martínez-Álvarez, O., López-Caballero, M., Gómez-Guillén, M., & Montero, P. (2017). Fermented seafood products and health. In *Fermented foods in health and disease prevention* (pp. 177–202). Elsevier.

Mason, T. J., Riera, E., Vercet, A., & Lopez-Buesa, P. (2005). Application of ultrasound. In *Emerging technologies for food processing* (pp. 323–351). Elsevier.

Meda, V., Orsat, V., & Raghavan, V. (2017). Microwave heating and the dielectric properties of foods. In *The microwave processing of foods* (pp. 23–43). Elsevier.

Meinlschmidt, P., Schweiggert-Weisz, U., & Eisner, P. (2016a). Soy protein hydrolysates fermentation: Effect of debittering and degradation of major soy allergens. *LWT-Food Science and Technology, 71*, 202–212.

Meinlschmidt, P., Ueberham, E., Lehmann, J., Reineke, K., Schlüter, O., Schweiggert-Weisz, U., & Eisner, P. (2016b). The effects of pulsed ultraviolet light, cold atmospheric pressure plasma, and gamma-irradiation on the immunoreactivity of soy protein isolate. *Innovative Food Science and Emerging Technologies, 38*, 374–383.

Mills, C., Moreno, F. J., Sancho, A. I., Jenkins, J. A., & Wichers, H. J. (2004). *Processing approaches to reducing allergenicity in proteins.*

Mills, E. C., & Shewry, P. R. (2004). *Plant food allergens*. Wiley Online Library.

Mondoulet, L., Paty, E., Drumare, M., Ah-Leung, S., Scheinmann, P., Willemot, R., … Bernard, H. (2005). Influence of thermal processing on the allergenicity of peanut proteins. *Journal of Agricultural and Food Chemistry, 53*(11), 4547–4553.

Moutiq, R., Misra, N. N., Mendonca, A., & Keener, K. (2019). *In-package decontamination of chicken breast using cold plasma technology: Microbial, quality and storage studies*. Meat Science. https://doi.org/10.1016/j.meatsci.2019.107942

Mueller, G. A., Maleki, S. J., & Pedersen, L. C. (2014). The molecular basis of peanut allergy. *Current Allergy Asthma Reports, 14*(5), 1–9.

Nooji, J. K. (2011). *Reduction of wheat allergen potency by pulsed ultraviolet light, high hydrostatic pressure, and non-thermal plasma*. University of Florida.

Norton, T., & Sun, D.-W. (2008). Recent advances in the use of high pressure as an effective processing technique in the food industry. *Food and Bioprocess Technology, 1*(1), 2–34.

Novembre, E., Bernardini, R., Bertini, G., Massai, G., & Vierucci, A. (1995). Skin-prick-test-induced anaphylaxis. *Allergy, 50*(6), 511–513.

Oh, S., Jang, D.-I., Lee, J.-W., Kim, J.-H., Byun, M.-W., & Lee, S.-Y. (2009). Evaluation of reduced allergenicity of irradiated peanut extract using splenocytes from peanut-sensitized mice. *Radiation Physics and Chemistry, 78*(7–8), 615–617.

Ojha, K. S., Kerry, J. P., Duffy, G., Beresford, T., & Tiwari, B. K. (2015). Technological advances for enhancing quality and safety of fermented meat products. *Trends in Food Science & Technology, 44*(1), 105–116.

Olatunde, O. O., & Benjakul, S. (2018). Nonthermal processes for shelf-life extension of Seafoods: A revisit. *Comprehensive Reviews in Food Science and Food Safety, 17*(4), 892–904. https://doi.org/10.1111/1541-4337.12354

Olatunde, O. O., Benjakul, S., & Vongkamjan, K. (2019a). Combined effects of high voltage cold atmospheric plasma and antioxidants on the qualities and shelf-life of Asian sea bass slices. *Innovative Food Science & Emerging Technologies, 54*, 113–122.

Olatunde, O. O., Benjakul, S., & Vongkamjan, K. (2019b). High voltage cold atmospheric plasma: Antibacterial properties and its effect on quality of Asian sea bass slices. *Innovative Food Science & Emerging Technologies, 52*, 305–312.

Olatunde, O. O., Shiekh, K. A., & Benjakul, S. (2021). Pros and cons of cold plasma technology as an alternative non-thermal processing technology in seafood industry. *Trends in food science and technology.*

Olatunde, O. O., Singh, A., Shiekh, K. A., Nuthong, P., & Benjakul, S. (2021). Effect of high voltage cold plasma on oxidation, Physiochemical, and gelling properties of myofibrillar protein isolate from Asian sea bass (*Lates calcarifer*). *Foods, 10*(2), 326.

O'Sullivan, J. J., Park, M., Beevers, J., Greenwood, R. W., & Norton, I. T. (2017). Applications of ultrasound for the functional modification of proteins and nanoemulsion formation: A review. *Food Hydrocolloids, 71*, 299–310.

Paschke, A. (2009). Aspects of food processing and its effect on allergen structure. *Molecular Nutrition Food Research, 53*(8), 959–962.

Peñas, E., Gomez, R., Frias, J., Baeza, M. L., & Vidal-Valverde, C. (2011). High hydrostatic pressure effects on immunoreactivity and nutritional quality of soybean products. *Food Chemistry, 125*(2), 423–429.

Peñas, E., Préstamo, G., Polo, F., & Gomez, R. (2006a). Enzymatic proteolysis, under high pressure of soybean whey: Analysis of peptides and the allergen Gly m 1 in the hydrolysates. *Food Chemistry, 99*(3), 569–573.

Peñas, E., Restani, P., Ballabio, C., Prestamo, G., Fiocchi, A., & Gomez, R. (2006b). Assessment of the residual immunoreactivity of soybean whey hydrolysates obtained by combined enzymatic proteolysis and high pressure. *European Food Research and Technology, 222*(3/4), 286–290.

Pereira, R., & Vicente, A. (2010). Environmental impact of novel thermal and non-thermal technologies in food processing. *Food Research International, 43*(7), 1936–1943.

Perry, T. T., & Pesek, R. D. (2013). Clinical manifestations of food allergy. *Pediatric Annals, 42*(6), e106–e111.

Rahaman, T., Vasiljevic, T., & Ramchandran, L. (2016). Effect of processing on conformational changes of food proteins related to allergenicity. *Trends in Food Science and Technology, 49*, 24–34.

Rahman, M. M., Byanju, B., Grewell, D., & Lamsal, B. P. (2020). High-power sonication of soy proteins: Hydroxyl radicals and their effects on protein structure. *Ultrasonics Sonochemistry, 64*, 105019.

Rajauria, G., Sharma, S., Emerald, M., & Jaiswal, A. K. (2016). Novel fermented marine-based products. In *Novel food fermentation technologies* (pp. 235–262). Springer.

Rizzello, C. G., De Angelis, M., Coda, R., & Gobbetti, M. (2006). Use of selected sourdough lactic acid bacteria to hydrolyze wheat and rye proteins responsible for cereal allergy. *European Food Research and Technology, 223*(3), 405–411.

Saleem, R., & Ahmad, R. (2016). Effect of ultrasonication on secondary structure and heat induced gelation of chicken myofibrils. *Journal of Food Science and Technology, 53*(8), 3340–3348.

Sampson, H. A. J. (2004). Update on food allergy. *Journal of Allergy and Clinical Immunology, 113*(5), 805–819.

Schwager, C., Kull, S., Behrends, J., Röckendorf, N., Schocker, F., Frey, A., ... Jappe, U. (2017). Peanut oleosins associated with severe peanut allergy—importance of lipophilic allergens for comprehensive allergy diagnostics. *Journal of Allergy Clinical Immunology, 140*(5), 1331–1338. e1338.

Shahali, Y., & Dadar, M. (2018). Plant food allergy: Influence of chemicals on plant allergens. *Food and Chemical Toxicology, 115*, 365–374.

Sharma, G. M., Roux, K. H., & Sathe, S. K. (2009). A sensitive and robust competitive enzyme-linked immunosorbent assay for Brazil nut (*Bertholletia excelsa* L.) detection. *Journal of Agricultural Food Chemistry, 57*(2), 769–776.

Shaw, A., Shama, G., & Iza, F. (2015). Emerging applications of low temperature gas plasmas in the food industry. *Biointerphases, 10*(2), 029402. https://doi.org/10.1116/1.4914029

Shewry, P. R., & Hey, S. J. (2015). The contribution of wheat to human diet and health. *Food Energy Security, 4*(3), 178–202.

Shiekh, K. A., & Benjakul, S. (2020a). Effect of high voltage cold atmospheric plasma processing on the quality and shelf-life of Pacific white shrimp treated with Chamuang leaf extract. *Innovative Food Science & Emerging Technologies, 64*, 102435.

Shiekh, K. A., & Benjakul, S. (2020b). Effect of pulsed electric field treatments on melanosis and quality changes of Pacific white shrimp during refrigerated storage. *Journal of Food Processing and Preservation, 44*(1), e14292.

Shi, X., Guo, R., White, B. L., Yancey, A., Sanders, T. H., Davis, J. P., ... Kulis, M. (2013). Allergenic properties of enzymatically hydrolyzed peanut flour extracts. *International Archives of Allergy and Immunology, 162*(2), 123–130.

Shriver, S. K., & Yang, W. W. (2011). Thermal and nonthermal methods for food allergen control. *Food Engineering Reviews, 3*(1), 26–43.

Sicherer, S. H., Furlong, T. J., Muñoz-Furlong, A., Burks, A. W., & Sampson, H. A. (2001). A voluntary registry for peanut and tree nut allergy: Characteristics of the first 5149 registrants. *Journal of Allergy Clinical Immunology, 108*(1), 128–132.

Sicherer, S. H., & Sampson, H. A. (2018). Food allergy: A review and update on epidemiology, pathogenesis, diagnosis, prevention, and management. *Journal of Allergy Clinical Immunology, 141*(1), 41–58.

Sicherer, S. H., & Wood, R. A. (2013). Advances in diagnosing peanut allergy. *The Journal of Allergy and Clinical Immunology: In Practice, 1*(1), 1–13.

Šotkovský, P., Sklenář, J., Halada, P., Cinová, J., Šetinová, I., Kainarová, A., ... Tučková, L. (2011). A new approach to the isolation and characterization of wheat flour allergens. *Clinical Experimental Allergy, 41*(7), 1031–1043.

Sun, D.-W. (2014). *Emerging technologies for food processing*. Elsevier.

Su, M., Venkatachalam, M., Teuber, S. S., Roux, K. H., & Sathe, S. K. (2004). Impact of γ-irradiation and thermal processing on the antigenicity of almond, cashew nut and walnut proteins. *Journal of the Science of Food Agriculture, 84*(10), 1119–1125.

Taheri-Kafrani, A., Gaudin, J.-C., Rabesona, H., Nioi, C., Agarwal, D., Drouet, M., ... Haertle, T. (2009). Effects of heating and glycation of β-lactoglobulin on its recognition by IgE of sera from cow milk allergy patients. *Journal of Agricultural and Food Chemistry, 57*(11), 4974–4982.

Tao, Y., & Sun, D.-W. (2015). Enhancement of food processes by ultrasound: A review. *Critical Reviews in Food Science and Nutrition, 55*(4), 570–594.

Thomas, K., Herouet-Guicheney, C., Ladics, G., Bannon, G., Cockburn, A., Crevel, R., ... Vieths, S. (2007). Evaluating the effect of food processing on the potential human allergenicity of novel proteins: International workshop report. *Food Chemical Toxicology, 45*(7), 1116–1122.

Tobajas, A. P., Agulló-García, A., Cubero, J. L., Colás, C., Segura-Gil, I., Sánchez, L., ... Pérez, M. D. (2020). Effect of high pressure and pulsed electric field on denaturation and allergenicity of Pru p 3 protein from peach. *Food Chemistry, 321*, 126745.

Torley, P. J., Truong, T. T., & Bhandari, B. (2020). Ultrasound in food processing and preservation. In *Handbook of food preservation* (p. 1072). Boca Raton: CRC Press.

Vanga, S. K., Singh, A., & Raghavan, V. (2017). Review of conventional and novel food processing methods on food allergens. *Critical Reviews in Food Science & Nutrition, 57*(10), 2077–2094.

Vaz, A. F., Costa, R. M., Coelho, L. C., Oliva, M. L., Santana, L. A., Melo, A. M., & Correia, M. T. (2011). Gamma irradiation as an alternative treatment to abolish allergenicity of lectins in food. *Food Chemistry, 124*(4), 1289–1295.

Vaz, A. F., Souza, M. P., Carneiro-da-Cunha, M. G., Medeiros, P. L., Melo, A. M., Aguiar, J. S., ... Correia, M. T. (2012). Molecular fragmentation of wheat-germ agglutinin induced by food irradiation reduces its allergenicity in sensitised mice. *Food Chemistry, 132*(2), 1033–1039.

Venkatachalam, M., Monaghan, E. K., Kshirsagar, H. H., Robotham, J. M., O'Donnell, S. E., Gerber, M. S., ... Sathe, S. K. (2008). Effects of processing on immunoreactivity of cashew nut (*Anacardium occidentale* L.) seed flour proteins. *Journal of Agricultural Food Chemistry, 56*(19), 8998–9005.

Venkatachalam, M., Teuber, S. S., Peterson, W. R., Roux, K. H., & Sathe, S. K. (2006). Antigenic stability of pecan [*Carya illinoinensis* (Wangenh.) K. Koch] proteins: Effects of thermal treatments and in vitro digestion. *Journal of Agricultural Food Chemistry, 54*(4), 1449–1458.

Venkatachalam, M., Teuber, S. S., Roux, K., & Sathe, S. (2002). Effects of roasting, blanching, autoclaving, and microwave heating on antigenicity of almond (*Prunus dulcis* L.) proteins. *Journal of Agricultural Food Chemistry, 50*(12), 3544–3548.

Venkataratnam, H., Cahill, O., Sarangapani, C., Cullen, P., & Barry-Ryan, C. (2020). Impact of cold plasma processing on major peanut allergens. *Scientific Reports, 10*(1), 1–11.

Venkataratnam, H., Sarangapani, C., Cahill, O., & Ryan, C. B. (2019). Effect of cold plasma treatment on the antigenicity of peanut allergen Ara h 1. *Innovative Food Science and Emerging Technologies, 52*, 368–375.

Verhoeckx, K. C., Vissers, Y. M., Baumert, J. L., Faludi, R., Feys, M., Flanagan, S., ... van der Bolt, N. (2015). Food processing and allergenicity. *Food and Chemical Toxicology, 80*, 223–240.

Vidal, A. R., Cechin, C.d. F., Cansian, R. L., Mello, R.d. O., Schmidt, M. M., Demiate, I. M., … Dornelles, R. C. P. (2018). Enzymatic hydrolysis (pepsin) assisted by ultrasound in the functional Properties of hydrolyzates from different collagens. *Ciência Rural, 48*(3).

Vieths, S., Zagon, J., Fischer, K., Dehne, L. I., & Bögl, K. W. (1995). Verarbeitungsbedingte Einflüsse auf das allergene Potential von Lebensmitteln. *Die Sojaallergie als Modellbeispiel Lebensmittelchemie, 49*, 29−31.

Villemin, C., Tranquet, O., Solé-Jamault, V., Smit, J. J., Pieters, R. H., Denery-Papini, S., & Bouchaud, G. (2019). Deamidation and enzymatic hydrolysis of gliadins alter their processing by dendritic cells in vitro. *Journal of Agricultural and Food Chemistry, 68*(5), 1447−1456.

Wal, J. M. (2003). *Thermal processing and allergenicity of foods*. Wiley Online Library.

Wigotzki, M., Steinhart, H., & Paschke, A. (2000). Influence of varieties, storage and heat treatment on IgE-binding proteins in hazelnuts (*Corylus avellana*). *Food Agricultural Immunology, 12*(3), 217−229.

von Woedtke, T., Reuter, S., Masur, K., & Weltmann, K. D. (2013). Plasmas for medicine. *Physics Reports, 530*(4), 291−320. https://doi.org/10.1016/j.physrep.2013.05.005

Worm, M., Jappe, U., Kleine-Tebbe, J., Schäfer, C., Reese, I., Saloga, J., … Fuchs, T. (2014). Food allergies resulting from immunological cross-reactivity with inhalant allergens. *Allergo Journal International, 23*(1), 1−16.

Xu, J., Chen, Z., Han, D., Li, Y., Sun, X., Wang, Z., & Jin, H. (2017). Structural and functional properties changes of β-conglycinin exposed to hydroxyl radical-generating systems. *Molecules, 22*(11), 1893.

Yang, X., Sun, J., Tao, J., Ma, Y., Wei, J., & Long, F. (2017). The allergenic potential of walnuts treated with high pressure and heat in a mouse model of allergy. *Innovative Food Science and Emerging Technologies, 39*, 165−170.

Yang, A., Zuo, L., Cheng, Y., Wu, Z., Li, X., Tong, P., & Chen, H. (2018). Degradation of major allergens and allergenicity reduction of soybean meal through solid-state fermentation with microorganisms. *Food and Function, 9*(3), 1899−1909.

Yu, J., Ahmedna, M., Goktepe, I., Cheng, H., & Maleki, S. (2011). Enzymatic treatment of peanut kernels to reduce allergen levels. *Food Chemistry, 127*(3), 1014−1022.

Yuanqing, H., Min, C., Lingling, S., Quancai, S., Pengyao, Y., Rui, G., … Haile, M. (2020). Ultrasound Pretreatment increases the bioavailability of dietary proteins by dissociating protein structure and composition. *Food Biophysics, 15*, 409−415.

Yu, J., Goktepe, I., & Ahmedna, M. (2013). Enzymatic treatment of peanut butter to reduce the concentration of major peanut allergens. *International Journal of Food Science and Technology, 48*(6), 1224−1234.

Yu, J., Hernandez, M., Li, H., Goktepe, I., Robinette, C., Auerbach, A., … Ahmedna, M. (2015). Allergenicity of roasted peanuts treated with a non-human digestive protease. *Food Research International, 69*, 341−347.

Zhang, Q., Cheng, Z., Zhang, J., Nasiru, M. M., Wang, Y., & Fu, L. (2021). Atmospheric cold plasma treatment of soybean protein isolate: Insights into the structural, physicochemical, and allergenic characteristics. *Journal of Food Science, 86*(1), 68−77.

Zhang, T., Shi, Y., Zhao, Y., Wang, J., Wang, M., Niu, B., & Chen, Q. (2019). Different thermal processing effects on peanut allergenicity. *Journal of the Science of Food Agriculture, 99*(5), 2321−2328.

Zhang, W., Xiao, S., & Ahn, D. U. (2013). Protein oxidation: Basic principles and implications for meat quality. *Critical Reviews in Food Science and Nutrition, 53*(11), 1191−1201.

Chapter 5

Functionality of plant-based proteins

Yong Wang, Woojeong Kim, Rishi Ravindra Naik and Cordelia Selomulya
School of Chemical Engineering, UNSW Sydney, Sydney, NSW, Australia

5.1 Introduction

Functionality of proteins can be ascribed to those physical and chemical properties that affect their behavior while interacting with different molecules during processing, storage, as well as consumption and digestion (Muhammad & Pauls, 2018). These functional properties determine different characteristics of the final product, including rheological, textural, and other sensorial aspects of the food. Both intrinsic and extrinsic factors can influence the functionality of proteins—these include composition of amino acid, molecular weight, net charge, isoelectric point, pH, and temperature (Sikorski, 2001). A number of plant proteins from cereals, pseudo cereals, pulses, and legumes have been identified for human consumption. The successful utilization of these plant proteins for human consumption depends upon their digestibility, functional, organoleptic, and nutritive properties, which vary among plant proteins. Modifications of plant proteins are often required to enhance their functional, nutritional, and organoleptic properties, while reducing their tendency to degrade (Shukla, 1982). To date, the study of functional properties of food proteins has been limited to emulsions, gels, foam formation, and other nonnutritional characteristics. However, protein modifications can affect or alter not only the nonnutritional attributes but also the nutritional quality of the foods. Both the nutritional (bioaccessibility in particular) and nonnutritional functional properties may be explained based on the structure of individual protein biomolecule and their interactions at nano scale. The knowledge of composition, structure, conformation, and different physical properties of plant proteins is useful to better understand their functionality (Foegeding, 2015).

5.2 Functionality of plant-based proteins

Recently, plant protein research has shifted its focus from purification of protein to improving its functionality. The functionality of these plant proteins can be improved via synergistic interaction of plant proteins with other components in the plants. Minimal processing is preferable to enhance the functionality of the whole plant proteins rather than their purified/isolated forms (Kornet et al., 2020). The following sections provide an overview of the major functional attributes of plant proteins.

5.2.1 Organoleptic and kinesthetic properties

While the major purpose of human diet is to fulfill their nutritional requirements, its consumption is also based on positive sensory experience. The sensory profile of food originates from its organoleptic and kinesthetic properties, which includes flavor, color, odor or aroma, texture, and others. These properties are perceived by the complex human—food interactions that involve chemosense of olfaction, sight, auditory, chemesthesis, and taste, which, in turn, provide a specific sensory profile to the food. These sensory profiles of food impact their acceptance and, thereby, the overall psychological response of the consumer (Damodaran & Arora, 2013; Hollis, 2018).

5.2.1.1 Color

Color is one of the important criteria used to evaluate the quality and safety of the food. If the plant proteins are to be used extensively, then the color they impart on the food must also be considered an important aspect of consumer acceptability. Plant proteins do not have their own color as such, but they can widely interact with different color pigments, including

Engineering Plant-Based Food Systems. https://doi.org/10.1016/B978-0-323-89842-3.00010-5

anthocyanins, carotenoids, and chlorophyll pigments. These interactions are mainly based on the molecular size, conformation, and hydrophobicity of the plant protein fragments. These phenolic compounds are formed by different enzymatic and nonenzymatic browning reactions and are attached to the proteins via oxidative reactions. Gluten, soybean, pea, and lentil proteins have demonstrated pH-dependent affinity to interact with proanthocyanadins, catechins, epicatechins, and other phenols by competitive adsorption or co-operative binding (Blouin et al., 1981; Granato et al., 2018). If used as an ingredient, during thermal processing, plant proteins will show significant color changes. This change is due to the Maillard reaction in which the ε- or terminal α-amino group reacts with the carbonyl groups of the food that can further lead to formation of different Maillard reaction products (Naik et al., 2021). Most of the protein recovery procedures from seed flours require multiple steps (simultaneous deoiling and protein extraction) or the use of cation-exchange resins to reduce/remove the content of glucosinolates and phenolic components from the final isolate or concentrates. However, this extraction process has not been upscaled for industrial application due to relatively high costs and lower yields (Von Der Haar et al., 2014).

5.2.1.2 Flavor

In food, the flavor is generated due to the combination of volatile and nonvolatile compounds in different concentrations, which are recognized by the taste (receptors on tongue and oral cavity) and olfactory organs, that are produced, released, or retained from the food matrix. These diverse volatile components mainly include aldehydes, ketones, alcohols, esters, hydrocarbons, furans, and sulfated compounds. Flavor is one of the crucial attributes that determine the consumer acceptance of the food product. Plant proteins do not have their own flavor, but they are able to bind or absorb the flavor compounds, which provide a flavor when consumed (Sharan et al., 2021; Yu et al., 2018; Zhang, Yan, et al., 2021, Zhang, Kang, et al., 2021). In an ideal scenario, the isolated plant protein should be flavorless. However, most of the natural plant proteins have a distinct flavor profile, which can be detected even in a commercially available plant protein isolate. One of the major examples of such is observed in legume proteins with a beany flavor. Many protein—flavor interaction studies have demonstrated that the binding of flavor molecules to plant proteins basically depends on the moisture content, pH, temperature, and the side chain structures that are sites wherein multiple covalent and noncovalent forces (van der Waals, hydrogen and ionic bonding, hydrophobic interactions) are active in the food system (Heng et al., 2004). Plant proteins, especially raw, unmodified legumes and pulses, have a distinct off-flavor, which is perceived immediately once they are ingested. This off-flavor has been one of the major reasons for its reduced acceptability as an ingredient for direct human consumption (Damodaran & Arora, 2013). Flavors in legumes, nuts, and pulses are produced partly during farm to fork unit operations and some are partly inherent. During farm to fork operations, these flavors can be formed by the changes in temperature leading to chemical reaction of amino acids. The inherent flavor components can arise from the oxidative degradation of PUFA to produce hydroperoxides and furans, hexanal, and ketones during its growth period that contribute to the grassy, green, or beany flavors (Roland et al., 2017). During most functional product development using plant proteins, a higher percentage of the formulation is made up using these proteins, which impact the final texture and flavor. In order to overcome these intrinsic flavors, two major approaches are utilized. First approach is to test different combinations so as to nullify or reduce the off-flavor. For example, beany or earthy flavors can be used with nutty flavors during the development of peanut butter- or other chocolate-based products, while bitter- or astringent-flavored plant proteins can be combined with the citrus food developments or its intensity can be reduced using natural plant-based sweeteners (AIDP, 2022). Next approach is to mask the inherent flavors. This approach is discussed exhaustively by (Bangratz, 2020), in which they described a process of creating a flavor masking solution to reduce the harsh taste of plant proteins.

5.2.1.3 Texture

Texture is a multiparameter attribute derived from the structure of the food and detected by sensory organs. Apart from flavor and color, which can also act as an indicator of food safety, texture is crucial as an indicator of food quality for the consumer la (Lawless & Heymann, 2010). This attribute has been described extensively in the rheology section of this chapter.

5.2.2 Hydration properties

Hydration properties of proteins are particularly based on their amino acid composition and to the polar: nonpolar and ionic: neutral amino acids ratio (Kuntz & Kauzmann, 1974). In terms of rehydration process of protein powders, the sequence starts from wetting (first contact of particles with liquid), dispersion, and ends with solubilization (release of materials from the particle surface into the liquid) (Ji et al., 2016). If used as an ingredient in product development, inappropriate hydration of the plant proteins can have a drastic effect on the emulsification, stability, water holding capacity, and the integrity of the final product, which can directly affect its shelf life (Tarté et al., 2009).

5.2.2.1 Wettability and water absorption capacity

When protein flour/concentrate/isolate is added to water, it might be easily able to cross the air—water interface or it might accumulate above it; this attribute of the powders is related to its wettability. Wettability is calculated as the time required for complete wetting and immersion of powders on a liquid surface (Nascimento et al., 2020). Wettability of plant protein is dependent on its water absorption capacity. Water absorption capacity is the amount of water absorbed per gram of the protein and is based on the proteins hydrophilic or hydrophobic characteristics. This property is a crucial index that explains the ability of the protein to prevent fluid leakage from the food product during processing and/or storage (Jarpa-Parra, 2018). Both of these properties are influenced by the physicochemical attributes of the food systems that include the composition (presence of fats or carbohydrates), surface topology, protein extraction method, protein charge, as well as the plant protein species (Schopf & Scherf, 2021; Sopade & Obekpa, 1990). Gelation can also affect the water absorption capacity if the network created by the proteins has larger pore sizes, which lower its retention capacity (Yada et al., 2018). Water absorption capacity of different plant proteins, including pea, lentils, chickpea, and fava bean, has been reported to range from 0.6 to 4.9 g/g, which is lower than animal proteins. However, their wettability and water absorption can be tailored to compete with animal proteins using different treatments (Singhal et al., 2016). These techniques include both physical (heat, ultrasound, pressure, cold plasma) and chemical treatments (glycation, conjugation, crosslinking, fermentation), which have been reviewed expansively by (Akharume et al., 2021; Avelar et al., 2021; Naik et al., 2021). Every treatment mentioned above can cause a structural rearrangement of the proteins that influences the hydrophobic: hydrophilic ratio of the surface, which ultimately impacts the water absorption capacity and wettability (Avelar et al., 2021).

5.2.2.2 Dispersibility and solubilization

Dispersibility is one of the major rehydration properties considered for protein powders and is defined as the capacity of the powder to disperse homogeneously when in contact with water. It is calculated using the dispersibility index, in which the powder is considered dispersible if the index is more than 85% (Le Roux et al., 2020). Most of the globular plant proteins are salt soluble and are slightly soluble in water, which results in their lower dispersibility index. Hence, these plant proteins require different chemical and physical processing that includes hydrolysis, pH shift, microfluidization, high pressure, ultrasound, etc. in order to improve their dispersibility (de Oliveira et al., 2020). Protein dispersibility is mainly influenced by the processing parameters, storage time, pH, composition of the medium, and the type of plant protein (Batal et al., 2000; Serrano et al., 2013; Volkert & Klein, 1979).

Protein solubility is a thermodynamic parameter related to the formation of intrinsic protein aggregates and protein solvation. It can be defined as the concentration of protein in the saturated solution under particular set of conditions, that is in equilibrium with the solid phase (crystalline/amorphous) (Garidel, 2013; Kramer et al., 2012). Solubility can be calculated as the ratio of nitrogen content of supernatant to the original plant protein suspension, and according to studies, majority of plant proteins have solubility less than 20% (Ebert et al., 2020). Nevertheless, broad generalization cannot be inferred among the solubilities of different plant protein families. Most of the legume, cucurbit, and mallow family exhibit higher solubility in NaOH, NaCl, and Na_2HPO_4, while the seed protein families including Iridaceae, Scrophulariaceae, Euphorbiaceae, Umbelliferae, Papaveraceae, and Chenopodiaceae exhibit lower solubilities in the same solvents (Smith et al., 1959).

Solubility of plant proteins is influenced by numerous intrinsic and extrinsic factors, including amino acid composition and its sequence, molecular structure, pH, ionic strength of the medium, surface charges, amino acid accessibility, as well as the temperature (Sathe, 2012). For example, on heating, the solubility of globular plant protein increases up to 40°C and then declines rapidly above this temperature leading to denaturation and precipitation. When the pH of the solution is above or below the isoelectric point of plant protein, they interact and solvate more efficiently increasing their solubility. Although, when the isoelectric point of protein and pH of the solution equilibrates, it results in decline of electrostatic repulsion leading to its precipitation. Protein solubility can also be influenced by the presence of salts, and its concentration can result in the salting-in (lower concentration of salts enhances the solubility) and salting-out phenomena (higher concentration of salts lead to precipitation) (Milewski, 2001). These phenomena are also influenced by the higher ratio of polar: nonpolar amino acid, which can increase the protein interactions, while lowering the solubility of proteins. However, the increased charges on the amino acids can enhance the interaction between solvent and protein resulting in improved solubility (Muhammad & Pauls, 2018).

5.2.2.3 Viscosity and gelation

Protein stability is one of the crucial factors that influence the quality and shelf life of the product as unstable proteins can result in unfolding and aggregation that leads to increased viscosity; hence viscosity is one of the major quality attributes. Viscosity of food system is directly influenced by the concentration of proteins, while the other components present in the food system have a negligible effect. In the case of plant proteins, the viscosity of cereal, pulses, oilseeds, and others varies significantly as different proteins change the protein matrix density of the microgel structure at different rates. For example, rice proteins show lower viscosity as compared to other plant proteins. Moreover, the increase in water absorption capacity can also increase the viscosity, which can be observed in the case of soy protein isolates (Akin & Ozcan, 2017; Kornet et al., 2020). High viscosities of plant proteins are sometimes considered as a limiting factor during processing and can also hinder solubilization. Hence, they are modified to achieve a wide range of tuned viscosities by varying the pH, temperature, mixing regime, and the slurry media (Fleming et al., 1974). Different approaches can be opted to modify the viscosity including the reduction of the steric repulsion between solutes, addition of salts ($\sim$50 mM NaCl) to reduce the electrostatic interaction or the hydrophobic and aromatic interactions between the proteins (Hong et al., 2018).

Proteins, when cooled after subsequent heating above the denaturation temperature, result in the formation of a three-dimensional network, which is related to its gelation attribute. This tendency of gelation is measured using minimum gelling concentration (MGC) and is defined as the minimum concentration of protein required for the formation of a stable gel. This network is usually stabilized by the hydrophobic interaction, weak forces, and the covalent disulfide bridging, particularly in the case of pulse proteins (Jarpa-Parra, 2018). The minimum gelling concentration varies significantly among different plant proteins and is affected by variety, extraction parameters, and degree of denaturation of proteins (Wang & Xiong, 2019). Protein gelation can be induced using physical, chemical, and enzymatic treatments, and each treatment has its influence on the three-dimensional structure and stability of gels. Furthermore, the gelling of plant proteins can be modified and improved by reacting them with polysaccharides, which was demonstrated using oat proteins by (Nieto-Nieto et al., 2016). The mechanism of heat-induced gelation of plant globulins (major protein in legumes) has been found to be analogous with that of the dairy proteins, but, further emphasis is required to explore its exact mode of action (Nicolai & Chassenieux, 2019).

5.2.3 Surface interactions

5.2.3.1 Binding properties

Understanding the interactions between plant protein and other food components, including water, oil, polysaccharide, flavor, etc. is important for designing food products with enhanced product quality. Water holding capacity (WHC) and oil holding capacity (OHC) of proteins are relevant to the ratio of hydrophilic and hydrophobic amino acids. Although the WHC of plant proteins is relatively inferior to those of animal proteins, denatured plant proteins can adsorb a large amount of water due to the exposure of hydrophilic groups (Lam et al., 2018). Heat and high-pressure treatments have improved the WHC of plant proteins such as soy and canola proteins (Puppo et al., 2005; Zhang et al., 2017). Improving the water adsorption capacity of plant protein shows its potential in expanding its use for food products such as meat, where the moisture preservation is prioritized (Li et al., 2021). The OHC depends on interactions between the aliphatic chains of lipid and the nonpolar side chains of amino acids (Lam et al., 2018). The increase in surface hydrophobicity of proteins leads to an increase in OHC. For example, these results were found in plant proteins such as canola protein and sunflower meal protein treated by ultrasound or pulsed electric field due to exposure of their hydrophobic groups (Malik et al., 2017; Zhang et al., 2017).

Interactions between protein and polysaccharide lead to protein—polysaccharide complex formation, which holds diverse industrial applicability. Plant protein—polysaccharide complex formation enables the improvement of protein functionality for emulsions and foams. A detailed review on protein—polysaccharide complex with particular attention on plant proteins was written by Warnakulasuriya and Nickerson (2018), covering the factors affecting complex formation, functional properties and application. Intermolecular and/or intramolecular interactions between plant protein and polysaccharide have been known to improve emulsifying and foaming properties, while the structures of the plant protein require modifications to bind strongly with polysaccharides. Disulfide bonds of proteins could be cleaved during processing such as ultrasound, microfluidization, and oxidization, making it easier to form covalent bonds with polysaccharides (Hu et al., 2013; Shen & Tang, 2012; Wang et al., 2016). Furthermore, the surface properties of plant proteins can be improved by conjugation with monosaccharides such as glucose, possessing loosened structure and increased flexibility (Tang et al., 2019). Naik et al. (2021) comprehensively reviewed the recent investigations on plant protein modifications by Maillard conjugation, including conjugation methods, functionalities, and processing technologies. The flavor binding properties of plant proteins can be considered as an essential aspect that can improve the acceptance of consumers. Generally, the binding sites of proteins for flavor compounds are arranged, and flavor ligands with reactive groups can be more attracted to unfolded proteins (Zhang, Yan, et al., 2021, Zhang, Kang, et al., 2021). Compared to animal proteins, there have been relatively few studies on interactions between plant proteins and flavor compounds. The oxidization of unsaturated fatty acids by lipoxygenase catalysis was found to cause beany and off-flavor of soy protein (Wang & Arntfield, 2017). However, more research on interactions between plant proteins and flavor compounds is needed.

5.2.3.2 Emulsifying properties

Emulsifying property of protein is a capability of protein to be involved in emulsion formation and stabilizing oil/water interface, which plays an important role in developing new plant-based food products. Although globular plant proteins with amphiphilicity can be suitable candidates as emulsifiers, rigid tertiary, and quaternary structures of plant proteins can lead to poor emulsifying properties due to low solubility and less surface activity (Kim et al., 2020). Properties of proteins such as solubility, surface hydrophobicity, and molecular flexibility affect the adsorption behavior, consequently impacting the emulsifying properties. The correlation between surface hydrophobicity and emulsifying properties has been established for many years (Jiang et al., 2015; Tang et al., 2019). Plant proteins commonly used in emulsion formation including legume proteins, rice proteins, and oilseed proteins are relatively hydrophilic (Yan et al., 2020). Therefore, physical and/or chemical treatments including chemical and enzyme hydrolysis, heat treatment, ultrasonication, etc, have been attempted. These treatments inhibit protein particle aggregation and promote disulfide bond formation between proteins in their secondary structures, resulting in denaturation, in which hidden hydrophobic groups are exposed, allowing protein molecules to readily stabilize the interface. In addition, interacting with other food components such as polysaccharide was suggested to be effective in enhancing emulsifying properties of plant proteins (Warnakulasuriya & Nickerson, 2018). For example, complexation or conjugation between protein and polysaccharide involves in covalent and/or noncovalent interactions, leading to steric interfacial layer formation in addition to electrostatic repulsive force on droplet surfaces (Liu et al., 2010). Furthermore, glycosylation with simple sugars including lactose, galactose, and glucose produced significant changes in the interfacial properties of proteins (Tang et al., 2019).

Plant protein-based colloidal particles for stabilizing the oil/water interface have been advanced in recent years. A Pickering emulsion is an emulsion that is stabilized by solid particles, which maintains the stability by wetting the particles partially and equally onto the oil/water interface. Properties such as hydrophobicity, shape, and size of the particle can remarkably impact the emulsion stability. A recent review paper by Shi et al. (2020) described plant proteins used as Pickering stabilizers and their preparation methods. Plant proteins that possess prolamin group, such as zein, kafirin, gliadin, and hordein, have been known to form excellent Pickering particles because they are alcohol-soluble and relatively hydrophobic, enabling particles to remain intact in emulsions during storage (Yan et al., 2020). However, highly hydrophilic or hydrophobic protein nanoparticles may slope to either oil or water phase, becoming unsuccessful for Pickering stabilizers, although properties including particle size, charge, and surface hydrophobicity can be improved to some extent by controlling pH, ionic strength, etc. Desirable Pickering particles from highly hydrophilic or hydrophobic proteins can be obtained by physicochemical treatment or incorporation of other protein, polysaccharide, and polyphenols. For example, forming complex with polysaccharides have been attempted to modify surface wettability of nanoparticles (Zhu et al., 2019). Furthermore, hydrophilic polyphenols including tannic acid altered the surface wettability close to neutral (Zou et al., 2015). Yan et al. (2020) have systemically reviewed the Pickering stabilization using protein microgels and fibrils in addition to nanoparticles.

5.2.3.3 Foaming properties

The foaming properties of protein are important for aerated foods, such as breads, cakes, ice creams, whipped creams, etc. Although emulsifying properties of plant proteins have been extensively investigated, utilization of plant proteins to foam is relatively scarce. Wheat proteins, including gluten, are the main ingredients used in bread foam, and thus significant studies have been done on physicochemical properties of wheat foams (Mita et al., 1977). The foaming properties of wheat, soybean, rice, and pea proteins were compared by (Zhao, Shen, et al., 2020, Zhao, Xiong, et al., 2020), showing the highest foam capacity of wheat protein, followed by similar foaming capacities of soybean and pea proteins and poor performance of rice protein. However, foam stability was found to be higher in soybean and pea proteins (82.44% and 89.74%, respectively) compared to wheat protein (68.03%), indicating the utilization of modified legume proteins as potential foaming agents. Although wheat protein shows relatively low foam stability, the foam stability of wheat gluten could be improved by enzyme hydrolysis, ultrasound, and microwave treatments due to the partial unfolding of wheat protein (Wang et al., 2006; Yalcin et al., 2008; Zhang et al., 2011).

Extraction methods can influence the foaming properties of legume proteins, suggested by Stone et al. (2015), showing that alkali extraction produced stable foam, while salt extraction showed better foam capacity in pea protein. Furthermore, enzyme hydrolysis and ultrasound treatment of soy and pea proteins have successfully enhanced the foaming properties by partial unfolding of proteins, in turn reducing surface tension (Liang et al., 2020; Morales et al., 2015; Xiong et al., 2018). Complex or conjugate formation with polysaccharide is generally known to enhance foam properties of plant proteins due to the changes in secondary structure of the proteins (Liu et al., 2010; Wang et al., 2016). Recently, (Zhang et al., 2020) found that foams with novel mechanical properties can be produced with Pickering emulsions using soy protein and bacterial cellulose nanofibril when the solvent is removed.

Although the foaming properties of rice protein have been assessed to be poor, recent attempts to improve foam capacity and stability through heat treatment in alkali condition and protein hydrolysis have been tested, showing potential for replacing dairy proteins (Amagliani et al., 2019; Zhao, Shen, et al., 2020, Zhao, Xiong, et al., 2020). In addition, the potential use from various sources, including rapeseed, quinoa, and amaranth, have been evaluated (Chmielewska et al., 2020; Dakhili et al., 2019).

5.2.4 Rheological and textural functionalities

5.2.4.1 Rheological functionalities of liquid/semi-solid system

Rheology can characterize the flow state of the plant-based protein under different mechanical forces and processing conditions and help understand its functionalities. Rheology is therefore essential to understand how plant proteins behave in liquid and semi-solid foods. Viscosity, as one of the most commonly used parameters, has been introduced in the Section 5.2.2.3 of this chapter. Therefore, the focus of this section is on dynamic rheology and texture analysis.

First, the classical rheology testing has been used to understand the flow behavior of plant-based protein by different formulation, including apparent viscosity, creep-recovery, stress relaxation, yield stress, thixotropy, etc. Additionally, small amplitude oscillatory shear (SAOS) has been used to compare the static viscosity/complex viscosity/elastic moduli/viscous moduli of plant-based proteins. Xia et al. (2021) used the elastic and viscous moduli during heating to investigate the gelling behavior of soy protein and found that they are good indicators to differentiate the degree of enzyme hydrolysation on the gelation of soy protein. Moreno et al. (2020) also reported that SAOS rheology is useful tool to evaluate the performance of different commercial pea proteins, especially in the gelation.

In addition, large amplitude oscillatory shear (LAOS) describes the elastic and viscous properties of plant-based protein at large deformations (outside the linear viscoelastic domain) (Wang & Selomulya, 2022), which are closer to the real processing and application conditions (Klost, Brzeski, & Drusch, 2020). With rapid development of their methodology in the last decade, LAOS can provide stress response up to 1000% strain, show detailed change with Lissajous curves, and analyze the strain-stiffening/strain-softening behaviors. The proposed new nonlinear rheological properties (elastic/viscous Chebyshev coefficients, higher harmonics, strain-stiffening/softening ratios) have been used to interpret physical and structural changes in the deformation behavior of complex structured materials in the nonlinear region (Joyner, 2021). Peanut, soy, lentil, and pea protein have been extensively studied using LAOS rheology. Bi et al. (2020) studied from Lissajous curve that the elastic/viscosity ratio changed significantly when the concentration of peanut protein increased from 23% to 27%; moreover, using higher harmonic intensity to distinct the critical modulus of peanut protein in LAOS regime was worth studying. Gallego et al. (2020) studied the modulus and transformation point of bean protein paste in LAOS regime by using strain curve; then found lentil protein was more significantly affected by cooking methods than soybeans or peas. Klost, Brzeski, and Drusch (2020) found that the Lissajous curve of pea protein gels can show its transition from elasticity to plasticity when the curve change from bone shape (pH = 6.0/6.5) to rectangular (pH = 8.0), and the strain-stiffening factor value from negative to positive (pH 6.0—8.0) can further proved, as shown in Fig. 5.1. From another study by Klost, Giménez-Ribes, and Drusch (2020), the hydrolyzed pea protein gel showed strain thinning (type I). The increased S factor indicated hardening in circulation, while T factor indicated softening/thinning behavior of pea protein gel at strain>160%. Elastic and viscous Lissajous curves can show significant differences of yellow pea with different molecular weights, while S and T factors can reflect density and interactions in protein network formation (Kornet et al., 2021).

5.2.4.2 Textural functionalities of solid system

There are two types of LAOS rheology experimental setup in studies on plant-source proteins. The first type is a normal shear rheometer similar to the animal source proteins used as summarized in the previous section. The second is an enclosed cavity geometry, in which the plant protein can simulate the behavior of extruded or substitute meat products at

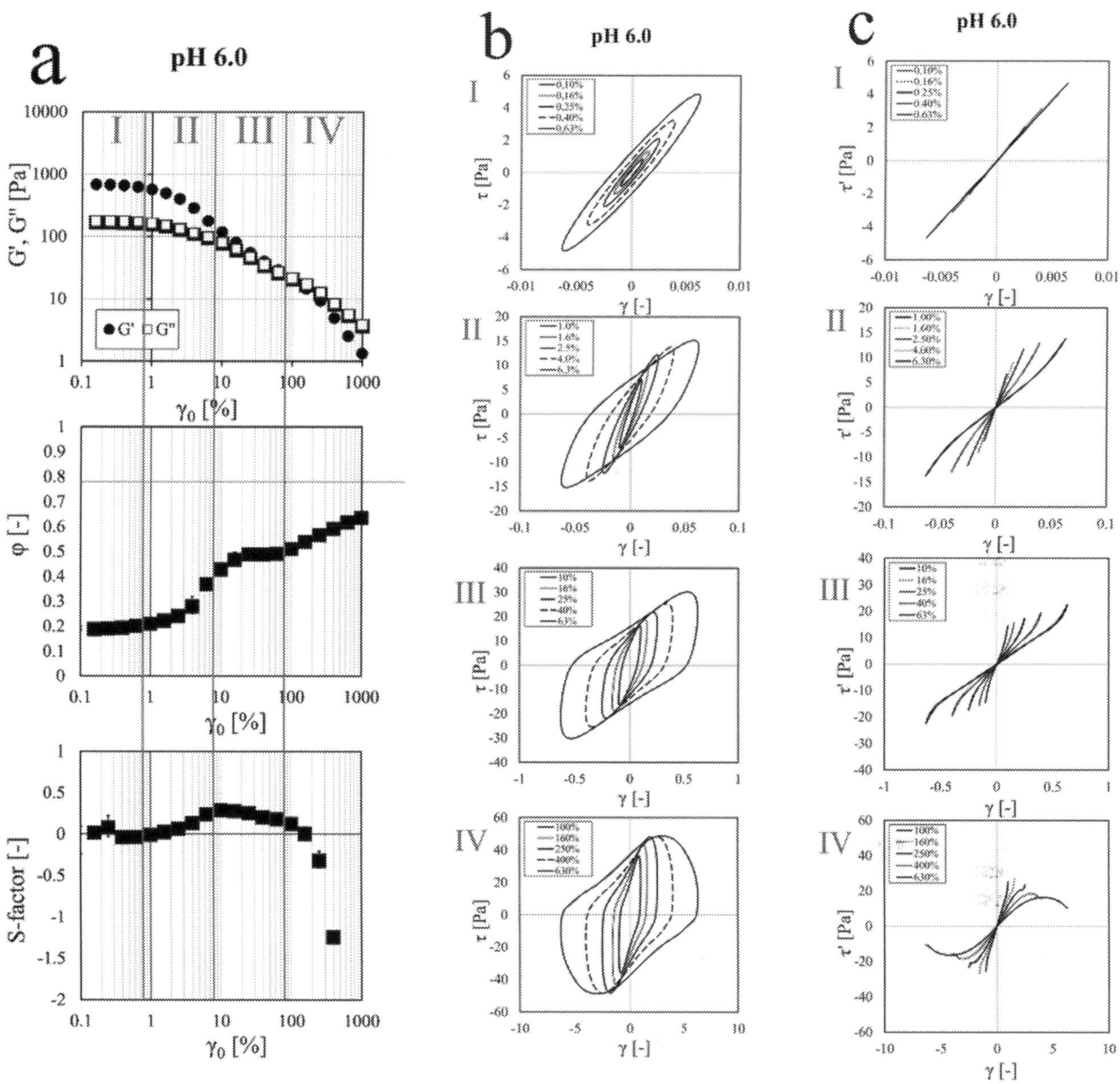

FIGURE 5.1 Amplitude sweeps of fermentation induced pea protein gels made from pea protein pre-treated at pH 6.0 at intercycle strain amplitudes $\gamma 0$ between 0.1% and 1010% and a frequency of 1 s−1. (A) top row: storage and loss modulus G′ and G″ over, $\gamma 0$, (A) middle row: dissipation ratio ϕ over $\gamma 0$, (A) bottom row: stiffening ratio (S-factor) over $\gamma 0$, (B): elastic Lissajous plots, (C) elastic stress. Red labels (I, II, III, IV) stand for different stages when the rheology testing emerge from linear viscoelastic region to the nonlinear region. *Adopted from Klost, M., Brzeski, C. & Drusch, S. (2020). Effect of protein aggregation on rheological properties of pea protein gels.* Food Hydrocolloids, 108. *https://doi.org/10.1016/j.foodhyd.2020.106036, with permission.*

higher concentrations. Using a closed cavity rheometer to cook samples onsite, Schreuders et al. (2021) investigated the LAOS behavior of pea protein, soy protein, and wheat gluten at high concentrations (30−60 wt%). The cooking conditions were reported to significantly affect the textural characteristics of different protein combination formulations. More interestingly, the combination of strain and stress analysis in LAOS range allows the prediction of the texture of meat analog made from different plant proteins, including brittle, mushy, tough, and rubbery. Witteck et al. (2020) also utilized the closed cavity rheometer to study soy protein with the addition of methylcellulose to mimic extrusion. An increase in elasticity through the addition of methylcellulose was detected in the LAOS regime, and the Lissajous curves reveal significant material behavior changes with increasing strain. These studies can provide valuable information, for example, in the production of textured plant source proteins via extrusion.

5.2.4.3 Rheology at the interface

Interfacial rheology is helpful to understand the mechanisms of emulsion formation and stabilization and is a common tool in food emulsion research because of the additional information at the interface. Similar to conventional shear rheology, interfacial rheology (dilatational rheology) can also be used for LAOS analysis through tools such as Lissajous curve and Chebyshev decomposition (Rühs et al., 2013). Particularly, the roles of plant-based protein in emulsions can be characterized with varying fat volume fraction/emulsifiers ratio/pH/homogenization pressure. Feng et al. (2021) used interfacial shear rheology to investigate the role of pea protein in the Pickering emulsion and found that there was a satisfactory relation between interfacial shear rheology and Pickering emulsion stability. Zhang, Yan, et al. (2021), Zhang, Kang, et al. (2021) reported that the interfacial shear rheology could be used to explain the mechanism of competitive displacement of soy proteins by Tween 20 in oil-in-water emulsions, and addition of Tween 20 improves the stability of the soy protein-stabilized emulsions.

Foam is a common form present in food products and is often used to enhance sensory properties, including beverage, milk (especially for coffee-making), and several baking products. This is one of the technical difficulties in replacing dairy milk with plant-based milk. It is challenging to quantify the behavior of food foams, as foams can change in a short amount of time during testing due to the high surface tension. Wang, Zhang, Zhang, Bak, et al. (2021) reported that interfacial rheology together with surface tension can quantify the behavior of foams from soy protein and found that partial hydrolyzation can help to improve the foaming capability and stability of soy protein. Völp et al. (2021) used interfacial rheology to analyze the foaming of lupin protein isolate and found that lupin proteins exhibited interfacial elasticity contributing to the foam rheology, which is one of the key reasons why lupin "plant milk" can yield comparable foam volume as cow milk but with smaller bubbles and thus higher yield stress.

5.3 Comparison between plant proteins and animal proteins

5.3.1 Behavior in emulsions, foams, and gels

Comparison of food proteins that forms food structures in emulsions, foams, and gels has been attempted in recent years. The functionality of plant proteins can be compared with animal proteins under same processing conditions and measurements. A considerable number of studies have been carried out on comparing plant protein such as soy protein, pea protein, wheat protein, and potato protein with animal proteins such as gelatin, casein, whey protein, and egg proteins (Table 5.1). For example, Stone et al. (2015) compared the functionality of egg, whey, soy, pea, and wheat proteins, reporting that whey protein showed exceptional performance in emulsions and foams followed by egg protein, while plant proteins showed poor functionalities. In their study, the solubility of food proteins was high in the order of egg, whey, soy,

TABLE 5.1 Recent studies on comparison of plant proteins with animal proteins in functionalities.

Form	Animal protein	Plant protein	Processing condition	References
Solution	WPI, egg white powder	SPI	Deamidation	Wong et al. (2009)
Emulsion	Casein, WPI	SPI	5% corn oil-in-water emulsion with 0.2%–1.5% protein, pH 3	Hu, McClements &Decker (2003)
Emulsion	WPC	SPI	Heat treatment, 20% soybean oil with 0.75%–3% protein, pH 3–7	Euston et al. (2009)
Emulsion	SC, WPI	Potato, soy, pea proteins	10% MCT oil-in-water emulsion with 0.1%–7.5% protein, pH 4,7, and 10	Amine et al. (2014)
Emulsion	Bovine gelatin, fish gelatin, egg white protein	pea, soy, rice proteins	0.1%–10% protein, ultrasound treatment (20 kHz, 95% amplitude for 2 min), pH 7	O'sullivan et al. (2016)
Emulsion	WPI	Rice glutelin	Rice glutelin hydrolysis by trypsin	Xu et al. (2017)
Emulsion	SC, WPI	PPI, SPI (soluble fractions)	Lycopene-loaded emulsion with 5–7 g/L protein, pH 7	Ho et al. (2017)
Emulsion	WPI, SC	SPI	Heat-treated Pickering nanoparticle	Zhu et al. (2017)

TABLE 5.1 Recent studies on comparison of plant proteins with animal proteins in functionalities.—cont'd

Form	Animal protein	Plant protein	Processing condition	References
Emulsion	Gelatin, WPI	SPI, zein, peanut protein	Ionic strength (0–300 mM NaCl), 10% rapeseed oil-in-water emulsion with 1% protein	Taha et al. (2019)
Emulsion	WPI	Pumpkin seed protein, rice protein	W/O/W emulsion with 10%–25% protein, cheese production at pH 6.3	Paximada et al. (2021)
Emulsion	SC	PPI	Maillard conjugation with dextran (60°C, 76.5% RH for 24 or 48 h), pH7	Caballero and Davidov-Pardo (2021)
Emulsion, Surface	WPI	PPI (soluble fraction)	10% sunflower oil with 1% protein	Hinderink, Berton-Carabin, et al. (2021), Hinderink, Sagis, et al. (2021)
Surface	WPI (soluble fraction)	PPI (soluble fraction)	0.1–100 mg/mL protein, heat treatment, pH 6.8	Zembyla et al. (2021)
Emulsion, Foam	SC, WPI	Deamidated wheat protein, SPI	Deamidation of wheat protein	Webb et al. (2002)
Emulsion, Foam	Egg, whey protein	Soy, pea, wheat proteins	Pea protein extraction methods	Stone et al. (2015)
Foam	ovalbumin, β-casein, β-lactoglobulin	Soy glycinin (pH3 and 6.7)	Neutral pH (except soy glycinin)	Martin et al. (2002)
Foam	β-casein, SC, WPI	β-conglycinin, glycinin	pH 5 and 7	Niño et al. (2005)
Emulsion gel	WPI	SPI	Alginate-emulsion gel beads	Lin et al. (2021)
Gel	Patatin, glycinin	β-lactoglobulin, ovalbumin	Heat-induced gelation, pH 7	Creusot et al. (2011)
Gel	Rubisco from spinach	Whey protein, egg white protein	Heat-induced gelation	Martin et al. (2014)

pea, and wheat proteins, suggesting increasing the solubility of plant proteins is imperative for improving their functionalities. Although plant protein functionalities are insufficient for food manufacturing, structural changes of plant proteins during processing have shown potential application as emulsifiers and foam stabilizers. Amine et al. (2014) found that the emulsifying properties of food proteins are pH-dependent in the study of comparing plant proteins with dairy proteins (sodium caseinate, whey protein, soy protein, pea protein, and potato protein). Furthermore, (Taha et al., 2019) produced stable emulsions using gelatin, whey protein isolate (WPI), and soy protein isolate (SPI), while zein and peanut protein did not produce emulsions by ultrasound treatment. Nevertheless, plant proteins such as soy protein were found to be vulnerable against high ionic strength and heat compared to animal proteins (Euston et al., 2009; Taha et al., 2019).

Since the functionalities of food proteins are related to their physicochemical nature, extraction of certain protein fractions and/or structural modifications of proteins can remarkably influence their functionalities. In the case of using the entire pea protein, pea protein is adsorbed as soluble aggregate without any conformational change, while whey protein adsorbed with their Trp-rich regions at the oil—water interface (Hinderink, Berton-Carabin, et al., 2021, Hinderink, Sagis, et al., 2021). However, chemical modifications including deamidation can result in higher emulsifying properties of deamidated wheat protein than SC, WPI, and SPI and comparable foaming properties to sodium caseinate (SC) (Webb et al., 2002). Deamidation leads to translation from ordered to disordered structures and exposure of tryptophan residues, indicating plant protein structures can be modified similar to those in whey protein when adsorbing at the oil/water interface (Wong et al., 2009). Moreover, emulsifying properties of soluble soy and pea protein fractions were compared with SC and whey protein by Ho et al. (2017), demonstrating that soluble pea protein had comparable performance to SC in terms of physical and chemical stability. In addition, hydrolysates of rice protein could form emulsions with small droplets (< 200 nm), and the emulsions were stable against pH changes, ionic strength, and thermal processing (Xu et al., 2017). Another strategy is chemical conjugation via Maillard reaction to improve functional properties of plant proteins. Recently, pea protein isolate (PPI) and SC were compared in Maillard reaction with dextran, resulting in the decrease in particle size of PPI, demonstrating a potential application of pea protein (Caballero & Davidov-Pardo, 2021).

Comparison of functionalities between dairy and legume proteins is the most widely studied, and thus these proteins have been employed in novel food systems, such as Pickering emulsions and emulsion gels. Lin et al. (2021) incorporated WPI or SPI to alginate-based emulsion gel beads, reporting that WPI leads to higher viscosity, smaller droplet size, and less flocculation due to higher solubility of WPI and stronger interactions with alginate than SPI. However, heated SPI and whey protein showed better freeze-thaw stability against coalescence and creaming than unheated proteins in Pickering emulsions, indicating SPI could perform close to whey protein as a Pickering stabilizer (Zhu et al., 2017). Interestingly, Zembyla et al. (2021) compared the surface adsorption and lubrication properties of WPI and PPI, reporting that pea protein at high concentrations offers inferior lubricity than whey protein, potentially resulting in unpleasant mouthfeel.

More attention has been given to common plant proteins, including soy, pea, wheat, and rice proteins, while plant proteins from unique sources have produced gels with a promising application to food products. Heat-induced gels were produced using Rubisco from spinach, showing that low critical gelling concentration and low denaturation temperature were required compared to whey and egg white protein, while a brittle character was observed upon deformation (Martin et al., 2014). Creusot et al. (2011) also found that minimum concentration to produce patatin gel (6%) was lower than β-lactoglobulin and ovalbumin (8%−11%) due to highly exposed structure of patitin by hydrophobic interaction. Moreover, low-fat cheddar cheese was produced using W/O/W emulsions stabilized by WPI or plant protein (pumpkin seed protein and rice protein) by Paximada et al. (2021), demonstrating that plant proteins can be utilized for fat-based emulsions with high encapsulation efficiencies.

5.3.2 Mixed plant-/animal-based protein systems

Recent health and environmental concerns have promoted the trends in increasing intake of plant-based products, while vegan products are often unsuccessfully accepted by consumers. Mixed animal/plant system in food products can enhance the functionalities of plant proteins and provide innovative diets. Mixing dairy and legume proteins are common combinations found in emulsions, foams and gels. Aoki et al. (1984) found that mixing SC and SPI negatively affected emulsion stability because SC inhibits adsorption of soy protein at the interface. However, in recent years, interactions in mixed protein system have been fundamentally investigated to approach reported synergistic effects in food systems. Molecular approach on the interactions and behaviors between heteroproteins was elucidated in detail in recent literature (Wu et al., 2021). Since interactions or network formation was not found by mixing two different proteins in aqueous phase, physicochemical treatments have been applied to modify secondary structures of proteins and to generate interactions. Heat treatment in mixed dairy/legume protein blends leads to distinctive results depending on protein structures. Interestingly, covalent disulfide bonds are formed between whey and legume proteins, while casein and legume proteins do not interact when heat is applied (Chihi et al., 2016; Mession et al., 2017). Rather, interactions between legume proteins are strengthened in this case. This is because whey and legume proteins with globular structures can interact readily, while casein with flexible structure makes legume protein difficult to interact. Therefore, enzyme cross-linking between casein and rice glutelin successfully bonded proteins, resulting in improved functional properties and digestibility (He et al., 2021).

Mixed protein system has been extensively employed in food systems such as emulsions, foams, and gels, as discussed in recent literature (Guyomarc'h et al., 2021; Kim et al., 2020; Wu et al., 2021). In emulsions, synergistic or antagonistic effect has been found depending on protein type and processing conditions, which mainly affect the emulsion stability (Hinderink et al., 2019; Ho et al., 2018). Thus, strategies are introduced to enhance the stability by varying processing methods, such as sequential adsorption (Hinderink, Berton-Carabin, et al., 2021, Hinderink, Sagis, et al., 2021). It was recently found that polysaccharide addition in mixed whey-/pea protein-stabilized emulsion enhances protein adsorption at the interface, thereby improving emulsion stability. Binary whey/pea protein emulsifiers formed thick membrane at the oil/water interface at pH 8.0, and small (~ 1 μm) and uniform droplets were formed at pH 11.0 (Fig. 5.2, unpublished data). Interestingly, in weak acidic condition, particle aggregation from pea protein in binary whey/pea protein-stabilized emulsions, which fails to encompass oil droplets and promotes destabilization, were found to be impeded by pea protein−polysaccharide interaction. Furthermore, both proteins from plant sources were utilized as an emulsifier, such as quinoa and lentil mixture, showing excellent physical stability (Alonso-Miravalles et al., 2022).

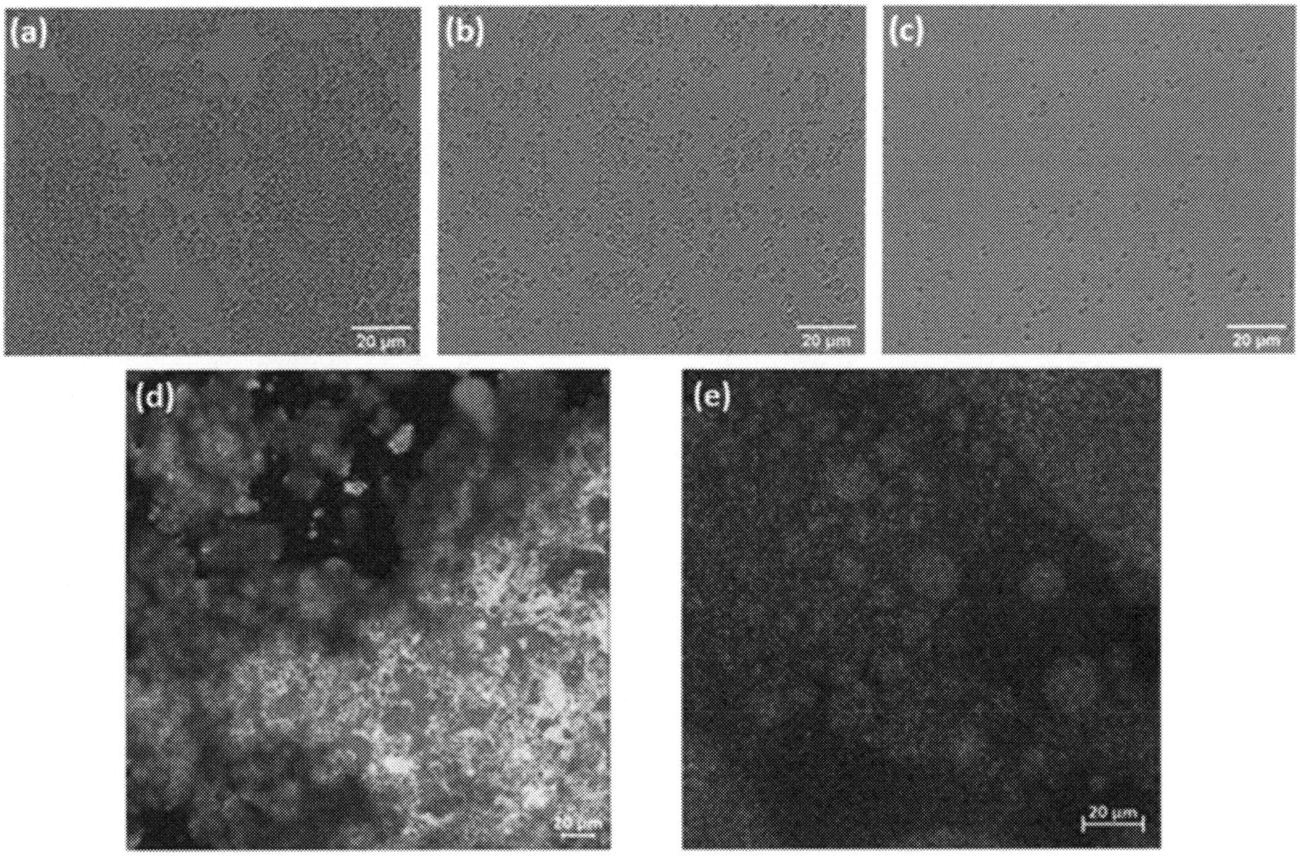

FIGURE 5.2 Microstructural images of binary protein-stabilized emulsions with/without sodium alginate using optical microscope with 40× objectives (A–C) and confocal laser scanning microscope (CLSM) with 20× and 40× objectives stained with Nile blue A (protein in the water phase) and Nile red (oil phase). Optical microscopic images of binary protein-stabilized emulsion at pH 6.6 (A), pH 8.0 (B), and pH 11.0 (C). CLSM images of binary protein-stabilized emulsion at pH 6.6 (D), and binary protein-sodium alginate-stabilized emulsion at pH 6.6 (E) (Kim et al., 2022).

Mixed protein system has also been applied to heat- or acid-induced gelation. Thermal aggregation was recommended due to less protein concentration required, rapid gelation, and higher gels strength when using pea protein mixed with casein or whey protein (Chihi et al., 2018; Mession et al., 2017). When gelatin or casein is mixed with legume proteins, independent network is formed by each protein without interactions between different proteins in the gel, and thus microphase separation is observed (Ersch et al., 2015; Silva et al., 2019). However, Jose et al. (2016) found that one network gel was formed by mixed whey/soy protein gel, and both proteins contribute to gel properties. With increase in the amount of legume protein in the mixed whey/legume protein gels, the gel strength is weakened (Chihi et al., 2018; Jose et al., 2016; McCann et al., 2018). Therefore, plant proteins, especially legume proteins, act rather as a filler in the gel network when mixed with whey protein.

5.4 Bioaccessibility of plant-based proteins

Bioavailability is a complex process including bio-accessibility, absorption, and transformation of nutrients (metabolism and bioactivity). Bioavailability can be defined as the ingested food fraction that can be absorbed in the gastrointestinal tract, metabolized, and distributed further to different organs and tissues (Parada & Aguilera, 2007). These processes are affected by both endogenous and exogenous factors and can influence the bioavailability of the plant proteins. These factors include the physicochemical properties of the protein, complexity of the food matrix, processing and storage condition, bio-accessibility within the intestine, solubility, and the interaction of proteins with other components inside the intestinal tract (Dima et al., 2020).

Bioaccessibility is the first phase of bioavailability and is defined as the fraction of an ingested biocomponent that becomes available for absorption through the epithelial membrane of the intestine (gastrointestinal tract). In terms of proteins, its bioaccessibility is expressed as the ratio of the protein solubilized in the micellar phase to the protein in raw digested form. The release of protein from the food matrix/carriers and its solubilization in the micelles is crucial for the protein to be absorbed (Dima et al., 2020).

In order to understand the bioaccessibility of proteins, the simulation that can predict the digestive transformation of the food into the simpler compounds absorbed by the gut (hepatic and intestinal absorption) is necessary. In vitro, this can be achieved by digesting the protein using static digestion INFOGEST protocol (Fig. 5.3) that can simulate the gastric, intestinal digestion, and epithelial uptake of protein (Brodkorb et al., 2019; Hayes, 2018). Moreover, different dynamic methods like TNO gastric-intestinal model to understand protein hydrolysis and overall bioavailability can also be employed as described by Verhoeckx et al. (2015).

In terms of plant proteins, their digestion is mainly influenced by the protein origin and the food matrix. During static in vitro studies, the release of α-amino acid and hydrolysis of protein was found to be higher in protein emulsions as compared to the protein isolates/concentrates. This particular trend was studied by Reynaud et al. (2020) where soya juice (51% hydrolyzed) and pea protein emulsion were hydrolyzed better than the fresh tofu (33% hydrolyzed), while soy protein and pea protein isolates (~18% hydrolyzed) were hydrolyzed the least when compared to tofu or emulsions. This also signifies that solid plant proteins exhibit lower degree of hydrolysis than their liquid counterparts. Different antinutritional factors (e.g., protease inhibitors, phytates, etc.) present in the plant-based sources can also hinder the interaction and absorption of proteins leading to reduced digestion and further metabolism, which is one of the shortcomings of plant proteins. Moreover, plant proteins also show inferior bioaccessibility compared to animal-based proteins due to the complexity of the structures, which can hinder the accessibility of proteins to the gastric enzymes (Maehre et al., 2016). Nevertheless, it has been found that some leaf proteins like amaranth, chia, alfalfa, and broad bean were able to demonstrate better bioaccessibility when compared to daily standard diet among the subjects (Ramírez-Rodrigues et al., 2022). It has also been found that plant proteins can alter the bioaccessibility of other components like carotenoids by modulating and stabilizing the digestive emulsion, or can hinder the process by interacting with the enzymes that are responsible for carotenoid breakdown and absorption. This effect is dependent on the concentration, type of protein, and carotenoid type as well (Iddir et al., 2020). Plant proteins can also influence the bioaccessibility and stability of nano-emulsions. For example, Tangeretin (flavone) had better absorption in presence of soy protein in the nanoemulsion compared to the standard emulsions without inclusion of proteins (Wan et al., 2017). Complexation of phenols with plant proteins has also shown to increase the bioaccessibility of phenols. For example, garden-cress protein or pea protein when complexed with curcumin have shown to increase the bioaccessibility of the phenol up to 72%–95% (Kadam et al., 2019; Kannamangalam Vijayan et al., 2021). Bioaccessibility of plant proteins can be improved by different techniques that include heat treatments to breakdown the complex proteins and sugars that interfere with absorption of the proteins or by using enzymatic pretreatment (using xylanase or cellulase) over the raw plant sources before protein extraction (Maehre et al., 2016; Mæhre et al., 2016).

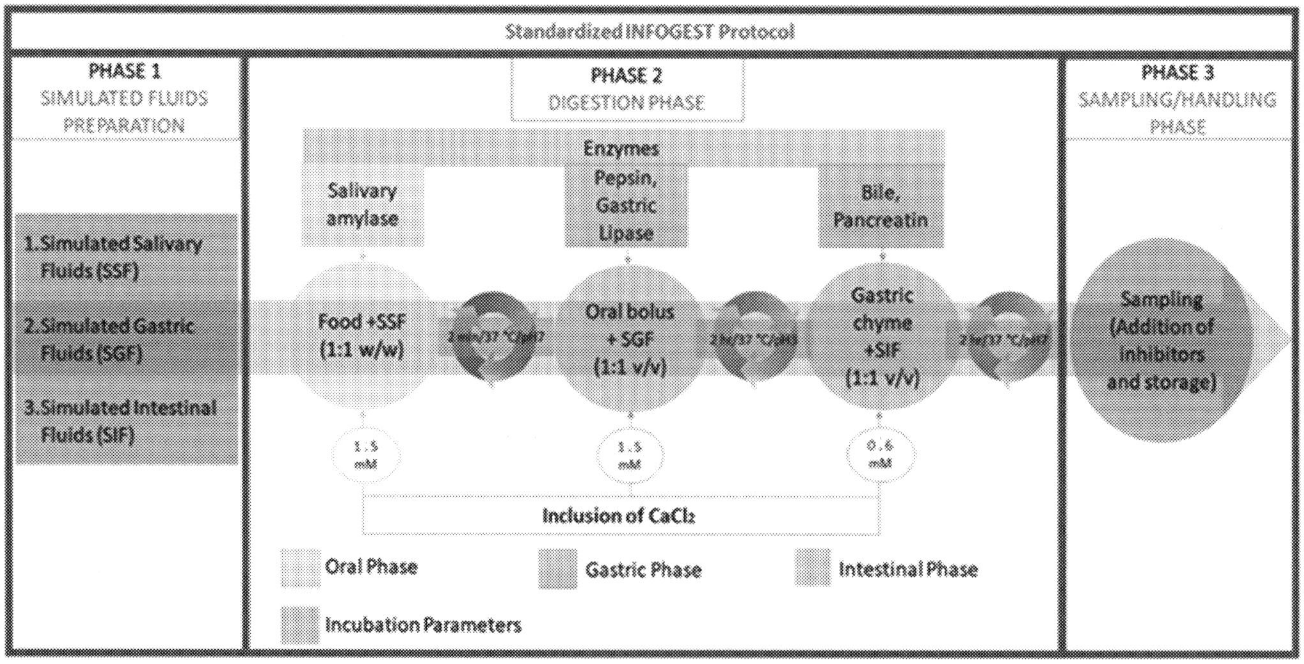

FIGURE 5.3 Standardized static digestion INFOGEST Protocol.

5.5 Functional limitations of plant-based proteins in food applications

Although the field of plant proteins has been increasingly researched, there are still many functionality problems to hinder further application of plant proteins.

- The stability of proteins is affected by a range of environmental factors (such as pH, temperature, and ionic strength), which limit their use in production. In the production and storage of plant protein beverages, instability such as protein precipitation and phase separation can occur, which can have a significant impact on the perceived quality of the product (Feng et al., 2021). When the pH of the system is close to the isoelectric point of the protein, the stability of the system will be severely affected (Wang, Zhang, Zhang, Bak, et al., 2021, Wang, Zhang, Zhang, Zeng, et al., 2021). However, the acidic pH of most plant protein solutions, combined with the fact that emulsion stability is not well controlled, limit the development, and application of plant proteins in food.
- The addition of large amounts of plant protein to replace animal protein can affect the texture of the product and introduce an unpleasant odor that might be difficult to remove later by processing. Therefore, food additives have been used to satisfy consumers' desire for flavor in meat products, but this has not resulted in an ideal solution to mask the unpleasant smell and/or taste from plant proteins. The current removal methods for beany flavor include biotechnological, physical, and chemical methods, which might increase production costs (Wang, Zhang, Zhang, Bak, et al., 2021, Wang, Zhang, Zhang, Zeng, et al., 2021) thus need further investigations.
- There are some disadvantages of plant protein powders in practical applications. Solubility is one of the most important functional properties of proteins and a major determinant of other functional properties (emulsification, foaming, and gelation), but there are factors affecting plant protein solubility, including the presence of high amounts of insoluble protein aggregates (especially from the glutelins), the sensitivity to heat- and acid-induced coagulation, the nonprotein components such as polysaccharide (Wang & Selomulya, 2020). These factors will directly determine the feasibility and convenience of plant proteins in food processing applications. The processing of large quantities of plant protein powders can result in low yields due to the complex processes that usually require fermentation, precipitation, ultrafiltration, and/or thermal treatment (Ebert et al., 2020) with challenges to achieve as high-purity product as animal-based proteins.

5.6 Summary and future direction

This chapter focuses on the functionalities of plant-based proteins, including color, flavor, texture, wettability, dispersibility, solubility, rheological properties, emulsifying properties, foaming properties, etc. Although there are still some gaps in the functionality of plant-based proteins limiting their applications compared to animal proteins, the overview presented in this chapter demonstrates that recent developments have progressed to overcome the functional limitations of plant-based proteins, showcasing their enormous potential. Looking ahead, plant-based proteins are bound to make a significant difference in replacing animal proteins due to their availability, variety (different functionalities) and relatively low production costs. The current commercialization trends of plant-based meat have demonstrated the consumer's demand, but research is still needed in the texturization of plant proteins to closely resemble the network structure and sensory properties (especially chewing) of animal proteins. Flavor masking of plant-based proteins also needs further research, especially in techniques to obtain flavor improvements by physical processing instead of flavorings. Regarding mixed plant-animal proteins, there is still a need to find the most suitable formulation combinations for specific food applications, especially in emulsion systems, in terms of structural compatibility and design processing techniques accordingly. In addition, plant protein powders will be an important factor for their wider applications as ingredients, and thus the development of highly soluble powders with improved rehydration functionalities will be crucial.

Acknowledgments

This project was supported by ARC Discovery grant (DP200100642).

References

AIDP Advanced ingredient's for dietary products.(2020). https://aidp.com/the-abc's-of-formulating-with-plant-proteins.html (Accessed 23 July 2021).

Akharume, F. U., Aluko, R. E., & Adedeji, A. A. (2021). Modification of plant proteins for improved functionality: A review. *Comprehensive Reviews in Food Science and Food Safety, 20*(1), 198–224. https://doi.org/10.1111/1541-4337.12688

Akin, Z., & Ozcan, T. (2017). Functional properties of fermented milk produced with plant proteins. *LWT, 86*, 25–30. https://doi.org/10.1016/j.lwt.2017.07.025

Alonso-Miravalles, L., Zannini, E., Bez, J., Arendt, E. K., & O'Mahony, J. A. (2022). Formation and thermal and colloidal stability of oil-in-water emulsions stabilized using quinoa and lentil protein blends. *Journal of the Science of Food and Agriculture, 102*(12), 5077–5085.

Amagliani, L., O'Regan, J., Schmitt, C., Kelly, A. L., & O'Mahony, J. A. (2019). Characterisation of the physicochemical properties of intact and hydrolysed rice protein ingredients. *Journal of Cereal Science, 88*, 16–23. https://doi.org/10.1016/j.jcs.2019.04.002

Amine, C., Dreher, J., Helgason, T., & Tadros, T. (2014). Investigation of emulsifying properties and emulsion stability of plant and milk proteins using interfacial tension and interfacial elasticity. *Food Hydrocolloids, 39*, 180–186. https://doi.org/10.1016/j.foodhyd.2014.01.001

Aoki, H., Shirase, Y., Kato, J., & Watanabe, Y. (1984). Emulsion stabilizing properties of soy protein isolates mixed with sodium caseinates. *Journal of Food Science, 49*(1), 212–216.

Avelar, Z., Vicente, A. A., Saraiva, J. A., & Rodrigues, R. M. (2021). The role of emergent processing technologies in tailoring plant protein functionality: New insights. *Trends in Food Science & Technology, 113*, 219–231. https://doi.org/10.1016/j.tifs.2021.05.004

Bangratz. (2020). Masking flavours for plant proteins. *Food Science and Technology, 34*(1), 29–31. https://doi.org/10.1002/fsat.3401_8.x

Batal, A. B., Douglas, M. W., Engram, A. E., & Parsons, C. M. (2000). Protein dispersibility index as an indicator of adequately processed soybean meal. *Poultry Science, 79*(11), 1592–1596. https://doi.org/10.1093/ps/79.11.1592

Bi, C., Chi, S., Hua, Z., Li, D., Huang, Z., & Liu, Y. (2020). Rheological properties and fractal-rheology analysis of peanut protein isolate suspension. *International Journal of Agricultural and Biological Engineering, 13*(6), 220–226. https://doi.org/10.25165/j.ijabe.20201306.5717

Blouin, F. A., Zarins, Z. M., & Cherry, J. P. (1981). Color. In *Protein Functionality in foods* (pp. 21–39). American Chemical Society. https://doi.org/10.1021/bk-1981-0147.ch002. Vol. 147.

Brodkorb, A., Egger, L., Alminger, M., Alvito, P., Assunção, R., Ballance, S., Bohn, T., Bourlieu-Lacanal, C., Boutrou, R., Carrière, F., Clemente, A., Corredig, M., Dupont, D., Dufour, C., Edwards, C., Golding, M., Karakaya, S., Kirkhus, B., Le Feunteun, S., … Recio, I. (2019). INFOGEST static in vitro simulation of gastrointestinal food digestion. *Nature Protocols, 14*(4), 991–1014. https://doi.org/10.1038/s41596-018-0119-1

Caballero, S., & Davidov-Pardo, G. (2021). Comparison of legume and dairy proteins for the impact of Maillard conjugation on nanoemulsion formation, stability, and lutein color retention. *Food Chemistry, 338*, 128083.

Chihi, M.-L., Mession, J., Sok, N., & Saurel, R. (2016). Heat-induced soluble protein aggregates from mixed pea globulins and β-lactoglobulin. *Journal of Agricultural and Food Chemistry, 64*(13), 2780–2791.

Chihi, M., Sok, N., & Saurel, R. (2018). Acid gelation of mixed thermal aggregates of pea globulins and β-lactoglobulin. *Food Hydrocolloids, 85*, 120–128.

Chmielewska, A., Kozłowska, M., Rachwał, D., Wnukowski, P., Amarowicz, R., Nebesny, E., & Rosicka-Kaczmarek, J. (2020). Canola/rapeseed protein–nutritional value, functionality and food application: A review. *Critical Reviews in Food Science and Nutrition*, 1–21.

Creusot, N., Wierenga, P. A., Laus, M. C., Giuseppin, M. L., & Gruppen, H. (2011). Rheological properties of patatin gels compared with β-lactoglobulin, ovalbumin, and glycinin. *Journal of the Science of Food and Agriculture, 91*(2), 253–261.

Dakhili, S., Abdolalizadeh, L., Hosseini, S. M., Shojaee-Aliabadi, S., & Mirmoghtadaie, L. (2019). Quinoa protein: Composition, structure and functional properties. *Food Chemistry, 299*, 125161.

Damodaran, S., & Arora, A. (2013). Off-flavor precursors in soy protein isolate and novel strategies for their removal. *Annual Review of Food Science and Technology, 4*(1), 327–346. https://doi.org/10.1146/annurev-food-030212-182650

Dima, C., Assadpour, E., Dima, S., & Jafari, S. M. (2020). Bioavailability and bioaccessibility of food bioactive compounds; overview and assessment by in vitro methods. *Comprehensive Reviews in Food Science and Food Safety, 19*(6), 2862–2884. https://doi.org/10.1111/1541-4337.12623

Ebert, S., Gibis, M., Terjung, N., & Weiss, J. (2020). Survey of aqueous solubility, appearance, and pH of plant protein powders from carbohydrate and vegetable oil production. *LWT, 133*, 110078. https://doi.org/10.1016/j.lwt.2020.110078

Ersch, C., ter Laak, I., van der Linden, E., Venema, P., & Martin, A. (2015). Modulating fracture properties of mixed protein systems. *Food Hydrocolloids, 44*, 59–65.

Euston, S. R., Al-Bakkush, A.-A., & Campbell, L. (2009). Comparing the heat stability of soya protein and milk whey protein emulsions. *Food Hydrocolloids, 23*(8), 2485–2492. https://doi.org/10.1016/j.foodhyd.2009.08.004

Feng, T., Wang, X., Wang, X., Xia, S., & Huang, Q. (2021). Plant protein-based antioxidant Pickering emulsions and high internal phase Pickering emulsions against broad pH range and high ionic strength: Effects of interfacial rheology and microstructure. *Lwt*, 150. https://doi.org/10.1016/j.lwt.2021.111953

Fleming, S., Sosulski, F., Kilara, A., & Humbert, E. (1974). Viscosity and water absorption charactercstics of slurries of sunflower and soybean flours, concentrates and isolates. *Journal of Food Science, 39*(1), 188–192.

Foegeding, E. A. (2015). Food protein functionality—a new model. *Journal of Food Science, 80*(12), C2670–C2677. https://doi.org/10.1111/1750-3841.13116

Gallego, M., Arnal, M., Barat, J. M., & Talens, P. (2020). Effect of cooking on protein digestion and antioxidant activity of different legume pastes. *Foods, 10*(1). https://doi.org/10.3390/foods10010047

Garidel, P. (2013). Protein solubility from a biochemical, physicochemical and colloidal perspective. *American Pharmarcy Reviews, 2*(5), 26–28.

Granato, T. M., Ferranti, P., Iametti, S., & Bonomi, F. (2018). Affinity and selectivity of plant proteins for red wine components relevant to color and aroma traits. *Food Chemistry, 256*, 235–243. https://doi.org/10.1016/j.foodchem.2018.02.085

Guyomarc'h, F., Arvisenet, G., Bouhallab, S., Canon, F., Deutsch, S.-M., Drigon, V., Dupont, D., Famelart, M.-H., Garric, G., & Guedon, E. (2021). Mixing milk, egg and plant resources to obtain safe and tasty foods with environmental and health benefits. *Trends in Food Science & Technology, 108*, 119–132.

Hayes, M. (2018). Bioavailability, bioaccessibility, and nutritional measurement of proteins. In *Novel Proteins for food, Pharmaceuticals and agriculture* (pp. 183–195). https://doi.org/10.1002/9781119385332.ch9

He, C., Hu, Y., Woo, M. W., Xiong, H., & Zhao, Q. (2021). Effect of microbial transglutaminase on the structural and rheological characteristics and in vitro digestion of rice glutelin—casein blends. *Food Research International, 139*, 109832.

Heng, L., van Koningsveld, G. A., Gruppen, H., van Boekel, M. A. J. S., Vincken, J. P., Roozen, J. P., & Voragen, A. G. J. (2004). Protein—flavour interactions in relation to development of novel protein foods. *Trends in Food Science & Technology, 15*(3), 217—224. https://doi.org/10.1016/j.tifs.2003.09.018

Hinderink, E. B. A., Berton-Carabin, C. C., Schroën, K., Riaublanc, A., Houinsou-Houssou, B., Boire, A., & Genot, C. (2021). Conformational changes of whey and pea proteins upon emulsification approached by front-surface fluorescence. *Journal of Agricultural and Food Chemistry, 69*(23), 6601—6612. https://doi.org/10.1021/acs.jafc.1c01005

Hinderink, E. B., Münch, K., Sagis, L., Schroën, K., & Berton-Carabin, C. C. (2019). Synergistic stabilisation of emulsions by blends of dairy and soluble pea proteins: Contribution of the interfacial composition. *Food Hydrocolloids, 97*, 105206.

Hinderink, E. B., Sagis, L., Schroën, K., & Berton-Carabin, C. C. (2021). Sequential adsorption and interfacial displacement in emulsions stabilized with plant-dairy protein blends. *Journal of Colloid and Interface Science, 583*, 704—713.

Hollis, J. (2018). In vivo foundations of sensory in vitro testing systems. In *Functional Foods and beverages* (pp. 53—85). https://doi.org/10.1002/9781118823309.ch3

Hong, T., Iwashita, K., & Shiraki, K. (2018). Viscosity control of protein solution by small solutes: A review. *Current Protein & Peptide Science, 19*(8), 746—758. https://doi.org/10.2174/1389203719666171213114919

Ho, K. K., Schroën, K., Martín-González, M. F., & Berton-Carabin, C. C. (2018). *Synergistic and antagonistic effects of plant and dairy protein blends on the physicochemical stability of lycopene-loaded emulsions.*

Ho, K. K., Schroën, K., San Martín-González, M. F., & Berton-Carabin, C. C. (2017). Physicochemical stability of lycopene-loaded emulsions stabilized by plant or dairy proteins. *Food Structure, 12*, 34—42.

Hu, H., Wu, J., Li-Chan, E. C., Zhu, L., Zhang, F., Xu, X., Fan, G., Wang, L., Huang, X., & Pan, S. (2013). Effects of ultrasound on structural and physical properties of soy protein isolate (SPI) dispersions. *Food Hydrocolloids, 30*(2), 647—655.

Iddir, M., Dingeo, G., Porras Yaruro, J. F., Hammaz, F., Borel, P., Schleeh, T., Desmarchelier, C., Larondelle, Y., & Bohn, T. (2020). Influence of soy and whey protein, gelatin and sodium caseinate on carotenoid bioaccessibility. *Food & Function, 11*(6), 5446—5459. https://doi.org/10.1039/D0FO00888E

Jarpa-Parra, M. (2018). Lentil protein: A review of functional properties and food application. An overview of lentil protein functionality. *International Journal of Food Science & Technology, 53*(4), 892—903. https://doi.org/10.1111/ijfs.13685

Jiang, L., Wang, Z., Li, Y., Meng, X., Sui, X., Qi, B., & Zhou, L. (2015). Relationship between surface hydrophobicity and structure of soy protein isolate subjected to different ionic strength. *International Journal of Food Properties, 18*(5), 1059—1074.

Ji, J., Fitzpatrick, J., Cronin, K., Maguire, P., Zhang, H., & Miao, S. (2016). Rehydration behaviours of high protein dairy powders: The influence of agglomeration on wettability, dispersibility and solubility. *Food Hydrocolloids, 58*, 194—203. https://doi.org/10.1016/j.foodhyd.2016.02.030

Jose, J., Pouvreau, L., & Martin, A. H. (2016). Mixing whey and soy proteins: Consequences for the gel mechanical response and water holding. *Food Hydrocolloids, 60*, 216—224.

Joyner, H. S. (2021). Nonlinear (Large-Amplitude oscillatory shear) rheological properties and their impact on food processing and quality. *Annual Reviews of Food and Science Technology, 12*, 591—609. https://doi.org/10.1146/annurev-food-061220-100714

Kadam, D., Palamthodi, S., & Lele, S. S. (2019). Complexation of curcumin with Lepidium sativum protein hydrolysate as a novel curcumin delivery system. *Food Chemistry, 298*, 125091. https://doi.org/10.1016/j.foodchem.2019.125091

Kannamangalam Vijayan, U., Shah, N. N., Muley, A. B., & Singhal, R. S. (2021). Complexation of curcumin using proteins to enhance aqueous solubility and bioaccessibility: Pea protein vis-à-vis whey protein. *Journal of Food Engineering, 292*, 110258. https://doi.org/10.1016/j.jfoodeng.2020.110258

Kim, W., Wang, Y., & Selomulya, C. (2020). *Dairy and plant proteins as natural food emulsifiers*. Trends in Food Science & Technology.

Kim, W., Wang, Y., & Selomulya, C. (2022). Impact of sodium alginate on binary whey/pea protein-stabilised emulsions. *Journal of Food Engineering, 321*, 110978.

Klost, M., Brzeski, C., & Drusch, S. (2020). Effect of protein aggregation on rheological properties of pea protein gels. *Food Hydrocolloids*, 108. https://doi.org/10.1016/j.foodhyd.2020.106036

Klost, M., Giménez-Ribes, G., & Drusch, S. (2020). Enzymatic hydrolysis of pea protein: Interactions and protein fractions involved in fermentation induced gels and their influence on rheological properties. *Food Hydrocolloids*, 105. https://doi.org/10.1016/j.foodhyd.2020.105793

Kornet, R., Veenemans, J., Venema, P., van der Goot, A. J., Meinders, M., Sagis, L., & van der Linden, E. (2021). Less is more: Limited fractionation yields stronger gels for pea proteins. *Food Hydrocolloids*, 112. https://doi.org/10.1016/j.foodhyd.2020.106285

Kornet, C., Venema, P., Nijsse, J., van der Linden, E., van der Goot, A. J., & Meinders, M. (2020). Yellow pea aqueous fractionation increases the specific volume fraction and viscosity of its dispersions. *Food Hydrocolloids, 99*, 105332. https://doi.org/10.1016/j.foodhyd.2019.105332

Kramer, R. M., Shende, V. R., Motl, N., Pace, C. N., & Scholtz, J. M. (2012). Toward a molecular understanding of protein solubility: Increased negative surface charge correlates with increased solubility. *Biophysical Journal, 102*(8), 1907—1915. https://doi.org/10.1016/j.bpj.2012.01.060

Kuntz, I., Jr., & Kauzmann, W. (1974). Hydration of proteins and polypeptides. *Advances in Protein Chemistry, 28*, 239—345.

Lam, A., Can Karaca, A., Tyler, R., & Nickerson, M. (2018). Pea protein isolates: Structure, extraction, and functionality. *Food Reviews International, 34*(2), 126—147.

Lawless, H. T., & Heymann, H. (2010). Texture evaluation. In *Sensory Evaluation of food: Principles and practices* (pp. 259—281). New York: Springer. https://doi.org/10.1007/978-1-4419-6488-5_11

Le Roux, L., Mejean, S., Chacon, R., Lopez, C., Dupont, D., Deglaire, A., Nau, F., & Jeantet, R. (2020). Plant proteins partially replacing dairy proteins greatly influence infant formula functionalities. *LWT, 120*, 108891. https://doi.org/10.1016/j.lwt.2019.108891

Liang, G., Chen, W., Qie, X., Zeng, M., Qin, F., He, Z., & Chen, J. (2020). Modification of soy protein isolates using combined pre-heat treatment and controlled enzymatic hydrolysis for improving foaming properties. *Food Hydrocolloids, 105*, 105764.

Li, Y., Kang, Z., Sukmanov, V., & Ma, H. (2021). Effects of soy protein isolate on gel properties and water holding capacity of low-salt pork myofibrillar protein under high pressure processing. *Meat Science, 176*, 108471. https://doi.org/10.1016/j.meatsci.2021.108471

Lin, D., Kelly, A. L., Maidannyk, V., & Miao, S. (2021). Effect of structuring emulsion gels by whey or soy protein isolate on the structure, mechanical properties, and in-vitro digestion of alginate-based emulsion gel beads. *Food Hydrocolloids, 110*, 106165.

Liu, S., Elmer, C., Low, N. H., & Nickerson, M. T. (2010). Effect of pH on the functional behaviour of pea protein isolate—gum Arabic complexes. *Food Research International, 43*(2), 489—495. https://doi.org/10.1016/j.foodres.2009.07.022

Maehre, H. K., Edvinsen, G. K., Eilertsen, K.-E., & Elvevoll, E. O. (2016). Heat treatment increases the protein bioaccessibility in the red seaweed dulse (Palmaria palmata), but not in the brown seaweed winged kelp (Alaria esculenta). *Journal of Applied Phycology, 28*(1), 581—590.

Mæhre, H. K., Jensen, I.-J., & Eilertsen, K.-E. (2016). Enzymatic pre-treatment increases the protein bioaccessibility and extractability in dulse (palmaria palmata). *Marine Drugs, 14*(11), 196. https://www.mdpi.com/1660-3397/14/11/196.

Malik, M. A., Sharma, H. K., & Saini, C. S. (2017). High intensity ultrasound treatment of protein isolate extracted from dephenolized sunflower meal: Effect on physicochemical and functional properties. *Ultrasonics Sonochemistry, 39*, 511—519.

Martin, A. H., Nieuwland, M., & de Jong, G. A. (2014). Characterization of heat-set gels from RuBisCO in comparison to those from other proteins. *Journal of Agricultural and Food Chemistry, 62*(44), 10783—10791.

Martin, A. H., Grolle, K., Bos, M. A., Stuart, M. A. C., & van Vliet, T. (2002). Network forming properties of various proteins adsorbed at the air/water interface in relation to foam stability. *Journal of Colloid and Interface Science, 254*(1), 175—183.

McCann, T. H., Guyon, L., Fischer, P., & Day, L. (2018). Rheological properties and microstructure of soy-whey protein. *Food Hydrocolloids, 82*, 434—441.

Mession, J.-L., Roustel, S., & Saurel, R. (2017). Interactions in casein micelle-Pea protein system (Part II): Mixture acid gelation with glucono-δ-lactone. *Food Hydrocolloids, 73*, 344—357.

Milewski, S. (2001). In Z. E. Sikorski (Ed.), *Protein structure and physicochemical properties. Chemical and Functional Properties of Food Proteins* (pp. 35—55). Technomic Publishing Company, Inc. Pennsylvania.

Mita, T., Nikai, K., Hiraoka, T., Matsuo, S., & Matsumoto, H. (1977). Physicochemical studies on wheat protein foams. *Journal of Colloid and Interface Science, 59*(1), 172—178.

Morales, R., Martínez, K. D., Pizones Ruiz-Henestrosa, V. M., & Pilosof, A. M. R. (2015). Modification of foaming properties of soy protein isolate by high ultrasound intensity: Particle size effect. *Ultrasonics Sonochemistry, 26*, 48—55. https://doi.org/10.1016/j.ultsonch.2015.01.011

Moreno, H. M., Domínguez-Timón, F., Díaz, M. T., Pedrosa, M. M., Borderías, A. J., & Tovar, C. A. (2020). Evaluation of gels made with different commercial pea protein isolate: Rheological, structural and functional properties. *Food Hydrocolloids*, 99. https://doi.org/10.1016/j.foodhyd.2019.105375

Muhammad, A., & Pauls, K. (2018). *Properties of plant proteins* (pp. 121—142). https://doi.org/10.1007/978-1-4939-8616-3_8

Naik, R. R., Wang, Y., & Selomulya, C. (2021). Improvements of plant protein functionalities by Maillard conjugation and Maillard reaction products. *Critical Reviews in Food Science and Nutrition*, 1—26.

Nascimento, R. F., Rosa, J. G., Andreola, K., & Taranto, O. P. (2020). Wettability improvement of pea protein isolate agglomerated in pulsed fluid bed. *Particulate Science and Technology, 38*(4), 511—521. https://doi.org/10.1080/02726351.2019.1574940

Nicolai, T., & Chassenieux, C. (2019). Heat-induced gelation of plant globulins. *Current Opinion in Food Science, 27*, 18—22. https://doi.org/10.1016/j.cofs.2019.04.005

Nieto-Nieto, T. V., Wang, Y., Ozimek, L., & Chen, L. (2016). Improved thermal gelation of oat protein with the formation of controlled phase-separated networks using dextrin and carrageenan polysaccharides. *Food Research International, 82*, 95—103. https://doi.org/10.1016/j.foodres.2016.01.027

Niño, M. R. R., Sánchez, C. C., Ruíz-Henestrosa, V. P., & Patino, J. M. R. (2005). Milk and soy protein films at the air—water interface. *Food Hydrocolloids, 19*(3), 417—428.

O'sullivan, J., Murray, B., Flynn, C., & Norton, I. (2016). The effect of ultrasound treatment on the structural, physical and emulsifying properties of animal and vegetable proteins. *Food hydrocolloids, 53*, 141—154.

de Oliveira, A. P. H., Omura, M. H., Barbosa, É. de A. A., Bressan, G. C., Vieira, É. N. R., Coimbra, J. S., dos, R., & de Oliveira, E. B. (2020). Combined adjustment of pH and ultrasound treatments modify techno-functionalities of pea protein concentrates. *Colloids and Surfaces A: Physicochemical and Engineering Aspects, 603*, 125156. https://doi.org/10.1016/j.colsurfa.2020.125156

Parada, J., & Aguilera, J. M. (2007). Food microstructure affects the bioavailability of several nutrients. *Journal of Food Science, 72*(2), R21—R32. https://doi.org/10.1111/j.1750-3841.2007.00274.x

Paximada, P., Howarth, M., & Dubey, B. N. (2021). Double emulsions fortified with plant and milk proteins as fat replacers in cheese. *Journal of Food Engineering, 288*, 110229.

Puppo, M. C., Speroni, F., Chapleau, N., de Lamballerie, M., Añón, M. C., & Anton, M. (2005). Effect of high-pressure treatment on emulsifying properties of soybean proteins. *Food Hydrocolloids, 19*(2), 289—296. https://doi.org/10.1016/j.foodhyd.2004.07.001

Ramírez-Rodrigues, M. M., Metri-Ojeda, J. C., González-Ávila, M., Ruiz-Álvarez, B. E., & Baigts-Allende, D. K. (2022). Digestibility and bioaccessibility of leaf protein concentrates and their impact on children gut microbiota. *Waste and Biomass Valorization, 13*(1), 299—314. https://doi.org/10.1007/s12649-021-01521-y

Reynaud, Y., Lopez, M., Riaublanc, A., Souchon, I., & Dupont, D. (2020). Hydrolysis of plant proteins at the molecular and supra-molecular scales during in vitro digestion. *Food Research International, 134*, 109204. https://doi.org/10.1016/j.foodres.2020.109204

Roland, W. S. U., Pouvreau, L., Curran, J., van de Velde, F., & de Kok, P. M. T. (2017). Flavor aspects of pulse ingredients. *Cereal Chemistry, 94*(1), 58–65. https://doi.org/10.1094/CCHEM-06-16-0161-FI

Rühs, P. A., Affolter, C., Windhab, E. J., & Fischer, P. (2013). Shear and dilatational linear and nonlinear subphase controlled interfacial rheology of β-lactoglobulin fibrils and their derivatives. *Journal of Rheology, 57*(3), 1003–1022. https://doi.org/10.1122/1.4802051

Sathe, S. K. (2012). Protein solubility and functionality. *Food Proteins and Peptides: Chemistry, Functionality, Interactions, and Commercialization*, 95–124.

Schopf, M., & Scherf, K. A. (2021). Water absorption capacity determines the functionality of vital gluten related to specific bread volume. *Foods, 10*(2), 228. https://www.mdpi.com/2304-8158/10/2/228.

Schreuders, F. K. G., Sagis, L. M. C., Bodnár, I., Erni, P., Boom, R. M., & van der Goot, A. J. (2021). Mapping the texture of plant protein blends for meat analogues. *Food Hydrocolloids*, 118. https://doi.org/10.1016/j.foodhyd.2021.106753

Serrano, M. P., Rebollar, P. G., Sueiro, S., Hermida, M., & Mateos, G. G. (2013). Influence of duration of storage on protein quality traits of soybean meals. *Journal of Applied Poultry Research, 22*(3), 423–429. https://doi.org/10.3382/japr.2012-00620

Sharan, S., Zanghelini, G., Zotzel, J., Bonerz, D., Aschoff, J., Saint-Eve, A., & Maillard, M.-N. (2021). Fava bean (Vicia faba L.) for food applications: From seed to ingredient processing and its effect on functional properties, antinutritional factors, flavor, and color. *Comprehensive Reviews in Food Science and Food Safety, 20*(1), 401–428. https://doi.org/10.1111/1541-4337.12687

Shen, L., & Tang, C.-H. (2012). Microfluidization as a potential technique to modify surface properties of soy protein isolate. *Food Research International, 48*(1), 108–118. https://doi.org/10.1016/j.foodres.2012.03.006

Shi, A., Feng, X., Wang, Q., & Adhikari, B. (2020). Pickering and high internal phase pickering emulsions stabilized by protein-based particles: A review of synthesis, application and prospective. *Food Hydrocolloids*, 106117.

Shukla, T. P. (1982). Chemical modification of food proteins. In *Food protein deterioration* (pp. 275–300). American Chemical Society. https://doi.org/10.1021/bk-1982-0206.ch011. Vol. 206.

Sikorski, Z. E. (2001). Functional properties of proteins in food systems. In *Chemical and functional properties of food proteins* (pp. 113–135). CRC Press. https://doi.org/10.1201/9781482279047

Silva, J. V., Cochereau, R., Schmitt, C., Chassenieux, C., & Nicolai, T. (2019). Heat-induced gelation of mixtures of micellar caseins and plant proteins in aqueous solution. *Food Research International, 116*, 1135–1143.

Singhal, A., Karaca, A. C., Tyler, R., & Nickerson, M. (2016). Pulse proteins: From processing to structure-function relationships. *Grain Legumes*, 55.

Smith, C. R., Earle, F. R., Wolff, I. A., & Jones, Q. (1959). Seed protein solubility, comparison of solubility characteristics of selected seed proteins. *Journal of Agricultural and Food Chemistry, 7*(2), 133–136. https://doi.org/10.1021/jf60096a011

Sopade, P. A., & Obekpa, J. A. (1990). Modelling water absorption in soybean, cowpea and peanuts at three temperatures using peleg's equation. *Journal of Food Science, 55*(4), 1084–1087. https://doi.org/10.1111/j.1365-2621.1990.tb01604.x

Stone, A. K., Karalash, A., Tyler, R. T., Warkentin, T. D., & Nickerson, M. T. (2015). Functional attributes of pea protein isolates prepared using different extraction methods and cultivars. *Food Research International, 76*, 31–38. https://doi.org/10.1016/j.foodres.2014.11.017

Taha, A., Ahmed, E., Hu, T., Xu, X., Pan, S., & Hu, H. (2019). Effects of different ionic strengths on the physicochemical properties of plant and animal proteins-stabilized emulsions fabricated using ultrasound emulsification. *Ultrasonics Sonochemistry, 58*, 104627. https://doi.org/10.1016/j.ultsonch.2019.104627

Tang, M., Zhu, Y., Li, D., Adhikari, B., & Wang, L. (2019). Rheological, thermal and microstructural properties of casein/κ-carrageenan mixed systems. *Lwt, 113*. https://doi.org/10.1016/j.lwt.2019.108296

Tarté, R., Egbert, W. R., & Payne, C. T. (2009). Plant proteins. In *Ingredients in meat products: Properties, Functionality and applications* (pp. 111–129). New York: Springer. https://doi.org/10.1007/978-0-387-71327-4_5

Verhoeckx, K., Cotter, P., López-Expósito, I., Kleiveland, C., Lea, T., Mackie, A., Requena, T., Swiatecka, D., Wichers, H., & Minekus, M. (2015). The TNO gastro-intestinal model (TIM). In *The Impact of food Bioactives on health: In vitro and ex vivo models* (pp. 37–46). Springer International Publishing. https://doi.org/10.1007/978-3-319-16104-4_5

Volkert, M. A., & Klein, B. P. (1979). Protein dispersibility and emulsion characteristics of four soy products. *Journal of Food Science, 44*(1), 93–96. https://doi.org/10.1111/j.1365-2621.1979.tb10014.x

Völp, A. R., Seitz, J., & Willenbacher, N. (2021). Structure and rheology of foams stabilized by lupin protein isolate of Lupinus.angustifolius. *Food Hydrocolloids*, 120. https://doi.org/10.1016/j.foodhyd.2021.106919

Von Der Haar, D., Müller, K., Bader-Mittermaier, S., & Eisner, P. (2014). Rapeseed proteins—Production methods and possible application ranges. *Ocl, 21*(1), D104.

Wang, K., & Arntfield, S. D. (2017). Effect of protein-flavour binding on flavour delivery and protein functional properties: A special emphasis on plant-based proteins. *Flavour and Fragrance Journal, 32*(2), 92–101.

Wang, W., Li, J., Yan, L., Huang, G., & Dong, Z. (2016). Effect of oxidization and chitosan on the surface activity of soy protein isolate. *Carbohydrate Polymers, 151*, 700–706.

Wang, Y., & Selomulya, C. (2020). Spray drying strategy for encapsulation of bioactive peptide powders for food applications. *Advanced Powder Technology, 31*(1), 409–415. https://doi.org/10.1016/j.apt.2019.10.034

Wang, Q., & Xiong, Y. L. (2019). Processing, nutrition, and functionality of hempseed protein: A review. *Comprehensive Reviews in Food Science and Food Safety, 18*(4), 936–952. https://doi.org/10.1111/1541-4337.12450

Wang, B., Zhang, Q., Zhang, N., Bak, K. H., Soladoye, O. P., Aluko, R. E., Fu, Y., & Zhang, Y. (2021). Insights into formation, detection and removal of the beany flavor in soybean protein. *Trends in Food Science & Technology, 112*, 336–347. https://doi.org/10.1016/j.tifs.2021.04.018

Wang, Z., Zhang, L., Zhang, X., Zeng, M., He, Z., & Chen, J. (2021). Interfacial rheology and foaming properties of soy protein and hydrolysates under acid condition. *Food Biophysics, 16*(4), 484–491. https://doi.org/10.1007/s11483-021-09685-9

Wang, J., Zhao, M., Yang, X., & Jiang, Y. (2006). Improvement on functional properties of wheat gluten by enzymatic hydrolysis and ultrafiltration. *Journal of Cereal Science, 44*(1), 93–100.

Wang, Y., & Selomulya, C. (2022). Food rheology applications of large amplitude oscillation shear (LAOS). *Trends in Food Science & Technology, 127*, 221–244. https://doi.org/10.1016/j.tifs.2022.05.018

Wan, J., Li, D., Song, R., Shah, B. R., Li, B., & Li, Y. (2017). Enhancement of physical stability and bioaccessibility of tangeretin by soy protein isolate addition. *Food Chemistry, 221*, 760–770. https://doi.org/10.1016/j.foodchem.2016.11.106

Warnakulasuriya, S. N., & Nickerson, M. T. (2018). Review on plant protein–polysaccharide complex coacervation, and the functionality and applicability of formed complexes. *Journal of the Science of Food and Agriculture, 98*(15), 5559–5571.

Webb, M., Naeem, H., & Schmidt, K. (2002). Food protein functionality in a liquid system: A comparison of deamidated wheat protein with dairy and soy proteins. *Journal of Food Science, 67*(8), 2896–2902.

Wittek, P., Zeiler, N., Karbstein, H. P., & Emin, M. A. (2020). Analysis of the complex rheological properties of highly concentrated proteins with a closed cavity rheometer. *Applied Rheology, 30*(1), 64–76. https://doi.org/10.1515/arh-2020-0107

Wong, H.-W., Choi, S.-M., Phillips, D. L., & Ma, C.-Y. (2009). Raman spectroscopic study of deamidated food proteins. *Food Chemistry, 113*(2), 363–370.

Wu, C., Wang, T., Ren, C., Ma, W., Wu, D., Xu, X., Wang, L., & Du, M. (2021). Advancement of food-derived mixed protein systems: Interactions, aggregations, and functional properties. *Comprehensive Reviews in Food Science and Food Safety, 20*(1), 627–651.

Xia, W., Siu, W. K., & Sagis, L. M. C. (2021). Linear and non-linear rheology of heat-set soy protein gels: Effects of selective proteolysis of β-conglycinin and glycinin. *Food Hydrocolloids*, 120. https://doi.org/10.1016/j.foodhyd.2021.106962

Xiong, T., Xiong, W., Ge, M., Xia, J., Li, B., & Chen, Y. (2018). Effect of high intensity ultrasound on structure and foaming properties of pea protein isolate. *Food Research International, 109*, 260–267.

Xu, X., Luo, L., Liu, C., & McClements, D. J. (2017). Utilization of anionic polysaccharides to improve the stability of rice glutelin emulsions: Impact of polysaccharide type, pH, salt, and temperature. *Food Hydrocolloids, 64*, 112–122. https://doi.org/10.1016/j.foodhyd.2016.11.005

Yada, R. Y., Aryee, A. N. A., Agyei, D., & Udenigwe, C. C. (2018). 2 - impact of processing on the chemistry and functionality of food proteins. In *Proteins in food processing* (2nd ed., pp. 27–45). Woodhead Publishing. https://doi.org/10.1016/B978-0-08-100722-8.00003-6

Yalcin, E., Sakiyan, O., Sumnu, G., Celik, S., & Koksel, H. (2008). Functional properties of microwave-treated wheat gluten. *European Food Research and Technology, 227*(5), 1411–1417.

Yan, X., Ma, C., Cui, F., McClements, D. J., Liu, X., & Liu, F. (2020). Protein-stabilized Pickering emulsions: Formation, stability, properties, and applications in foods. *Trends in Food Science & Technology, 103*, 293–303.

Yu, P., Low, M. Y., & Zhou, W. (2018). Design of experiments and regression modelling in food flavour and sensory analysis: A review. *Trends in Food Science & Technology, 71*, 202–215. https://doi.org/10.1016/j.tifs.2017.11.013

Zembyla, M., Liamas, E., Andablo-Reyes, E., Gu, K., Krop, E. M., Kew, B., & Sarkar, A. (2021). Surface adsorption and lubrication properties of plant and dairy proteins: A comparative study. *Food Hydrocolloids, 111*, 106364.

Zhang, H., Claver, I. P., Zhu, K.-X., & Zhou, H. (2011). The effect of ultrasound on the functional properties of wheat gluten. *Molecules, 16*(5), 4231–4240.

Zhang, J., Kang, D., Zhang, W., & Lorenzo, J. M. (2021). Recent advantage of interactions of protein-flavor in foods: Perspective of theoretical models, protein properties and extrinsic factors. *Trends in Food Science & Technology, 111*, 405–425. https://doi.org/10.1016/j.tifs.2021.02.060

Zhang, L., Wang, L.-J., Jiang, W., & Qian, J.-Y. (2017). Effect of pulsed electric field on functional and structural properties of canola protein by pretreating seeds to elevate oil yield. *LWT, 84*, 73–81. https://doi.org/10.1016/j.lwt.2017.05.048

Zhang, S. B., Yan, D. Q., Jiang, Y. S., & Ding, C. H. (2021). Competitive displacement of interfacial soy proteins by Tween 20 and its effect on the physical stability of emulsions. *Food Hydrocolloids*, 113. https://doi.org/10.1016/j.foodhyd.2020.106515

Zhang, X., Zhou, J., Chen, J., Li, B., Li, Y., & Liu, S. (2020). Edible foam based on pickering effect of bacterial cellulose nanofibrils and soy protein isolates featuring interfacial network stabilization. *Food Hydrocolloids, 100*, 105440. https://doi.org/10.1016/j.foodhyd.2019.105440

Zhao, H., Shen, C., Wu, Z., Zhang, Z., & Xu, C. (2020). Comparison of wheat, soybean, rice, and pea protein properties for effective applications in food products. *Journal of Food Biochemistry, 44*(4), e13157.

Zhao, M., Xiong, W., Chen, B., Zhu, J., & Wang, L. (2020). Enhancing the solubility and foam ability of rice glutelin by heat treatment at pH12: Insight into protein structure. *Food Hydrocolloids, 103*, 105626. https://doi.org/10.1016/j.foodhyd.2019.105626

Zhu, Q., Lu, H., Zhu, J., Zhang, M., & Yin, L. (2019). Development and characterization of pickering emulsion stabilized by zein/corn fiber gum (CFG) complex colloidal particles. *Food Hydrocolloids, 91*, 204–213.

Zhu, X.-F., Zhang, N., Lin, W.-F., & Tang, C.-H. (2017). Freeze-thaw stability of pickering emulsions stabilized by soy and whey protein particles. *Food Hydrocolloids, 69*, 173–184. https://doi.org/10.1016/j.foodhyd.2017.02.001

Zou, Y., Guo, J., Yin, S.-W., Wang, J.-M., & Yang, X.-Q. (2015). Pickering emulsion gels prepared by hydrogen-bonded zein/tannic acid complex colloidal particles. *Journal of Agricultural and Food Chemistry, 63*(33), 7405–7414.

Section II

Plant-based dairy alternatives

Chapter 6

Plant-based beverages

Malik Adil Nawaz[1], Roman Buckow[2], Lita Katopo[1] and Regine Stockmann[1]
[1]*Commonwealth Scientific and Industrial Research Organisation (CSIRO), Agriculture and Food, Werribee, VIC, Australia;* [2]*The University of Sydney, Centre for Advanced Food Engineering, School of Chemical and Biomolecular Engineering, Darlington, NSW, Australia*

6.1 Introduction

Human health and regulation of physiological needs are maintained by food (Eertmans et al., 2001), as it provides nutritional contents, minerals, and other bioactive compounds to the human body. Human food predominantly comes from animals such as chicken, beef, fish, lamb, and pork or from plants like fruits, beans, vegetables, grains, and nuts. The increased consumption of food products of animal origin, especially in developed countries, has significantly increased the global demand for cereals, which is one of the main ingredients of animal feed (Salami et al., 2019). However, plant proteins (65%–90%) fed to livestock are lost due to the low efficacy of feed conversion ratio (Mottet et al., 2017). Animal-based foods have a major share in the environmental impacts of the agro-food sector as animal proteins require approximately 100-fold more water compared to that required in the production of the same amount of plant proteins (Jeske et al., 2018a). Beef and dairy products are also implicated in greenhouse gas emissions due to the high methane emissions of *Bos taurus* (Tan et al., 2021).

Generally, bovine milk and associated milk products are associated with several concerns based on health, nutrition, safety, or other dietary requirements. Usually, these concerns are divided into distinctive groups such as chemical (contaminants from the environment), physical (improper handling of milk), biological (intolerance, e.g., lactose intolerance), allergens (cow's milk allergy), toxins (pesticide residues), and pathogenicity), and other issues (religious and ethical concerns) (Crittenden & Bennett, 2005; Ishaq & Nawaz, 2018; Lifschitz & Szajewska, 2015; Vesa et al., 2000). On the other hand, plant-based beverages are usually considered best suited for a group of the population having cow's milk allergy, lactose intolerance, and coronary cardiac diseases. A considerable increase in the consumption of plant-based beverages has been recorded recently with health and sustainability concerns being major drivers. The share of plant-based beverages has increased by 61% in the United States between 2012 and 2019 (Mcclements et al., 2019). According to a survey, commonly consumed plant-based beverages in the United States are almond (64% market share), soy (13% market share), and coconut (12% market share). It was also observed that the sale of new plant-based beverages also increased due to high customer satisfaction; for instance, the sale of pecan- and quinoa-based beverages in 2017 was 63% and 58% higher than the previous year, respectively (Mintel, 2018). The current global market of plant-based beverages is valued at USD 22.6 billion in 2020 and is projected to grow up to USD 40.6 billion by 2026, at a compound annual growth rate (CAGR) of 10.3%, as shown in Fig. 6.1 (Markets & Markets, 2021).

Several studies have been conducted in the past on the nutritional quality and colloidal stability of plant-based beverages; however, most of them were focused on soy-based beverages (Jiang et al., 2013). Exploration of other plant-based beverages due to the increasing popularity and displacement of soy products in the market is gaining research interest (Nawaz, Tan, et al., 2020b). Scientists have assessed the efficacy of various legumes, cereals, oilseeds, pseudocereals, and nuts sources for their functional attributes due to the presence of health-promoting nutrients and bioactive components, and these plant sources can be used to replace not only bovine milk but also soy-based beverages (Munekata et al., 2020). However, several technological challenges such as off-flavor and colloidal instability remain to be solved to satisfy increasingly discerning consumers. This chapter provides an overview of the current knowledge on fundamental processing steps required to convert plant material into plant-based beverages, what are the unique processing challenges for different plant sources, how to best overcome these challenges and potential quality deficiencies, and what are the opportunities to maximize nutritional and sensory aspects of plant-based beverages.

Engineering Plant-Based Food Systems. https://doi.org/10.1016/B978-0-323-89842-3.00015-4

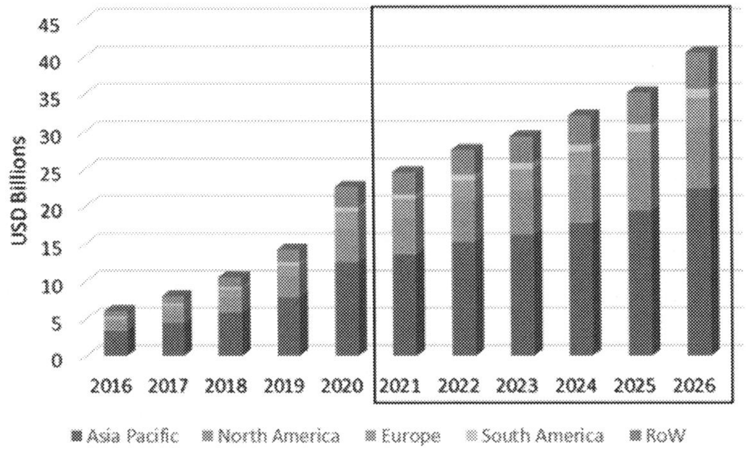

FIGURE 6.1 Plant-based beverages global market. The predicted market share for 2021–2026 is mentioned in box. *Adapted from Markets & Markets. (2021). Dairy alternatives market by source (soy, almond, coconut, oats, rice, hemp), application (milk, Yogurt, ice creams, cheese, creamers), distribution channel (supermarkets, health food stores, pharmacies), formulation, and region—Global forecast to 2026. https://www.marketsandmarkets.com/Market-Reports/dairy-alternative-plant-milk-beverages-market-677.html.*

6.2 Processing methods employed to manufacture plant-based beverages

Cow's milk has a complex structure and composition and contains distinct constituents viz., proteins, fats, carbohydrates and minerals, and colloidal assemblies viz., casein micelles and fat globules. This unique composition and structure of milk determine the functionality of milk and other dairy products. Therefore, the plant-based beverages should depict the desirable attributes of dairy such as appearance, texture, mouthfeel, and creaminess, which are usually stimulated by the milk fat globules (Jeske et al., 2018b). There are two different approaches used by industry to achieve this goal: (i) breakdown of plant material and extraction of natural oil bodies, which are further homogenized to achieve the particle size distribution ranging 5–20 μm to resemble bovine milk fat (Sethi et al., 2016) and (ii) preparation of O/W emulsions using vegetable oil and stabilization of this emulsion by plant proteins and addition of these emulsions to plant proteins to formulate the plant beverage (Felix et al., 2019).

6.2.1 Extraction of natural oil bodies from plant material

In this approach, oil bodies are extracted from plant material to formulate plant seed-based beverages, such as soy-, almond-, cashew-, or coconut-based beverages (Nikiforidis et al., 2014). Seeds are rich in fat, which is present in the form of oil bodies; therefore, beans and nuts are used to extract these oil bodies in legume- and nut-based beverages. A typical oil body has a triacylglycerol core, which is surrounded by a layer of protein and phospholipids (Tzen & Huang, 1992). The structure, composition, and particle size ($\sim$0.1 to a few μm) of these plant oil bodies are very similar to those of bovine milk fat globules (Gallier et al., 2012). These plant oil bodies can be easily extracted from seeds by soaking in water followed by wet grinding to disintegrate the surrounding tissues and release of oil bodies (Adams et al., 2012; Iwanaga et al., 2007). This isolation of oil bodies can be facilitated by modifying the pH and the addition of salts and enzymes. Isolated oil bodies are then dispersed in an adequate amount of water to prepare a beverage with viscosity similar to that of cow's milk. Also, the watery phase contains carbohydrates, proteins, fibers, and minerals that can also contribute to the sensory and nutritional attributes similar to the aqueous phase of milk.

Apart from physical similarities to bovine milk, plant seed-based beverages prepared with natural oil bodies have some important differences from milk. For instance, cow's milk is usually bland but has a distinct aroma and taste profile (Haug et al., 2007). On the other hand, plant seed-based beverages contain different types of fats and other molecules that result in a characteristic mouthfeel and flavor, which usually depends on the plant source; for example, legume-based beverages often have a beany flavor, whereas, nut-based beverages have a sweet nutty flavor (Aydar et al., 2020). The presence or absence of such distinct flavors of plant-based beverages is the major driving force for the product acceptability to consumers. Research has shown that consumers who consume plant-based beverages for reasons other than dairy intolerance often expect flavors and functionality similar to dairy milk. Therefore, dairy milk cannot always be replaced with plant-based milk analogs (Nawaz, Tan, et al., 2020b).

Specific steps of the industrial manufacturing process of plant seed-based beverages depend on the raw material and final product requirements. However, some of the basic steps frequently used in the creation of plant-based beverages from natural oil bodies from seeds are summarized in Fig. 6.2. Plant raw materials such as seeds or beans are usually soaked in water to soften the plant tissues, followed by wet grinding of soaked raw materials, inactivation of endogenous enzymes by blanching, centrifugation, and filtration to remove undissolved material, thermal treatment, fortification with additives viz., oil, thickeners, stabilizers, colors, flavors, other nutrients, etc., and packaging. It is noteworthy that compared with dairy milk where a large proportion of products sits in the refrigerated milk category, most of the plant-based beverages are sold in the nonrefrigerated long shelf-life category.

6.2.2 Preparation of O/W emulsions

Oil-in-water (O/W) emulsions are defined as thermodynamically unstable systems, which include two immiscible liquids (generally water and oil), in which oil is distributed into the water (Chen & Tao, 2005). Bovine milk is a typical example of O/W emulsion that contains protein-coated fat globules suspended in an aqueous fluid. Therefore, in the second approach, instead of extracting natural oil bodies from plant material, it is possible to prepare O/W emulsion using vegetable oil and plant proteins (Do et al., 2018). Vegetable oil and water are homogenized together in the presence of plant proteins (having emulsifying properties) and/or hydrophilic emulsifiers to prepare an O/W emulsion that resembles bovine milk in most of its physicochemical properties and sensory attributes (Amine et al., 2014). Important ingredients and main processing steps to prepare plant-based beverages using O/W emulsion technique are discussed in this section.

6.2.2.1 Ingredients

Vegetable oil, aqueous phase (water or buffers), plant proteins, and/or other emulsifiers, and other additives are the important ingredients of plant-based beverage formulation (McClements, 2020a). The functional properties of the final product depend on these ingredients; therefore, each ingredient must be carefully selected (Gumus et al., 2017).

6.2.2.1.1 Vegetable oil

A variety of vegetable oils are available that can be used to prepare O/W emulsions such as sunflower, canola, olive, coconut, sesame, linseed, corn, soybean, and palm oils. Every vegetable oil has its distinct fatty acid profile,

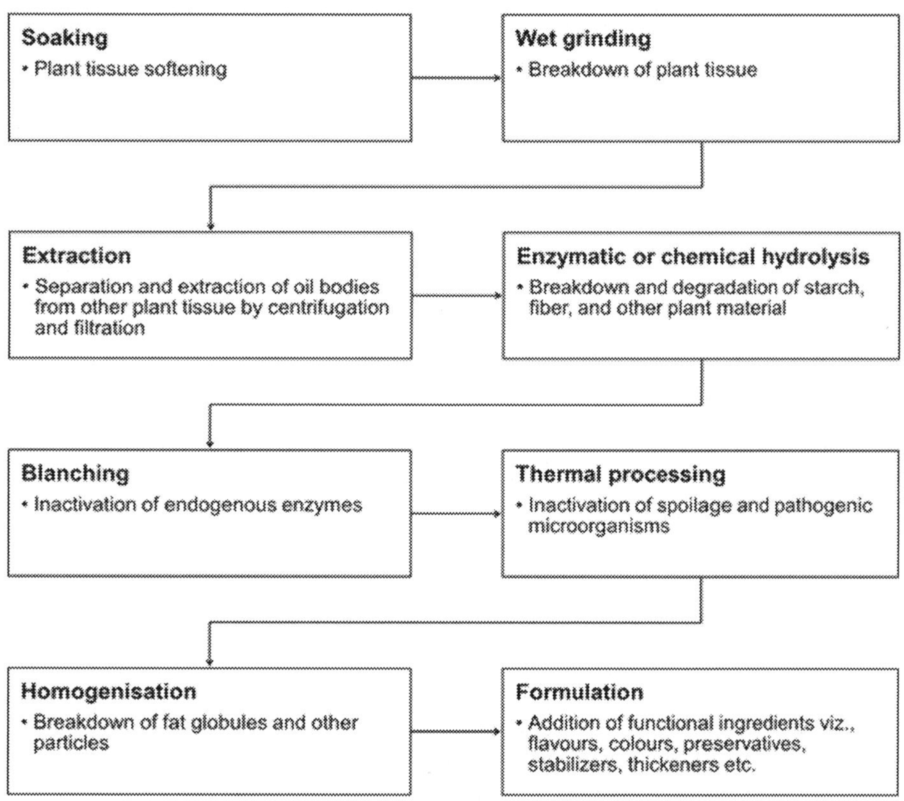

FIGURE 6.2 Some of the basic steps used in the creation of plant-based beverages from natural oil bodies. *Adapted from McHugh, T. (2018). How plant-based milks are processed.* Food Technology, *72, 63–64 and McClements, D. J., Newman, E., & McClements, I. F. (2019). Plant-based milks: A review of the science underpinning their design, fabrication, and performance.* Comprehensive Reviews in Food Science and Food Safety.

physiochemical properties, sensory, and nutritional attributes. For instance, palm and coconut oils are high in medium-chain saturated fatty acids; therefore, they are highly stable for lipid oxidation. However, there are a lot of health concerns associated with these oils that may promote cardiovascular diseases due to the high content of saturated fatty acids (Boateng et al., 2016). On the other hand, olive and linseed oils are rich in monounsaturated fatty acids such as oleic acid (ω-9) and polyunsaturated fatty acids mainly α-linoleic acid (ω-3), respectively, which make them more prone to lipid oxidation and thereby reduce the product stability but add more health benefits (Gorzynik-Debicka et al., 2018; Tripathi et al., 2013). Also, the physiochemical properties of vegetable oils such as glass transition properties, rheological properties, and interfacial tensions, etc. also impact the formation and colloidal stability of O/W emulsions. For example, the reduction in viscosity and interfacial tensions during homogenization results in reduced oil droplet size (Taha et al., 2018). Moreover, solid fats should be melted prior to homogenization to improve emulsion stability. Therefore, these attributes should be considered during emulsification.

6.2.2.1.2 Aqueous phase

The aqueous phase is also important as the quality of water used to formulate the O/W emulsions can affect the quality of plant-based drinks. Commercial sources of water usually vary in pH, minerals levels, and may contain organic matter, which can interact with other ingredients of the emulsion resulting in significant changes in the physicochemical properties of products (Navarini & Rivetti, 2010). Therefore, prior treatment of water viz., thermal processing, filtration, or reverse osmosis is important. In some high-quality products, buffers are used as the aqueous phase to improve the protein solubility, reduce coalescence and flocculation of fat globules, and improve the functional properties of the beverage.

6.2.2.1.3 Plant proteins and/or other emulsifiers

O/W emulsions of plant-based beverages are usually stabilized with plant protein and/or other emulsifiers including polysaccharides, phospholipids, and biosurfactants (Gumus et al., 2017). The selection of an appropriate emulsifier is important, as every emulsifier has different efficacy to form and stabilize emulsion, which determines the physicochemical properties of processed products (Dickinson, 2009). It is important for an emulsifier to form and stabilize the emulsion with the smallest mean droplet size at minimum concentration.

As far as emulsion stability is concerned, the efficacy of emulsifiers is usually rated on the ability to retard the droplet aggregation during emulsion storage when the product is exposed to environmental stresses viz., changes in pH, temperature, and ionic strength. For instance, legume proteins form emulsions that are unstable near isoelectric points (pH 4–5), high temperatures (>70°C), and high ionic strengths (>50 mM) (Ladjal Ettoumi et al., 2016). By contrast, emulsions stabilized by plant-based polysaccharides (such as gums and lecithin) and saponins are usually stable over most of these conditions (McClements & Gumus, 2016). For example, data on the change of droplet charge and mean particle diameter of O/W emulsions at different pH (2–8) of O/W emulsions stabilized by various plant-based emulsifiers showed that emulsions prepared by saponins, lecithin, and gum were more stable (Fig. 6.3).

Typically, plant-based emulsifiers are used individually; however, the functional performance of plant-based emulsifiers can be improved by using a combination of two or more emulsifiers in the emulsion system (Lv et al., 2020). This can be achieved either by simply adding and allowing selected emulsifiers to stabilize the droplet surfaces (Salminen et al., 2019) or by using various chemical and physical techniques to preform covalent or noncovalent mixed molecule entities prior to emulsion formation such as (i) creation of covalent conjugates by using Maillard reaction (Zha et al., 2019) and (ii) physical complexes formed by using electrostatic attraction (Li & de Vries, 2018). Multiple plant-based emulsifiers can be mixed together before homogenization or sequential adsorption after homogenization to form multilayer systems. For example, fat globules can be first coated with a charged plant-based emulsifier followed by adsorption of oppositely charged emulsifier on the top (Burgos-Díaz et al., 2016). This process should be carefully repeated multiple times to create multiple layers around the fat globules without aggregation. Also, the economic viability of the process for emulsion formulation and creation at a commercial scale should be ensured (Guzey & McClements, 2006).

6.2.2.1.4 Other additives

Usually, various types of other additives are also incorporated in O/W emulsions drinks to improve the physicochemical properties, nutritional value, and sensory properties (Nawaz, Tan, et al., 2020b). For instance, thickening agents such as plant-based gums (guar gum, gum arabic, etc.), plant hydrocolloids (starch, pectin, etc.), seaweed hydrocolloids (carrageenan, alginate, etc.), or microbial hydrocolloids (xanthan gum) are usually added to the products to improve the rheological properties and mouthfeel and retard the cream separation of the products during storage (Silva et al., 2020).

Studies showed that the incorporation of electrically charged hydrocolloids in O/W emulsion could stabilize the oil droplets by forming a protective coating around them and restricting the flocculation and coalescence (Chuah et al., 2009;

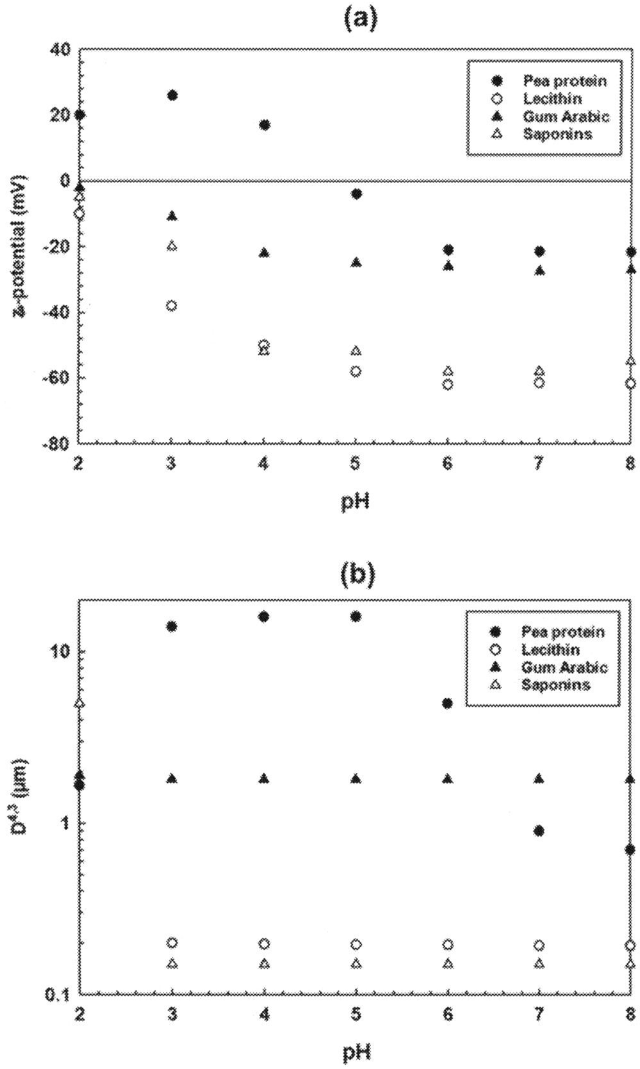

FIGURE 6.3 Change in (A) droplet charge (ʓ-potential) and (B) mean particle diameter (D4,3) with pH for different kinds of plant-based emulsifiers viz., pea proteins, lecithin, gum Arabic, and saponins. *Data adapted from McClements, D. J. & Gumus, C. E. (2016). Natural emulsifiers—Biosurfactants, phospholipids, biopolymers, and colloidal particles: Molecular and physicochemical basis of functional performance.* Advances in Colloid and Interface Science, 234, *3–26 and McClements, D. J., Newman, E. & McClements, I. F. (2019). Plant-based milks: A review of the science underpinning their design, fabrication, and performance.* Comprehensive Reviews in Food Science and Food Safety.

Lutz et al., 2009). Xu et al. (2017) found that the stability of O/W emulsions improved when they were stabilized with hydrolyzed rice glutelin and anionic polysaccharides such as xanthan gum and pectin. These anionic polysaccharides may be adsorbed to the cationic patches on the surface of emulsion hydrocolloids, thereby inhibiting them to aggregation near their isoelectric points by generating strong steric or electrostatic repulsive forces (Qiu et al., 2015).

6.2.2.2 Processing operations

Homogenization is the most widely used process in the preparation of O/W emulsions. Various mechanical devices viz., high shear mixers, colloid mills, high and ultra-high-pressure homogenizers, ultrasound homogenizers, and microfluidizers may be used in the industry to manufacture emulsions from vegetable oil, aqueous phase, and plant proteins/emulsifiers (Fig. 6.4). Traditionally, these devices have been used to homogenize raw bovine milk. Each of these mechanical homogenization devices has its pros and cons regarding fixed and running cost, maintenance, energy requirements, safety, rate of production, and capability to produce microemulsion with narrow particle size distribution (Stang et al., 2001). For instance, microfluidizers and ultrasound homogenizers are very efficient in producing microemulsion and more stable products; however, they are usually not preferred in the industry due to higher production and maintenance costs

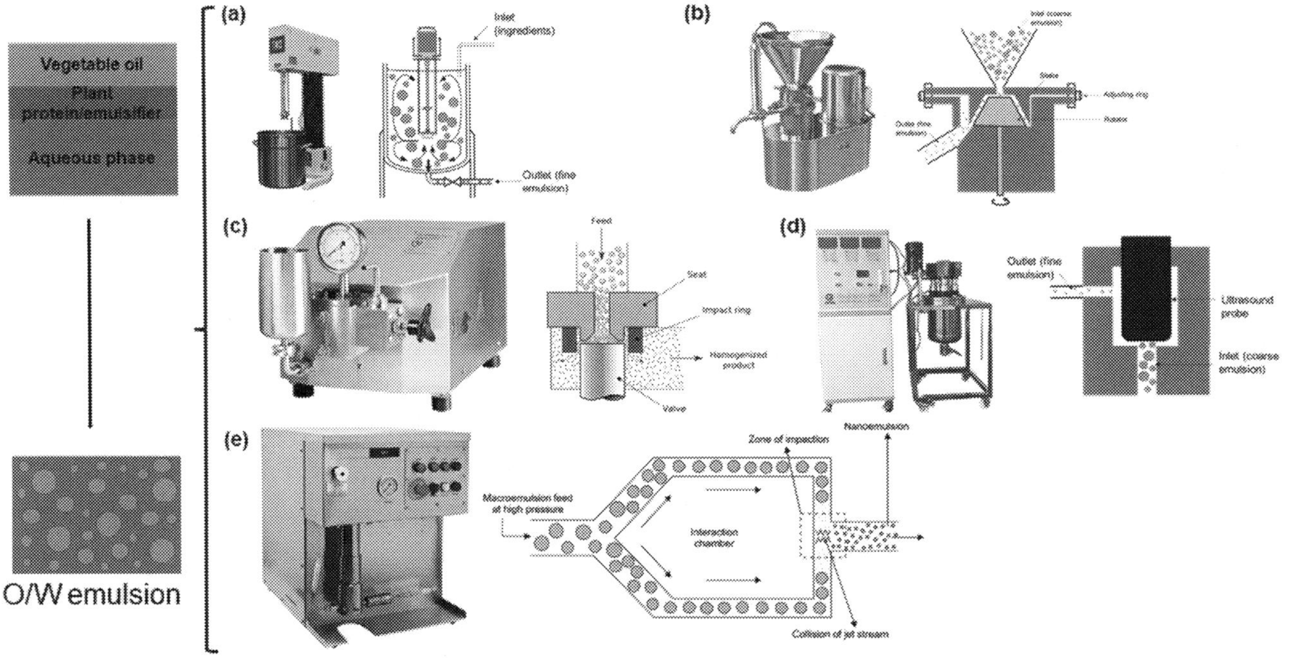

FIGURE 6.4 Schematic representation of some of the mechanical devices commonly used to prepare plant-based O/W emulsion; (A) high shear mixer, (B) colloidal mill, (C) high-pressure homogenizer, (D) ultrasound homogenizer, and (E) microfluidizer. *Adapted from Rayner, M. (2015). Scales and forces in emulsification. Engineering aspects of food emulsification and homogenization. In M. Rayner & P. Dejmek (Eds.), 3−32 and Narsimhan, G., Wang, Z., & Xiang, N. (2019). Guidelines for processing emulsion-based foods. In G. L. Hasenhuettl & R. W. Hartel (Eds.),* Food emulsifiers and their applications. *Springer International Publishing.*

(Pinnamaneni et al., 2003; Taha et al., 2020). Moreover, ultrasound homogenizers are often considered unsafe to use on a commercial scale. Therefore, high-pressure homogenizers are the most common homogenizers used in industry to prepare plant-based beverages due to less production and maintenance costs and are capable of producing fairly stable products at a homogenizer reasonable production rate (Lee & Norton, 2013).

After formulation and production, plant-based beverages are thermally sterilized to increase the shelf life by killing spoilage micro-organisms and inactivating enzymes (Quan et al., 2020). Oil droplets are coated with plant proteins and other emulsifiers, which are prone to aggregation during thermal processing resulting in reduced colloidal stability. Various studies have been conducted in the past to evaluate the efficacy of various plant proteins/emulsifiers and homogenization techniques on the colloidal stability of thermally treated plant-based O/W emulsion drinks (Nawaz, Singh, et al., 2020a; Qamar et al., 2019). These studies concluded that emulsifiers, as well as homogenization techniques, have a significant effect on emulsion stability. Therefore, to avoid these types of challenges, the appropriate selection of homogenization techniques is a crucial part of the production design for these products (Hebishy et al., 2015).

6.3 Plant-based beverages currently available

Plant-based beverages are generally classified into five broad categories based on their origin viz., legume-, nut-, cereal-, pseudocereal-, and seed-based beverages as shown in Fig. 6.5.

6.3.1 Legume-based beverages

6.3.1.1 Soy-based beverages

According to literature, people in China began making plant-based milks from soybean (*Glycine* max L.) around 2000 years ago (Jeske et al., 2018b). Among all legume-based milk beverages, the soy-based drink was the main avenue for obtaining nutrients wherever there was a shortage of cow's milk (Sethi et al., 2016). People with lactose intolerance and/or milk allergies also favor soy-based beverages (Hodges et al., 2019). The beverage contains many bioactive components such as essential fatty acids and isoflavones. Essential fatty acids, in the form of mono- and poly-unsaturated, are known to benefit cardiovascular health (Hasler, 2002). Isoflavones are categorized into two types, i.e., aglycones and glucosidic, with

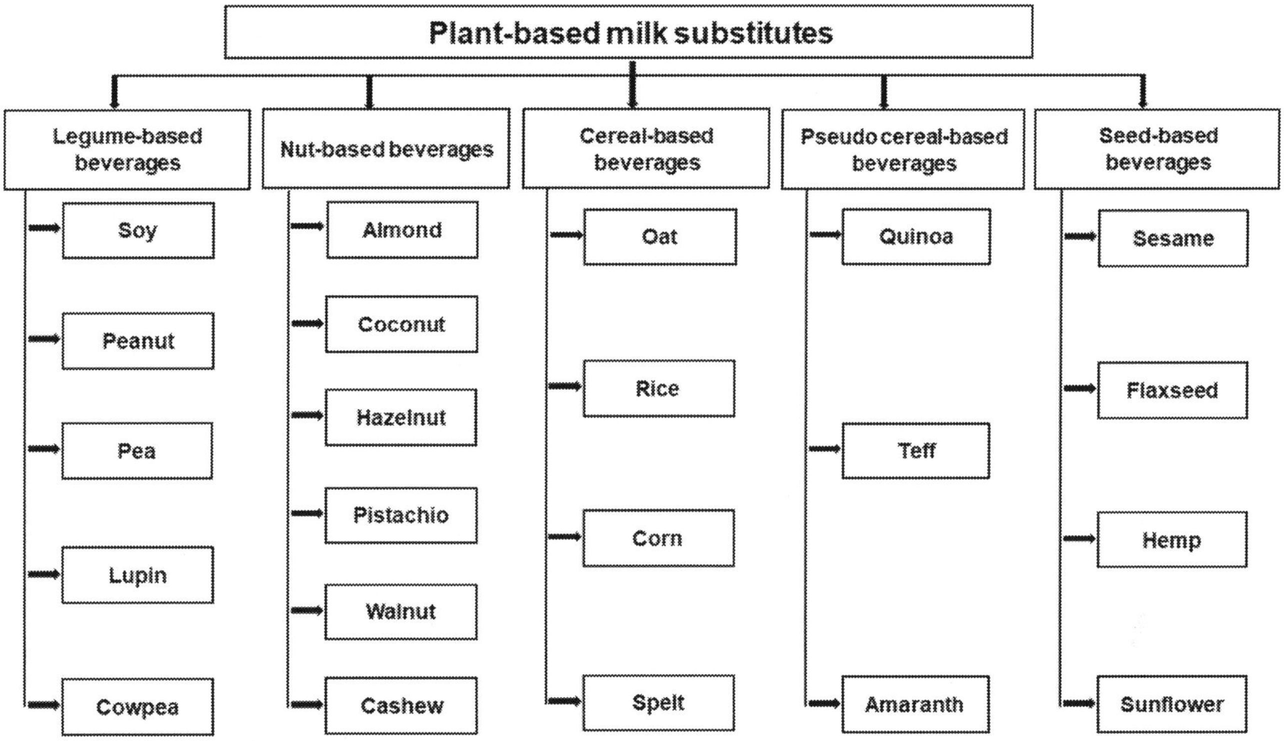

FIGURE 6.5 Various sources of plant-based beverages (Nawaz, Tan, et al., 2020b). *Data from Nawaz, M. A., Tan, M., Oiseth, S., & Buckow, R. (2020). An emerging segment of functional legume-based beverages: A review.* Food Reviews International. *https://doi.org/10.1080/87559129.2020.1762641.*

the former includes mainly daidzein and genistein and the latter comprised of primarily genistin and daidzin (Friedman, 1996). Genistein constitutes the largest fraction of isoflavones in soybean and is considered the most biologically active soy isoflavone (Cohen et al., 2000). Table 6.1 summarizes other potential health advantages of soybeans. Studies have documented the protective and therapeutic effects of isoflavones and proteins from soybean on age-related conditions, such as Alzheimer's disease and osteoporosis (Omoni & Aluko, 2005; Xiao, 2008). Moreover, phytochemicals, e.g., phytosterols, are found abundantly in soy-based foodstuff, and these compounds possess cholesterol-lowering properties (Khatoon et al., 2010). However, consumption of soy-based food may elicit allergic reactions in sensitive people due to the presence of certain proteins, namely, β-conglycinin and glycinin. Those reactions affect the intestinal tract, skin, and respiratory system in vulnerable individuals. Recent animal trials conducted by Peng et al. (2018) concluded that β-conglycinin and glycinin caused intestinal damage in piglet models. There are many varieties of soy-based beverages available on the market owing to their popularity and they can be sorted by: (i) solid concentration (light, milk-like, and rich); (ii) formulation (original, flavored, and sweetened); and (iii) fortification (regular, blended, and enriched) (Kumar et al., 2016). An undesirable beany flavor and a short shelf-life are typical drawbacks associated with soy-based beverages prepared in the traditional or conventional method viz., isolation of oil bodies from endosperm tissues. However, those limitations can be addressed with advanced processing techniques shown in Table 6.2. A number of approaches have been developed to remove off flavor in soy-based milk substitutes with the most commonly used methods include (i) application of vacuum at an elevated temperature to remove volatile compounds (e.g., sterols and short-chain fatty acids) (Yoo & Chang, 2016); (ii) Cornell hot grinding method, which involves soaking and crushing soybeans in steam or boiling water, and heating the hot slurry to inactivate lipoxygenase (Nawaz, 2018); and (iii) Illinois preblanching method, which involves soaking of soybeans followed by blanching (to inactivate lipoxygenase) and grinding (Nelson et al., 1976). Further, application of soy protein isolate (Damodaran & Arora, 2013), removing fat from soy flour (Hsieh et al., 1982), and soaking of soybeans in ethanol and alkaline environment (Ashraf & Snyder, 1981; Khaleque et al., 1970) are examples of method that can be used to partly mask or reduce the beany flavor in soy milk substitutes. In addition, new soybean varieties with the lipoxygenase gene knocked out have been developed and off-flavor development during the processing of such beans is markedly reduced. Finally, lot of soy-based drinks also incorporate other ingredients, for instance, flavors and food gums, to either imitate the taste of dairy milk or mask the undesirable beany flavor.

TABLE 6.1 Functional/bioactive components, potential health benefits of various plant sources.

Plant sources	Functional/bioactive components	Potential health benefits
Soybean	Isoflavones phytosterols	• Protection against age-related diseases such as osteoporosis (Wei et al., 2012) and Alzheimer's disease (Hsieh et al., 2009). • Cardiovascular disease (Sathyapalan et al., 2018). • Hormone-dependent cancer (Amaral et al., 2017). • Cognitive function (Cui et al., 2018). • Cholesterol reduction (Isanga & Zhang, 2008).
Peanut	Phenolic compounds	• Natural source of resveratrol (Wang et al., 2005). • Anti-inflammatory (Chang et al., 2006). • Coronary heart disease, strokes, and various cancers (Toomer, 2018; Wien et al., 2014).
Pea	Prebiotic oligosaccharides phenolic compounds	• Improved gastrointestinal function (Dahl et al., 2012). • Anticarcinogenic and anti-inflammatory (Dueñas et al., 2004).
Almonds	Phenolic and flavonoids compounds Arabinose	• Hypocholesterolemic benefits. • Glucose regulation (Kamil & Chen, 2012). • Cell wall pectic substances and has potential prebiotic properties by lowering blood serum cholesterol level (Grundy et al., 2016).
Coconut	Medium-chain triglycerides and medium-chain fatty acids (MCFAs)	• Improved digestibility. • Potential treatment of brain impairment conditions such as Alzheimer's disease (Chatterjee et al., 2020; Karunasiri et al., 2020).
Oat	Antioxidants and polyphenols viz., avenanthramides, saponins (avenacoside A and avenacoside B) β-glucan	• Anticancerous potential by reducing the causative agents of colon cancer along with significant reduction in blood cholesterol level blood pressure (Mathews et al., 2020).
Rice	Hydroxycinnamic acids viz., ferulic acid, sinapic acid, and p-coumaric acid	• Rice hydroxycinnamic acids have antihyperglycemic and antioxidant properties (Haldipur & Srividya, 2021).
Quinoa	High-quality protein with essential amino acids, dietary fiber, and polyphenols.	• Low glycemic index. • Blood pressure regulation. • Reduced risk of coronary heart diseases. • Anticarcinogenic properties (Navruz-Varli & Sanlier, 2016).
Sesame	Sesamine Sesamolin	Bio-functional component (CBDA) provides health benefits against chronic diseases such as; • Obesity • Inflammatory bowel disease • Rheumatoid arthritis • Alzheimer's disease (Vigli et al., 2021).
Hemp	Cannabidiolic acid (CBDA) Lignanamides viz., cannabisin M, cannabisin N, cannabisin O, and 3,3′-demethyl-heliotropamodes	• Anti-neuroinflammatory properties due to inhibition of nuclear factor kappa-light-chain-enhancer of the activated B cells (NF-κB) signaling pathway (Leonard et al., 2020).

Apart from the organoleptic obstacle, soybeans contain trypsin inhibitors that could adversely affect trypsin in digesting proteins (Liu, 2019). Trypsin inhibitors can be inactivated with heat treatment such as blanching (Yuan et al., 2008) and ultra-high-temperature process (Kwok et al., 1993). However, such methods could also pose other processing challenges as they could induce protein denaturation and aggregation leading to sedimentation and reduced solubility (Zhang et al., 2005)]. As a result, nonthermal processes, e.g., pulsed electric field (PEF), have emerged as an attractive option in preparing soy-based beverages with minimal heat-induced changes (Vagadia et al., 2017).

6.3.1.2 Peanut-based beverages

Peanuts are one of the most promising sources of legumes or oilseeds for preparing nondairy drinks (Massey, 2003). In developing countries, peanut-based beverages are popular milk alternatives among vegans, low-income communities, malnourished infants, and populations with cow's milk allergies (De Albuquerque et al., 2015). Peanuts are a good source

TABLE 6.2 Limitation in consumption and technological interventions for increased acceptability of plant-based beverages.

Plant sources	Factors affecting product quality and overall acceptability	Technological interventions to resolve the problem
Soybean	• Sedimentation and product stability • Breakdown of unsaturated fatty acids due to LOX activity resulting in beany off-flavor • presence of inhibitors	• Ultra-high-pressure homogenization (UHPH) at 300–400 MPa can improve product stability (Floury et al., 2004; Poliseli-Scopel et al., 2012). • Enzymatic treatment to remove beany flavor viz., Aldehyde dehydrogenase (Chiba et al., 1979). • Vacuum treatment at high temperature (Yoo & Chang, 2016). • Hot grinding (Zhang et al., 2012). • Blanching in boiling water (Yuan et al., 2008). • Effect of soaking conditions viz., alkaline (Khaleque et al., 1970), and ethanolic (Ashraf & Snyder, 1981). • Use of soy protein isolates and/or addition of flavoring compounds (Damodaran & Arora, 2013). • Elimination of inhibitor such as trypsin inhibitor by blanching (Yuan et al., 2008). • Using ultra-high heat treatment to inactivate trypsin inhibitors (Kwok et al., 1993). • Use of nonthermal techniques such as PEF to inactivate trypsin inhibitors (Vagadia et al., 2017; Vanga et al., 2017).
Peanut	• Beany off-flavour due to LOX activity • Chalkiness and viscosity issues	• Defatting, roasting, alkali soaking, steaming (Diarra et al., 2005; Isanga & Zhang, 2009; Jeske et al., 2018b; Lee & Beuchat, 1991). • Homogenization at 20.68 MPa and sterilization at 110°C resulted in reduced chalkiness and optimum viscosity of the final peanut milk substitute (Diarra et al., 2005).
Pea	• Sedimentation and product stability	• Homogenization at 50 MPa followed by two-stage ultra-high temperature treatment viz., preheating at 95°C for 15 s and high heat treatment at 140°C for 2 s resulted in more stable pea protein emulsions (Qamar et al., 2019).
Almonds	• Sedimentation and product stability	• Ultra-high-pressure homogenization (UHPH) at 200 MPa at 55 and 75°C improved the colloidal stability in stored almond-based beverages (Ferragut et al., 2015).
Coconut	• Beverage yield	• Yield and quality of beverage depend on various factors viz., grinding time, viscozyme-L enzyme, and incubation period. • Combination of grinding time (2–4 min), amount of Viscozyme-L (60–180 FBGU), and incubation time (2–7 h) produced the best yield and quality of coconut milk alternative (Agarwal & Bosco, 2014).
Oat	• increased viscosity due to starch gelatinization	• Enzymatic liquefication of oat starch (Deswal et al., 2014a).
Rice	• Increased viscosity due to starch gelatinization • Chalkiness	• Enzymatic liquefication of rice starch (Munera-Picazo et al., 2014). • Chalky mouthfeel (Banyavongsa et al., 2019) due to insoluble large particles can be minimized by using ultra high-pressure processing.
Quinoa	• Increased viscosity due to starch gelatinization • Reduced colloidal stability due to poor protein solubility	• Enzymatic liquefication of quinoa starch (Li et al., 2018). • Improved protein solubility by the combination of pH and ionic strength. Saline solution (NaCl ~ 0.03 mol/L) and acidic conditions (pH 5) (Pineli et al., 2015).

Continued

TABLE 6.2 Limitation in consumption and technological interventions for increased acceptability of plant-based beverages.—cont'd

Plant sources	Factors affecting product quality and overall acceptability	Technological interventions to resolve the problem
Sesame	• Colloidal stability • Poor protein solubility and thermal-induced gelation • Bitterness and chalkiness	• Pasteurization (<85°C) (Ahmadian-Kouchaksaraei et al., 2014) or nonthermal sterilization such as pulsed electric field (PEF) (Poojary et al., 2020). • Roasting and alkali soaking can improve solubility and reduce bitterness and chalkiness (Afaneh et al., 2011; El-Adawy et al., 2000).
Hemp	• Colloidal stability	• Hemp-based beverages have poor colloidal stability due to high flocculation and coalescence of oil droplets. It can be improved by co-stabilization using vegetable emulsifiers. • High-pressure processing (>60 MPa) can minimize flocculation and coalescence and improve product stability (Wang et al., 2018).

of macro-nutrients, other components present in peanuts include minerals, vitamins, volatiles, functional, and bioactive components (Arya et al., 2016). Phenolics in peanuts are noted for their antioxidant properties and play a protective function in the prevention of diseases, namely strokes, coronary heart disease (CHD), and various cancers (Toomer, 2018; Wien et al., 2014). The potential health benefits of peanuts are addressed in Table 6.1.

Despite the positive benefits, consumption of peanuts elicits allergic reactions in susceptible individuals owing to the existence of different protein fractions (Ara h 1 to Ara h 8), with Ara h 1 and Ara h 2 being regarded as the primary allergens. The allergenicity of these allergens is not affected by the thermal treatment utilized in the preparation of peanut-based drinks (FSANZ, 2016). In contrast, enzymatic treatment (peroxidase hydrolysis) (Deshpande et al., 2008), activated charcoal (Mazumder & Hongsprabhas, 2016), and phytic acid (Mensah-Brown et al., 2014) has the capability to diminish the allergenic potency of the proteins.

To improve the properties of peanut-based beverages, researchers developed several approaches, i.e., defatting, roasting, steaming, and soaking in alkali solutions (Diarra et al., 2005; Galvez et al., 1990; Jeske et al., 2018a). The defatting process has been shown to improve the physical stability of pasteurized peanut-based drinks (Zhao et al., 2005). Spray-dried peanut protein isolates have been utilized in recent years to manufacture high-quality peanut-based beverages with improved stability, but the high cost associated with spray drying restricts the commercialization of this method. Furthermore, other problems such as chalkiness, sedimentation-induced viscosity, and lipoxygenases (LOX)-induced off flavors generally reduce the quality, including consumer acceptability, of the beverages, as shown in Table 6.2 (Mäkinen et al., 2016). Studies have suggested different methods, e.g., homogenization (Diarra et al., 2005), sodium bicarbonate combined with homogenization (Lee and Beuchat (1991), thermal treatment (Isanga & Zhang, 2009), use of emulsifying and stabilizing agents (Yu et al., 2009), and incorporation of flavorings, to resolve the aforementioned issues. Nutrients can also be added in the form of additives to attain the desired nutritional balance (Deshpande et al., 2008).

6.3.1.3 Pea-based beverages

Garden peas (*Pisum sativum* ssp. *hortense*) and field peas (*P. sativum* ssp. *arvense*) are two different types of commercially cultivated peas. Literature has suggested that people began growing field pea as a crop since the Neolithic era (Smýkal et al., 2012). The crop is currently grown as animal feed and human food in various parts of the world. Peas have promising health benefits but could also pose a risk due to the presence of allergens. Among all plant proteins containing more than 70% (w/w) protein, pea protein isolate (PPI) is regarded as one of the most functional protein products (Qamar et al., 2019). The excellent quality of PPI can be attributed to its nutritional properties, presence of antioxidants (Zhang et al., 2013), and low allergenic potency (Tamm et al., 2016). The potential health benefits of peas are addressed in Table 6.1.

With reference to techno-functionality, pea proteins have similar emulsifying properties to that of other top-quality food proteins. Emulsions made with pea proteins have good stability and their fat or oil droplet sizes are small with a narrow range of particle distribution. Furthermore, superb protein solubility and foaming capacity at a wide range of pH are characteristics of pea proteins (Adebiyi & Aluko, 2011). Distinct extraction methods, e.g., isoelectric precipitation, salt extraction, and ultrafiltration, also influence the PPI's functional properties (Boye et al., 2010; Karaca et al., 2011). Isoelectric precipitation was reported to produce PPI with a higher surface charge and solubility than the other two methods.

Research has demonstrated that oil-in-water emulsions prepared with pea protein have a higher degree of stability as compared to that made with other plant proteins (Lam et al., 2018; Liang & Tang, 2013). While pea protein emulsions have good stability, the key processing challenge faced by food manufacturers is sedimentation, as outlined in Table 6.2. The stability of pea-based drinks and related products is influenced by the methods used to prepare or process the products. Enzymatic hydrolysis of PPIs using proteases has been shown to increase the solubility and lower the viscosity of the pea protein emulsions (Bajaj et al. (2017). Common food processes may also amend the structural and functional properties of pea-based beverages. Recent work by Qamar et al. (2019) showed that homogenization and heating were able to stabilize pea protein emulsions by decreasing the size of the fat particles.

Physicochemical properties of pea-based beverages can also be influenced by the maturity stage of the crop as its nutrient composition changes during maturation (Johnson, 1968; Kluczkovski et al., 2015). Felberg et al. (2009) concluded that matured pea seeds had a higher starch content than their young counterparts; hence, thermally processed beverages prepared from the matured seeds could have a viscosity issue due to starch gelatinization. As for the nutritional aspect, the levels of amino acids (arginine, alanine, methionine, glutamic acid, and threonine) in peas decreased as the legume reached maturity.

6.3.2 Nut-based beverages

6.3.2.1 Almond-based beverages

Nuts are very nutritious and considered an essential part of a healthy diet. Among nuts consumption globally, almond (*Prunus dulcis*) has the largest share due to its good quality of protein and other bio-functional ingredients. Almonds contain around 20%–25% protein, which is usually in the form of amandin (a legumin-type protein) (Lipan et al., 2020). Among the bio-functional ingredients, almonds are a good source of essential minerals such as manganese, calcium, magnesium, selenium, potassium, zinc, phosphorus, and copper and also rich in vitamins such as vitamin E (25.9 mg/100 g), which is present in the form of α-tocopherol, vitamin B complex, comprising of B_1 (0.3 mg/100 g), B_2 (0.8 mg/100 g), B_3 (4 mg/100 g), B_5 (0.3 mg/100 g), and B_6 (0.1 mg/100 g) (Jeske et al., 2018b). In addition, almonds also contain 30%–35% monounsaturated fat with oleic acid as a dominating fatty acid, total dietary fibers (10%–15%), and total phenolic compounds (26–350 mg/100 g) (Barreca et al., 2020).

Apart from nutritional benefits, almonds also contain arabinose, which is present in the cell wall pectic substances and has potential prebiotic properties and lowering blood serum cholesterol levels (Kalita et al., 2018). Almonds are a rich source of fatty acids, which can result in the growth of microbes especially pathogenic *Escherichia coli* and *Listeria monocytogenes* (Iorio et al., 2019). Although sterilization can be done to minimize the presence of these pathogens, still both these pathogens can cause a considerable reduction in nutritive value (Van Impe et al., 2018). Moreover, being a tree nut, almond allergy is common and ranked as the fourth most common tree nut allergy in the USA (Mcwilliam et al., 2015). The almond allergic responses can vary from mild such as simple oral allergy to very complex such as fatal anaphylaxis. Studies showed that the major storage protein (approximately 65%) of almonds viz., amandin, legumin (11S), prunin, and prunin (Pru du 6) are the main allergens in almonds (Chhabra et al., 2017). Almond allergens, especially Pru du 6, are resistant to thermal processing; however, enzymatic hydrolysis using pepsin showed promising results in the removal of these allergens (Mandalari & Mackie, 2018). The potential health benefits of almonds are mentioned in Table 6.1.

Commercially, almond-based milk beverages are prepared by heating an aqueous dispersion (8% W/W) of almond powder (partially defatted) with an emulsifying agent (0.1%) at 90°C for sufficient time to solubilize components (Topcuoglu & Yilmaz-Ersan, 2020). The slurry is then ground and centrifuged to remove the undissolved particles. The product is usually thermally treated (pasteurization or ultra-high-temperature treatment) and homogenized (18 MPa), followed by aseptic packaging (Berger et al., 1997). By the virtue of processing efficiency, almond-based milk beverages have the second largest market share after soy-based milk beverages globally (Sethi et al., 2016). Potential processing challenges of almond-based beverages are mentioned in Table 6.2.

6.3.2.2 Coconut-based beverages

Coconut (*Cocos nucifera* L.) is one of the most important and extensively grown palm trees around the globe (Prades et al., 2012a). Coconut is a fruit and not a nut but is usually referred to as a nut due to the hard outer shell of the endosperm. The inner part of the endosperm is divided into two edible parts viz., a white kernel and clear water or coconut water. Both parts are usually processed in several ways and lead to different products such as copra, virgin coconut oil, coconut cream, desiccated coconut, coconut water, nata de coco, and coconut milk (Prades et al., 2012b).

Commercially, coconut-based beverages and creams are usually used as ingredients in confectionaries, bakeries, and ice-creams. Domestically, they are essential ingredients in various Asian countries such as India, Sri Lanka, Thailand, Vietnam, etc (Karunasiri et al., 2020). Coconut beverage is rich in medium-chain triglycerides, which makes it an easily digestible plant-based milk alternative (Zentek et al., 2011). In addition, it contains medium-chain fatty acids (MCFAs) that are readily metabolized by the liver into ketone compounds, having potential benefits in brain functioning and relief from memory impairment like Alzheimer's disease (Chatterjee et al., 2020). Along with MCFAs, soluble and insoluble fiber content increases the nutritional value of coconut milk with various antioxidant properties. Also, it contains a fairly good amount of micro-nutrients viz., vitamins, and minerals (Chalupa-Krebzdak et al., 2018). The potential health benefits of coconut are mentioned in Table 6.1.

Apart from the nutritional benefits of the coconut milk alternative, it contains the highest amount of fat and the least amount of protein among all commercially available plant-based beverages. Moreover, approximately 87% of this fat is saturated with lauric acid (44%) followed by caprylic and capric acids (13%), which can adversely increase the low-density lipoprotein (LDL) cholesterol (Amarasiri & Dissanayake, 2006). Also, coconut milk contains very low levels of calcium (4% of the total recommended level) making it a calcium-deficient plant-based beverage (Balogun et al., 2016).

Coconut milk alternative is manufactured by mechanically deshelling the nut and separating the edible endosperm, which is cleaned and grated, and mixed with warm water followed by grinding to extract the milk alternative (Belewu & Belewu, 2007). Various factors affect the yield and quality of coconut milk; for instance, grinding time, incubation period, and the viscozyme-L enzyme (Table 6.2). Precisely, coconut milk alternative yield is directly proportional to enzyme concentration and incubation period and indirectly proportional to grinding time (Agarwal & Bosco, 2014).

6.3.3 Cereal-based beverages

6.3.3.1 Oat-based beverages

Oat (*Avena sativa* L.) is the sixth-largest cereal crop produced in the world following wheat, maize, rice, barley, and sorghum (Ahmad et al., 2014). Oats are rich source of micro-nutrients, starch (60%), protein (11%−15%), lipids (5% −9%), dietary fibers (2.5%−8.5%), and calcium (0.54%) (Rasane et al., 2015). Oats lipids are mainly composed of mono- and polyunsaturated fatty acids. Besides this, oats are also dense in antioxidants and polyphenols viz., avenanthramides (2−289 mg/kg), saponins (avenacoside A-290 mg/kg, and avenacoside B-110 mg/kg) (Sang & Chu, 2017). Also, oats β-glucans have known anticancerous potential by reducing the causative agents of colon cancer along with significant reduction in blood cholesterol level blood pressure. The recommended dose of β-glucan for a single food is 0.75 g/serving (Daou & Zhang, 2012). The potential health benefits of oats are mentioned in Table 6.1.

Oat-based beverages occupy an important position in the global market of plant-based beverages (Deswal et al., 2014b). Despite many potential health benefits of whole oats, oat-based beverages are usually deficient in bioavailable calcium, which is commercially fortified in oat-based drinks. Moreover, fermentation is considered an important technique to improve the health-promoting properties of oat-based beverages. Studies showed that the fermentation of oats using *Monascus anka* could increase the phenolic compounds (free, conjugated, and bound) along with the rise in glucosamine content, resulting in the enhancement of free radical scavenging properties of oat-based drinks (Bei et al., 2017; Chen et al., 2020). However, it is not always beneficial as some major phytochemicals may get utilized as a substrate for the process. Also, concurrent production of some important secondary metabolites, i.e., phenolic compounds like chlorogenic acid and quercetin, were detected in oats after fermentation by *M. anka*. On the other hand, some compounds may be lost in the process, e.g., sinapic acid a major component of unfermented seeds was not detected in fermented oats (Bei et al., 2020). Potential processing challenges of oat-based beverages are mentioned in Table 6.2.

6.3.3.2 Rice-based beverages

Rice (*Oryza sativa* L.) is the second most important cereal crop after wheat and is considered a staple food for almost half of the world's population (Nawaz et al., 2016). It is primarily grown/consumed in south-east Asian countries; however, western countries have seen increased consumption of rice during the last 2 decades, and hence, an escalation in production has been recorded (Muthayya et al., 2014). Being a cereal, rice is rich in starch (~90%) and contains around 6%−8% proteins, which are deficient in threonine and lysine (Jiang et al., 2016). Rice contains appreciable amounts of hydroxycinnamic acids (class of aromatic acids or phenylpropanoids having a typical skeleton of C_6-C_3) viz., ferulic acid ($C_{10}H_{10}O_4$), sinapic acid ($C_{11}H_{12}O_5$), and p-coumaric acid ($C_9H_8O_3$), sinapic acid with CO_2 reduced along with p-hydroxybenzoic acid ($C_7H_6O_3$) (Goufo et al., 2015; Shao & Bao, 2015). These rice hydroxycinnamic acids have anti-hyperglycemic and antioxidant properties (Jung et al., 2010).

Apart from these health benefits, rice is deficient in β-carotene (precursor of vitamin A). This is synthesized by the activity of plant enzymes viz., phytoene synthase (PSY), phytoene desaturase (PDS), ζ-carotene desaturase (ZDS), carotene *cis-trans*-isomerase (CRTISO), and lycopene-β-carotene (LCYB), which convert geranylgeranyl diphosphate into β-carotene (Zhao et al., 2020). Although, these enzymes are functional in rice stems, leaves, and immature grains, mature grains are lacking in these enzyme systems (Datta & Datta, 2020). Also, the rice bran portion is rich in health-promoting components such as polyphenols (phenolic acids, anthocyanins, proanthocyanidins consisting of catechin and epicatechin) and minerals such as iron (Chaudhari et al., 2018; Khosravi & Razavi, 2020). However, rice-based beverages produced from brown or unpolished rice usually lack sensory quality and overall acceptability by the consumers (Paul et al., 2020); therefore, rice-based beverages are traditionally prepared from polished rice and fortified with iron (Mehansho, 2006). The potential health benefits of oats are mentioned in Table 6.1.

Nowadays, the food industry is focusing on producing rice bran-based beverages due to its underutilization and health benefits (Ardali et al., 2013). Previous studies conducted on the physicochemical properties of rice bran-based beverages showed that the bran beverages had a lower ratio of total soluble solids and viscosity than soy-based beverages (Faccin et al., 2009; Issara & Rawdkuen, 2016). In another study conducted by Sharma and Gayathri (2018), it was reported that the overnight soaking of rice before preparing beverage resulted in significantly increased contents of minerals, water-soluble vitamins especially B_6 and B_{12}, insoluble fiber, and other functional components. The authors further allowed the soaked rice to ferment naturally by lactic acid bacteria, which broke down the antinutritional factors and enhanced the bioavailability of essential minerals such as calcium, magnesium, and iron. These minerals are not only beneficial to the immune system of humans but also increase the activity of beneficiary bacteria in the digestive tract. Potential processing challenges of rice-based beverages are mentioned in Table 6.2.

6.3.4 Pseudocereal-based beverages

6.3.4.1 Quinoa-based beverages

Quinoa (*Chenopodium quinoa*) is a pseudocereal of high nutritional importance; it contains a good quality of protein (~17%) with approximately 80% bioavailability (Elsohaimy et al., 2015) and contains a balanced content of essential amino acids with high levels of methionine, cysteine, and lysine, which are usually limiting in most cereals (Dakhili et al., 2019). Quinoa proteins have the ability to form gel structures on acidification and can be used as an ingredient in a range of processed products (Mäkinen et al., 2015). Also, gluten is absent in quinoa and thus can be used for the production of gluten-free foods (Dakhili et al., 2019).

Among other macro-nutrients, carbohydrate is the most abundant nutrient, being composed of mainly starch (~32%−69%) (Vilcacundo & Hernández-Ledesma, 2017). Quinoa also contains around 5% lipids, which is higher than other pseudocereals. Quinoa lipids are mainly composed of unsaturated fatty acids (~85%), with linoleic (18:2) acid being the most abundant, followed by oleic (18:1) acid (Vera et al., 2019). Interestingly, the fatty acid profile of quinoa is very similar to that of soybean oil (Nowak et al., 2016). Quinoa is rich in micro-nutrients viz., minerals (potassium, iron, magnesium, copper, calcium, and manganese) and vitamins (vitamin E, folic acid, carotenes, and vitamin C and the B complex) (Jancurová et al., 2009). The bitterness in quinoa is related to the presence of saponins. Saponins content depends mainly on the variety, the crop, and the processing of quinoa, such as washing or peeling. Quinoa is generally classified as "sweet" or "bitter" depending on its saponin content (El Hazzam et al., 2020). The potential health benefits of quinoa are mentioned in Table 6.1.

Quinoa is rich in starch, which results in increased viscosity of quinoa-based beverages leading to reduced colloidal stability and sedimentation of large particles during storage (Ujiroghene et al., 2019). Therefore, it is functionally and nutritionally important to liquefy starch prior to preparing quinoa-based beverages (Table 6.2). Usually, amylases are used to liquefy the native quinoa starch, which not only improves the flowability of the drink but also increases the sweetness (Pineli et al., 2015). Quinoa-based beverages have different textural attributes than that of bovine milk, and therefore, it is used as an ingredient in various food products such as shakes, pasta, confectionaries, and coffee (Panzolini et al., 2017). Previous studies have reported that a combination of ionic strength and pH could be used to increase the protein solubility in quinoa drinks. For instance, Pineli et al. (2015) developed a quinoa drink with relatively higher protein contents (0.8%−1.7%) by using a saline solution (NaCl ~ 0.03 mol/L) and acidified to pH 5.

6.3.5 Seed-based beverages

6.3.5.1 Sesame-based beverages

Sesame (*Sesamum indicum* L.) is an ancient crop and is thought to have originated in Africa. Sesame seeds and sesame oil are popular ingredients in Asian, African, and Middle Eastern dishes due to their high nutritive value and pleasant flavor (Nagendra Prasad et al., 2012). It is a dense source of protein (~25%), and oil (~55%) (Onsaard, 2012), also noted for its high antioxidant activity (Medina-Vera et al., 2021). Sesame protein is rich in sulfur-containing amino acids (methionine and cysteine); however, it is deficient in lysine (Anilakumar et al., 2010). Also, sesame oil has a nutritional quality and is mainly composed of palmitic (16:0), stearic (18:0), oleic (18:1), and linoleic (18:2) fatty acids (Were et al., 2006). Sesame seeds are also rich in minerals (calcium, iron, phosphorus, copper, magnesium, manganese, and zinc) and vitamins; however, they are mostly concentrated in husk and barks, respectively. Therefore, dehusking results in a significant reduction in minerals and vitamins, but still milled sesame seeds have a decent amount of vitamin B complex (Nandi & Ghosh, 2015). Moreover, sesame seeds contain two unique components viz., sesamine and sesamolin; these components can be converted to antioxidants viz., sesamol and sesamolin by using heating or refining operations such as acid-activated clay bleaching (Huang et al., 2012). Both sesamol and sesamolin have known cholesterol-reducing effects in humans (Andargie et al., 2021). The potential health benefits of sesame are mentioned in Table 6.1.

Sesame-based beverages are usually prepared by blending the soaked seeds to extract the oil bodies from seed tissues (Afaneh et al., 2011). Various pretreatments such as roasting and blanching can be done to improve the flavor. Extraction of oil bodies from roasted seeds gives roasted flavor and blanching results in reduced enzymatic off-flavor (Ahmadian-Kouchaksaraei et al., 2014). Sesame-based beverages are a dense source of lipids and contain sterols, saturated short-chain fatty acids, proteins, and calcium. In comparison with soy-based beverages, sesame drinks contain less amount raffinose; hence, there are fewer chances of flatulence (Fitrotin et al., 2015). Ahmadian-Kouchaksaraei et al. (2014) reported that the roasting pretreatment of sesame seeds resulted in a significant reduction in total soluble solids and available protein in the sesame-based beverages. This reduction of soluble solids especially protein may be attributed to protein denaturation during roasting, which resulted in decreased solubility of denatured protein (Lee & Beuchat, 1992; Obizoba & Amaechi, 1993).

Among the processing challenges, colloidal stability is the most important challenge (Table 6.2). Studies showed that the roasting and bleaching pretreatments usually resulted in reduced colloidal stability of sesame-based beverages due to protein denaturation and exposure of hydrophobic residues of globular proteins (Ahmadian-Kouchaksaraei et al., 2014). However, immersion of seeds in water with baking soda ($NaHCO_3$) during grinding and extraction of oil bodies can improve the colloidal stability, possibly due to the increased protein solubility as a result of alkaline conditions that increase the development of protein—lipid hydrophilic complexes (El-Adawy et al., 2000).

As discussed earlier, sesame proteins are poorly soluble in water and readily thermally denatured during UHT processing; therefore, pasteurization and/or nonthermal sterilization techniques such as pulsed electric field (PEF) processing are usually recommended (Nawaz, Tan, et al., 2020b; Sethi et al., 2016). For instance, sesame-based beverages can sustain the pasteurization temperature but are unstable at UHT condition; this is an indication of sesame protein heat sensitivity (Quasem et al., 2009). Afaneh et al. (2011) worked on various pasteurization conditions of sesame-based beverages and reported the highest colloidal stability in samples pasteurized at 75°C/5 min when compared to other pasteurization conditions viz., 65°C/30 min and 85°C/5 min. Also, colloidal stability improved from 10% to 14% with an initial increase in the original sesame seed percentage. The other issue with the sesame-based drinks is the bitterness and chalkiness, which can be minimized by prior roasting or alkali soaking of seeds (McClements, 2020b). Sesame milk alternatives have an advantage over traditional soymilk due to lower tendency of flatulence production, relatively lesser allergens, and no beany off-flavor (Fitrotin et al., 2015).

6.3.5.2 Hemp-based beverages

Hemp (*Cannabis sativa* L.) or industrial hemp contains negligible levels of cannabinoids especially Δ^9-tetrahydrocannabinol (THC) unlike marijuana (Small, 2016). During the last 2 decades, hemp has been processed into various food products in Europe and North America (Cherney & Small, 2016). Hemp is a rich source of macro and micronutrients viz., protein (20%—25%), carbohydrates (20%—30%), lipids (25%—35%), insoluble fibers (10%—15%), vitamin E (90 mg/100 g), and minerals (phosphorus, potassium, sodium, magnesium, sulfur, calcium, iron, and zinc) (Frassinetti et al., 2018). Hemp fat is a rich source of polyunsaturated fatty acids (80—90 g/100 g) viz., α-linolenic acid and linoleic acid. These fatty acids provide prostanoids and leukotriene with antithrombotic, antivasoconstrictive, and anti-inflammatory properties, making it a prominent ingredient of new commercially available health food products (Andre et al., 2016). Moreover, hemp seeds contain a dense composition of bioactive components with cannabidiolic acid (CBDA) being the most predominant bio-functional component. CBDA provides health benefits against chronic diseases such as obesity, inflammatory bowel disease, rheumatoid arthritis, and Alzheimer's disease (Crescente et al., 2018). Hemp seeds also exhibit antineuroinflammatory properties due to the native lignanamides such as cannabisin M, cannabisin N, cannabisin O, and 3,3′-demethyl-heliotropamodes (Yan et al., 2015), which inhibit the nuclear factor kappa-light-chain-enhancer of the activated B cells (NF-κB) signaling pathway (Luo et al., 2017). Although total phenylpropionamides (TPA) have antineuroinflammatory properties, high dosages of TPA might cause toxicity which counters their effect on inhibiting inflammatory cytokines production (Zhou et al., 2018). Apart from the health benefits, there are various antinutritional factors associated with hemp seeds such as phytic acid (inositol hexaphosphate, IP6) and trypsin inhibitors (Schultz et al., 2020). The potential health benefits of hemp are mentioned in Table 6.1.

Hemp's low allergenicity and high nutritional value make it a potential alternative for bovine milk. Usually, hemp-based milk beverages are an O/W emulsion and colloidal stability is poor due to flocculation and coalescence of oil droplets, which is the most crucial challenge faced by industry because it results in reduced quality (Aydar et al., 2020). This colloidal stability can be improved by the addition of emulsifiers; however, it results in increased production costs. In addition, added emulsifiers may have some health concerns viz., obesity-associated diseases, chronic inflammatory diseases, and metabolic ailments. Therefore, high-pressure homogenization techniques are preferred commercially to maintain the colloidal stability of hemp-based beverages. In these techniques, oil droplets are mechanically disrupted, and protein is adsorbed on the O/W interface, which not only results in better colloidal stability but also reduces rancidity (Wang et al., 2018). The concentration of bio-functional components of hemp seeds, such as polyphenols, can be extracted by using ultrasonic treatment. This technique is more economical and environmentally friendly than chemical extraction (Teh & Birch, 2014). Potential processing challenges of hemp-based beverages are mentioned in Table 6.2.

6.4 Physicochemical, nutritional, and sensory

6.4.1 Physicochemical properties

The physicochemical properties of plant-based beverages mainly depend on the raw material and the specifications of the individual disintegration, homogenization, formulation, emulsification, and storage processes (Nawaz et al., 2021). Various studies have been conducted in the past to develop a process of manufacturing plant-based beverages with physiochemical properties especially stability resembling bovine milk, which is a natural emulsion (Paul et al., 2020). For instance, cereal- and pseudocereal-based beverages are rich in starch and therefore can easily gelate during thermal sterilization viz., UHT, or pasteurization, which can cause processing problems downstream (Mäkinen et al., 2016). On the other hand, seed- and nut-based have excessive lipid, which may lead to undesirable phase separation and colloidal instability (Briviba et al., 2016).

6.4.2 Nutritional value and bioactive components

Plants can be used as an ingredient in processed food with certain properties (Fig. 6.6). For instance, legumes and seeds contain protein comparable to bovine milk, though the amino acid profile is not as good as bovine milk (Chalupa-Krebzdak et al., 2018). Moreover, nuts and pseudocereals are rich in micronutrients (vitamins and minerals), bioactive components such as antioxidants and dietary fibers (Dreher et al., 1996; Mir et al., 2018). Plant foods are also rich in phytoestrogens, which has lot of health benefits viz., protective role against osteoporosis, cardiovascular diseases, breast cancer, and menopausal complications, such as vulvovaginal atrophy, periodontal disease, and urinary incontinence (Desmawati & Sulastri, 2019; Kargozar et al., 2017; Tham et al., 1998). Apart from the potential health benefits, phytohormones may have health concerns to susceptible individuals (Patisaul & Jefferson, 2010).

Recent advances in the understanding of the health benefits and molecular mechanisms associated with plant-based foods led to several studies based on the presence and impact of macro- and micro-nutrients, and bioactive compounds in plant materials (Zhang et al., 2020). For instance, nuts are a good source of vitamins especially vitamin E and C, having antioxidant properties (Alasalvar & Bolling, 2015). Legumes are rich in essential mono- and polyunsaturated fatty acids, minerals such as iron, zinc, and magnesium, and phytoestrogens such as isoflavones (Kamboj & Nanda, 2018). Furthermore, cereals and pseudocereals are dense in certain health-promoting components such as proanthocyanidins and β-glucans, which not only help in lowering cholesterol levels but also improve the sensorial attributes of final products (Ahmad et al., 2012; Zhu, 2019).

Even with these undisputed beneficial properties of plant raw materials, studies showed that commercial plant-based beverages are lacking in balanced nutrition and therefore cannot be comparable to bovine milk (Khandpur et al., 2021). For instance, more than 50% of commercially available plant-based beverages contain no or very little protein (<0.5%); there are only a few commercial soy-based beverages having protein levels similar to commercial bovine milk (~3.7%) (Jeske et al., 2017). Generally, vegan proteins have poor digestibility and reduced bioavailability and also lack essential amino acids such as L-lysine, L-methionine, L-cysteine, and L-tryptophan (Lynch et al., 2018). In addition to low protein quality, edible plant sources are also deficient in some essential vitamins mainly vitamin D and B_{12}, which is one of the reasons for vitamin deficiency in populations having a vegetarian lifestyle (Herrmann & Geisel, 2002). Also, vitamins are sensitive to food processing operations and are easily washed or degraded by washing or thermal processing, respectively. Moreover, certain phytochemicals are present in such a form that makes their low bioavailability. For instance, isoflavones

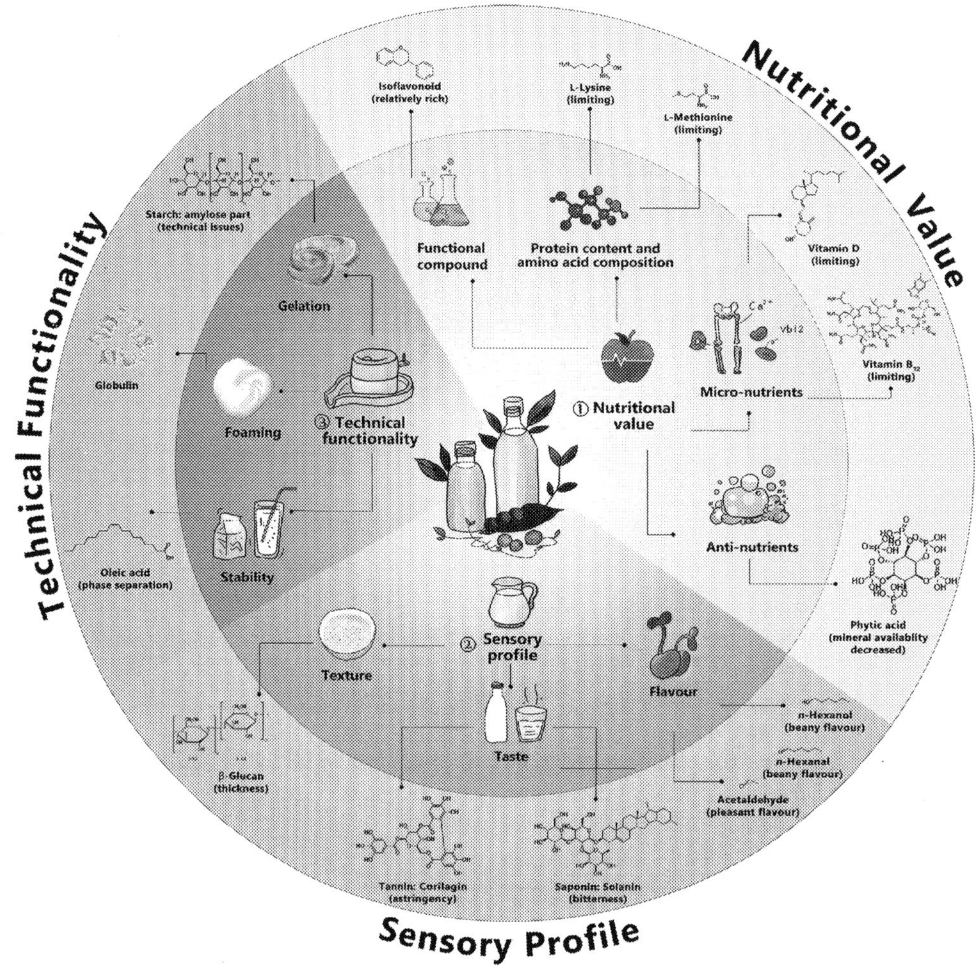

FIGURE 6.6 Quality criteria of plant-based beverages (Tangyu et al., 2019). *From Tangyu, M., Muller, J., Bolten, C., & Wittmann, C. (2019). Fermentation of plant-based milk alternatives for improved flavor and nutritional value.* Applied Microbiology and Biotechnology, 103, *9263–9275. https://doi.org/10.1007/s00253-019-10175-9.*

in legumes especially soybean mainly exist as genistin and daidzin, which are glucosides of genistein and daidzein and have much lower bioavailability than the other corresponding aglycone forms (Hsiao et al., 2020).

Plant-based food ingredients are also dense in antinutritional components such as lectins, urease, phytates, saponins, protease inhibitors, and oligosaccharides (Lal et al., 2017). Phytates and saponins can form insoluble complexes with certain essential minerals viz., Ca^{+2}, Mg^{+2}, Fe^{+2}, and Zn^{+2}, which result in decreased bioavailability of these essential minerals (Amos et al., 2020). Among protease inhibitors, trypsin inhibitor is considered as one of the most important antinutritional factors in plant-based food ingredients. The presence of trypsin inhibitors in plant-based beverages and associated products can significantly reduce the digestion and absorption of protein and cause adverse health effects to humans such as pancreatic hypertrophy (Weizman, 2004). Furthermore, legumes are rich in oligosaccharides such as raffinose, stachyose, and verbascose (Fan et al., 2015). The digestive systems of some individuals are unable to break down these compounds because of α-galactosidase deficiency. Consequently, they can accumulate in the colon and act as substrates for anaerobic fermentation by the microbial flora, causing flatulence and abdominal complications (Kurbel et al., 2006).

6.4.3 Sensory attributes

Taste is considered as a key factor that impacts food sensorial attributes and selection by consumers, and therefore, it is considered a critical indicator in all consumer and marketing surveys and studies of food products (Clark, 1998). In relation to the natural taste of plant-based beverages, unfortunately, products have low acceptability (Tso et al., 2021). Though high contents of dietary fibers in plant-based beverages can positively influence the texture and mouthfeel (Vasquez-Orejarena et al., 2018), yet their taste is generally perceived by consumers as displeasing, possibly due to due to prior experiences of inferior taste products in the market (Litt & Shiv, 2012). Regarding aroma or smell profiles, legume- and nut-based beverages have a typical beany or grassy smell and nutty smell, respectively, which is generally considered as off-flavor in cultures, where consumption of legume-based products is not common (Keast & Lau, 2006). As discussed earlier, legumes and tree nuts are rich in mono- and polyunsaturated fatty acids, which can easily oxidize during the processing and storage of beverages and other associated products. This oxidation of unsaturated fatty acids produces compounds such as n-hexanal and n-hexanol, which are mainly responsible for these types of off-flavors (Bott & Chambers Iv, 2006). Also, certain phytochemicals such as polyphenols (including antinutrients viz., saponins and tannins), flavonoids, terpenes, and glucosinolates can induce bitterness and astringent taste (Axten et al., 2008). Unfortunately, functional components such as isoflavonoids are responsible for distinct aftertaste (Ntukidem et al., 2019). Furthermore, plant ingredients also negatively induce greenish, greyish, or brownish coloration; chalky or sandy texture; and watery mouthfeel in beverage due to the presence of undissolved material and colloidal instability during processing and storage, resulting in reduced consumer acceptability (Kyriakopoulou et al., 2021).

6.4.4 Bioavailability of plant-based beverages

The bioavailability of nutrients is usually defined by their accessibility to normal metabolic and physiologic processes (Hambidge, 2010). The bioavailability of functional ingredients of plant-based beverages depends on several factors viz., structure, dosage, contextual diet, the origin of the food, and gut microflora (Ribas-Agustí et al., 2018). The bioavailability of vitamins and minerals is usually low in plant-based beverages due to the presence of antinutrients such as tannins, phytic acids, trypsin inhibitors, saponins, etc. Also, calcium bioavailability is low by its reduced absorption due to phytate and oxalate, which are predominant antinutrients in seeds especially sesame (Abebe et al., 2007). Studies showed that the adequate absorption of calcium occurred when the calcium and phosphate were in 1:2 or 2:1. However, a higher ratio of phosphate can result in an increased quantity of insoluble calcium phosphate and decreased calcium bioavailability (Dubey & Patel, 2018). In addition to calcium, the bioavailability of iron, zinc, and magnesium is often impaired by phytate due to insoluble compounds produced as a consequence of reactions with mineral cations; thus, absorption of these minerals reduces (Sandberg, 2002), for example, reduced absorption of iron and zinc was recorded as a consequence of phytic acid in oat-, soy-, and cashew-based beverages (Scholz-Ahrens et al., 2020).

6.4.4.1 Effects of polyphenols plant proteins and other micro-nutrients

Apart from the health benefits of polyphenols, studies showed that some of the polyphenols restricted the transport of some bioactive compounds (organic cations, glucose, and certain vitamins viz., B_1 and B_9) at the intestinal epithelia and the placenta (Martel et al., 2010). Polyphenols can also bind with proteins and reduce digestibility (Gibson et al., 2006). However, food processing such as thermal application can influence the bioavailability of vitamins and minerals. Studies showed that the thermal treatment can break the bonds of formed complexes and antinutrients (Lola, 2009; Martel et al., 2010). For example, oxalate in legumes and seeds can be degraded by boiling, which can improve calcium bioavailability (Kamchan et al., 2004).

6.4.4.2 Bioavailability of genistein

Genistein is one of the naturally occurring isoflavones and has known health benefits such as maintenance of normal arterial elasticity and hence help to prevent hypertension (Hemati et al., 2020). However, various studies have indicated the low bioavailability of genistein (Kwon et al., 2007; Lu et al., 2019; Tang et al., 2011). While the desirable absorption potential of genistein in the intestine has been observed, low water solubility inhibits the absorption of a higher concentration (Jaiswal et al., 2019). Yang et al. (2012) reported that the metabolism of genistein involves several extensive processes viz., oxidation, reduction, and conjugation reactions (glucuronidation and sulfation). This extensive genistein's metabolism is the most critical factor of its low oral bioavailability. Moreover, genistein and its glucuronide and sulfate conjugates are the substrates of breast cancer—resistant protein (BCRP) and ATP-binding cassette transporters (responsible for the intestinal and biliary excretion of numerous compounds and their metabolites). Studies showed that the high levels of efflux transporters mainly BCRP might be the most important factor responsible for genistein's reduced oral bioavailability (Yang et al., 2012). Interestingly, genistein's low bioavailability may have desirable effects on some vulnerable consumers especially pregnant women (Aydar et al., 2020; Pande et al., 2019). Pistollato et al. (2015) reported that the consumption of genistein by pregnant women might cause one type of leukemia in infants, MLL + acute myeloid leukemia. However, further large-scale clinical and epidemiological studies are needed to validate these risks.

6.5 Advantages and limitations

Plant-based alternative milk consumption is rapidly expanding around the world driven by health- and ethically conscious consumers that are shifting toward flexitarian diets where animal-derived protein is partially or wholly replaced by plant-based protein products. Milk and other dairy products provide the majority of saturated fat in western diets and, therefore, are believed to be the main contributor to public health challenges such as chronic heart disease, type 2 diabetes, and Alzheimer's disease (Chen et al., 2016). At a time of potentially devastating climatic shifts, a growing population, and threats to food and water security in many parts of the world, there is also the need to drastically reduce the environmental footprint of our diets (Willett et al., 2019). There is strong evidence that the environmental footprint of plant-based beverages is significantly better than that of bovine milk, especially those derived from grains and pseudo-grains (Mcauliffe et al., 2020).

Despite the clear rationale for plant-based beverages as healthy, ethical, and sustainable milk alternative options, the majority of consumers do not fully embrace animal-free milk alternatives. The slow growth of the plant-based milk alternative market underlines the fact that the mainstream consumer continues to prefer bovine milk products (Haas et al., 2019). Thus, a viable path to human and planetary health must involve alternative beverages that mimic the familiar appearance, tastes, textures, and mouthfeel of bovine milk. However, mimicking bovine milk using plant-based ingredients only has not been achieved yet at commercially viable costs. The main challenges include a lack of varied and affordable raw ingredients for the development of plant-based dairy; a lack of sustainable technologies for creating dairy-like textures and flavors; limited understanding of the safety and nutritional value of alternative ingredients; a lack of regulations concerning the use of alternative protein and other novel food ingredients; and a limited understanding of, and strategies to address, consumer preferences for, and acceptance of these novel alternatives (Tso et al., 2021).

The current plant-based dairy alternatives often feature distinct flavors and the presence of antinutritionals and allergens derived from the origin of their main protein ingredients, such as soy or almond milks (Vogelsang-O'dwyer et al., 2021). Almost all milk alternatives use relatively cheap vegetable oils and thickening agents to enhance a creamy mouthfeel and often require the use of stabilizers to improve colloidal stability. However, all of these ingredients tend to negatively impact the flavor and nutritional characteristics of these plant-based beverages (Oduro et al., 2021). A major constraint in developing more authentic alternative milks is the dearth of access to suitable base ingredients, especially proteins. Mammal milk proteins, especially caseins, nearly taste and odorless and have unique functional and nutritional properties that are rarely matched by plant proteins. Consequently, significant innovation in the development of plant proteins and other ingredients is critical, especially those that can be produced at low cost and at scale.

The opportunities start on-farm in selecting suitable crop varieties that can grow in a given climate, are high in protein yield, and low in antinutrients and other components that are causing distinctive off-flavors and colors during processing. There is also an even bigger opportunity to develop new processing technologies and techniques that can safely convert these crops into alternative milk beverages at scale and in a cost-effective manner, for example, dry milling, fractionation, etc (Vogelsang-O'dwyer et al., 2021).

Apart from sensory and economic improvements, more evidence and efforts are needed to clearly demonstrate and improve the nutritional value and safety of plant-based alternative milk products. Plant protein and mineral bioavailability and digestibility should also better match those attributes of bovine milk (Vogelsang-O'dwyer et al., 2021). This can be achieved through enhanced fractionation and/or fermentation of plant proteins or processing and conjugation with other ingredients to tailor the release of amino acids and minerals at oral, gastric, and intestinal phases. The release of potential antinutritional factors and their interaction with other food components is still poorly understood and merits further investigation.

6.6 Summary and future directions

Nowadays, a large population is interested in reducing animal-based food products in their diet for health, ethical, and environmental reasons. In response, the food industry is developing a wide range of plant-based milk alternatives viz., legume-, nut-, cereal-, pseudocereal-, and seed-based beverages. The success of these plant-based beverages depends on the factors that are usually ranked by consumers as important, such as sensory attributes, convenience, and cost. In this chapter, the current knowledge on fundamental processing steps to convert plant material into plant-based beverages, what are processing challenges for different plant sources, how to overcome these challenges and potential quality deficiencies, and what are the opportunities to maximize textural, nutritional, and sensory aspects of plant-based beverages. It is predicted that more and more consumers will incorporate plant-based beverages into their daily diet. Therefore, it is important to develop high-quality plant-based beverages that have better physicochemical properties especially colloidal stability and higher sensorial attributes. Also, it is also important to process beverages by using innovative minimal processing methods, which can not only maintain the bioavailability of health-promoting bioactive ingredients but also have low environmental impacts and are more sustainable. In fact, standardized techniques for quantification of nutritional benefits and environmental pressures of different plant-based beverages should be developed, resulting in a better comparison of the pros and cons of various plant-based drinks.

References

Abebe, Y., Bogale, A., Hambidge, K. M., Stoecker, B. J., Bailey, K., & Gibson, R. S. (2007). Phytate, zinc, iron and calcium content of selected raw and prepared foods consumed in rural Sidama, Southern Ethiopia, and implications for bioavailability. *Journal of Food Composition and Analysis, 20*, 161–168.

Adams, G. G., Imran, S., Wang, S., Mohammad, A., Kok, M. S., Gray, D. A., Channell, G. A., & Harding, S. E. (2012). Extraction, isolation and characterisation of oil bodies from pumpkin seeds for therapeutic use. *Food Chemistry, 134*, 1919–1925.

Adebiyi, A. P., & Aluko, R. E. (2011). Functional properties of protein fractions obtained from commercial yellow field pea (*Pisum sativum L.*) seed protein isolate. *Food Chemistry, 128*, 902–908.

Afaneh, I., Abu-Alruz, K., Quasem, J. M., Sundookah, A., Abbadi, J., Alloussi, S., & Sawalha, S. (2011). Effect of critical processing variables on sesame milk quality. *American Journal of Agricultural and Biological Sciences, 6*, 517–520.

Agarwal, R. K., & Bosco, S. (2014). Optimization of viscozyme-L assisted extraction of coconut milk and virgin coconut oil. *Asian Journal of Dairy and Food Research, 33*, 276–284.

Ahmad, A., Anjum, F. M., Zahoor, T., Nawaz, H., & Dilshad, S. M. R. (2012). Beta glucan: A valuable functional ingredient in foods. *Critical Reviews in Food Science and Nutrition, 52*, 201–212.

Ahmad, M., Dar, Z. A., & Habib, M. (2014). A review on oat (*Avena sativa L.*) as a dual-purpose crop. *Scientific Research and Essays, 9*, 52–59.

Ahmadian-Kouchaksaraei, Z., Varidi, M., Varidi, M. J., & Pourazarang, H. (2014). Influence of processing conditions on the physicochemical and sensory properties of sesame milk: A novel nutritional beverage. *LWT-Food Science and Technology, 57*, 299–305.

Alasalvar, C., & Bolling, B. W. (2015). Review of nut phytochemicals, fat-soluble bioactives, antioxidant components and health effects. *British Journal of Nutrition, 113*, S68–S78.

de Albuquerque, E. M. B., Almeida, F.d. A. C., Gomes, J. P., Alves, N. M. C., & da Silva, W. P. (2015). Production of "peanut milk" based beverages enriched with umbu and guava pulps. *Journal of the Saudi Society of Agricultural Sciences, 14*, 61–67.

Amaral, C., Toloi, M. R. T., Vasconcelos, L. D., Fonseca, M. J. V., Correia-da-Silva, G., & Teixeira, N. (2017). The role of soybean extracts and isoflavones in hormone-dependent breast cancer: Aromatase activity and biological effects. *Food & Function, 8*, 3064–3074.

Amarasiri, W. A., & Dissanayake, A. S. (2006). Coconut fats. *The Ceylon Medical Journal, 51*, 47–51.

Amine, C., Dreher, J., Helgason, T., & Tadros, T. (2014). Investigation of emulsifying properties and emulsion stability of plant and milk proteins using interfacial tension and interfacial elasticity. *Food Hydrocolloids, 39*, 180–186.

Amos, A., Alvan, A., & Florence, A. (2020). The anti-nutritional effect of phytate on zinc, iron and calcium bioavailabilities of some cereals staple foods in Zaria, Nigeria. *European Journal of Nutrition & Food Safety*, 1–6.

Andargie, M., Vinas, M., Rathgeb, A., Möller, E., & Karlovsky, P. (2021). Lignans of sesame (*Sesamum indicum L.*): A comprehensive review. *Molecules, 26*.

Andre, C. M., Hausman, J.-F., & Guerriero, G. (2016). Cannabis sativa: The plant of the thousand and one molecules. *Frontiers in Plant Science, 7*, 19.

Anilakumar, K. R., Pal, A., Khanum, F., & Bawa, A. S. (2010). Nutritional, medicinal and industrial uses of sesame (*Sesamum indicum L.*) seeds-an overview. *Agriculturae Conspectus Scientificus, 75*, 159–168.

Ardali, F., Hojjatoleslamy, M., & Shariaty, M. (2013). Production of a new functional rice bran beverage. *African Journal Science Research, 2*, 20–23.

Arya, S. S., Salve, A. R., & Chauhan, S. (2016). Peanuts as functional food: A review. *Journal of Food Science and Technology, 53*, 31–41.

Ashraf, H. R. L., & Snyder, H. (1981). Influence of ethanolic soaking of soybeans on flavor and lipoxygenase activity of soymilk. *Journal of Food Science, 46*, 1201–1204.

Axten, L. G., Wohlers, M. W., & Wegrzyn, T. (2008). Using phytochemicals to enhance health benefits of milk: Impact of polyphenols on flavor profile. *Journal of Food Science, 73*, H122–H126.

Aydar, E. F., Tutuncu, S., & Ozcelik, B. (2020). Plant-based milk substitutes: Bioactive compounds, conventional and novel processes, bioavailability studies, and health effects. *Journal of Functional Foods, 70*, 103975.

Bajaj, P. R., Bhunia, K., Kleiner, L., Joyner, H. S., Smith, D., Ganjyal, G., & Sablani, S. S. (2017). Improving functional properties of pea protein isolate for microencapsulation of flaxseed oil. *Journal of Microencapsulation, 34*, 218–230.

Balogun, M. A., Kolawole, F. L., Joseph, J. K., Adebisi, T. T., & Ogunleye, O. T. (2016). Effect of fortification of fresh cow milk with coconut milk on the proximate composition and yield of warankashi, a traditional cheese. *Croatian Journal of Food Science and Technology, 8*, 10–14.

Banyavongsa, A., Linh, T. T. T., Van Tai, N., & Thuy, N. M. (2019). Response surface optimization of enzymatic hydrolysis of purple rice (cultivated in Soc Trang, Vietnam) using amylases and formulation of healthy rice milk. *International Journal of Advances in Agricultural Science and Technology, 6*, 17–28.

Barreca, D., Nabavi, S. M., Sureda, A., Rasekhian, M., Raciti, R., Silva, A. S., Annunziata, G., Arnone, A., Tenore, G. C., & Süntar, İ. (2020). Almonds (Prunus dulcis mill. DA Webb): A source of nutrients and health-promoting compounds. *Nutrients, 12*, 672.

Bei, Q., Liu, Y., Wang, L., Chen, G., & Wu, Z. (2017). Improving free, conjugated, and bound phenolic fractions in fermented oats (*Avena sativa L.*) with Monascus anka and their antioxidant activity. *Journal of Functional Foods, 32*, 185–194.

Bei, Q., Wu, Z., & Chen, G. (2020). Dynamic changes in the phenolic composition and antioxidant activity of oats during simultaneous hydrolysis and fermentation. *Food Chemistry,* 305, 125269.

Belewu, M. A., & Belewu, K. Y. (2007). Comparative physico-chemical evaluation of tiger-nut, soybean and coconut milk sources. *International Journal of Agriculture and Biology,* 5, e787.

Berger, J., Bravay, G., & Berger, M. (1997). *Almond milk preparation process and products obtained.* Google Patents.

Boateng, L., Ansong, R., Owusu, W., & Steiner-Asiedu, M. (2016). Coconut oil and palm oil's role in nutrition, health and national development: A review. *Ghana Medical Journal,* 50, 189–196.

Bott, L., & Chambers Iv, E. (2006). Sensory characteristics of combinations of chemicals potentially associated with beany aroma in foods. *Journal of Sensory Studies,* 21, 308–321.

Boye, J. I., Aksay, S., Roufik, S., Ribéreau, S., Mondor, M., Farnworth, E., & Rajamohamed, S. H. (2010). Comparison of the functional properties of pea, chickpea and lentil protein concentrates processed using ultrafiltration and isoelectric precipitation techniques. *Food Research International,* 43, 537–546.

Briviba, K., Gräf, V., Walz, E., Guamis, B., & Butz, P. (2016). Ultra high pressure homogenization of almond milk: Physico-chemical and physiological effects. *Food Chemistry,* 192, 82–89.

Burgos-Díaz, C., Wandersleben, T., Marqués, A. M., & Rubilar, M. (2016). Multilayer emulsions stabilized by vegetable proteins and polysaccharides. *Current Opinion in Colloid & Interface Science,* 25, 51–57.

Chalupa-Krebzdak, S., Long, C. J., & Bohrer, B. M. (2018). Nutrient density and nutritional value of milk and plant-based milk alternatives. *International Dairy Journal,* 87, 84–92.

Chang, J.-C., Lai, Y.-H., Djoko, B., Wu, P.-L., Liu, C.-D., Liu, Y.-W., & Chiou, R. Y.-Y. (2006). Biosynthesis enhancement and antioxidant and anti-inflammatory activities of peanut (*Arachis hypogaea L.*) arachidin-1, arachidin-3, and isopentadienylresveratrol. *Journal of Agricultural and Food Chemistry,* 54, 10281–10287.

Chatterjee, P., Fernando, M., Fernando, B., Dias, C. B., Shah, T., Silva, R., Williams, S., Pedrini, S., Hillebrandt, H., & Goozee, K. (2020). Potential of coconut oil and medium chain triglycerides in the prevention and treatment of Alzheimer's disease. *Mechanisms of Ageing and Development,* 186, 111209.

Chaudhari, P. R., Tamrakar, N., Singh, L., Tandon, A., & Sharma, D. (2018). Rice nutritional and medicinal properties: A. *Journal of Pharmacognosy and Phytochemistry,* 7, 150–156.

Chen, M., Li, Y., Sun, Q., Pan, A., Manson, J. E., Rexrode, K. M., Willett, W. C., Rimm, E. B., & Hu, F. B. (2016). Dairy fat and risk of cardiovascular disease in 3 cohorts of US adults. *The* American Journal of Clinical Nutrition, 104, 1209–1217.

Chen, G., Liu, Y., Zeng, J., Tian, X., Bei, Q., & Wu, Z. (2020). Enhancing three phenolic fractions of oats (Avena sativa L.) and their antioxidant activities by solid-state fermentation with Monascus anka and Bacillus subtilis. *Journal of Cereal Science,* 93, 102940.

Chen, G., & Tao, D. (2005). An experimental study of stability of oil–water emulsion. *Fuel Processing Technology, 86*, 499–508.

Cherney, J. H., & Small, E. (2016). Industrial hemp in North America: Production, politics and potential. *Agronomy,* 6, 58.

Chhabra, G. S., Liu, C., Su, M., Venkatachalam, M., Roux, K. H., & Sathe, S. K. (2017). Effects of the Maillard reaction on the immunoreactivity of amandin in food matrices. *Journal of Food Science,* 82, 2495–2503.

Chiba, H., Takahashi, N., & Sasaki, R. (1979). Enzymatic improvement of food flavor II. Removal of Beany flavor from soybean products by aldehyde dehydrogenase. *Agricultural and Biological Chemistry,* 43, 1883–1889.

Chuah, A. M., Kuroiwa, T., Kobayashi, I., & Nakajima, M. (2009). Effect of chitosan on the stability and properties of modified lecithin stabilized oil-in-water monodisperse emulsion prepared by microchannel emulsification. *Food Hydrocolloids,* 23, 600–610.

Clark, J. E. (1998). Taste and flavour: Their importance in food choice and acceptance. *Proceedings of the Nutrition Society,* 57, 639–643.

Cohen, L., Zhao, Z., Pittman, B., & Scimeca, J. (2000). Effect of intact and isoflavone-depleted soy protein on NMU-induced rat mammary tumorigenesis. *Carcinogenesis,* 21, 929–935.

Crescente, G., Piccolella, S., Esposito, A., Scognamiglio, M., Fiorentino, A., & Pacifico, S. (2018). Chemical composition and nutraceutical properties of hempseed: An ancient food with actual functional value. *Phytochemistry Reviews,* 17, 733–749.

Crittenden, R. G., & Bennett, L. E. (2005). Cow's milk allergy: A complex disorder. *Journal of the American College of Nutrition,* 24, 582S–591S.

Cui, C., Birru, R., Snitz, B. E., Ihara, M., Lopresti, B. J., Aizenstein, H. J., Lopez, O. L., Mathis, C., Miyamoto, Y., & Kuller, L. H. (2018). Effects of soy isoflavones on cognitive function: A systematic review and meta-analysis of randomized controlled trials. Alzheimer's & dementia. *The Journal of the Alzheimer's Association,* 14, P1365–P1366.

Dahl, W. J., Foster, L. M., & Tyler, R. T. (2012). Review of the health benefits of peas (Pisum sativum L.). *British Journal of Nutrition,* 108, S3–S10.

Dakhili, S., Abdolalizadeh, L., Hosseini, S. M., Shojaee-Aliabadi, S., & Mirmoghtadaie, L. (2019). Quinoa protein: Composition, structure and functional properties. *Food Chemistry,* 299, 125161.

Damodaran, S., & Arora, A. (2013). Off-flavor precursors in soy protein isolate and novel strategies for their removal. *Annual Review of Food Science and Technology,* 4, 327–346.

Daou, C., & Zhang, H. (2012). Oat beta-glucan: Its role in health promotion and prevention of diseases. *Comprehensive Reviews in Food Science and Food Safety,* 11, 355–365.

Datta, S. K., & Datta, K. (2020). *Golden rice. The Future of rice demand: Quality beyond productivity*. Springer.

Deshpande, R. P., Chinnan, M. S., & Phillips, R. D. (2008). Process development of a chocolate-flavoured peanut–soy beverage. *International Journal of Food Science & Technology,* 43, 886–894.

Desmawati, D., & Sulastri, D. (2019). Phytoestrogens and their health effect. *Open Access Macedonian Journal of Medical Sciences,* 7, 495.

Deswal, A., Deora, N. S., & Mishra, H. N. (2014a). Effect of concentration and temperature on the rheological properties of oat milk. *Food and Bioprocess Technology, 7*, 2451–2459.

Deswal, A., Deora, N. S., & Mishra, H. N. (2014b). Optimization of enzymatic production process of oat milk using response surface methodology. *Food and Bioprocess Technology, 7*, 610–618.

Diarra, K., Nong, Z. G., & Jie, C. (2005). Peanut milk and peanut milk based products production: A review. *Critical Reviews in Food Science and Nutrition, 45*, 405–423.

Dickinson, E. (2009). Hydrocolloids as emulsifiers and emulsion stabilizers. *Food Hydrocolloids, 23*, 1473–1482.

Do, D. T., Singh, J., Oey, I., & Singh, H. (2018). Biomimetic plant foods: Structural design and functionality. *Trends in Food Science & Technology, 82*, 46–59.

Dreher, M. L., Maher, C. V., & Kearney, P. (1996). The traditional and emerging role of nuts in healthful diets. *Nutrition Reviews, 54*, 241–245.

Dubey, M. R., & Patel, V. P. (2018). Probiotics: A promising tool for calcium absorption. *The Open Nutrition Journal*, 12.

Dueñas, M., Estrella, I., & Hernández, T. (2004). Occurrence of phenolic compounds in the seed coat and the cotyledon of peas (*Pisum sativum L.*). *European Food Research and Technology, 219*, 116–123.

Eertmans, A., Baeyens, F., & Van den Bergh, O. (2001). Food likes and their relative importance in human eating behavior: Review and preliminary suggestions for health promotion. *Health Education Research, 16*, 443–456.

El Hazzam, K., Hafsa, J., Sobeh, M., Mhada, M., Taourirte, M., El Kacimi, K., & Yasri, A. (2020). An insight into saponins from quinoa (Chenopodium quinoa Willd): A review. *Molecules, 25*, 1059.

Elsohaimy, S. A., Refaay, T. M., & Zaytoun, M. A. M. (2015). Physicochemical and functional properties of quinoa protein isolate. *Annals of Agricultural Sciences, 60*, 297–305.

El-Adawy, T. A., Rahma, E. H., El-Bedawy, A. A., & Sobihah, T. Y. (2000). Effect of soaking process on nutritional quality and protein solubility of some legume seeds. *Food/Nahrung, 44*, 339–343.

Faccin, G. L., Miotto, L. A., do Nascimento Vieira, L., Barreto, P. L. M., & Amante, E. R. (2009). Chemical, sensorial and rheological properties of a new organic rice bran beverage. *Rice Science, 16*, 226–234.

Fan, P.-H., Zang, M.-T., & Xing, J. (2015). Oligosaccharides composition in eight food legumes species as detected by high-resolution mass spectrometry. *Journal of the Science of Food and Agriculture, 95*, 2228–2236.

Felberg, I., Antoniassi, R., Deliza, R., Freitas, S. C.d., & Modesta, R. C. D. (2009). Soy and Brazil nut beverage: Processing, composition, sensory, and color evaluation. *Food Science and Technology, 29*, 609–617.

Felix, M., Cermeño, M., & FitzGerald, R. J. (2019). Assessment of the microstructural characteristics and the in vitro bioactive properties of sunflower oil-based emulsions stabilized by fava bean (vicia faba) protein. *Food Hydrocolloids, 97*, 105220.

Ferragut, V., Valencia-Flores, D. C., Pérez-González, M., Gallardo, J., & Hernández-Herrero, M. (2015). Quality characteristics and shelf-life of ultra-high pressure homogenized (UHPH) almond beverage. *Foods, 4*, 159–172.

Fitrotin, U., Utami, T., Hastuti, P., & Santoso, U. (2015). Antioxidant properties of fermented sesame milk using Lactobacillus plantarum Dad 13. *International Research Journal of Biological Sciences, 4*, 56–61.

Floury, J., Legrand, J., & Desrumaux, A. (2004). Analysis of a new type of high pressure homogeniser. Part B. study of droplet break-up and recoalescence phenomena. *Chemical Engineering Science, 59*, 1285–1294.

Frassinetti, S., Moccia, E., Caltavuturo, L., Gabriele, M., Longo, V., Bellani, L., Giorgi, G., & Giorgetti, L. (2018). Nutraceutical potential of hemp (*Cannabis sativa L.*) seeds and sprouts. *Food Chemistry, 262*, 56–66.

Friedman, M. (1996). Nutritional value of proteins from different food sources. A review. *Journal of Agricultural and Food Chemistry, 44*, 6–29.

FSANZ. (2016). *Plant-based milk alternatives*.

Gallier, S., Gordon, K. C., & Singh, H. (2012). Chemical and structural characterisation of almond oil bodies and bovine milk fat globules. *Food Chemistry, 132*, 1996–2006.

Galvez, F. C. F., Resurreccion, A. V. A., & Koehler, P. E. (1990). Optimization of processing of peanut beverage 1. *Journal of Sensory Studies, 5*, 1–17.

Gibson, R. S., Perlas, L., & Hotz, C. (2006). Improving the bioavailability of nutrients in plant foods at the household level. *Proceedings of the Nutrition Society, 65*, 160–168.

Gorzynik-Debicka, M., Przychodzen, P., Cappello, F., Kuban-Jankowska, A., Marino Gammazza, A., Knap, N., Wozniak, M., & Gorska-Ponikowska, M. (2018). Potential health benefits of olive oil and plant polyphenols. *International Journal of Molecular Sciences, 19*, 686.

Goufo, P., Ferreira, L. M. M., Trindade, H., & Rosa, E. A. S. (2015). Distribution of antioxidant compounds in the grain of the Mediterranean rice variety 'Ariete. *CyTA-Journal of Food, 13*, 140–150.

Grundy, M. M. L., Lapsley, K., & Ellis, P. R. (2016). A review of the impact of processing on nutrient bioaccessibility and digestion of almonds. *International Journal of Food Science & Technology, 51*, 1937–1946.

Gumus, C. E., Decker, E. A., & McClements, D. J. (2017). Formation and stability of ω-3 oil emulsion-based delivery systems using plant proteins as emulsifiers: Lentil, pea, and faba bean proteins. *Food Biophysics, 12*, 186–197.

Guzey, D., & McClements, D. J. (2006). Formation, stability and properties of multilayer emulsions for application in the food industry. *Advances in Colloid and Interface Science, 128*, 227–248.

Haas, R., Schnepps, A., Pichler, A., & Meixner, O. (2019). Cow milk versus plant-based milk substitutes: A comparison of product image and motivational structure of consumption. *Sustainability, 11*, 5046.

Haldipur, A. C., & Srividya, N. (2021). A comparative evaluation of in vitro antihyperglycemic potential of Bamboo seed rice (*Bambusa arundinacea*) and Garudan samba (*Oryza sativa*): An integrated metabolomics, enzymatic and molecular docking approach. *Journal of Cereal Science*, 103200.

Hambidge, K. M. (2010). Micronutrient bioavailability: Dietary reference intakes and a future perspective. *The American Journal of Clinical Nutrition*, 91, 1430S–1432S.

Hasler, C. M. (2002). The cardiovascular effects of soy products. *Journal of Cardiovascular Nursing*, 16, 50–63.

Haug, A., Høstmark, A. T., & Harstad, O. M. (2007). Bovine milk in human nutrition – a review. *Lipids in Health and Disease*, 6, 25.

Hebishy, E., Buffa, M., Guamis, B., Blasco-Moreno, A., & Trujillo, A.-J. (2015). Physical and oxidative stability of whey protein oil-in-water emulsions produced by conventional and ultra high-pressure homogenization: Effects of pressure and protein concentration on emulsion characteristics. *Innovative Food Science & Emerging Technologies*, 32, 79–90.

Hemati, N., Asis, M., Moradi, S., Mollica, A., Stefanucci, A., Nikfar, S., Mohammadi, E., Farzaei, M. H., & Abdollahi, M. (2020). Effects of genistein on blood pressure: A systematic review and meta-analysis. *Food Research International*, 128, 108764.

Herrmann, W., & Geisel, J. (2002). Vegetarian lifestyle and monitoring of vitamin B-12 status. *Clinica Chimica Acta*, 326, 47–59.

Hodges, J. K., Cao, S., Cladis, D. P., & Weaver, C. M. (2019). Lactose intolerance and bone health: The challenge of ensuring adequate calcium intake. *Nutrients*, 11, 718.

Hsiao, Y.-H., Ho, C.-T., & Pan, M.-H. (2020). Bioavailability and health benefits of major isoflavone aglycones and their metabolites. *Journal of Functional Foods*, 74, 104164.

Hsieh, O. A.-L., Huang, A.-S., & Chang, S. S. (1982). Isolation and identification of objectionable volatile flavor compounds in defatted soybean flour. *Journal of Food Science*, 47, 16–18.

Hsieh, H.-M., Wu, W.-M., & Hu, M.-L. (2009). Soy isoflavones attenuate oxidative stress and improve parameters related to aging and Alzheimer's disease in C57BL/6J mice treated with d-galactose. *Food and Chemical Toxicology*, 47, 625–632.

Huang, J., Song, G., Zhang, L., Sun, Q., & Lu, X. (2012). A novel conversion of sesamolin to sesaminol by acidic cation exchange resin. *European Journal of Lipid Science and Technology*, 114, 842–848.

Iorio, M. C., Bevilacqua, A., Corbo, M. R., Campaniello, D., Sinigaglia, M., & Altieri, C. (2019). A case study on the use of ultrasound for the inhibition of *Escherichia coli* O157: H7 and Listeria monocytogenes in almond milk. *Ultrasonics Sonochemistry*, 52, 477–483.

Isanga, J., & Zhang, G.-N. (2008). Soybean bioactive components and their implications to health—a review. *Food Reviews International*, 24, 252–276.

Isanga, J., & Zhang, G. (2009). Production and evaluation of some physicochemical parameters of peanut milk yoghurt. *LWT - Food Science and Technology*, 42, 1132–1138.

Ishaq, Z., & Nawaz, M. A. (2018). Analysis of contaminated milk with organochlorine pesticide residues using gas chromatography. *International Journal of Food Properties*, 21, 879–891.

Issara, U., & Rawdkuen, S. (2016). Rice bran: A potential of main ingredient in healthy beverage. *International Food Research Journal*, 23.

Iwanaga, D., Gray, D. A., Fisk, I. D., Decker, E. A., Weiss, J., & McClements, D. J. (2007). Extraction and characterization of oil bodies from soy beans: A natural source of pre-emulsified soybean oil. *Journal of Agricultural and Food Chemistry*, 55, 8711–8716.

Jaiswal, N., Akhtar, J., Singh, S. P., & Ahsan, F. (2019). An overview on genistein and its various formulations. *Drug Research*, 69, 305–313.

Jancurová, M., Minarovičová, L., & Dandar, A. (2009). Quinoa–a rewiev. *Czech Journal of Food Sciences*, 27, 71–79.

Jeske, S., Zannini, E., & Arendt, E. K. (2017). Evaluation of physicochemical and glycaemic properties of commercial plant-based milk substitutes. *Plant Foods for Human Nutrition*, 72, 26–33.

Jeske, S., Zannini, E., & Arendt, E. K. (2018a). Past, present and future: The strength of plant-based dairy substitutes based on gluten-free raw materials. *Food Research International*, 110, 42–51.

Jeske, S., Zannini, E., Cronin, M. F., & Arendt, E. K. (2018b). Impact of protease and amylase treatment on proteins and the product quality of a quinoa-based milk substitute. *Food & Function*, 9, 3500–3508.

Jiang, S., Cai, W., & Xu, B. (2013). Food quality improvement of soy milk made from short-time germinated soybeans. *Foods*, 2, 198–212.

Jiang, S.-Y., Ma, A., Xie, L., & Ramachandran, S. (2016). Improving protein content and quality by over-expressing artificially synthetic fusion proteins with high lysine and threonine constituent in rice plants. *Scientific Reports*, 6, 34427.

Johnson, G. W. (1968). Simulated milk comprising soy bean flour, sesame seed flour and coconut meal. *Google Patents*, Article US3386833A.

Jung, E.-H., Ha, T.-Y., & Hwang, I.-K. (2010). Anti-hyperglycemic and antioxidative activities of phenolic acid concentrates of rice bran and hydroxycinnamic acids in cell assays. *The Korean Journal of Food and Nutrition*, 23, 233–239.

Kalita, S., Khandelwal, S., Madan, J., Pandya, H., Sesikeran, B., & Krishnaswamy, K. (2018). Almonds and cardiovascular health: A review. *Nutrients, 10*, 468.

Kamboj, R., & Nanda, V. (2018). Proximate composition, nutritional profile and health benefits of legumes-A review. *Legume Research-An International Journal*, 41, 325–332.

Kamchan, A., Puwastien, P., Sirichakwal, P. P., & Kongkachuichai, R. (2004). In vitro calcium bioavailability of vegetables, legumes and seeds. *Journal of Food Composition and Analysis*, 17, 311–320.

Kamil, A., & Chen, C. Y. O. (2012). Health benefits of almonds beyond cholesterol reduction. *Journal of Agricultural and Food Chemistry*, 60, 6694–6702.

Karaca, A. C., Low, N., & Nickerson, M. (2011). Emulsifying properties of chickpea, faba bean, lentil and pea proteins produced by isoelectric precipitation and salt extraction. *Food Research International*, 44, 2742–2750.

Kargozar, R., Azizi, H., & Salari, R. (2017). A review of effective herbal medicines in controlling menopausal symptoms. *Electronic Physician*, 9, 5826.

Karunasiri, A. N., Gunawardane, M., Senanayake, C. M., Jayathilaka, N., & Seneviratne, K. N. (2020). Antioxidant and nutritional properties of domestic and commercial coconut milk preparations. *International Journal of Food Science, 2020*, Article 3489605.

Keast, R. S. J., & Lau, J. J. (2006). Culture-specific variation in the flavor profile of soymilks. *Journal of Food Science*, 71, S567–S572.

Khaleque, A., Bannatyne, W. R., & Wallace, G. M. (1970). Studies on the processing and properties of soymilk I.—Effect of preprocessing conditions on the flavour and compositions of soymilks. *Journal of the Science of Food and Agriculture,* 21, 579—583.

Khandpur, N., Martinez-Steele, E., & Sun, Q. (2021). Plant-based meat and dairy substitutes as appropriate alternatives to animal-based products? *The Journal of Nutrition,* 151, 3—4.

Khatoon, S., Raja Rajan, R. G., & Gopala Krishna, A. G. (2010). Physicochemical characteristics and composition of Indian soybean oil deodorizer distillate and the recovery of phytosterols. *Journal of the American Oil Chemists' Society,* 87, 321—326.

Khosravi, A., & Razavi, S. H. (2020). The role of bioconversion processes to enhance polyphenol bioaccessibility in rice bioaccessibility of polyphenols in rice. *Food Bioscience*, 100605.

Kluczkovski, A. M., Martins, M., Mundim, S. M., Simoes, R. H., Nascimento, K. S., Marinho, H. A., & Junior, A. K. (2015). Properties of Brazil nuts: A review. *African Journal of Biotechnology,* 14, 642—648.

Kumar, V., Rani, A., & Hussain, L. (2016). Essential amino acids profile of differentially processed soy products and their efficiency in meeting daily requirement. *Nutrition & Food Science,* 46, 237—245.

Kurbel, S., Kurbel, B., & Včev, A. (2006). Intestinal gases and flatulence: Possible causes of occurrence. *Medical Hypotheses,* 67, 235—239.

Kwok, K. C., Qin, W. H., & Tsang, J. C. (1993). Heat inactivation of trypsin inhibitors in soymilk at ultra-high temperatures. *Journal of Food Science,* 58, 859—862.

Kwon, S. H., Kang, M. J., Huh, J. S., Ha, K. W., Lee, J. R., Lee, S. K., Lee, B. S., Han, I. H., Lee, M. S., & Lee, M. W. (2007). Comparison of oral bioavailability of genistein and genistin in rats. *International Journal of Pharmaceutics,* 337, 148—154.

Kyriakopoulou, K., Keppler, J. K., van der Goot, A. J., & Boom, R. M. (2021). Alternatives to meat and dairy. *Annual Review of Food Science and Technology*, 12.

Ladjal Ettoumi, Y., Chibane, M., & Romero, A. (2016). Emulsifying properties of legume proteins at acidic conditions: Effect of protein concentration and ionic strength. *LWT—Food Science and Technology,* 66, 260—266.

Lal, N., Barcchiya, J., Raypuriya, N., & Shiurkar, G. (2017). Anti-nutrition in legumes: Effect in human health and its elimination. *Innovative Farming,* 2, 32—36.

Lam, A. C. Y., Can Karaca, A., Tyler, R. T., & Nickerson, M. T. (2018). Pea protein isolates: Structure, extraction, and functionality. *Food Reviews International,* 34, 126—147.

Lee, C., & Beuchat, L. R. (1991). Changes in chemical composition and sensory qualities of peanut milk fermented with lactic acid bacteria. *International Journal of Food Microbiology,* 13, 273—283.

Lee, C., & Beuchat, L. R. (1992). Chemical, physical and sensory characteristics of peanut milk as affected by processing conditions. *Journal of Food Science,* 57, 401—405.

Lee, L., & Norton, I. T. (2013). Comparing droplet breakup for a high-pressure valve homogeniser and a microfluidizer for the potential production of food-grade nanoemulsions. *Journal of Food Engineering,* 114, 158—163.

Leonard, W., Zhang, P., Ying, D., & Fang, Z. (2020). Lignanamides: Sources, biosynthesis and potential health benefits—a minireview. *Critical Reviews in Food Science and Nutrition,* 1—11.

Liang, H.-N., & Tang, C.-H. (2013). pH-dependent emulsifying properties of pea [Pisum sativum (L.)] proteins. *Food Hydrocolloids,* 33, 309—319.

Li, X., & de Vries, R. (2018). Interfacial stabilization using complexes of plant proteins and polysaccharides. *Current Opinion in Food Science,* 21, 51—56.

Lifschitz, C., & Szajewska, H. (2015). Cow's milk allergy: Evidence-based diagnosis and management for the practitioner. *European Journal of Pediatrics,* 174, 141—150.

Lipan, L., Cano-Lamadrid, M., Hernández, F., Sendra, E., Corell, M., Vázquez-Araújo, L., Moriana, A., & Carbonell-Barrachina, Á. A. (2020). Long-term correlation between water deficit and quality markers in HydroSOStainable almonds. *Agronomy,* 10, 1470.

Litt, A., & Shiv, B. (2012). Manipulating basic taste perception to explore how product information affects experience. *Journal of Consumer Psychology,* 22, 55—66.

Liu, K. (2019). Soybean trypsin inhibitor Assay: The sequence effect of adding reagents, factors involved, and mechanistic explanations. *Journal of the American Oil Chemists' Society,* 96, 619—633.

Li, Z.-j., Xue, Y.-b., & Zhang, L. (2018). Study on liquefaction and saccharification of quinoa beverage. *Journal of Anhui Agricultural Sciences*, 18, 2018.

Lola, A. (2009). The effect of boiling on the nutrients and anti-nutrients in two non-conventional vegetables. *Pakistan Journal of Nutrition, 8*, 1430—1433.

Luo, Q., Yan, X., Bobrovskaya, L., Ji, M., Yuan, H., Lou, H., & Fan, P. (2017). Anti-neuroinflammatory effects of grossamide from hemp seed via suppression of TLR-4-mediated NF-κB signaling pathways in lipopolysaccharide-stimulated BV2 microglia cells. *Molecular and Cellular Biochemistry,* 428, 129—137.

Lutz, R., Aserin, A., Wicker, L., & Garti, N. (2009). Double emulsions stabilized by a charged complex of modified pectin and whey protein isolate. *Colloids and Surfaces B: Biointerfaces,* 72, 121—127.

Lu, Y., Zhao, A., Wu, Y., Zhao, Y., & Yang, X. (2019). Soybean soluble polysaccharides enhance bioavailability of genistein and its prevention against obesity and metabolic syndrome of mice with chronic high fat consumption. *Food & Function, 10*, 4153—4165.

Lv, S., Zhou, H., Bai, L., Rojas, O. J., & McClements, D. J. (2020). Development of food-grade Pickering emulsions stabilized by a mixture of cellulose nanofibrils and nanochitin. *Food Hydrocolloids*, 106451.

Lynch, H., Johnston, C., & Wharton, C. (2018). Plant-based diets: Considerations for environmental impact, protein quality, and exercise performance. *Nutrients,* 10, 1841.

Mäkinen, O. E., Uniacke-Lowe, T., O'Mahony, J. A., & Arendt, E. K. (2015). Physicochemical and acid gelation properties of commercial UHT-treated plant-based milk substitutes and lactose free bovine milk. *Food Chemistry,* 168, 630–638.

Mäkinen, O. E., Wanhalinna, V., Zannini, E., & Arendt, E. K. (2016). Foods for special dietary needs: Non-dairy plant-based milk substitutes and fermented dairy-type products. *Critical Reviews in Food Science and Nutrition,* 56, 339–349.

Mandalari, G., & Mackie, A. R. (2018). Almond allergy: An overview on prevalence, thresholds, regulations and allergen detection. *Nutrients, 10,* 1706.

Markets, & Markets. (2021). *Dairy alternatives market by source (soy, almond, coconut, oats, rice, hemp), application (milk, Yogurt, ice creams, cheese, creamers), distribution channel (supermarkets, health food stores, pharmacies), formulation, and region—Global Forecast to 2026.*

Martel, F., Monteiro, R., & Calhau, C. (2010). Effect of polyphenols on the intestinal and placental transport of some bioactive compounds. *Nutrition Research Reviews,* 23, 47–64.

Massey, L. K. (2003). Dietary animal and plant protein and human bone health: A whole foods approach. *The Journal of Nutrition,* 133, 862S–865S.

Mathews, R., Kamil, A., & Chu, Y. (2020). Global review of heart health claims for oat beta-glucan products. *Nutrition Reviews,* 78, 78–97.

Mazumder, M. A., & Hongsprabhas, P. (2016). A review on nutrient quality of soymilk powder for malnourished population. *Pakistan Journal of Nutrition,* 15, 600–606.

McAuliffe, G. A., Takahashi, T., & Lee, M. R. F. (2020). Applications of nutritional functional units in commodity-level life cycle assessment (LCA) of agri-food systems. *The international Journal of Life Cycle Assessment,* 25, 208–221.

McClements, D. J. (2020a). Development of next-generation nutritionally fortified plant-based milk substitutes: Structural design principles. *Foods,* 9, 421.

McClements, D. J. (2020b). Development of next-generation nutritionally fortified plant-based milk substitutes: Structural design principles. *Foods*, 9.

McClements, D. J., & Gumus, C. E. (2016). Natural emulsifiers—Biosurfactants, phospholipids, biopolymers, and colloidal particles: Molecular and physicochemical basis of functional performance. *Advances in Colloid and Interface Science,* 234, 3–26.

McClements, D. J., Newman, E., & McClements, I. F. (2019). Plant-based milks: A review of the science underpinning their design, fabrication, and performance. *Comprehensive Reviews in Food Science and Food Safety, 18*(6), 2047–2067.

McHugh, T. (2018). How plant-based milks are processed. *Food Technology,* 72, 63–64.

McWilliam, V., Koplin, J., Lodge, C., Tang, M., Dharmage, S., & Allen, K. (2015). The prevalence of tree nut allergy: A systematic review. *Current Allergy and Asthma Reports,* 15, 54.

Medina-Vera, I., Gómez-de-Regil, L., Gutiérrez-Solis, A. L., Lugo, R., Guevara-Cruz, M., Pedraza-Chaverri, J., & Avila-Nava, A. (2021). Dietary strategies by foods with antioxidant effect on nutritional management of dyslipidemias: A systematic review. *Antioxidants*, 10.

Mehansho, H. (2006). Iron fortification technology development: New approaches. *The Journal of Nutrition,* 136, 1059–1063.

Mensah-Brown, H., Afoakwa, E. O., & Wadie, W. B. (2014). Optimization of the production of a chocolate-flavoured, soy-peanut beverage with acceptable chemical and physicochemical properties using a three-component constrained extreme lattice mixture design. *Optimization,* 23, 46–54.

Mintel. (2018). *US non-dairy milk sales grow 61% over the last five years.*

Mir, N. A., Riar, C. S., & Singh, S. (2018). Nutritional constituents of pseudo cereals and their potential use in food systems: A review. *Trends in Food Science & Technology,* 75, 170–180.

Mottet, A., de Haan, C., Falcucci, A., Tempio, G., Opio, C., & Gerber, P. (2017). Livestock: On our plates or eating at our table? A new analysis of the feed/food debate. *Global Food Security,* 14, 1–8.

Munekata, P. E. S., Domínguez, R., Budaraju, S., Roselló-Soto, E., Barba, F. J., Mallikarjunan, K., Roohinejad, S., & Lorenzo, J. M. (2020). Effect of innovative food processing technologies on the physicochemical and nutritional properties and quality of non-dairy plant-based beverages. *Foods,* 9, 288.

Munera-Picazo, S., Ramírez-Gandolfo, A., Cascio, C., Castaño-Iglesias, C., Signes-Pastor, A. J., Burló, F., Haris, P. I., & Carbonell-Barrachina, Á. A. (2014). Chapter 29—Arsenic in rice-based infant foods. In R. R. Watson, V. R. Preedy, & S. Zibadi (Eds.), *Wheat and rice in disease prevention and health.* Academic Press. San Diego.

Muthayya, S., Sugimoto, J. D., Montgomery, S., & Maberly, G. F. (2014). An overview of global rice production, supply, trade, and consumption. *Annals of the New York Academy of Sciences,* 1324, 7–14.

Nagendra Prasad, M. N., Sanjay, K. R., Prasad, D. S., Vijay, N., Kothari, R., & Nanjunda Swamy, S. (2012). A review on nutritional and nutraceutical properties of sesame. *Journal of Nutrition and Food Sciences,* 2, 1–6.

Nandi, I., & Ghosh, M. (2015). Studies on functional and antioxidant property of dietary fibre extracted from defatted sesame husk, rice bran and flaxseed. *Bioactive Carbohydrates and Dietary Fibre, 5,* 129–136.

Narsimhan, G., Wang, Z., & Xiang, N. (2019). Guidelines for processing emulsion-based foods. In G. L. Hasenhuettl, & R. W. Hartel (Eds.), *Food emulsifiers and their applications.* Springer International Publishing. Cham.

Navarini, L., & Rivetti, D. (2010). Water quality for Espresso coffee. *Food Chemistry,* 122, 424–428.

Navruz-Varli, S., & Sanlier, N. (2016). Nutritional and health benefits of quinoa (Chenopodium quinoa Willd.). *Journal of Cereal Science,* 69, 371–376.

Nawaz, M. A., Fukai, S., & Bhandari, B. (2016). Effect of alkali treatment on the milled grain surface protein and physicochemical properties of two contrasting rice varieties. *Journal of Cereal Science,* 72, 16–23.

Nawaz, M. A. (2018). *Processing and quality of glutinous rice. PhD.* The University of Queensland.

Nawaz, M. A., Singh, T. K., Jegasothy, H., & Buckow, R. (2020a). UHT treatment on the stability of faba bean protein emulsion. In *1st international electronic conference on food science and functional foods.* MDPI.

Nawaz, M. A., Tan, M., Øiseth, S., & Buckow, R. (2020b). An emerging segment of functional legume-based beverages: A review. *Food Reviews International*, 1–39.

Nawaz, M. A., Singh, T. K., Stockmann, R., Jegasothy, H., & Buckow, R. (2021). Quality attributes of ultra-high temperature-treated model beverages prepared with faba bean protein concentrates. *Foods,* 10(6), 1244.

Nelson, A. I., Steinberg, M. I., & Wei, L. S. (1976). Illinois process for preparation of soymilk. *Journal of Food Science,* 41, 57–61.

Nikiforidis, C. V., Matsakidou, A., & Kiosseoglou, V. (2014). Composition, properties and potential food applications of natural emulsions and cream materials based on oil bodies. *RSC Advances,* 4, 25067–25078.

Nowak, V., Du, J., & Charrondière, U. R. (2016). Assessment of the nutritional composition of quinoa (Chenopodium quinoa Willd.). *Food Chemistry,* 193, 47–54.

Ntukidem, V. E., Ukwo, P. S., Udoh, I. E., & Umoinyang, E. U. (2019). Influence of different pre-treatments on the nutritional and organoleptic properties of vegetable milk produced locally from tiger-nut (Cyperus esculentus) tubers. *IOSR Journal of Environmental Science, Toxicology and Food Technology,* 13, 55–61.

Obizoba, I. C., & Amaechi, N. A. (1993). The effect of processing methods on the chemical composition of baobab (Adansonia digitata. L) pulp and seed. *Ecology of Food and Nutrition,* 29, 199–205.

Oduro, A. F., Saalia, F. K., & Adjei, M. Y. B. (2021). Sensory acceptability and proximate composition of 3-blend plant-based dairy alternatives. *Foods,* 10, 482.

Omoni, A. O., & Aluko, R. E. (2005). Soybean foods and their benefits: Potential mechanisms of action. *Nutrition Reviews,* 63, 272–283.

Onsaard, E. (2012). Sesame proteins. *International Food Research Journal, 19.*

Pande, P., Fleck, S. C., Twaddle, N. C., Churchwell, M. I., Doerge, D. R., & Teeguarden, J. G. (2019). Comparative estrogenicity of endogenous, environmental and dietary estrogens in pregnant women II: Total estrogenicity calculations accounting for competitive protein and receptor binding and potency. *Food and Chemical Toxicology,* 125, 341–353.

Panzolini, C. R. L. D., Lima, J. P., Nascimento, P. G. B. D., & Ghesti, G. F. (2017). Estudo Prospectivo sobre Produtos A base de quinoa Para A indústria Alimenticia. *Cadernos de Prospecção,* 10, 765.

Patisaul, H. B., & Jefferson, W. (2010). The pros and cons of phytoestrogens. *Frontiers in Neuroendocrinology,* 31, 400–419.

Paul, A. A., Kumar, S., Kumar, V., & Sharma, R. (2020). Milk Analog: Plant based alternatives to conventional milk, production, potential and health concerns. *Critical Reviews in Food Science and Nutrition,* 60, 3005–3023.

Peng, C., Cao, C., He, M., Shu, Y., Tang, X., Wang, Y., Zhang, Y., Xia, X., Li, Y., & Wu, J. (2018). Soybean glycinin- and β-conglycinin-induced intestinal damage in piglets via the p38/JNK/NF-κB signaling pathway. *Journal of Agricultural and Food Chemistry,* 66, 9534–9541.

Pineli, L.d. L.d. O., Botelho, R. B. A., Zandonadi, R. P., Solorzano, J. L., de Oliveira, G. T., Reis, C. E. G., & Teixeira, D.d. S. (2015). Low glycemic index and increased protein content in a novel quinoa milk. *LWT-Food Science and Technology,* 63, 1261–1267.

Pinnamaneni, S., Das, N. G., & Das, S. K. (2003). Comparison of oil-in-water emulsions manufactured by microfluidization and homogenization. *Die Pharmazie-An International Journal of Pharmaceutical Sciences,* 58, 554–558.

Pistollato, F., Sumalla Cano, S., Elio, I., Masias Vergara, M., Giampieri, F., & Battino, M. (2015). Plant-based and plant-rich diet patterns during gestation: Beneficial effects and possible shortcomings. *Advances in Nutrition,* 6, 581–591.

Poliseli-Scopel, F. H., Hernández-Herrero, M., Guamis, B., & Ferragut, V. (2012). Comparison of ultra high pressure homogenization and conventional thermal treatments on the microbiological, physical and chemical quality of soymilk. *LWT—Food Science and Technology,* 46, 42–48.

Poojary, M. M., Lund, M. N., & Barba, F. J. (2020). *Pulsed electric field (PEF) as an efficient technology for food additives and nutraceuticals development. Pulsed Electric Fields to Obtain Healthier and Sustainable Food for Tomorrow.* Elsevier.

Prades, A., Dornier, M., Diop, N., & Pain, J.-P. (2012a). Coconut water preservation and processing: A review. *Fruits,* 67, 157–171.

Prades, A., Dornier, M., Diop, N., & Pain, J.-P. (2012b). Coconut water uses, composition and properties: A review. *Fruits,* 67, 87–107.

Qamar, S., Bhandari, B., & Prakash, S. (2019). Effect of different homogenisation methods and UHT processing on the stability of pea protein emulsion. *Food Research International,* 116, 1374–1385.

Qiu, C., Zhao, M., & McClements, D. J. (2015). Improving the stability of wheat protein-stabilized emulsions: Effect of pectin and xanthan gum addition. *Food Hydrocolloids,* 43, 377–387.

Quan, W., Tao, Y., Qie, X., Zeng, M., Qin, F., Chen, J., & He, Z. (2020). Effects of high-pressure homogenization, thermal processing, and milk matrix on the in vitro bioaccessibility of phenolic compounds in pomelo and kiwi juices. *Journal of Functional Foods,* 64, 103633.

Quasem, J. M., Mazahreh, A. S., & Abu-Alruz, K. (2009). Development of vegetable based milk from decorticated sesame (Sesamum indicum). *American Journal of Applied Sciences,* 6, 888.

Rasane, P., Jha, A., Sabikhi, L., Kumar, A., & Unnikrishnan, V. S. (2015). Nutritional advantages of oats and opportunities for its processing as value added foods-a review. *Journal of Food Science and Technology,* 52, 662–675.

Rayner, M. (2015). In M. Rayner, & P. Dejmek (Eds.), *Scales and forces in emulsification. Engineering aspects of food emulsification and homogenization* (pp. 3–32).

Ribas-Agustí, A., Martín-Belloso, O., Soliva-Fortuny, R., & Elez-Martínez, P. (2018). Food processing strategies to enhance phenolic compounds bioaccessibility and bioavailability in plant-based foods. *Critical Reviews in Food Science and Nutrition,* 58, 2531–2548.

Salami, S. A., Luciano, G., O'Grady, M. N., Biondi, L., Newbold, C. J., Kerry, J. P., & Priolo, A. (2019). Sustainability of feeding plant by-products: A review of the implications for ruminant meat production. *Animal Feed Science and Technology,* 251, 37–55.

Salminen, H., Bischoff, S., & Weiss, J. (2019). Impact of concentration ratio on the formation and stability of emulsions stabilized by Quillaja saponin–sodium caseinate mixtures. *Food Biophysics,* 14, 109–119.

Sandberg, A.-S. (2002). Bioavailability of minerals in legumes. *British Journal of Nutrition,* 88, 281–285.

Sang, S., & Chu, Y. (2017). Whole grain oats, more than just a fiber: Role of unique phytochemicals. *Molecular Nutrition & Food Research,* 61, 1600715.

Sathyapalan, T., Aye, M., Rigby, A. S., Thatcher, N. J., Dargham, S. R., Kilpatrick, E. S., & Atkin, S. L. (2018). Soy isoflavones improve cardiovascular disease risk markers in women during the early menopause. *Nutrition, Metabolism and Cardiovascular Diseases,* 28, 691–697.

Scholz-Ahrens, K. E., Ahrens, F., & Barth, C. A. (2020). Nutritional and health attributes of milk and milk imitations. *European Journal of Nutrition,* 59, 19–34.

Schultz, C. J., Lim, W. L., Khor, S. F., Neumann, K. A., Schulz, J. M., Ansari, O., Skewes, M. A., & Burton, R. A. (2020). Consumer and health-related traits of seed from selected commercial and breeding lines of industrial hemp, Cannabis sativa L. *Journal of Agriculture and Food Research,* 2, 100025.

Sethi, S., Tyagi, S. K., & Anurag, R. K. (2016). Plant-based milk alternatives an emerging segment of functional beverages: A review. *Journal of Food Science and Technology,* 53, 3408–3423.

Shao, Y., & Bao, J. (2015). Polyphenols in whole rice grain: Genetic diversity and health benefits. *Food Chemistry,* 180, 86–97.

Sharma, N., & Gayathri, R. (2018). Assessment of nutritional value of overnight soaked cooked rice over un-soaked cooked rice. *International Journal of Research in Pharmaceutical Sciences*, 9.

Silva, K., Machado, A., Cardoso, C., Silva, F., & Freitas, F. (2020). Rheological behavior of plant-based beverages. *Food Science and Technology,* 40, 258–263.

Small, E. (2016). *Cannabis: A complete guide*. CRC Press.

Smýkal, P., Aubert, G., Burstin, J., Coyne, C. J., Ellis, N. T. H., Flavell, A. J., Ford, R., Hýbl, M., Macas, J., Neumann, P., McPhee, K. E., Redden, R. J., Rubiales, D., Weller, J. L., & Warkentin, T. D. (2012). Pea (Pisum sativum L.) in the genomic era. *Agronomy,* 2, 74–115.

Stang, M., Schuchmann, H., & Schubert, H. (2001). Emulsification in high-pressure homogenizers. *Engineering in Life Sciences,* 1, 151–157.

Taha, A., Hu, T., Zhang, Z., Bakry, A. M., Khalifa, I., Pan, S., & Hu, H. (2018). Effect of different oils and ultrasound emulsification conditions on the physicochemical properties of emulsions stabilized by soy protein isolate. *Ultrasonics Sonochemistry,* 49, 283–293.

Taha, A., Ahmed, E., Ismaiel, A., Ashokkumar, M., Xu, X., Pan, S., & Hu, H. (2020). Ultrasonic emulsification: An overview on the preparation of different emulsifiers-stabilized emulsions. *Trends in Food Science & Technology, 105*, 363–377.

Tamm, F., Herbst, S., Brodkorb, A., & Drusch, S. (2016). Functional properties of pea protein hydrolysates in emulsions and spray-dried microcapsules. *Food Hydrocolloids,* 58, 204–214.

Tang, J., Xu, N., Ji, H., Liu, H., Wang, Z., & Wu, L. (2011). Eudragit nanoparticles containing genistein: Formulation, development, and bioavailability assessment. *International Journal of Nanomedicine,* 6, 2429.

Tangyu, M., Muller, J., Bolten, C. J., & Wittmann, C. (2019). Fermentation of plant-based milk alternatives for improved flavour and nutritional value. *Applied Microbiology and Biotechnology,* 103, 9263–9275.

Tan, M., Nawaz, M. A., & Buckow, R. (2021). Functional and food application of plant proteins—A review. *Food Reviews International,* 1–29.

Teh, S.-S., & Birch, E. J. (2014). Effect of ultrasonic treatment on the polyphenol content and antioxidant capacity of extract from defatted hemp, flax and canola seed cakes. *Ultrasonics Sonochemistry,* 21, 346–353.

Tham, D. M., Gardner, C. D., & Haskell, W. L. (1998). Potential health benefits of dietary phytoestrogens: A review of the clinical, epidemiological, and mechanistic evidence. *The Journal of Clinical Endocrinology & Metabolism,* 83, 2223–2235.

Toomer, O. T. (2018). Nutritional chemistry of the peanut (*Arachis hypogaea*). *Critical Reviews in Food Science and Nutrition,* 58, 3042–3053.

Topcuoglu, E., & Yilmaz-Ersan, L. (2020). Effect of fortification with almond milk on quality characteristics of probiotic yoghurt. *Journal of Food Processing and Preservation*, e14943.

Tripathi, V., Abidi, A. B., Markerb, S., & Bilal, S. (2013). Linseed and linseed oil: Health benefits-a review. *International Journal of Pharmacy and Biological Sciences, 3*, 434–442.

Tso, R., Lim, A. J., & Forde, C. G. (2021). A critical Appraisal of the evidence supporting consumer motivations for alternative proteins. *Foods*, 10.

Tzen, J. T., & Huang, A. H. (1992). Surface structure and properties of plant seed oil bodies. *Journal of Cell Biology,* 117, 327–335.

Ujiroghene, O. J., Liu, L., Zhang, S., Lu, J., Zhang, C., Pang, X., & Lv, J. (2019). Potent α-amylase inhibitory activity of sprouted quinoa-based yoghurt beverages fermented with selected anti-diabetic strains of lactic acid bacteria. *RSC Advances,* 9, 9486–9493.

Vagadia, B. H., Vanga, S. K., & Raghavan, V. (2017). Inactivation methods of soybean trypsin inhibitor – a review. *Trends in Food Science & Technology,* 64, 115–125.

Van Impe, J., Smet, C., Tiwari, B., Greiner, R., Ojha, S., Stulić, V., Vukušić, T., & Režek Jambrak, A. (2018). State of the art of nonthermal and thermal processing for inactivation of micro-organisms. *Journal of Applied Microbiology,* 125, 16–35.

Vanga, S. K., Singh, A., & Raghavan, V. (2017). Review of conventional and novel food processing methods on food allergens. *Critical Reviews in Food Science and Nutrition,* 57, 2077–2094.

Vasquez-Orejarena, E., Simons, C. T., Litchfield, J. H., & Alvarez, V. B. (2018). Functional properties of a high protein beverage stabilized with oat-β-glucan. *Journal of Food Science,* 83, 1360–1365.

Vera, E. P., Alca, J. J., Saravia, G. R., Campioni, N. C., & Alpuy, I. J. (2019). Comparison of the lipid profile and tocopherol content of four Peruvian quinoa (Chenopodium quinoa Willd.) cultivars ('Amarilla de Maranganí','Blanca de Juli', INIA 415 'Roja Pasankalla', INIA 420 'Negra Collana') during germination. *Journal of Cereal Science,* 88, 132–137.

Vesa, T. H., Marteau, P., & Korpela, R. (2000). Lactose intolerance. *Journal of the American College of Nutrition,* 19, 165S–175S.

Vigli, D., Cosentino, L., Pellas, M., & De Filippis, B. (2021). Chronic treatment with cannabidiolic acid (CBDA) reduces thermal Pain sensitivity in male mice and rescues the Hyperalgesia in a mouse model of rett syndrome. *Neuroscience,* 453, 113–123.

Vilcacundo, R., & Hernández-Ledesma, B. (2017). Nutritional and biological value of quinoa (Chenopodium quinoa Willd.). *Current Opinion in Food Science,* 14, 1–6.

Vogelsang-O'Dwyer, M., Sahin, A. W., Zannini, E., & Arendt, E. K. (2022). Physicochemical and nutritional properties of high protein emulsion-type lupin-based model milk alternatives: Effect of protein source and homogenisation pressure. *Journal of the Science of Food and Agriculture, 102*(12), 5086–5097 (n/a).

Vogelsang-O'Dwyer, M., Zannini, E., & Arendt, E. K. (2021). Production of pulse protein ingredients and their application in plant-based milk alternatives. *Trends in Food Science & Technology,* 110, 364–374.

Wang, Q., Jiang, J., & Xiong, Y. L. (2018). High pressure homogenization combined with pH shift treatment: A process to produce physically and oxidatively stable hemp milk. *Food Research International,* 106, 487–494.

Wang, K.-H., Lai, Y.-H., Chang, J.-C., Ko, T.-F., Shyu, S.-L., & Chiou, R. Y. Y. (2005). Germination of peanut kernels to enhance resveratrol biosynthesis and prepare sprouts as a functional vegetable. *Journal of Agricultural and Food Chemistry,* 53, 242–246.

Wei, P., Liu, M., Chen, Y., & Chen, D.-C. (2012). Systematic review of soy isoflavone supplements on osteoporosis in women. *Asian Pacific Journal of Tropical Medicine,* 5, 243–248.

Weizman, Z. (2004). Vegetarian diet and exocrine pancreatic function using fecal tests. *Journal of Pediatric Gastroenterology and Nutrition,* 39, 212.

Were, B. A., Onkware, A. O., Gudu, S., Welander, M., & Carlsson, A. S. (2006). Seed oil content and fatty acid composition in East African sesame (Sesamum indicum L.) accessions evaluated over 3 years. *Field Crops Research,* 97, 254–260.

Wien, M., Oda, K., & Sabaté, J. (2014). A randomized controlled trial to evaluate the effect of incorporating peanuts into an American Diabetes Association meal plan on the nutrient profile of the total diet and cardiometabolic parameters of adults with type 2 diabetes. *Nutrition Journal,* 13, 10.

Willett, W., Rockström, J., Loken, B., Springmann, M., Lang, T., Vermeulen, S., Garnett, T., Tilman, D., DeClerck, F., & Wood, A. (2019). Food in the Anthropocene: The EAT–lancet commission on healthy diets from sustainable food systems. *The Lancet,* 393, 447–492.

Xiao, C. W. (2008). Health effects of soy protein and isoflavones in humans. *The Journal of Nutrition,* 138, 1244S–1249S.

Xu, X., Luo, L., Liu, C., & McClements, D. J. (2017). Utilization of anionic polysaccharides to improve the stability of rice glutelin emulsions: Impact of polysaccharide type, pH, salt, and temperature. *Food Hydrocolloids,* 64, 112–122.

Yang, Z., Zhu, W., Gao, S., Yin, T., Jiang, W., & Hu, M. (2012). Breast cancer resistance protein (ABCG2) determines distribution of genistein phase II metabolites: Reevaluation of the roles of ABCG2 in the disposition of genistein. *Drug Metabolism and Disposition,* 40, 1883–1893.

Yan, X., Tang, J., dos Santos Passos, C., Nurisso, A., Simões-Pires, C. A., Ji, M., Lou, H., & Fan, P. (2015). Characterization of lignanamides from hemp (cannabis sativa L.) seed and their antioxidant and Acetylcholinesterase inhibitory activities. *Journal of Agricultural and Food Chemistry,* 63, 10611–10619.

Yoo, S.-H., & Chang, Y. H. (2016). Volatile compound, physicochemical, and antioxidant properties of beany flavor-removed soy protein isolate Hydrolyzates obtained from combined high temperature pre-treatment and enzymatic hydrolysis. *Preventive Nutrition and Food Science,* 21, 338–347.

Yuan, S., Chang, S. K. C., Liu, Z., & Xu, B. (2008). Elimination of trypsin inhibitor activity and beany flavor in soy milk by consecutive blanching and ultrahigh-temperature (UHT) processing. *Journal of Agricultural and Food Chemistry,* 56, 7957–7963.

Yu, Q., Zhao, Q., & Zhao, M. (2009). Effects of ratio and dosage of emulsifiers on stability of peanut milk. *Modern Food Science and Technology,* 25, 903–906.

Zentek, J., Buchheit-Renko, S., Ferrara, F., Vahjen, W., Van Kessel, A. G., & Pieper, R. (2011). Nutritional and physiological role of medium-chain triglycerides and medium-chain fatty acids in piglets. *Animal Health Research Reviews,* 12, 83.

Zha, F., Dong, S., Rao, J., & Chen, B. (2019). Pea protein isolate-gum Arabic Maillard conjugates improves physical and oxidative stability of oil-in-water emulsions. *Food Chemistry,* 285, 130–138.

Zhang, Y., Guo, S., Liu, Z., & Chang, S. K. C. (2012). Off-Flavor related volatiles in soymilk as affected by soybean variety, grinding, and heat-processing methods. *Journal of Agricultural and Food Chemistry,* 60, 7457–7462.

Zhang, H., Li, L., Tatsumi, E., & Isobe, S. (2005). High-pressure treatment effects on proteins in soy milk. *LWT—Food Science and Technology,* 38, 7–14.

Zhang, X., Xiong, Y. L., Chen, J., & Zhou, L. (2013). Synergistic inhibition of lipid oxidation by pea protein hydrolysate coupled with licorice extract in a liposomal model system. *Journal of Agricultural and Food Chemistry,* 61, 8452–8461.

Zhang, B., Zhang, Y., Li, H., Deng, Z., & Tsao, R. (2020). A review on insoluble-bound phenolics in plant-based food matrix and their contribution to human health with future perspectives. *Trends in Food Science & Technology, 105,* 347–362.

Zhao, Y., Martin, B. R., & Weaver, C. M. (2005). Calcium bioavailability of calcium carbonate fortified soymilk is equivalent to cow's milk in young women. *The Journal of Nutrition,* 135, 2379–2382.

Zhao, M., Lin, Y., & Chen, H. (2020). Improving nutritional quality of rice for human health. *Theoretical and Applied Genetics,* 133, 1397–1413.

Zhou, Y., Wang, S., Lou, H., & Fan, P. (2018). Chemical constituents of hemp (*Cannabis sativa L.*) seed with potential anti-neuroinflammatory activity. *Phytochemistry Letters,* 23, 57–61.

Zhu, F. (2019). Proanthocyanidins in cereals and pseudocereals. *Critical Reviews in Food Science and Nutrition,* 59, 1521–1533.

Further reading

Abraham, E. M., Ganopoulos, I., Madesis, P., Mavromatis, A., Mylona, P., Nianiou-Obeidat, I., Parissi, Z., Polidoros, A., Tani, E., & Vlachostergios, D. (2019). The use of lupin as a source of protein in animal feeding: Genomic tools and breeding approaches. *International Journal of Molecular Sciences,* 20, 851.

Abudu, I. A., & Akinyele, I. O. (1990). The effect of germination on the oligosaccharides, trypsin inhibitors and nutrient content of cowpea milk. *Food Chemistry,* 35, 161–166.

Adeleke, B. S., & Babalola, O. O. (2020). Oilseed crop sunflower (*Helianthus annuus*) as a source of food: Nutritional and health benefits. *Food Science & Nutrition, 8*, 4666–4684.

Aisa, H. A., Gao, Y., Yili, A., Ma, Q., & Cheng, Z. (2019). *Beneficial role of chickpea (*Cicer arietinum L.*) functional factors in the intervention of metabolic syndrome and diabetes mellitus.* Bioactive Food as dietary Interventions for diabetes. Elsevier.

Alasalvar, C., Pelvan, E., & Topal, B. (2010). Effects of roasting on oil and fatty acid composition of Turkish hazelnut varieties (*Corylus avellana L.*). *International Journal of Food Sciences and Nutrition, 61*, 630–642.

Anastasio, M., Pepe, O., Cirillo, T., Palomba, S., Blaiotta, G., & Villani, F. (2010). Selection and use of phytate-degrading LAB to improve cereal-based products by mineral solubilization during dough fermentation. *Journal of Food Science, 75*, M28–M35.

Arnoldi, A., Boschin, G., Zanoni, C., & Lammi, C. (2015). The health benefits of sweet lupin seed flours and isolated proteins. *Journal of Functional Foods, 18*, 550–563.

Arzoumanidis, I., Petti, L., Raggi, A., & Zamagni, A. (2013). *Life cycle assessment for the agri-food sector. Product-Oriented Environmental Management Systems (POEMS)*. Springer.

Asif, M., Rooney, L. W., Ali, R., & Riaz, M. N. (2013). Application and opportunities of pulses in food system: A review. *Critical Reviews in Food Science and Nutrition, 53*, 1168–1179.

Austria, J. A., Richard, M. N., Chahine, M. N., Edel, A. L., Malcolmson, L. J., Dupasquier, C. M. C., & Pierce, G. N. (2008). Bioavailability of alpha-linolenic acid in subjects after ingestion of three different forms of flaxseed. *Journal of the American College of Nutrition, 27*, 214–221.

Awika, J. M., & Duodu, K. G. J. J.o. F. F. (2017). Bioactive polyphenols and peptides in cowpea (*Vigna unguiculata*) and their health promoting properties: A review. *Journal of Functional Foods, 38*, 686–697.

Bader, S., Oviedo, J. P., Pickardt, C., & Eisner, P. (2011). Influence of different organic solvents on the functional and sensory properties of lupin (*Lupinus angustifolius L.*) proteins. *LWT—Food Science and Technology, 44*, 1396–1404.

Bashir, T., Zahara, K., Haider, S., & Tabassum, S. (2015). Chemistry, pharmacology and ethnomedicinal uses of Helianthus annuus (sunflower): A review. *Pure and Applied Biology, 4*, 226.

Bekhit, A. E.-D. A., Shavandi, A., Jodjaja, T., Birch, J., Teh, S., Ahmed, I. A. M., Al-Juhaimi, F. Y., Saeedi, P., & Bekhit, A. A. (2018). Flaxseed: Composition, detoxification, utilization, and opportunities. *Biocatalysis and Agricultural Biotechnology, 13*, 129–152.

Bernacchia, R., Preti, R., & Vinci, G. (2014). Chemical composition and health benefits of flaxseed. *Austin Journal of Nutritients and Food Science, 2*, 1045.

Bernat, N., Cháfer, M., Rodríguez-García, J., Chiralt, A., & González-Martínez, C. (2015). Effect of high pressure homogenisation and heat treatment on physical properties and stability of almond and hazelnut milks. *LWT—Food Science and Technology, 62*, 488–496.

Blum, J. E. S., Ramoni, E. O., & Balbi, M. E. (2016). Extract of preparing soluble (milk) from sunflower seeds germinated (*Helianthus Annus L., Asteraceae*) and evaluation of their nutritional composition. *Visão Acadêmica*, 17.

Cabanillas, B., Jappe, U., & Novak, N. (2018). Allergy to peanut, soybean, and other legumes: Recent advances in allergen characterization, stability to processing and IgE cross-reactivity. *Molecular Nutrition & Food Research, 62*, 1700446.

Cai, R., Hettiarachchy, N. S., & Jalaluddin, M. (2003). High-performance liquid chromatography determination of phenolic constituents in 17 varieties of cowpeas. *Journal of Agricultural and Food Chemistry, 51*, 1623–1627.

Camacho, L., Vásquez, M., Leiva, M., & Vargas, E. (1988). Effect of processing and methionine addition on the sensory quality and nutritive value of spray-dried lupin milk. *International Journal of Food Science & Technology, 23*, 233–240.

Câmara, C. R. S., & Schlegel, V. (2016). A review on the potential human health benefits of the black walnut: A comparison with the English walnuts and other tree nuts. *International Journal of Food Properties, 19*, 2175–2189.

Carvajal-Larenas, F., Linnemann, A., Nout, M., Koziol, M., & Van Boekel, M. (2016). Lupinus mutabilis: Composition, uses, toxicology, and debittering. *Critical Reviews in Food Science and Nutrition, 56*, 1454–1487.

Chandrasekara, N., & Shahidi, F. (2011). Oxidative stability of cashew oils from raw and roasted nuts. *Journal of the American Oil Chemists' Society, 88*, 1197–1202.

Costa, J., Mafra, I., Carrapatoso, I., & Oliveira, M. B. P. P. (2016). Hazelnut allergens: Molecular characterization, detection, and clinical relevance. *Critical Reviews in Food Science and Nutrition, 56*, 2579–2605.

Cramer, D. W., Xu, H., & Sahi, T. (1994). Adult hypolactasia, milk consumption, and age-specific fertility. *American Journal of Epidemiology, 139*, 282–289.

Creusot, N., & Gruppen, H. (2007). Enzyme-induced aggregation and gelation of proteins. *Biotechnology Advances, 25*, 597–601.

Dalgleish, D. G., & Corredig, M. (2012). The structure of the casein micelle of milk and its changes during processing. *Annual Review of Food Science and Technology, 3*, 449–467.

Dendena, B., & Corsi, S. (2014). Cashew, from seed to market: A review. *Agronomy for Sustainable Development, 34*, 753–772.

Downs, M. L., Baumert, J. L., Taylor, S. L., & Mills, E. N. C. (2016). Mass spectrometric analysis of allergens in roasted walnuts. *Journal of Proteomics, 142*, 62–69.

Dubovis, M., & Muneyyirci-Delale, O. (2020). Effects of marijuana on human reproduction. *Reproductive Toxicology, 94*, 22–30.

Dueñas, M., Fernández, D., Hernández, T., Estrella, I., & Muñoz, R. (2005). Bioactive phenolic compounds of cowpeas (*Vigna sinensis L*). Modifications by fermentation with natural microflora and with Lactobacillus plantarum ATCC 14917. *Journal of the Science of Food and Agriculture, 85*, 297–304.

Elsamani, M. O., Habbani, S. S., Babiker, E. E., & Mohamed Ahmed, I. A. (2014). Biochemical, microbial and sensory evaluation of white soft cheese made from cow and lupin milk. *LWT—Food Science and Technology, 59*, 553–559.

Erdemoglu, N., Ozkan, S., & Tosun, F. (2007). Alkaloid profile and antimicrobial activity of Lupinus angustifolius L. alkaloid extract. *Phytochemistry Reviews, 6*, 197–201.

FAO. (2017). In U. N. FAO (Ed.), *Statistical databases*. Rome.

Feldheim, W. (2000). The use of lupins in human nutrition. *Lupin an Ancient Crop for the New Millennium: Proceedings of the 9th International Lupin Conference*, 434–437. In this issue.

Fleming, S. E., & Sosulski, F. W. (1977). The utilization of sunflower protein in milk-like beverages. *Canadian Institute of Food Science and Technology Journal*, 10, 229–232.

Forges, T., & Monnier-Barbarino, P. (2003). Premature ovarian failure in galactosaemia: Pathophysiology and clinical management. *Pathologie-Biologie*, 51, 47–56.

Fridovich-Keil, J. L. (2006). Galactosemia: The good, the bad, and the unknown. *Journal of Cellular Physiology*, 209, 701–705.

Galves, C., Stone, A. K., Szarko, J., Liu, S., Shafer, K., Hargreaves, J., Siarkowski, M., & Nickerson, M. T. (2019). Effect of pH and defatting on the functional attributes of safflower, sunflower, canola, and hemp protein concentrates. *Cereal Chemistry*, 96, 1036–1047.

González-Pérez, S., & Vereijken, J. M. (2007). Sunflower proteins: Overview of their physicochemical, structural and functional properties. *J*ournal of the Science of Food and Agriculture, 87, 2173–2191.

Goyal, A., Sharma, V., Upadhyay, N., Gill, S., & Sihag, M. (2014). Flax and flaxseed oil: An ancient medicine & modern functional food. *Journal of Food Science and Technology*, 51, 1633–1653.

Grant, C. A., & Hicks, A. L. (2018). Comparative life cycle assessment of milk and plant-based alternatives. *Environmental Engineering Science*, 35, 1235–1247.

Gul, O., Saricaoglu, F. T., Besir, A., Atalar, I., & Yazici, F. (2018). Effect of ultrasound treatment on the properties of nano-emulsion films obtained from hazelnut meal protein and clove essential oil. *Ultrasonics Sonochemistry*, 41, 466–474.

Gültekin-Özgüven, M., Davarcõ, F., Paslõ, A. A., Demir, N., & Özçelik, B. (2015). Determination of phenolic compounds by ultra high liquid chromatography-tandem mass spectrometry: Applications in nuts. *LWT—Food Science and Technology*, 64, 42–49.

Gupta, R. K., Gangoliya, S. S., & Singh, N. K. (2015). Reduction of phytic acid and enhancement of bioavailable micronutrients in food grains. *Journal of Food Science and Technology*, 52, 676–684.

Gupta, R. K., Gupta, K., Sharma, A., Das, M., Ansari, I. A., & Dwivedi, P. D. (2017). Health risks and benefits of chickpea (cicer arietinum) consumption. *Journal of Agricultural and Food Chemistry*, 65, 6–22.

Han, Y., Ni, Z., Li, S., Qu, M., Tang, F., Mo, R., Ye, C., & Liu, Y. (2018). Distribution, relationship, and risk assessment of toxic heavy metals in walnuts and growth soil. *Environmental Science and Pollution Research*, 25, 17434–17443.

Heredia, N., & García, S. (2018). Animals as sources of food-borne pathogens: A review. *Animal Nutrition*, 4, 250–255.

Høst, A. (2002). Frequency of cow's milk allergy in childhood. *Annals of Allergy, Asthma & Immunology*, 89, 33–37.

Hyvärinen, H. K., Pihlava, J.-M., Hiidenhovi, J. A., Hietaniemi, V., Korhonen, H. J. T., & Ryhänen, E.-L. (2006). Effect of processing and storage on the stability of flaxseed lignan added to dairy products. *Journal of Agricultural and Food Chemistry*, 54, 8788–8792.

Jan, C. C., & Seiler, G. J. (2007). *Sunflower. Genetic Resources, Chromosome Engineering, and Crop Improvement*, 4, 103–165.

Jayachandran, M., & Xu, B. (2019). An insight into the health benefits of fermented soy products. *Food Chemistry*, 271, 362–371.

Jayasena, V., Khu, W. S., & Nasar-Abbas, S. M. (2010). The development and sensory acceptability of lupin-based tofu. *Journal of Food Quality*, 33, 85–97.

Jayathilake, C., Visvanathan, R., Deen, A., Bangamuwage, R., Jayawardana, B. C., Nammi, S., & Liyanage, R. (2018). Cowpea: An overview on its nutritional facts and health benefits. *Journal of the Science of Food and Agriculture*, 98, 4793–4806.

Jha, P. K., Materia, S., Zizzi, G., Costa-Saura, J. M., Trabucco, A., Evans, J., & Bregaglio, S. (2021). Climate change impacts on phenology and yield of hazelnut in Australia. *Agricultural Systems*, 186, 102982.

Kaur, P., Waghmare, R., Kumar, V., Rasane, P., Kaur, S., & Gat, Y. (2018). Recent advances in utilization of flaxseed as potential source for value addition. *OCL*, 25.

Khan, M. K., Karnpanit, W., Nasar-Abbas, S. M., Huma, Z.-e.-., & Jayasena, V. (2015a). Phytochemical composition and bioactivities of lupin: A review. *International Journal of Food Science & Technology*, 50, 2004–2012.

Kohajdova, Z., Karovicova, J., & Schmidt, Š. (2011). Lupin composition and possible use in bakery—A review. *Czech Journal of Food Sciences*, 29, 203–211.

Köksal, A.İ., Artik, N., Şimşek, A., & Güneş, N. (2006). Nutrient composition of hazelnut (*Corylus avellana L.*) varieties cultivated in Turkey. *Food Chemistry*, 99, 509–515.

Koyuncu, M. A., Islam, A., & Küçük, M. (2005). Fat and fatty acid composition of hazelnut kernels in vacuum packages during storage. *Grasas Y Aceites*, 56, 263–266.

Kyle, W. (1994). *The current and potential uses of lupins for human food. Procedings of first Australian Lupin Technical Symposium* (pp. 89–97). Perth, Australia: Department of Agriculture.

Leitzmann, C. (2005). *Vegetarian diets: What are the advantages? Diet diversification and health promotion*. Karger Publishers.

Lima-Cabello, E., Alche, V., Foley, R. C., Andrikopoulos, S., Morahan, G., Singh, K. B., Alche, J. D., & Jimenez-Lopez, J. C. (2017). Narrow-leafed lupin (*Lupinus angustifolius L.*) β-conglutin proteins modulate the insulin signaling pathway as potential type 2 diabetes treatment and inflammatory-related disease amelioration. *Molecular Nutrition & Food Research*, 61, 1600819.

Lima, J. R., Bruno, L. M., Wurlitzer, N. J., Sousa, P. H. M.d., & Holanda, S. A.d. M. (2020). Cashew nut-based beverage: Development, characteristics and stability during refrigerated storage. *Food Science and Technology, 41*(1), 60–64.

Lima, J. R., de Souza, A. C. R., Pinto, C. O., Araujo, I. M.d. S., Bruno, L. M., Goiana, M. L., Wurlitzer, N. J., & Tajra, T. F. (2018). *Estabilidade durante armazenamento em temperatura ambiente de extrato hidrossolúvel de amêndoa de castanha-de-caju*. Embrapa Agroindústria Tropical-Boletim de Pesquisa e Desenvolvimento (INFOTECA-E).

Lundin, K. E. A., Nilsen, E. M., Scott, H. G., Løberg, E. M., Gjøen, A., Bratlie, J., Skar, V., Mendez, E., Løvik, A., & Kett, K. (2003). Oats induced villous atrophy in coeliac disease. *Gut,* 52, 1649.

Mahmood, T., Iqbal, Z., Akhtar, M., Ali, R., & M, Y. (2014). Exploring the potential of domesticating lupins in Punjab, Pakistan. *Soil and Environment,* 33, 59–66.

Martinchik, A. N., Baturin, A. K., Zubtsov, V. V., & Molofeev, V. (2012). [Nutritional value and functional properties of flaxseed]. *Vopr Pitan,* 81, 4–10.

Oak, S. J., & Jha, R. (2019). The effects of probiotics in lactose intolerance: A systematic review. *Critical Reviews in Food Science and Nutrition,* 59, 1675–1683.

Oliver, S. P., Jayarao, B. M., & Almeida, R. A. (2005). Foodborne pathogens in milk and the dairy farm environment: Food safety and public health implications. *Foodbourne Pathogens & Disease,* 2, 115–129.

Onyesom, I., Enaholo, A., & Mordi, J. (2005). Effect of processing techniques on the contents of flatulence factors and emulsion properties of cowpea (*Vigna unguiculata*). *Journal of Applied Sciences and Environmental Management,* 9, 65–72.

Parmar, P. T., Singh, A. K., & Borad, S. G. (2021). *Coconut (cocos nucifera). Oilseeds: Health attributes and food applications*. Springer.

Petchiammal, C., Hopper, W. J. I. J.o. P., & Sciences, P. (2014). Antioxidant activity of proteins from fifteen varieties of legume seeds commonly consumed in India. *International Journal of Pharmacy and Pharmaceutical Sciences,* 6, 476–479.

Pham, L. B., Wang, B., Zisu, B., Truong, T., & Adhikari, B. (2020). Microencapsulation of flaxseed oil using polyphenol-adducted flaxseed protein isolate-flaxseed gum complex coacervates. *Food Hydrocolloids,* 107, 105944.

Poore, J., & Nemecek, T. (2018). Reducing food's environmental impacts through producers and consumers. *Science,* 360, 987.

Queirós, R. P., Saraiva, J. A., & da Silva, J. A. L. (2018). Tailoring structure and technological properties of plant proteins using high hydrostatic pressure. *Critical Reviews in Food Science and Nutrition,* 58, 1538–1556.

Ramya, K. B., & Thaakur, S. (2007). Herbs containing L- Dopa: An update. *Ancient Science of Life,* 27, 50–55.

Rao, T. R., Rao, D. N., Kotilingam, K., & Athota, R. R. (2000). Isolation and characterization of allergens from the seeds of Vigna sinensis. *Asian Pacific Journal of Allergy and Immunology,* 18, 9–14.

Rekha, C. R., & Vijayalakshmi, G. (2010). Bioconversion of isoflavone glycosides to aglycones, mineral bioavailability and vitamin B complex in fermented soymilk by probiotic bacteria and yeast. *Journal of Applied Microbiology,* 109, 1198–1208.

Rifna, E. J., Ratish Ramanan, K., & Mahendran, R. (2019). Emerging technology applications for improving seed germination. *Trends in Food Science & Technology,* 86, 95–108.

Röös, E., Patel, M., & Spångberg, J. (2016). Producing oat drink or cow's milk on a Swedish farm—environmental impacts considering the service of grazing, the opportunity cost of land and the demand for beef and protein. *Agricultural Systems,* 142, 23–32.

Sánchez-González, C., Ciudad, C. J., Noé, V., & Izquierdo-Pulido, M. (2017). Health benefits of walnut polyphenols: An exploration beyond their lipid profile. *Critical Reviews in Food Science and Nutrition,* 57, 3373–3383.

Sanchez-Monge, R., Lopez-Torrejón, G., Pascual, C. Y., Varela, J., Martin-Esteban, M., & Salcedo, G. (2004). Vicilin and convicilin are potential major allergens from pea. *Clinical & Experimental Allergy, 34*, 1747–1753.

Saricaoglu, F. T., Gul, O., Besir, A., & Atalar, I. (2018). Effect of high pressure homogenization (HPH) on functional and rheological properties of hazelnut meal proteins obtained from hazelnut oil industry by-products. *Journal of Food Engineering,* 233, 98–108.

Sellick, C. A., Campbell, R. N., & Reece, R. J. (2008). Galactose metabolism in yeast—structure and regulation of the Leloir pathway enzymes and the genes encoding them. *International Review of Cell and Molecular Biology,* 269, 111–150.

Sen, D., & Kahveci, D. (2020). Production of a protein concentrate from hazelnut meal obtained as a hazelnut oil industry by-product and its application in a functional beverage. *Waste and Biomass Valorization*, 1–9.

Shahidi, F., Alasalvar, C., & Liyana-Pathirana, C. M. (2007). Antioxidant phytochemicals in hazelnut kernel (*Corylus avellana L.*) and hazelnut byproducts. *Journal of Agricultural and Food Chemistry,* 55, 1212–1220.

Shevkani, K., Singh, N., Chen, Y., Kaur, A., & Yu, L. (2019). Pulse proteins: Secondary structure, functionality and applications. *Journal of Food Science and Technology*, 1–12.

Siger, A., Czubinski, J., Kachlicki, P., Dwiecki, K., Lampart-Szczapa, E., & Nogala-Kalucka, M. (2012). Antioxidant activity and phenolic content in three lupin species. *Journal of Food Composition and Analysis,* 25, 190–197.

Singh, K. K., Mridula, D., Rehal, J., & Barnwal, P. (2011). Flaxseed: A potential source of food, feed and fiber. *Critical Reviews in Food Science and Nutrition,* 51, 210–222.

Singh, K. V., & Singh, R. (2016). Quinoa (Chenopodium quinoa Willd), functional superfood for today's world: A review. *world Scientific News,* 58, 84–96.

Slatnar, A., Mikulic-Petkovsek, M., Stampar, F., Veberic, R., & Solar, A. (2014). HPLC-MSn identification and quantification of phenolic compounds in hazelnut kernels, oil and bagasse pellets. *Food Research International,* 64, 783–789.

Snowden, J., Sipsas, S., & John, C. S. (2007). Method to produce lupin protein-based dairy substitutes. *Google Patents*, Article US20070154611A1.

Stambolian, D. (1988). Galactose and cataract. *Survey of Ophthalmology,* 32, 333–349.

Syngai, G. G., Gopi, R., Bharali, R., Dey, S., Lakshmanan, G. M. A., & Ahmed, G. (2016). Probiotics-the versatile functional food ingredients. *Journal of Food Science and Technology,* 53, 921–933.

Taghizadeh, S. F., Badibostan, H., Hayes, A. W., Giesy, J. P., & Karimi, G. (2021). Residues levels of pesticides in walnuts of Iran and associated health risks. *Human and Ecological Risk Assessment: An International Journal, 27*, 191–204.

Tano-Debrah, K., Asiamah, K., Sakyi-Dawson, E., & Budu, A. S. (2005). *Effect of malt enzyme treatment on the nutritional and physicochemical characteristics of cowpea-peanut milk*. Pretoria: University of Pretoria.

Tatar, F., Tunç, M. T., & Kahyaoglu, T. (2015). Turkish tombul hazelnut (*Corylus avellana L.*) protein concentrates: Functional and rheological properties. *Journal of Food Science and Technology,* 52, 1024–1031.

Thambiraj, S. R., Phillips, M., Koyyalamudi, S. R., & Reddy, N. (2018). Yellow lupin (*Lupinus luteus L.*) polysaccharides: Antioxidant, immunomodulatory and prebiotic activities and their structural characterisation. *Food Chemistry,* 267, 319–328.

Tola, J., & Mazengia, Y. (2019). Cashew production benefits and opportunities in Ethiopia: A review. *Journal of Agricultural and Crop Research,* 7, 18–25.

Touma, J., La Vieille, S., Guillier, L., Barrere, V., Manny, E., Théolier, J., Dominguez, S., & Godefroy, S. B. (2021). Occurrence and risk assessment of sesame as an allergen in selected Middle Eastern foods available in Montreal, Canada. *Food Additives & Contaminants: A,* 1–13.

Trehan, I., Benzoni, N. S., Wang, A. Z., Bollinger, L. B., Ngoma, T. N., Chimimba, U. K., Stephenson, K. B., Agapova, S. E., Maleta, K. M., & Manary, M. J. J. T. (2015). Common beans and cowpeas as complementary foods to reduce environmental enteric dysfunction and stunting in Malawian children: Study protocol for two randomized controlled trials. *Trials,* 16, 520.

Ugidos-Rodríguez, S., Matallana-González, M. C., & Sánchez-Mata, M. C. (2018). Lactose malabsorption and intolerance: A review. *Food & Function,* 9, 4056–4068.

Ursini, T., Moro, L., Requena-Méndez, A., Bertoli, G., & Buonfrate, D. (2020). A review of outbreaks of cryptosporidiosis due to unpasteurized milk. *Infection, 48,* 659–663.

Vamanu, E., & Gatea, F. (2020). Correlations between microbiota bioactivity and bioavailability of functional compounds: A mini-review. *Biomedicines,* 8, 39.

Vanga, S. K., & Raghavan, V. (2018). How well do plant based alternatives fare nutritionally compared to cow's milk? *Journal of Food Science and Technology,* 55, 10–20.

Vargas-Guerrero, B., Garcia-Lopez, P. M., Martinez-Ayala, A. L., Dominguez-Rosales, J. A., & Gurrola-Diaz, C. M. (2014). Administration of Lupinus albus gamma conglutin (Cgamma) to n5 STZ rats augmented Ins-1 gene expression and pancreatic insulin content. *Plant Foods for Human Nutrition,* 69, 241–247.

Venkidasamy, B., Selvaraj, D., Nile, A. S., Ramalingam, S., Kai, G., & Nile, S. H. (2019). Indian pulses: A review on nutritional, functional and biochemical properties with future perspectives. *Trends in Food Science & Technology,* 88, 228–242.

Wäsche, A., Müller, K., & Knauf, U. (2001). New processing of lupin protein isolates and functional properties. *Food/Nahrung,* 45, 393–395.

Wenzel, P., & Jungbluth, N. (2017). *The environmental impact of vegan drinks compared to whole milk*. Switzerland: ESU-Services Ltd.: Schaffhausen.

Wu, X., Lu, Y., Xu, H., Lv, M., Hu, D., He, Z., Liu, L., Wang, Z., & Feng, Y. (2018). Challenges to improve the safety of dairy products in China. *Trends in Food Science & Technology,* 76, 6–14.

Xiao, M., Du, G., Zhong, G., Yan, D., Zeng, H., & Cai, W. (2016). Gas chromatography/mass spectrometry-based metabolomic profiling reveals Alterations in mouse Plasma and liver in response to fava beans. *PLOS ONE,* 11, e0151103.

Xu, Y., & Hanna, M. A. (2011). Nutritional and anti-nutritional compositions of defatted Nebraska hybrid hazelnut meal. *International Journal of Food Science & Technology,* 46, 2022–2029.

Yakubu, B. L., Mbonu, O. A., & Nda, A. J. (2012). Cowpea (Vigna unguiculata) pest control methods in storage and recommended practices for efficiency: A review. *Journal of Biology, Agriculture and Healthcare,* 2, 27–33.

Yorgancilar, M., & Bilgiçli, N. (2014). Chemical and nutritional changes in bitter and sweet lupin seeds (*Lupinus albus L.*) during bulgur production. *Journal of Food Science and Technology,* 51, 1384–1389.

Yu, Q., Ma, Y.-X., Qin, Z., Luo, X.-R., Liu, H.-M., & Wang, X.-D. (2021). Using solid acid catalysts to improve the oxidative stability of cold-pressed sesame oil. *LWT,* 141, 110928.

Zaffran, V. D., Gupta, S., Sathe, S. K., & Roper, M. G. (2021). Effect of deglycosylation on immunoreactivity and in vitro pepsin digestibility of major cashew (*Anacardium occidentale L.*) allergen, Ana o 1. *Journal of Food Science, 86*(3), 1144–1152 (n/a).

Chapter 7

Plant-based gels

Nitya Sharma[1], Vasudha Bansal[2] and Jatindra K. Sahu[1]
[1]*Food Customization Research Lab, Centre for Rural Development and Technology, Indian Institute of Technology Delhi, New Delhi, India;*
[2]*Department of Food and Nutrition, Government Home Science College, Chandigarh, India*

7.1 Introduction

Gels can be defined as a crosslinked 3D network that is swelled to a certain extent without dissolution due to solvent absorption. They hold an intermediate state that is between a solid and a liquid with both elastic and flow characteristics. A food gel is a similar 3D polymeric network with properties such as (a) high moisture content, (b) flow resistance (even under mild pressures), and (c) mechanical rigidity (Banerjee & Bhattacharya, 2012). The process of gelation is a very common phenomena associated with the processing of foods. Food gels pre-exists in the market in various forms and have been gaining further popularity due to its low calorie, appealing taste, and satiety-enhancing property. Edible food gels are usually made up of polysaccharide- and protein-based ingredients that are crosslinked either using covalent bonds (for protein gels) or a combination of weak intermolecular forces like hydrogen bonds, electrostatic forces, van der Waals forces, and hydrophobic interactions. These ingredients are basically a range of food hydrocolloids that are derived from natural sources including plants. With an increasing shift of diet trend toward plant-based food products, various researchers are now targeting plant-based ingredients for applications like fat replacement, thickening and gelling agents, stabilizing agents (for foams, dispersions and emulsions), reducing syneresis, and increasing water retention, therefore, creating an increased economical demand for plant-based gels due to its low price, sustainable production, high stability, worldwide availability, nontoxicity, and diverse potential.

7.2 Classification of food gels based on plant ingredients

Gels manufactured from food biopolymers like proteins and polysaccharides have diverse applications in food, pharmaceutical, and cosmetic sectors, with added advantages such as affordability, biodegradability, biocompatibility, and edibility. Food systems are usually a complex combination of several types of biopolymers that has the ability to form variety of gel structures, depending upon the type of interaction between the polymers. Most of the food polysaccharide and protein gels are physically crosslinked and are induced by either temperature, pH, or salts. Although, some food gels are found to be chemically crosslinked by either intermolecular disulfide bonds, enzymatic reactions (like transglutaminase), Millard reactions, or presence of some food-grade crosslinkers that do not required purification (Yang et al., 2021). Food gels can be produced from various plant-based ingredients that have been categorized based on the polymer source: polysaccharide gels, protein gels, and binary gels.

7.2.1 Polysaccharide gels

These gels can only be formed with certain selected polysaccharides at a certain concentration of gelling agent (known as "critical concentration"), with the help of either temperatures, pH, salts, or freeze-thaw cycles. Gelation occurs when the selected polysaccharides hydrate completely, and the polymer strands start interacting with each other and form junction zones. These categories of food gels are thermoreversible, although their thermal hysteresis can also be modulated with ionic strength, molecular weight, and abundance of side residues of the polymer interaction (Nazir et al., 2017). Polysaccharide gels are considered to be nontoxic, hypoallergenic, and biodegradable, with higher bioavailability and biocompatibility. Some examples for plant-based ingredients used to develop biocompatible polysaccharide gels are plant

Engineering Plant-Based Food Systems. https://doi.org/10.1016/B978-0-323-89842-3.00003-8

exudates (pectin, inulin, acacia gum, cashew gum, konjac mannan, gum karaya, gum tragacanth, gum arabic), seed gums (guar gum, locust bean gum, fenugreek gum, cassia seed gum, basil seed gum, mesquite seed gum, oat gum, rye gum, tamarind seed gum, psyllium, starch, amylase, cellulose), and seaweeds (agar—agar, carrageenan, alginic acid, sodium alginate, furcellaran, ulvan, laminarin, red alga, xylan) (Manzoor et al., 2020).

7.2.2 Protein gels

These gels are usually formed by either globular proteins, that are water soluble coiled-shaped structures, or fibrous proteins, that are water insoluble parallel long chain peptides linked by disulfide (-S-S-) cross-bridges. Here, for precise regulation of gelation, various physical (temperature, pressure) and chemical (pH, ionic strength, enzymes, etc.) parameters have to be regulated to achieve required physico-chemical properties (Nazir et al., 2017). For instance, heating of globular proteins helps in achieving hydrophobic interactions, disulfide bonds and gelation by molecular unfolding, and exposure of the hydrophobic patches and thiol moieties. Most studied plant-based proteins for the formation of food gels are pea and soy proteins, due to their high nutritional value and nonallergenicity. The predominance of legumin (11S) and vicilin (7S) in pea protein, and β-conglycinin (7S) and glycinin (11S) in soy protein, makes them a suitable candidate to form gels (Munialo et al., 2018).

7.2.3 Binary gels

These gels are composed of one or more biopolymers, usually a combination of polysaccharide—polysaccharide, protein—protein, and polysaccharide—protein. Although, polysaccharide—polysaccharide mixtures and protein—protein mixtures are effective in optimizing or designing food gels with modified textural properties, but polysaccharide—protein mixtures are more likely to produce binary food gels with diversified gel structure design. This is due to the different chemical structures, physical properties, and functionalities of polysaccharides and proteins, that can produce more extensive range of phase behaviors. These phase behaviors result from long- or short-range interactions between polysaccharides and proteins and from different affinities between the polymers and solvent. Another added potential for such mixed gels is that functional food gels with multistructures and high nutritional values can be easily obtained by combining the individual advantages of polysaccharides (remarkable rheological properties) with proteins (nutritional function). For instance, soy protein isolate (SPI) has become a model plant to produce protein gels formulated with a variety of food polysaccharides. This is because of the ability of this formulation to produce mixed gels with an extensive range of microstructures and physical properties (Yang et al., 2021). Table 7.1provides some recent examples of plant-based food gels and their gelation conditions categorized based on the polymer sources.

7.3 Fabrication techniques of plant-based food gels

Gels developed using plant-based ingredients can be designed into varying size, shape, and form in order to meet a wide range of applications in the food industry. Food gels with desired properties and functionalities can be fabricated using various approaches (Fig. 7.1).

7.3.1 Fabrication of hydrogels

Hydrogels can be identified by a polymeric 3D network that are randomly crosslinked linearly or in branches by physical or chemical bonds. They have a characteristic ability of absorbing large amounts of water or biological fluids and yet retain their structure without dissolving (Batista et al., 2019). Plant-based hydrogels of any shape, size, and form can be easily fabricated by the process of molding. Apart from this, micro- and nano-hydrogels can be produced using a wide range of different techniques that can broadly be classified as top-down and bottom-up methods (Fig. 7.2) (Cao and Mezzenga, 2020).

The **top-down approach** comprises of various techniques like grinding, spray-drying, extrusion, and emulsification. Although grinding is the simplest technique to produce micro- or nano-hydrogels, but they result in highly irregular and polydisperse shaped gel particles. While spray-drying techniques is a commonly used technique at industrial-scale, which involves spray-drying of biopolymer solution, followed by formation of hydrogel by rehydrating these spray-dried particles in an appropriate medium. The dried intermediates produced in this method can be advantageous in minimizing the transport and storage costs, while its disadvantages may be limited accessibility to high-viscosity polymers and possible breakdown of gel particles during rehydration. Another top-down technique involves formation of precursor droplets using

TABLE 7.1 Recent examples of plant-based food gels.

Source polymer	Plant-based ingredient	Gelation condition	References
Polysaccharide	*Mesona chinensis* polysaccharide with rice/corn starch	Freezing	Ren, Jiang et al. (2020), Ren, Rong et al. (2020), Luo et al. (2020)
	Mesona chinensis polysaccharide with maize/potato/pea starch	Cooling	Liu et al. (2020)
	Plantago asiatica L. seed polysaccharide with ascorbic acid	Cooling	Li, Wu, et al. (2020)
	Mesona chinensis Benth polysaccharide with sweet potato/tapioca starch	Heating	Ren, Jiang et al. (2020), Ren, Rong et al. (2020); Xiao, Li, et al. (2020), Xiao, Chen, et al. (2020)
	Eucheuma spinosum polysaccharide with kappa carrageenan	Heating	Yang and Yang (2020)
	Tamarind seed polysaccharide with corn starch	Heating	Xie et al. (2020)
	Okara polysaccharides with soy oil	High shear homogenization	Yang, Li and Tang (2020)
	Mesona chinensis polysaccharide with tapioca starch	Ionization (NaCl)	Xie et al. (2021)
Protein	Pea protein concentrate hydrolyzed using enzymes (trypsin and endo-proteases)	Fermentation using lactic acid culture (*S. thermophilus and L. bulgaricus*)	Klost, Giménez-Ribes and Drusch (2020)
	Soy protein isolate with glucose	Fermentation using lactic acid culture (*L. plantarum*)	Yang, Su and Li (2020)
	Soybean protein isolate with soybean oil	High-pressure homogenization	Bi et al. (2020)
	Peanut protein isolate with pH-shifting treatment	Heating	Li, He et al. (2020), Li, Zhang et al. (2020); Wang, Yang et al. (2020)
	Pea protein isolates hydrolyzed using enzyme (microbial transglutaminase)	Ionization (NaCl) → heating	Moreno et al. (2020)
	Bambara protein isolate with pH adjustment	Ionization (NaCl) → heating	Ruzengwe, Amonsou and Kudanga (2020)
	Zein with water/ethanol mixture	Slow stirring	Gagliardi et al. (2020)
Binary	Soy protein isolate with *Mesona blumes* polysaccharide/wheat bran cellulose/soy soluble polysaccharide/native corn starch/acetylated corn starch/acetylated distarch phosphate corn starch	Heating	Wang, Shen et al. (2020); Xiao, Li, et al. (2020), Xiao, Chen, et al. (2020) Yang et al. (2020), Yu et al. (2020)
	Peanut arachin with basil seed gum	Heating	Yang et al. (2021)
	Soy protein isolate with konjac gum/gellan gum/curdlan gum	Ionization ($CaSO4$)	Zhao et al. (2020)
	Peanut oil processing residue complexes with peanut oil	High shear homogenization	Ye et al. (2020)

Continued

TABLE 7.1 Recent examples of plant-based food gels.—cont'd

Source polymer	Plant-based ingredient	Gelation condition	References
	Soy glycinin with soy soluble polysaccharide and soy oil	High shear homogenization	Hao, Peng and Tang (2020)
	Pea protein with carrageenan/xanthan gum/gum Arabic/sodium alginate/pectin/gellan gum/locust bean gum/tara gum	High shear homogenization	Vélez-Erazo et al. (2020)
	Soy milk with kappa carrageenan	High shear homogenization → heating	Pang et al. (2020)
	Soy protein isolate with pectin, glycyrrhizic acid and sunflower oil	High shear homogenization → ultrasonic homogenization	Li, Wang, et al. (2020)
	Soy protein isolate with sodium alginate and sunflower oil	High shear homogenization	Lin, Kelly and Miao (2021)
	Zein with sodium alginate and corn oil	Enzyme activity (transglutaminase) → high shear homogenization → ionization ($CaCl_2$)	Yan et al. (2021)

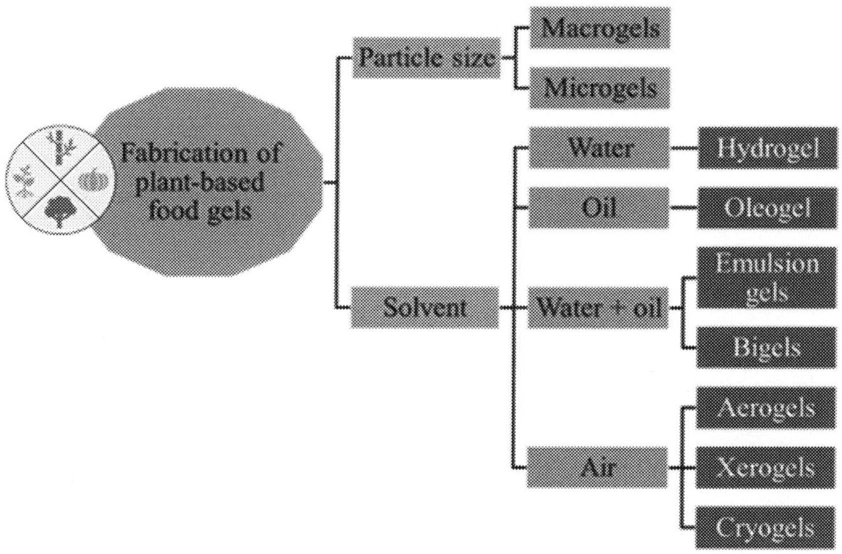

FIGURE 7.1 Fabrication of plant-based food gels based on their configuration and solvent.

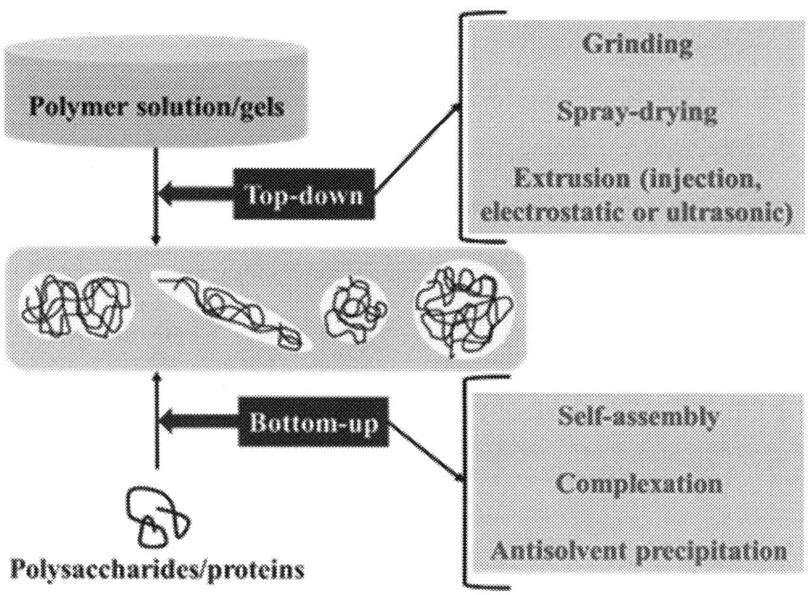

FIGURE 7.2 Top-down and bottom-up approach for fabricating plant-based hydrogels. *Modified from Cao, Y., & Mezzenga, R. (2020). Design principles of food gels.* Nature Food, 1*(2), 106–118. https://doi.org/10.1038/s43016-019-0009-x.*

extrusion (injection, electrostatic, ultrasonic) and thereafter causing controlled gelation of these liquid droplets in a specific size and shape. In this case, the biopolymer solution is simply injected through a needle into a gel causing medium. This injection method can be further modified where either electric current or ultrasonic vibrations are used to produce separate biopolymer droplets forming gels (Cao and Mezzenga, 2020).

The **bottom-up approach** constitutes the process of self-assembly of polysaccharides and/or proteins for the development of hydrogels. The commonly used routes to promote self-association are-thermal denaturation, crosslinking, desolvation, and simple coacervation. Heating globular proteins above their denaturation temperatures increases the protein surface hydrophobicity causing protein aggregation via hydrophobic interactions and intermolecular disulfide bonds, thus forming heat-induced micro-hydrogels. The size and charge of these microgels depend upon the initial biopolymer concentration, heating temperature, heating time, pH value, and ionic strength. In addition to this, cold-set gelation of preheated globular proteins also helps in fabricating micro-hydrogels. This involves heating of protein solution for protein unfolding and filament formation, followed by manipulation of the protein solution conditions (e.g., pH shifting or addition of salt) for the association of protein filaments. Another bottom-up self-associating approach to fabricate hydrogels is crosslinking polysaccharides and random coil proteins. This can either be done by addition of crosslinking substances like chemical reagents, e.g., glutaraldehyde, enzymes, e.g., transglutaminase and mineral counterions, e.g., calcium or tripolyphosphate or, by heat treatment of polysaccharides that causes swelling of the starch granules due to imbibing water, which ends to gelatinization. Other two self-assembling methods for hydrogel formation, i.e., desolvation and simple coacervation depends upon various polar and nonpolar interactions that determines the net effect of an organic solvent on protein structure. Here, protein denaturation occurs when apolar patches are significantly solvated at high organic solvent contents, while at low contents of organic solvents, protein structure may stabilize and protected against denaturing agents (Farjami and Madadlou, 2017). The difference between both the approaches to fabricate food hydrogels is that bottom-up methods are more suitable for preparing nano- and micro-hydrogel particles, while top-down methods are suitable for preparing micro- or milli-hydrogel particles.

7.3.2 Fabrication of aerogels

Aerogels are uniformly porous solid material characterized by low densities and high specific surface areas. Fabrication of plant-based aerogels can be done using two main routes: (a) using molecular precursors to form colloids, e.g., alginate, pectin or (b) using nanoscale particle-based precursors to form colloids, e.g., protein nanofiber aggregates, nanofibrous cellulose. The properties of the aerogels produced using any of these approaches is primarily evaluated based on the concentration and functional groups of precursors, sol pH value, type, and concentration of crosslinkers, post-gelation modification of aerogels (if any) and most importantly, the method of drying applied to fabricate aerogels (Fig. 7.3) (Zhao et al., 2018).

7.3.3 Fabrication of oleogels

Structuring of oil produces a self-standing thermo-reversible viscoelastic edifice, leading to the formation of a gelator network. Oleogels are formed by structuring selected oil media comprising of a certain molecule or a combination of molecules (termed as oleogelators) with characteristics like hydrophobic nature and crystallinity. Depending upon the nature of oleogelator, a myriad of colloidal architectures can be exhibited by oleogels, including the networks of crystalline species, self-assembled fibrillar conformations, polymeric arrangements, and indirect templated systems. Oleogels can be fabricated directly by mixing an oleogelator with the oil medium at temperatures above its melting point (like, waxes, ethylcellulose, monoglycerides). Besides this, oleogels can also be fabricated using a multistep solvent exchange procedure, where modification of the polarity of organic solvent is carried out by replacing the previous continuous phase for oil (Okuro et al., 2020). This section discusses oil structuring by direct or indirect dispersion of oleogelators in the oil medium (Fig. 7.4).

7.3.4 Fabrication of emulsion gels

Emulsion gels are complex colloidal materials, that structurally exist either as emulsion droplets embedded in continuous phase polymeric gel matrix (emulsion-filled gels) or as a gel network structure of aggregated emulsion droplets (emulsion particulate gels). In emulsion-filled gels, the oil droplets act as filler particles that effect the gel properties based on their type, i.e., active or inactive fillers. Active fillers are bound fillers that are attached to the gel network and impart gel strength, while inactive fillers and unbound fillers that have minimal or no interaction with the gel matrix with low chemical affinity. However, emulsion particulate gels comprise of aggregated particles occupying the enclosing volume and forming a continuous three-dimensional network (Farjami and Madadlou, 2019). Plant-based emulsion gels can be fabricated into three different forms-bulk emulsion gel, emulsion gel particles, and liquid emulsion gels. Alginate-based emulsion gels are shown in Fig. 7.5A, B and C, along with its fabrication process in Fig. 7.6. Bulk emulsion gel matrix is prepared either using protein (e.g., soy protein), polysaccharides (e.g., carrageenan, agar, alginate, inulin), and LMW (low molecular weight) compounds (e.g., glycyrrhizic acid and mixture of β-sitosterol and γ-oryzanol). Based on these matrix materials used and the end application of the emulsion gel to be fabricated, preparation methods are selected, that includes heat-set and cold-set gelation methods like one-step cold-set, cold-set after heat treatment, enzyme treatment, acidification treatment, addition of ions, and self-assembly. Another form of emulsion gel is the emulsion gel particles or gel beads that are further classified into macrogel particles (>1 mm) and microgel particles (0.2−1000 μm) for food applications, with alginate being the most frequently used matrix material. Most recent methods to develop emulsion gel beads are simple dripping, jet back-up extrusion, electrostatic extrusion, spinning disk, atomization, impinging aerosol method, emulsification technique, microfluidics, and templating method. The third type of emulsion gel is the liquid

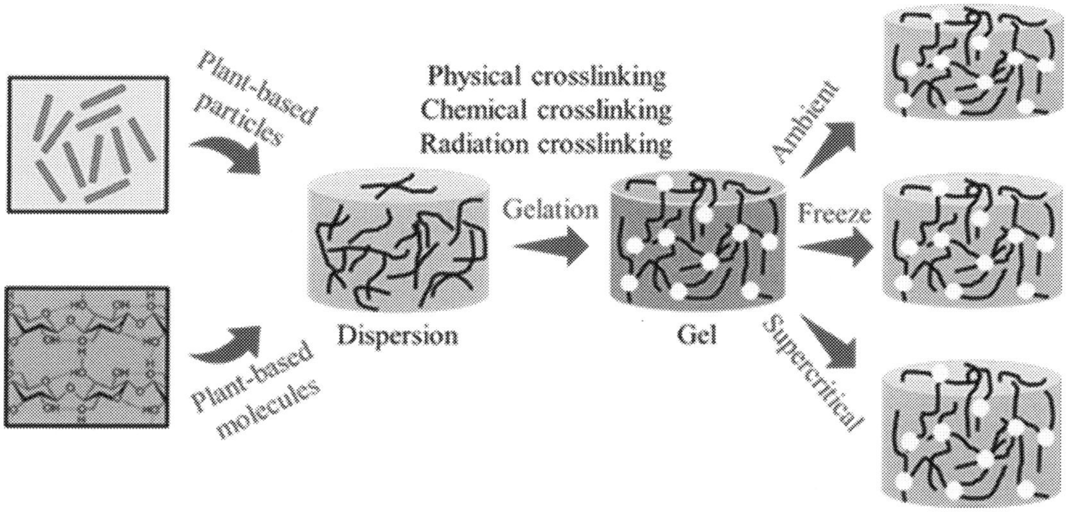

FIGURE 7.3 Fabrication approach of aerogels.

- In *molecular approach* for food aerogel fabrication, the plant-based biopolymer molecules are dissolved in a solvent, that is usually water, but can also be an organic solvent or mixtures thereof. In this method of fabrication, the gelation kinetics and final structure of the plant-based aerogel produced is dependent upon the type of crosslinking between the biopolymeric chains. The crosslinking can be done in three ways based on the nature and type of biopolymer used-physical, chemical, and radiation crosslinking. Physical crosslinks can be formed either with weak forces like hydrogen bonding, ionic interactions, and physical entanglement or with stronger forces like covalent bonding and multivalent functionalization. Chemical crosslinking produces the strongest aerogels due to covalent bonding, even though physically crosslinked aerogels are preferred due to their ease of implementation, biocompatibility, and nontoxic nature to be used as food. Furthermore, application of gamma radiation also results in gelation of plant-based polysaccharides like starch, cellulose, and alginate, forming radiation crosslinks in the aerogel structure (Zhao et al., 2018).
- In particle-based approach, controlled aggregation of nano- and micro-particles forms a network with the help of physical or chemical crosslinking. This fabrication approach uses predefined particle building blocks with a desired micro or nanostructure and functionality, which leads to the formation of aerogels with direction-dependent physical properties, including mechanical strength, thermal conduction, and molecular diffusion. However, inappropriate mixing many cause inhomogenous particle distribution in the aerogel network, thereby reducing its functionality. For instance, accumulation of large and polydisperse pores increases the thermal conductivity leading to scattering of light from larger pores, along with uneven distribution of mechanical stresses in the aerogel matrix (Zhao et al., 2018). Drying of the pore fluid with maximum retention of structural properties constitutes an important step in the process of fabrication of aerogels. The process of drying for a plant-based food hydrogel and organogel can be carried out in different ways. The most common method is vacuum drying of ambient pressure drying, which usually leads to the disruption of the soft structure due to the involvement of strong capillary forces. Therefore, as an alternative to these capillary forces at the liquid–gas interphase, controlled sublimation of the frozen pore fluid can be carried out under isobaric conditions through freeze drying. Another alternative is to use increased temperature and pressure though supercritical drying, which is an efficient technique to fabricate aerogels with minimal damage to the microstructure (Zhao et al., 2018). *From Zhao, S., Malfait, W. J., Guerrero-Alburquerque, N., Koebel, M. M., & Nyström, G. (2018). Biopolymer Aerogels and Foams: Chemistry, Properties, and Applications.* Angewandte Chemie - International Edition, *57(26), 7580–7608. https://doi.org/10.1002/anie.201709014.*

emulsion gel that is characterized as the emulsion gel with higher viscoelastic properties than the conventional ones. Based on the preparation method, fluid emulsion gel can be further classified into gel-like emulsions and disrupted emulsion gel systems. Gel-like emulsions are fabricated using pickering emulsions that are stabilized by amphiphilic solid particles like proteins, polysaccharides or a combination of both. These pickering emulsions are then converted into gel-like emulsions under specific conditions like, proper solid particle type, solid particle concentration, oil phase concentration, pH, and ionic strength. However, disrupted gel systems are fabricated by breaking down bulk emulsion gels that may lead to disruption of gel matrix-covered structure, thereby separating gel matrix and oil droplets during homogenization (Lin, Kelly and Miao, 2020).

7.3.5 Fabrication of bigels

Bigels are soft-solid formulations of two immiscible gelled systems, in which the density and viscosity of dispersed and continuous phase would match during internal phase gelation, resulting in improved compatibility between the two phases. Bigels are fabricated mostly by high shear mixing of hydrogels and organogels (oleogels/emulsion gels) at an appropriate

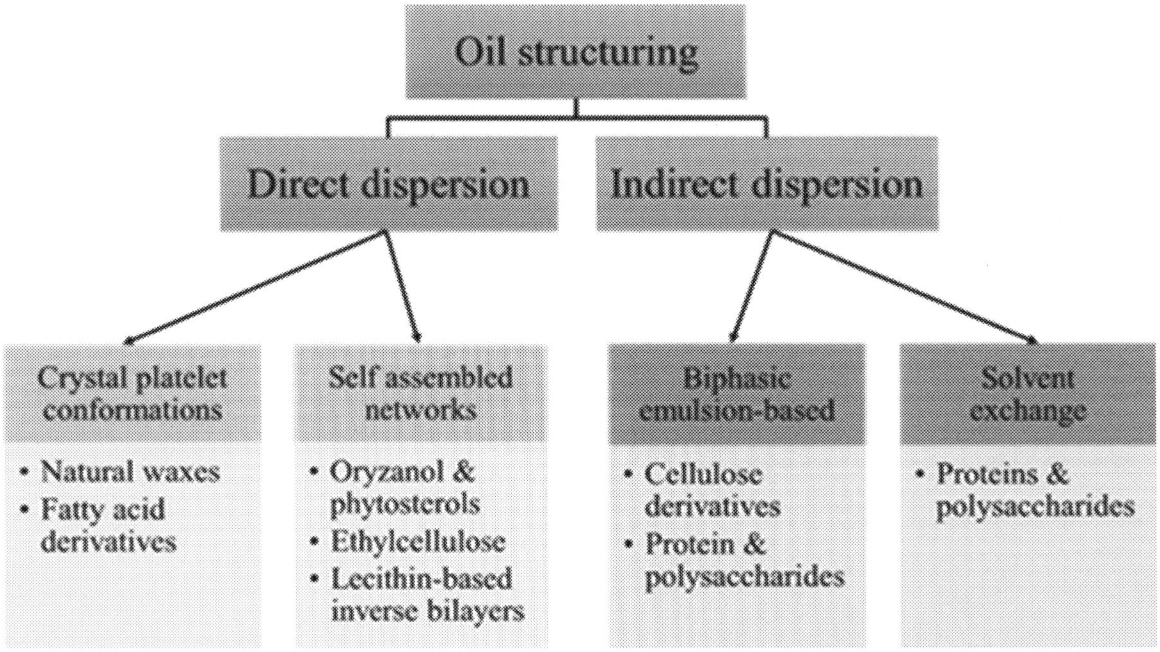

FIGURE 7.4 Direct and indirect fabrication strategies for plant-based food oleogels.

- In the process of *direct dispersion,* when the oleogelator is mixed in liquid edible oil and heated above melting point, a gelator network is formed during the cooling period that promotes entrapping of the oil in a solid structure, thus forming an edible oleogel. The food matrix and the type of oleogelator incorporated decide the crystallization kinetics of the oleogel. In direct dispersion method of oleogel fabrication, crystal platelet conformation can be formed when wax crystallization takes place at high temperatures in the presence of oil medium. This adaptable structural conformation produced by high- and low-melting wax oleogelators can easily lead to edible oil structuring, thus producing oleogels with distinct characteristics. Apart from this, combining phytosterols and γ-oryzanol (composed by esters and phytosterols) induce the formation of self-assembled tubular conformations, thus forming oleogels with better mechanical properties (Martins et al., 2020).
- *Indirect dispersion* method of oleogel fabrication uses amphiphilic molecules as building blocks for oil structuring. The plant-based emulsion complexes that form a template for oleogel fabrication can be cellulose with carboxyl methyl cellulose, proteins and polysaccharides, polysaccharides, and hydrocolloids like carrageenan and xanthan gum. Natural antioxidant molecules like polyphenols can also be used to stabilize emulsions due to their ability to form covalent or noncovalent interactions with carbohydrates and proteins. The crystals thus formed due to these interactions can act as reinforcement agents for water-in-oil as well as oil-in-water emulsions, leading to oil structuring. Another approach to fabricate oleogels is the solvent exchange method that can form complexes between protein aggregates and polysaccharides. The first step in this method is the formation of hydrogels with protein network in aqueous medium. The aqueous phase is then replaced with an organic solvent followed by incorporation of oil phase by frequent dipping or immersing the polymeric network in the edible oil. Furthermore, in a different approach to form oleogels, these series of steps were transformed to the addition of alcohol instead of oil, followed by supercritical drying with carbon dioxide and forming a high-porosity surface gel, i.e., aerogel, that could be used to fabricated oleogel by absorbing up to 80% of edible oil (Martins et al., 2020). *From Okuro, P. K., Martins, A. J., Vicente, A. A., & Cunha, R. L. (2020). Perspective on oleogelator mixtures, structure design and behavior toward digestibility of oleogels.* Current Opinion in Food Science, 35, *27–35. https://doi.org/10.1016/j.cofs.2020.01.001.*

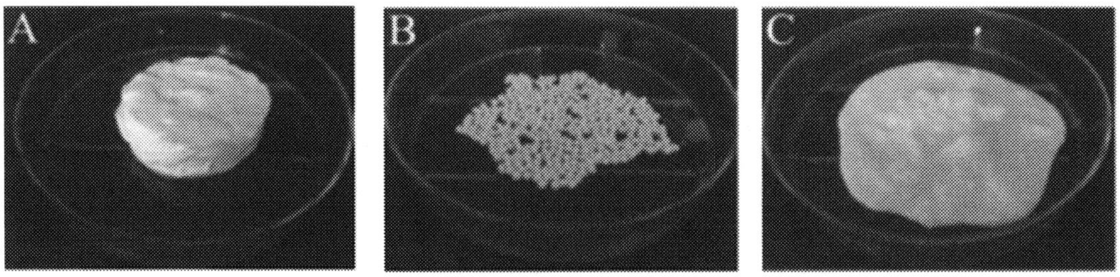

FIGURE 7.5 (A) Bulk emulsion gels, (B) emulsion gel particles, and (C) fluid emulsion gels fabricated using alginate as matrix material. From *Lin, D., Kelly, A. L., & Miao, S. (2020). Preparation, structure-property relationships and applications of different emulsion gels: Bulk emulsion gels, emulsion gel particles, and fluid emulsion gels.* Trends in Food Science and Technology, 102, *123–137. https://doi.org/10.1016/j.tifs.2020.05.024.*

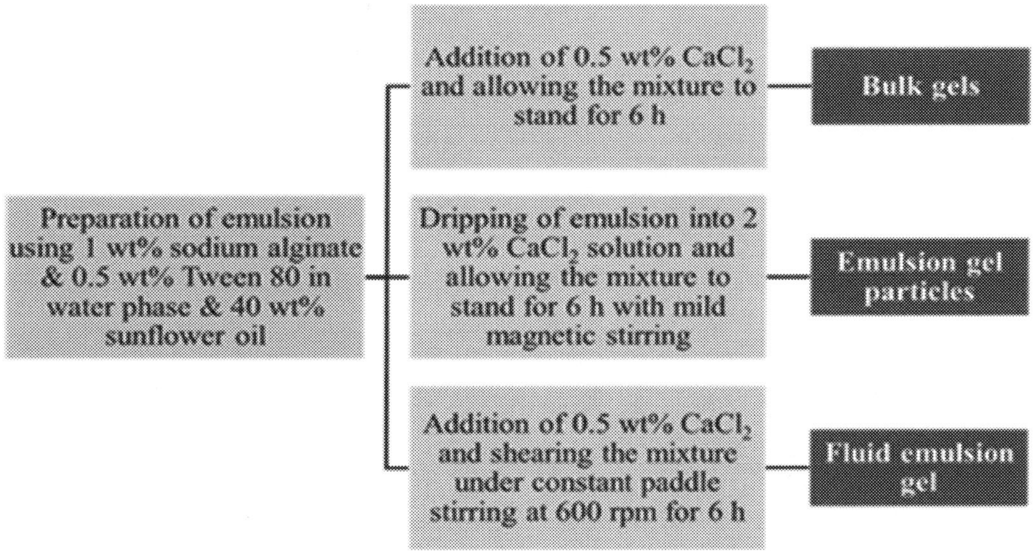

FIGURE 7.6 Process of preparation for alginate-based bulk emulsion gels, emulsion gel particles, and fluid emulsion gel. From *Lin, D., Kelly, A. L., & Miao, S. (2020). Preparation, structure-property relationships and applications of different emulsion gels: Bulk emulsion gels, emulsion gel particles, and fluid emulsion gels.* Trends in Food Science and Technology, 102*, 123–137. https://doi.org/10.1016/j.tifs.2020.05.024.*

temperature (based on the matrix material). Thus, based on its structural organization and disposition of organogels and hydrogels, bigels can be fabricated into organogel dispersed in hydrogel, hydrogel dispersed in organogel, bicontinuous bigel, and complex bigel. Fig. 7.7 shows a schematic illustration of the process for fabrication of organogel dispersed in hydrogel and hydrogel dispersed in organogel. The process involves: (a) preparation of organogels using organogelators and hydrogels using biopolymers, (b) combining both organogel and hydrogel at a definite shear speed and temperature, while retaining the characteristic properties of both the components, followed by (c) formation of a stable bigel, depending

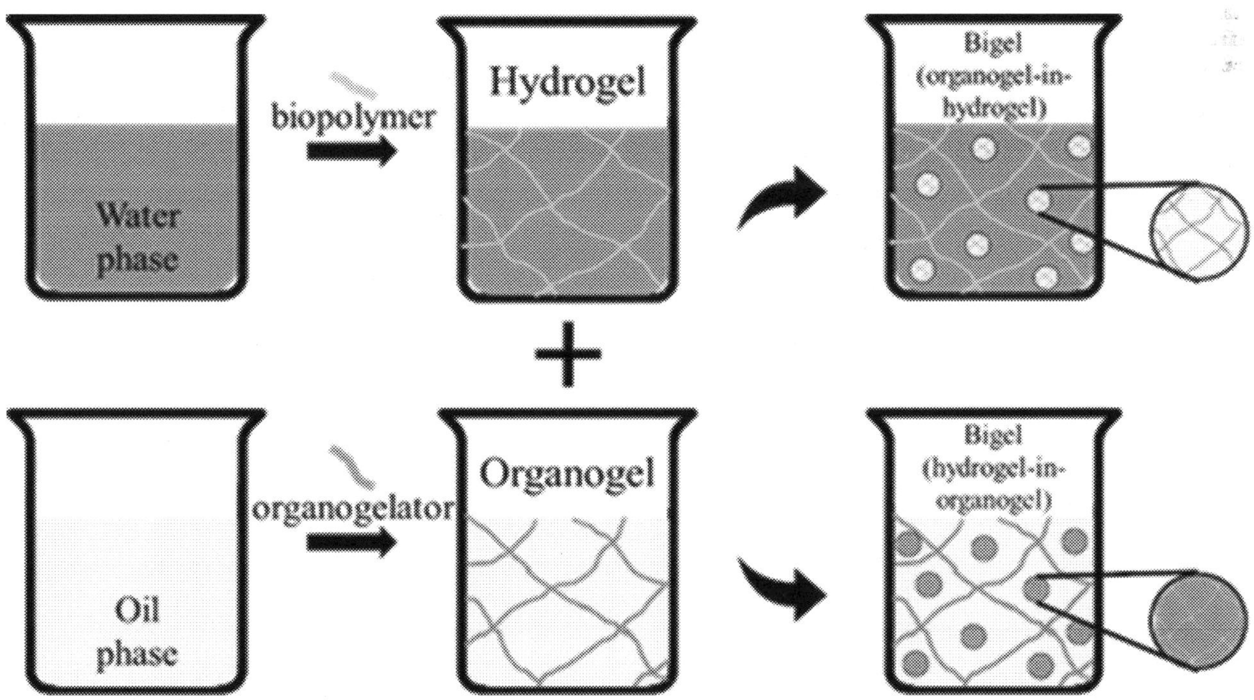

FIGURE 7.7 Schematic representation of processes for fabrication of bigels. *From Mao, L., Lu, Y., Cui, M., Miao, S., & Gao, Y. (2020). Design of gel structures in water and oil phases for improved delivery of bioactive food ingredients.* Critical Reviews in Food Science and Nutrition, 60*(10), 1651–1666. https://doi.org/10.1080/10408398.2019.1587737.*

upon the composition of both the phases. Alternatively, fabrication of a bicontinuous bigel involves dispersion of higher proportions of hydrogel-in-organogel in lower proportions of organogel-in-hydrogel. However, complex bigels can be fabricated by adding organogel-in-hydrogel to an oil-in-water/water-in-oil structured emulsion (Singh et al., 2018). Furthermore, based on these approaches for food gel fabrication that have been outlined and discussed above, some recent examples of plant-based food gels developed using different fabrication techniques is mentioned in Table 7.2.

TABLE 7.2 Recent examples of plant-based food gels obtained using different fabrication techniques.

Gel type	Plant-based ingredient	Fabrication	References
Hydrogel	Starch, xanthan gum	Extrusion and thermo-pressing	Simões et al. (2020)
	κ-carrageenan, agar, thiamine (encapsulated)	Semi-solid extrusion process (3D printing)	Kamlow et al. (2021)
	Plant cell lines isolated from carrot callus samples, alginate	Callus-based food ink was printed with 3D lattice scaffolds in gelatin slurry support bath	Park et al. (2020)
	Celery cellulose, short-chain fatty acids (encapsulated)	Mixing of cellulose solution in Lithium chloride (LiCl)/N,N-dimethylacetamide (DMAc) solution	Yan et al. (2020)
	Maleic anhydride esterified tapioca starch (MS), sodium trimetaphosphate (STMP), curcumin (encapsulated)	Crosslinking MS and STMP at 40°C	Meng et al. (2020)
	Coconut water, gellan gum, xanthan gum	High speed mixing	Seow et al. (2021)
Oleogel	*Schisandra chinensis*fruit oil	Using oleogelators potato starch, candelilla wax	Gao and Wu (2020)
	Camellia oil	Using oleogelators glycerol monolaurate	Pan et al. (2021)
	Sunflower oil	Using oleogelators ceramide, soybean lecithin	Guoc et al. (2020)
	Soybean oil	Using oleogelators monopalmitate, carnauba wax	Yang et al. (2020)
	Soybean oil	Using oleogelator unsaturated 2-monoglycerides	Chen et al. (2021)
	Olive oil	Using oleogelator soybean fatty acid distillate wax esters	Papadaki et al. (2020)
	Canola oil	Using oleogelators soy lecithin, stearic acid	Zhuang et al. (2021)
Emulsion gel	Corn oil, octenylsuccinate quinoa starch granule, lutein (encapsulated)	High speed homogenization	Li, He, et al. (2020), Li, Zhang, et al. (2020)
	Pea protein isolate, inulin	High-pressure homogenization and Ca2+-induced gelation	Xu et al. (2021)
	Olive oil, chia mucilage	High speed homogenization with gelling agents	Câmara et al. (2020)
	Sunflower oil, κ-carrageenan, ι-carrageenan, λ-carrageenan	High speed homogenization with KCl	Fontes-Candia et al. (2020)
	Sunflower oil	High shear homogenization using glycyrrhizic acid nanofibrils and soy protein isolate-pectin colloidal nanoparticles as complex stabilizers, followed by ultrasonic homogenization	Li, He, et al. (2020), Li, Zhang, et al. (2020)
	Sunflower oil	Ultrasonication with soybean lipophilic protein and hydroxypropyl methylcellulose, followed by Ca + induced gelation	Zhang et al. (2020)
	Soybean oil	High pressure homogenization with soybean protein isolate	Bi et al. (2020)
	Sunflower oil	High shear mixing with soy protein isolate and sodium alginate	Lin et al. (2020)

TABLE 7.2 Recent examples of plant-based food gels obtained using different fabrication techniques.—cont'd

Gel type	Plant-based ingredient	Fabrication	References
Bigel	κ-carrageenan hydrogel, monoglycerides oleogel, β-carotene (encapsulated)	High speed mixing of hydrogel and oleogel	Zheng et al. (2020)
	Soybean oil and dual modified rice starch emulsion, alginate hydrogel beads, lycopene (encapsulated)	Injecting lycopene loaded emulsion/alginate mixture in calcium chloride solution using a syringe to form lycopene loaded alginate beads	Jain et al. (2020)
	Medium-chain triglyceride (MCT) oil and cumin essential oil emulsion, sodium alginate hydrogel	Emulsion was prepared using ultrasonication and encapsulated in sodium alginate hydrogels	Gholamian et al. (2021)
	Cottonseed and cannabis oil organogel, alginate and ferula gum hydrogel, quercetin (encapsulated)	High speed mixing of cottonseed oil organogel and alginate/ferula gum hydrogel, followed by dropwise addition of mixture to cold cannabis oil to form organogel beads	Mousavi et al. (2021)
	High oleic sunflower oil and glycerol monostearate organogel, gellan gum hydrogel	Heating with constant stirring	Fasolin et al. (2021)
Aerogel	Germinated, nongerminated wheat starch, poly-ethylene oxide	Physical cross-linking using cycles of freezing and thawing	da Silva et al. (2020)
	Yerba-mate cellulose/nanocellulose	Physical cross-linking using cycles of freezing and thawing	de Oliveira et al. (2020)
	Medium chain triglycerides, ethylcellulose	Supercritical-carbon dioxide extraction of oil from oleogels	Manzocco et al. (2021)
	Potato protein isolate	Solvent exchange and supercritical-carbon dioxide drying of potato protein isolate hydrogel	Andlinger et al. (2021)
	Camelina seed mucilage	Solvent exchange and supercritical-carbon dioxide drying of mucilage hydrogel	Ubeyitogullari et al. (2020)
	Corn starch, agar, microcrystalline cellulose	Thermal gelation in aqueous media, retrogradation, solvent exchange, followed by supercritical-carbon dioxide drying	Dogenski et al. (2020)
	Canola seed meal	pH adjustment, heat treatment, followed by freeze drying	Fitzpatrick et al. (2020)
Aerogel/cryogel	Waxy maize, potato and pea starch	Starch dissolution, retrogradation, solvent exchange, followed by supercritical-carbon dioxide drying (for aerogel)/freeze drying (for cryogel)	Zou and Budtova (2021)

7.4 Functions of plant-based gels in food industry

Food gels developed with plant-based ingredients have demonstrated various functions based on their peculiar characteristics.

7.4.1 Water absorption, water retention, and moisture absorption

The water absorption and retention capability of gels depend on crosslink density and the types of monomers used. Water absorption is higher at lower crosslink density and the presence of polar groups in the monomer, thus attracting water due to static potential energy. Water absorptivity within the gel can be controlled by two parameters (a) compatibility between a polyelectrolyte and water and (b) osmotic pressure originated due to high ionic concentration. Following Eq. (7.1) can be used to determine the ability of a gel to absorb water.

$$\text{Water holding capacity} = \frac{\text{Osmotic pressure of ions} + \text{Compatibility of polyelectrolyte and water}}{\text{Crosslink density}} \quad (7.1)$$

In case of food gels, the structure of gel network determines the water holding capacity, which is affected by biopolymer concentration and composition, pH, ionic strength, heat treatment, and coagulator. For example, addition of sugar beet pectin increased the water holding capacity of soy protein gels, due to a decrease in pH. As it approached to isoelectric point of soy protein isolate, there was a change in protein spatial structure. pH also affected the ionization and electrostatic charge of protein molecules, thereby decreasing the attraction between protein molecules and increasing the rigidity of proteins (Chen et al., 2019). Similarly, the water holding capacity of quinoa protein hydrolysate gel was directly related to its microstructure, as a less interconnected protein network and large size pore (formed at different hydrolysis times) lead to reduced water retention in gel mesh (Galante et al., 2020).

7.4.2 Sustained release

Gastric acid is responsible for the denaturation of digested food and degradation of food structures, which involve breakdown of cell wall structure of plant-based foods in different extent. Previous studies have provided evidence that as gastric juices diffuse, the pectin present in the cell walls disintegrates, thus reducing their cohesiveness and increasing the intercellular spaces of cell wall matrix of plant-based gels. The possible reasons for pectin destabilization could be acid hydrolysis of glycosidic linkages within pectin strands in the gastric environment; displacement of Ca^{2+} cross-links by H^{+}; or solubilization of calcium ions in the middle lamella (Somaratane et al., 2020). Thus, lower binding of calcium prevented erosive degradation and fragmentation and promoted swelling, thereby yielding high values of diffusion coefficient indicating sustained release profiles of the encapsulated compound from food gels, specifically caffeine encapsulated in calcium induced alginate gels (Morrish et al., 2020). In addition to this, protein hydrogel microstructure also restricts the proteolysis process during the gastric digestion. For example, higher porosity of pea protein isolate gel promoted hydrolysis at a fast rate as compared to soy protein isolate gel (Opazo-Navarrete et al., 2018). Furthermore, even the texture of gel particles affected the diffusion process, as soft particles of food gels disintegrated and released more nutrients at a considerably faster rate than the harder ones (Somaratane et al., 2020).

7.4.3 Emulsion stabilization

The negative attribute of low solubility of plant structural proteins in both aqueous media and edible oils is now considered as a valued functional attribute for pickering systems. The pickering stabilization behavior of plant-based gels is largely attributed to the detachment energy (ΔG_d) associated with dislodging spherical colloidal particles from the fluid–fluid interface Eq. (7.2).

$$\Delta G_{d\ sphere} = \gamma_{\alpha\beta}\ \pi r^2 (1 - |\cos\theta|)^2 \quad (7.2)$$

Here, r is the radius of the spherical colloidal particle, $\gamma_{\alpha\beta}$ is the surface free energy or fluid–fluid interfacial tension, andθ is the equilibrium three-phase contact angle reflecting the partial wetting of the particles by the two fluids. It is evident from the equation that particle size, shape, and the detailed surface chemistry of the particles further affect their orientation at the interface. Some of the most popular classes of materials used for making plant-based Pickering emulsions are prolamin-based particles, e.g., zein, soy protein particles/aggregates, globular proteins behaving as soft nanoparticles, and thermo-sheared protein microgel particles/aggregates. Apart from protein particles, insoluble polysaccharide particles possessing semicrystalline structures over mesoscopic length scales, e.g., cellulose and starch, along with some secondary metabolites in the form of organic crystals (flavonoids) also forms plant-based pickering stabilizers (Sarkar and Dickinson, 2020). Development of gels using naturally available biopolymers, especially plant-based biopolymers, have recently gained a lot of research attention due to its biocompatibility and digestibility. Since biopolymers like polysaccharides and protein are commonly used as natural hydrogelation agents; therefore, the food-grade gels obtained with these plant-based ingredients are bestowed with improved properties and reduced risk of toxicity in foods. Owing to these distinct characteristics and the diverse functions discussed above, plant-based hydrogels have found various applications in the food sector. Fig. 7.8 represents the diverse applications of plant-based gels in the food sectors.

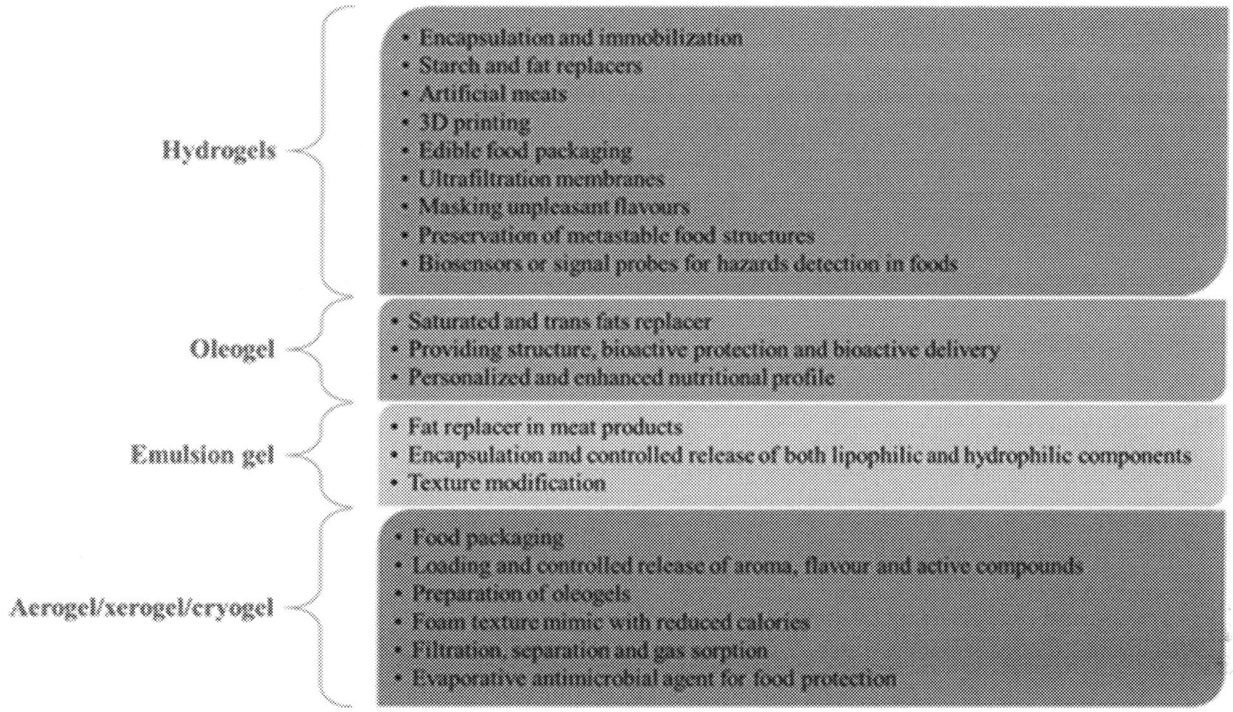

FIGURE 7.8 Applications of plant-based food gels.

7.5 Physico-chemical and sensory of plant-based gels

Gels (either polymer or protein based) are usually defined as the 3D fibrillar networks that get swollen by absorbing or trapping large amount of water or solvent. The pattern of gelation is essential in order to obtain the prudent sensory and textural structures in foods. The thorough understanding of the structure gel is essential in order to achieve the clear perception of sensory. The different properties of these comprise of thermal resistance, nature of polarity, surface activity, and molecular sizes. The peculiar features corresponding to each are different owing to the vivid parameters in terms of protein concentration, degree of denaturation caused by factors *viz.* pH, temperature, pressure, ionic strength, etc. The process of gelation requires an external medium or the driving force to unfold the protein structure followed by retainment of order of matrix between the protein strands. The driving force can be applied physically or chemically to bring out alterations in the protein—protein interaction. Physical treatments include heat and pressure, whereas chemical treatment leading to the use of acids and salts by varying their conditions of concentration and pH, respectively (Nazir et al., 2017). Gels possess the intermediate state of matter (something between solid and liquid) owing to which they are viscoelastic in nature. Depending upon the trapping agent, e.g., solvent in organogels and water in hydrogels, their usage varies in multiple applications. For example, organogels have porosity inside them; thereby, they can be employed under the involvement of long chain polymers as well as small molecules. The entrapped medium in the organogels can be organic solvent, vegetable oil, and mineral oil, respectively. Furthermore, the type of medium renders the properties like firmness and elasticity. However, hydrogels, involving the building blocks of small molecules opening the avenues of multiple applications particularly in the field of biomedical (drug delivery due to their sustainable release quality, tissue engineering, etc). Hydrogels are usually attained from natural resources and owing to which they fall under the huge range of polysaccharides and proteins (Banerjee and Bhattacharya, 2012). Due to their higher water content, their inbuilt flexibility is more similar to the natural tissues. The natural materials in the form of cellulose and agarose are recognized as hydrogels. Similarly, there is another category of aerogels, whose density is very low owing to the dispersion of air between the networks of colloidal gel and therefore, the presence of air the aerogel does not gets shrink. Concurrently, the surface area of the gels gets increased. On contrary, xerogel is formed without hampering of the shrinkage of the internal molecular network along with retainment of remarkable surface area and porosity.

7.6 Bioavailable components from plant-based gels

The significance of bioactive compounds (polyphenolds, carotenoids, vitamins, flavonoids, functional lipids) has been burgeoning owing to their functional role inside the body. Therefore, they are clinically beneficial for quality maintenance of the body owing to their defensive action against numerous lifestyle disorders in terms of preventing CVDs, diabetes, growth of cancerous cells, fastening of aging, etc. The bioactive components such as polyphenolic compounds and flavonoids are further usually the crosslinking agents that render stability to the plant-based gels. In addition, variety of polysaccharides (natural gums, agar, alginate, starches, pectin, glucomannan, etc) is also derived from these gels (Mao et al., 2020; Nazir et al., 2017). It was also observed if these components are added in the gel-based structures, then in return these components render stability and protection to the entire structure of the gel. The structure of the gel comprises of the fine network of organic solvents and liquid, and these fine networks hold the bioactive compounds (Da Pieve et al., 2011). In addition, the network of gels also makes the dispersion and solubility of the bioactive components orderly, which, in turn, result in their appropriate delivery. Thereby, gels acquire nutritional and functional properties with capacity of high loading and good stability. In the current scenario, the products enriched with bioactive components are attracting more attention owing to their multiple grades of clinical significance. Further, the addition of plant-based compounds to the gels results in the increase of the water holding capacity of the gels owing to decrease of the syneresis (Ozdemir and Ozcan, 2020). Various studies have supported the concept of ameliorating the stability (by preventing the crystallization of the gels) as well the bioavailability of bioactive compounds by incorporating them via gel-based delivery. However, in releasing of bioactives from the gels, it is the pH which plays the major role (Qiu and Park et al., 2001). In one of the supported studies, the presence of $CaCl_2$ when added to the protein-based whey gel, it stimulated the release of caffeine (from caffeine loaded whey protein) (Zand-Rajabi and Madadlou, 2016). Similarly, the controlled release of tocopherol was also investigated on $CaCl_2$ laddened protein gel. Thereby, we can produce pH-based gels in order to achieve the optimum release of bioactive components (Somchue et al., 2009). There are numbers of ways through which bioactive components are inserted in the form of their (aqueous or solvent dispersed) microstructures (*viz.* micelles, gels, particles, droplets, etc.). The binding blocks under the networks of the gels act as efficient carriers for the functional components like bioactives owing to their prominent characteristic of lower mobility (Shakeel et al., 2018; Zheng et al., 2017). The different types of gels have their peculiar properties for their suitability of delivering the bioactive compounds. Likewise, one of the alginate gels which is also based on the type of polysaccharide hydrogels have been studied widely for the transport of functional components owing to their enrichment with metal cations, e.g., calcium salts. Wang et al. (2019) explored the optimization of the delivery of hydrophobic bioactive compounds using the rapeseed protein-based nanogel. Concurrently, emulsion type gels (oil and water based) have also been explored for their efficacious suitability and stability for delivering lipophilic-based bioactive components or neutraceuticals (~ hydrophobic compounds in terms of curcumin, lycopene, β-carotene, quercetin) owing to their peculiar feature of flexibility in controlling and directing the structure as well as the composition of ingredients (McClements et al., 2016; Zou et al., 2015). However, in the case of organogels, the oil is usually entrapped inside the gel-network and bioactives get dispersed in it and thereby protected by the network from their degradation. Various studies have reported the efficacy of organogels with the fat-soluble components as these gels prevent the precipitation. The compounds of potent clinical significance in terms of major antioxidants like β-carotene, curcumin, and quercetin were found to have increased bio-accessibility on their stuffing to organogels (Cui et a., 2019; Xu; Yu et al., 2012). The entrapment of bioactive components within the lipid leads to retardation in their oxidation and thereby and improves the dissolvability as well as solubility of these functional metabolites. However, among the categorization of the gels, there also exists a combination of organogel and hydrogel, which is termed as bigel. They are basically the emulsions, which involve both the states of water and oil. Their significant characteristic lies in their compatibility for both lipophilic and hydrophilic bioactive metabolites (Shakeel et al., 2018). Owing to the robustness of the bigels, they attain the higher strength (or hardness) and resistance to temperature. The positive benefits of applications of bigels have been seen more on medicinal compounds due to their advantage of optimized slow release of the bioactive ingredient because of the tanginess of the bigel structure. Nevertheless, there is a scarcity of work done in the area of food using bigels.

7.7 Recent trends and future for improving the quality-based gels

Development of gels using plant-based food polymers makes them a versatile and ubiquitous form of matter with a unique set of features like source renewability, biocompatibility, biodegradability, and affordability. This chapter focuses on the food gels only with plant-based ingredients that have been classified as protein, polysaccharide, and binary gels, along with a detailed discussion on its fabrication methods, functions, applications, and properties. Particular attention has been paid to understand the process of fabrication of food gels with plant-based biopolymers and its effect on the desired functionality, that is yet to be exploited for the development of innovative food gels in near future. Apart from this, in-spite of the multiple advantages of the various types of gels, still there are certain nonclarities on part of their digestibility aspect, which can be improved in future by throwing more light on them through applications-based experiments. However, future trends exist in the development of gels in the micro-scale and nano-scale by extensive research on their multiple structures in terms of beads, coatings, edible films, and their potential applications in the area of food technology. Furthermore, gaps to be filled in the areas of optimized releasing of bioactive ingredients and their protection from various kinds of degradation during their loading or incorporation to the gels. In addition, more improvement is required in the properties of gels in terms of their loading strength, hardness, porosity, stability, noncrystallization, water holding capacity, firmness by settling down their functionalities and compatibilities with bioactive metabolites. Therefore, there exist a lot of potential in exploring innovative approaches toward an improved understanding of the fundamental properties of food gels based on their fabrication and designing for already available as well as other unexplored applications.

References

Andlinger, D. J., Bornkeßel, A. C., Jung, I., Schroeter, B., Smirnova, I., & Kulozik, U. (2021). Microstructures of potato protein hydrogels and aerogels produced by thermal crosslinking and supercritical drying. *Food Hydrocolloids, 112*, 106305. https://doi.org/10.1016/j.foodhyd.2020.106305

Banerjee, S., & Bhattacharya, S. (2012). Food gels: Gelling process and new applications. *Critical Reviews in Food Science and Nutrition, 52*(4), 334–346. https://doi.org/10.1080/10408398.2010.500234

Batista, R. A., Espitia, P. J. P., Quintans, J. de S. S., Freitas, M. M., Cerqueira, M.Â., Teixeira, J. A., & Cardoso, J. C. (2019). Hydrogel as an alternative structure for food packaging systems. *Carbohydrate Polymers, 205*, 106–116. https://doi.org/10.1016/j.carbpol.2018.10.006

Bi, C., Wang, P., Sun, D., Yan, Z., Liu, Y., Huang, Z., & Gao, F. (2020). Effect of high-pressure homogenization on gelling and rheological properties of soybean protein isolate emulsion gel. *Journal of Food Engineering, 277*, 109923. https://doi.org/10.1016/j.jfoodeng.2020.109923

Câmara, A. K. F. I., Okuro, P. K., Santos, M., Paglarini, C. de S., da Cunha, R. L., Ruiz-Capillas, C., Herrero, A. M., & Pollonio, M. A. R. (2020). Understanding the role of chia (Salvia Hispanica L.) mucilage on olive oil-based emulsion gels as a new fat substitute in emulsified meat products. *European Food Research and Technology, 246*(5), 909–922. https://doi.org/10.1007/s00217-020-03457-4 C

Cao, Y., & Mezzenga, R. (2020). Design principles of food gels. *Nature Food, 1*(2), 106–118. https://doi.org/10.1038/s43016-019-0009-x

Chen, H., Gan, J., Ji, A., Song, S., & Yin, L. (2019). Development of double network gels based on soy protein isolate and sugar beet pectin induced by thermal treatment and laccase catalysis. *Food Chemistry, 292*, 188–196. https://doi.org/10.1016/j.foodchem.2019.04.059

Chen, C., Zhang, C., Zhang, Q., Ju, X., Wang, Z., & He, R. (2021). Study of monoglycerides enriched with unsaturated fatty acids at sn-2 position as oleogelators for oleogel preparation. *Food Chemistry, 354*, 129534. https://doi.org/10.1016/j.foodchem.2021.129534

Cui, M., Mao, L., Lu, Y., Yuan, F., & Gao, Y. (2019). Effect of monoglyceride content on the solubility and chemical stability of β-carotene in organogels. *LWT, 106*, 83–91. https://doi.org/10.1016/j.lwt.2019.02.042

Da Pieve, S., Calligaris, S., Panozzo, A., Arrighetti, G., & Nicoli, M. C. (2011). Effect of monoglyceride organogel structure on cod liver oil stability. *Food Research International, 44*(9), 2978–2983. https://doi.org/10.1016/j.foodres.2011.07.011

da Silva, F. T., de Oliveira, J. P., Fonseca, L. M., Bruni, G. P., da Rosa Zavareze, E., & Dias, A. R. G. (2020). Physically cross-linked aerogels based on germinated and non-germinated wheat starch and PEO for application as water absorbers for food packaging. *International Journal of Biological Macromolecules, 155*, 6–13. https://doi.org/10.1016/j.ijbiomac.2020.03.123

de Oliveira, J. P., Bruni, G. P., Fonseca, L. M., da Silva, F. T., da Rocha, J. C., & da Rosa Zavareze, E. (2020). Characterization of aerogels as bioactive delivery vehicles produced through the valorization of yerba-mate (Illex paraguariensis). *Food Hydrocolloids, 107*, 105931. https://doi.org/10.1016/j.foodhyd.2020.105931

Dogenski, M., Navarro-Díaz, H. J., de Oliveira, J. V., & Ferreira, S. R. S. (2020). Properties of starch-based aerogels incorporated with agar or microcrystalline cellulose. *Food Hydrocolloids, 108*, 106033. https://doi.org/10.1016/j.foodhyd.2020.106033

Farjami, T., & Madadlou, A. (2017). Fabrication methods of biopolymeric microgels and microgel-based hydrogels. *Food Hydrocolloids, 62*, 262–272. https://doi.org/10.1016/j.foodhyd.2016.08.017

Farjami, T., & Madadlou, A. (2019). An overview on preparation of emulsion-filled gels and emulsion particulate gels. *Trends in Food Science & Technology, 86*, 85–94. https://doi.org/10.1016/j.tifs.2019.02.043

Fasolin, L. H., Martins, A. J., Cerqueira, M. A., & Vicente, A. A. (2021). Modulating process parameters to change physical properties of bigels for food applications. *Food Structure, 28*, 100173. https://doi.org/10.1016/j.foostr.2020.100173

Fitzpatrick, S. E., Deb-Choudhury, S., Ranford, S., & Staiger, M. P. (2020). Novel protein-based bio-aerogels derived from canola seed meal. *Journal of Materials Science, 55*(11), 4848–4863. https://doi.org/10.1007/s10853-019-04330-w

Fontes-Candia, C., Ström, A., Lopez-Sanchez, P., López-Rubio, A., & Martínez-Sanz, M. (2020). Rheological and structural characterization of carrageenan emulsion gels. *Algal Research, 47*, 101873. https://doi.org/10.1016/j.algal.2020.101873

Gagliardi, A., Froiio, F., Salvatici, M. C., Paolino, D., Fresta, M., & Cosco, D. (2020). Characterization and refinement of zein-based gels. *Food Hydrocolloids, 101*, 105555. https://doi.org/10.1016/j.foodhyd.2019.105555

Galante, M., De Flaviis, R., Boeris, V., & Spelzini, D. (2020). Effects of the enzymatic hydrolysis treatment on functional and antioxidant properties of quinoa protein acid-induced gels. *LWT, 118*, 108845. https://doi.org/10.1016/j.lwt.2019.108845

Gao, Y., & Wu, S. (2020). Development and evaluation of a novel oleogel system based on starch—water—wax—oil. *Food & Function, 11*(9), 7727—7735. https://doi.org/10.1039/D0FO01785J

Gholamian, S., Nourani, M., & Bakhshi, N. (2021). Formation and characterization of calcium alginate hydrogel beads filled with cumin seeds essential oil. *Food Chemistry, 338*, 128143. https://doi.org/10.1016/j.foodchem.2020.128143

Guo, S., Song, M., Gao, X., Dong, L., Hou, T., Lin, X., Tan, W., Cao, Y., Rogers, M., & Lan, Y. (2020). Assembly pattern of multicomponent supramolecular oleogel composed of ceramide and lecithin in sunflower oil: Self-assembly or self-sorting? *Food & Function, 11*(9), 7651—7660. https://doi.org/10.1039/D0FO00635A

Hao, Z.-Z., Peng, X.-Q., & Tang, C.-H. (2020). Edible pickering high internal phase emulsions stabilized by soy glycinin: Improvement of emulsification performance and pickering stabilization by glycation with soy polysaccharide. *Food Hydrocolloids, 103*, 105672. https://doi.org/10.1016/j.foodhyd.2020.105672

Jain, S., Winuprasith, T., & Suphantharika, M. (2020). Encapsulation of lycopene in emulsions and hydrogel beads using dual modified rice starch: Characterization, stability analysis and release behaviour during in-vitro digestion. *Food Hydrocolloids, 104*, 105730. https://doi.org/10.1016/j.foodhyd.2020.105730

Kamlow, M.-A., Vadodaria, S., Gholamipour-Shirazi, A., Spyropoulos, F., & Mills, T. (2021). 3D printing of edible hydrogels containing thiamine and their comparison to cast gels. *Food Hydrocolloids, 116*, 106550. https://doi.org/10.1016/j.foodhyd.2020.106550

Klost, M., Giménez-Ribes, G., & Drusch, S. (2020). Enzymatic hydrolysis of pea protein: Interactions and protein fractions involved in fermentation induced gels and their influence on rheological properties. *Food Hydrocolloids, 105*, 105793. https://doi.org/10.1016/j.foodhyd.2020.105793

Li, J., Wu, M., Wang, Y., Li, K., Du, J., & Bai, Y. (2020). Effect of pH-shifting treatment on structural and heat induced gel properties of peanut protein isolate. *Food Chemistry, 325*, 126921. https://doi.org/10.1016/j.foodchem.2020.126921

Li, O.-Y., Wang, L., Liu, X.-Y., Yin, J.-Y., & Nie, S.-P. (2020). Interactions between ascorbic acid and water soluble polysaccharide from the seeds of Plantago asiatica L.: Effects on polysaccharide physicochemical properties and stability. *Food Hydrocolloids, 99*, 105351. https://doi.org/10.1016/j.foodhyd.2019.105351

Li, Q., He, Q., Xu, M., Li, J., Liu, X., Wan, Z., & Yang, X. (2020). Food-grade emulsions and emulsion gels prepared by soy protein—pectin complex nanoparticles and glycyrrhizic acid nanofibrils. *Journal of Agricultural and Food Chemistry, 68*(4), 1051—1063. https://doi.org/10.1021/acs.jafc.9b04957

Li, S., Zhang, B., Li, C., Fu, X., & Huang, Q. (2020). Pickering emulsion gel stabilized by octenylsuccinate quinoa starch granule as lutein carrier: Role of the gel network. *Food Chemistry, 305*, 125476. https://doi.org/10.1016/j.foodchem.2019.125476

Lin, D., Kelly, A. L., & Miao, S. (2020). Preparation, structure-property relationships and applications of different emulsion gels: Bulk emulsion gels, emulsion gel particles, and fluid emulsion gels. *Trends in Food Science & Technology, 102*, 123—137. https://doi.org/10.1016/j.tifs.2020.05.024

Lin, D., Kelly, A. L., & Miao, S. (2021). The role of mixing sequence in structuring O/W emulsions and emulsion gels produced by electrostatic protein-polysaccharide interactions between soy protein isolate-coated droplets and alginate molecules. *Food Hydrocolloids, 113*, 106537. https://doi.org/10.1016/j.foodhyd.2020.106537

Liu, S., Shen, M., Xiao, Y., Luo, Y., & Xie, J. (2020). Effect of maize, potato, and pea starches with Mesona chinensis polysaccharide on pasting, gelatinization properties, granular morphology and digestion. *Food Hydrocolloids, 108*, 106047. https://doi.org/10.1016/j.foodhyd.2020.106047

Luo, Y., Shen, M., Li, E., Xiao, Y., Wen, H., Ren, Y., & Xie, J. (2020). Effect of Mesona chinensis polysaccharide on pasting, rheological and structural properties of corn starches varying in amylose contents. *Carbohydrate Polymers, 230*, 115713. https://doi.org/10.1016/j.carbpol.2019.115713

Manzocco, L., Basso, F., Plazzotta, S., & Calligaris, S. (2021). Study on the possibility of developing food-grade hydrophobic bio-aerogels by using an oleogel template approach. *Current Research in Food Science, 4*, 115—120. https://doi.org/10.1016/j.crfs.2021.02.005

Manzoor, M., Singh, J., Bandral, J. D., Gani, A., & Shams, R. (2020). Food hydrocolloids: Functional, nutraceutical and novel applications for delivery of bioactive compounds. *International Journal of Biological Macromolecules, 165*, 554—567. https://doi.org/10.1016/j.ijbiomac.2020.09.182

Mao, L., Lu, Y., Cui, M., Miao, S., & Gao, Y. (2020). Design of gel structures in water and oil phases for improved delivery of bioactive food ingredients. *Critical Reviews in Food Science and Nutrition, 60*(10), 1651—1666. https://doi.org/10.1080/10408398.2019.1587737

Martins, A. J., Vicente, A. A., Pastrana, L. M., & Cerqueira, M. A. (2020). Oleogels for development of health-promoting food products. *Food Science and Human Wellness, 9*(1), 31—39. https://doi.org/10.1016/j.fshw.2019.12.001

McClements, D. J., Saliva-Trujillo, L., Zhang, R., Zhang, Z., Zou, L., Yao, M., & Xiao, H. (2016). Boosting the bioavailability of hydrophobic nutrients, vitamins, and nutraceuticals in natural products using excipient emulsions. *Food Research International, 88*, 140—152. https://doi.org/10.1016/j.foodres.2015.11.017

Meng, R., Wu, Z., Xie, H.-Q., Xu, G.-X., Cheng, J.-S., & Zhang, B. (2020). Preparation, characterization, and encapsulation capability of the hydrogel cross-linked by esterified tapioca starch. *International Journal of Biological Macromolecules, 155*, 1—5. https://doi.org/10.1016/j.ijbiomac.2020.03.141

Moreno, H. M., Domínguez-Timón, F., Díaz, M. T., Pedrosa, M. M., Borderías, A. J., & Tovar, C. A. (2020). Evaluation of gels made with different commercial pea protein isolate: Rheological, structural and functional properties. *Food Hydrocolloids, 99*, 105375. https://doi.org/10.1016/j.foodhyd.2019.105375

Morrish, C., Teimouri, S., Istivan, T., & Kasapis, S. (2020). Molecular characterisation of hot moulded alginate gels as a delivery vehicle for the release of entrapped caffeine. *Food Hydrocolloids, 109*, 106142. https://doi.org/10.1016/j.foodhyd.2020.106142

Mousavi, S. N., Hosseini, E., Seyed Dorraji, M. S., Sheikh Mohammadi, S., Pourmansouri, Z., Rasoulifard, M. H., Doosti, M., & Chiti, H. (2021). Synthesis of a green bigel using cottonseed oil/cannabis oil/alginate/ferula gum for quercetin release: Synergistic effects for treating infertility in rats. *International Journal of Biological Macromolecules, 177*, 157—165. https://doi.org/10.1016/j.ijbiomac.2021.02.121

Munialo, C. D., Euston, S. R., & de Jongh, H. H. J. (2018). 19 - protein gels. In F. P. Yada (Ed.), *Woodhead publishing Series in Food Science, Technology and Nutrition* (pp. 501–521). Woodhead Publishing. https://doi.org/10.1016/B978-0-08-100722-8.00020-6

Nazir, A., Asghar, A., & Aslam Maan, A. (2017). Chapter 13 - food gels: Gelling process and new applications. In J. Ahmed, P. Ptaszek, S. B, T. A., F. R., & I. A. Basu (Eds.), *Woodhead publishing Series in Food Science, Technology and Nutrition* (pp. 335–353). Woodhead Publishing. https://doi.org/10.1016/B978-0-08-100431-9.00013-9

Okuro, P. K., Martins, A. J., Vicente, A. A., & Cunha, R. L. (2020). Perspective on oleogelator mixtures, structure design and behaviour towards digestibility of oleogels. *Current Opinion in Food Science, 35*, 27–35. https://doi.org/10.1016/j.cofs.2020.01.001

Opazo-Navarrete, M., Altenburg, M. D., Boom, R. M., & Janssen, A. E. M. (2018). The effect of gel microstructure on simulated gastric digestion of protein gels. *Food Biophysics, 13*(2), 124–138. https://doi.org/10.1007/s11483-018-9518-7

Ozdemir, T., & Ozcan, T. (2020). Effect of steviol glycosides as sugar substitute on the probiotic fermentation in milk gels enriched with red beetroot (Beta vulgaris L.) bioactive compounds. *LWT, 134*, 109851. https://doi.org/10.1016/j.lwt.2020.109851

Pan, J., Tang, L., Dong, Q., Li, Y., & Zhang, H. (2021). Effect of oleogelation on physical properties and oxidative stability of camellia oil-based oleogels and oleogel emulsions. *Food Research International, 140*, 110057. https://doi.org/10.1016/j.foodres.2020.110057

Pang, Z., Luo, Y., Li, B., Zhang, M., & Liu, X. (2020). Effect of different hydrocolloids on tribological and rheological behaviors of soymilk gels. *Food Hydrocolloids, 101*, 105558. https://doi.org/10.1016/j.foodhyd.2019.105558

Papadaki, A., Kopsahelis, N., Freire, D. M. G., Mandala, I., & Koutinas, A. A. (2020). Olive oil oleogel formulation using wax esters derived from soybean fatty acid distillate. *Biomolecules, 10*(1). https://doi.org/10.3390/biom10010106

Park, S. M., Kim, H. W., & Park, H. J. (2020). Callus-based 3D printing for food exemplified with carrot tissues and its potential for innovative food production. *Journal of Food Engineering, 271*, 109781. https://doi.org/10.1016/j.jfoodeng.2019.109781

Qiu, Y., & Park, K. (2001). Environment-sensitive hydrogels for drug delivery. *Advanced Drug Delivery Reviews, 53*(3), 321–339. https://doi.org/10.1016/S0169-409X(01)00203-4

Ren, Y., Jiang, L., Wang, W., Xiao, Y., Liu, S., Luo, Y., Shen, M., & Xie, J. (2020). Effects of Mesona chinensis Benth polysaccharide on physicochemical and rheological properties of sweet potato starch and its interactions. *Food Hydrocolloids, 99*, 105371. https://doi.org/10.1016/j.foodhyd.2019.105371

Ren, Y., Rong, L., Shen, M., Liu, W., Xiao, W., Luo, Y., & Xie, J. (2020). Interaction between rice starch and Mesona chinensis Benth polysaccharide gels: Pasting and gelling properties. *Carbohydrate Polymers, 240*, 116316. https://doi.org/10.1016/j.carbpol.2020.116316

Ruzengwe, F. M., Amonsou, E. O., & Kudanga, T. (2020). Rheological and microstructural properties of Bambara groundnut protein gels. *LWT, 123*, 109070. https://doi.org/10.1016/j.lwt.2020.109070

Sarkar, A., & Dickinson, E. (2020). Sustainable food-grade Pickering emulsions stabilized by plant-based particles. *Current Opinion in Colloid & Interface Science, 49*, 69–81. https://doi.org/10.1016/j.cocis.2020.04.004

Seow, E.-K., Muthu, V., Che Muhamed, A. M., Ooi, C. H., Singh, R., Mat Easa, A., & Tan, T.-C. (2021). Influence of low-level gelling agents on the dissolution and in-vitro nutrient release study of coconut water-based hydrogel. *Journal of Food Science and Technology*. https://doi.org/10.1007/s13197-021-05045-0

Shakeel, A., Lupi, F. R., Gabriele, D., Baldino, N., & De Cindio, B. (2018). Bigels: A unique class of materials for drug delivery applications. *Soft Materials, 16*(2), 77–93. https://doi.org/10.1080/1539445X.2018.1424638

Simões, B. M., Cagnin, C., Yamashita, F., Olivato, J. B., Garcia, P. S., de Oliveira, S. M., & Eiras Grossmann, M. V. (2020). Citric acid as crosslinking agent in starch/xanthan gum hydrogels produced by extrusion and thermopressing. *LWT, 125*, 108950. https://doi.org/10.1016/j.lwt.2019.108950

Singh, V. K., Qureshi, D., Nayak, S. K., & Pal, K. (2018). 10 - bigels. In K. Pal, & I. B. T. Banerjee (Eds.), *Woodhead publishing Series in Biomaterials* (pp. 265–282). Woodhead Publishing. https://doi.org/10.1016/B978-0-08-102179-8.00010-7

Somaratne, G., Ferrua, M. J., Ye, A., Nau, F., Floury, J., Dupont, D., & Singh, J. (2020). Food material properties as determining factors in nutrient release during human gastric digestion: A review. *Critical Reviews in Food Science and Nutrition, 60*(22), 3753–3769. https://doi.org/10.1080/10408398.2019.1707770

Somchue, W., Sermsri, W., Shiowatana, J., & Siripinyanond, A. (2009). Encapsulation of α-tocopherol in protein-based delivery particles. *Food Research International, 42*(8), 909–914. https://doi.org/10.1016/j.foodres.2009.04.021

Ubeyitogullari, A., & Ciftci, O. N. (2020). Fabrication of bioaerogels from camelina seed mucilage for food applications. *Food Hydrocolloids, 102*, 105597. https://doi.org/10.1016/j.foodhyd.2019.105597

Vélez-Erazo, E. M., Bosqui, K., Rabelo, R. S., Kurozawa, L. E., & Hubinger, M. D. (2020). High internal phase emulsions (HIPE) using pea protein and different polysaccharides as stabilizers. *Food Hydrocolloids, 105*, 105775. https://doi.org/10.1016/j.foodhyd.2020.105775

Wang, Z., Zhang, R. X., Zhang, C., Dai, C., Ju, X., & He, R. (2019). Fabrication of stable and self-assembling rapeseed protein nanogel for hydrophobic curcumin delivery. *Journal of Agricultural and Food Chemistry, 67*(3), 887–894. https://doi.org/10.1021/acs.jafc.8b05572

Wang, W., Shen, M., Jiang, L., Song, Q., Liu, S., & Xie, J. (2020). Influence of Mesona blumes polysaccharide on the gel properties and microstructure of acid-induced soy protein isolate gels. *Food Chemistry, 313*, 126125. https://doi.org/10.1016/j.foodchem.2019.126125

Wang, Y., Yang, F., Wu, M., Li, J., Bai, Y., Xu, W., & Qiu, S. (2020). Synergistic effect of pH shifting and mild heating in improving heat induced gel properties of peanut protein isolate. *LWT, 131*, 109812. https://doi.org/10.1016/j.lwt.2020.109812

Xiao, Y., Li, J., Liu, Y., Peng, F., Wang, X., Wang, C., Li, M., & Xu, H. (2020). Gel properties and formation mechanism of soy protein isolate gels improved by wheat bran cellulose. *Food Chemistry, 324*, 126876. https://doi.org/10.1016/j.foodchem.2020.126876

Xiao, Y., Shen, M., Luo, Y., Ren, Y., Han, X., & Xie, J. (2020). Effect of Mesona chinensis polysaccharide on the pasting, rheological, and structural properties of tapioca starch varying in gelatinization temperatures. *International Journal of Biological Macromolecules, 156*, 137–143. https://doi.org/10.1016/j.ijbiomac.2020.04.041

Xie, F., Zhang, H., Xia, Y., & Ai, L. (2020). Effects of tamarind seed polysaccharide on gelatinization, rheological, and structural properties of corn starch with different amylose/amylopectin ratios. *Food Hydrocolloids, 105*, 105854. https://doi.org/10.1016/j.foodhyd.2020.105854

Xie, J., Ren, Y., Xiao, Y., Luo, Y., & Shen, M. (2021). Interactions between tapioca starch and Mesona chinensis polysaccharide: Effects of urea and NaCl. *Food Hydrocolloids, 111*, 106268. https://doi.org/10.1016/j.foodhyd.2020.106268

Xu, Q., Qi, B., Han, L., Wang, D., Zhang, S., Jiang, L., Xie, F., & Li, Y. (2021). Study on the gel properties, interactions, and pH stability of pea protein isolate emulsion gels as influenced by inulin. *LWT, 137*, 110421. https://doi.org/10.1016/j.lwt.2020.110421

Xu, X. Improve bioaccessibility of quercetin using pseudo-organogel based nanoemulsions.Retrieved from https://doi.org/10.7282/T3FJ2DVK.

Yan, L., Wang, L., Gao, S., Liu, C., Zhang, Z., Ma, A., & Zheng, L. (2020). Celery cellulose hydrogel as carriers for controlled release of short-chain fatty acid by ultrasound. *Food Chemistry, 309*, 125717. https://doi.org/10.1016/j.foodchem.2019.125717

Yan, J., Liang, X., Ma, C., McClements, D. J., Liu, X., & Liu, F. (2021). Design and characterization of double-cross-linked emulsion gels using mixed biopolymers: Zein and sodium alginate. *Food Hydrocolloids, 113*, 106473. https://doi.org/10.1016/j.foodhyd.2020.106473

Yang, S., Yang, G., Chen, X., Chen, J., & Liu, W. (2020). Interaction of monopalmitate and carnauba wax on the properties and crystallization behavior of soybean oleogel. *Grain & Oil Science and Technology, 3*(2), 49–56. https://doi.org/10.1016/j.gaost.2020.05.001

Yang, Q., Wang, Y.-R., Li-Sha, Y.-J., & Chen, H.-Q. (2021). The effects of basil seed gum on the physicochemical and structural properties of arachin gel. *Food Hydrocolloids, 110*, Article 106189. https://doi.org/10.1016/j.foodhyd.2020.106189

Ye, J., Hua, X., Zhao, Q., Dong, Z., Li, Z., Zhang, W., & Yang, R. (2020). Characteristics of alkali-extracted peanut polysaccharide-protein complexes and their ability as Pickering emulsifiers. *International Journal of Biological Macromolecules, 162*, 1178–1186. https://doi.org/10.1016/j.ijbiomac.2020.06.245

Yu, H., & Huang, Q. (2012). Improving the oral bioavailability of curcumin using novel organogel-based nanoemulsions. *Journal of Agricultural and Food Chemistry, 60*(21), 5373–5379. https://doi.org/10.1021/jf300609p

Yu, B., Ren, F., Zhao, H., Cui, B., & Liu, P. (2020). Effects of native starch and modified starches on the textural, rheological and microstructural characteristics of soybean protein gel. *International Journal of Biological Macromolecules, 142*, 237–243. https://doi.org/10.1016/j.ijbiomac.2019.09.095

Zand-Rajabi, H., & Madadlou, A. (2016). Caffeine-loaded whey protein hydrogels reinforced with gellan and enriched with calcium chloride. *International Dairy Journal, 56*, 38–44. https://doi.org/10.1016/j.idairyj.2015.12.011

Zhang, S., Holmes, M., Ettelaie, R., & Sarkar, A. (2020). Pea protein microgel particles as Pickering stabilisers of oil-in-water emulsions: Responsiveness to pH and ionic strength. *Food Hydrocolloids, 102*, 105583. https://doi.org/10.1016/j.foodhyd.2019.105583

Zhao, S., Malfait, W. J., Guerrero-Alburquerque, N., Koebel, M. M., & Nyström, G. (2018). Biopolymer aerogels and foams: Chemistry, properties, and applications. *Angewandte Chemie International Edition, 57*(26), 7580–7608. https://doi.org/10.1002/anie.201709014

Zhao, H., Chen, J., Hemar, Y., & Cui, B. (2020). Improvement of the rheological and textural properties of calcium sulfate-induced soy protein isolate gels by the incorporation of different polysaccharides. *Food Chemistry, 310*, 125983. https://doi.org/10.1016/j.foodchem.2019.125983

Zheng, B., Zhang, Z., Chen, F., Luo, X., & McClements, D. J. (2017). Impact of delivery system type on curcumin stability: Comparison of curcumin degradation in aqueous solutions, emulsions, and hydrogel beads. *Food Hydrocolloids, 71*, 187–197. https://doi.org/10.1016/j.foodhyd.2017.05.022

Zheng, H., Mao, L., Cui, M., Liu, J., & Gao, Y. (2020). Development of food-grade bigels based on κ-carrageenan hydrogel and monoglyceride oleogels as carriers for β-carotene: Roles of oleogel fraction. *Food Hydrocolloids, 105*, 105855. https://doi.org/10.1016/j.foodhyd.2020.105855

Zhuang, X., Gaudino, N., Clark, S., & Acevedo, N. C. (2021). Novel lecithin-based oleogels and oleogel emulsions delay lipid oxidation and extend probiotic bacteria survival. *LWT, 136*, 110353. https://doi.org/10.1016/j.lwt.2020.110353

Zou, F., & Budtova, T. (2021). Tailoring the morphology and properties of starch aerogels and cryogels via starch source and process parameter. *Carbohydrate Polymers, 255*, 117344. https://doi.org/10.1016/j.carbpol.2020.117344

Zou, L., Zheng, B., Liu, W., Liu, C., Xiao, H., & McClements, D. J. (2015). Enhancing nutraceutical bioavailability using excipient emulsions: Influence of lipid droplet size on solubility and bioaccessibility of powdered curcumin. *Journal of Functional Foods, 15*, 72–83. https://doi.org/10.1016/j.jff.2015.02.044

Further reading

Belščak-Cvitanović, A., Komes, D., Karlović, S., Djaković, S., Špoljarić, I., Mršić, G., & Ježek, D. (2015). Improving the controlled delivery formulations of caffeine in alginate hydrogel beads combined with pectin, carrageenan, chitosan and psyllium. *Food Chemistry, 167*, 378–386. https://doi.org/10.1016/j.foodchem.2014.07.011

Benetti, J. V. M., do Prado Silva, J. T., & Nicoletti, V. R. (2019). SPI microgels applied to pickering stabilization of O/W emulsions by ultrasound and high-pressure homogenization: Rheology and spray drying. *Food Research International, 122*, 383–391. https://doi.org/10.1016/j.foodres.2019.04.020

Chen, K., & Zhang, H. (2019). Alginate/pectin aerogel microspheres for controlled release of proanthocyanidins. *International Journal of Biological Macromolecules, 136*, 936–943. https://doi.org/10.1016/j.ijbiomac.2019.06.138

Dafe, A., Etemadi, H., Dilmaghani, A., & Mahdavinia, G. R. (2017). Investigation of pectin/starch hydrogel as a carrier for oral delivery of probiotic bacteria. *International Journal of Biological Macromolecules, 97*, 536–543. https://doi.org/10.1016/j.ijbiomac.2017.01.060

Dai, H., & Huang, H. (2017). Enhanced swelling and responsive properties of pineapple peel carboxymethyl cellulose-g-poly(acrylic acid-co-acrylamide) superabsorbent hydrogel by the introduction of carclazyte. *Journal of Agricultural and Food Chemistry, 65*(3), 565–574. https://doi.org/10.1021/acs.jafc.6b04899

Gautam, M., & Santhiya, D. (2019). Pectin/PEG food grade hydrogel blend for the targeted oral co-delivery of nutrients. *Colloids and Surfaces A: Physicochemical and Engineering Aspects, 577*, 637–644. https://doi.org/10.1016/j.colsurfa.2019.06.027

Ghafar, A., Gurikov, P., Subrahmanyam, R., Parikka, K., Tenkanen, M., Smirnova, I., & Mikkonen, K. S. (2017). Mesoporous guar galactomannan based biocomposite aerogels through enzymatic crosslinking. *Composites Part A: Applied Science and Manufacturing, 94*, 93–103. https://doi.org/10.1016/j.compositesa.2016.12.013

Groult, S., & Budtova, T. (2018). Tuning structure and properties of pectin aerogels. *European Polymer Journal, 108*, 250–261. https://doi.org/10.1016/j.eurpolymj.2018.08.048

Gul, K., Gan, R.-Y., Sun, C.-X., Jiao, G., Wu, D.-T., Li, H.-B., Kenaan, A., Corke, H., & Fang, Y.-P. (2021). Recent advances in the structure, synthesis, and applications of natural polymeric hydrogels. *Critical Reviews in Food Science and Nutrition*, 1–16. https://doi.org/10.1080/10408398.2020.1870034

Guo, J., Giusti, M. M., & Kaletunç, G. (2018). Encapsulation of purple corn and blueberry extracts in alginate-pectin hydrogel particles: Impact of processing and storage parameters on encapsulation efficiency. *Food Research International, 107*, 414–422. https://doi.org/10.1016/j.foodres.2018.02.035

Guo, J., Zhou, Q., Liu, Y.-C., Yang, X.-Q., Wang, J.-M., Yin, S.-W., & Qi, J.-R. (2016). Preparation of soy protein-based microgel particles using a hydrogel homogenizing strategy and their interfacial properties. *Food Hydrocolloids, 58*, 324–334. https://doi.org/10.1016/j.foodhyd.2016.03.008

Gupta, V., & M., R. (2019). Development of programmed shape-shifting wheat xerogel discs and characterization of its hydromorphic behavior. *Colloids and Surfaces A: Physicochemical and Engineering Aspects, 583*, 123928. https://doi.org/10.1016/j.colsurfa.2019.123928

Gupta, V., R. G., T. K., Stephen, J., & Radhakrishnan, M. (2020). Cold plasma-assisted shape-shifting of a flat two-dimensional wheat xerogel and its morphological behavior. *Journal of Food Process Engineering, 43*(9), e13456. https://doi.org/10.1111/jfpe.13456

Hou, J.-J., Guo, J., Wang, J.-M., He, X.-T., Yuan, Y., Yin, S.-W., & Yang, X.-Q. (2015). Edible double-network gels based on soy protein and sugar beet pectin with hierarchical microstructure. *Food Hydrocolloids, 50*, 94–101. https://doi.org/10.1016/j.foodhyd.2015.04.012

Hu, H., Zhu, X., Hu, T., Cheung, I. W. Y., Pan, S., & Li-Chan, E. C. Y. (2015). Effect of ultrasound pre-treatment on formation of transglutaminase-catalysed soy protein hydrogel as a riboflavin vehicle for functional foods. *Journal of Functional Foods, 19*, 182–193. https://doi.org/10.1016/j.jff.2015.09.023

K, R. R., & R, M. (2020). Morphogenesis and characterization of wheat xerogel structure and insights into its 4D transformation. *Food Structure*, 100170. https://doi.org/10.1016/j.foostr.2020.100170

Kavousi, H. R., Fathi, M., & Goli, S. A. H. (2017). Stability enhancement of fish oil by its encapsulation using a novel hydrogel of cress seed mucilage/chitosan. *International Journal of Food Properties, 20*(Suppl. 2), 1890–1900. https://doi.org/10.1080/10942912.2017.1357042

Kemas, C. U., Ngwuluka, N. C., Ochekpe, N. A., & Nep, E. I. (2017). Starch-based xerogels: Effect of acetylation on Physicochemical and rheological properties. *International Journal of Biological Macromolecules, 98*, 94–102. https://doi.org/10.1016/j.ijbiomac.2017.01.101

Klein, M., & Poverenov, E. (2020). Natural biopolymer-based hydrogels for use in food and agriculture. *Journal of the Science of Food and Agriculture, 100*(6), 2337–2347. https://doi.org/10.1002/jsfa.10274

Larrea-Wachtendorff, D., Sousa, I., & Ferrari, G. (2020). Starch-based hydrogels produced by high-pressure processing (HPP): Effect of the starch source and processing time. *Food Engineering Reviews*. https://doi.org/10.1007/s12393-020-09264-7

Liu, F., Li, R., Mao, L., & Gao, Y. (2018). Ethanol-induced composite hydrogel based on propylene glycol alginate and zein: Formation, characterization and application. *Food Chemistry, 255*, 390–398. https://doi.org/10.1016/j.foodchem.2018.02.072

Luu, T. P., Do, N. H. N., Chau, N. D. Q., Lai, D. Q., Nguyen, S. T., Le, D. K., Thai, Q. B., Le, P. T. K., & Duong, H. M. (2020). Morphology control and advanced properties of bio-aerogels from pineapple leaf waste. *Chemical Engineering Transactions, 78*, 433–438. https://doi.org/10.3303/CET2078073. SE-Research Articles.

Mohammadinejad, R., Maleki, H., Larrañeta, E., Fajardo, A. R., Nik, A. B., Shavandi, A., Sheikhi, A., Ghorbanpour, M., Farokhi, M., Govindh, P., Cabane, E., Azizi, S., Aref, A. R., Mozafari, M., Mehrali, M., Thomas, S., Mano, J. F., Mishra, Y. K., & Thakur, V. K. (2019). Status and future scope of plant-based green hydrogels in biomedical engineering. *Applied Materials Today, 16*, 213–246. https://doi.org/10.1016/j.apmt.2019.04.010

Mu, R.-J., Yuan, Y., Wang, L., Ni, Y., Li, M., Chen, H., & Pang, J. (2018). Microencapsulation of Lactobacillus acidophilus with konjac glucomannan hydrogel. *Food Hydrocolloids, 76*, 42–48. https://doi.org/10.1016/j.foodhyd.2017.07.009

Ni, X., Ke, F., Xiao, M., Wu, K., Kuang, Y., Corke, H., & Jiang, F. (2016). The control of ice crystal growth and effect on porous structure of konjac glucomannan-based aerogels. *International Journal of Biological Macromolecules, 92*, 1130–1135. https://doi.org/10.1016/j.ijbiomac.2016.08.020

Nivala, O., Nordlund, E., Kruus, K., & Ercili-Cura, D. (2021). The effect of heat and transglutaminase treatment on emulsifying and gelling properties of faba bean protein isolate. *LWT, 139*, 110517. https://doi.org/10.1016/j.lwt.2020.110517

Parikka, K., Nikkilä, I., Pitkänen, L., Ghafar, A., Sontag-Strohm, T., & Tenkanen, M. (2017). Laccase/TEMPO oxidation in the production of mechanically strong arabinoxylan and glucomannan aerogels. *Carbohydrate Polymers, 175*, 377–386. https://doi.org/10.1016/j.carbpol.2017.07.074

Ratish Ramanan, K., Rifna, E. J., & Mahendran, R. (2018). Effect of concentration and temperature on the formation of wheat hydrogel and xerogel pattern. *Colloids and Surfaces A: Physicochemical and Engineering Aspects, 559*, 385–391. https://doi.org/10.1016/j.colsurfa.2018.09.075

Slavutsky, A. M., & Bertuzzi, M. A. (2019). Formulation and characterization of hydrogel based on pectin and brea gum. *International Journal of Biological Macromolecules, 123*, 784–791. https://doi.org/10.1016/j.ijbiomac.2018.11.038

Ubeyitogullari, A., & Ciftci, O. N. (2016). Formation of nanoporous aerogels from wheat starch. *Carbohydrate Polymers, 147*, 125–132. https://doi.org/10.1016/j.carbpol.2016.03.086

Wu, B., & McClements, D. J. (2015). Functional hydrogel microspheres: Parameters affecting electrostatic assembly of biopolymer particles fabricated from gelatin and pectin. *Food Research International, 72*, 231–240. https://doi.org/10.1016/j.foodres.2015.02.028

Xu, M., & Dumont, M.-J. (2015). Evaluation of the stability of pea and canola protein-based hydrogels in simulated gastrointestinal fluids. *Journal of Food Engineering, 165*, 52–59. https://doi.org/10.1016/j.jfoodeng.2015.04.033

Yang, D., & Yang, H. (2020). The temperature dependent extraction of polysaccharides from eucheuma and the rheological synergistic effect in their mixtures with kappa carrageenan. *LWT, 129*, 109515. https://doi.org/10.1016/j.lwt.2020.109515

Yang, Y., He, S., Ye, Y., Cao, X., Liu, H., Wu, Z., Yue, J., & Sun, H. (2020). Enhanced hydrophobicity of soybean protein isolate by low-pH shifting treatment for the sub-micron gel particles preparation. *Industrial Crops and Products, 151*, 112475. https://doi.org/10.1016/j.indcrop.2020.112475

Yang, T., Li, X.-T., & Tang, C.-H. (2020). Novel edible pickering high-internal-phase-emulsion gels efficiently stabilized by unique polysaccharide-protein hybrid nanoparticles from Okara. *Food Hydrocolloids, 98*, 105285. https://doi.org/10.1016/j.foodhyd.2019.105285

Yang, X., Su, Y., & Li, L. (2020). Study of soybean gel induced by Lactobacillus plantarum: Protein structure and intermolecular interaction. *LWT, 119*, 108794. https://doi.org/10.1016/j.lwt.2019.108794

Chapter 8

Plant-based butter like spreads

Dian Widya Ningtyas
The University of Queensland, School of Agriculture and Food Sciences, Brisbane, Australia, Brawijaya University, Faculty of Agricultural Technology, Department of Food Science and Technology, Malang, Indonesia

8.1 Introduction

Butter-like spread is an emulsion product containing fat either from dairy or nondairy sources. According to Kumar (2015), the terms "butter" and "spread" have a slightly different meaning from the percentage of the main ingredients. Spread emulsions are differentiated into oil-in-water type emulsion (contain <15% fat) and water-in-oil type emulsion (contain >15% fat). Butter spread should contain a minimum of 80% fat from milk or cream. It is generally characterized by poor spreadability and is available in a solid state at refrigeration temperature. Consumption of conventional butter (dairy-based) has widely been associated with obesity, heart disease, and other degenerative health problems due to significant amounts of saturated fatty acids and cholesterol. Margarine has been an alternative butter-like spread from plant source oil/fat, but this is also characterized by high amount of trans fatty acids. Due to the above reasons, the consumption of margarine or butter as spreads is being discouraged by health practitioners. In addition, the use of fruit-based spreads such as jam and jelly is also considered unhealthy due to high amounts of sugar and inadequate nutritional components such as protein, fat, iron, and calcium contents. These health concerns have led to the substitution of animal-based spreads with plant-based spreads that are palatable, healthy, and possess good textural properties (Fig. 8.1).

In the past, margarine and peanut butter were the only alternatives to dairy butter, but nowadays, numerous kinds of spreads have been developed worldwide. Some examples are spreads that contain a high amount of omega-3 (ω-3) and omega-6 (ω-6) fatty acids, low-fat spread, and trans-free margarine spread (Gorrepati et al., 2015). Besides peanut butter, different varieties of plant-based butter/spreads originating from almond, pistachio, macadamia, cashew, sesame, sunflower, and olive are now available in the market. These spread products are not only nutrient-dense (rich in protein, fiber, essential fatty acids) but also easy to consume in various ways. In addition, some of these spreads have superior nutritional value than dairy butter, as presented in Table 8.1.

FIGURE 8.1 Cashewnut butter spread on a bread.

Engineering Plant-Based Food Systems. https://doi.org/10.1016/B978-0-323-89842-3.00013-0

TABLE 8.1 Nutritional information of dairy butter and plant-based spreads per 100g.

Products	Protein (g)	Fat (g)	Calorie (cal)	Calcium (mg)	Zinc (mg)
Dairy butter	0.6	82.2		18	0.1
Nut-based butter-like spread					
Almond	16.9	66.9	711.8	303	3.5
Cashew	19.7	56.4	655.4	49.3	5.6
Hazelnut	14.1	66.9	662.4	–	–
Peanut—natural	26.8	56.4	662.4	49.3	2.8
Peanut—reduced fat	28.2	42.3	669.5	–	2.8
Seed-based butter-like spread					
Sunflower	21.1	49.3	563.8	–	–
Sesame	18.3	56.4	627.2	451	4.9
Bean-based butter-like spread					
Soy (sweetened)	28.2	38.8	599.0	352.4	–
Soy-peanut (sweetened)	14.1	8.5	352.4	281.9	–

Sources: Mangels, R. (2001). Guide to nut and nut butters. *Vegetarian J XXI:20–23* and Gorrepati, K., Balasubramanian, S., & Chandra, P. (2015). Plant based butters. *Journal of Food Science and Technology, 52*(7), 3965–3976. https://doi.org/10.1007/s13197-014-1572-7

In general, nuts, seeds, oilseeds, and fruits that contain high-fat levels are used to produce plant-based butter. The USDA (2011) mandates that nut butter comprises at least 90% nuts, while nut spread includes a minimum of 40% nuts dispersed in the continuous oil phase of nut oil. Emphasizing the importance of healthy diets, the researchers have developed a healthier version of plant-based butter: pumpkin seed oil spread with a high amount of ω-3 and ω-6 fatty acids (Radočaj et al., 2012), trans-free spread (Hwang et al., 2014; Lumor et al., 2010), and low-fat spreads such as low-fat chocolate spread (Kumar, 2015), low-fat pumpkin seed spread (Devi et al., 2019; Nikolić et al., 2014), low-fat sesame paste (Razavi et al., 2007), and low-calorie pistachio butter (Emadzadeh et al., 2011, 2012) by adding fat replacers in the formulation to maintain their textural properties. Using up to 30% of olive oil as a stabilizer in a chocolate spread has also been found to improve its textural quality (hardness, stickiness, spreadability) and consumer acceptability (Kumar, 2015).

Plant-based spreads' textural, color, and flavor properties play a significant role in consumer appeal, buying decisions, and eventual consumption. The overall quality of plant-based spread is related to the quality of the paste. It is highly associated with the milling process as it determines the particle size, consistency, viscosity, and mouthfeel. Numerous authors have emphasized the importance of the structure—rheology interface (Nikolić et al., 2014; Razavi et al., 2007; Shakerardekani et al., 2013b). Stabilizers are typically added to the production of oil/fat spreads to control oil migration. However, an excessive amount of stabilizers in the product can result in a hard butter that is difficult to spread, while an insufficient amount can result in unacceptable oil separation (Lima & Guraya, 2005). Although *trans* fatty acids in hydrogenated vegetable oil have been linked to coronary heart disease, this type of oil is frequently utilized to stabilize the spread (Lumor et al., 2010). In their study, it has been suggested the usage of nonhydrogenated vegetable oil, such as palm mid-fraction and cottonseed oil, which do not contain *trans* fatty acids, for margarine spread production. Meanwhile, Hwang et al. (2014) discussed the possibility of using sunflower wax organogels to replace saturated and trans fats in margarine and spread products.

Numerous researchers have examined the relationship between the particle size distribution of spreads (such as from peanuts) prior to swallowing (Flynn et al., 2011) and the structural changes in products after mastication using temporal dominance of sensations (TDS) (Rosenthal & Share, 2014). Similar to other food products with high oil content and high unsaturated fatty acid composition, plant-based spreads are also sensitive to autoxidation. According to Miraliakbari & Shahidi (2008), the oil of pecans and pistachios was more stable than that of pine nuts and walnuts due to their higher unsaturated fatty acid content. Their oxidation, however, can be regulated by the addition of antioxidants and processes that limit tocopherol and other natural antioxidant losses.

8.2 Type of plant-based butter-like spreads

8.2.1 Nut-based butter

Nut-based butter spread is made mostly of nuts (peanut, almond, cashew, macadamia, hazelnut, pecan, pistachio, or walnut), sweeteners, protein sources, with or without the addition of vegetable oil (Liedl & Rowe, 2007; Shakerardekani et al., 2013b). One of the most famous nut spreads is made from peanuts, which is widely used as a sandwich spread throughout the world. Typical peanut butter contains a minimum of 90% peanut paste with the addition of other ingredients such as hydrogenated vegetable oil (1%–5%), sweeteners (1%–6%), and salts (1%–1.5%) to enhance flavor, while small amounts (0.5%–1.5%) of stabilizers (hydrogenated vegetable oil) and emulsifiers (lecithin or hydroxylated lecithin) are required to prevent oil separation (Akhtar et al., 2014). In general, peanut butter has about 50% fat and according to USDA (2011), peanut butter that contains 27.8% protein and 49.4% fat, is classified as a high-energy diet (581 kcal/100 g). Additionally, the high nutritional value of peanut butter-based ready-to-use therapeutic food (RUTF) has been utilized in diet programs to alleviate malnutrition among children in African countries (Humphrey, 2008). However, the food sector has faced significant obstacles in developing reduced-fat peanut butter. Currently, commercially available reduced-fat peanut butter contains at least 25% less fat (Akhtar et al., 2014). Singh et al. (2000) developed reduced-fat peanut pastes (with a fat reduction of more than 50%) by combining protein and starch-based fat substitutes (whey protein concentrate and microcrystalline cellulose, respectively).

Pistachio butter is a nutritious pasty-like product made using pistachio, sugar, soy protein isolate, and flavor enhancer (Ardakani, 2006). In contrast, Shakerardekani et al. (2013a) developed pistachio spread using pistachio paste, sugar, soy protein isolate, and red palm oil. Due to the high unsaturated fat content, it is essential to incorporate an adequate number of emulsifying and antioxidant chemicals into its composition (Gorrepati et al., 2015). Additionally, Shakerardekani et al. (2013a) showed that adding an emulsifier mixture (1% mono-diglycerides and lecithin) to a pistachio butter formulation (73%–83% roasted pistachio kernels and 15%–25% sugar) results in butter with a consistent texture and appropriate consistency.

Cashew nuts are a significant tropical nut due to their high nutritional content, which includes protein (21%), fat (46%), carbohydrates (29%), and fiber (1.13%). The fat and oil content of cashew nuts contribute significantly to their energy content and are primarily composed of oleic acid glycerides (73.8%) and linoleic acid glycerides (7.7%) (Olaleye et al., 2021). Lima et al. (2012) prepared cashew butter using various grades of kernels and added sugar (8%), soybean lecithin (2%), and salt (0.1%). The sensory evaluation results from their study indicate that the shiny, nutty aroma, and flavor of cashew butter, which are desirable characteristics, were associated with cashew nut kernels from butts, pieces, and splits. While the graininess and consistency were associated with cashew butter made from cashew nut kernel pieces and granules.

8.2.2 Beans and legumes-based butter

Soy butter is a new product that is only commercially available in limited quantities (Agrahar-Murugkar et al., 2013). It offers a healthier alternative to peanut butter for consumers allergic to peanuts (Gorrepati et al., 2015). In general, soy butter is made by removing the skins from soybeans, roasting them to a dark brown color, reducing them to a fine powder, and then combining the powder with edible vegetable oil to form a paste. However, although it has similar characteristics as peanut butters, consumer in America found that soy butter is lack of creamy texture and rich flavor (Dubost et al., 2003). Previously published research has detailed ways for enhancing the quality of the resulting soy butters (Abd-Elsattar & Abdel-Haleem, 2016; Murugkar et al., 2014). For instance, to minimize the grassy or beany flavor, the moisturizing stage during sprouting should be regulated by applying just enough moisture to soak some of the beans completely (Agrahar-Murugkar et al., 2013). Processing of soybeans such as blanching and roasting of the kernels should be properly controlled as it contributes to development of flavor, aroma, and color of the final nut butters (Abd-Elsattar & Abdel-Haleem, 2016; Agrahar-Murugkar et al., 2013). Additionally, it has been reported that adding low-calorie sweeteners to soy butter increases the firmness and decreases the water activity, resulting in a product of incomparable quality to that created with sugar (Gorrepati et al., 2015). Moreover, Agrahar-Murugkar et al. (2013) successfully developed butter from sprouted soybeans with better quality, such as decreased particle size, apparent viscosity, and flow behavior index, compared to nonsprouted beans.

Meanwhile, relatively limited research has been conducted on the creation of butter made from beans. Commercially available pulse-based spreads are a relatively recent notion, but hummus, the primary spread-like vegetable protein product, is an old meal. Hummus, a spread made from cooked-mashed chickpeas, is a popular dish dip in the Middle East, North Africa, Mediterranean, and Middle East (Marks, 2010). Unlike other type of plant-based butter, hummus spread is high in carbohydrate (14.29%), protein (7.9%), fiber (6%), polyunsaturated fatty acids, vitamins A, E, and C, folate, and minerals (Mg, K, and Fe) but relatively low amount of fat (9.6%) (Wallace et al., 2016).

8.2.3 Seed-based butter

Seed butters are spreadable products made from milling seeds into a paste. Oilseed spreads are usually good sources of protein and have a significant market presence nowadays as a substitute for margarine due to their high fatty acid (ω-3 and ω-6), γ-tocopherols, copper, folic acid, potassium, and zinc (Mangels, 2001; Radočaj et al., 2012). Similar to method in the production of nut butter (Fig. 8.3), the seed butter processing also include roasting, grinding, chilling, and tempering (Gorrepati et al., 2015). Additionally, homogenization is used in the final stages of commercial production to give the seed butter a fine and creamy texture. The seed butter that are popular are made from sesame seeds, pumpkin seeds, and sunflower seeds.

Sesame (*Sesamum indicum*) seed contains about 55% highly stable oil and 20% protein, rich in methionine, tryptophan, and valine. In Egypt, as suggested by El-Adawy & Mansour (2000), dehulled roasted sesame seeds at 130°C for 1 h make a high-quality sesame butter called Tahina. In addition, Zahedi & Mazaheri-Tehrani (2012) also stated that sesame butter has a high amount of lipids (54%—65%), proteins (23%—27%), carbohydrates (6.4%—9%), dietary fiber (9.3%), and minerals (Mg, K, Zn, Fe), which contribute to its nutritional value. Furthermore, sesame seeds also contain natural antioxidants (sesamol, sesaminol, and tocopherols) that cause the sesame paste to have a longer shelf life and high oxidation resistance.

Pumpkin seeds are known for their distinct flavor and nutty taste and have been used as a snack in a variety of regions around the world, including Mexico, Canada, the United States of America, China, and Europe. The seeds contained a high concentration of oil (41.59%), protein (25.4%), and carbohydrate (25.19%), making them a viable source of alternative butter. Additionally, the seeds also contain significant amounts of iron, protein, manganese, magnesium, zinc, potassium, copper, and phosphorus, as well as polyunsaturated fatty acids (PUFA), c-tocopherol, and carotenoids. Due to the above reasons, pumpkin seeds and seed oil have been used in the treatment and management of diabetes, inflammation, hyperlipidemia, hypertension, and cancer (Syed, 2019). Radoĉaj et al. (2011) optimized the formula for manufacturing pumpkin seed butter from hull-less pumpkin seed press-cake. According to their investigation, stabilizer (a blend of hydrogenated canola and coconut oils) and hemp oil were added in the formulation to create a product that resembled peanut butter in appearance and texture. Pumpkin seed butter with the closest texture features to peanut butter is made with a mix of 1% stabilizer and 40% hemp oil or 1.25% stabilizer and 20% hemp oil.

Sunflower (*Helianthus annuus*) seed butter was first introduced in the early 1980s (Lima & Guraya, 2005). Still, they failed in the market due to the sunflower seed's fibrous outer layer, which causes moisture retention when improperly roasted. In addition, these fibrous substances result in a product with unacceptable texture and appearance, including a green color and a bitter under-roast flavor, which leads to poor flavor and contributed to the product's unacceptability (Lima & Guraya, 2005). Apart from its poor texture, the sunflower seed butter has more monounsaturated fat, magnesium, phosphorus, zinc, copper, iron, manganese, vitamin E, and selenium than peanut butter as well as less saturated fat (Gorrepati et al., 2015). Additionally, their investigation found that the level of roasting and stabilizer had a substantial effect on the hardness and oil separation of the butter, with addition of 1.8% stabilizer and a low level of roasting produce the most acceptable butter, according to the panellist.

8.2.4 Fruits and vegetables-based butter

Avocado (*Persea americana*) or butter fruit is a tropical fruit with a buttery texture containing high nutritional value and several health benefits as it is termed "functional food." According to Arachchige et al. (2017), the avocado pulp consists of 67%—78% moisture, 13.5%—24% lipids, 0.8%—4.8% carbohydrates, 1.0%—3.0% proteins, 1.4%—3.0% fiber, and energy density between 140 and 228 kcal. In addition, numerous phytochemicals such as anthocyanins, gallic acid, chlorogenic acid, lutein, zeaxanthin, β-cryptoxanthin, α-carotene, β-carotene (pro-vitamin A, retinol), tocopherol, vanillic acid, glutathione, and beta-sitosterol also contributes to the health-promoting function (Bhuyan et al., 2019). According to Sethi & Balasubramanyam (2018), avocado is a potential fruit used in plant-based butter due to its creamy texture and high monounsaturated fatty acids (MUFAs). Arachchige et al. (2017) prepared avocado butter using 35% dairy butter, honey (sweetener), and cinnamon as antioxidants. While in another study, skim milk powder, vegetable oil, and an emulsifier (xanthan gum) were added to the manufacturing process of avocado spread (Fig. 8.2) in order to improve the final product's textural and sensory properties (Sethi & Balasubramanyam, 2018). The optimal formulation for avocado butter fruit spread is 8.61% skim milk powder, 0.2% xanthan gum, and 15% avocado pulp.

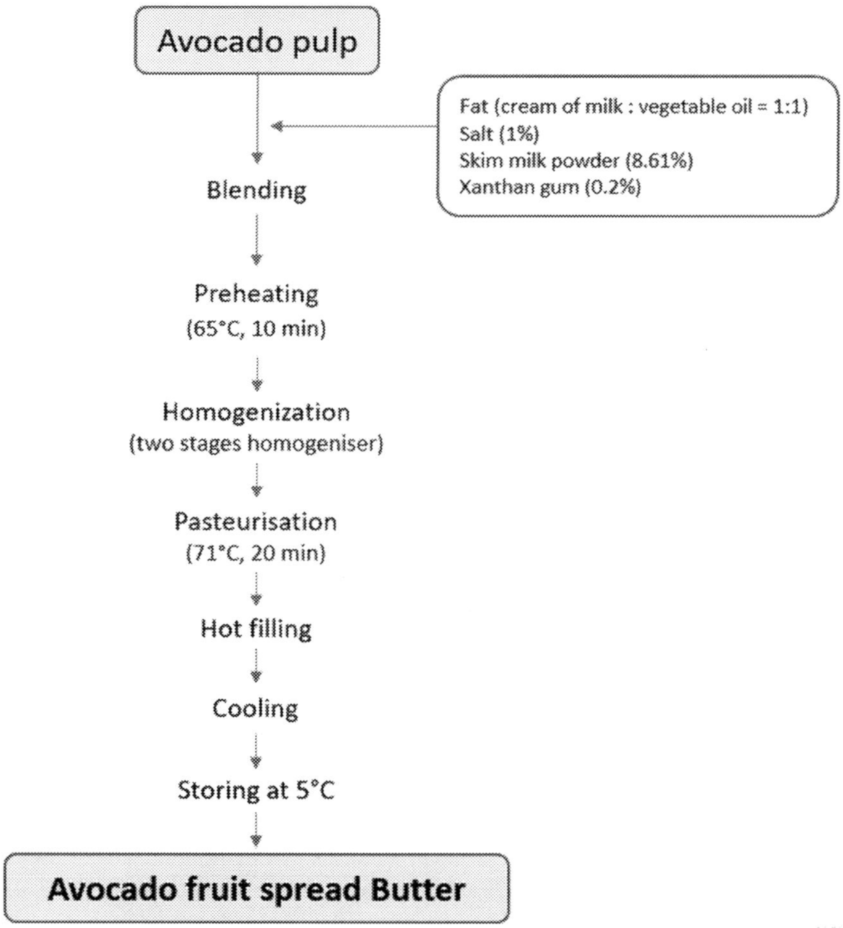

FIGURE 8.2 Processing steps of fruits-based butter made from Avocado.

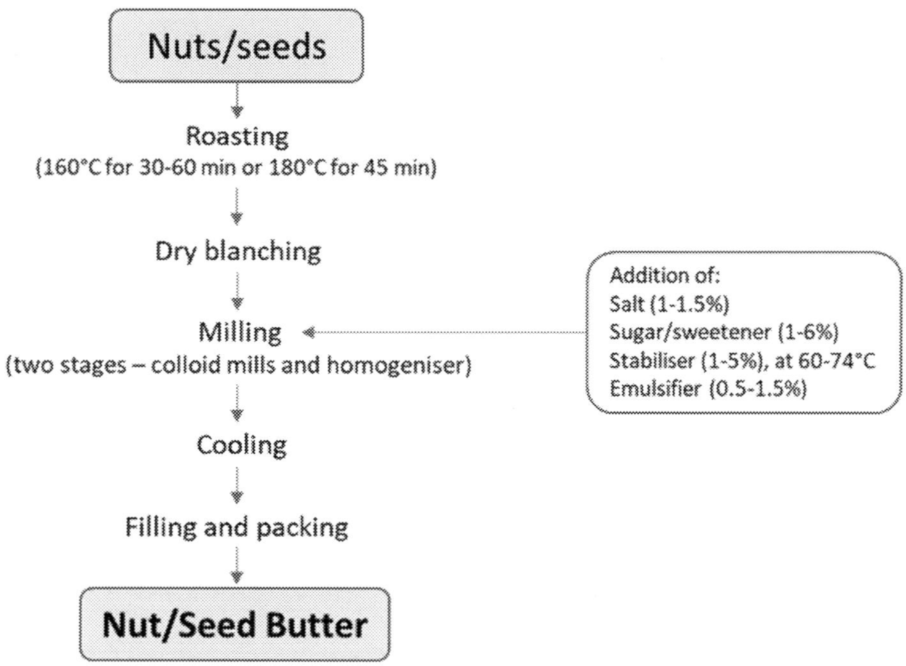

FIGURE 8.3 Processing steps of plant-based butter from nuts and seeds.

8.3 Factors affecting the textural properties of plant-based butter

The textural qualities of plant-based butters, such as nut butter, are determined by the nut variety, processing conditions, and the inclusion of additional ingredients. The plant-based butter must maintain an optimal amount of firmness, a high degree of cohesion and spreadability, and adequate adhesiveness to impart a sticky texture. Nut and seed butter are generally prepared by roasting and milling, as described in Fig. 8.3, followed by refrigeration to prolong the shelf-life. However, due to its short self-life mainly due to oxidation potential, the butter needs to be consumed within the period of few months.

8.3.1 Effect of processing

8.3.1.1 Particle size

The textural, rheological characteristic, and overall quality of the plant-based butter are influenced by its particle size (Çiftçi et al., 2008; Genovese et al., 2007). A study by Matsunobu et al. (1984) showed that almond paste containing a substantial number of particles with a diameter larger than 105 μm resulted in precipitation, indicating instability of the paste. Therefore, the roasting and milling process in nut-based butter processing are aimed to create smaller particle sizes.

Carbohydrates, protein, and other nonfat components of the nuts/seeds will be in a continuous phase prior to grinding. Meanwhile, fat cells entrapped in nonfat substances will enter a phase of discontinuity. In the manufacturing of peanut butter, a two-step milling procedure is used to reduce the particle size of the nuts. After the second milling step using a high-speed comminutor (at 9600 rpm), the ultrafine nuts particles are achieved, with 98% of the particles having a diameter less than 73.7 mm and a maximum diameter of 254 mm (Mohd Rozalli et al., 2015). After grinding into a paste, fat cells break and become continuous, whereas nonfat materials form a discontinuous phase. Following the formation of the paste, the continuous phase (which contains the fat/oil) will separate from the nonfat particles. As Marangoni (2012) further describe, peanut butter is a paste-like dispersion composed of a large number of small (but not colloidal) solid peanut particles suspended in a continuous oil phase.

According to previous work by Lima et al. (2000), the oil separation of reduced-fat peanut butter is highly dependent on the fat distribution surrounding the solid particles. When large particles (>300 mm) are present, the overall surface area of the particles decreases, resulting in less oil adhering to the solid particles and more oil available for separation. Thus, it may be concluded that the stability of peanut butter is attributable to the compact packing of peanut particles inside the continuous phase, not the any crystal network as the peanut oil does not crystallize at ambient temperature (Marangoni, 2012). At the same time, adding other ingredients such as vegetable oil, protein sources, stabilizers, salt, and sweeteners helps produce a stable nut-based butter.

8.3.1.2 Moisture content

The overall quality of plant-based spread is determined by the paste used as the primary ingredient. Roasting reduces the water content of peanuts to around 1%, followed by the release of oil from the cytoplasm of the cells extending the shelf life of peanuts and assisting in the development of peanut butter flavor. Low water activity will prevent microbial growth. However, as the roasting temperature increased, the moisture loss is accelerated, resulting in the changes in peanut texture becoming more rigid, fragile, and crumbly. According to Di Monaco et al. (2008), the ideal roasting conditions for peanuts were 180°C for 45 min. In contrast, Ogunsanwo et al. (2005) discovered that peanut butter prepared by roasting at 160°C for 30 min was similar to commercial samples. Further, the roasting condition of kernels should be carefully controlled because it affects not only textural characteristics but also the development of flavor, aroma, and color of peanut butter.

Formulations incorporating water or substances with a high moisture content may present significant difficulties. According to prior research, increasing the moisture level of peanut butter by adding water or high-moisture substances such as honey or corn syrup affected the flavor and viscosity of the butter, resulting in a much stiffer butter (Abegaz & Kerr, 2006). Additionally, Felland & Koehler (1997) reported quality faults in peanut butter as the water content increased (darker color, decreased roasted peanut scent and flavor, off-odor, and off-flavors). In comparison, Abegaz et al. (2004) discovered that adding moisture at 2% and 5% prevented lipid oxidation but resulted in the loss of volatile pyrazines linked with roasted peanut flavor. Although the mechanisms on how water interact with other ingredients, it is speculated that adding different quantities of water to peanut butter may operate as a prooxidant or an antioxidant.

8.3.2 Effect of ingredients

8.3.2.1 Oil content

In the case of oil content, products containing a higher amount of oil are less sticky (lower adhesion force) and more spreadable (higher work of adhesion) than samples having a lower oil content (Shakerardekani et al., 2013b). However, Wagener & Kerr (2018) found that the inherent high oil content of pecan (65%–75%) results in a pecan butter that is more runny than spreadable and also tends to have higher oil separation. As a result, it is critical to maintain the total solids to oil ratio in the formulation in order to achieve an acceptable pecan butter.

Different types of oil also influence the spreadability characteristics of the products. Because crystals of fat are not present in the natural oil of peanut butter at ambient temperatures, the peanut paste is uniformly blended with hydrogenated vegetable oil, which crystallizes at these temperatures. Generally, the mixing process is accomplished above the melting point of the hydrogenated peanut oil to insure more complete dispersion of the hydrogenated oil (Sanders, 2003). A study by Gills & Resurreccion (2000) showed that peanut butter containing hydrogenated vegetable oil was more difficult to spread on crackers than the nonstabilized counterpart due to solidification of the oil retained the peanut solids in suspension. Further Aryana et al. (2003) and Gills & Resurreccion (2000) also found that using mixed hydrogenated rapeseed and cottonseed oils as a stabilizer in peanut butter were preferable to palm oil. Therefore, the addition of suitable emulsifiers, such as lecithin and mono- and diglycerides, into peanut butter help to reduce the stickiness and enhance its spreadability properties.

8.3.2.2 Stabilizer and emulsifier

Ingredients such as salt, sugar, stabilizers, and emulsifiers are added during the milling of nut butter. The stabilizers (such as hydrogenated canola and cottonseed oils) are added to the formulation to prevent the separation of less viscous oil from solid particles during storage at room temperature. The addition of stabilizers reduces the oil separation by creating a strong network between protein and fat (Aryana et al., 2000) and improve the texture to more solid-like behavior where the spread begins to hold its' shape (Tanti et al., 2016). Without stabilizer, the texture of peanut butter is less dense (more liquid-like sample) with >2% oil separation after 12 weeks of storage (Galvez et al., 2006). It cannot hold its shape but instead spreads outward on the flat surface due to oil separation (Tanti et al., 2016). These separations on peanut butter where the paste settling at the bottom and forming a hard layer, while the oil remaining on top was also observed by Aryana et al. (2000). Further, this can lead to lipid peroxidation, which develops the off-flavor and rancidity. During stabilizer addition to produce a more homogenized product, the temperature of the paste should be more than the melting point of stabilizer that is about 60–74°C. Regarding the desired consistency and mouthfeel of peanut butter as well as oil content and particle size, Woodroof (1983) emphasized the critical factors on type and amount of stabilizer used in the formulation. According to Totlani & Chinnan (2007), adding 1%–2% stabilizer to peanut butter was enough to keep it stable for 3 months at 35°C. While Aryana et al. (2000) discovered that 2%–2.5% palm oil is efficient for stabilizing peanut butter for more than a year at 21–24°C.

Emulsifier is another ingredient that is frequently used in the manufacture of butters to prevent oil separation. It is believed that emulsifiers retard the settling of the dispersed phase by improving the dispersability of the protein particles in the oil (Marangoni, 2012). Mono-diglycerides and lecithin are frequently added as emulsifiers in nut butters to increase their long-term storage stability (Haghighi et al., 2008) and have been successfully used in chocolate spread emulsions (Said et al., 2019). It has been found that hazelnut butter with 2% lecithin and 1% mono-diglycerides increases the hardness and spreadability without impairing the sensory features (Pourfarzad & Kisomi, 2020). Similarly, among the other concentrations, pistachio butter with 1% mono-diglyceride combinations and lecithin had the most consistent, coherent, and homogeneous texture (Taghizadeh & Razavi, 2009).

Due to the health risks associated with the consumption of *trans* and saturated fatty acids found in hydrogenated vegetable oils used in butter, margarine, and spreads, oleogels have been investigated as a possible fat structuring agent (Hwang et al., 2014). Oleogel is a self-supporting, thermoreversible, anhydrous, and viscoelastic material composed of a three-dimensional supramolecular network of self-assembled molecules with a low solubility in organic solvents (Marangoni, 2012; Patel & Dewettinck, 2016). According to Hwang et al. (2014), before oleogels can be used in place of saturated and *trans* fats in margarine, butter, or spread products, it must be established that the oleogel is stable in the emulsion system and does not react unexpectedly with any other ingredient in the product. Further, margarine with up to 80% oleogels (Hwang et al., 2012) and butter containing 4.5% oleogels (Yõlmaz & Öğütcü, 2015) exhibit textural properties comparable to their commercial counterparts.

According to Yõlmaz & Öğütcü (2014b), beeswax and saturated monoglyceride have suitable thermal, textural, and stability properties, allowing them to be successfully used in spread/margarine-type products. (Hwang et al. (2012) discovered a promising gel made with soybean oil and sunflower wax, candelilla wax, and rice bran wax (at concentrations of 0.5%−1%). Additionally, the firmness and melting properties of margarine made with the oleogels are comparable to those of commercial margarine. Hwang et al. (2014) also reported in another study that oleogel made from sunflower wax in soybean oil has a firmer texture and a higher firmness than rice bran wax and candelilla wax. The results of the preceding studies indicate that oleogel-based emulsions can be used as easily spreading fat materials. However, the oleogel's application as spread or butter was characterized in terms of microstructure and rheological properties. According to Patel et al. (2020), there are several critical issues with the addition of wax-based oleogels to spreads and margarine/butter-like products. These include residual waxy taste due to high melting points of waxes, low-firmness due to the lack of strain-hardening properties in wax gels, and poor long-term emulsion stability. Therefore, Yõlmaz & Öğütcü (2014a) suggest that using oleogels with a neutral taste and aroma or a well-gelling oleogel at very low addition levels may help improve the sensory quality and consumer acceptability of spreadable products.

8.3.2.3 Fat replacer

To meet consumer demand for a low-calorie or reduced-fat version of plant-based butter, fat replacers are incorporated into the composition (Fig. 8.4). Numerous studies on low-fat spreads based on nuts, seeds, and fat substitutes have been published (Emadzadeh et al., 2011; Nikolić et al., 2014; Singh et al., 2000; Taghizadeh & Razavi, 2009). In their study, they have used fat substitutes derived from fat (salatrim), protein (whey protein concentrate), starch (microcrystalline cellulose, Xanthan gum, Reihan seed gum, and Balangu seed gum, wheat fibers, maltodextrin), or a combination of plant- and starch-based fat substitutes. According to Emadzadeh et al. (2012), all samples exhibit excellent stability during storage, and there is no discernible effect on the concentrations of the gums used in low-calorie pistachio butter, except for the Balangu seed gum. Meanwhile, Singh et al. (2000) recommend combining a protein-based fat replacer (SimplesseTM-500 made from whey protein concentrate) with a starch-based fat replacer (PCR-352-1, a microcrystalline cellulose) at a ratio of 5:1.25 and 10:2.5 to produce a peanut paste that is >50% reduced in fat but retains acceptable textural and rheological characteristics. It can be inferred from their study that in comparison to the original butter, the reduced-fat peanut butter produced had increased spreadability, decreased hardness, and was less gritty.

8.4 Physical and sensory characteristics of the plant-based butter-like spreads

8.4.1 Physical characteristics

The textural, color, and flavor qualities of plant-based spreads contribute significantly to consumer decisions and subsequent demand. Textural characteristics such as hardness, shear work, and stickiness are frequently used to characterize the spread's quality and forecast consumer acceptance. The hardness of the material is directly proportional to its spreadability qualities. The lower the hardness, the better for oil-based spreads, as evidenced by the ease with which products spread uniformly on a flat surface (more spreadability) (Di Monaco et al., 2008). A study on pumpkin spread (Radočaj et al., 2012) and another on sunflower butter (Lima & Guraya, 2005) demonstrated that the hardness of fat-based spreads is significantly dependent on the kind and content of the stabiliser. It has been reported by Tanti et al. (2016) that incorporation of polymer additives, freeze-dried hydroxypropyl methylcellulose (HPMC) and methylcellulose (MC), in peanut butter increases the amount of small fragments formation during spreading (Fig. 8.5B). These small fragments hinder the aggregation of protein bodies, thus resulting in a less oily product because more oils will be adsorbed in the protein. Further, as the stabilizer concentration increases, the spreads become more difficult to develop due to establishing a dense crystal network (Fig. 8.5B) that makes it as a solid-state.

Moreover Gills & Resurreccion (2000) found that cohesiveness and adhesiveness rose linearly with an increasing amount of stabilizer (palm oil and hydrogenated vegetable oil), resulting in a greater concentration of peanut particles suspended in oil. Excess stabilizers also resulted in a less oily product, reducing slipperiness, as evaluated by the texture profile analyser. Further, to avoid breaking the bread or crumbling the crackers, high-quality spreadable products such as table spread and butter should have a soft feel and be easy to spread (Shakerardekani et al., 2013b). The structural features of spreads products can be modified by adjusting the fat content, including the percentage of solids fat obtained through blending (Fig. 8.6), crystallization and melting rates, and other processing parameters (Rønholt et al., 2013).

Based on the particle size of the nuts, Wagener & Kerr (2018) classified peanut butter into two types: smooth (fine consistency, even texture with few noticeable particles) and chunky/crunchy (partially fine and partially grainy particles with significant amounts greater than 1.58 mm in diameter). According to Liedl & Rowe (2007), nut spread can have

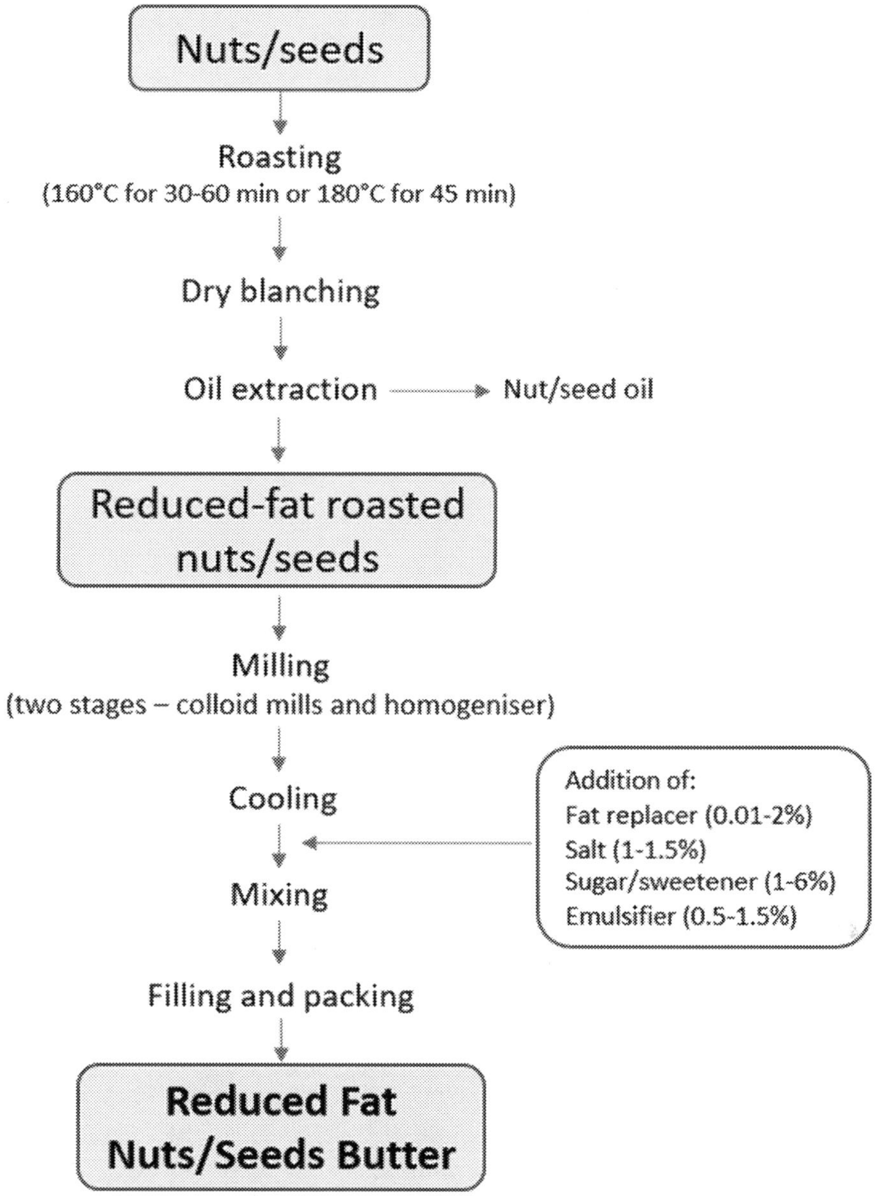

FIGURE 8.4 Flowchart of reduced fat plant-based butter from nuts or seeds.

acceptable sensory and textural features if at least 90% of the particles are smaller than 40 μm, 50% are smaller than 10 μm, and 10% are smaller than 3 μm. Optimizing particle size distribution in suspensions also resulted in 50-fold reductions in the shear viscosity. In addition, a higher degree of homogenization resulted in a complete breakdown of the nut cell wall (w), protein bodies (p), and starch grains (s), which are uniformly distributed in the oil matrix (o), as shown in Fig. 8.7 (1). In contrast, a clump of proteins (indicated by an arrow in Fig. 8.7 (2)) was observed due to aggregation of protein bodies and cell wall fragments resulting from the inefficient milling process of the peanut butter (Aryana et al., 2000).

The elastic behaviors increase with grinding time, suggesting that the particle size has a pronounced effect on peanut butter's viscosity and gel strength (Mohd Rozalli et al., 2015). In addition, previous research on low-calorie pistachio butter (Emadzadeh et al., 2012), pistachio spread (Shakerardekani et al., 2013b), and sesame paste (Çiftçi et al., 2008) has shown that the degree of milling affects the particle distribution, which alters their textural and rheological properties. According to their findings, the resulting butter acts as a non-Newtonian pseudoplastic fluid under yield stress. With the sesame paste butter, Çiftçi et al. (2008) suggested that the consistency index and thixotropic areas decreased with increase in temperature

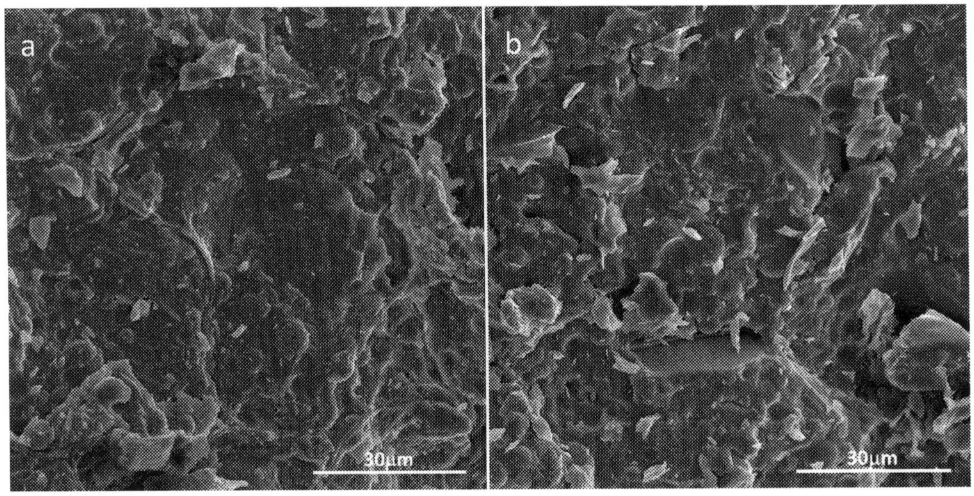

FIGURE 8.5 Cryo-scanning electron microscopy images of (A) natural peanut butter, and (B) peanut butter with 1.5% (w/w) freeze-dried hydroxypropyl methylcellulose (HPMC). *Sources: Adapted from Tanti, R., Barbut, S., & Marangoni, A. G. (2016). Oil stabilization of natural peanut butter using food grade polymers.* Food Hydrocolloids, 61, *399–408. https://doi.org/10.1016/j.foodhyd.2016.05.034*

from 15 to 40°C. This is possibly due to irreversible, shear-induced, and permanent damage that could affect the structure of food biopolymer including protein and change the flow behavior of sesame paste (Fig. 8.8A). Further, as depicted from Fig. 8.8B, the dynamic viscoelasticity structure of sesame paste transformed regularly to a viscous behavior with decreasing particle size. At the same time, the magnitudes of the storage (G') and loss (G'') moduli increased with frequency. Sesame pastes with smaller particle size showed a liquid-like behavior indicated by higher G″ (viscous-like behavior) than G′ (solid-like behavior) (Çiftçi et al., 2008). Hence, it can be inferred that small particle size results in a strong particle–particle interaction and produce more stable spread products.

Previous research on hazelnuts (Özdemir & Devres, 2000), peanuts (Cämmerer & Kroh, 2009), sesame seeds (Kahyaoglu, 2008), and macadamia nuts (Wall & Gentry, 2007) has demonstrated that browning and caramelization occur during roasting. As a result, the brown pigments are formed, affecting the volatile flavor characteristics (nutty, sweet, salty, roasted, and rancidity). Roasting was also observed to boost antioxidant capacity through the generation of Maillard reaction products, affecting the storage stability of nuts-based butter (Talcott et al., 2005). Even after being processed into peanut butter, the high protein and unsaturated fatty acid content of peanuts make them an excellent and cost-effective source of protein and lipids.

8.4.2 Sensory characteristics

The overall sensory quality of plant-based butter is related to the quality of the paste used as the main ingredient. Only a few studies can be found that focusing on the sensory properties of a plant-based butter. Gills & Resurreccion (2000) conducted a descriptive analysis of peanut butter. They identified eight oral textural attributes: stickiness and graininess (prior to mastication), hardness of the first bite, adhesiveness and gumminess during mastication, and residual sensations such as oiliness, mouth-coating, and mouth-dryness. Meanwhile, Lima et al. (2012) found that in the case of cashew nut butter, graininess and consistency are two attributes directly associated with the particles perceived visually or in the mouth, as well as the force required to remove the butter from a spoon or between tongue and mouth. Because the desirable characteristic of nut butter/spread is to have a smooth and creamy texture, the consumer preference ratings decreased as the particle size increased. According to Giboreau et al. (2007), shiny is an essential attribute for a creamy foods product such as butter, peanut butter, and other fat-based spread associated with an oily surface (not dry). From the Temporal Dominance of Sensations (TDS) study on peanut butter conducted by Rosenthal & Share (2014), peanut paste elicits only two dominant sensations: soft (early consumption) and stick to the palate (at the end of consumption), as shown in Fig. 8.9.

The peanut butter and peanut paste are also perceived to have a moist appearance that primarily comes from the high oil instead of water content. Toward to the high amount of oil in the product, peanut butter and peanut paste are expected to be well lubricated and easy to swallow. Interestingly, as shown in Fig. 8.8, peanut paste's sticky sensation and difficulty swallowing are caused by the solid matter suspended within the oil becoming hydrated by the saliva, resulting in particle aggregation and an increase in the suspension's viscosity (Rosenthal & Share, 2014). In addition, a previous study from Abegaz & Kerr (2006) has reported that the addition of a small amount of water to a peanut butter spread formulation increased the hardness, adhesiveness, and chewiness of the paste.

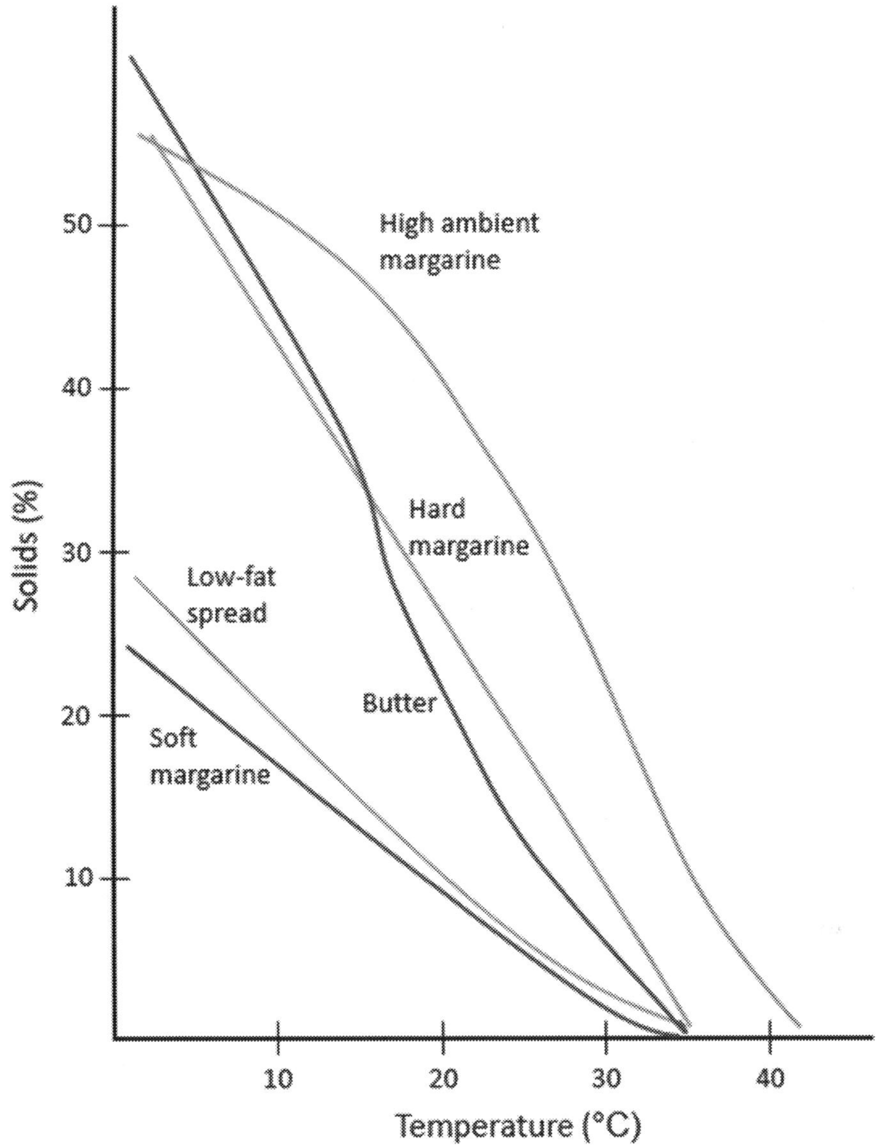

FIGURE 8.6 The typical percentage of fat solids for different spreads products. *Sources: Redrawn from Arellano, M., Norton, I. T., & Smith, P. (2015).* Specialty oils and fats in margarines and low-fat spreads *(pp. 241–270). Elsevier BV. https://doi.org/10.1016/b978-1-78242-376-8.00010-7*

8.4.3 Flavor characteristics

The shelf life of peanut butter, soy butter, and other spreads resembling butter is significantly reduced by rancidity, which occurs due to the oxidation of primarily unsaturated fatty acids (Nepote et al., 2006). The oxidative and/or hydrolytic rancidity significantly reduces plant-based butters' shelf life that includes a high concentration of polyunsaturated fatty acids. The combination of high temperature, oxygen, and enzymes accelerates this process. A sensory evaluation of commercial fat spreads made from oilseeds and walnuts conducted by Dimić et al. (2013) revealed that the most pleasant flavor was noticed in pumpkin seed spreads. Additionally, sunflower and walnut spreads have a mild rancid taste. Gills & Resurreccion (2000) also noted that panelists who identified peanut butter with hydrogenated vegetable oil reported a lower intensity of oxidized flavor than panelists who identified peanut butter with palm oil as a stabilizer or peanut butter without a stabiliser. Thus, oxidized flavor is a critical sensory characteristic of fat-based spreads and ultimately determines shelf life.

When it comes to acceptability, Dubost et al. (2003) discovered that the most distinguishing characteristic of peanut butter/spreads is their aroma. Their study demonstrates that the highest rating for the aroma of roasted peanuts was obtained from commercial peanut butter, with decreasing intensity as the amount of peanuts decreased. Hathorn & Sanders (2012)

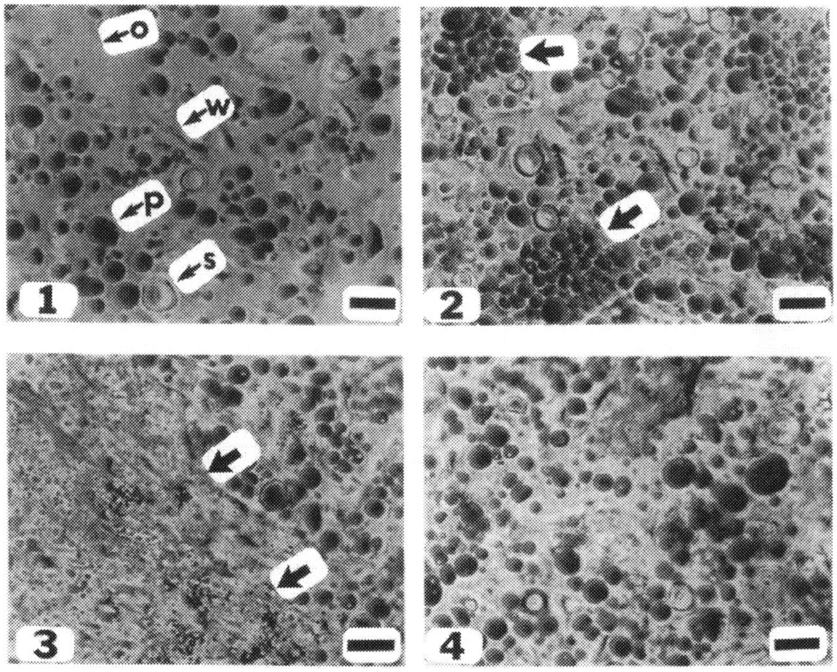

FIGURE 8.7 The microstructure of stabilized (1) and nonstabilized (2) peanut butter. The cell wall fragments (w), protein bodies (p), and starch grains (s) are dispersed in oil (o). (Bar size = 10 μm). *Sources: Adapted from Young, C. T., & Schadel, W. E. (1991). Comparison of the effects of three different grinding procedures on the microstructure of old-fashioned non-stabilised peanut butter.* Food Structure, 10*(3), 213–216.*

also reported a decrease in the roast peanuty intensity and an increase in the woody/hulls/skins, bitter, and astringency intensities of peanut paste and peanut butter when peanut skins are added at a rate greater than 5%.

8.5 Advantages and limitations of plant-based spreads

Due to the high oil content of most plant-based spreads, oil separation is the main challenge for this industry. The separated oil contaminates the packaging material, resulting in a decrease in the quality and appearance of the spreads. Additionally, the oil that serves as a lubricant in the emulsion suspension system separates from the main solid mass of the product, resulting in the hardening of the paste solids layer (solidification). As a result, natural spreadability reduces over time, and its stiffness increases. Therefore, limiting oil separation is crucial for increasing these products' acceptance and marketability. Many research on plant-based spreads have attempted to fix the oil separation problem by utilizing a range of commercially available food additives or combinations. Galvez et al. (2006) detected the separation of oil (more than 2%)

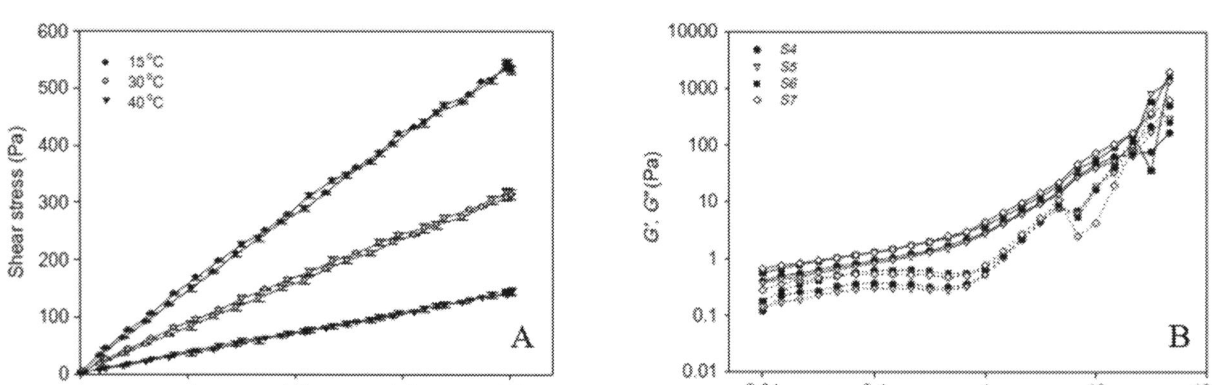

FIGURE 8.8 The effect of temperature (15, 30, and 45°C) on flow curves (A) and mean particle size (S4: 34.77 μm; S5: 27.63 μm; S6: 19.04 μm; S7: 14.23 μm) on viscoelastic behavior (B) of sesame paste. *Sources: Adapted from Çiftçi, D., Kahyaoglu, T., Kapucu, S. and Kaya, S. (2008). Colloidal stability and rheological properties of sesame paste.* Journal of Food Engineering, 87*(3), 428–435. https://doi.org/10.1016/j.jfoodeng.2007.12.026*

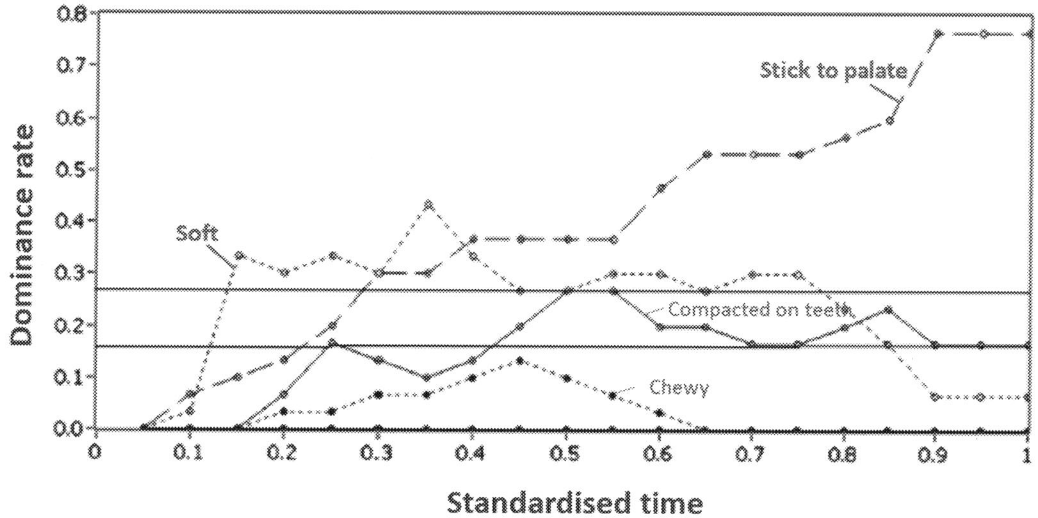

FIGURE 8.9 Temporal dominance of sensations (TDS) curve from peanut butter. *Sources: Adapted from Rosenthal, A. J., & Share, C. (2014). Temporal dominance of sensations of peanuts and peanut products in relation to Hutchings and Lillford's breakdown path.* Food Quality and Preference, 32, *311–316. https://doi.org/10.1016/j.foodqual.2013.09.004*

from peanut butter after 12 weeks of storage in the absence of a stabilizer. Ereifej et al. (2005) investigated the effects of several additives on the oil separation of tahini halva, including palm oil, soy protein isolate, gelatine, lecithin, gum arabic, pectin, and powdered sugar (sesame spread). Their investigation discovered that powdered sugar, gum arabic, and pectin significantly reduced oil leakage compared to control samples.

The primary impediment to consumer acceptance of plant-based butter is its shelf life. Rancidity is the primary sign of plant-based butter's lipid oxidation during storage. Due to high amount in polyunsaturated fatty acids (PUFAs), these oils are prone to lipid oxidation, which results in rancidity and off-flavor. Spreadability and hardness of natural peanut butter are significantly modified by storage temperature and time. In contrast to the hardness attribute, spreadability increases with temperature and decreases with storage duration. In addition, plant-based spread with a high moisture level deteriorates more quickly and develops off flavors than those with low moisture content. Based on previous study by Abegaz et al. (2004), the addition of moisture to peanut butter resulted in a less pronounced 'roasted peanutty' flavor, as evidenced by the decreased pyrazine concentrations. On contrary, the oxidative qualities such as 'rancid'," 'painty,' and 'cardboard' were more prevalent.

Monounsaturated, polyunsaturated, and saturated fatty acids are found in plant-based butter. Saturated, hydrogenated and trans fatty acid fats contribute to the flavor, palatability, and texture of several spreads. However, because these fats are closely related to cardiovascular disease, the World Health Organization (WHO) recommends a total fat intake of fewer than 30%, 10% for saturated fats, and 1% for trans fats. Hence, in order to address these difficulties, organogels have been identified as one of the most promising solutions for structuring food products such as margarine and table spreads (Pușcaș et al., 2020).

8.6 Summary and future direction

Natural food products have increased in popularity and demand in this decade. Certain customers will pick minimally processed foods over processed goods, and when they do choose processed foods, they will choose those with fewer additives and/or natural additives (Carocho et al., 2014). In addition, consumer attention has recently shifted to people who consume fewer carbohydrates and fats or consume protein- and vitamin-fortified table spreads such as low-calorie pistachio butter, milk-based fortified spreads, peanut spreads fortified with soy flour, and protein-fortified peanut butter. Unfortunately, consumers are not willing to sacrifice organoleptic qualities over health benefits. As a result, reduced carbohydrate products should have comparable organoleptic qualities to their high carbohydrate table spread counterparts. In light of the recent focus on carbohydrate and fat reduction, there is still a demand for cleanly packaged and easily applied spreads with lower fat and carbohydrate content. Additionally, interest in functional lipids and nutritional enhancement of lipid-containing foods has increased. Thus, supplementing spread/butter products with seeds (pumpkin seeds, chia seeds, watermelon seeds, and cucumber seeds) or olive oil has many potentials to develop. In addition, oleogelation, a novel technology for converting liquid plant-based oil into a solid-like gel, is gaining prominence in the structure of plant-based spreads. More exploratory research required that focuses on the synergy of certain organogelators or blends to create perfect oleogels that are stable and offer extra benefits for human health or the environment.

References

Abd-Elsattar, H. H., & Abdel-Haleem, A. M. H. (2016). Production of soybean butter using different technological treatments. *LWT—Food Science and Technology, 69*, 40–46. https://doi.org/10.1016/j.lwt.2016.01.029

Abegaz, E. G., & Kerr, W. L. (2006). Effect of moisture, sugar and tertiary butylhydroquinone on color, texture and microstructure of peanut paste. *Journal of Food Quality, 29*(6), 643–657. https://doi.org/10.1111/j.1745-4557.2006.00102.x

Abegaz, E. G., Kerr, W. L., & Koehler, P. E. (2004). The role of moisture in flavor changes of model peanut confections during storage. *LWT—Food Science and Technology, 37*(2), 215–225. https://doi.org/10.1016/j.lwt.2003.07.007

Agrahar-Murugkar, D., Kotwaliwale, N., Kumar, M., & Gupta, C. (2013). Effect of sprouting on rheological properties of soy-butter. *LWT—Food Science and Technology, 54*(1), 95–100. https://doi.org/10.1016/j.lwt.2013.05.010

Agrahar Murugkar, D., Kotwaliwale, N., Kumar, M., & Gupta, C. (2013). Effect of roasting parameters on soy-butter product quality. *International Journal of Food Science and Technology, 48*(7), 1359–1365. https://doi.org/10.1111/ijfs.12096

Akhtar, S., Khalid, N., Ahmed, I., Shahzad, A., & Suleria, H. A. R. (2014). Physicochemical characteristics, functional properties, and nutritional benefits of peanut oil: A review. *Critical Reviews in Food Science and Nutrition, 54*(12), 1562–1575. https://doi.org/10.1080/10408398.2011.644353

Arachchige, U., Ranaweera, S., Ampemohotti, T., Nayanakanthi, P. K. T., & Edirisinghe, D. S. P. (2017). Preparation of avocado butter. *International Journal of Scientific & Technology Research, 8*(10), 3527–3530.

Ardakani, A. S. (2006). The vital role of pistachio processing industries in development of Iran non-oil exports. In *Acta horticulturae* (Vol. 726, pp. 579–581). International Society for Horticultural Science. https://doi.org/10.17660/actahortic.2006.726.97

Arellano, M., Norton, I. T., & Smith, P. (2015). *Specialty oils and fats in margarines and low-fat spreads* (pp. 241–270). Elsevier BV. https://doi.org/10.1016/b978-1-78242-376-8.00010-7

Aryana, K. J., Resurreccion, A. V. A., Chinnan, M. S., & Beuchat, L. R. (2000). Microstructure of peanut butter stabilized with palm oil. *Journal of Food Processing and Preservation, 24*(3), 229–241. https://doi.org/10.1111/j.1745-4549.2000.tb00415.x

Aryana, K. J., Resurreccion, A. V. A., Chinnan, M. S., & Beuchat, L. R. (2003). Functionality of palm oil as a stabilizer in peanut butter. *Journal of Food Science, 68*(4), 1301–1307. https://doi.org/10.1111/j.1365-2621.2003.tb09643.x

Bhuyan, D. J., Alsherbiny, M. A., Perera, S., Low, M., Basu, A., Devi, O. A., Barooah, M. S., Li, C. G., & Papoutsis, K. (2019). The odyssey of bioactive compounds in Avocado (Persea Americana) and their health benefits. *Antioxidants, 8*(10). https://doi.org/10.3390/antiox8100426

Cämmerer, B., & Kroh, L. W. (2009). Shelf life of linseeds and peanuts in relation to roasting. *LWT—Food Science and Technology, 42*(2), 545–549. https://doi.org/10.1016/j.lwt.2008.08.003

Carocho, M., Barreiro, M. F., Morales, P., & Ferreira, I. C. F. R. (2014). Adding molecules to food, pros and cons: A review on synthetic and natural food additives. *Comprehensive Reviews in Food Science and Food Safety, 13*(4), 377–399. https://doi.org/10.1111/1541-4337.12065

Çiftçi, D., Kahyaoglu, T., Kapucu, S., & Kaya, S. (2008). Colloidal stability and rheological properties of sesame paste. *Journal of Food Engineering, 87*(3), 428–435. https://doi.org/10.1016/j.jfoodeng.2007.12.026

Devi, A., Narayanan, R., Masilamoni, R. S., & Manikkavasagan. (2019). Optimization and development of fat spread from pumpkin seeds. *International Journal of Chemical Studies, 7*(5), 2508–2512.

Di Monaco, R., Giancone, T., Cavella, S., & Masi, P. (2008). Predicting texture attributes from microstructural, rheological and thermal properties of hazelnut spreads. *Journal of Texture Studies, 39*(5), 460–479. https://doi.org/10.1111/j.1745-4603.2008.00154.x

Dimić, E. B., Vujasinović, V. B., Radočaj, O. F., & Borić, B. D. (2013). Sensory evaluation of commercial fat spreads based on oilseeds and walnut. *Acta Periodica Technologica, 44*, 21–30. https://doi.org/10.2298/APT1344021D

Dubost, N. J., Shewfelt, R. L., & Eitenmiller, R. R. (2003). Consumer acceptability, sensory and instrumental analysis of peanut soy spreads. *Journal of Food Quality, 26*(1), 27–42. https://doi.org/10.1111/j.1745-4557.2003.tb00224.x

El-Adawy, T. A., & Mansour, E. H. (2000). Nutritional and physicochemical evaluations of tahina (sesame butter) prepared from heat-treated sesame seeds. *Journal of the Science of Food and Agriculture, 80*(14), 2005–2011. https://doi.org/10.1002/1097-0010(200011)80:14<2005::AID-JSFA740>3.0.CO;2-J

Emadzadeh, B., Razavi, S., & Hashemi, M. (2011). Viscous flow behavior of low-calorie pistachio butter: A response surface methodology. *International Journal of Nuts and Related Sciences, 2*(1), 37–47.

Emadzadeh, B., Razavi, S. M. A., & Mahallati, M. N. (2012). Effects of fat replacers and sweeteners on the time-dependent rheological characteristics and emulsion stability of low-calorie pistachio butter: A response surface methodology. *Food and Bioprocess Technology, 5*(5), 1581–1591. https://doi.org/10.1007/s11947-010-0490-6

Ereifej, K. I., Rababah, T. M., & Al-Rababah, M. A. (2005). Quality attributes of halva by utilization of proteins, non-hydrogenated palm oil, emulsifiers, gum Arabic, sucrose, and calcium chloride. *International Journal of Food Properties, 8*(3), 415–422. https://doi.org/10.1080/10942910500267323

Felland, S. L., & Koehler, P. E. (1997). Sensory, chemical, and physical changes in increased water activity peanut butter products. *Journal of Food Quality, 20*(2), 145–156. https://doi.org/10.1111/j.1745-4557.1997.tb00459.x

Flynn, C. S., Foster, K. D., Bronlund, J. E., Lentle, R. G., Jones, J. R., & Morgenstern, M. P. (2011). Identification of multiple compartments present during the mastication of solid food. *Archives of Oral Biology, 56*(4), 345–352. https://doi.org/10.1016/j.archoralbio.2010.10.010

Galvez, F. C. F., Aquino, M. B., Villarino, B. J., Lustre, A. O., Francisco, M. L. L., & Resurreccion, A. V. A. (2006). Development and optimization of choco-peanut spread. In *Peanut butter and spreads*.

Genovese, D. B., Lozano, J. E., & Rao, M. A. (2007). The rheology of colloidal and noncolloidal food dispersions. *Journal of Food Science, 72*(2), R11−R20. https://doi.org/10.1111/j.1750-3841.2006.00253.x

Giboreau, A., Dacremont, C., Egoroff, C., Guerrand, S., Urdapilleta, I., Candel, D., & Dubois, D. (2007). Defining sensory descriptors: Towards writing guidelines based on terminology. *Food Quality and Preference, 18*(2), 265−274. https://doi.org/10.1016/j.foodqual.2005.12.003

Gills, L. A., & Resurreccion, A. V. A. (2000). Sensory and physical properties of peanut butter treated with palm oil and hydrogenated vegetable oil to prevent oil separation. *Journal of Food Science, 65*(1), 173−180. https://doi.org/10.1111/j.1365-2621.2000.tb15975.x

Gorrepati, K., Balasubramanian, S., & Chandra, P. (2015). Plant based butters. *Journal of Food Science and Technology, 52*(7), 3965−3976. https://doi.org/10.1007/s13197-014-1572-7

Hahighi, Azar, M., Mazloumi, M., & Fonoud. (2008). Survey of formulation. Production and sensory evaluation of pistachio butter. *Iranian Journal of Food Science and Technology, 5*(3), 19−26.

Hathorn, C. S., & Sanders, T. H. (2012). Flavor and antioxidant capacity of peanut paste and peanut butter supplemented with peanut skins. *Journal of Food Science, 77*(11), S407−S411. https://doi.org/10.1111/j.1750-3841.2012.02953.x

Humphrey, J. H. (2008). Underweight malnutrition in infants in developing countries: An intractable problem. *Archives of Pediatrics and Adolescent Medicine, 162*(7), 692−694. https://doi.org/10.1001/archpedi.162.7.692

Hwang, H.-S., Kim, S., Singh, M., Winkler-Moser, J. K., & Liu, S. X. (2012). Organogel formation of soybean oil with waxes. *Journal of the American Oil Chemists' Society, 89*(4), 639−647. https://doi.org/10.1007/s11746-011-1953-2

Hwang, H. S., Singh, M., Winkler-Moser, J. K., Bakota, E. L., & Liu, S. X. (2014). Preparation of margarines from organogels of sunflower wax and vegetable oils. *Journal of Food Science, 79*(10), C1926−C1932. https://doi.org/10.1111/1750-3841.12596

Kahyaoglu, T. (2008). Optimization of the pistachio nut roasting process using response surface methodology and gene expression programming. *LWT—Food Science and Technology, 41*(1), 26−33. https://doi.org/10.1016/j.lwt.2007.03.026

Kumar, P. (2015). Process optimization for the preparation of chocolate spread incorporating whey protein concentrate, cocoa powder, olive oil and butterfat using response surface methodology. *Journal of Food Processing and Preservation, 39*(6), 745−757. https://doi.org/10.1111/jfpp.12284

Liedl, F. G., & Rowe, K. F. (2007). *Nut butter and related products and method of making same.*

Lima, J. R., Garruti, D. S., & Bruno, L. M. (2012). Physicochemical, microbiological and sensory characteristics of cashew nut butter made from different kernel grades-quality. *LWT—Food Science and Technology, 45*(2), 180−185. https://doi.org/10.1016/j.lwt.2011.08.018

Lima, I. M., & Guraya, H. S. (2005). Optimization analysis of sunflower butter. *Journal of Food Science, 70*(6), s365−s370. https://doi.org/10.1111/j.1365-2621.2005.tb11457.x

Lima, I. M., Guraya, H. S., & Champagne, E. T. (2000). Improved peanut flour for a reduced-fat peanut butter product. *Journal of Food Science, 65*(5), 854−861. https://doi.org/10.1111/j.1365-2621.2000.tb13600.x

Lumor, S. E., Pina-Rodriguez, A. M., Shewfelt, R. L., & Akoh, C. C. (2010). Physical and sensory attributes of a trans-free spread formulated with a blend containing a structured lipid, palm mid-fraction, and cottonseed oil. *Journal of the American Oil Chemists' Society, 87*(1), 69−74. https://doi.org/10.1007/s11746-009-1470-8

Mangels, R. (2001). Guide to nut and nut butters. *Vegetarian J, XXI*, 20−23.

Marangoni, A. G. (2012). Organogels: An alternative edible oil-structuring method. *Journal of the American Oil Chemists' Society, 89*(5), 749−780. https://doi.org/10.1007/s11746-012-2049-3

Marks, G. (2010). *Encyclopedia of Jewish food.* New York: John Wiley and Son.

Matsunobu, Horishita, S., & Yamada, T. (1984). *Almond nut paste for beverages and desserts.* US Patent. Publication.

Miraliakbari, H., & Shahidi, F. (2008). Oxidative stability of tree nut oils. *Journal of Agricultural and Food Chemistry, 56*(12), 4751−4759. https://doi.org/10.1021/jf8000982

Mohd Rozalli, N. H., Chin, N. L., & Yusof, Y. A. (2015). Particle size distribution of natural peanut butter and its dynamic rheological properties. *International Journal of Food Properties, 18*(9), 1888−1894. https://doi.org/10.1080/10942912.2014.971184

Murugkar, D. A., Kotwaliwale, N., Gupta, C., Gulati, P., & Kumar, M. (2014). Optimization of soy-butter from sprouted soybean and its quality parameters. *Journal of Food Quality, 37*(1), 63−72. https://doi.org/10.1111/jfq.12069

Nepote, V., Mestrallet, M. G., Accietto, R. H., Galizzi, M., & Grosso, N. R. (2006). Chemical and sensory stability of roasted high-oleic peanuts from Argentina. *Journal of the Science of Food and Agriculture, 86*(6), 944−952. https://doi.org/10.1002/jsfa.2442

Nikolić, I., Dokić, L., Krstonošić, V., Šereš, Z., & Šoronja-Simović, D. (2014). Possibility of the production of functional low-fat food spread of hull-less pumpkin seed flour from rheological and textural aspect. *Journal of Texture Studies, 45*(4), 324−333. https://doi.org/10.1111/jtxs.12078

Ogunsanwo, B. M., Faboya, O. O. P., Idowu, O. R., & Adewuyi, G. O. (2005). Using rating to evaluate quality of peanut products. *African Journal of Biotechnology, 4*(12), 1469−1471. http://www.academicjournals.org/AJB/PDF/Pdf2005/Dec/Ogunsanwo%20et%20al.pdf.

Olaleye, H. T., Oresanya, T. O., & Bello, B. Z. (2021). Physicochemical, textural, rheological and sensory properties of chia seed-cashew nut spread. *Journal of Culinary Science and Technology*, 1−11. https://doi.org/10.1080/15428052.2021.1929635

Özdemir, M., & Devres, O. (2000). Kinetics of color changes of hazelnuts during roasting. *Journal of Food Engineering, 44*(1), 31−38. https://doi.org/10.1016/S0260-8774(99)00162-4

Patel, A. R., & Dewettinck, K. (2016). Edible oil structuring: An overview and recent updates. *Food and Function, 7*(1), 20−29. https://doi.org/10.1039/c5fo01006c

Patel, A. R., Nicholson, R. A., & Marangoni, A. G. (2020). Applications of fat mimetics for the replacement of saturated and hydrogenated fat in food products. *Current Opinion in Food Science, 33*, 61–68. https://doi.org/10.1016/j.cofs.2019.12.008

Pourfarzad, A., & Kisomi, R. S. (2020). Effect of lecithin and mono- and di-glycerides on quality and shelf life of hazelnut butter: Chemometric approach. *Polish Journal of Food and Nutrition Sciences, 70*(4), 399–408. https://doi.org/10.31883/pjfns/129719

Puşcaş, A., Mureşan, V., Socaciu, C., & Muste, S. (2020). Oleogels in food: A review of current and potential applications. *Foods, 9*(1), 70. https://doi.org/10.3390/foods9010070

Radoĉaj, O. F., Dimić, E. B., & Vujasinović, V. B. (2011). Optimization of the texture of fat-based spread containing hullless pumpkin (*Cucurbita pepo L.*) seed press-cake. *Acta Periodica Technologica, 42*, 131–143. https://doi.org/10.2298/APT1142131R

Radočaj, O., Dimić, E., & Vujasinović, V. (2012). Development of a hull-less pumpkin (cucurbita pepo L.) seed oil press-cake spread. *Journal of Food Science, 77*(9), C1011–C1017. https://doi.org/10.1111/j.1750-3841.2012.02874.x

Razavi, S. M. A., Habibi Najafi, M. B., & Alaee, Z. (2007). The time independent rheological properties of low fat sesame paste/date syrup blends as a function of fat substitutes and temperature. *Food Hydrocolloids, 21*(2), 198–202. https://doi.org/10.1016/j.foodhyd.2006.03.008

Rønholt, S., Mortensen, K., & Knudsen, J. C. (2013). The effective factors on the structure of butter and other milk fat-based products. *Comprehensive Reviews in Food Science and Food Safety, 12*(5), 468–482. https://doi.org/10.1111/1541-4337.12022

Rosenthal, A. J., & Share, C. (2014). Temporal Dominance of Sensations of peanuts and peanut products in relation to Hutchings and Lillford's breakdown path. *Food Quality and Preference, 32*, 311–316. https://doi.org/10.1016/j.foodqual.2013.09.004

Said, A., Nasir, N. A. M., Bakar, C. A. A., & Mohamad, W. A. F. W. (2019). Chocolate spread emulsion: Effects of varying oil types on physico-chemical properties, sensory qualities and storage stability. *Journal of Agrobiotechnology, 10*(2), 32–42.

Sanders, T. H. (2003). *Peanuts*. Elsevier BV. https://doi.org/10.1016/b0-12-227055-x/00897-x

Sethi, N., & Balasubramanyam, B. V. (2018). Optimization of butter fruit incorporated fat spread using instrumental textural analysis: A response surface methodology. *Journal of Food Measurement and Characterization, 12*(2), 859–866. https://doi.org/10.1007/s11694-017-9700-5

Shakerardekani, A., Karim, R., Ghazali, H. M., & Chin, N. L. (2013a). Development of pistachio (Pistacia vera L.) spread. *Journal of Food Science, 78*(3), S484–S489. https://doi.org/10.1111/1750-3841.12045

Shakerardekani, A., Karim, R., Ghazali, H. M., & Chin, N. L. (2013b). Textural, rheological and sensory properties and oxidative stability of nut spreads-a review. *International Journal of Molecular Sciences, 14*(2), 4223–4241. https://doi.org/10.3390/ijms14024223

Singh, S. K., Castell-Perez, M. E., & Moreira, R. G. (2000). Viscosity and textural attributes of reduced-fat peanut pastes. *Journal of Food Science, 65*(5), 849–853. https://doi.org/10.1111/j.1365-2621.2000.tb13599.x

Syed, Q. A. (2019). Nutritional and therapeutic importance of the pumpkin seeds. *Biomedical Journal of Scientific & Technical Research, 21*(2). https://doi.org/10.26717/bjstr.2019.21.003586

Taghizadeh, M., & Razavi, S. M. A. (2009). Modeling time-independent rheological behavior of pistachio butter. *International Journal of Food Properties, 12*(2), 331–340. https://doi.org/10.1080/10942910701772048

Talcott, S. T., Duncan, C. E., Pozo-Insfran, D. D., & Gorbet, D. W. (2005). Polyphenolic and antioxidant changes during storage of normal, mid, and high oleic acid peanuts. *Food Chemistry, 89*(1), 77–84. https://doi.org/10.1016/j.foodchem.2004.02.020

Tanti, R., Barbut, S., & Marangoni, A. G. (2016). Oil stabilization of natural peanut butter using food grade polymers. *Food Hydrocolloids, 61*, 399–408. https://doi.org/10.1016/j.foodhyd.2016.05.034

Totlani, V. M., & Chinnan, M. S. (2007). Effect of stabilizer levels and storage conditions on texture and viscosity of peanut butter. *Peanut Science, 34*(1), 1–9. https://doi.org/10.3146/0095-3679(2007)34[1:EOSLAS]2.0.CO;2

USDA. (2011). *Commercial item description: nut butters and nut spreads*. Retrieved from https://www.ams.usda.gov/sites/default/files/media/NutButtersNutSpreads.pdf.

Wagener, E. A., & Kerr, W. L. (2018). Effects of oil content on the sensory, textural, and physical properties of pecan butter (Carya illinoinensis). *Journal of Texture Studies, 49*(3), 286–292. https://doi.org/10.1111/jtxs.12304

Wallace, T., Murray, R., & Zelman, K. (2016). The nutritional value and health benefits of chickpeas and hummus. *Nutrients, 8*(12), 766. https://doi.org/10.3390/nu8120766

Wall, M. M., & Gentry, T. S. (2007). Carbohydrate composition and color development during drying and roasting of macadamia nuts (Macadamia integrifolia). *LWT—Food Science and Technology, 40*(4), 587–593. https://doi.org/10.1016/j.lwt.2006.03.015

Woodroof, J. G. (1983). *Peanut production, processing, products* (pp. 181–227). AVI Publishing Co.

Yõlmaz, E., & Öğütcü, M. (2014a). Comparative analysis of olive oil organogels containing beeswax and sunflower wax with breakfast margarine. *Journal of Food Science, 79*(9), E1732–E1738. https://doi.org/10.1111/1750-3841.12561

Yõlmaz, E., & Öğütcü, M. (2014b). Properties and stability of hazelnut oil organogels with beeswax and monoglyceride. *Journal of the American Oil Chemists' Society, 91*(6), 1007–1017. https://doi.org/10.1007/s11746-014-2434-1

Yõlmaz, E., & Öğütcü, M. (2015). Oleogels as spreadable fat and butter alternatives: Sensory description and consumer perception. *RSC Advances, 5*(62), 50259–50267. https://doi.org/10.1039/C5RA06689A

Young, C. T., & Schadel, W. E. (1991). Comparison of the effects of three different grinding procedures on the microstructure of old-fashioned non-stabilised peanut butter. *Food Structure, 10*(3), 213–216.

Zahedi, Y., & Mazaheri-Tehrani, M. (2012). Development of spreadable halva fortified with soy flour and optimization of formulation using mixture design. *Journal of Food Quality, 35*(6), 390–400. https://doi.org/10.1111/jfq.12003

Section III

Plant-based meat alternatives

Chapter 9

Plant-based meat analogue

Oni Yuliarti, Muhammad Faiz Bin Muhd Faizal Abdullah Tan and Jun Kiat Kovis Tay
School of Chemical & Life Sciences, Singapore Polytechnic, Singapore

9.1 Introduction

Plant-based meat analogues are not a new food category in Asia, as they have been consumed widely in the community where soybeans are used as the primary protein source (Bohrer, 2019). Recently, the growing interest in consuming plant-based meat analogue is in a rise worldwide as growing awareness of healthy foods and the sustainability issues of animal protein. Concerns over personal health, the environment, and animal ethics are the driving factors in this growing interest. High levels of meat consumption for long term have been associated with the development of health conditions such as cardiovascular disease, type 2 diabetes, obesity, and cancer (Raphaely & Marinova, 2015). Livestock activities have an enormous impact on the environment; the livestock sector is reported to be one of the most significant contributors to environmental problems globally. Based on the United Nations Food and Agriculture Organisation's (2006) report, "Livestock's Long Shadow," the livestock sector is accountable for approximately 18% of the world's greenhouse gas emissions (Steinfeld et al., 2006). Animal agriculture is also greatly associated with large-scale land-based pollution, loss of biodiversity, and deforestation (Hallström et al., 2014).

In the livestock sector, animals are raised in confined animal feeding operations (CAFOs) and are often subjected to inhumane treatment, as a means for maximizing efficiency and profits. According to a 2007 US Department of Agriculture statistic, it is reported that approximately over 10 billion land animals were slaughtered in farms all over the United States of America (Freeman, 2011).

The meat analogue market is the fastest growing industry and is estimated to be valued at USD 1.9 billion in 2021 and is projected to reach USD 4.0 billion by 2027 (Market & Market, 2021). Therefore, innovative, nutritious, and sustainable plant-based products are then required to fill the growing market demands. Meat analogues are defined as plant-based products that mimic the appearance and the fibrous texture of traditional meat products (Pietsch et al., 2017). They are made from non-animal-related ingredients, in order to derive and obtain a product that is aesthetically (taste, appearance and texture) as well as nutritionally similar to that of animal meat (Joshi & Kumar, 2015). Meat analogues can be derived from a wide variety of protein sources such as vegetable proteins (texturized vegetable proteins) and fungal-derived proteins such as QuornTM (Jones, 2016). There are also meat analogues that are made through the use of traditional high protein preparations, which includes products such as tempeh (fermented soybean cake), seitan (cooked wheat gluten), and "Kori tofu" (freeze-texturized soy protein) (Consolacion & Jelen, 1986).

Despite having the potential to mimic the texture and structure of meat, meat analogues are plagued with issues regarding their taste, texture, and nutrition. A truly fibrous meat structure is challenging to obtain in meat analogues, and there is also the issue of distinctive off-flavors associated with certain types of plant proteins. Furthermore, meat analogues are lacking in essential nutrients commonly found in animal proteins. The use of processed plant proteins with varying nutritional makeup, along with other ingredients such as hydrocolloids and flavor compounds, have the potential to improve the nutritional and sensory attributes of meat analogues. In this chapter, the importance of structuring meat analogue by using certain techniques is reviewed.

9.2 Structuring techniques of plant-based meat

Meat is characterized by its unique texture and structure. It is a dynamic tissue made up of cross-linked and aligned muscle fibers forming macro and micro muscle fibers and together with connective tissues. These alignments affect the tenderness of

Engineering Plant-Based Food Systems. https://doi.org/10.1016/B978-0-323-89842-3.00009-9

meat (Gonzalez et al., 2018). Meat analogue and other meat alternatives are developed with the aim of mimicking the distinct texture of meats, undergoing specific structuring processes in order to achieve a fibrous structure. Different techniques can be used to make fibrous products that mimic muscle meats. It can either be the bottom-up strategy, which refers to assembly of structural elements that are combined such as cell culturing and electrospinning or the top-down strategy, which refers to structuring of biopolymer blends using an overall force field such extrusion and freeze structuring techniques (Dekkers et al., 2016). The most current production techniques for meat analogues are described in this section.

Table 9.1 indicates an overview of comparative structuring technique used and also the main protein sources for the development of plant-based meat analogue with respective techniques. Equipment types with its specific action to the protein are also provided. Although several approaches are available to resemble the fibrous structure, the downside of cultured meat and electrospinning depends on the scalability, labor intensiveness, and requirement of higher level of resources (Dekkers et al., 2018). Methods such as freeze structuring and mechanical elongation are still preferred by many small commercial-scale industries because they are less expensive than other strategies.

9.2.1 Extrusion

This technique has recently gained growing attention among the manufacturers and researchers for the production of meat analogues. However, this processing method can be an expensive investment for many small-scale manufacturers due the expensive processing equipment as well as the requirement of high material input for consistent and uniform operation (Yuliarti et al., 2021). These techniques utilize a disruptive force to facilitate the aggregation of proteins within an emulsion where the unique structure of the products is formed. The two main components in a top-down strategy are a multiphase biopolymer blend and a force field. The biopolymer blend is made up of two immiscible distinct phases, which consist of a dispersed phase and a continuous phase, while the force field refers to a disruptive force used in the structuring process. An example of a force field would be shear force, while a possible biopolymer blend would be water dispersed into a continuous phase of protein. The application of the force field leads to the deformation and alignment of the dispersed phase within the continuous phase. After solidification, a fibrous structure is obtained from the disrupted continuous phase (Dekkers et al., 2018).

TABLE 9.1 Techniques for structuring meat analogues.

Structuring technique	Equipment types and functions	Protein sources	Product structure	References
Extrusion	Extruder with twin screw. This is for the protein to undergo hydrothermal treatment, to unfold protein chains and to align the protein molecules in the fibrous direction	Soy protein isolate, wheat gluten	Sponge like, fibrous, and layered structure	Chiang et al. (2019), Lin and Huang (2008), Lin et al. (2000)
Mechanical elongation	Mixer, high rotational speed of a knife blade, incubator. This is for protein network formation, stretching and elongation	Soy protein isolate, zein, wheat gluten	Fibrillar structure	Chiang et al. (2021), Mattice and Marangoni (2020)
Freeze structuring	Uni-directional freezer, mixer and freezer to form ice crystals and to facilitate cross-linking between protein molecules by compacting the proteins in the network	Pea protein, wheat protein, soy protein isolate	Fibrous, layered structure, fibers/ sheet-like material	Guan et al. (2010), Yuliarti et al. (2021)
Cultured meat	Bioreactor. The cell from animals will grow and multiply to create muscle tissues in the bioreactor	Precursor skeletal muscle cells, growth medium	Fibrous as muscle fibers	Benjaminson et al. (2002), Kadim et al. (2015)
Electrospinning	Electrospinning set-up which consists of barrel, high voltage supply, ground elector to form of very thin fibers or nanofibers	Protein isolate, zein	Fibers/nanofibers	Dekkers et al. (2018), Lim et al. (2019), Mattice and Marangoni (2020)

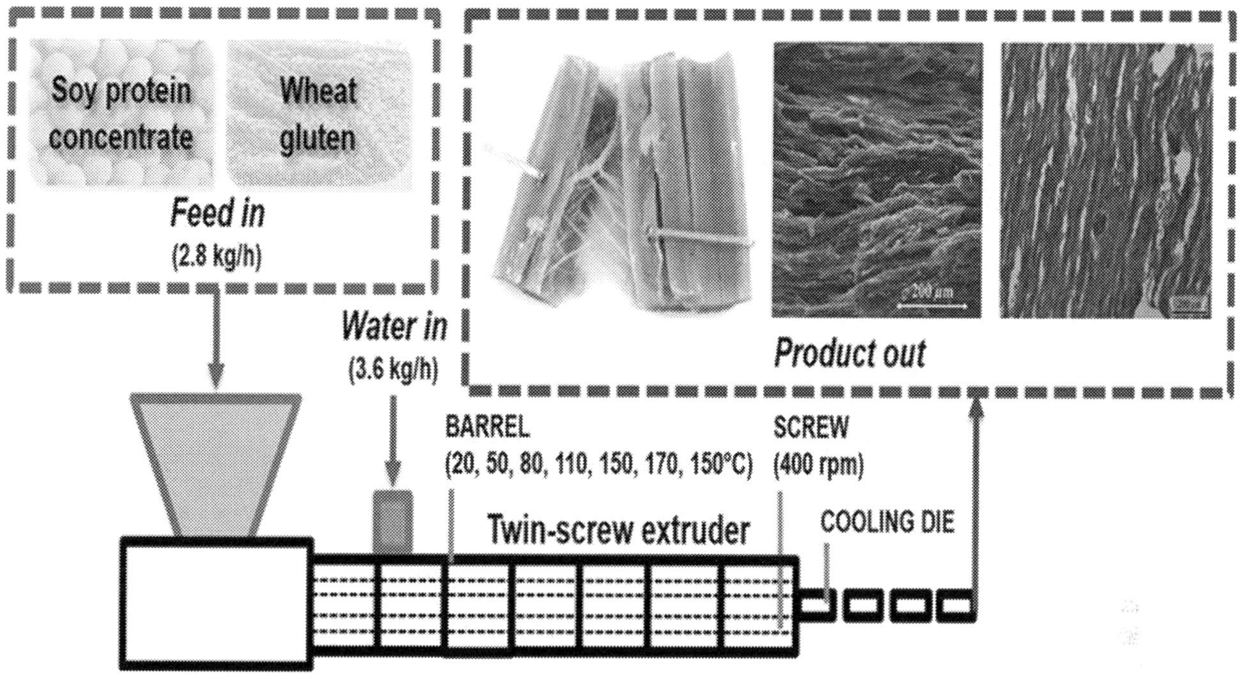

FIGURE 9.1 Extrusion technique of extruded meat analogues using combination of soy protein concentrate (SPC) and wheat gluten. *From Chiang, J. H., Loveday, S. M., Hardacre, A. K., & Parker, M. E. (2019). Effects of soy protein to wheat gluten ratio on the physicochemical properties of extruded meat analogues.* Food Structure, 19. *https://doi.org/10.1016/j.foostr.2018.11.002.*

Extrusion is split into two broad categories, low-moisture and high moisture extrusion, which are similar processes that differ in terms of their feed properties and products. In low-moisture extrusion, flours or concentrates are mechanically processed into textured vegetable proteins (TVP), which are dry, slightly expanded products that are hydrated later. On the other hand, as for high moisture extrusion, fibrous products are produced with moisture contents of at least 50% w/w. During extrusion processing, proteins undergo hydrothermal treatment in the extruder barrel, which unfolds long protein chains, effectively plasticizing the proteins into a melt (Fig. 9.1). This melted protein is then passed through a die, which cools and aligns the protein molecules in the direction of flow via drag and shear flow, forming a highly fibrous structure (Cheftel et al., 1992).

9.2.2 Mechanical elongation method

This method is considered as a less expensive and a lower energy intensive method to produce meat analogues compared to the extrusion method where the resulted products have similar texture as animal meat (Mattice & Marangoni, 2020). It involves thermal treatment followed by elongating and orientating the protein fine fibrils with a diameter of 0.5–2 μm in the self-assembled protein network (Fig. 9.2). These authors further stated that the formation of fiber structure such the fiber width and number of fibers can easily be customised by adjusting the amount of stretching during the formation.

The formation of meat analogue by this technique had yielded meat analogues with fibrillar structure as described by some studies (Chiang et al., 2021; Mattice & Marangoni, 2020) with the summarized method as below:

1. Formation of protein network by mixing of the protein ingredients with water to form a dough.
2. Incubation in an oven at a certain temperature of 40–60°C to allow the gluten to be fully hydrated and allow gluten network formation.
3. Mechanical elongation method, which consists of cutting the incubated dough into smaller pieces using a food processor. This process will stretch and pull small protein fragments to form a secondary dough via the high rotational speed allowing the proteins to agglomerate and form stronger networks. Further elongation is required to form cohesive protein network.

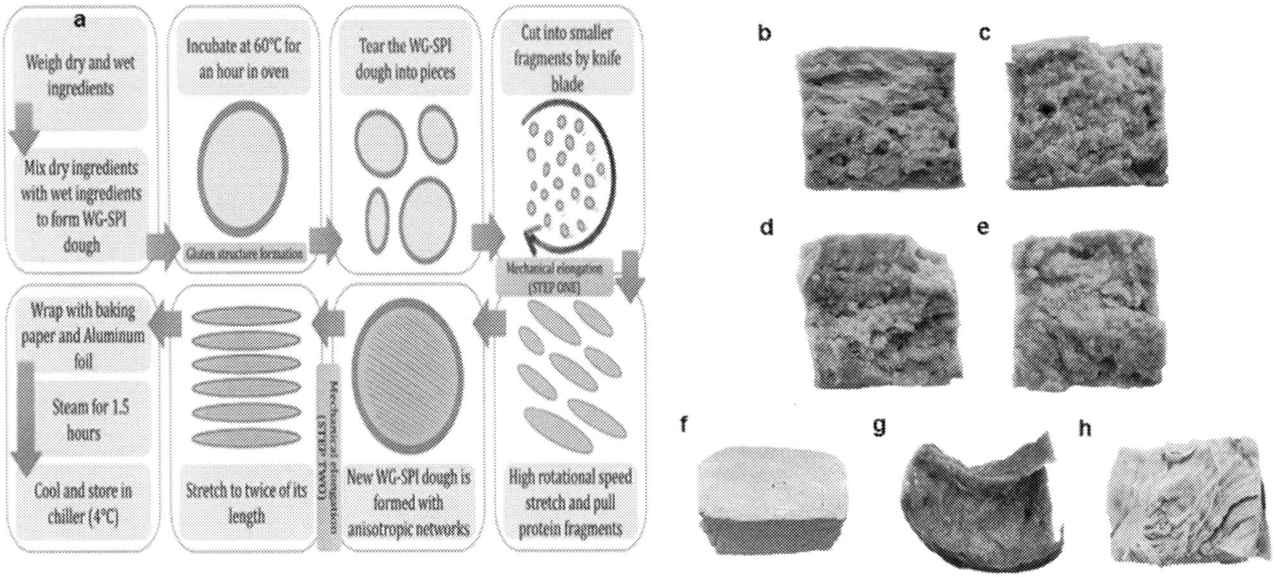

FIGURE 9.2 (A) Schematic illustration of the mechanical elongation method to produce meat analogues and meat analogues produced with different wheat gluten to soy protein isolate (WG-SPI) ratio (B) 100 %WG, (C) 80 %WG, (D) 60 %WG, (E) 40 %WG, (F) firm tofu, (G) mock chicken, and (H) steamed chicken (Chiang et al., 2021). *From Chiang, J. H., Tay, W., Ong, D. S. M., Liebl, D., Ng, C. P., & Henry, C. J. (2021). Physicochemical, textural and structural characteristics of wheat gluten-soy protein composited meat analogues prepared with the mechanical elongation method.* Food Structure, 28. *https://doi.org/10.1016/j.foostr.2021.100183.*

9.2.3 Freeze structuring

The freeze structuring technique utilizes freezing as a means to deform and align the dispersed phase of an aqueous solution of proteins to form a fibrous structure. Unidirectional removal of heat from a well-mixed aqueous solution of protein results in the freezing and alignment of water (dispersed phase) within the slurry, forming ice crystal needles. The formation of ice crystal serves to facilitate cross-linking between protein molecules by compacting the proteins (continuous phase) together. Upon removal of the ice crystals, a porous structure with sheet-like parallel orientation of proteins is formed, resulting in a cohesive and fibrous product (Lugay & Kim, 1978). In the case of freeze structuring technique used in plant-based chicken nugget development, the protein texturization is achieved by the following steps (Yuliarti et al., 2021), and the illustration is shown in Fig. 9.3:

1. Oil in water emulsion formation, which is achieved by mixing the binding agent, water and vegetable oil. This formation helps to disperse the oil droplets in water phase.
2. Mixing of hydration of protein to form a uniform parallel zones of gluten network.
3. Hydrated protein mixing with emulsion to disperse the oil droplets in gluten network.
4. Heat processing of the mixture by steaming to firm up the network structure and to facilitate the entrapment of air released by leavening agents in the gluten network. This is followed by immediate freezing of fibrous protein network to convert the trapped water to produce small ice crystals.

The freeze structuring method is further enhanced and facilitated via denaturation of globular vegetable proteins into random coils before freezing. This denaturation process facilitates cross-linking and protein polymerization between protein chains, enhancing the structure of the final product (Consolacion & Jelen, 1986). The structure of plant-based nugget using freeze structuring technique indicated fibrous and layered arrangement (Yuliarti et al., 2021) demonstrating that the freeze structuring technique applied had governed the final analogue structure. It is important to note that the presence of a binder or fat replacer is important for the success of meat analogue cohesive structure.

9.2.4 Cell culture

This technique is classified under bottom-up strategies, which involve the production of individual structural elements, which are then assembled to form a larger product. It is also known as cultured meat or in vitro meat and considered as a safe and disease-free way forward to meet increasing meat requirement without involving animal sacrifices and at the same

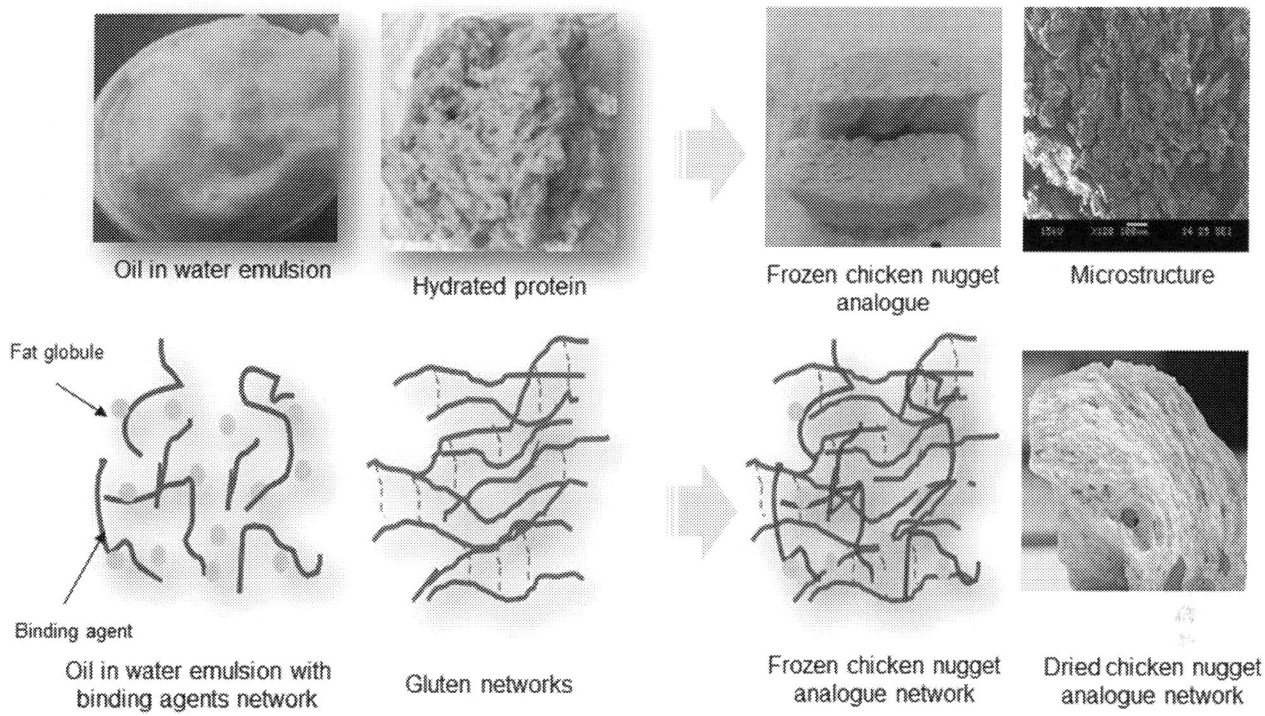

FIGURE 9.3 Schematic illustration of the mechanical elongation method to produce chicken nugget analogues produced by freezing structuring technique. *Adapted from Yuliarti, O., Kiat Kovis, T. J., & Yi, N. J. (2021). Structuring the meat analogue by using plant-based derived composites.* Journal of Food Engineering, 288. *https://doi.org/10.1016/j.jfoodeng.2020.110138.*

time, reducing greenhouse emissions (Gaydhane et al., 2018). However, as reported by these authors, this method has limited commercial feasibility due it is costly, difficult to scale-up, public neophobia and technophobia, and incomplete understanding of health benefits.

This method utilizes tissue-engineering techniques to facilitate meat production in vitro, and the study in this area is still in its initial stages (Froggatt & Wellesley, 2019; Post, 2012). As shown in Fig. 9.4, to culture meat fibers, myoblast cells (muscle cells that develop into muscle fibers) are harvested and isolated from the skeletal muscle of animals. These myoblast cells are then grown in a nutrient-rich controlled medium, to facilitate cell division and cell growth. Once a sufficient number of cells have been obtained, they are then attached to a scaffold/supporting structure with pre-existing anchor points that allow for alignment and connection between myoblast cells with the support of electric fields, forming a multicellular tissue. After a maturation period of 3 weeks, the muscle fibers can then be harvested and used to make meat or meat products such as hamburger, sausages, and other processed meats (Langelaan et al., 2010).

9.2.5 Electrospinning

This technique is regarded as a sensitive and expensive process, which requires high concentration of acid/alkaline solvents to produce thin fibers, in which a concentrated high plant protein is spun at high speeds to produce simulated meat (Singh et al., 2021). In electrospinning, technically a plant-based biopolymer/protein solution is extruded through a hollow needle or a spinneret that has an electric potential relative to a ground electrode, with the solution accumulating charge along the process (Fig. 9.5). The resultant attraction of the solution to the ground electrode ultimately leads to the formation of extremely thin protein fibers (Schiffman & Schauer, 2008). The solvent will evaporate and is collected on the collector, which is connected to the ground electrode (Nieuwland et al., 2013).

Despite its potential in meat analogue application, electrospinning requires highly soluble, adequately concentrated, and randomly coiled proteins for efficient usage (Dekkers et al., 2016). As the plant proteins are usually globular in their native state, these authors suggested the protein to be unfolded by heating prior to electrospinning to produce thin fibers and to avoid the formation of insoluble aggregates. To date, zein was the only plant-based protein that reported (Nieuwland et al., 2013) to produce thin fibrils with the help of non-plant-based protein such gelatine to yield regular fibrils of homogeneous thickness that mimic animal meat texture.

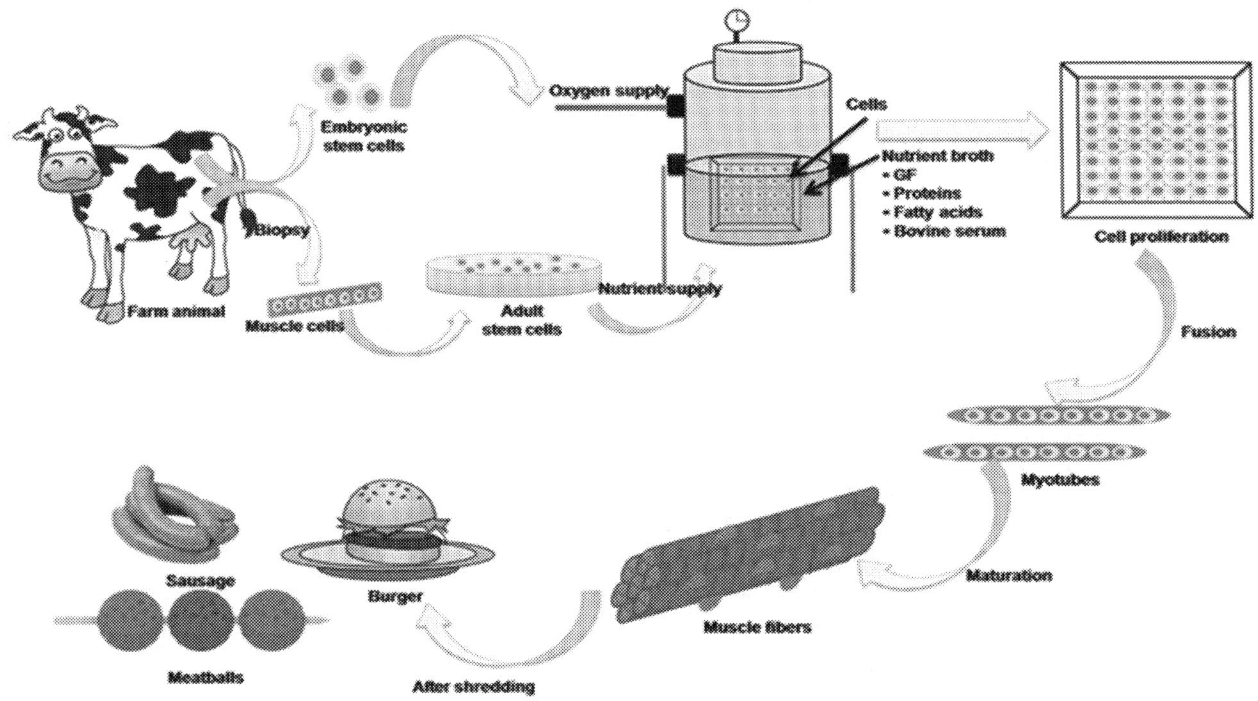

FIGURE 9.4 Illustration of cultured meat production. *From Zhang, L., Hu, Y., Badar, I. H., Xia, X., Kong, B., & Chen, Q. (2021). Prospects of artificial meat: Opportunities and challenges around consumer acceptance.* Trends in Food Science and Technology, 116, *434–444. https://doi.org/10.1016/j.tifs.2021.07.010.*

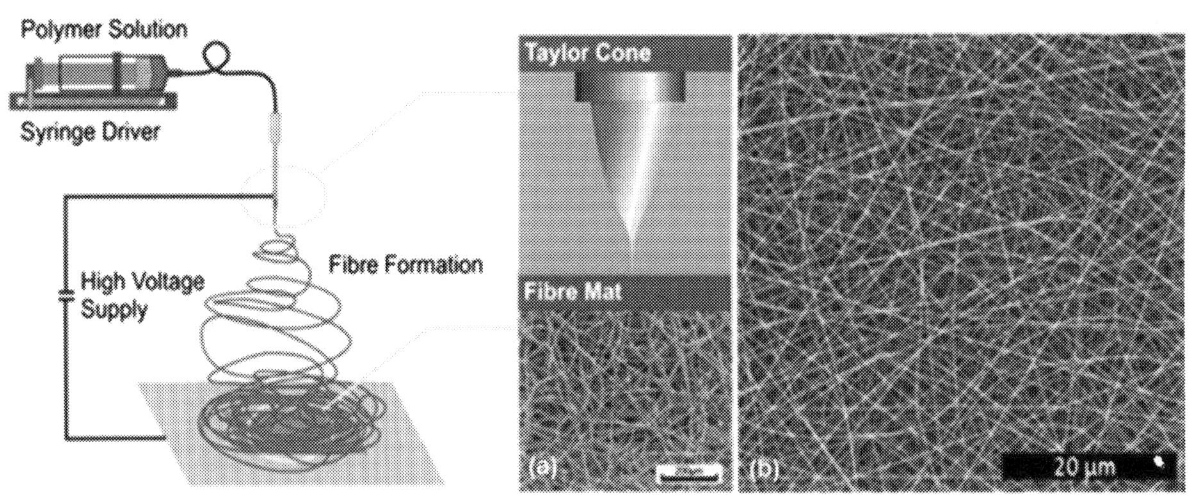

FIGURE 9.5 (A) Schematic illustration of principle and electrospinning setup to produce thin fibrils of meat analogues and (B) scanning electron microscopy (SEM) pictures of zein fine fibers. *Adapted from Nieuwland, M., Geerdink, P., Brier, P., Van Den Eijnden, P., Henket, J. T. M. M., Langelaan, M. L. P., Stroeks, N., Van Deventer, H. C., & Martin, A. H. (2013). Food-grade electrospinning of proteins.* Innovative Food Science and Emerging Technologies, 20, *269–275. https://doi.org/10.1016/j.ifset.2013.09.004.*

9.2.6 3D printing

The 3D printing (3DP) has been gaining attention as an emerging food processing technology recently, and many studies have been focusing on incorporation of 3D printing to create innovative food products to meet the consumer needs. It is a sustainable and promising novel technology in meat analogue development where it involves digital modeling of food formulations and texture to mimic a very complex animal meat matrix (Ramachandraiah, 2021; Singh et al., 2021).

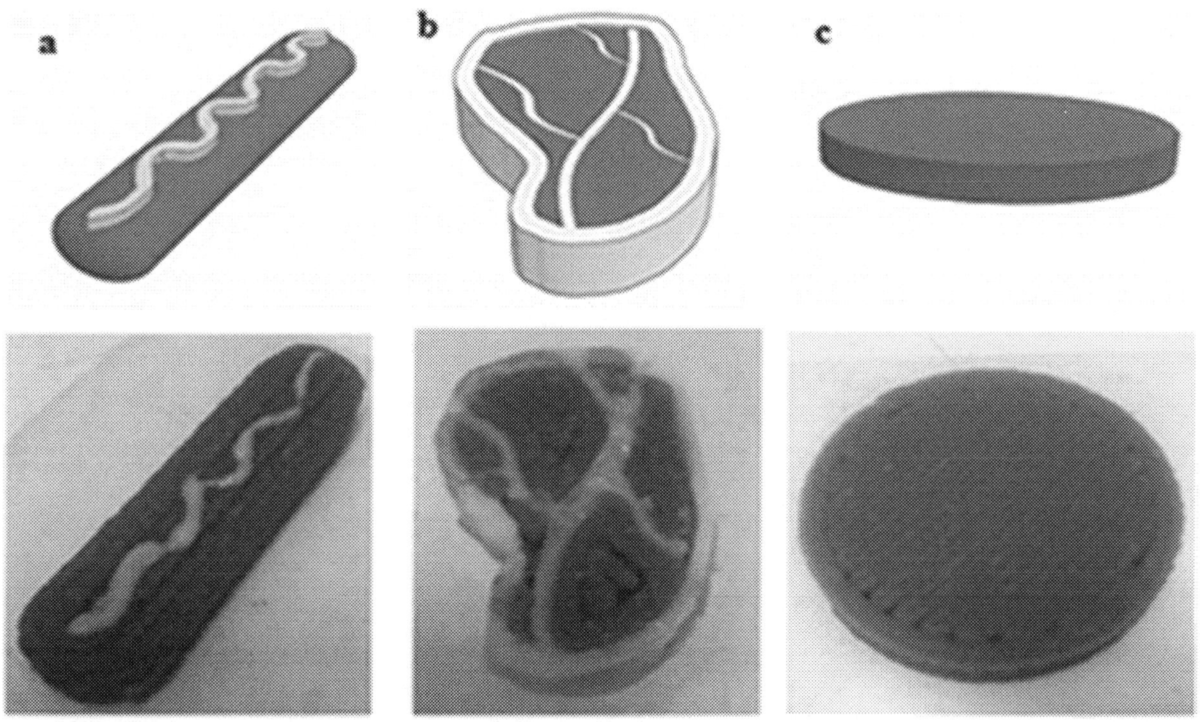

FIGURE 9.6 Hypothetical food designs for age care homes: (A) sausage, (B) steak "recombined meat," and (C) patty. *From Dick, A., Bhandari, B., & Prakash, S. (2019). 3D printing of meat.* Meat Science, 153, *35–44. https://doi.org/10.1016/j.meatsci.2019.03.005.*

However, this method is considered difficult and challenging as it is strongly dependent on the material properties and binding mechanisms (Godoi et al., 2016). A study (Lipton et al., 2010) suggested that optimizing the particle size of the materials, that is, particle size reduction, as well as addition of plasticiser such as plant-based gelling agents in the formulation can be done to improve the material printability.

For the plant-based meat analogue, the meat products are typically printed by an extrusion process, in which fibrous plant-based meat liquid materials are extruded in a layer-by-layer from a printing nozzle in order to form 3D structures (Ramachandraiah, 2021). Fig. 9.6 illustrates three 3D-printed designs of recombined meat products, such as sausage, steaks where the liquid printing materials are customized to suit the texture, flavors, and nutrient of respective products (Dick et al., 2019).

9.3 Plant-based meat ingredients

9.3.1 Protein source

The protein source is the key ingredient in the production of meat analogues. It is not only to supplement protein content but also sought after in the meat replacers or alternatives. The type of the protein is largely responsible in determining the texture and structure of meat analogues (Egbert & Borders, 2006). Different proteins could have distinctive structure–function interactions in terms of their hydration and solubility, gelation, and texture formation, which govern the meat analogue final structure (Meade et al., 2005). Typical plant-based sources of proteins incorporated into meat analogues include soy, pea, wheat, and fungi protein (Joshi & Kumar, 2015). Out of all the plant-based proteins, soybean and wheat proteins are the most commonly used proteins in the food industry due to their suitable rheological and viscoelastic properties while considering their availability, cost, and processing functionality (Sha & Xiong, 2020; Yuliarti et al., 2021).

Table 9.2 shows an overview of the selected commercialized meat analogues using plant proteins currently available in the market, pioneering to mimic the sensorial attributes of meat. They are mostly in the form of burger and chicken nuggets, as they are considered as the most popular ready to eat meat products due to the conveniences and widely used as a Ready-to-Eat Meal (Heck et al., 2017). Soy, pea, and wheat proteins are the most commonly used proteins in their formulation as they are economical and superior due to the presence of high amounts of essential nutrients and possess properties that promote the formation of meat-like structures (Asgar et al., 2010). Among the wide variety of meat

TABLE 9.2 Commercialized beef and chicken meat analogues made by using various plant protein sources.

Product name	Type of meat product simulated	Type of plant-based proteins	Protein content in NIP (%)	Website
Beyond Burger	Beef burger	Pea protein isolate, rice protein and mung bean protein	17.70 (113 g serving size)	https://www.beyondmeat.com/products/the-beyond-burger
Impossible burger	Beef burger	Soy protein concentrate, soy protein isolate and potato protein	16.80 (113 g serving size)	https://impossiblefoods.com/products
MorningStar farms grillers original burger	Beef burger	Wheat protein, soy protein, and hydrolyzed soy protein	16.60 (120 g serving size)	https://www.incogmeato.com/en_US/home.html#products
BOCA Original Vegan Veggie Burgers	Beef burger	Wheat protein and soy protein isolate	18.30 (71 g serving size)	https://www.bocaburger.com/products
Gardein Meatless Meat balls	Beef Meat balls	Textured vegetable protein, vital wheat gluten, soy protein isolate and pea protein	16.67 (90 g serving size)	https://gardein.com/beefless-and-porkless/classics/meatless-meatballs
Quorn chicken nuggets	Quorn Meatless Nuggets	Mycoprotein, textured wheat protein and wheat protein	11.0 (118 g serving size)	https://www.quorn.us/products/quorn-meatless-chicken-nuggets
Rebellyous Foods	Rebellyous Nuggets	Soy Protein Concentrate, textured wheat protein, isolated soy protein	14.10 (85 g serving size)	https://rebellyous.com/products/nuggets/
Field Roast	Burger	Vital wheat gluten, wheat protein isolate	22.8 (92 g serving size)	https://fieldroast.com/product/fieldburger/

analogues in the market, Quorn meat analogue is a novel plant-based chicken nugget, where mycoprotein is used as one of the protein sources. Mycoprotein is fungus obtained from a fermentation technology to create a fibrous texture that resembles meat (Ismail et al., 2020). The protein content has been reported to be comparable to animal-derived sources of protein with a Protein Digestibility-Corrected Amino Acid Score (PDCAAS) of 1.00 (Joshi & Kumar, 2015). The focus of the section will be on four specific protein groups, oilseed proteins, legume proteins, cereal protein and microalgae proteins as they are widely used in the manufacturing meat analogues. In addition, the specific functional properties of protein are summarized in Table 9.3.

9.3.1.1 Oil seed proteins

Oilseeds are oil-producing plants such as soybean, rapeseed, cotton seeds, etc., which have been found to show good potential for utilization in meat analogues due to their relatively high in protein content, various health benefits, and good gelling properties. Soybean is regarded as an economic, high-quality protein source for humans, and is regularly touted for its health benefits and excellent gelling properties (Asgar et al., 2010). Soybeans have two main protein fractions, β-conglycinin and glycinin, which contain eight essential amino acids at even greater quantities than in animal products (Hidayat et al., 2011). According to Protein Digestibility-Corrected Amino Acid Score (PDCAAS) studies, soybean protein had been evaluated to be comparable to animal proteins, with a PDCAAS score of 0.95−1.00, almost identical to that of animal proteins (Singh et al., 2008). Despite this, soybean protein suffers from low-levels of sulfur containing amino acids such as cysteine and methionine.

In terms of health benefits, increased soybean consumption is associated with a lowered risk of coronary heart disease as well reduced low-density lipoprotein (LDL) cholesterol concentrations (Sacks et al., 2006). Soy protein is the predominant protein used in the alternative meat industry, largely due to its favorable gelling properties and its ability to produce highly interconnected fibrous matrices (Banerjee & Bhattacharya, 2012). This capability stems from the gelling and emulsification properties of the protein fractions (Table 9.3), which thus enables its use in a variety of different applications.

Several soy-based ingredients such as soy protein isolate (SPI), textured soy protein (TSP), and soy protein concentrate (SPC) have been utilized in producing meat analogues as shown in Table 9.3. However, to achieve the desired texture and

TABLE 9.3 Summary of protein types and the functional properties.

Protein types	Composition (%w/w)	Functionality and meat analogue types
Soy protein isolate	~90% protein	Good solubility, gelling, and emulsification properties Products: Burger patties, minced meat, sausages
Soy protein concentrate	~70% protein	Good texturization properties. Products: Burger patties, mincedmeat, sausages, muscle-type products
Soy flour (defatted)	~43%–56% protein	Water binding capacity and fat retention. Products: Burger patties, minced meat, sausages, muscle-type products
Pea protein isolate	~85% protein	Water and fat binding, emulsification, and firm texture after thermal processing Products: Burger patties, minced meat, sausages, muscle-type products
Wheat gluten	~75%–80% protein	Binding, dough forming, cross-linking capacity via S-S bridges, low solubility Products: Burger patties, muscle-type products
Microalgae protein	~58%–69% protein	Good solubility, emulsification and nutritional properties. Contributes to product color. Product: Burger patties, nuggets, and muscle type products

Adapted from Kyriakopoulou, K., Keppler, J. K., & van der Goot, A. J. (2021). Functionality of ingredients and additives in plant-based meat analogues. *Foods, 10*(3). https://doi.org/10.3390/foods10030600; Böcker, L., Bertsch, P., Wenner, D., Teixeira, S., Bergfreund, J., Eder, S., Fischer, P., & Mathys, A. (2021). Effect of Arthrospira platensis microalgae protein purification on emulsification mechanism and efficiency. *Journal of Colloid and Interface Science, 584*, 344–353. https://doi.org/10.1016/j.jcis.2020.09.067.

nutritional benefits, the use of soy protein combinations is recommended than a single type of soy protein. A study (Malav et al., 2015) reported that a combination of textured and nontextured soy protein could enhance the meat-like texture, its amino acid composition, and the flavor retention. The use of wheat gluten and SPI combination had produced satisfactory meat analogue with fibrillar structure when formed using the mechanical elongation method (Chiang et al., 2021). The use of wheat gluten in the wheat gluten-SPI blend had enhanced the elastic behavior of the meat analogue as SPI was responsible for a brittle and resilient gel.

In addition to soy protein, tempeh, a fermented soybean product, is a functional food that is a plant-based protein food and recently has gained popularity in Asia and worldwide (Nicole et al., 2021). This is due to it having a complete source of protein and contains vitamin B12 (Erkan et al., 2020). Moreover, due to its versatility, nowadays, this fermented soybean has been incorporated extensively as a plant-based protein for vegan, and many other types of modern dishes such as chips, meat balls, and burgers (Yuliarti, 2020).

9.3.1.2 Legume proteins

Legume proteins have risen in popularity in recent years due to their high nutritive quality, good techno-functional properties, and low cost. This has led to an increase in conducted research on the nutritional and functional characteristics of legume proteins. A few examples of increasingly utilized proteins are pea and chickpea proteins. Pea protein in particular is seeing an increase in its use as a replacement or an alternative to soy protein in meat analogue production. Pea protein has a similar nutritional quality to soy protein, is not listed as allergenic, contains less antinutrients such as proteinase inhibitors, and contain good quality starch and fibers (Lam et al., 2018). However, the gelling capability of pea proteins is not comparable to that of soy proteins, as it produces weaker and less elastic gels (Bildstein et al., 2008). The main protein fractions of pea proteins are Legumin, Vicilin, and Convicilin, each with their own specific nutritional, physicochemical, and functional properties (Barac et al., 2015). In the case of chickpea protein, research has shown that it shows good potential for use in the production of meat analogues. Studies have indicated good nutritional properties, minimal off-flavor, and the ability to form stable emulsions and gels (Jones, 2016).

9.3.1.3 Wheat protein

Gluten is the main protein fraction found in cereals such as wheat and rye. It is highly utilized in the food industry and is regarded as key elements in manufacturing meat analogue due to its optimal rheological and viscoelastic properties (Agyare et al., 2007). Gluten makes up 80%–85% of the total wheat protein and consists of monomeric gliadins and polymeric glutenin proteins. Gluten proteins determine the viscoelastic properties of the protein as it is able to form a network upon hydrating and mixing. Gluten functionality is determined by the ratio of glutenin: gliadins (Kyriakopoulou et al., 2021). Glutenin proteins build up the polymeric protein network that provides cohesiveness and elasticity to mixture, whereas gliadins act as plasticisers of the glutenin network and contribute to dough viscosity and extensibility (Ooms & Delcour, 2019).

Gluten is mainly used as a meat extender or as a structurant in protein blends due to its strength and viscoelastic capabilities, which allows for unique textures to be achieved (Grabowska et al., 2014); therefore, it enhances overall texture and mouth-feel of meat products. During hydration, gluten forms thin protein films upon simple deformation and elongation, transforming the meat analogue dough into a fibrous material (Kyriakopoulou et al., 2021). However, one of the main setbacks of wheat gluten is its insolubility in water, which greatly limits its potential and functionality as a protein source for meat alternatives (Kong et al., 2007).

9.3.1.4 Microalgae protein

Microalgae have demonstrated potential to be a protein source because their protein content can be up to 71%, where Arthrospira sp. and Chlorella sp. are currently the most commonly sold microalgae in the market with the protein content of over 50% (Becker, 2007). Microalgae are biologically a more efficient source for protein production as it have low land requirements compared to animal-based proteins and the protein content exceeds the protein content of soybean 30%–40% protein content in soybeans (Caporgno et al., 2020). In term of amino acid composition, microalgae is superior compared to other sources and contains balanced amounts of essential amino acids (EAAs) (Fu et al., 2021).

Microalgae proteins have outstanding solubility properties; hence, the protein exhibits excellent functional properties such as gelation, emulsification, and foaming (Böcker et al., 2021; Grossmann et al., 2020). Therefore, it can be utilized in a wide range of meat analogue applications, for example to supplement the extrusion materials for texturizing meat analogue. Incorporation of microalgae in high moisture extrusion had increased the tenderness and improves the nutritional profiles of the soy protein based-meat analogue (Caporgno et al., 2020). Although, it has demonstrated as a suitable protein source for meat analogues, the production of microalgae does however present some drawbacks where the cultivation and processing is still economically expensive (Caporgno et al., 2020; Mattice & Marangoni, 2020).

9.3.2 Binding agents

The current formulation of meat analogue involves ingredients, which act as binding agents to develop a cohesive network structure. These binding agents are normally obtained from plant origin, which can aid the manufacturing process of the analogue, i.e., to improve the interaction of protein, water, and lipid in the system (Bohrer, 2019), therefore, can lead to a better product mouthfeel and juiciness. They are grouped into starches, to improve product texture and consistency, and binding agents to improve the stability and cohesiveness of the products such as guar gum, methylcellulose, xanthan gums, zein, alginates, and others (Bohrer, 2019; Singh et al., 2021). These polysaccharides play a crucial role as a textural and structural component and can be incorporated before or during the thermomechanical processing. They improve the water-binding capacity of meat analogues as well as the rheological properties of mixture, as was observed in the case of soy-/pectin-based meat analogues produced in the shear cell (Dekkers et al., 2018; Kyriakopoulou et al., 2021).

Methylcelluose or modified cellulose dietary fiber (Ashland, 2019; Yuliarti et al., 2021) is known to be an effective binder because the gel form can be thermally induced at lower temperature, hence, can maintain the firmness of the plant-based chicken nuggets at lower temperature. The incorporation of methylcellulose into a protein mixture could results in hydrocolloid–protein interactions therefore strengthen the product structure. In addition, it has a good binding ability and moisture retention ability. During its application, as more water can be added, a lesser amount of fat is required resulting in healthier meat analogue products.

Konjac glucomannan (KG), a polysaccharide produced by Amorphophallus konjac plant, offers a great potential for application in food technology, due to it is ability to perform as a binding ingredient and could improve the meat analogue stability. It comprises of β –1, 4-linked glucose and mannose units, with the glucomannan backbone of KG consists 5%–10% acetyl-substituted residues (Katsuraya et al., 2003). The KG was used as fat analogues in the formulation of reduced/low-fat animal meat products such as frankfurters and sausages due to the gel forms could mimic animal fat textural properties (Jiménez-Colmenero et al., 2012). Konjac gel was further reported by these authors to be thermally stable and exhibits mainly elastic during heating, similar behaviors as animal fat.

9.3.3 Miscellaneous plant-based ingredients

9.3.3.1 *Agricultural wastes*

The effort to use agricultural wastes as food ingredients to create plant-based meat products is considered systematic and sustainable ways as the problem of food wastage is rapidly increasing (Smetana et al., 2018, pp. 145–155). Many of these agricultural wastes such as okara, spent grains, potato peels etc. are a valuable source of dietary fibers, proteins, lipids, starch, and micronutrients. Okara, obtained from soybeans during the production of soymilk of tofu, is known for its high fiber, protein content, and isoflavones (Li et al., 2013). In addition, brewer's spent grains, derived from the brewing industry, are rich in fibers, lipid, and protein (Rachwał et al., 2020).

Despite being rich in the nutritional properties, direct incorporation of these by-products as food ingredients into food products could face several limitations and strongly influence the physicochemical and sensorial attributes of the foods. For instance, okara has been known for the presence of enzyme inhibitors and contributes to "beany" flavor (Vong & Liu, 2016). Therefore, pretreatments such as drying, heat treatment, or reduction of particle size are required to improve their functional and nutritional characteristics. There are many studies that have been conducted in this area, for example, the use of okara in the production of sausages (Su et al., 2013), peanut protein biomass in high-moisture extrusion process of fibrous meat analogue (Zhang et al., 2019) and incorporation of spent grains in beef frankfurters. Incorporation of these by products in the analogue have strongly imparted the product water holding capacity and the textural properties (Su et al., 2013; Zhang et al., 2019).

9.3.3.2 *Color and flavor*

Color is the most important physical attribute that influences consumer choice of meat analogue (Bohrer, 2019). The meat analogue should mimic the original color of the animal meat products that they are simulating before, during, and after cooking; therefore, additional of desired color are required (Kyriakopoulou et al., 2019). Heat stable coloring agents are normally added to obtain a bloody color of fresh meat or match the Maillard effect (browning) of cooked meat. The commercial products such as Beyond burger- and MorningStar Farm burger-incorporated beet juice extract and tomato paste as the color ingredients, respectively (Bohrer, 2019).

In order to enhance the consumer plant-based eating experience, intended final product flavor is added in the formulation. Addition of flavor ingredients is also envisioned to increase the acceptability of the products as the plant protein ingredients use could impede several issues of their off-flavor and off-taste (Kyriakopoulou et al., 2021). This off-flavor was attributed to the presence of volatile off-flavor compounds, such as aldehydes, alcohols, ketones, acids, pyrazines, sulfur compounds, and for the off-taste is strongly correlated to the presence of saponins, phenolic compounds, and sometimes alkaloids (Ismail et al., 2020). Addition of salt as sodium chloride is not only enhancing the overall analogue flavor and taste, but more importantly, it could alter the structure of proteins as it solubilizes and unfolds the protein, which leads to strengthen the protein network structure (Kyriakopoulou et al., 2021).

9.4 Processing factors

In this section, some crucial factors influencing the structuring meat analogue such as processing temperature, protein combination, and pH are taken into consideration and discussed briefly.

9.4.1 Temperature

The processing temperature can greatly affect the structure and eating quality of meat analogues. Heat is used to denature the vegetable proteins and facilitate cross-linking forming intra- and intermolecular bonds between protein molecules that stabilize the protein network structure. Hydrogen bonds are weakened as the temperature increases, while hydrophobic interactions can increase with temperature (Cordier & Grzesiek, 2002; Cornet et al., 2020). Furthermore, high heat treatment is used to set and coagulate the vegetable proteins in the meat analogue to form the fibrous structure. For example, when wheat proteins are heated at above 85°C, hydrated gluten will coagulate irreversibly, yielding a firm, nonsticky, moist, and resilient gel without losing the original structure (Kalin, 1979). Its adhesive and film forming property binds the system's particles together resulting in a distinctive structure.

During structuring of fibrous protein using thermomechanical techniques such as high-moisture extrusion cooking (HMEC) and shear cell technologies, the heating treatment steps are applied simultaneously with shear at elevated pressure. This combination of treatments is crucial as applying heat or shear alone does not result in a fibrous structure (Cornet et al., 2020). In addition to high temperature, these authors also reported that cooling steps are also is highly considered as the "structuring zone" where the extruded protein is cooled to a temperature of approximately 50°C die in order to reduce product expansion and to induce possible fiber formation (Caporgno & Mathys, 2018). Studies reported (Cornet et al., 2020; Lin & Huang, 2008; Osen et al., 2014), that lowering temperature in the cooling stage can increase protein—protein interactions and crosslinking thereby strengthening of the noncovalent interactions of the melt protein.

In the freeze structuring technique, freezing temperature is crucial to develop a sheet-like parallel orientation of the proteins. The size of the ice crystal needles can be modified by varying the freezing temperature and rate and therefore to obtain distinct fibrous structure, the proteins should have relatively good solubility prior to freezing, as during the freezing process, these proteins become solidify (Dekkers et al., 2018; Lugay & Kim, 1978).

9.4.2 Composite of proteins

Research has been done to evaluate the effect of incorporating different proteins at different ratios on the textural and sensorial properties of meat analogues to acquire desirable functional attributes. This is because of different vegetable proteins have different physicochemical properties such as color, water holding capacity, and gelling properties as well as their bond formation capabilities (Schreuders et al., 2019). Table 9.1 indicates the functional properties of the protein and the analogue types. Modifying the functional properties of the protein, by combining protein ingredients as a composite protein, could enhance overall protein functionality and therefore can improve the final structure and textural properties of meat analogues (Beniwal et al., 2021; Schreuders et al., 2019).

A study on the effects of soy protein to wheat gluten ratio found that the incorporation of higher amounts of wheat gluten resulted in the development of a more fibrous structure with greater degree of molecular interlinkage and bond formation between protein chains. Wheat gluten had played an essential role in the development of disulfide bonds, which was the key force in the formation of a fibrous structure in high moisture extrusion (Chiang et al., 2019). Incorporation of pea protein in wheat gluten plant-based nuggets appeared to increase the sensorial qualities and acceptance of the analogue. The use of wheat protein alone decreases the overall nuggets elasticity (Yuliarti et al., 2021).

9.4.3 pH

pH is commonly associated with the net charge value of protein, i.e., proteins are poly-ampholytes; they can carry both positive and negative charges; hence protein stability is strongly pH dependence (Klost & Drusch, 2019; Yang & Honig, 1993). In the structuring of meat analogues, the formulation pH can affect the texture of analogues. The fibrous structure of meat analogues is obtained by the cross-linking and bond formation between protein chains, which forms the protein gel structure. In order to facilitate cross-linking and bond formation between protein chains, a relatively high pH is required in order to denature and unfold the globular vegetable proteins into a random protein coil arrangement. Therefore, the initial pH of the protein blend/emulsion can greatly influence the obtained structure of the meat analogue (Consolacion & Jelen, 1986).

Alkaline conditions in the thermomechanical processing were reported to increase intermolecular association of the proteins, which strengthen the formation of the gluten network and textural characteristics of the gluten extrudates (Li et al., 2018; Shiau & Yeh, 2001). It was possible that protein in the analogues with a higher pH have higher overall negative charge groups hence exhibited a higher intramolecular repulsion, which contributed to an increase in gel strength of the meat analogue (Yuliarti et al., 2021).

9.5 Summary and future outlook

The market of plant-based meat products is growing rapidly as the demand of these products upsurge steadily. This has brought about the development of new products that utilizes a wider range of ingredients that is able to appeal to a larger consumer base, which has led to an increase in the availability of meat alternatives in the market (Sadler, 2004). Food manufacturers face the challenge of providing healthier and economical food products while at the same time ensuring that the product has an appealing taste and texture (Asgar et al., 2010). Therefore, many manufacturers and research studies are not only focusing on the development of plant-based meat product varieties but also toward improving the quality by diversifying the plant-based protein alternative sources and techniques.

Important to highlight also that the consumer's demand of healthier plant-based proteins has been increasing, as growing concern of the ultra-process nature of current plant-based alternative proteins in the market. The next generation of plant-based protein sources with a lesser ultra-process is a challenge that industry needs to address and more research is required to overcome this challenge.

Acknowledgment

Authors would like to thank Ms Toh Hui Kheng for her assistance in this chapter work.

References

Agyare, K., Xiong, Y., & Addo, K. (2007). Influence of salt and pH on the solubility and structural characteristics of transglutaminase-treated wheat gluten hydrolysate. *Food Chemistry, 107*(3). https://doi.org/10.1016/j.foodchem.2007.09.039

Asgar, M. A., Fazilah, A., Huda, N., Bhat, R., & Karim, A. A. (2010). Nonmeat protein alternatives as meat extenders and meat analogs. *Comprehensive Reviews in Food Science and Food Safety, 9*(5), 513–529. https://doi.org/10.1111/j.1541-4337.2010.00124.x

Ashland. (2019). *Benecel MX modified cellulose for plant based foods.* Ashland Asia Pacific — Singapore.

Banerjee, S., & Bhattacharya, S. (2012). Food gels: Gelling process and new applications. *Critical Reviews in Food Science and Nutrition, 52*(4), 334–346. https://doi.org/10.1080/10408398.2010.500234

Barac, M. B., Pesic, M. B., Stanojevic, S. P., Kostic, A. Z., & Bivolarevic, V. (2015). Comparative study of the functional properties of three legume seed isolates: Adzuki, pea and soy bean. *Journal of Food Science and Technology, 52*(5), 2779–2787. https://doi.org/10.1007/s13197-014-1298-6

Becker, E. W. (2007). Micro-algae as a source of protein. *Biotechnology Advances, 25*(2), 207–210. https://doi.org/10.1016/j.biotechadv.2006.11.002

Beniwal, A. S., Singh, J., Kaur, L., Hardacre, A., & Singh, H. (2021). Meat analogs: Protein restructuring during thermomechanical processing. *Comprehensive Reviews in Food Science and Food Safety, 20*(2), 1221–1249. https://doi.org/10.1111/1541-4337.12721

Benjaminson, M., Gilchriest, J., & Lorenz, M. (2002). In vitro edible muscle protein production system (mpps): stage 1, fish. *Acta Astronautica, 51*(12), 879–889. https://doi.org/:10.1016/S0094-5765(02)00033-4

Bildstein, M., Lohmann, M., Hennigs, C., Krause, A., & Hilz, H. (2008). An enzyme-based extraction process for the purification and enrichment of vegetable proteins to be applied in bakery products. *European Food Research and Technology, 228*(2), 177–186. https://doi.org/10.1007/s00217-008-0921-z

Böcker, L., Bertsch, P., Wenner, D., Teixeira, S., Bergfreund, J., Eder, S., Fischer, P., & Mathys, A. (2021). Effect of Arthrospira platensis microalgae protein purification on emulsification mechanism and efficiency. *Journal of Colloid and Interface Science, 584*, 344–353. https://doi.org/10.1016/j.jcis.2020.09.067

Bohrer, B. M. (2019). An investigation of the formulation and nutritional composition of modern meat analogue products. *Food Science and Human Wellness, 8*(4), 320–329. https://doi.org/10.1016/j.fshw.2019.11.006

Caporgno, M. P., Böcker, L., Müssner, C., Stirnemann, E., Haberkorn, I., Adelmann, H., Handschin, S., Windhab, E. J., & Mathys, A. (2020). Extruded meat analogues based on yellow, heterotrophically cultivated Auxenochlorella protothecoides microalgae. *Innovative Food Science and Emerging Technologies, 59*, 102275. https://doi.org/10.1016/j.ifset.2019.102275

Caporgno, M. P., & Mathys, A. (2018). Trends in microalgae incorporation into innovative food products with potential health benefits. *Frontiers in Nutrition, 5*. https://doi.org/10.3389/fnut.2018.00058

Cheftel, J. C., Kitagawa, M., & Quéguiner, C. (1992). New protein texturization processes by extrusion cooking at high moisture levels. *Food Reviews International, 8*(2), 235–275. https://doi.org/10.1080/87559129209540940

Chiang, J. H., Loveday, S. M., Hardacre, A. K., & Parker, M. E. (2019). Effects of soy protein to wheat gluten ratio on the physicochemical properties of extruded meat analogues. *Food Structure, 19*. https://doi.org/10.1016/j.foostr.2018.11.002

Chiang, J. H., Tay, W., Ong, D. S. M., Liebl, D., Ng, C. P., & Henry, C. J. (2021). Physicochemical, textural and structural characteristics of wheat gluten-soy protein composited meat analogues prepared with the mechanical elongation method. *Food Structure, 28*. https://doi.org/10.1016/j.foostr.2021.100183

Consolacion, F., & Jelen, P. (1986). Freeze texturation of proteins: Effect of the alkali, acid and freezing treatments on texture formation. *Food Structure, 5*.

Cordier, F., & Grzesiek, S. (2002). Temperature-dependence of protein hydrogen bond properties as studied by high-resolution NMR. *Journal of Molecular Biology, 317*(5), 739–752. https://doi.org/10.1006/jmbi.2002.5446

Cornet, S. H. V., Snel, S. J. E., Schreuders, F. K. G., van der Sman, R. G. M., Beyrer, M., & van der Goot, A. J. (2020). Thermo-mechanical processing of plant proteins using shear cell and high-moisture extrusion cooking. *Critical Reviews in Food Science and Nutrition*. https://doi.org/10.1080/10408398.2020.1864618

Dekkers, B. L., Boom, R. M., & van der Goot, A. J. (2018). Structuring processes for meat analogues. *Trends in Food Science and Technology, 81*, 25–36. https://doi.org/10.1016/j.tifs.2018.08.011

Dekkers, B. L., Nikiforidis, C. V., & van der Goot, A. J. (2016). Shear-induced fibrous structure formation from a pectin/SPI blend. *Innovative Food Science and Emerging Technologies, 36*, 193–200. https://doi.org/10.1016/j.ifset.2016.07.003

Dick, A., Bhandari, B., & Prakash, S. (2019). 3D printing of meat. *Meat Science, 153*, 35–44. https://doi.org/10.1016/j.meatsci.2019.03.005

Egbert, R., & Borders, C. (2006). Achieving success with meat analogs. *Food Technology, 60*(1), 28–34.

Erkan, S. B., Gürler, H. N., Bilgin, D. G., Germec, M., & Turhan, I. (2020). Production and characterization of tempehs from different sources of legume by Rhizopus oligosporus. *LWT, 119*. https://doi.org/10.1016/j.lwt.2019.108880

Freeman. (2011). *Green food: An a-to-z guide*. SAGE Publications, Inc.

Froggatt, A., & Wellesley, L. (2019). *Meat analogues: Considerations for the EU.*

Fu, Y., Chen, T., Chen, S. H. Y., Liu, B., Sun, P., Sun, H., & Chen, F. (2021). The potentials and challenges of using microalgae as an ingredient to produce meat analogues. *Trends in Food Science and Technology, 112*, 188–200. https://doi.org/10.1016/j.tifs.2021.03.050

Gaydhane, M. K., Mahanta, U., Sharma, C. S., Khandelwal, M., & Ramakrishna, S. (2018). Cultured meat: State of the art and future. *Biomanufacturing Reviews, 3*(1). https://doi.org/10.1007/s40898-018-0005-1

Godoi, F. C., Prakash, S., & Bhandari, B. R. (2016). 3d printing technologies applied for food design: Status and prospects. *Journal of Food Engineering, 179*, 44–54. https://doi.org/10.1016/j.jfoodeng.2016.01.025

Gonzalez, J., Drouillard, J., O'Quinn, T., & Phelps, K. (2018). 493 the influence of muscle fiber structure on meat palatability. *Journal of Animal Science, 96*, 273–274. https://doi.org/10.1093/jas/sky404.600. Suppl. 1_3.

Grabowska, K. J., Tekidou, S., Boom, R. M., & van der Goot, A. J. (2014). Shear structuring as a new method to make anisotropic structures from soy-gluten blends. *Food Research International, 64*, 743–751. https://doi.org/10.1016/j.foodres.2014.08.010

Grossmann, L., Hinrichs, J., & Weiss, J. (2020). Cultivation and downstream processing of microalgae and cyanobacteria to generate protein-based technofunctional food ingredients. *Critical Reviews in Food Science and Nutrition, 60*(17), 2961–2989. https://doi.org/10.1080/10408398.2019.1672137

Guan, J., Porter, D., Tian, K., Shao, Z., & Chen, X. (2010). Morphology and mechanical properties of soy protein scaffolds made by directional freezing. *Journal of Applied Polymer Science*, n/a-n/a. https://doi.org/10.1002/app.32579

Hallström, E., Röös, E., & Börjesson, P. (2014). Sustainable meat consumption: A quantitative analysis of nutritional intake, greenhouse gas emissions and land use from a Swedish perspective. *Food Policy, 47*, 81–90. https://doi.org/10.1016/j.foodpol.2014.04.002

Heck, R. T., Vendruscolo, R. G., de Araújo Etchepare, M., Cichoski, A. J., de Menezes, C. R., Barin, J. S., Lorenzo, J. M., Wagner, R., & Campagnol, P. C. B. (2017). Is it possible to produce a low-fat burger with a healthy n − 6/n − 3 PUFA ratio without affecting the technological and sensory properties? *Meat Science, 130*, 16–25. https://doi.org/10.1016/j.meatsci.2017.03.010

Hidayat, M., Sujatno, M., Sutadipura, N., & Faried, A. (2011). β-Conglycinin content obtained from two soybean varieties using different preparation and extraction methods. *HAYATI Journal of Biosciences, 18*(1), 37–42. https://doi.org/10.4308/hjb.18.1.37

Ismail, I., Hwang, Y. H., & Joo, S. T. (2020). Low-temperature and long-time heating regimes on non-volatile compound and taste traits of beef assessed by the electronic tongue system. *Food Chemistry, 320*. https://doi.org/10.1016/j.foodchem.2020.126656

Jiménez-Colmenero, F., Cofrades, S., Herrero, A. M., Fernández-Martín, F., Rodríguez-Salas, L., & Ruiz-Capillas, C. (2012). Konjac gel fat analogue for use in meat products: Comparison with pork fats. *Food Hydrocolloids, 26*(1), 63–72. https://doi.org/10.1016/j.foodhyd.2011.04.007

Jones, O. G. (2016). Recent advances in the functionality of non-animal-sourced proteins contributing to their use in meat analogs. *Current Opinion in Food Science, 7*, 7–13. https://doi.org/10.1016/j.cofs.2015.08.002

Joshi, V., & Kumar, S. (2015). Meat Analogues: Plant based alternatives to meat products- A review. *International Journal of Food and Fermentation Technology, 5*(2), 107. https://doi.org/10.5958/2277-9396.2016.00001.5

Kadim, I. T., Mahgoub, O., Baqir, S., Faye, B., & Purchas, R. (2015). Cultured meat from muscle stem cells: a review of challenges and prospects. *Journal of Integrative Agriculture, 14*(2), 222–233. https://doi.org/10.1016/S2095-3119(14)60881-9

Kalin, F. (1979). Wheat gluten applications in food products. *Journal of the American Oil Chemists' Society, 56*(3Part3), 477–479. https://doi.org/10.1007/bf02671549

Katsuraya, K., Okuyama, K., Hatanaka, K., Oshima, R., Sato, T., & Matsuzaki, K. (2003). Constitution of konjac glucomannan: chemical analysis and 13C NMR spectroscopy. *Carbohydrate Polymers, 53*(2), 183–189. https://doi.org/10.1016/S0144-8617(03)00039-0

Klost, M., & Drusch, S. (2019). Functionalisation of pea protein by tryptic hydrolysis — characterisation of interfacial and functional properties. *Food Hydrocolloids, 86*, 134–140. https://doi.org/10.1016/j.foodhyd.2018.03.013

Kong, X., Zhou, H., & Qian, H. (2007). Enzymatic hydrolysis of wheat gluten by proteases and properties of the resulting hydrolysates. *Food Chemistry, 102*(3), 759–763. https://doi.org/10.1016/j.foodchem.2006.06.062

Kyriakopoulou, K., Dekkers, B., & van der Goot, A. J. (2019). *Plant-based meat analogues* (pp. 103–126). Elsevier BV. https://doi.org/10.1016/b978-0-12-814874-7.00006-7

Kyriakopoulou, K., Keppler, J. K., & van der Goot, A. J. (2021). Functionality of ingredients and additives in plant-based meat analogues. *Foods, 10*(3). https://doi.org/10.3390/foods10030600

Lam, A. C. Y., Can Karaca, A., Tyler, R. T., & Nickerson, M. T. (2018). Pea protein isolates: Structure, extraction, and functionality. *Food Reviews International, 34*(2), 126–147. https://doi.org/10.1080/87559129.2016.1242135

Langelaan, M. L. P., Boonen, K. J. M., Polak, R. B., Baaijens, F. P. T., Post, M. J., & van der Schaft, D. W. J. (2010). Meet the new meat: Tissue engineered skeletal muscle. *Trends in Food Science and Technology, 21*(2), 59–66. https://doi.org/10.1016/j.tifs.2009.11.001

Li, T., Guo, X. N., Zhu, K. X., & Zhou, H. M. (2018). Effects of alkali on protein polymerization and textural characteristics of textured wheat protein. *Food Chemistry, 239*, 579–587. https://doi.org/10.1016/j.foodchem.2017.06.155

Lim, L.-T., Mendes, A. C., & Chronakis, I. S. (2019). *Electrospinning and electrospraying technologies for food applications* (pp. 167–234). https://doi.org/10.1016/bs.afnr.2019.02.005

Lin, K. W., & Huang, C. Y. (2008). Physicochemical and textural properties of ultrasound-degraded konjac flour and their influences on the quality of low-fat Chinese-style sausage. *Meat Science, 79*(4), 615–622. https://doi.org/10.1016/j.meatsci.2007.10.026

Lin, S., Huff, H. E., & Hsieh, F. (2000). Texture and chemical characteristics of soy protein meat analog extruded at high moisture. *Journal of Food Science, 65*(2), 264–269. https://doi.org/10.1111/j.1365-2621.2000.tb15991.x

Lipton, J., Arnold, D., Nigl, F., Norén, N., & Lipson, H. (2010). *Multi-material food printing with complex internal structure suitable for \n conventional post-processing*.

Li, S., Zhu, D., Li, K., Yang, Y., Lei, Z., & Zhang, Z. (2013). Soybean curd residue: Composition, utilization, and related limiting factors. *ISRN Industrial Engineering*, 1–8. https://doi.org/10.1155/2013/423590

Lugay, J. C., & Kim, M. K. (1978). *Freeze alignment: A novel method for protein texturization*.

Malav, O. P., Talukder, S., Gokulakrishnan, P., & Chand, S. (2015). Meat analog: A review. *Critical Reviews in Food Science and Nutrition, 55*(9), 1241–1245. https://doi.org/10.1080/10408398.2012.689381

Market, & Market. (2021). *Retrieved on 14*. https://www.reportlinker.com/p05779291/Plant-based-Meat-Market-by-Source-Product-Type-Process-And-Region-Global-Forecast-to.html?utm_source=GNW.

Mattice, K. D., & Marangoni, A. G. (2020). Comparing methods to produce fibrous material from zein. *Food Research International, 128*. https://doi.org/10.1016/j.foodres.2019.108804

Meade, S. J., Reid, E. A., & Gerrard, J. A. (2005). The impact of processing on the nutritional quality of food proteins. *Journal of AOAC INTERNATIONAL, 88*(3), 904–922. https://doi.org/10.1093/jaoac/88.3.904

Nicole, T. Z. H., Nichelle, T. S., Elizabeth, T. E., & Yuliarti, O. (2021). Formulation of functional crackers enriched with fermented soybean (tempeh) paste: Rheological and microstructural properties. *Future Foods, 4*. https://doi.org/10.1016/j.fufo.2021.100050

Nieuwland, M., Geerdink, P., Brier, P., Van Den Eijnden, P., Henket, J. T. M. M., Langelaan, M. L. P., Stroeks, N., Van Deventer, H. C., & Martin, A. H. (2013). Food-grade electrospinning of proteins. *Innovative Food Science and Emerging Technologies, 20*, 269–275. https://doi.org/10.1016/j.ifset.2013.09.004

Ooms, N., & Delcour, J. A. (2019). How to impact gluten protein network formation during wheat flour dough making. *Current Opinion in Food Science, 25*, 88–97. https://doi.org/10.1016/j.cofs.2019.04.001

Osen, R., Toelstede, S., Wild, F., Eisner, P., & Schweiggert-Weisz, U. (2014). High moisture extrusion cooking of pea protein isolates: Raw material characteristics, extruder responses, and texture properties. *Journal of Food Engineering, 127*, 67–74. https://doi.org/10.1016/j.jfoodeng.2013.11.023

Pietsch, V. L., Emin, M. A., & Schuchmann, H. P. (2017). Process conditions influencing wheat gluten polymerization during high moisture extrusion of meat analog products. *Journal of Food Engineering, 198*, 28–35. https://doi.org/10.1016/j.jfoodeng.2016.10.027

Post, M. J. (2012). Cultured meat from stem cells: Challenges and prospects. *Meat Science, 92*(3), 297–301. https://doi.org/10.1016/j.meatsci.2012.04.008

Rachwał, K., Waśko, A., Gustaw, K., & Polak-Berecka, M. (2020). Utilization of brewery wastes in food industry. *PeerJ, 8*, e9427. https://doi.org/10.7717/peerj.9427

Ramachandraiah, K. (2021). Potential development of sustainable 3D-printed meat analogues: A review. *Sustainability, 13*(2), 938. https://doi.org/10.3390/su13020938

Raphaely, T., & Marinova, D. (2015). Impact of meat consumption on health and environmental sustainability. In *Impact of meat consumption on health and environmental sustainability* (pp. 1–410). https://doi.org/10.4018/978-1-4666-9553-5. IGI Global.

Sacks, F. M., Lichtenstein, A., Van Horn, L., Harris, W., Kris-Etherton, P., & Winston, M. (2006). Soy protein, isoflavones, and cardiovascular health. *Circulation, 113*(7), 1034–1044. https://doi.org/10.1161/circulationaha.106.171052

Sadler, M. J. (2004). Meat alternatives-market developments and health benefits. *Trends in Food Science and Technology, 15*(5), 250–260. https://doi.org/10.1016/j.tifs.2003.09.003

Schiffman, J. D., & Schauer, C. L. (2008). A review: Electrospinning of biopolymer nanofibers and their applications. *Polymer Reviews, 48*(2), 317–352. https://doi.org/10.1080/15583720802022182

Schreuders, F., Dekkers, B., Bodnár, I., Erni, P., Boom, R., & Goot, A. J. (2019). Comparing structuring potential of pea and soy protein with gluten for meat analogue preparation. *Journal of Food Engineering, 261*, 32–39.

Sha, L., & Xiong, Y. L. (2020). Plant protein-based alternatives of reconstructed meat: Science, technology, and challenges. *Trends in Food Science and Technology, 102*, 51–61. https://doi.org/10.1016/j.tifs.2020.05.022

Shiau, S. Y., & Yeh, A. I. (2001). Effects of alkali and acid on dough rheological properties and characteristics of extruded noodles. *Journal of Cereal Science, 33*(1), 27–37. https://doi.org/10.1006/jcrs.2000.0344

Singh, P., Kumar, R., Sabapathy, S. N., & Bawa, A. S. (2008). Functional and edible uses of soy protein products. *Comprehensive Reviews in Food Science and Food Safety, 7*(1), 14–28. https://doi.org/10.1111/j.1541-4337.2007.00025.x

Singh, M., Trivedi, N., Enamala, M. K., Kuppam, C., Parikh, P., Nikolova, M. P., & Chavali, M. (2021). Plant-based meat analogue (PBMA) as a sustainable food: A concise review. *European Food Research and Technology, 247*(10), 2499–2526. https://doi.org/10.1007/s00217-021-03810-1

Smetana, S., Aganovic, K., Irmscher, S., & Heinz, V. (2018). *Agri-food waste streams utilization for development of more sustainable food substitutes*. Springer Science and Business Media LLC. https://doi.org/10.1007/978-3-319-66981-6_17

Steinfeld, H., Gerber, P., Wassenaar, T., Castel, V., Rosales, M., & de Haan, C. (2006). Livestock's long shadow: Environmental issues and options. *Renewable Resources Journal, 24*(4), 15–17.

Su, S. I. T., Yoshida, C. M. P., Contreras-Castillo, C. J., Quiñones, E. M., & Venturini, A. C. (2013). Okara, a soymilk industry by-product, as a non-meat protein source in reduced fat beef burgers. *Food Science and Technology, 33*(1), 52–56. https://doi.org/10.1590/S0101-20612013000500009

Vong, W. C., & Liu, S. Q. (2016). Biovalorisation of okara (soybean residue) for food and nutrition. *Trends in Food Science and Technology, 52*, 139–147. https://doi.org/10.1016/j.tifs.2016.04.011

Yang, A. S., & Honig, B. (1993). On the pH dependence of protein stability. *Journal of Molecular Biology, 231*(2), 459–474. https://doi.org/10.1006/jmbi.1993.1294

Yuliarti, O. (2020). Textural characteristics of Indonesian foods. In *Textural characteristics of world foods* (pp. 137–150). Wiley. https://doi.org/10.1002/9781119430902.ch10

Yuliarti, O., Kiat Kovis, T. J., & Yi, N. J. (2021). Structuring the meat analogue by using plant-based derived composites. *Journal of Food Engineering, 288*. https://doi.org/10.1016/j.jfoodeng.2020.110138

Zhang, J., Liu, L., Jiang, Y., Faisal, S., Wei, L., Cao, C., Yan, W., & Wang, Q. (2019). Converting peanut protein biomass waste into \double green\ meat substitutes using a high-moisture extrusion process: A multiscale method to explore a process for forming a meat-like fibrous structure. *Journal of Agricultural and Food Chemistry, 67*(38), 10713–10725. https://doi.org/10.1021/acs.jafc.9b02711

Chapter 10

Plant-based imitated fish

Lankatillake C., Dias D. and Huynh T.
STEM School of Science, RMIT University, Melbourne, VIC, Australia

10.1 Introduction

Fish has long been a component of the human diet. Over the years, the importance of fish consumption to human nutrition and health has become increasingly evident with research linking diets high in fish with reduced risk of lifestyle diseases such as cardiovascular disease (CVD) (He et al., 2004; Hosomi et al., 2012; Zhang et al., 2020), metabolic syndrome (Liaset et al., 2019; Patel et al., 2009), and certain cancers (Huang et al., 2015; Song et al., 2014; Zheng et al., 2013). The dietary value and the health benefits of fish stem primarily from its high protein and polyunsaturated fat content. Oily fish (e.g., salmon and mackerel) in particular are rich in the omega-3 fatty acids docosahexaenoic acid (DHA) and eicosapentaenoic acid (EPA). Omega-3 fatty acids protect against CVD and age-related cognitive decline, are involved in multiple physiological processes including inflammation and anticoagulation, and play an important role in fetal development (Swanson et al., 2012). Fish also provide essential micronutrients such as vitamin B12, vitamin D, iodine, and selenium (Bourre & Paquotte, 2008).

While health is the number one motivator of fish consumption, consumers report taste and convenience as other major drivers. On the other hand, high cost, concerns regarding quality, and unfamiliarity with selecting and cooking fish are the main barriers to fish consumption among consumers (Christenson et al., 2017; Lamy & Szejda, 2020). In recent years, increased awareness of the plight of the oceans, sustainability issues surrounding seafood, and safety concerns regarding contaminants in fish, have been driving consumers to seek plant-based alternatives to fish. This, together with increasing consumer preference toward more plant-based diets, have created an inviting yet challenging opportunity for the food industry to innovate plant-based alternatives to fish and seafood (Lam et al., 2016; The Good Food Institute, 2019).

This chapter discusses plant-based alternatives to fish that are currently available for consumer purchase in many developed countries. The major plant-based proteins and other ingredients used in the manufacture of plant-based imitation fish are discussed with an overview of the processing technologies and sensory characteristics. The chapter focuses on the nutritional composition and dietary value of the different plant-based products in comparison to a representative product made from real fish. The advantages and limitations of different available imitation fish are assessed to provide consumers better knowledge based on nutrition and sustainability.

10.2 Currently available plant-based alternatives to fish

Plant-based fish analogs or imitation fish represent a small but rapidly growing share of the plant-based meat alternatives market. Manufacturers of plant-based imitation fish are global with prominent representatives from the United States, the United Kingdom, Canada, and South Africa.

The major types of products currently available for consumer purchase can be categorized as those mimicking fish fillets, steaks and fish fingers; tinned tuna; and smoked salmon (Table 10.1). The main ingredients, additives, and nutritional value of these products are summarized. Many of the products evaluated were only available for purchase in developed countries including the United States, the United Kingdom, Canada, Australia, New Zealand, and parts of Europe with the exception of Fry's Fish-Style Fillets, which were exported to 30 countries around the globe including the United States, the United Kingdom, Australia, parts of Europe, Asia, Africa, and the Middle East.

Engineering Plant-Based Food Systems. https://doi.org/10.1016/B978-0-323-89842-3.00008-7

TABLE 10.1 Currently available plant-based alternatives to fish.

	Product name	Main ingredients	Additives	Manufacturer	Reference
Fillet and steak-style fish	Fry's fish-style fillets	Vegetable protein (22%) (soybean, wheat)	Wheat fiber, thickener (methyl cellulose, gum arabic), pea fiber, flaxseed oil (6%), wheat flour, flavorings, smoke flavor, maltodextrin, emulsifier (E1518), garlic	Fry Group Foods	(Fry Family Food, n.d.)
	VBites™ fsh steaks	Konjac flour, tapioca starch, wheat starch	Water, breadcrumb coating (wheat flour (flour, calcium carbonate, niacin, iron, thiamine), canola oil, water, salt, yeast), konjac flour, tapioca starch, wheat starch, salt, natural flavoring, sugar, wheat fiber, yeast extract, thickener: carrageenan, preservative: potassium sorbate, color: titanium dioxide	VBites™ Foods	(VBites Foods Ltd, n.d.)
	Gardein™ golden fishless filets	Water, soybean protein concentrate	Expeller pressed canola oil, potato starch, modified vegetable gum, soybean protein isolate, vital wheat gluten, sea salt, natural plant favors, yeast extract, organic cane sugar, citric acid, algal oil (omega 3 DHA, EPA), onion powder, pea protein, carrot fiber, color, paprika, rosemary, turmeric extract	Conagra Brands	(Conagra Brands Inc, 2021)
	Quorn™ crumbed fish free fillets	Rice flakes, wheat flour (calcium carbonate, iron, niacin, thiamin), mycoprotein (14%)	Water, canola oil, flavoring, wheat starch, emulsifier (401), gelling agent (461), salt, yeast, black pepper, citric acid (330), natural lemon flavoring	Marlow Foods	(Marlow Foods Ltd, n.d.)
	Sophie's Kitchen™ vegan fish fillet	Textured vegetable protein (pea protein, pea starch)	Canola oil, brown rice flakes, konjac powder, seaweed powder, potato starch, powdered cellulose, organic agave nectar, turmeric, white pepper, sea salt, ginger	Sophie's Kitchen™	(Sophie's Kitchen, n.d.)
	Vivera® plant fish fillet	Rehydrated wheat protein (65%)	Breadcrumbs (wheat flour, salt, parsley, yeast), sunflower oil, water, vinegar, wheat flour, thickener (methylcellulose), natural flavourings, salt, wheat fiber, wheat starch, flaxseed oil, potato fiber, maltodextrin, herbs, spices, maize starch, preservative (sodium diacetate), onion powder, garlic powder, vitamin B12, iron	Vivera®	(Vegan Food UK, n.d.)
Tinned tuna	Good Catch ® fish-free tuna	Water, pea protein isolate, soybean protein concentrate, chickpea flour, lentil protein, faba protein, navy bean flour	Algal oil, sea salt, sunflower oil, seaweed powder, citric acid, onion powder, yeast	Gathered Foods	(Gathered Foods, n.d.)

TABLE 10.1 Currently available plant-based alternatives to fish.—cont'd

	Product name	Main ingredients	Additives	Manufacturer	Reference
			extract, garlic powder, soy lecithin.		
	Loma Linda® Tuno™	Textured soybean protein (63%),	Spring water, vegan fish flavor (yeast extract, maltodextrin, natural flavors), thickener (gum arabic), sunflower oil, coconut oil), sea salt (0.2%), seaweed powder, potassium chloride	Atlantic Natural Foods	(Woolworths Group Ltd, n.d.)
	Sophie's Kitchen™ plant-based toona sea salt	Water, pea protein	Safflower oil (expeller pressed high oleic), vegan spice mix (potato starch, sea salt, vinegar powder, umami flavor [yeast extract, natural flavor], kombu powder, beet powder, torula yeast, canola oil [processing aid])	Sophie's Kitchen™	(Sophie's Kitchen Inc, n.d.-b)
Smoked salmon	Sophie's Kitchen™ Vegan smoked salmon	Water, olive oil, konjac powder, pea starch, potato starch, pea protein	Sea salt, organic agave nectar, seaweed powder, fenugreek, seaweed alginate, paprika, calcium hydroxide	Sophie's Kitchen™	(Sophie's Kitchen Inc, n.d.-a)
	VBites™ smokey slmn	Water, canola oil, soya protein	Thickeners: carrageenan, konjac flour; potato starch, sea salt, onion powder, sugar, natural flavoring, preservative: potassium sorbate, smoke flavoring, colors: annatto, iron oxide	VBites™ Foods	(VBites Foods Ltd, n.d.-b)

Keywords: Imitation fish, plant-based fish, ingredients, additives, nutrition.

10.3 Ingredients used for the manufacture of plant-based fish

The ingredients used in the manufacture of plant-based fish (Table 10.1) determine the organoleptic characteristics (e.g., flavor, texture, appearance, and mouthfeel), nutritional value, and the cost of the end product. The main ingredients present in plant-based imitation fish include water, vegetable proteins, fat or oils, and additives such as thickeners, binding agents, flavors, and preservatives (Egbert & Borders, 2006).

10.3.1 Water

Water is a major ingredient in the product formulation of plant-based fish. The high water content lowers overall ingredient costs, assists with ingredient hydration, distribution, and emulsification, and provides juiciness (Egbert & Borders, 2006). Further, the interaction of water with plant proteins is important for their functional properties such as gelation, swelling, viscosity, foaming, and the ability to stabilize emulsions (Cornet et al., 2020; Kyriakopoulou et al., 2021; Peters et al., 2017).

10.3.2 Protein

A combination of plant-based protein sources, mostly from legumes and cereals, are used in the manufacture of imitation fish. Plant-based proteins, which form the solid bulk of the product in most cases, were key ingredients as they contributed significantly to the organoleptic properties, functional characteristics, and the nutritional composition (the protein content in particular) of the end product. Therefore, selecting the correct type and combination of plant-based proteins is critical in the development of plant-based fish. In the products reviewed, the sources of plant-based proteins were predominantly legumes (Fabaceae family) and cereals (Poaceae family). These ingredients were usually incorporated as protein isolates or concentrates but could also appear in the form of flour, powder, starch, or fiber. The product formulations can include one source of plant-based protein (e.g., soybean) or a combination of up to six sources (e.g., pea protein, soybean protein, chickpea flour, lentil protein, faba protein, and navy bean flour). The most commonly used were soybean (*Glycine max*), peas (*Pisum sativum* L.), or wheat (*Triticum aestivum* L.) (Fig. 10.1A).

Soybean (*Glycine max*) was a preferred ingredient for the manufacture of plant-based fish due to its functional flexibility and nutritional properties (Thrane et al., 2017). Soybean has a fibrous structure which lends itself favorably to the texturization processes. Soybean (Fabaceae) can be grown with low fertilizer use as it readily fixes nitrogen from the soil; is widely cultivated (mainly in the United States, Brazil, and Argentina), has an established supply chain, and is therefore a low cost ingredient (Givaudan, 2020; Thrane et al., 2017). In terms of nutrition, soybean has the highest protein content (36 g/100 g raw beans) among cereals (wheat, rice, corn, oats, barley) and legumes (lentils, peas, chickpeas) and provides a complete protein containing all the essential amino acids required for human nutrition (Michelfelder, 2009). Importantly, soybean produces high-quality proteins with a protein digestibility-corrected amino acid score (PDCAAS) equal to milk and eggs, and the soybean protein PDCAAS score is the highest among all plant proteins (Lefranc-Millot & Teichman-Dubois, 2018; Thrane et al., 2017). The disadvantage of soybean is its status as one of the eight major allergens, which means that soy-based products are not suitable for all consumers (Food and Drug Administration, 2020; Givaudan, 2020).

Peas (*Pisum sativum L.*) were the second-most common source of plant protein utilized in the manufacture of plant-based fish. Peas are also a widely cultivated, nitrogen fixing, high-yield legume mainly grown in Canada, Russia, and China (Lefranc-Millot & Teichman-Dubois, 2018). Nutritionally, peas have a protein content similar to soybean and a PDCAAS score of 70 (soybean PDCAAS score is 100) (Lefranc-Millot & Teichman-Dubois, 2018). However, pea protein is functionally inferior to soybean, has a stronger off-taste, and is slightly more expensive (Givaudan, 2020).

Wheat (*Triticum aestivum L.*) provides plant proteins in the form of gluten and is the most common cereal ingredient. Wheat is a popular ingredient as it is inexpensive, has good sourcing options and functionality, and imparts a clean taste (Givaudan, 2020). However, wheat is of lower nutritional value (Givaudan, 2020) and like soybean, is among the big eight food allergens (Teuber et al., 2006) and is associated with several other gastrointestinal disorders; the most important of which is celiac disease (Cabanillas, 2020). Soybean and wheat can be substituted with other plant or fungal protein sources such as mycoprotein from the fungus *Fusarium venenatum* to make products appropriate for a wider range of consumers (Marlow Foods, 2021; Wiebe, 2004).

10.3.3 Fats and oils

Plant fats and oils can be a major component of the product (e.g., vegan smoked salmon) or a minor additive depending on the amount used. Fats and oils commonly incorporated into plant-based imitation fish products included canola oil, sunflower oil, and flaxseed oil. Fats are an essential component which improves overall palatability and contribute to the

FIGURE 10.1 Word cloud representing the most common main ingredients (A) and additives (B) used in the manufacture of plant-based imitation fish.

flavor, tenderness, mouthfeel, and give succulence to products which can otherwise be dry (Egbert & Borders, 2006; Kyriakopoulou et al., 2021). Algal oil is added to some products in small amounts as a source of plant-based omega-3 fatty acids DHA and EPA, and to impart a fish-like taste (Cision, 2018; The Good Food Institute, 2019). Fats and oils must be carefully moderated as high fat contents not only affect the nutritional composition of the final product but too much oil in the product formulation (>15%) also impede the texturization process. This is due to increased lubrication, which negatively impacts extrusion shear forces and prevents proper macromolecule alignment required for the formation of well-aligned fibrous structures during extrusion (Gwiazda et al., 1987).

10.3.4 Additives

Several additives are employed in the formulation of plant-based fish to achieve the desired sensorial properties including mimicking the fishy taste of real fish (Fig. 10.1B) and raise valid health concerns. Salt (sodium chloride) plays multiple roles in food processing and was a prominent additive. Salt imparts a saline taste to the product, improves water-binding capacity, improves handling and functionality of ingredients, and acts as a flavor enhancer and a preservative by lowering water activity which prevents microbial growth (Man, 2007). Flavoring agents and colors were also common additives. Many of the products contained yeast extract as a flavoring agent, and seaweed powder was used to impart a fishy taste. Annatto (E169b) and iron oxide (E172) were added to imitation salmon to mimic the pink-red color of real salmon, and titanium dioxide (E171) whitened products to mimic white fish (Boutillier et al., 2020; Green, 2020). Some products also contained unspecified natural flavorings and colors.

Binding agents (e.g., carrageenan E407) and gelling or thickening agents (e.g., methyl cellulose E461, konjac powder E425, and gum arabic E414) were added to improve the rheological properties and the water binding capacity of manufactured products (Egbert & Borders, 2006; Kyriakopoulou et al., 2019). Preservatives such as potassium sorbate (E202) retard the growth of yeasts, molds, and bacteria to extend the shelf-life.

10.4 Processing and manufacture

The manufacture of plant-based imitation fish products employs modern food processing technologies to mold plant protein ingredients into fibrous structures that simulate the texture and physicochemical properties of fish meat (Kazir & Livney, 2021; Kyriakopoulou et al., 2019). These technologies can be classified as bottom-up or top-down approaches. In bottom-up approaches, such as wet spinning and electrospinning, production starts with the creation of individual anisotropic elements (e.g., individual protein fibers), which are then assembled into macrostructural products using binding agents or enzymatic cross-linking (Dekkers et al., 2018). Top-down structuring technologies include extrusion cooking, and shear cell technology where fibrous structures are formed by the application of shear forces to protein blends. Bottom-up approaches can create products that closely resemble the ordered structure of real fish meat, whereas top-down approaches only achieve the overall fibrous texture. However, top-down approaches are more prominent on a commercial scale due to better resource efficiency, scalability, and robustness (Dekkers et al., 2018).

Extrusion is a well-developed technology as it has been widely used in the food industry for manufacturing breakfast cereal, confectionary products, extruded snacks, and baby food (Choton et al., 2020; Sun et al., 2021). It is capable of producing large quantities of products in a short time in an energy efficient manner and is the most common technique used in the manufacture of plant-based meat and seafood analogs (Dekkers et al., 2018). Extrusion cooking utilizes a drag flow device consisting of one or two (single vs. twin screw extruder) rotating screws inside a barrel attached to a cooling die. A preconditioned ingredient mixture is fed into the barrel, heated under pressure, and the rotating screws shears the mixture and pushes it through the cooling die (Kazir & Livney, 2021; Kinney et al., 2019; Kyriakopoulou et al., 2019; Sha & Xiong, 2020). Meat and seafood analog manufacture mostly uses high moisture extrusion cooking (HMEC) (>50% moisture content) in twin-screw extruders. Inside the barrel, the plant protein mixture is plasticized by the application of high heat (>140 °C) and shear force for a resident time of around 150 s, followed by cooling inside a long cooling die. This process facilitates protein fibrillation and produces the fibrous texture seen in textured vegetable protein (TVP) (Akdogan, 1999; Cheftel et al., 2009; Manski et al., 2007). The extruded product is then cut into desired lengths and may undergo additional processing steps such as hydration, cooking, and flavor coating before being packaged (Kinney et al., 2019).

Shear cell processing is an example of flow-induced structuring. The processing equipment can either be a shear cell, which is a cone-in-cone device based on a cone-plate rheometer consisting of a stationary top cone and a rotating bottom cone or a cylindrical Couette cell (Cornet et al., 2021; Krintiras et al., 2015, 2016; Kyriakopoulou et al., 2019). In both systems, the unit operations are similar to HMEC; a hydrated protein ingredient mixture is subjected to heat and simple flow shear, followed by cooling, which causes the alignment of proteins to form fibrillar structures (Cornet et al., 2021). However, there are number of important differences between HMEC and shear cell processing. Shear cell processing operates at a narrower temperature range (95—140 °C, compared to 100—175 °C for HMEC) and a longer residence time (20 min *vs*. 2—5 min for HMEC) and mixing of ingredients takes place outside the shearing device. In addition, shear cell processing is a batch process, unlike HMEC which can operate continuously (Cornet et al., 2021; Krintiras et al., 2015).

10.5 Physicochemical and sensory properties

Physicochemical and sensory properties such as water holding capacity (WHC), cooking loss, texture and appearance are important as they influence product quality, shelf-life, stability and consumer acceptance. These properties depend on the ingredients and additives used, and processing parameters such as moisture content during extrusion, extrusion temperature, and screw rotation speed (Palanisamy et al., 2019; Samard et al., 2019; Wi et al., 2020; nullS).

WHC is related to the entrapment of water within protein networks and has a direct influence on the juiciness of the product. WHC is impacted by factors such as ionic strength and pH, and the WHC capacity of plant-based protein products can be favorably modified by marination and different cooking methods (Cornet et al., 2021; Gómez et al., 2019). Cooking loss refers to the shrinkage of a product during cooking and is linked with final product yield, as well as juiciness. Reducing cooking loss increases yield and gives a juicier product. The composite ingredients and additives used in the formulation of a product have been shown to affect cooking loss. For example, the addition of oil to TVP lowers cooking loss due to the formation of hydrophobic oil—protein or protein—protein interactions, which surround and trap the water within the gel matrix (Wi et al., 2020).

Texture is an important sensory attribute that can determine the success of a product at market. How plant-based fish analogs are created to simulate the fibrous texture of fish meat by different structuring processes was described previously. In addition to the fibrous texture, fish analogs must remain tender when cooked. Bite tenderness is measured as shear force, while work of shearing is an indicator of chewiness of a product. In real fish, heat-induced solubilization of connective tissue softens meat, while denaturation of myofibrillar proteins causes toughening (Gómez et al., 2019; Kong et al., 2008). Other factors that affect texture include the composition and water content of product formulations (Kitcharoenthawornchai & Harnsilawat, 2015; Wi et al., 2020). For instance, increasing the proportion of TVP lowers shear force and work of shearing, which correlates to increased tenderness of the product (Kitcharoenthawornchai & Harnsilawat, 2015).

The color of fish analogs is another important determinant of consumer acceptance and can also affect the perceived flavor of the product. Color parameters are determined instrumentally and include lightness, L*; redness, a*; yellowness b*; chroma, C*; and hue angle H*. The plant-based proteins included in the product formulation greatly influences the color of the final product. For example, formulations with higher TVP content are lighter and redder, with lower yellowness and chroma, and has a hue angle in the yellow-orange region. Color can be enhanced using additives such as annatto (E160b) and iron oxide (E172) (added to plant-based salmon products to provide a pinkish color), and titanium dioxide (E171) which has been used to whiten products that mimic white fish (Boutillier et al., 2020; Green, 2020).

10.6 Nutritional composition

Fish plays an important role in the diet as a source of protein and heart-healthy omega-3 fatty acids (Tilami & Sampels, 2018). The nutritional value of fish due to its high protein and unsaturated fatty acid content and low saturated fat content is reported by consumers as the primary motivator for the consumption of fish (Christenson et al., 2017). As more people adopt flexitarian, vegetarian, or vegan diets, it is important to assess and understand the nutritional profiles of plant-based fish products in comparison to similar products manufactured from real fish in order to inform consumers.

Currently available plant-based fish products were grouped by product type (Table 10.1), and the nutritional composition was analyzed for crumbed fillet- and steak-style fish (Table 10.2), tinned tuna (Table 10.3), and smoked salmon (Table 10.4). All three product types varied greatly in their nutritional composition, and the majority of the plant-based fish were nutritionally different to the real fish, which was likely related to the ingredients, protein source, as well as the additives in the product formulations.

Energy was variable with lower (1 out of 6), similar (3 out of 6) or higher (2 out of 6) values for the fillet- and steak-style plant-based fish compared with real fish (Table 10.2). Sophie's Kitchen plant-based fish fillet was the highest in energy at 1158.6 kJ/100 g. The high energy content of Sophie's Kitchen plant-based fish fillet can be attributed to the higher macronutrient content of the product, particularly the high carbohydrates, which provided 523.6 kJ (30.8 g of carbohydrate x 17 kJ of energy/g of carbohydrate) of energy per 100 g of the product. The VBites plant-based fish steaks provided the lowest amount of energy as it contained the lowest amount of fat (the most energy-rich macronutrient

TABLE 10.2 Comparison of nutrition information per 100 g for real fish versus plant-based fillet and steak-style fish. Nutritional information was based on Nutrition Information Panels.

	Real Fish	Plant-Based Fish					
Per 100 g	Woolworths fish fillets crumbed	Fry's fish-style fillets	Gardein™ golden fish-less filets	Quorn™ crumbed fish free fillets	Sophie's Kitchen™ vegan fish fillets	VBites™ fsh fteaks	Vivera® plant fish fillet
Energy (kJ)	898.0	1073.1	784.4	761.1	1158.6	568.0	866.5
Protein (g)*	11.8	8.2	9.4	4.4	12.3	2.0	14.0
Fat, total (g)	9.3	20.0	10.4	7.2	12.3	5.7	8.9
Saturated (g)	1.0	3.0	0.5	1.6	1.5	0.4	1.1
Monounsaturated (g)*	-	3.0	6.3	-	-	-	-
Polyunsaturated (g)*	-	14.0	3.1	-	-	-	-
Carbohydrate (g)	20.2	10.0	14.6	21.4	30.8	17.8	16.0
Sugars (g)	3.8	0.3	0.0	1.2	4.6	1.1	0.8
Fibre (g)*	-	6.0	3.1	6.8	4.6	2.7	3.4
Sodium (g)	0.3	0.6	0.3	0.3	0.3	1.2	1.6

considered healthy.

TABLE 10.3 Comparison of nutrition information per 100 g for real tinned tuna versus plant-based tinned tuna. Nutritional information is based on Nutrition Information Panels. *considered healthy.

	Real tinned tuna	Plant-based tinned tuna		
Per 100 g	**Woolworths tuna in springwater**	**Good Catch® fish-free tuna**	**Loma Linda® Tuno™ ⁱn spring water**	**Sophie's Kitchen™ plant-based toona sea salt**
Energy (kJ)	484.0	401.1	296.0	590.6
Protein (g)*	26.6	14.9	12.3	11.8
Fat, total (g)	0.9	2.7	0.0	9.4
Saturated (g)	0.3	0.0	0.0	0.6
Monounsaturated (g)*	-	0.0	-	-
Polyunsaturated (g)*	-	1.1	-	-
Carbohydrate (g)	< 0.1	5.3	3.5	4.7
Sugars (g)	< 0.1	0.0	0.0	0.0
Fiber (g)*	-	1.1	3.5	1.2
Sodium (g)	0.1	0.4	0.7	0.6

TABLE 10.4 Comparison of nutrition information per 100 g for real smoked salmon versus plant-based fake smoked salmon. Nutritional information is based on Nutrition Information Panels. *considered healthy.

	Real smoked salmon	Plant-based smoked salmon	
Per 100 g	Tassal smoked salmon	Sophie's Kitchen™ vegan smoked salmon	VBites™ smokey slmn
Energy (kJ)	879.0	719.7	523.2
Protein (g)*	23.7	0.4	5.7
Fat, total (g)	12.8	4.7	9.2
Saturated (g)	2.8	0.7	0.6
Monounsaturated (g)*	-	-	-
Polyunsaturated (g)*	-	-	-
Carbohydrate (g)	<0.8	32.0	4.5
Sugars (g)	<0.5	0.0	0.5
Fiber (g)*	-	-	1.0
Sodium (g)	1.1	0.5	1.2

providing 37 kJ/1 g of fat), protein, and carbohydrates. Water was listed as the first ingredient in this product, and it is therefore likely to contain a high amount of water/100 g, which reduced the energy content. In comparison, Sophie's Kitchen plant-based fish contained more protein-rich ingredients (textured vegetable protein), more fats from canola oil which was the second ingredient in the product, and more sources of refined carbohydrates such as pea starch, konjac powder, and potato starch.

Protein was variable with lower (4 out of 6) or higher (2 out of 6) values for the plant-based fish compared with real fish. Vivera plant-based fish fillet contained 65% wheat protein, and this contributed to the high protein value (14.0 g/100 g) that exceeded the real fish fillet (11.8 g/100 g). Other products did not disclose the percentage of added protein and this could account for the variations in final protein value.

Fat content was variable with lower (2 out of 6), similar (3 out of 6), or higher (1 out of 6) values for the plant-based fish compared with real fish. Fry's fish-style fillets contained the highest total fat, saturated, and polyunsaturated fat content (20.0 g/100 g) which was more than double that of the real fish fillet (9.3 g/100 g). This was likely due to the high fat content of soybeans (20.0 g/100 g) which was the main ingredient. In comparison, VBites plant-based fish steaks were the lowest in total fat (5.7 g/100 g, which was 71% less than the Fry's product) and saturated fat (0.4 g/100 g), and the main ingredients were from konjac, tapioca and wheat which have lower fat content.

Sodium was variable with lower (2 out of 6), same (3 out of 6), or higher (2 out of 6) values for the plant-based fish compared with real fish. The two saltier plant-based products contained over 4 times the amount of sodium than the real fish fillets. Vivera plant-based fish fillet was the highest in sodium at 1.6 g/100 g, and VBites plant-based fish steaks were the second highest containing 1.2 g/100 g. The high sodium content of these products can be attribute not only to the addition of salt during product formulation but also to the amount of salt contained in individual ingredients (e.g., flavorings mixes) and salt used in the extraction and processing of plant protein ingredients (Lam et al., 2016; Man, 2007).

High sodium intake is associated with deleterious effects on the kidney, heart, brain, and blood vessels (Farquhar et al., 2015) often with multiorgan dysfunction and health complications. Therefore, reducing salt levels in these products at each stage of processing requires careful consideration and checking. Further, the sodium content of some products (1600 mg/100 g) exceeded the recommended adequate intake level for adults and the upper limit for children under the age of nine (National Health & Medical Research Council, 2017). Therefore, these plant-based products cannot be recommended for young children, or susceptible groups with cardiovascular disease.

A comparison of the tinned tuna products (Table 10.3) showed that the plant-based tuna was inferior to real tuna, contributing less nutrition but adding more salt to the diet. Energy was variable with lower (2 out of 3) or higher (1 out of 3) values for the plant-based tuna compared with real tuna. The plant-based tuna contained significantly less protein than real tuna. Total fats were variable with one plant-based tuna by Sophie's Kitchen containing more than nine times that of real tuna - likely due to the high addition of safflower oil. All of the plant-based tuna products contained more carbohydrates than real tuna (which is naturally low in carbohydrates) due to the inclusion of carbohydrate-containing plant protein ingredients such as pea protein and chickpea flour. The higher carbohydrates in plant-based tuna may not appeal to weight-conscious consumers. The level of sodium in the plant-based tuna was significantly higher (up to 7 times) than real tinned tuna and was one of the contributing factors for its inferiority rating.

Comparisons between smoked salmon products indicated that plant-based salmon contained lower energy, protein, fats, and salt; and higher carbohydrates, but was similar in other categories compared to real salmon (Table 10.4). Salmon is valued for its heart-healthy omega-3 fats and unsaturated fat content and the plant-based salmon contained less fat than the real salmon making them nutritionally inferior. The higher carbohydrate content of plant-based salmon was likely due to starch from konjac, pea, and potato.

Real smoked salmon undergoes a brining process before smoking. Therefore, smoked salmon is usually high in sodium. Out of the two plant-based smoked salmon products evaluated, one contained sodium levels similar to that of real smoked salmon (1.2 g/100 g for real smoked salmon vs. 1.2 g/100 g for VBites fake smoked salmon), and the other contained less than half the sodium (0.5 g/100 g for Sophie's Kitchen fake smoked salmon). As plant-based salmon products do not require brine curing, the lower salt content would be the main advantage over real smoked salmon.

10.7 Value for money

The cost to consumers is a key element that influences consumer purchasing and can determine the success of a product. The most expensive product by unit weight was Sophie's Kitchen plant-based smoked salmon, which was more than double the price of its plant-based competitor and slightly more expensive than the real smoked salmon that was nutritionally superior (Table 10.5).

Comparing the prices and the nutritional value of real fish against that of plant-based imitation fish was difficult. However, as plant-based fish products are produced to imitate real fish, consumers would expect the imitation fish products to offer a similar nutritional profile for a similar price. As the value of fish in the diet is its protein content, plant-based

TABLE 10.5 Cost comparisons of plant-based fish analogues with similar products based on real fish. (Cost in US dollars)

Product	Cost/100 g (USD)	Protein/100 g	Cost/ 10 g protein (USD)
Fish fillet, and steak-style products			
Woolworths Crumbed Fish Fillet (real fish)	**0.80**	**11.8**	**0.68**
Fry's Fish-Style Fillets	1.85	8.2	2.25
Gardein™ golden fishless filets	2.47	9.4	2.63
Quorn™ Crumbed Fish Free Fillets	2.58	4.4	5.87
Sophie's Kitchen™ Vegan Fish Fillets	3.41	12.3	2.77
VBites™ Fsh Steaks	2.56	2.0	12.8
Vivera® Plant Fish Fillet	2.35	14.0	1.68
Tinned tuna-style products			
Woolworths Tuna in Springwater (real fish)	**0.74**	**26.6**	**0.28**
GOOD CATCH® fish-free TUNA	2.13	14.9	2.13
Loma Linda® Tuno™	1.24	12.3	1.24
Sophie's Kitchen™ Plant-Based Toona Sea Salt	2.94	11.8	2.49
Smoked salmon-style products			
Tassal smoked salmon (real fish)	**5.54**	**23.7**	**2.34**
Sophie's Kitchen™ Vegan Smoked Salmon	6.86	0.4	171.50
VBites™ Smokey Slmn	2.86	5.7	5.02

imitation fish can be judged for value based on their protein content and the cost per 10 g of protein can be used to assess whether products offer good "protein value for money".

A comparison of the protein cost of plant-based fish to real fish (Table 10.5) showed that the cost of the imitation fish products were much higher regardless of product category, sometimes as high as 18 times (e.g., Woolworths crumbed real fish fillet *vs.* VBites imitation fish steak). The protein cost of tinned tuna was also pricier for plant-based products compared to real tuna. The plant-based salmon was pricier than the real smoked salmon with Sophie's Kitchen imitated salmon, which had a negligible protein content, costing almost 73 times more for the same amount of protein compared to real salmon. This shows that judging the value for money is not always straightforward with these products and consumers can be at a disadvantage if they rely solely on the sale price of the product to determine protein value for money.

10.8 Gourmet plant-based fish for the food service industry

In addition to consumer products, there are also a small number of gourmet plant-based imitation fish products developed especially for the high-end food service industry. Ocean Hugger Foods created the world's first alternative to raw tuna, Ahimi, from tomatoes (Ocean Hugger, 2019a). A proprietary manufacturing processes is utilized to give tomatoes a texture similar to raw fish. Glutamic acid, which is responsible for the savoury taste of meat, is naturally present in tomatoes. The flavor is further enhanced with a marinade consisting of simple ingredients (gluten-free soy sauce, sugar, water, and sesame oil) (Chaker, 2018). Ocean Hugger Foods also developed UnamiTM, a plant-based alternative to eel made from eggplant, gluten-free soy sauce, mirin, sugar, rice bran oil, algae oil, and konjac powder (Ocean Hugger, 2019b). Ahimi and UnamiTM are used by chefs in the US and Europe in sushi dishes (Chaker, 2018; Southey, 2019).

10.9 Conclusion and future recommendations

Consumers turn to plant-based alternatives for fish due to concerns about sustainability, and the impact of commercial fishing on marine ecosystems. Over 90% of wild fisheries are overfished or harvested at maximum capacity. Overfishing, by-catch and unsustainable fishing practices harm marine ecosystems and leads to the loss of biodiversity (Coleman & Williams, 2002; The Good Food Institute, 2019). Seafood safety concerns due to the bioaccumulation of microplastics, industrial chemicals, chemicals in sunscreen products, and heavy metals such as mercury are also factors driving consumer interest in plant-based alternatives (Lamy & Szejda, 2020; The Good Food Institute, 2019). Currently available plant-based imitation fish products offer consumers sustainable food choices that look and taste like fish. The advantages of these products are that they are easier to prepare than conventional fish and have a longer shelf life (Lamy & Szejda, 2020; Yadav et al., 2020) which adds to their convenience.

These products may be perceived by consumers as a healthier option due to their plant-based origins. However, many of these products were ultra-processed and under-nutritional. Many products were very high in sodium and have significantly lower levels of protein than would be obtained from similar fish-based products. Further, the products cost a lot more per 10 g of protein than real fish. Therefore, these plant-based alternatives require further development to provide consumers with value for money and be justified as alternatives to conventional fish.

The raw materials used in the manufacture of plant-based fish analogs are inexpensive, easily available in abundance and without seasonal variation in supply. Therefore, plant-based imitation fish products can be manufactured cost-effectively. Combined with advancements in food technology and increasing investment in the plant-based food market, there is opportunity for creating products which are sustainable, affordable and mirror the nutritional profile of fish more closely. Improvements are required in the nutritional quality of current products and there are promising options. As the primary purpose of fish in the diet is to provide protein and omega-3 fatty acids, fish analogs must provide these nutrients in comparable amounts. None of the current products are fortified with vitamin B12 and iron which are important micronutrients often deficient in vegan and vegetarian diets and need to be addressed. Algal oil can be formulated into more products as a plant-based source of omega-3 fatty acids. Using enrichedt ingredients for manufacture and fortifying products with micronutrients would enhance their nutritional value and benefit consumer health.

10.10 Abbreviations

DHA docosahexaenoic acid
EPA eicosapentaenoic acid
HMEC high moisture extrusion cooking
PDCAAS Protein Digestibility-Corrected Amino Acid Score

NIP Nutrition Information Panel
TVP textured vegetable protein
WHC water holding capacity

References

Akdogan, H. (1999). High moisture food extrusion. *International Journal of Food Science and Technology, 34*(3), 195–207. https://doi.org/10.1046/j.1365-2621.1999.00256.x

Bourre, J. M., & Paquotte, P. (2008). Seafood (wild and farmed) for the elderly: Contribution to the dietary intakes of iodine, selenium, DHA and vitamins B12 and D. *Journal of Nutrition, Health and Aging, 12*(3), 186–192. https://doi.org/10.1007/BF02982617

Boutillier, S., Fourmentin, S., & Laperche, B. (2020). Food additives and the future of health: An analysis of the ongoing controversy on titanium dioxide. *Futures, 122.* https://doi.org/10.1016/j.futures.2020.102598

Cabanillas, B. (2020). Gluten-related disorders: Celiac disease, wheat allergy, and nonceliac gluten sensitivity. *Critical Reviews in Food Science and Nutrition, 60*(15), 2606–2621. https://doi.org/10.1080/10408398.2019.1651689

Chaker, A. M. (2018). Fish: The final frontier in fake meat. *The Wall Street Journal*, Article 1539610151. https://www.wsj.com/articles/fish-the-final-frontier-in-fake-meat-1539610151.

Cheftel, J. C., Kitagawa, M., & Queguiner, C. (2009). New protein texturization processes by extrusion cooking at high moisture levels. *Food Reviews International, 8*, 235–275.

Choton, S., Gupta, N., Bandral, J. D., Anjum, N., & Choudary, A. (2020). Extrusion technology and its application in food processing: A review. *The Pharma Innovation Journal, 9*, 162–168.

Christenson, J. K., O'Kane, G. M., Farmery, A. K., & McManus, A. (2017). The barriers and drivers of seafood consumption in Australia: A narrative literature review. *International Journal of Consumer Studies, 41*(3), 299–311. https://doi.org/10.1111/ijcs.12342

Cision. (2018). *Algarithm to supply Good Catch with wild-sourced plant-based algal omega-3 DHA for their fishless seafood line: Will debut new O3OmegaTM Smoothie flavors at SupplySide West.* Canada NewsWire.

Coleman, F. C., & Williams, S. L. (2002). Overexploiting marine ecosystem engineers: Potential consequences for biodiversity. *Trends in Ecology and Evolution, 17*(1), 40–44. https://doi.org/10.1016/S0169-5347(01)02330-8

Conagra Brands Inc. (2021). *Golden fishless filet.* https://www.gardein.com/fishless/golden-fishless-filet.

Cornet, S. H. V., Snel, S. J. E., Schreuders, F. K. G., Van Der Sman, R. G. M., Beyrer, M., & Van Der Goot, A. J. (2021). Thermo-mechanical processing of plant proteins using shear cell and high-moisture extrusion cooking. *Critical Reviews in Food Science and Nutrition*, 1–18. https://doi.org/10.1080/10408398.2020.1864618

Cornet, S. H. V., van der Goot, A. J., & van der Sman, R. G. M. (2020). Effect of mechanical interaction on the hydration of mixed soy protein and gluten gels. *Current Research in Food Science, 3*, 134–145. https://doi.org/10.1016/j.crfs.2020.03.007

Dekkers, B. L., Boom, R. M., & van der Goot, A. J. (2018). Structuring processes for meat analogues. *Trends in Food Science and Technology, 81*, 25–36. https://doi.org/10.1016/j.tifs.2018.08.011

Egbert, R., & Borders, C. (2006). Achieving success with meat analogs. *Food Technology, 60*(1), 28–34.

Farquhar, W. B., Edwards, D. G., Jurkovitz, C. T., & Weintraub, W. S. (2015). Dietary sodium and health: More than just blood pressure. *Journal of the American College of Cardiology, 65*(10), 1042–1050. https://doi.org/10.1016/j.jacc.2014.12.039

Gathered Foods, Plant-based tuna, n.d., https://goodcatchfoods.com/product-cat/plant-based-tuna/.

Givaudan. (2020). *Plant-based fish and seafood on the rise.* https://www.givaudan.com/taste-wellbeing/health-wellbeing/protein/plant-based-fish.

Gómez, I., Ibañez, F. C., & Beriain, M. J. (2019). Physicochemical and sensory properties of sous vide meat and meat analog products marinated and cooked at different temperature-time combinations. *International Journal of Food Properties, 22*(1), 1693–1708. https://doi.org/10.1080/10942912.2019.1666869

Green, E. (2020). *Givaudan tips plant-based fish to make waves in alternative proteins.* https://www.foodingredientsfirst.com/news/givaudan-tips-plant-based-fish-to-make-waves-in-alternative-proteins-space.html.

Gwiazda, S., Noguchi, A., & Saio, K. (1987). Microstructural studies of texturized vegetable protein products: Effects of oil addition and transformation of raw materials in various sections of a twin screw extruder. *Food Structure, 6*, 57–61.

He, K., Song, Y., Daviglus, M. L., Liu, K., Van Horn, L., Dyer, A. R., & Greenland, P. (2004). Accumulated evidence on fish consumption and coronary heart disease mortality: A meta-analysis of cohort studies. *Circulation, 109*(22), 2705–2711. https://doi.org/10.1161/01.CIR.0000132503.19410.6B

Hosomi, R., Yoshida, M., & Fukunaga, K. (2012). Seafood consumption and components for health. *Global Journal of Health Science, 4*(3), 72–86. https://doi.org/10.5539/gjhs.v4n3p72

Huang, R. X., Duan, Y. Y., & Hu, J. A. (2015). Fish intake and risk of liver cancer: A meta-analysis. *PLoS ONE, 10*(1). https://doi.org/10.1371/journal.pone.0096102

Kazir, M., & Livney, Y. D. (2021). Plant-based seafood analogs. *Molecules, 26*(6). https://doi.org/10.3390/molecules26061559

Kinney, M. J., Weston, Z., & Bauman, J. D. (2019). *Plant-based meat manufacturing by extrusion.*

Kitcharoenthawornchai, N., & Harnsilawat, H. (2015). Characterization of meat analogue nugget: Effect of textured vegetable protein. *Food and Applied Bioscience Journal, 3*, 121–129.

Kong, F., Tang, J., Lin, M., & Rasco, B. (2008). Thermal effects on chicken and salmon muscles: Tenderness, cook loss, area shrinkage, collagen solubility and microstructure. *LWT—Food Science and Technology, 41*(7), 1210–1222. https://doi.org/10.1016/j.lwt.2007.07.020

Krintiras, G. A., Gadea Diaz, J., Van Der Goot, A. J., Stankiewicz, A. I., & Stefanidis, G. D. (2016). On the use of the Couette Cell technology for large scale production of textured soy-based meat replacers. *Journal of Food Engineering, 169*, 205–213. https://doi.org/10.1016/j.jfoodeng.2015.08.021

Krintiras, G. A., Göbel, J., Van Der Goot, A. J., & Stefanidis, G. D. (2015). Production of structured soy-based meat analogues using simple shear and heat in a Couette cell. *Journal of Food Engineering, 160*, 34–41. https://doi.org/10.1016/j.jfoodeng.2015.02.015

Kyriakopoulou, K., Dekkers, B., & Van Der Goot, A. J. (2019). Plant-based meat analogues. In C. M. Galanakis (Ed.), *Sustainable meat production and processing* (pp. 103–126). Elsevier. https://www.sciencedirect.com/science/article/pii/B9780128148747000067.

Kyriakopoulou, K., Keppler, J. K., & van der Goot, A. J. (2021). Functionality of ingredients and additives in plant-based meat analogues. *Foods, 10*(3). https://doi.org/10.3390/foods10030600

Lam, A. C. Y., Can Karaca, A., Tyler, R. T., & Nickerson, M. T. (2016). Pea protein isolates: Structure, extraction, and functionality. *Food Reviews International, 34*(2), 126–147. https://doi.org/10.1080/87559129.2016.1242135

Lamy, J., & Szejda, K. (2020). *Literature review: Consumer preferences for seafood and applications to plant-based and cultivated seafood*. https://www.gfi.org/files/sci-tech/seafood/SeafoodConsumerResearchLiteratureReview2020.pdf.

Lefranc-Millot, C., & Teichman-Dubois, V. (2018). Protein from vegetable sources: A focus on pea protein. In M. Hayes (Ed.), *Novel proteins for food, pharmaceuticals and agriculture* (pp. 197–216). John Wiley & Sons. https://doi.org/10.1002/9781119385332.ch10

Liaset, B., Øyen, J., Jacques, H., Kristiansen, K., & Madsen, L. (2019). Seafood intake and the development of obesity, insulin resistance and type 2 diabetes. *Nutrition Research Reviews, 32*(1), 146–167. https://doi.org/10.1017/s0954422418000240

Man, C. M. D. (2007). Technological functions of salt in food products. In *Reducing salt in foods: Practical strategies* (pp. 157–173). Elsevier Ltd. https://doi.org/10.1533/9781845693046.2.157

Manski, J. M., Van Der Goot, A. J., & Boom, R. M. (2007). Advances in structure formation of anisotropic protein-rich foods through novel processing concepts. *Trends in Food Science and Technology, 18*(11), 546–557. https://doi.org/10.1016/j.tifs.2007.05.002

Marlow Foods. (2021). *Mycoprotein*. https://www.quorn.com.au/mycoprotein.

Marlow Foods Ltd, Quorn Vegan Crumbed Fish Free Fillets, n.d., https://www.quorn.com.au/products/quorn-vegan-crumbed-fish-free-fillets.

Michelfelder, A. J. (2009). Soy: A complete source of protein. *American Family Physician, 79*(1), 43–47.

National Health, & Medical Research Council. (2017). *Nutrient reference values*. https://www.nhmrc.gov.au/about-us/publications/nutrient-reference-values-australia-and-new-zealand-including-recommended-dietary-intakes.

Ocean Hugger. (2019a). *Ahimi*. https://oceanhuggerfoods.com/ahimi.

Ocean Hugger. (2019b). *Unami*. https://oceanhuggerfoods.com/unami.

Palanisamy, M., Töpfl, S., Berger, R. G., & Hertel, C. (2019). Physico-chemical and nutritional properties of meat analogues based on Spirulina/lupin protein mixtures. *European Food Research and Technology, 245*(9), 1889–1898. https://doi.org/10.1007/s00217-019-03298-w

Patel, P. S., Sharp, S. J., Luben, R. N., Khaw, K. T., Bingham, S. A., Wareham, N. J., & Forouhi, N. G. (2009). Association between type of dietary fish and seafood intake and the risk of incident type 2 diabetes: The European Prospective Investigation of Cancer (EPIC)-Norfolk cohort study. *Diabetes Care, 32*(10), 1857–1863. https://doi.org/10.2337/dc09-0116

Peters, J. P. C. M., Vergeldt, F. J., Boom, R. M., & van der Goot, A. J. (2017). Water-binding capacity of protein-rich particles and their pellets. *Food Hydrocolloids, 65*, 144–156. https://doi.org/10.1016/j.foodhyd.2016.11.015

Samard, S., Gu, B. Y., & Ryu, G. H. (2019). Effects of extrusion types, screw speed and addition of wheat gluten on physicochemical characteristics and cooking stability of meat analogues. *Journal of the Science of Food and Agriculture, 99*(11), 4922–4931. https://doi.org/10.1002/jsfa.9722

Sha, L., & Xiong, Y. L. (2020). Plant protein-based alternatives of reconstructed meat: Science, technology, and challenges. *Trends in Food Science and Technology, 102*, 51–61. https://doi.org/10.1016/j.tifs.2020.05.022

Song, J., Su, H., Wang, B. L., Zhou, Y. Y., & Guo, L. L. (2014). Fish consumption and lung cancer risk: Systematic review and meta-analysis. *Nutrition and Cancer, 66*(4), 539–549. https://doi.org/10.1080/01635581.2014.894102

Sophie's Kitchen Inc, Plant-based fish fillets, n.d., https://www.sophieskitchen.com/fish-fillets.

Sophie's Kitchen Inc, Plant-based smoked salmon, n.d., https://www.sophieskitchen.com/smoked-salmon.

Sophie's Kitchen, Toona sea salt, n.d., https://www.sophieskitchen.com/sea-salt-toona.

Southey, F. (2019). *Plant based raw tuna and freshwater eel with fishlike texture come to Europe*. https://www.foodnavigator.com/Article/2019/07/10/Plant-based-raw-tuna-and-freshwater-eel-with-fishlike-texture-come-to-Europe.

Sun, C., Ge, J., He, J., Gan, R., & Fang, Y. (2021). Processing, quality, safety, and acceptance of meat analogue products. *Engineering, 7*(5), 674–678. https://doi.org/10.1016/j.eng.2020.10.011

Swanson, D., Block, R., & Mousa, S. A. (2012). Omega-3 fatty acids EPA and DHA: Health benefits throughout life. *Advances in Nutrition, 3*(1), 1–7. https://doi.org/10.3945/an.111.000893

Teuber, S. S., Beyer, K., Comstock, S., & Wallowitz, M. (2006). The big eight foods: Clinical and epidemiological overview. In S. J. Maleki, A. W. Burks, & R. M. Helm (Eds.), *Food allergy*. USA: ASM Press. https://onlinelibrary.wiley.com/doi/abs/10.1128/9781555815721.ch3.

The Good Food Institute. (2019). *An ocean of opportunity: Plant-based and cell-based seafood for sustainable oceans without sacrifice*. https://www.gfi.org/files/sci-tech/seafood/GFI_An_Ocean_of_Opportunity.pdf.

The Fry Family Food Co, Golden crispy fish-style fillets, n.d., https://www.fryfamilyfood.com/au/our-food/4-golden-crispy-vegan-fish-fillet/.

Thrane, M., Paulsen, P. V., Orcutt, M. W., & Krieger, T. M. (2017). Chapter 2 - soy protein: Impacts, production, and applications. In *Sustainable protein sources* (pp. 23–45). Elsevier. https://www.sciencedirect.com/science/article/pii/B9780128027783000020.

Tilami, S. K., & Sampels, S. (2018). Nutritional value of fish: Lipids, proteins, vitamins, and minerals. *Reviews in Fisheries Science & Aquaculture, 26*(2), 243–253. https://doi.org/10.1080/23308249.2017.1399104

U. S. Food and Drug Administration. (2020). *Food allergies*. https://www.fda.gov/food/food-labeling-nutrition/food-allergies.

VBites Foods Limited, VBites - Fsh Steaks, the Delicious Vegan Version of Fish "n" Chips, n.d., https://www.vbites.com/product/fsh-steaks/.

VBites Foods Limited, Vegan Plant-Based Fish Free Salmon Slices, n.d., https://vbites.com/products/vegan-plant-based-salmon-slices.

Vegan Food UK, Vivera Veggie Fishless Fillet 200g, n.d., https://veganfooduk.co.uk/pin/vivera-veggie-fishless-fillet-200g/.

Wi, G., Bae, J., Kim, H., Cho, Y., & Choi, M. J. (2020). Evaluation of the physicochemical and structural properties and the sensory characteristics of meat analogues prepared with various non-animal based liquid additives. *Foods, 9*(4). https://doi.org/10.3390/foods9040461

Wiebe, M. G. (2004). QuornTM myco-protein - overview of a successful fungal product. *Mycologist, 18*(1), 17–20. https://doi.org/10.1017/S0269915X04001089

Woolworths Group Limited, Loma Linda Tuno In Spring Water 142g, n.d., https://www.woolworths.com.au/shop/productdetails/67685/loma-linda-tuno-in-spring-water.

Yadav, R., Shaikh, A., & Modi, S. (2020). Plant based seafood. *Food and Agriculture Spectrum Journal, 1*.

Zhang, B., Xiong, K., Cai, J., & Ma, A. (2020). Fish consumption and coronary heart disease: A meta-analysis. *Nutrients, 12*(8), 1–22. https://doi.org/10.3390/nu12082278

Zheng, J.-S., Hu, X.-J., Zhao, Y.-M., Yang, J., & Li, D. (2013). Intake of fish and marine n-3 polyunsaturated fatty acids and risk of breast cancer: meta-analysis of data from 21 independent prospective cohort studies. *BMJ, 346*(f3706), 1–10. https://doi.org/10.1136/bmj.f3706

Zhou, X., Jiang, S., Zhao, D., Zhang, J., Gu, S., Pan, Z., & Ding, Y. (2017). Changes in physicochemical properties and protein structure of surimi enhanced with camellia tea oil. *LWT, 84*, 562–571. https://doi.org/10.1016/j.lwt.2017.03.026

Chapter 11

Plant-based imitated seafood

William Leonard and Zhongxiang Fang
School of Agriculture and Food, University of Melbourne, Parkville, VIC, Australia

11.1 Introduction

Seafood products are the primary protein source for a third of the world's population, and its perceived health properties compared to "red" meat have contributed to its high demand over the past decade (Link & Watson, 2019; OECD/FAO, 2019). Nevertheless, the surge in consumption of seafood products has led to marine over-exploitation and ecological challenge. Another growing safety concern among consumers is the high concentration of heavy metals (e.g., mercury, lead) and microplastics contamination in seafood (Bosch et al., 2016; Smith et al., 2018). Imitation meat or meat analog is described as food products that do not contain meat but resemble the taste, texture, appearance, and nutrition of meat products (Wild, 2016). In recent years, several manufacturers, particularly those based in developed western countries, have commenced the commercialization of imitation seafood products from plant ingredients, such as fishless-fillets, imitation canned tuna and sushi, crab cakes, and smoked salmon (Fig. 11.1). This chapter reviews the raw ingredients,

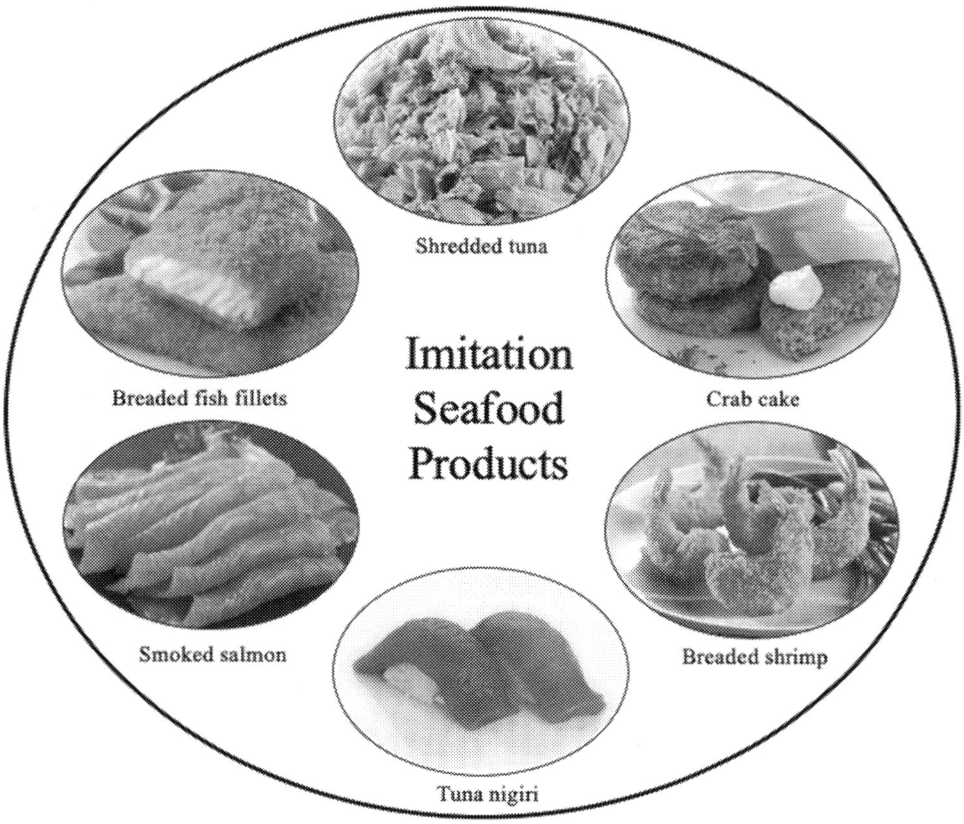

FIGURE 11.1 Common imitated seafood products in the market.

Engineering Plant-Based Food Systems. https://doi.org/10.1016/B978-0-323-89842-3.00007-5

processing technologies, postprocessing aspects, and the safety concern of plant-based seafood alternatives. The nutritional value of available imitation seafood in the market is compared to traditional seafood products. Finally, the environmental impact, market demand, and the future challenges of plant-based imitation seafood are discussed.

11.2 Raw ingredients

Mimicking the texture of meat by using plant-based materials has been a challenge for food technologists and manufacturers. Unlike myofibrillar proteins, the native globular structure of plant proteins does not allow the development of fibrous meat texture, along with its gelling and fat-emulsifying activity (Sha & Xiong, 2020). This necessitates the use of food processing technologies to restructure and reorder the protein assembly and thus its functionality. Even though imitation meat is one of the fastest growing research areas in food science, there is still a lack of peer-reviewed articles specifically developing fully plant-based imitation seafood products. Therefore, the ingredients discussed in this section are based on the existing popular commercial products in the market (Table 11.1).

11.2.1 Protein sources

Protein is a major component in most imitation meat systems due to its effect on the product's hydration, gelation, flavor binding, and texturization (Dekkers et al., 2018). Selection of the primary plant protein source for imitation seafood requires a balancing act in multiple aspects, such as functional, nutritional and sensory properties, cost, availability, and allergenicity. Considering these aspects, soy, pea, and wheat have been the most extensively utilized protein sources in imitation seafood industry (Table 11.1). Soy and pea proteins are widely available with low production cost. The protein digestibility—corrected amino acid score for soy protein, in particular, at a range of 0.95—1 is similar to animal proteins such as beef and milk proteins (Asgar et al., 2010). Both proteins are dominated by 7S (e.g., β-conglycinin, vicilin) and 11S (e.g., glycinin, legumin) globulins that readily unfold and aggregate when subjected to appropriate temperature and ionic concentration (Sha & Xiong, 2020). These components contribute to the functionality of these proteins in terms of solubility, gelation, and emulsification properties. Gelation property is particularly important as it determines the protein's capacity to trap water in the matrix and formation of a stable protein network. However, soy protein emanates bitter and beany off-flavors from the presence of lipooxygenases, saponins, and isoflavones. Similarly, the aldehyde and ketone compounds in pea protein have been linked to its undesirable aroma and flavor profile (Roland et al., 2017). Despite its weaker functionality and apparently stronger off-flavors compared, pea protein may provide an alternative to soy protein as it contains no allergen, is not associated to genetic modification, and its growth adaptable to more diverse climate conditions.

Plant globulins alone are rarely sufficient to satisfy the typical water binding capacity and fibrous texture of meat muscle, thus other proteins are often added in the production of imitation seafood. Wheat glutens, due to its glutenins and gliadins, may be added to improve the meaty chewiness, elasticity, and extensibility of the product (Sha & Xiong, 2020). In breaded fish-less fillet and stick products, the interaction between wheat gluten and starch is critical for Maillard reaction, thus the formation of golden-brown color, baked aroma, and hardness of the outer "crust" layer (Schreuders et al., 2019). The authors demonstrated that the tensile strength of pea protein-wheat gluten blends is in the range to the reference chicken meat but significantly weaker to soy-gluten blends. This may imply the higher suitability of pea-wheat blends in imitation seafood products with relatively weaker tensile strength such as shredded canned tuna and fish fingers. Manufacture of imitation canned tuna has also involved the addition of other legume proteins into the ingredient blend, such as faba, lentil, navy bean, and chickpeas, some of which have comparable nutrition and gelation property to soy protein (Asgar et al., 2010; Withana-Gamage et al., 2011). Nevertheless, much like soybean and peas, product developers should take into account the presence of antinutritional factors (e.g., protease inhibitors, phytic acid, polyphenols) in legumes and the less investigated interaction effect between pulse proteins. In addition to pulse proteins, the integration of mycoprotein from the filamentous fungus *Fusarium venenatum* together with wheat constituents has been commercially applied for Quorn's fish-less sticks and breaded imitation fish (Quorn, Stokesley, UK). Mycoprotein integration has been commercially established over the past few decades but requires an intensive use of energy and raw materials demanding (Dekkers et al., 2018).

11.2.2 Other nonprotein sources

In some cases, protein is not the most important ingredient in the development of imitation seafood, as demonstrated by the use of tomato, eggplant, and jackfruit as the dominant ingredients of manufacturing tuna sushi, eel unagi, and crab

TABLE 11.1 Common ingredients found in commercial plant-based imitation seafood products.

Category	Ingredients	Function
Proteins	Soy, pea, wheat, rice, faba, lentil, navy bean, chickpea	Gelation and aggregation during high heat and shear treatment, major texture and structure of product, fat binding capacity, elasticity and extensibility.
Carbohydrates	Tomato, eggplant, jackfruit, wheat flour	Major texture and structure of product, fiber formation, natural color to traditional seafood products, breading.
Fats	Algal, canola, sunflower, safflower, soy, olive and palm oil	Seafood taste, aroma and mouthfeel. Nutrition comparable to marine fishes (omega-3 and omega-6 fatty acids).
Micronutrients	Iron, sodium chloride, potassium chloride, calcium chloride, B-group vitamins (e.g., cobalamin, thiamine, folic acid).	Supplementation purposes.
Adhering and thickening agents	Wheat flour and starch, brown rice flour, potato starch, pea starch, methylcellulose, gum Arabic, Konjac glucomannan, calcium hydroxide, maltodextrin, carrot fiber, calcium alginate (seaweed)	Emulsification, binding of fats, protein and water. Texture and rheology enhancement.
Coloring and flavoring agents	Sodium chloride, potassium chloride, soy sauce, barley powder, celery powder, onion powder, garlic powder, kombu powder, beet powder, paprika, yeast extract, citric acid, konjac powder, other spices and herbs	Improve seafood-like color, flavor and taste of product.
Antioxidants	Tocopherols, spices, herbs and similar extracts	Reduce risk of lipid oxidation and rancidity.
Antimicrobials	Spices, herbs and similar extracts	Maintain the shelf life of product.

Note: Adapted from Sha, L., & Xiong, Y. L., (2020). Plant protein-based alternatives of reconstructed meat: Science, technology, and challenges. *Trends in Food Science & Technology, 102*, 51–61 and several commercial imitation seafood product labels (Gardein, Richmond, Canada; Good Catch, Newtown, USA; Loma Linda, Nashville, USA; Ocean Hugger, Brooklyn, USA; Quorn, Stokesley, UK; Sophie's Kitchen, Sebastopol, USA; Worthington, Collegedale, USA).

imitation, respectively (Ocean Hugger Foods, Brooklyn, USA). However, nonprotein sources generally occupy a minor composition to improve the impression of seafood-like feel in the imitation meat products. Lipid influences the sensory (e.g., mouthfeel, lubricant, texture, juiciness) and nutritional properties of any food system (Talbot, 2015). Algal oil imparts the "sea" flavor and is one of the richest nonanimal sources for the seafood characteristic, such as polyunsaturated omega-3 fatty acids of docosapentaenoic acid (DPA), docosahexaenoic acid (DHA), and eicosapentaenoic acid (EPA). The oil-producing microalgae can be sustainably grown in industrial fermenters without contact to sea environment, with significantly higher growth rate and lipid production than land-based oil crops. Other unsaturated oils in imitation seafood are derived from canola, sunflower, safflower, soy, and olive oils, although the fatty acid profile of these plant oils is not as similar to the oil of typical marine fishes (Table 11.1). Similar to seafood, the high content of unsaturated fats of imitation seafood exposes the product to the risk of oxidation and rancidity, negatively impacting consumer acceptance. Thus, incorporation of antioxidants (e.g., mixed tocopherols) and precautionary measures (e.g., limit exposure to air and metals, addition of oil at the end stage of production process) may reduce the effects of oxidative degradation of these oils from the high heat and shear treatment during processing.

Complex carbohydrates, especially water-soluble polysaccharides, are commonly utilized as an emulsifying-thickening agent that holds the protein and fat component together in the imitation meat system. The glucomannan-rich konjac powder, partly due to its "fishy" aroma, has been included in the production of fish fillet, crab cakes, eel unagi, and shrimp alternatives (Ocean Hugger Foods, Brooklyn, USA; Sophie's Kitchen, Sebastopol, USA). Reaction of konjac powder with an alkali coagulant (e.g., calcium hydroxide) led to the deacetylation and a thermally stable gel to form, thus enhancing the fatty texture and chewiness of imitation seafood (Yang et al., 2017). Seaweed alginate and its sodium salt, cellulose derivatives (methylcellulose), and gum Arabic are frequently applied in imitation seafoods as additives to maintain product stability through its emulsifying, gelling, and binding properties (Sha & Xiong, 2020). Simpler carbohydrates, such as potatoes, rice, and wheat starch, are also added into imitation seafood due to its strong water-holding capacity, ability to control gelation, and stability of emulsion systems. These carbohydrates are often added together in the imitation meat system; hence, the possible synergistic, antagonistic, or absence of interactions should be considered to tailor the intended seafood product. For instance, while konjac glucomannan may show positive synergy with plant gums on textural attributes of many food systems, it appears to exert no effect on aroma retention and sensory properties when combined with potato starch (Lafarge et al., 2014). Unlike other tested gums (e.g., xanthan, carrageenan), gum Arabic showed no effect on the gelation and water mobilization properties of starch (Sikora et al., 2010).

Imitation seafood manufacturers use different seasonings and flavor enhancers that help mask the after-taste from plant proteins and complement the expected seafood flavor. Manufacturers may isolate individual "meat-like" aromatic compounds and integrate it into the imitation meat at appropriate concentration after several testings (Bohrer, 2019). In addition, developing imitation seafood with a color realistically similar to seafood, including its changes during cooking, is crucial in consumer decision-making. This issue is not as significant in breaded or battered imitation fish products, where the meat is covered by the crust, and deep frying is the general cooking method, which have changed the original seafood flavor. Product developers may resort to the use of coloring agents or primary plant ingredient with natural color, as in the case of Ocean Hugger's eel (eggplant) or tuna filler (tomato), and shredded tuna (beetroot) by Sophie's Kitchen (Table 11.2).

11.2.3 Application of plant ingredients as extenders in seafood products

Meat extenders are ingredients that partially replace the proportion of meat in a product. Far more peer-reviewed articles have reported the role of plant-based materials as an extender in seafood products, which may provide useful insights in manufacturing fully plant-based imitation seafood. Surimi is a minced fish paste that can be used to mimic the appearance and texture of crab, lobster, and other shellfish. Although not as effectual as animal-based additives, the general literature shows an improvement in surimi gel properties after minor (<5%) addition of legumes (e.g., soybean, mung bean, black bean), konjac glucomannan, starch, gums, and coconut husk extracts (Bashir et al., 2017). The main seed storage proteins in legumes, legumin, and vicilin are suitable gelling and binding agents for surimi gels. Legumes also contain inhibitors of endogenous protease in fish pastes, which limit protein denaturation and promote formation of gel with increased deformability. Incorporation of legumes in surimi's is occasionally supplemented with microbial transglutaminase, which provides additional support for cross-linking of myofibrillar proteins and gel stability from nondisulfide covalent bonds. Similarly, konjac glucomannan has been added in squid and carp surimis due to its ability to protect against freezing damage, or cryoprotective effect, and improve gelation properties, though this appears to depend on the molecular size of the glucomannan (Jian et al., 2016). Larger molecular size (Rz 108.5 nm) and its random coil conformation may prevent a thorough integration of konjac glucomannan with myofibrillar protein due to steric hindrance.

TABLE 11.2 Nutritional profiles of several commercial plant-based imitation (no shading) and traditional (gray shading) seafood.

Product	Manufacturer	Energy (kJ)	Protein (g)	Fat (g)	Saturated fats (g)	Carbohydrate (g)	Sugars (g)	Dietary fiber (g)	Sodium (mg)	Iron (mg)	Calcium (mg)	Potassium (mg)
Fish-less imitation fillets												
Gardein breaded fishless fillets	Gardein, Richmond, Canada	784	9.38	10.42	0.52	14.58	0	3.13	354	1.13	81.25	n.s.
Quorn's battered fishless fillets	Quorn, Stokesley, UK	855	4.2	9.3	1.3	25.2	3.9	1.6	500	Fortified	Fortified	n.s
Quorn's crumbed fishless fillets	Quorn, Stokesley, UK	761	4.4	7.2	1.6	21.4	1.2	6.8	327	Fortified	Fortified	n.s.
Sophie's Kitchen plant-based fish fillets	Sophie's Kitchen, Sebastopol, USA	1158	12.31	12.31	1.54	30.77	4.62	4.62	369	1.54	32.31	84.62
Quorn's fishless sticks	Quorn, Stokesley, UK	837	5	8	2	30	<1	6	340	Fortified	Fortified	n.s.
Great value fish fillets	Walmart, Bentonville, USA	814	7.41	11.11	2.78	21.3	1.85	<0.93	509	0.67	0	92.59
Processed imitation tuna												
Good Catch plant-based tuna (Naked in water)	Good Catch, Newtown, USA	401	14.89	2.66	0	5.32	0	1.06	479	1.06	31.91	106
Loma Linda Tuno (Spring water)	Loma Linda, Nashville, USA	294	12.28	0	0	3.51	0	3.51	298	1.89	91.23	165
Sophie's Kitchen Toona (sea salt flavor)	Sophie's Kitchen, Sebastopol, USA	591	11.76	9.41	0.59	4.71	0	1.18	600	2.12	30.59	111
Worthington tuna roll	Worthington, Collegedale, USA	738	17.65	7.94	1.76	7.06	0	1.76	635	3.53	90	259
StarKist chunk light tuna in water	StarKist, Pittsburgh, USA	188	10	0.5	0	0	0	0	180	0.72	0	100
Other imitation seafood products												
Good Catch plant-based crab cakes	Good Catch, Newtown, USA	592	17.7	4.42	0.44	7.96	0.88	0.88	540	3.01	35.4	168
Sophie's Kitchen plant-based crab cakes	Sophie's Kitchen, Sebastopol, USA	772	4.62	9.23	1.54	20	1.54	3.08	262	1.08	56.92	108

Continued

TABLE 11.2 Nutritional profiles of several commercial plant-based imitation (no shading) and traditional (gray shading) seafood.—cont'd

Product	Manufacturer	Energy (kJ)	Protein (g)	Fat (g)	Saturated fats (g)	Carbohydrate (g)	Sugars (g)	Dietary fiber (g)	Sodium (mg)	Iron (mg)	Calcium (mg)	Potassium (mg)
Crab cake (USDA database 783039)	–	858	14.57	14.75	2.4	2.58	0.51	0.2	416	0.7	78	215
Good Catch plant-based fish burgers	Good Catch, Newtown, USA	592	18.58	4.42	0.44	7.96	0.88	1.77	487	3.45	44.25	204
Fish cake or patty (USDA 783040)	–	854	13.9	10.66	2.77	12.63	0.84	0.9	285	0.7	28	443
Sophie Kitchen's plant-based breaded shrimp	Sophie's Kitchen, Sebastopol, USA	795	4	11	2	31	2	4	270	0.5	38	90
ICYBAY breaded shrimp	–	854	9.73	1.33	0	37.17	3.54	1.8	850	1.27	35	0
Sophie Kitchen's Vegan Smoked Salmon	Sophie's Kitchen, Sebastopol, USA	670	<2	4	0	32	0	6	520	0	24	60
Smoked salmon (USDA 782636)	–	490	18.28	4.32	0.93	0	0	0	672	0.85	11	175
Ocean Hugger's Ahimi Tuna Nigiri	Ocean Hugger foods, Brooklyn, USA	73.84	1.18	0	0.12	3.53	0	0	565	0	55.29	94.12
Fish, tuna, fresh, bluefin, raw (USDA 173706)	–	602	23.33	4.9	1.26	0	0	0	39	1.02	8	252
Ocean Hugger's Unami eel Nigiri top	Ocean Hugger foods, Brooklyn, USA	126	9.38	0	0	7	4	2	540	0.4	60	70
Eel, cooked (USDA 782842)	–	967	23.19	14.66	2.97	0	0	0	380	0.63	25	291

n.s. not specified in the label information. Compositions for traditional seafood products are highlighted in gray.

Starch may be used to replace part of the fish proteins, due to its low cost and high-water binding capacity. It has been reported that addition of either 4% native potato starch or hydroxypropylated potato starch enhanced the sensory evaluation scores (i.e., texture, juiciness) of shrimp surimi analog, although the latter showed a better performance (Remya et al., 2015). The presence of hydroxypropyl groups may interfere with the association of amylose/amylopectin chains and its moisture holding capacity. Mi et al. (2021) demonstrated that the combination of hydroxypropylated cassava starch and gums (κ-carrageenan) led to higher textural analysis (hardness, cohesiveness) and sensory values (appearance, texture, taste) than separate individual treatments or untreated control in silver carp surimi gel. The gumstarch interaction may promote cross-linking of hydroxyl groups or glucose residues with myofibrillar protein, and gums shift the surimi protein conformation away from the gel-disrupting α-helices. Moreover, an increasing number of studies have integrated fiber-dominated plant processing by-products, particularly coconut husk extracts, in seafood surimi due to its perceived health benefits. The improved cohesiveness in 0.075% coconut husk-incorporated *Tilapia* shrimp analogs indicates its cross-linking and complementary action in the emulsion system (Priyadarshini et al., 2020). Pre-emulsification of fish oil with coconut husk extract improved the surimi gel strength and lowered the incidence of lipid oxidation (Buamard & Benjakul, 2019). As with other plant-based materials, excessive amount of plant fiber may adversely impact the color (e.g., whiteness) and texture of products; thus, preliminary trials are needed to determine the upper ceiling for adding fiber in the imitation seafood.

11.3 Processing technologies in the manufacture of imitation seafood

Manufacturing of imitation seafood and meat could be categorized into bottom-top and top-bottom strategies as suggested by Dekkers et al. (2018). The bottom-top strategy is a highly detailed methodology that gradually assembles smaller meat fibers into a whole meat product, such as cell culture and spinning methods. On the other hand, the top-bottom strategy refers to a large scale-size production of imitation meat through a field force, of which minor adjustments or ingredients can be included during the final stages to improve the meat-like texture and taste. While the bottom-top strategy may result in imitation products with greater resemblance to meat, the top-bottom strategy has been established for a longer period of time, with proven commercial scalability, reduced costs, and environmental impact (Dekkers et al., 2018). Extrusion and shear cell, a part of the top-bottom strategy, are among the most utilized technologies in structuring and manufacturing imitation meat.

Extrusion merges the process of mixing, shearing, cooking, and compressing a material through a narrow opening (Leonard et al., 2020). The sudden pressure drop as the material leaves the narrow opening led to changes in its physicality and functionality, and thus this principle has been used to transform plant-based molten material into a fibrous, meat-like structure. Extrusion can be easily adapted to the manufacturer's needs as it allows the control of various processing parameters, including number of screw (single or twin extruder), moisture content (high or low), screw speed, and barrel temperature at different extruder sections. There is an increasing enthusiasm in premium imitation meat derived from high moisture or wet extrusion cooking, which processes materials with >50% moisture content (Wild, 2016). Imitation meat products from wet extrusion are claimed to have improved appearance, texture, and mouth-feel, in comparison to low-moisture extrusion products. Protein-rich material is sheared and heated at around 120−160°C in the barrel to allow denaturation and cross-linking, as the pressure builds up and pushes the mass in a laminar flow. Therefore, the elongated barrel allows sufficient heat transfer to the mass, and the shear flow of the channel allows the formation of an anisotropic, fibrous structure. The core temperature of the material should be around 100°C at the end of the cooling channel to prevent expansion and rupture of the product. Wet packaging, such as cans or pouches, is generally prepared for products of high moisture extrusion. The major criticism of extrusion is its energy-intensive process, and that its operation is often deduced from empiricism, rather than scientific knowledge (Grabrowska et al., 2016).

Shear cell technology was developed in the past decade to manufacture imitation meat and reportedly requires significantly less mechanical energy input than extrusion cooking (Grabrowska et al., 2016). This allows a smoother commercial scalability and control of additional parameters, such as residence time and deformation rate. The shearing devices, or cells, can be employed in Couette or cone-in-cone geometry. As in extrusion, the final structure of the product depends on the composition (e.g., polysaccharides, protein) of the raw ingredients and process parameters. Fibrous products have been developed from various protein blends in past research, including soy protein isolate-wheat gluten blends and soy protein-pectin blends (He et al., 2020). Extrusion and shear cell technology are not necessary in the formulation of imitation fresh seafood products, such as nigiri, in which the top-bottom mixing method of plant materials is sufficient. Other less common top-bottom strategy includes freeze structuring, whereas bottom-top strategies include cell culture and spinning (Dekkers et al., 2018). Although still in its infancy, these strategies have the potential to replace extrusion in the manufacture of commercial scale imitation meat in the future.

11.4 Postprocessing of imitation seafood: packaging, shelf life, and safety

Seafood is one of the most perishable food products ($a_w > 0.95$) due to its susceptibility to lipid oxidation, autolysis, and bacterial spoilage (Kerry, 2012). The presence of unsaturated fatty acids, nonprotein nitrogen sources (e.g., trimethylamine oxide), and native microflora considerably shortens its shelf life if appropriate handling practices and packaging are not applied. Most imitation seafood products have been processed under high temperature (e.g., extrusion), which confers the advantage over fresh seafood in terms of food safety, shelf life, and packaging requirements. Nonetheless, the high water activity ($a_w > 0.90$) in imitation seafood, particularly products derived from wet extrusion, facilitates bacteria and mold growth. This effect is exacerbated in products with neutral pH and low salt concentration (less than 2%). Plant-based food products are prone to spoilage from high population of *Clostridium botulinum* and *Enterococcus faecalis* (Geeraerts et al., 2020; Pernu et al., 2020). The heat-involving extrusion process itself is sufficient for microbial control of imitation meat products, with an average of less than 100 colony forming units in 1 gram of extruded products (He et al., 2020). However, some heat-resistant spores, mainly of the *Bacillus* and *Clostridium* species, may persist and reactivate in plant-based products when conditions are suitable, such as high level of water activity ($a_w > 0.90$) and the absence of refrigeration or freezing for storage. Contamination of product surface during handling, breading, and unsterile additives inclusion is another risk associated to extruded products (Wild, 2016). Therefore, additional postextrusion heat treatment may further extend its shelf life.

Given the risks of microbial spoilage, plant-based imitation seafoods are typically stored and packed in a similar manner as seafood products. Imitation seafoods are often sold frozen in pouches or flexible bags under vacuum or modified atmosphere environment. These products may last 6–18 months under freezing (−20°C) conditions, depending on added ingredients, or for several days to weeks when chilled at 4°C (Wild, 2016). The color of imitation seafood products is generally stable and rarely linked as a freshness indicator, which is not the case after extensive lipid oxidation in some fishes. For canned imitation tuna, products can be stored at room temperature up to several years due to the harsh heat sterilization and complete isolation of product from the environment (Kerry, 2012). As in typical canned seafood products, it should be consumed within days and chilled after exposure to open air. With further innovations in food packaging, application of active packaging, such as moisture scavenging and antimicrobial systems, or intelligent packaging, such as time, temperature, and toxin indicators, may improve the shelf life and help maintain the product safety for consumers (Fang et al., 2017).

Allergenicity is the major safety issue in plant-based imitation meat. Allergy is described as severe reactions from immune-related sensitivity to allergens, which occur more often in individuals with higher levels of immunoglobulin E (IgE) antibodies. Numerous IgE binding and allergenic proteins have been detected in legumes, such as soybean, pea, and lentil, although it appears that allergies linked to soybean are not as severe as peanuts (Asgar et al., 2010; Cabanillas et al., 2018). In wheat-containing imitation meat products, there is a health risk for gluten intolerance and celiac disease. Therefore, proper labeling and declaration of allergenic ingredients are imperative to provide relevant information and ensure the safety of consumers. Manufacturers may also opt to incorporate plant ingredients with lesser allergens (e.g., pea) or apply food processing methods (i.e., heat treatment) with the aim of altering protein functionality, to produce allergy-friendly imitation seafood. In addition, the high temperature in the manufacturing plant-based seafood alternatives may theoretically risk the formation of potentially carcinogenic compounds, such as polycyclic aromatic hydrocarbons and heterocyclic aromatic amines, as in meat products (He et al., 2020). Pesticides and heavy metals may be present in legume products, although the concentration of heavy metals (e.g., mercury, lead) appears to be significantly lower than in marine fishes and shellfishes (Bosch et al., 2016). Another advantage of imitation seafood is the lack of microplastics contamination, which is a growing safety concern for consumers over the past decade (Smith et al., 2018). Extrusion processing may significantly reduce the antinutritional factors (e.g., phytate, lectins) in legume-based imitation meat products but still risks unknown degree of adverse impact on nutrient bioavailability and human digestive system, particularly when testing of these factors in food products is rarely conducted (Leonard et al., 2020). With the lack of imitation seafood being investigated on the basis of these aspects, there is a need to assess and establish the control measures of these compounds in plant-based imitation seafood.

11.5 Nutritional profile of plant-based imitation seafood

Plant-based imitation seafood often prioritizes the sensory and organoleptic properties before the nutritional value of products. From a macronutrient perspective, the composition of protein, lipid, and water of imitation seafood is more likely to be similar to traditional seafood for nutrition and functionality. In contrast, carbohydrate and dietary fiber are less or even absent in marine fish and shellfish. The nutritional information of some imitation seafood outlined in Table 11.2 is publicly available from product labels and web databases as of November 2020. However, it does not provide a comprehensive picture on the nutritional value of a product, particularly of protein quality (i.e. amino acid composition), unsaturated fatty acids content, and several micronutrients (vitamins and minerals). The imitation seafood products listed in Table 11.2 are among the most well-known at the time of this chapter's writing, with manufacturers primarily targeting the fish fillet and processed tuna sector. To compare with traditional seafood products, the nutritional information of selected products was obtained from U.S. Department of Agriculture's FoodData Central database (U.S. Department of Agriculture, 2020) or the best-selling product of a top supermarket chain in the seafood imitator's country.

11.5.1 Fish-less imitation fillets

The breaded fish fillet is currently the most mimicked seafood product in the market. This could be due to the lack of the need to form a proper fish structure (as most fillets have been comminuted), and the crunchiness-crispness from the outer crust layer may mask the overall flavor and sensory value of the product. Comparison of nutritional information was carried out using five fillet imitation meat products (three of Quorn fishless fillets and sticks, Gardein breaded fishless fillets, and Sophie's Kitchen plant-based fish fillets) and one traditional crumbed fish fillet (Great Value fish fillets) (Quorn, Stokesley, UK; Gardein, Richmond, Canada; Walmart, Bentonville, USA) (Table 11.2). With the exception of Sophie's Kitchen, all fish-less fillets are similarly as energy-dense as the Great Value's fish fillets (761−855 vs. 814 kJ per 100 g). Most of the imitation seafood (i.e., Quorn) have lower protein and fat content, but all breaded fish-less products contain less saturated fatty acids compared to their traditional counterparts. This could be due to the addition of crop oils (e.g., canola and sunflower) or algal oil in imitation seafood that are less saturated than the soybean oil added in the Great Value's fish fillets. The majority of imitation fish fillets showed higher total carbohydrates, sugars, and dietary fiber than traditional breaded fillet, which is expected to originate from the legumes or rice flakes, in addition to the breading ingredients. However, it is difficult to determine the net added health value from this information due to the contrasting consensus on dietary fiber and sugar (Table 11.2). The sodium level in most imitation fish fillets is generally lower (327−369 vs. 509 mg/100 g) than the Great Value's fillets, and a single serving is still considerably below the recommended sodium intake at 2300 mg per day. Moreover, in fish-less fillet products (i.e., Sophie's Kitchen) with the available labeling information, the levels of iron, calcium, and potassium are greater than Great Value's fish fillets. Overall, plant-based fish-less fillets may provide a slightly healthier option than traditional breaded fish fillets (e.g., the Great Value fish fillet in Table 11.2) on the basis of its lower saturated fats and sodium. High intake of saturated fats and sodium has been linked to elevated risk of cardiovascular diseases and hypertension, respectively (Briggs et al., 2017; Grillo et al., 2019).

11.5.2 Processed imitation tuna

Processed imitation tuna in shredded and canned form is another popular product in the imitation seafood market, which could be due to its structurally less-demanding production process. Four processed imitation tuna products (Good Catch plant-based tuna, Loma Linda Tuno, Sophie's Kitchen Toona and Worthington's tuna roll) and one traditional processed tuna (StarKist chunk light tuna in water) were compared in terms of nutritional specification (Good Catch, Newtown, USA; Loma Linda, Nashville, USA; Worthington, Collegedale, USA; StarKist, Pittsburgh, USA). All processed imitation tuna exhibited significantly higher energy content (294−738 vs. 188 kJ/100 g), protein (11.76−17.65 vs. 10 g/100 g), carbohydrate (3.53−7.06 vs. 0 g/100 g), and sodium content (298−635 vs. 180 mg/100 g) than StarKist canned tuna. Only Loma Linda claimed no fat and no saturated fats content in its product, which are lower than StarKist tuna. Although the carbohydrate content in imitation processed tuna is greater than the traditional tuna product, it is comprised of dietary fiber with the absence of sugar. Past systematic review has indicated the possible health properties of higher fiber consumption against the risk of cardiovascular diseases and its related mortality and several cancers (e.g., pancreatic, gastric) (Veronese et al., 2018). These associations may not be of high significance in the context of shredded tuna, of which the product is often consumed with other foods high in simple carbohydrates such as white bread and pasta. In addition, the imitation processed tuna products are notably higher in calcium and iron, which are nearly absent in StarKist tuna. Assuming a balanced Western diet, the levels of calcium (30−90 mg/100 g) and iron (1−3.5 mg/100 g) in imitation processed tuna are adequate for a single food item to meet the recommended dietary intake for adults at 1000 mg/day for calcium and 8 mg/day for iron. However, sodium levels of some imitation processed tuna products (at over 600 mg/100 g) may be excessively high for a single serving, particularly to some individuals with risk of cardiovascular diseases. Aside from its dietary fiber and calcium content, imitation processed tuna does not appear to offer higher nutritional value compared to traditional processed tuna (StarKist's canned tuna chunk).

11.5.3 Other imitation seafood products

Crab and fish cakes are popular snacks made from mixing shredded seafood meat and other ingredients, commonly bread crumbs, egg, and parsley. Two imitation crab cake products (Good Catch and Sophie's Kitchen plant-based crab cakes) were compared for its nutritional specifications to the crab cake data provided by USDA FoodData Central database (U.S. Department of Agriculture, 2020). Despite the different chemical composition, the two imitation products recorded lower energy (592−772 vs. 858 kJ/100 g), fat (4.42−9.23 vs. 14.74 g/100 g), and calcium-potassium content but higher carbohydrate (7.96−20 vs. 2.58 mg/100 mg) and iron content, compared to traditional crab cakes. Imitation crab meat may therefore provide a more nutritious option as established on its lower energy density, saturated fatty acids, sugar, and higher dietary fiber. Nevertheless, the variability in protein and sodium content between the imitation crab meat products should be considered and adjusted to specific needs of the each individual. Similarly, Good Catch's plant-based burger patty may have enhanced nutritional properties as indicated by its lower energy and saturated fats content, higher protein, and dietary fiber compared to traditional fish patty from the USDA's database. The higher sodium content in imitation fish burger (487 vs. 285 mg/100 g) should be noted and, as with the issue of all product labeling, the quality (e.g., digestibility) of protein in imitation meat analogues may not correspond to its high measured content. Compared to traditional breaded shrimp (ICYBAY brand breaded shrimp, xx city, xx country), imitation breaded shrimp by Sophie's Kitchen contains significantly higher fat (11 vs. 1.33 g/100 g) and saturated fats (2 vs. 0 g/100 g), which may be derived from the use of canola oil. However, the imitation breaded shrimp is higher in dietary fiber and lower in sugar and sodium (270 vs. 850 mg/100 g) levels than the traditional breaded shrimp. Despite its lower sodium content, the currently available imitation smoked salmon (Sophie's Kitchen) may not be significantly healthier than the traditional smoked salmon due to its lower protein, higher carbohydrates, unclear proportion of marine-characteristic polyunsaturated fatty acids, and the absence of iron (Table 11.2).

Fewer manufacturers have attempted to mimic whole seafood products due to its structuring difficulty and elevated safety risk when consumed raw. For nutritional comparison, two imitation nigiris (Ocean Hugger's Ahimi imitation tuna and Unami imitation eel) were placed against their respective traditional meats as provided by USDA FoodData Central database. Ahimi imitation tuna is significantly lower in energy (73.84 vs. 602 kJ/100 g), protein (1.18 vs. 23.33 g/100 g), and fat (0 vs. 4.9 mg/100 g) content but higher in carbohydrate (3.53 vs. 0 g/100 g) and sodium (565 vs. 39 mg/100 g) content, than fresh bluefin tuna. The high sodium content could be a result of soy sauce addition to target the preferences in the food's country of origin. The same nutritional pattern was repeated for Unami imitation eel when compared to traditional cooked eel, with notably lower energy (126 vs. 967 kJ/100 g), protein (9.38 vs. 23.19 g/100 g), and higher sodium (540 vs. 380 mg/100 g). Even though these imitation products may share similar visual and organoleptic properties, they have a distinctive nutritional character compared to their traditional counterparts: rich carbohydrate/fiber, low protein, absence of fats in imitation product, and fiber-absent, high protein, and unsaturated fats in traditional seafood meats. Hence, the nutritional worth of these products depends on the consumer's preferences and dietary needs.

11.6 Environmental impact from the rise of imitation seafood

The fisheries industry holds a colossal economic and nutritional impact on humanity, with around 57 million workers employed in the sector and 35% of the world's population relying on its products as the primary protein source (Link & Watson, 2019). Nevertheless, the endless pursuit of economic goals requires an ever-increasing fishery harvest levels, thus ecosystem destruction and weakened social impact are very likely. This is manifested with overfishing and marine "defaunation," establishment of unsustainable aquaculture farms, water pollution, soil acidification, eutrophication, mangrove deforestation, and biological contamination from introducing exotic species (Martinez-Porchas & Martinez-Cordova, 2012). These adverse impacts could be mitigated with consumption of plant-based imitation seafood products. In terms of several metrics of environmental footprint, the greenhouse gas (i.e., carbon dioxide), acidifying (i.e., sulfur dioxide) and eutrophying (i.e., phosphate) emissions for "farmed fish and crustaceans" were up to 45-, 24-, and 45-fold the emissions of "peas and other pulses" group to produce 100 g protein, respectively (Poore & Nemecek, 2018). Switching from a seafood-based to legume-based diet, however, may increase the use of land on the same protein and weight basis. Ernstoff et al. (2019) demonstrated that a single meat-less meal (containing grains, fruits, legumes) may reduce the carbon footprint by up to 80%, as compared to fish-containing meat mixture meal. Soy-based imitation meat products also showed the least environmental impact (e.g., climate change, marine/freshwater ecotoxicity, freshwater eutrophication) on the same weight and protein basis than chicken, dairy-based or lab-grown meats, which could be due to effective processing, stream, and utilization of its by-products (Smetana et al., 2015). The measured environmental footprints between foods drastically change with the functional units (e.g., the basis of weight, protein, energy) used in these studies. Unlike red meats, consumer's perception on fishes is generally more positive due to the presence of unsaturated fatty acids, and this component is rarely included as a functional unit in comparison of environmental impacts. Most imitation seafood has also undergone processing (e.g., extrusion) steps, which increases the environmental footprint of a product and this processing aspect was rarely taken into account in impact assessments. Moreover, further investigation is needed to specifically compare the environmental footprint of a fully fish-based meal and legume-based seafood.

11.7 Market demand, consumer attitudes, and regulatory challenges for imitation seafood products

Consumption of fish per capita is expected to increase between the period of 2016–18 and 2028 in all continents, except Africa (OECD/FAO, 2019). In developed countries, such as Australia, the largest increases have been derived from canned and processed fish sectors, which are the main target sectors for seafood imitators (Bogard et al., 2019). The market for plant-based meat alternatives is predicted to grow over ten-fold over 2019–29 period and may seize around 10% of 2029s $ 1.4 trillion meat market in several countries, including the United States, the United Kingdom, and Germany (Ismail et al., 2020). Compared to other established imitation meat products (e.g., beef burgers, chicken, mince products), imitation seafood appears to occupy the smallest portion of the meat-less market. However, the novelty and lack of competitors may attract an increasing number of seafood imitators over the next few years. In the future, flexitarians, or semi-vegetarians, are expected to propel the meat alternatives market, with environmental sustainability and animal ethics as the primary drivers in purchasing decision. Salmon (smoked or sliced), tuna (canned or sliced), shrimp, crab (meat and cakes), and whitefish (breaded fillets and fingers) are predicted to be the top imitation seafood products (Givaudan, 2020).

Nearly all imitation seafood manufacturers are based in developed western regions, such as the United States and Europe. This could be due to the duration of exposure and degree of familiarity to a given product, rather than the culture or region factor. A survey conducted by Bryant et al. (2019) showed that China and India recorded higher acceptance for plant-based meat alternatives, with around 62%–63% of respondents very or extremely likely to purchase, compared to 33% of respondents expressing the same interest in the United States. However, the factors influencing consumer's purchasing intent may differ between countries. While healthiness and nutritional value were the most critical predictors in China, Indian respondents perceived sustainability and animal welfare issues as the more important factors when accepting meat alternatives. Therefore, there is a potential to introduce plant-based meat alternatives in developing countries with adapted marketing strategies and the possible improved acceptance with higher familiarity and longer exposure to meat alternatives. Several approaches to improve the acceptance for meat alternatives have been identified, including sensory stimulation, incorporating meat-less days in a week, and adjusting meal portion (He et al., 2020).

Development of plant-based imitation seafood is less likely to be met with the same degree of legal and food regulation issues compared to other meat alternatives (e.g., insects, cultured meat) due to the familiar ingredients used (van der Weele et al., 2019). Opposition is expected to come from traditional seafood industry and fisheries. The currently available imitation seafood products might be too insignificant to threaten the traditional seafood market. Nevertheless, these groups have begun demanding a proper classification and labeling requirements for imitation seafood products, so as not to provide misleading information for consumers. They argued that plant-based products should not be categorized as real meat and seafood, and that the terms for traditional meat products should not be included in the labels of plant-based products, such as "burger," "fillet," "fish," or "tuna" (Sha & Xiong, 2020). This will be the main issue of discussion for food regulation authorities.

11.8 Future outlooks and conclusion

The rising consumption of seafood products is not accompanied by the same degree of environmental sustainability. Development of imitation seafood products may mitigate the ecological destruction that have been brought by the seafood industry. Currently, plant-based imitation products are one of the most acceptable form of meat alternatives that balance nutritional, social-institutional, technological transformation, environmental sustainability, and regulation framework, as compared to cell cultured, insect-based, and algae-derived imitation meat (van der Weele et al., 2019). Cultured meat will likely pose the greatest competition to plant-based meat alternatives in the future. Although its sensory properties (e.g., texture, mouthfeel, taste) may be inferior to cultured meat, plant-based imitation meat presents several advantages including the familiarity of product, absence of ethical concerns as the use of cells from slaughtered animals is not required, lower manufacturing cost, and environmental footprint (Boukid, 2020). However, further optimization and introduction of novel-improved technologies to enhance the seafood-like taste and texture is still needed to consolidate its place in the market. The nutritional value of imitation seafood products may not be significantly superior to traditional seafood products due to its distinctive chemical composition. Application of intensive processing technologies should be considered as it may diminish the environment sustainability advantage of plant-based meat alternatives. Increased exposure and familiarity could be the key to market plant-based imitation products. Furthermore, food authorities and industry should deliberate on the regulations and policies of plant-based imitation seafood, including labeling and nutrition claims, to provide as much transparency to consumers.

References

Asgar, M. A., Fazilah, A., Huda, N., Bhat, R., & Karim, A. A. (2010). Nonmeat protein alternatives as meat extenders and meat analogs. *Comprehensive Reviews in Food Science and Food Safety, 9*(5), 513–529.

Bashir, K. M. I., Kim, J., An, J. H., Sohn, J. H., & Choi, J. (2017). Natural food additives and preservatives for fish-paste products: A review of the past, present, and future states of research. *Journal of Food Quality, 2017*, 9675469.

Bogard, J. R., Farmery, A. K., Baird, D. L., Hendrie, G. A., & Zhou, S. (2019). Linking production and consumption: The role for fish and seafood in a healthy and sustainable Australian diet. *Nutrients, 11*(8), 1766.

Bohrer, B. M. (2019). An investigation of the formulation and nutritional composition of modern meat analogue products. *Food Science and Human Wellness, 8*(4), 320–329.

Bosch, A. C., O'Neill, B., Sigge, G. O., Kerwath, S. E., & Hoffman, L. C. (2016). Heavy metals in marine fish meat and consumer health: A review. *Journal of the Science of Food and Agriculture, 96*(1), 32–48.

Boukid, F. (2020). Plant-based meat analogues: From niche to mainstream. *European Food Research and Technology, 247*, 297–308.

Briggs, M. A., Petersen, K. S., & Kris-Etherton, P. M. (2017). Saturated fatty acids and cardiovascular disease: Replacements for saturated fat to reduce cardiovascular risk. *Healthcare, 5*(2), 29.

Bryant, C., Szejda, K., Parekh, N., Deshpande, V., & Tse, B. (2019). A survey of consumer perceptions of plant-based and clean meat in the USA, India, and China. *Frontiers in Sustainable Food Systems, 3*.

Buamard, N., & Benjakul, S. (2019). Effect of ethanolic coconut husk extract and pre-emulsification on properties and stability of surimi gel fortified with seabass oil during refrigerated storage. *LWT-Food Science and Technology, 108*, 160–167.

Cabanillas, B., Jappe, U., & Novak, N. (2018). Allergy to peanut, soybean, and other legumes: Recent advances in allergen characterization, stability to processing and IgE cross-reactivity. *Molecular Nutrition and Food Research, 62*, 1700446.

Dekkers, B. L., Boom, R. M., & van der Goot, A. J. (2018). Structuring processes for meat analogues. *Trends in Food Science & Technology, 81*, 25–36.

Ernstoff, A., Tu, Q., Faist, M., Del Duce, A., Mandlebaum, S., & Dettling, J. (2019). Comparing the environmental impacts of meatless and meat-containing meals in the United States. *Sustainability, 11*(22), 6235.

Fang, Z., Zhao, Y., Warner, R. D., & Johnson, S. K. (2017). Active and intelligent packaging in meat industry. *Trends in Food Science and Technology, 61*, 60–71.

Food and Agriculture Organization of the United Nations. (2019). *OECD-FAO agricultural outlook 2019–2028*. Rome: OECD/FAO.

Geeraerts, W., Vuyst, L. D., & Leroy, F. (2020). Ready-to-eat meat alternatives, a study of their associated bacterial communities. *Food Bioscience, 37*, 100681.

Givaudan. (2020). *New research on plant-based fish and seafood in collaboration with the University of California*. Berkeley. Available from: Givaudan. [15 October 2020].

Grabowska, K. J., Zhu, S., Dekkers, B. L., de Ruijter, N. C. A., Gieteling, J., & van der Goot, A. J. (2016). Shear-induced structuring as a tool to make anisotropic materials using soy protein concentrate. *Journal of Food Engineering, 188*, 77–86.

Grillo, A., Salvi, L., Coruzzi, P., Salvi, P., & Parati, G. (2019). Sodium intake and hypertension. *Nutrients, 11*(9), 1970.

He, J., Evans, N. M., Liu, H., & Shao, S. (2020). A review of research on plant-based meat alternatives: Driving forces, history, manufacturing, and consumer attitudes. *Comprehensive Reviews in Food Science and Food Safety, 19*(5), 2639.

Ismail, I., Hwang, Y. H., & Joo, S. T. (2020). Meat analog as future food: A review. *Journal of Animal Science and Technology, 62*(2), 111–120.

Jian, W., Wu, H., Wu, L., Wu, Y., Jia, L., Pang, J., & Sun, Y. (2016). Effect of molecular characteristics of Konjac glucomannan on gelling and rheological properties of Tilapia myofibrillar protein. *Carbohydrate Polymers, 150*, 21–31.

Kerry, J. P. (2012). *Advances in meat, poultry and seafood packaging*. Amsterdam: Elsevier. Available from: Elsevier. [22 June 2012].

Lafarge, C., Cayot, N., Hory, C., Goncalves, L., Chassemont, C., & Bail, L. P. (2014). Effect of konjac glucomannan addition on aroma release in gels containing potato starch. *Food Research International, 64*, 412–419.

Leonard, W., Zhang, P., Ying, D., & Fang, Z. (2020). Application of extrusion technology in plant food processing byproducts: An overview. *Comprehensive Reviews in Food Science and Food Safety, 19*(1), 218–246.

Link, J. S., & Watson, R. A. (2019). Global ecosystem overfishing: Clear delineation within real limits to production. *Science Advances, 5*(6). eaav0474.

Martinez-Porchas, M., & Martinez-Cordova, L. R. (2012). World aquaculture: Environmental impacts and troubleshooting alternatives. *The Scientific World Journal, 2012*, 389623.

Mi, H., Li, Y., Wang, C., Yi, S., Li, X., & Li, J. (2021). The interaction of starch-gums and their effect on gel properties and protein conformation of silver carp surimi. *Food Hydrocolloids, 112*, 106290.

Pernu, N., Keto-Timonen, R., Lindström, M., & Korkeala, H. (2020). High prevalence of Clostridium botulinum in vegetarian sausages. *Food Microbiology, 91*, 103512.

Poore, J., & Nemecek, T. (2018). Reducing food's environmental impacts through producers and consumers. *Science, 360*(6392), 987–992.

Priyadarshini, M. B., Xavier, K. A. M., Dhanabalan, V., Nayak, B. B., & Balange, A. K. (2020). Development of ready-to-cook shrimp analogue from surimi: Effect of natural plant extracts on the chemical quality during refrigerated storage. *LWT—Food Science and Technology, 135*, 110239.

Remya, S., Basu, S., Venkateshwarlu, G., & Mohan, C. O. (2015). Quality of shrimp analogue product as affected by addition of modified potato starch. *Journal of Food Science and Technology, 52*(7), 4432–4440.

Roland, W. S. U., Pouvreau, L., Curran, J., van de Velde, F., & de Kok, P. M. T. (2017). Flavor aspects of pulse ingredients. *Cereal Chemistry, 94*(1), 58–65.

Schreuders, F. K. G., Dekkers, B. L., Bodnar, I., Erni, P., Boom, R. M., & van der Goot, A. J. (2019). Comparing structuring potential of pea and soy protein with gluten for meat analogue preparation. *Journal of Food Engineering, 261*, 32–39.

Sha, L., & Xiong, Y. L. (2020). Plant protein-based alternatives of reconstructed meat: Science, technology, and challenges. *Trends in Food Science & Technology, 102*, 51–61.

Sikora, M., Tomasik, P., & Krystyjan, M. (2010). Characterization of potato starch fractions and their interaction with hydrocolloids. *Starch/Stärke, 62*(7), 341–349.

Smetana, S., Mathys, A., Knoch, A., & Heinz, V. (2015). Meat alternatives: Life cycle assessment of most known meat substitutes. *The International Journal of Life Cycle Assessment, 20*(9), 1254–1267.

Smith, M., Love, D. C., Rochman, C. M., & Neff, R. A. (2018). Microplastics in seafood and the implications for human health. *Current Environmental Health Reports, 5*(3), 375–386.

Talbot, G. (2015). *Specialty oils and fats in food and nutrition : Properties, processing and applications*. Amsterdam: Elsevier. Available from: Elsevier. [07 July 2015].

U.S. Department of Agriculture. (2020). *Food composition database, USDA food composition database*. Washington: U.S. Department of Agriculture. Available from: https://ndb.nal.usda.gov/ndb [5 October 2020].

Van der Weele, C., Feindt, P., van der Goot, A. J., van Mierlo, B., & van Boekel, M. (2019). Meat alternatives: An integrative comparison. *Trends in Food Science & Technology, 88*, 505–512.

Veronese, N., Solmi, M., Caruso, M. G., Giannelli, G., Osella, A. R., Evangelou, E., Maggi, S., Fontana, L., Stubbs, B., & Tzoulaki, I. (2018). Dietary fiber and health outcomes: An umbrella review of systematic reviews and meta-analyses. *American Journal of Clinical Nutrition, 107*(3), 436–444.

Wild, F. (2016). *Reference module in food sciences*. Amsterdam: Elsevier. Available from: Elsevier. [22 March 2017].

Withana-Gamage, T. S., Wanasundara, J. P., Pietrasik, Z., & Shand, P. J. (2011). Physicochemical, thermal and functional characterisation of protein isolates from Kabuli and Desi chickpea (Cicer arietinum L.): A comparative study with soy (Glycine max) and pea (Pisum sativum L.). *Journal of the Science of Food and Agriculture, 91*(6), 1022–1031.

Yang, D., Yuan, Y., Wang, L., Wang, X., Mu, R., Pang, J., Xiao, J., & Zheng, Y. (2017). A review on konjac glucomannan gels: Microstructure and application. *International Journal of Molecular Sciences, 18*(11), 2250.

Section IV

Fermented plant-based beverages and foods

Chapter 12

Fermented plant-based beverage: kombucha

Elok Zubaidah[1], Ignatius Srianta[2] and Ihab Tewfik[3]

[1]Department of Food Science and Technology, Faculty of Agricultural Technology, Brawijaya University, Jawa Timur, Indonesia; [2]Department of Food Technology, Faculty of Agricultural Technology, Widya Mandala Catholic University Surabaya, Surabaya, Indonesia; [3]School of Life Sciences, University of Westminster, London, United Kingdom

12.1 Introduction

Fermentation is one of the most traditional methods of food preservation. Various fermented food products were originally known for their benefits to increase shelf-life, product safety, and improve organoleptic properties. Today more than ever, fermented foods are fast developing and have much higher functional health benefits. Fermentation results in the transformation of key substrates by microbes into bioactive metabolites. Many biochemical changes occur during fermentation and may improve bioavailability of essential nutrients and consequently enhance the properties of the final product, such as digestibility, utilization, and the bioactivity. Recently, this bioprocess has been applied for the production and extraction of bioactive ingredients from plants in food and beverage industries (Hur et al., 2014). Clinical trials have shown that fermented foods provide health benefits far beyond the starting ingredients. In addition, many fermented foods are rich in enzymes produced by microbes that assist with the digestive process and contain live microbes, which function as probiotics.

Some fermented foods have evidenced therapeutic benefits by virtue of the presence of live microbiome in them imparting probiotics or synbiotics properties. They are also termed: bio-foods, therapeutic foods, or nutraceutical foods and have been the subject of extensive research in the past few decades. The fermentation process generates new metabolites owing to the microbial activity, e.g., antioxidants and B vitamins [folic acid, riboflavin and vitamin B12, which is formed from nonvitamin compounds]. Some essential amino acids and their derivatives that either act as neurotransmitters such as gamma-aminobutyric acid or play an immunomodulatory function could also be produced during fermentation. Moreover, some of the proteins and exopolysaccharides created during fermentation are able to impair the attachment of disease-causing microbes to the intestinal wall and serve as cholesterol-lowering (hypocholesterolemic) agent. One of the fermented foods that have high functional properties is kombucha. The word "kombucha" comes from the Japanese words "seaweed" (Kombu) and "tea" (cha). Kombucha tea is also known as red tea fungus, Champignon de longue vie, Chainii grib, Ling zhi, kocha kinoko, and Chainii kvass (Amarasinghee et al., 2018).

12.2 Kombucha and its properties

Kombucha is a traditional tea beverage produced by fermenting sweet tea using Symbiotic Consortia of Bacteria and Yeast (SCOBY) (Filippis et al., 2018). SCOBY is a symbiotic community of acetic acid bacteria (AAB) Acetobacteraceae and osmophilic yeast (De Filippis et al., 2018). In general, kombucha is made using black or green tea as a raw material by symbiotic cultures of AAB (*Gluconobacter, Acetobacter*), lactic acid bacteria (*Lactobacillus, Lactococcus*), and yeast (*Saccharomycodes ludwiga, Zygosaccharomyces bailii,* etc.) (Villarreal-Soto et al., 2011). As a result of fermentation, the taste of kombucha will change from a fruity, sour, and carbonated taste to a mild vinegary flavor, thereby increasing consumer acceptance of the taste and other sensory aspects. The resulting kombucha tea will taste like apple cider, which is sour and slightly carbonated (Amarasinghee et al., 2018). Kombucha fermentation has traditionally been carried out from

Engineering Plant-Based Food Systems. https://doi.org/10.1016/B978-0-323-89842-3.00011-7

10 to 14 days (Zubaidah et al., 2018a, 2018b), but currently fermentation is generally carried out from 7 to 21 days (Ahmed et al., 2020; Amarasinghee et al., 2018).

Kombucha fermentation is a combination of three components: alcoholic, lactic, and acetic acid; this is because of the coexistence of several yeasts and bacteria in the medium, being initiated by osmotolerant microorganisms and ultimately dominated by acid-tolerant species (Villarreal-Soto et al., 2018). It also originated in China where the "Divine Che" was prized 220 BCE during the Tsin Dynasty for its detoxifying and energizing properties (Roche, 1998).

Various in vivo tests have proven the positive effects of kombucha on some, degenerative diseases, for instance, it is included in diabetes therapy, enhancing liver and pancreatic functions. Kombucha has been used as a therapeutic agent for hyperglycemia and dyslipidemia in diabetic animal models (Aloulou et al., 2012; Srihari et al., 2013). Kombucha is able to provide an inhibitory effect on alpha amylase and lipase activities in blood plasma and pancreas, so that it suppresses the surge in blood glucose levels (Aloulou et al., 2012). Thus, the health benefits thought to be associated with kombucha consumption are numerous and are attributed to the phenolic compounds, which rise during fermentation. In addition, there are organic acids, vitamins, amino acids, antibiotics, and various micronutrients produced during fermentation of kombucha, which are thought to have a role in health benefits (Vijayaraghavan et al., 2000). The literature outlined the multimechanisms through which kombucha assists in diabetes therapy through reduction of damage to pancreatic beta cells caused by oxidative stress so as to increase insulin production, lessen glucose uptake from the digestive system, and improve cellular glucose uptake (Alouluło et al., 2012; Bhattacharya et al., 2011; Srihari et al., 2013). In addition, kombucha is rich in enzymes that are beneficial for digestion, as well as potential as a probiotic drink. Kombucha was illustrated in Fig. 12.1.

Kombucha is known to have several health benefits such as immune response boosting and liver detoxification (Chakravorty et al., 2016) due to the genesis of key metabolites during the fermentation process (Dufresne & Farnworth, 2000) such as acetic acid, gluconic acid, and other compounds such as sugar, ethyl gluconate, oxalate, saccharic, lactate, keto-gluconate, amino acids, water soluble vitamins, tea components (catechins, teaflavins, flavonols), and hydrolytic enzymes (invertase, amylase) (Velićanski et al., 2014). In vitro study by Vázquez-Cabral et al. (2017) showed that kombucha fermented for 28 days has remarkable antioxidant properties that can fight oxidative damage by H_2O_2 in macrophages and can significantly decrease the levels of proinflammatory cytokines IL-6, tumor necrosis factor (TNF)-α, and nitric oxide (NO). (Zubaidah et al., 2020a, 2020b) reported that kombucha tea and snake fruit kombucha were able to significantly enhance the immune response in mice infected with *Salmonella typhi*.

12.3 Microbiology of kombucha

Microbial colonies in kombucha depend on fermentation conditions such as temperature, time, type of microbe selected, raw materials, and the source of sugar used. Hence, there are no unique or standardized kombucha microbes, but some of them are identified as species of AAB, lactic acid bacteria or LAB, and yeast (Morales, 2020). Kombucha consists of two components, namely, the cellulose layer and liquid kombucha. The cellulose layer is also called a biofilm, where the diameter corresponds to the diameter of the fermentation vessel. Microbes that produce cellulose, namely, the type of

FIGURE 12.1 Liquid kombucha and cellulose layer (Zubaidah et al., 2018a, 2018b).

Acetobacter with the dominant species *Acetobacter xylinum*. The latter was reclassified as *Gluconacetobacter xylinus*, then reclassified again as *Komagataeibacter xylinus* (Yamada et al., 2012; Villarreal-Soto et al., 2018). Particularly that the biochemical activity of these bacteria is the oxidation of glucose to gluconic acid (*found in the liquid phase*), then another specific metabolic pathway synthesizes microbial cellulose into biofilms on the surface of the liquid. The composition of microbial colonies in kombucha tea is yeast and AAB, which grow symbiotically, have synergistic activity, and complement each other in the fermentation process.

12.3.1 Yeast

Yeast is facultative anaerobic, which means it can grow well, with or without oxygen. During kombucha fermentation, yeast plays a role in producing alcohol. There are many yeasts genus and species in kombucha culture; a broad spectrum has been reported including species of *Zygosaccharomyces, Candida, Kloeckera/Hanseniaspora, Torulaspora, Pichia, Brettanomyces/Dekkera, Saccharomyces, Lachancea, Saccharomycoides, Schizosaccharomyces,* and *Kluyveromyces* (Chakravorty et al., 2016; Coton et al., 2017; Marsh et al., 2014). *Saccharomyces* sp. being one of the most common types of yeast is found in kombucha. Saccharomyces strains found in kombucha include *Saccharomycodes apiculatus*, *Saccharomycodes ludwigii*, *Schizosaccharomyces pombe*, *Zygosaccharomyces*, and *Saccharomyces cerevisiae*. Another strain is Brettanomyces sp., which also grows in both aerobic and anaerobic conditions and produces alcohol or acetic acid. *Zygosaccharomyces kombuchaensis* is a yeast strain in kombucha that produces alcohol, undergoes carbonation, and also contributes to the formation of SCOBY (Harler, 1964).

12.3.2 Acetic acid bacteria

The acetic acid bacteria that grow on kombucha tea are aerobic, which means that their growth requires oxygen. The main AAB identified in kombucha tea is *Acetobacter*, which produces acetic acid and gluconic acid. The Acetobacter strain is almost found in kombucha and contributes to the formation of the SCOBY fungus. The strains of Acetobacter that are most commonly found in kombucha are *Acetobacter xylinoides* and *Acetobacter ketogenum* (Harler, 1964). *Gluconacetobacter kombuchae* is an anaerobic bacterial strain that was identified from kombucha tea; the bacteria utilizes nitrogen in tea to produces acetic acid and gluconic acid and contributes to the formation of SCOBY (Harler, 1964). *Komagataeibacter xylinus* is also one of the strains of aerobic bacteria, which is known for its ability to produce cellulose (Römling & Galperin, 2015).

12.3.3 Lactic acid bacteria

In some kombucha tea, it was identified that lactic acid bacteria came from the Lactobacillus (aerobic) strain, a type of strain that produces lactic acid and mucus (Harler, 1964). Some lactic acid bacterial strains could be found in kombucha tea or be added to kombucha to increase its antioxidant and antimicrobial capacity as well as increasing the production of glucuronic acid. There are many strains of AAB and yeast, which have been identified in kombucha.

The identified microbial species in kombucha are shown Table 12.1.

During fermentation, yeast hydrolyzes sucrose to glucose and fructose. The emerged glucose and fructose will be fermented by yeast into ethanol, carbon dioxide, and glycerol (Gaggia et al., 2019). Ethanol is then oxidized by AAB to acetic acid so that the pH in kombucha will be eventually decreased (Saichana et al., 2015). These AAB are obligate aerobes (i.e., requiring oxygen for the fermentation process) (Saichana et al., 2015). In static conditions, a cellulose biofilm will be produced on the kombucha surface (Mao et al., 2019). This cellulose has a strong structure and can be used as protection from pathogenic bacteria (Mao et al., 2019).

During the fermentation process, further metabolites are also produced. AAB (*Gluconobacter species*) converts glucose into gluconic acid and glucuronic acid (Jayabalan et al., 2017, pp. 965–978). Meanwhile, yeast will stimulate the production of glucuronic acid by lactic acid bacteria (Nguyen et al., 2015). Other metabolites e.g., 1,4-lactone D-saccharide acid, lactic acid, quinic acid, oxalic acid, malic acid, and citric acid (Laureys et al., 2020) are also generated in the medium.

During kombucha fermentation, a symbiotic relationship between microbes and a selective environment is formed. Ethanol produced by yeast (Pfeiffer & Morley, 2014) and production of organic acids by yeast and bacteria will reduce the pH of kombucha to 2.0–4.0 (Chakravorty et al., 2016; Neffe-Skocinska et al., 2017); such selective medium inhibits the growth of certain microbes.

TABLE 12.1 Microbial species identified in kombucha beverages.

Microorganism	Genus	Species	Source
AAB	*Acetobacter*	*Acetobacter lovaniensis*	Coton et al. (2017)
		Acetobacter peroxydans	Coton et al. (2017)
	Gluconobacter	*Gluconobacter liquefaciens*	Coton et al. (2017)
		Gluconobacter europaeus	Coton et al. (2017)
	Tanticharoenia	*Tanticharoenia sakaeratensis*	Coton et al. (2017)
LAB	*Lactobacillus*	*Lactobacillus nagelli*	Coton et al. (2017)
Yeast	*Candida*	*Candida Boidinii*	Coton et al. (2017)
	Dekkera	*Dekkera anomala*	Coton et al. (2017)
	Hanseniaspora	*Hanseniaspora valbyensis*	Coton et al. (2017)
	Pichia	*Pichia anomala*	Coton et al. (2017)
		Pichia membranifaciens	Coton et al. (2017)
	Torulaspora	*Torulaspora microellipsoides*	Coton et al. (2017)
	Saccharomyces	*Saccharomyces uvarum*	Coton et al. (2017)
	Zygotorulaspora	*Zygotorulaspora florentina*	Coton et al. (2017)
	Zygosaccharomyces	*Zygosaccharomyces bailii*	Coton et al. (2017)
Yeast	*Candida*	*Candida parapsilosis*	Chakravorty et al. (2016)
	Eremothecium	*Eremothecium ashbyii*	Chakravorty et al. (2016)
	Hansemiaspora	*Hanseniaspora vineae*	Chakravorty et al. (2016)
	Kluyveromyces	*Kluyveromyces marxianus*	Chakravorty et al. (2016)
	Pichia	*Pichia mexicana*	Chakravorty et al. (2016)
	Saccharomyces	*Saccharomyces cerevisiae*	Chakravorty et al. (2016)
	Sporopachydermia	*Sporopachydermia lactativora*	Chakravorty et al. (2016)
	Zygowilliopsis	*Zygowilliopsis californica*	Chakravorty et al. (2016)

AAB = Acetic acid bacteria; *LAB* = Lactic acid bacteria.
Source: Morales (2020).

12.4 Kombucha processing

(Zubaidah et al., 2018a, 2018b), explained the process of making kombucha (Fig. 12.2). It starts by boiling water, adding 10% sugar, and 0.8%−1% of tea leaves. After the tea and sugar are evenly mixed, they are allowed to boil for 10 min and then poured into a jar, which is then tightly closed and cooled to room temperature. The jar used is presterilized to prevent contamination from other microbes that can hinder the fermentation process. The tea is then inoculated with the kombucha starter 10% and covered with a cloth to form an aerobic atmosphere. The fermentation process is carried out at a temperature of 20 to 30°C with optimal temperatures in the range of 23 to 28°C. The fermentation vessel should be placed on a level surface to maximize the formation of new SCOBY. Several days later, SCOBY will form and begin to float forming a thin, clear gel-like membrane on the entire surface, located above the parent SCOBY. The fermentation process takes between 10 and 14 days.

The tea leave used as a raw material for making kombucha could be replaced with other ingredients. Some of the ingredients developed for making kombucha include Jerusalem artichoke, echinacea flowers, mint leaves, eucalyptus leaves (Ayed et al., 2017; Vázquez-Cabral et al., 2016; Jayabalan et al., 2014; Lobanova et al., 2016; Yavari et al., 2010, 2011), and other leaves, which are high in antioxidants, e.g., guava leaves, tailings leaves, coffee leaves, and betel leaves can also be used (Zubaidah & Wistiani, 2015). Other ingredients for making kombucha include cherries, grapes, oranges, black currants, apples, dragon fruit and snake fruit (Zubaidah et al., 2018a, 2018b), dates and turmeric (Zubaidah, et al., 2020a, 2020b).

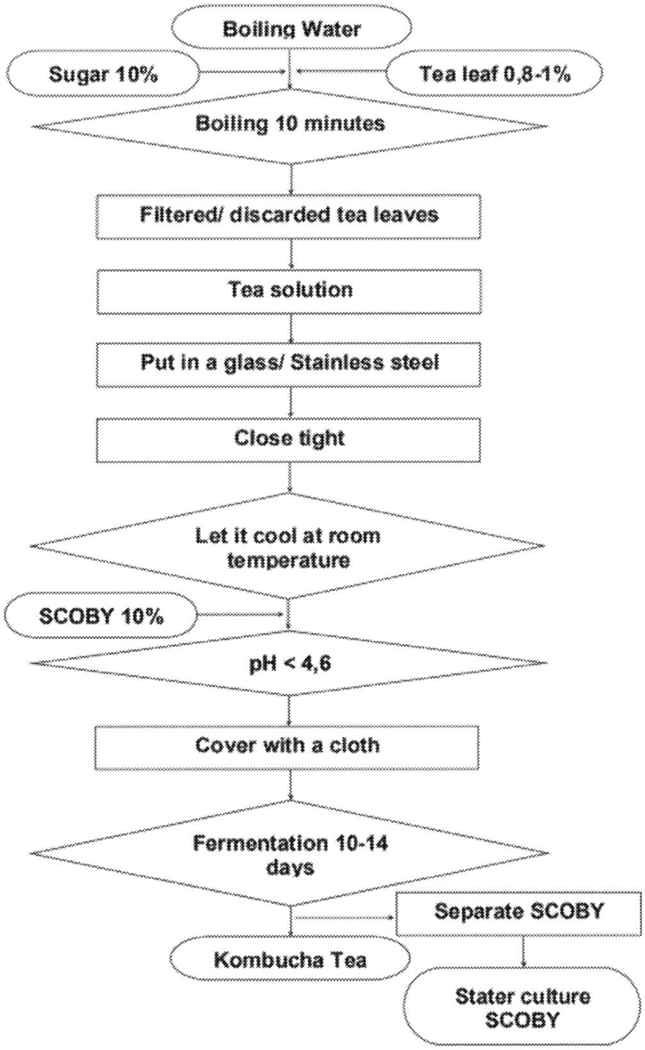

FIGURE 12.2 The process of making kombucha tea (Zubaidah et al., 2018a, 2018b).

12.4.1 Preparation of fermentation medium

In a liter of boiling water, 8–10 g of tea and 10% sucrose sugar are added, and then the tea is allowed to boil for 10 min. Afterward the tea is poured into a fermentation container. The fermentation container used could be made from glass or stainless steel that has been presterilized, tightly closed, and is at room temperature.

12.4.2 Inoculation

The inoculation process is carried out after the tea has cooled to room temperature. Kombucha culture, which is a symbiosis of bacteria and yeast (SCOBY), is added and as much as 10% (v/v) of the kombucha liquid from the previous fermentation. Sheet-shaped "tea mushrooms" can be added as much as 3% (w/v) (Jayabalan et al., 2008). Malbaša et al. (2006) revealed that the kombucha fermentation is faster with the addition of liquid starter compared to simply adding "tea mushroom." According to Ma et al. (2008), adding pure culture can be done to replace mixed cultures, with a combination of *Saccharomyces cerevisiae* 108 CFU/mL, *Gluconacetobacter* sp 108 CFU/mL, and *Lactobacillus plantarum* 108 CFU/mL with a ratio of 1: 1: 1.

12.4.3 Fermentation

The fermentation container is tightly closed with a breathable cloth, then left for 10–14 days. The length of kombucha fermentation depends on the ambient temperature and also the desired acidity level. The fermentation temperature ranges from 20 to 30°C. The longer the fermentation process, the higher the organic acids and phenols content (Leal et al., 2018). However, if the fermentation is continued for a longer time, the acidity level will increase and could cause potential harm if consumed. Fermentation can be stopped by separating the kombucha culture from the medium and then transferring the kombucha liquid into a closed bottle container (Greenwalt et al., 2000). During the fermentation process, microbes will produce primary and secondary metabolites, including acetic acid, glucuronic acid, alcohol, and nata/cellulose (Villarreal et al., 2018). The main acids present are acetic, gluconic, tartaric, malic, and in less proportion, citric acid. All these acids are responsible for its characteristic sour taste (Jayabalan et al., 2007). The composition and concentration of kombucha metabolite compounds are influenced by the source of the inoculum (Nguyen et al., 2015), the concentration of tea and sugar (Fu et al., 2014; Watawana et al., 2017), and the length of fermentation time (Zubaidah et al., 2018a, 2018b).

As outlined earlier, in static conditions, a cellulose biofilm will be produced on the kombucha surface (May et al., 2019). AAB synthesize and polymerize cellulose chains to form fibril tissue in SCOBY. The following is a reaction to the formation of glucuronic acid in kombucha (Prescott & Dunn, 1959; Wood, 1998). Synthesis and polymerization of cellulose chains to form fibril networks begins with the synthesis of uridine diphospho-glucose (UDPGlc), which is a precursor to cellulose and polymerization of β-1,4-glucan chains by single cell acetobacter bacteria. The production of cellulose in this way provides the advantage that bacteria can grow rapidly under controlled conditions and can produce cellulose from a variety of carbon sources including glucose, ethanol, sucrose, and glycerol (Villarreal-Soto et al., 2018).

In the first stage, cellulose-producing bacteria increase the population by utilizing dissolved oxygen and synthesizing cellulose membranes in the liquid phase. The thickness of the membrane continues to increase to form a solid structure in the culture medium. The development of the membrane together with hydrogen and C—H bonds will continue and then stop when the membrane begins to grow covering the entire surface and trapping all bacteria, causing the bacteria to no longer be active due to insufficient oxygen supply (Villarreal-Soto et al., 2018). These bacteria will reactivate when used as an inoculum in fermentation (Villarreal-Soto et al., 2018). The cellulose formed varies depending on many factors: the strain used, culture time, and chemical additives added to the fermentation medium (Villarreal-Soto et al., 2018). Several factors that need to be controlled to maximize cellulose yield, e.g., volume of inoculated media, incubation time, and surface area (Villarreal-Soto et al., 2018).

12.4.4 Composition of kombucha

Kombucha consists of two components, namely, the cellulose layer and liquid kombucha. Some of the compounds, which have been identified in kombucha, include organic acids, sugars (sucrose, glucose, and fructose), water-soluble vitamins, amino acids, biogenic amines, purines, pigments, lipids, proteins, hydrolytic enzymes, ethanol, carbon dioxide, polyphenols, minerals (Mn, Fe, Ni, Co, Zn, Cu, Cd, Pb), anions (fluorides, chlorides, bromides, iodides, nitrates, phosphates, and sulfates), D-saccharic acid-1,4-lactone, and metabolic products from yeast and bacteria (Leal et al., 2018). During the fermentation process of sugared tea infusion, disaccharides undergo decomposition into monosaccharides under the influence of enzymes and acids, i.e., simple sugars. Afterward, the role of yeasts in the fermentation process is to convert sugar into alcohol and carbon dioxide. Concurrently, AAB lives on the cellulose part of the kombucha starter culture. They play a vital role in the fermentation process because they are responsible for creating new cellulose layers, and moreover, they metabolize alcohol produced by yeasts into organic acids. Apart from organic acids, the basic metabolites in the fermentation process of kombucha are simple sugars, carbon dioxide, and ethanol (Chakravorty et al., 2016). According to Chen and Liu (2000), Lončar, et al. (2000), Malbaša et al. (2002) and Jayabalan et al. (2007), acetic, glucuronic and gluconic acids are the main healthy products of kombucha drink, which is produced from sucrose and black tea in a controlled and optimized fermentation process. According to Nguyen et al. (2015), a right choice of fermentation conditions and selection of kombucha starter culture enables production of beverages with the highest content of healthy organic acids, including glucuronic acid. Apart from acetic acid, it is one of the main organic acids resulting from fermentation (Jayabalan et al., 2007).

Kombucha has a total organic acid content of 2.5 g/L, dominated by acetic acid (1.55 g/L) tartaric (0.23 g/L) and citric acid (0.05 g/L) (Ivanišova et al., 2019). Acetic acid is a chemical compound that plays a role in the smell and sour taste of kombucha. This compound is produced by AAB through the conversion of sucrose to ethanol, glucose and fructose (Spedding, 2015). The amount of acetic acid is strongly influenced by the length of fermentation.

Jayabalan et al. (2007) showed that the acetic acid value in kombucha black tea ranged from 0.22 g/L (fermentation on day 3) to 9.51 g/L (fermentation on day 15). Oxalic, malic, lactic, tartaric and glucuronic acids are also found in kombucha. Glucuronic acid is considered as one of the key components in kombucha tea because it has the ability to detoxify by conjugation (Jayabalan et al., 2007).

Organic acids have the ability to capture reactive oxygen species (ROS) (Sentjurc et al., 2003) and increase the biological availability of phenolic compounds (van den Berg et al., 2003). Apart from that, organic acids can also be used as antimicrobials (Mau et al., 2002). Several studies have shown that organic acids can capture free radical compounds and have an antioxidant effect because they have the ability to chelate so they can deactivate reduced cations (Mau et al., 2004).

12.5 Functionality of kombucha

There are many reports that kombucha may help prevent chronic diseases and has beneficial properties for human health such as antihyperglycemic, antimicrobial, antioxidant, anti-carcinogenic and antihyperlipidemic; however, most of the benefits were stated in models (Chakravorty et al., 2016; Dufresne & Farnworth, 2000; Mo et al., 2008). Kombucha consists of the following favourable components: organic acids, vitamins, polyphenols, amino acids, antibiotics and microelements (Jayabalan et al., 2014; Marsh et al., 2014). Some scientific publications confirm induction of synthesis of vitamin C, B vitamins (e.g., folic acid) during the fermentation process of kombucha (Jayabalan et al., 2014; Malbaša et al., 2002). The most important organic acids produced during fermentation are glucuronic, gluconic, lactic, malic, citric, tartaric, folic, malonic, oxalic, succinic, pyruvic and usnic acids. Glucuronic acid deserves special attention: It is the result of a microbiological process of glucose oxidation and it is one of the most valuable, healthy kombucha components. Being produced by the liver in a human body, it exhibits detoxifying effects. Glucuronic acid has an ability to bind with xenobiotics, including phenols present in the liver, allowing these substances to be excreted by kidneys more efficiently. Glucuronic acid is also a precursor in the biosynthesis of vitamin C (Jayabalan et al., 2014; Nguyen et al., 2015).

12.5.1 Antioxidant activity

During fermentation there is an increase in antioxidant activity. The increase in antioxidant activity is followed by an increase in phenol levels. The increase in polyphenol compounds is also triggered by several enzymes produced during fermentation by yeast such as vinyl phenol reductase, ferulic acid reductase, vinyl phenol reductase enzymes, which are enzymes that catalyze the reaction of 4-vynilphenol + NAD + + 3H + to 4-ethylphenol + NADH and are commonly found in *Brettanimyces bruxellensis* (Tchobanov et al., 2008). The ferulic acid reductase enzyme is known as a phenol-forming enzyme through the decarboxylation of cinnamic acid and ferulic acid produced by bacteria. Cinnamic acid was identified as a phenol compound while firulic acid was identified as a derivative of cinnamic hydroxy acids. Both are able to act as antioxidants. The production of compounds that have radical scavenging properties depends on the culture period and the origin of the starter from which they determine which metabolites to produce (Watawana et al., 2015). The total phenol value in kombucha increases linearly during the fermentation process. According to research by Ivanišova et al. (2019), the total polyphenols in kombucha are 412 mg GAE/L, this value is higher than sweet tea (180.17 mg GAE/L). This is in accordance with Aidoo's research (2015) which shows that the total phenol content in kombucha is higher than unfermented tea. During the kombucha fermentation process, thearubigin depolymerization occurs which causes an increase in the total amount of phenol (Chu & Chen, 2006), besides that, polyphenols also undergo transformation and structural damage by microorganisms, which causes the color change in kombucha to become brighter (Chu & Chen, 2006).

Kombucha has a higher antioxidant content than unfermented tea (Fu et al., 2017) because it contains polyphones and metabolite compounds such as vitamins and organic acids (Essawet et al., 2015). Research by Fu et al. (2013) using green tea, black tea, and commercial tea showed that black tea had the highest DPPH binding ratio (95.30%), while green tea was 38.7%. However, the DPPH binding ratio in black tea can increase to 70% within 15 days of fermentation (Chu & Chen, 2006). Research by Jayabalan et al. (2008) showed that among the kombucha samples of green tea, black tea, and commercial teas analyzed, green tea had the highest ability to bind DPPH (88%) on fermentation on the 18th day.

The ability of kombucha to scavenge free radicals is also very diverse, depending on the microorganism and its metabolic process (Liu et al., 1996; Mayser et al., 1995) and its value increase 1.7 times higher after fermentation (Chu & Chen, 2006). This increase in ability to fight against DPPH radicals is ascribed to kombucha significant capacity to reverse chromate (VI) or compounds that cause oxidative damage (Dipti et al., 2003; Ram et al., 2000).

Chu and Chen's research (2006) showed that kombucha that has been fermented for 15 days has 1.4 times higher antioxidant level compared to regular tea, where the length of fermentation time has a positive correlation with the antioxidant yield. The effect of scavenging or free radical binding by catechins in tea varies widely, depending on the type of radical compounds, as well as antioxidant properties such as polarity, ionization level, and steric inhibition (Rice-Evans et al., 1996).

The difference in the composition of kombucha can be caused by the use of different starters, resulting in variability of microflora in each sample (Chen & Liu, 2000; Jayasekera et al., 2011; Reiss, 1994); this causes differences in antioxidant activity on each substrate (Malbaša et al., 2011).

12.5.2 Antibacterial activity

Kombucha is a fermented drink that has antibacterial activity against pathogenic bacteria. Various studies have demonstrated this antibacterial property against *Staphylococcus aureus, Pseudomonas aeruginosa, Escherichia coli, Bacillus cereus, Agrobacterium tumefaciens, Salmonella choleraesuis, and Helicobacter pylori* (Rinihapsari & Catur, 2008). By virtue of the fermentation process, the produced acetic acid destroys a number of Gram-positive and Gram-negative bacteria. Acetic acid acts as an antibacterial by penetrating the cell wall and denature cell's plasma proteins (Sreeramulu et al., 2000). During the fermentation process, kombucha culture, which is a symbiosis between bacteria and yeast, will metabolize the substrates resulting antibacterial activity, which increases with the longer kombucha fermentation time. According to Dufresne and Farnworth (2000), kombucha has the activity of inhibiting the growth of pathogenic microbes such as *Escherichia coli* (causes diarrhea), *Helicobacter pylori* (causes stomach ulcers), *Entamoeba cloacae, Pseudomonas aeruginosa, Staphylococcus aureus, Staphylococcus epidermis, Agrobacterium tumefaciens, Bacillus cereus, Aeromonas salmonella enter, Salmonella, Shigella sonnei, Leuconostoc monocytogenes, Yersinia enterocolitica, Campylobacter jejuni, and Candida albicans*. According to Jayabalan et al. (2014), kombucha's antibacterial activity is supported by low pH because kombucha contains acetic acid and several types of organic acids and catechins that support bacterial inhibitory properties.

Frank (1991) reported that kombucha tea is antibacterial against *Helicobacter pylori*, a bacteria that can cause gastritis. This drink contains vitamins B1, B2, B6, B12, folic acid, and glucuronic acid. These acidic compounds form bonds with unused metabolic waste, with drugs and poisons, thus helping the detoxification process. In general, the kombucha fermentation process occurs symbiosis between *Acetobacter xylinum* and the yeast *Saccharomyces cerevisiae.* This symbiosis produces acids and alcohols that block the growth of foreign microbes that do not come from the kombucha tea fungus (Blanc, 1996; Cvetkovic & Markov, 2002).

Kombucha contains 7 g/L (0.7%) acetic acid. Acetic acid or ethanoic acid (CH 3COOH) is a colorless liquid that has a sharp aroma and a sour taste. Acetic acid ionizes weakly in dilute solutions and has long been used as a preservative because of its antimicrobial effect. Acetic acid in kombucha has in vitro antibacterial activity against *Salmonella choleraesius*, serotype typhimurium, and *Agrobacterium tumifaciens* (Naidu et al., 2000).

12.5.3 Antidiabetes activity

Various in vivo tests have proven the positive effects of kombucha on degenerative diseases, including diabetes therapy, improving liver, and pancreatic function. kombucha was used as a therapeutic agent for hyperglycemia and dyslipidemia in diabetic animal models (Aloulou et al., 2012; Srihari et al., 2013). Kombucha has the capacity to provide an inhibitory effect on alpha amylase and lipase activity in blood plasma and pancreas so that it can suppress the increase in blood glucose levels (Aloulou et al., 2012). The health benefits thought to be associated with kombucha consumption are numerous and are attributed to the phenolic components, which increase during fermentation. In addition, there are organic acids, vitamins, amino acids, antibiotics, and various micronutrients produced during fermentation of kombucha, which are thought to have a role in health benefits (Vijayaraghavan et al., 2000). Several studies have shown that kombucha can be used as a diabetes therapy agent through several mechanisms to reduce damage to pancreatic beta cells caused by oxidative stress so as to increase insulin production, reduce glucose uptake from the digestive system, and increase cellular glucose uptake (Aloululo et al., 2012; Bhattacharya et al., 2011; Srihari et al., 2013).

12.5.3.1 Fasting plasma glucose level reduction

Zubaidah et al. (2019) compared snake fruit and black tea kombucha, and metformin as diabetes therapy agents. The research investigated the antidiabetic activity of snake fruit kombucha, black tea kombucha, and metformin in streptozotocin-induced diabetic rats. Snake fruit kombucha, black tea kombucha, and metformin were orally administered to the diabetic rats daily for 28-day experiment. The blood plasma was analyzed, and pancreas immunohistochemical study and β-cells quantification were carried out.

Changes in the FPG levels before and after the treatment are presented in Table 12.2. The FPG levels among the DM rat groups (P1, P2, P3, and P4) on day 0 were not significantly different ($P > .05$), while there were significant differences

TABLE 12.2 Effect of the kombucha and metformin administration on indices of the fasting plasma glucose (FPG) levels in the rats (Zubaidah et al., 2019).

	Fasting plasma glucose (FPG) levels (mg/dL)	
Treatment	**Day 0**	**Day 28**
P0 (normal rat)	119.5 ± 8.9 a	104.2 ± 3.4 a
P1 (diabetes mellitus)	463.7 ± 36.6 b	413.2 ± 3.8 c
P2(DM + kombucha tea)	473.7 ± 58.7 b	154.0 ± 54.4 b
P3(DM + snake fruit kombucha)	453.0 ± 12.9 b	110.2 ± 2.8 a
P4 (DR + metformin)	417.2 ± 59.3 b	104.0 ± 3.5 a

Note: 1. Different notations show significant differences ($\alpha = 0.05$).

($P < .05$) within the normal rats (P0) group. At the end of the 28-day treatment, the normal and DM rats appeared to show constant FPG levels, whereas the FPG levels of the DM groups P2, P3, and P4 decreased. The DM group with the KS (P3) and metformin (P4) administration showed lower FPG levels than the DM group with the KT (P2).

Metformin is an antidiabetic drug, and it works by increasing the sensitivity of the liver and peripheral tissues to insulin without affecting insulin secretion and increasing glucose uptake in the peripheral tissues to reduce insulin resistance (Iida et al., 2003). The snake fruit kombucha decreased the level of FPG because of its high content of antioxidant compounds such as phenolic, tannin, and other bioactives, as well as some other organic acids, such as citric, lactic, butyric, and propionic acids (Östman et al., 2005; Zubaidah et al., 2018a). Antioxidant compounds decrease FPG levels through increasing cellular glucose uptake. Increased insulin secretion will have implications for the body's ability to utilize blood glucose for normal metabolism. Glycolysis, glycogenesis, and lipogenesis are some of the metabolisms of insulin-regulated glucose (Dufresne & Farnworth, 2000; Aloulou et al., 2012). Antioxidant compounds can also inhibit glucose absorption in the small intestine by decreasing glucose-conducting activity, such as sodium-glucose transport protein 1 (GLUT1), glucose transporter 5 (GLUT 5), and glucose transporter 2 (GLUT 2). The gross effect of these is to reduce the glucose that enters the bloodstream (Kwon et al., 2007). Consequently, it can be inferred that the snake fruit kombucha had a comparable effect as the metformin and a better effect on FPG than the black tea kombucha.

12.5.3.2 Pancreas immunohistochemistry study

The result of IHC staining is shown in Figs. 12.3 and 12.4 (Zubaidah et al., 2019). The results showed increasing Langerhans island structures and insulin secretions in the three treated (kombucha and metformin) groups (Fig. 12.3). The size and shape of the structures from the DM group were irregular and smaller than those of the normal group (P0) and the three groups of KT, KS, and metformin. In addition, the DM group showed a very low immunoreactive response (brown color) to anti-insulin, which indicated low levels of insulin production. The DM rat groups with the kombucha and metformin improved in the Langerhans island structures, and in the size, shape, distributions, and numbers of the β-cells, as well as the high intensity of brown color when compared with the DM rats (Fig. 12.3).

The DM group with metformin (P4) had a high number of pancreatic β-cells, which is not significantly ($P > .05$) different from the snake fruit kombucha. Metformin has adenosine monophosphate-activated protein kinase (AMPK) enzyme that plays a role in repairing HbA1c, the main parameter of blood glucose. In addition, it can reduce hepatic glucose production, lower LDL and triglyceride levels, increase HDL levels, decrease platelet aggregation, increase fibrinolytic activity and improve weight, reduce the risk of hypoglycemia, and increase insulin sensitivity with a concomitant improvement of pancreatic performance (Gunton et al., 2003). The DM group with the snake fruit kombucha treatment had a pancreatic β-cells higher than the black tea kombucha, possibly due to its higher antioxidant compounds as discussed above that can protect and repair pancreatic β-cells and increase insulin secretions (Zubaidah et al., 2018a, 2018b).

12.5.4 Potential of kombucha as an immunomodulator

The immune system is a vital mechanism used to protect the body against foreign objects, microbes (e.g., bacteria, fungi, and parasites), viruses, cancer cells, and toxins that can cause a decrease in immune system and cause disease (Mao et al., 2019).

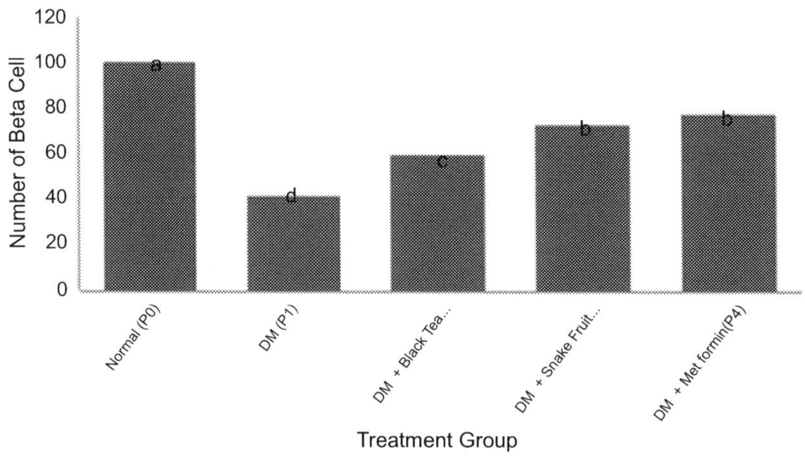

FIGURE 12.3 Effect of the kombucha and metformin administration on the number of pancreatic β-cells (Zubaidah et al., 2019) Values are mean ± SD of β = cells of five Langerhans islands. The same letter indicates no significant difference ($P > .05$). DM: diabetes mellitus.

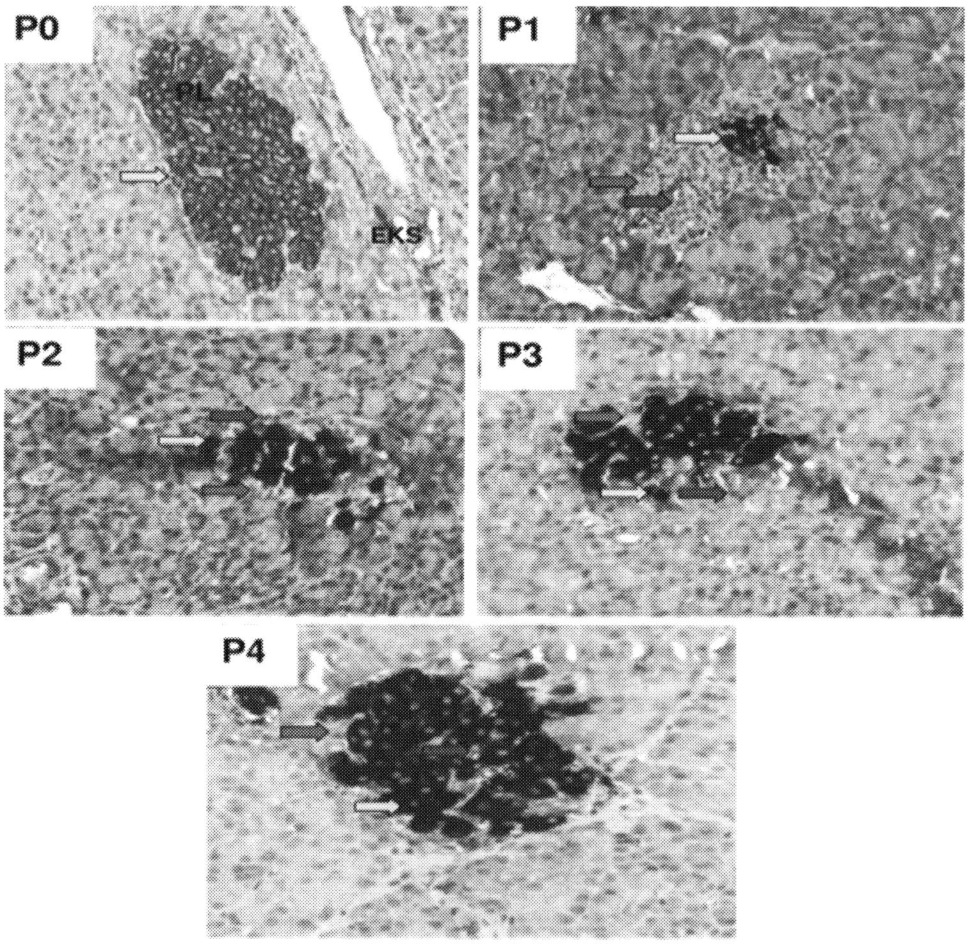

FIGURE 12.4 Effect of the kombucha and metformin administration on the pancreatic cells in rats evaluated by the IHC staining (400× magnification) (Zubaidah et al., 2019). PL: Langerhans Island, EKS: Exocrine glands (acini). Yellow arrow (light gray in print): Pancreatic β-cells, which have immunoreactivity to anti-insulin. Green arrow (gray in print): Endocrine cells, which are not immunoreactive to anti-insulin. Red arrow (dark gray in print): Empty space by necrosis, DM: Diabetes Mellitus, KT: black tea Kombucha, KS: snake fruit Kombucha, BW: Body Weight.

The immune system consists of two [2] types, namely.

1. the innate immune system (innate). It is the main defense against attack by foreign microorganisms (e.g., pathogens), has a fast response and does not show antigen in its working system. Innate immune cells include granulocytes, mast cells, macrophages, and dendritic cells (DCs)
2. the adaptive immune system has a slow response for days but develops to form immunological memory (Turvey & Broide, 2010; Bonilla & Oettgen, 2010; Murphy et al., 2007).

Immunomodulatory mechanisms are enhancing the immune system by means of stimulation (immunostimulants) and normalizing abnormal immune reactions (immunosuppressants). Immunomodulators are associated with increased nonspecific (innate) and specific (adaptive) immune responses, which are expected to restore a compromised immune system (Mao et al., 2019).

Several natural immunomodulatory compounds can enhance the immune system by affecting immune cell function or influencing the secretion of antibodies to control infection and maintain homeostatic conditions (Ortuño-Sahagún et al., 2017).

Research by Sofiakmi et al. (2015) showed that kombucha has immunomodulatory activity due to the presence of polyphenols, flavonoids, and organic acids. The most immunomodulatory activity is shown by acetic acid, glucuronic acid, lactic acid, and ascorbic acid. The increase in these compounds can induce the immune system as it escalates the response of T lymphocytes, improves the immune system, and stimulates lymphocyte cell proliferation (Arnas, 2009). In addition to some of the compounds above, probiotic bacteria in kombucha can also prevent infection because of the presence of T helper cells (Th1, Th2, Th17, and T regulator cells), affecting the production of various cytokines by DCs and macrophages, especially interleukin-12 (Kawashima et al., 2011).

Immunostimulants are substances that raise the body's immune system against foreign bodies, infectious diseases, tumors, and immunodeficiency. Immunostimulants can improve the immune system by using substances that can stimulate the immune system (Bascones-Martinez et al., 2014). Immunosuppressants are used to suppress the immune response, especially in the organ transplant process to prevent rejection reactions, and to treat autoimmune diseases such as pemphigus, lupus, or allergies (Lee et al., 2010).

Flavonoids such as quercetin and epigallocatechin 3-gallate (EGCG) in black tea increase the body's defense system through the increased mechanism of Interleukin-6 (IL-6) and IL-10, which affects the activation of T and B lymphocytes (Neiman et al., 2009). In addition, flavonoids can also modulate the immune system by inhibiting Th1, Th2, TNF-α, IL-2, IFN-γ, IL-12, and IL-6 cells. The inhibition of TNF-α and IL-12 can affect the activity of phagocytes and NK cells (Peluso et al., 2015).

Vitamin C in kombucha supports innate and adaptive immunity by supporting the function of the epithelial barrier against pathogens so that it can prevent pathogens. Vitamin C increases the differentiation and proliferation of B cells and T cells. Vitamin C deficiency results in impaired immunity and susceptibility to infection (Carr & Maggini, 2017).

Innate immunity is a specific immune system response due to the entry of pathogenic microorganisms and other foreign substances through phagocytosis by neutrophils and monocytes (macrophages), chemical barriers through internal and external secretions (lysozymes in mucus tissue, tears, lactoperxidase in saliva) and blood proteins (interferon, kinin system, complement), and natural killer (NK) cells (Fitzsimmons et al., 2014).

(Zubaidah et al., 2020a) conducted an investigation to compare the potential of kombucha tea Arabica coffee leaves and kombucha black tea as immunomodulators in mice infected with *Salmonella typhi*. Immune response analyses were carried out on the adaptive immune response of T cells with CD4 +, CD8 +, IFN-γ +, and TNF-α + cytokines are presented in Table 12.3.

Mice infected with *S. typhi* bacteria triggered an innate and adaptive immune response. Upon entering the body, the bacteria will phagocyte so that it stimulates macrophages so that it will increase interleukin 12 to activate NK cells. With the activation of NK cells, interferon-gamma is produced, which functions as a bacterial eliminator (Agnello et al., 2003; Mosser, 2003)

The mechanism of increasing CD4 +, CD8 +, CD4 + IFNɣ, CD4 + TNFα, CD8 + IFNɣ, and CD8 + TNFα. Cells took place due to the presence of phenolic compounds found in kombucha arabica coffee leaf tea and kombuchas black tea such as flavonoids and tannins. These bioactive compounds can increase the production of interleukin-2, which can stimulate a Th1 response by activating CD4 cell differentiation. The Th1 response increases the production of gamma interferon, which activates macrophages by releasing reactive O_2 compounds, which are toxic to microbes (Abbas & Lichman, 2011, pp. 113–121; Cheeke, 2000; Hoffmann et al., 2006; Lyu & Park, 2005; Qureshi et al., 2012).

Flavonoid compounds have been shown to increase IL-2 and lymphocyte proliferation (Nopitasari, 2006a, 2006b). The administration of kombucha black tea can increase the proliferation of lymphocyte cells of Falur BALB/c mice in vitro. In

TABLE 12.3 Immunomodulatory activity of kombucha black tea and kombucha tea arabica coffee leaves (Zubaidah et al., 2020a).

Group	$CD4^{+}IFN\ \gamma^{+}$ (%)	$CD4^{+}TNF\ \alpha^{+}$ (%)	$CD8^{+}IFN\ \gamma^{+}$ (%)	$CD8^{+}TNF\ \alpha^{+}$ (%)
Negative control - healthy mice (P0)	$0.26^{c} \pm 0.04$	$0.47^{c} \pm 0.01$	$0.43^{d} \pm 0.17$	$0.70^{d} \pm 0.17$
Healthy mice with kombucha arabica coffee leaf tea (P1)	$0.42^{c} \pm 0.10$	$0.65^{bc} \pm 0.05$	$0.67^{cd} \pm 0.27$	$1,19^{d} \pm 0.28$
Healthy mice with black tea kombucha (P2)	$0.53^{c} \pm 0.06$	$0.75^{bc} \pm 0.05$	$0.78^{c} \pm 0.20$	$1.34^{cd} \pm 0.51$
Positive control - infection mice (P3)	$0.62^{bc} \pm 0.10$	$0.79^{bc} \pm 0.17$	$1.39^{b} \pm 0.06$	$1.96^{bc} \pm 0.03$
Mice infection with kombucha arabica coffee leaf tea (P4)	$0.97^{b} \pm 0.34$	$1.75^{ab} \pm 0.22$	$1.89^{a} \pm 0.01$	$2.49^{ab} \pm 0.51$
Mice infection with kombucha black tea (P5)	$1.84^{a} \pm 0.22$	$2.13^{a} \pm 1.30$	$2.07^{a} \pm 0.02$	$3,15^{a} \pm 0.03$

Note: Different notations show significant differences ($\alpha = 0.05$) 2. Data obtained from the mean of four replications ± SD.

this study, the researchers outlined the effect of giving kombucha arabica coffee leaf tea and kombucha black tea on the increase in CD4 +, CD8 +, CD4 + IFNɣ, CD4 + TNFα, CD8 + IFNɣ, and CD8 + TNFα. cell. A significant increase occurred in the black kombucha group. The increase in TNF alpha is due to an ongoing infectious process (Sofiakmi et al., 2015).

The kombucha mechanism provides immunomodulatory effects such as mitogens that attach to the surface of lymphocyte cells. Bioactive compounds in this case such as flavonoids are thought to be ligands that will bind to the surface of the T and B cell receptors so that signal transduction activation occurs through second messengers including IP3 (Inositol triphosphate), which results in stimulation of Ca2 + to the cytoplasm so that the concentration increases. This increase will stimulate the activation of protein kinase C. This activation of protein kinase C will trigger cell-level gene expression such as the expression of interleukin 2 products, which causes the activation of B cells and T cells for proliferation (Erniati, 2018; Gomez-Florez et al., 2008; Schechter, 2010; Subaryono et al., 2017).

12.6 Summary and future direction

Currently there is a lot of information about the health benefits of kombucha based on people's experiences. However, research focusing on the in vivo effects of kombucha on human health is still unfolding. Along with the development of kombucha, which is currently not only made from tea leaves, but also based on various fruits.

In vivo research opportunities for the health effects of kombucha, both tea-based and fruit-based, are still possible. Nevertheless, research on possible side effects of consumption of kombucha has not been reported.

References

Abbas, A., & Lichman, A. H. (2011). *Basic immunology 3 updated edition*. Elsevier.

Agnello, D., Lankford, C. S. R., Bream, J., Morinobu, A., Gadina, M., O'Shea, J. J., & Frucht, D. M. (2003). Cytokines and transcription factors that regulate T helper cell differentiation: New players and new insights. *Journal of Clinical Immunology,* 23(3), 147–161. https://doi.org/10.1023/A:1023381027062

Ahmed, R. F., Hikal, M. S., & Abou-Taleb, K. A. (2020). Biological, chemical and antioxidant activities of different types Kombucha. *Annals of Agricultural Sciences,* 65(1), 35–41. https://doi.org/10.1016/j.aoas.2020.04.001

Aidoo, E. (2015). *Studies on the Cytotoxicity and antioxidant Activity of tea kombucha. Doctoral Dissertation.*

Aloulou, A., Hamden, K., Elloumi, D., Ali, M. B., Hargafi, K., Jaouadi, B., Ayadi, F., Elfeki, A., & Ammar, E. (2012). Hypoglycemic and antilipidemic properties of kombucha tea in alloxan-induced diabetic rats. *BMC Complementary and Alternative Medicine, 12*. https://doi.org/10.1186/1472-6882-12-63

Amarasinghe, H., Weerakkody, N. S., & Waisundara, V. Y. (2018). Evaluation of physicochemical properties and antioxidant activities of kombucha "Tea Fungus" during extended periods of fermentation. *Food Science and Nutrition,* 6(3), 659–665. https://doi.org/10.1002/fsn3.605

Arnas, Y. (2009). *Pengaruh pemberian seduhan teh hitam (Camellia sinensis) dengan dosis Bertingkat terhadap proliferasi limfosit mencit balb/C yang diinokulasi Salmonella typhimurium. Skripsi.*

Ayed, L., Ben Abid, S., & Hamdi, M. (2017). Development of a beverage from red grape juice fermented with the Kombucha consortium. *Annals of Microbiology, 67*(1), 111–121. https://doi.org/10.1007/s13213-016-1242-2

Bascones-Martinez, A., Mattila, R., Gomez-Font, R., & Meurman, J. H. (2014). Immunomodulatory drugs: Oral and systemic adverse effects. *Medicina Oral, Patologia Oral Y Cirugia Bucal, 19*(1), e24–e31. https://doi.org/10.4317/medoral.19087

Blanc, P. J. (1996). Characterization of the tea fungus metabolites. *Biotechnology Letters, 18*, 139–142. https://doi.org/10.1007/BF00128667

van den Berg, P. M., Korkmaz, E., & Abubakar, A. (2003). A constrained conjugate gradient method for solving the magnetic field boundary integral equation. *IEEE Transactions on Antennas and Propagation*, 51.

Bhattacharya, S., Gachhui, R., & Sil, P. C. (2011). Hepatoprotective properties of kombucha tea against TBHP-induced oxidative stress via suppression of mitochondria dependent apoptosis. *Pathophysiology, 18*(3), 221–234. https://doi.org/10.1016/j.pathophys.2011.02.001

Bonilla, F. A., & Oettgen, H. C. (2010). Adaptive immunity. *Journal of Allergy and Clinical Immunology, 125*(2), S33–S40. https://doi.org/10.1016/j.jaci.2009.09.017

Carr, A. C., & Maggini, S. (2017). Vitamin C and immune function. *Nutrients, 9*(11). https://doi.org/10.3390/nu9111211

Chakravorty, S., Bhattacharya, S., Chatzinotas, A., Chakraborty, W., Bhattacharya, D., & Gachhui, R. (2016). Kombucha tea fermentation: Microbial and biochemical dynamics. *International Journal of Food Microbiology, 220*, 63–72. https://doi.org/10.1016/j.ijfoodmicro.2015.12.015

Cheeke, P. R. (2000). Actual and potential application of yucca schidigera and quillaja saponaria saponin in human and animal nutrition. *Journal of Animal Science, 77*, 1–10.

Chen, C., & Liu, B. Y. (2000). Changes in major components of tea fungus metabolites during prolonged fermentation. *Journal of Applied Microbiology, 89*(5), 834–839. https://doi.org/10.1046/j.1365-2672.2000.01188.x

Chu, S. C., & Chen, C. (2006). Effects of origins and fermentation time on the antioxidant activities of kombucha. *Food Chemistry, 98*(3), 502–507. https://doi.org/10.1016/j.foodchem.2005.05.080

Coton, M., Pawtowski, A., Taminiau, B., Burgaud, G., Deniel, F., Coulloumme-Labarthe, L., Fall, A., Daube, G., & Coton, E. (2017). Unraveling microbial ecology of industrial-scale Kombucha fermentations by metabarcoding and culture-based methods. *FEMS Microbiology Ecology, 93*(5). https://doi.org/10.1093/femsec/fix048

Cvetkovic, D. D., & Markov, S. L. (2002). Cultivation of tea fungus on malt extract medium. *Acta Periodica Technologica, 33*, 117–124. https://doi.org/10.2298/APT0233117C

De Filippis, F., Troise, A. D., Vitaglione, P., & Ercolini, D. (2018). Different temperatures select distinctive acetic acid bacteria species and promotes organic acids production during Kombucha tea fermentation. *Food Microbiology, 73*, 11–16. https://doi.org/10.1016/j.fm.2018.01.008

Dipti, P., Yogesh, B., Kain, A. K., Pauline, T., Anju, B., Sairam, M., Singh, B., Mongia, S. S., Ilavazhagan Devendra Kumar, G., & Selvamurthy, W. (2003). Lead induced oxidative stress: Beneficial effects of kombucha tea. *Biomedical and Environmental Sciences, 16*(3), 276–282.

Dufresne, C., & Farnworth, E. (2000). Tea, kombucha, and health: A review. *Food Research International, 33*(6), 409–421. https://doi.org/10.1016/S0963-9969(00)00067-3

Erniati. (2018). *Aktivitas imunomodulator produk Geluring Dari rumput laut Gelidium sp.*

Essawet, N., Cvetkovic, D., Velicanski, A., Canadanovic-Brunet, J., Vulic, J., Maksimovic, V., & Markov, S. (2015). Polyphenols and antioxidant activities of Kombucha beverage enriched with Coffeeberry extract. *Chemical Industry and Chemical Engineering Quarterly, 21*(3), 399–409. https://doi.org/10.2298/CICEQ140528042E

Fitzsimmons, C., Falcone, M., & Dunne, H. (2014). Helminth allergens, parasite specific IgE, and its protective role in human immunity. *Frontiers in Immunology, 5*.

Frank, G. W. (1991). Kombucha. In *Healthy Beverage and natural Remedy from the far east. Wilhelm Ennsthaller.*

Fu, Y., Foden, J. A., Khayter, C., Maeder, M. L., Reyon, D., Joung, J. K., & Sander, J. D. (2013). High-frequency off-target mutagenesis induced by CRISPR-Cas nucleases in human cells. *Nature Biotechnology, 31*(9), 822–826. https://doi.org/10.1038/nbt.2623

Fu, Q. Y., Li, Q. S., Lin, X. M., Qiao, R. Y., Yang, R., Li, X. M., Dong, Z. B., Xiang, L. P., Zheng, X. Q., Lu, J. L., Yuan, C. B., Ye, J. H., & Liang, Y. R. (2017). Antidiabetic effects of tea. *Molecules, 22*(5). https://doi.org/10.3390/molecules22050849

Fu, C., Yan, F., Cao, Z., Xie, F., & Lin, J. (2014). Antioxidant activities of kombucha prepared from three different substrates and changes in content of probiotics during storage. *Food Science and Technology, 34*(1), 123–126. https://doi.org/10.1590/S0101-20612014005000012

Gaggìa, F., Baffoni, L., Galiano, M., Nielsen, D. S., Jakobsen, R. R., Castro-Mejía, J. L., Bosi, S., Truzzi, F., Musumeci, F., Dinelli, G., & Di Gioia, D. (2019). Kombucha beverage from green, black and rooibos teas: A comparative study looking at microbiology, chemistry and antioxidant activity. *Nutrients, 11*(1). https://doi.org/10.3390/nu11010001

Gomez-Flores, R., Verástegui-Rodríguez, L., Quintanilla-Licea, R., Guerra, P., Tamez-Guerra, R., & Rodríguez-Padilla, C. (2008). In vitro rat lymphocyte proliferation induced by Ocinum basilicum, Persea americana, Plantago virginica and Rosa spp. Extracts. *Journal of Medicinal Plants Research, 2*(1), 5–10.

Greenwalt, C. J., Steinkraus, K. H., & Ledford, R. A. (2000). Kombucha, the fermented tea: Microbiology, composition, and claimed health effects. *Journal of Food Protection, 63*(7), 976–981. https://doi.org/10.4315/0362-028X-63.7.976

Gunton, J. E., Delhanty, P. J. D., Takahashi, S. I., & Baxter, R. C. (2003). Metformin rapidly increases insulin receptor activation in human liver and signals preferentially through insulin-receptor substrate-2. *Journal of Clinical Endocrinology and Metabolism, 88*(3), 1323–1332. https://doi.org/10.1210/jc.2002-021394

Harler, C. R. (1964). *The culture and marketing of tea.* Oxford University Press.

Hoffmann, O., Zweigner, J., Smith, S. H., Freyer, D., Mahrhofer, C., Dagand, E., Tuomanen, E. I., & Weber, J. R. (2006). Interplay of pneumococcal hydrogen peroxide and host-derived nitric oxide. *Infection and Immunity, 74*(9), 5058–5066. https://doi.org/10.1128/IAI.01932-05

Hur, S. J., Lee, S. Y., Kim, Y. C., Choi, I., & Kim, G. B. (2014). Effect of fermentation on the antioxidant activity in plant-based foods. *Food Chemistry*, 160, 346–356. https://doi.org/10.1016/j.foodchem.2014.03.112

Iida, K. T., Kawakami, Y., Suzuki, M., Shimano, H., Toyoshima, H., Sone, H., Shimada, K., Iwama, Y., Watanabe, Y., Mokuno, H., Kamata, K., & Yamada, N. (2003). Effect of thiazolidinediones and metformin on LDL oxidation and aortic endothelium relaxation in diabetic GK rats. *American Journal of Physiology - Endocrinology and Metabolism*, 284(6), E1125–E1130. https://doi.org/10.1152/ajpendo.00430.2002

Ivanišová, E., Meňhartová, K., Terentjeva, M., Godočíková, L., Árvay, J., & Kačániová, M. (2019). Kombucha tea beverage: Microbiological characteristic, antioxidant activity, and phytochemical composition. *Acta Alimentaria*, 48(3), 324–331. https://doi.org/10.1556/066.2019.48.3.7

Jayabalan, R., Malbaša, R. V., Lončar, E. S., Vitas, J. S., & Sathishkumar, M. (2014). A review on kombucha tea-microbiology, composition, fermentation, beneficial effects, toxicity, and tea fungus. *Comprehensive Reviews in Food Science and Food Safety*, 13(4), 538–550. https://doi.org/10.1111/1541-4337.12073

Jayabalan, Rasu, Malbaša, R. V., & Sathishkumar, M. (2017). *Kombucha tea: Metabolites*. Springer Nature. https://doi.org/10.1007/978-3-319-25001-4_12

Jayabalan, R., Marimuthu, S., & Swaminathan, K. (2007). Changes in content of organic acids and tea polyphenols during kombucha tea fermentation. *Food Chemistry*, 102(1), 392–398. https://doi.org/10.1016/j.foodchem.2006.05.032

Jayabalan, R., Subathradevi, P., Marimuthu, S., Sathishkumar, M., & Swaminathan, K. (2008). Changes in free-radical scavenging ability of kombucha tea during fermentation. *Food Chemistry*, 109(1), 227–234. https://doi.org/10.1016/j.foodchem.2007.12.037

Jayasekera, S., Molan, A. L., Garg, M., & Moughan, P. J. (2011). Variation in antioxidant potential and total polyphenol content of fresh and fully-fermented Sri Lankan tea. *Food Chemistry*, 125(2), 536–541. https://doi.org/10.1016/j.foodchem.2010.09.045

Kawashima, T., Hayashi, K., Kosaka, A., Kawashima, M., Igarashi, T., Tsutsui, H., Tsuji, N. M., Nishimura, I., Hayashi, T., & Obata, A. (2011). Lactobacillus plantarum strain YU from fermented foods activates Th1 and protective immune responses. *International Immunopharmacology*, 11(12), 2017–2024. https://doi.org/10.1016/j.intimp.2011.08.013

Kwon, O., Eck, P., Chen, S., Corpe, C. P., Lee, J. H., Kruhlak, M., & Levine, M. (2007). Inhibition of the intestinal glucose transporter GLUT2 by flavonoids. *FASEB Journal*, 21(2), 366–377. https://doi.org/10.1096/fj.06-6620com

Laureys, D., Britton, S. J., & Clippeleer, D. (2020).

Leal, J. M., Suárez, L. V., Jayabalan, R., Oros, J. H., & Escalante-Aburto, A. (2018). A review on health benefits of kombucha nutritional compounds and metabolites. *CYTA - Journal of Food*, 16(1), 390–399. https://doi.org/10.1080/19476337.2017.1410499

Lee, S. J., Chinen, J., & Kavanaugh, A. (2010). Immunomodulator therapy: Monoclonal antibodies, fusion proteins, cytokines, and immunoglobulins. *Journal of Allergy and Clinical Immunology*, 125(2), S314–S323. https://doi.org/10.1016/j.jaci.2009.08.018

Liu, C. H., Hsu, W. H., Lee, F. L., & Liao, C. C. (1996). The isolation and identification of microbes from a fermented tea beverage, Haipao, and their interactions during Haipao fermentation. *Food Microbiology*, 13(6), 407–415. https://doi.org/10.1006/fmic.1996.0047

Lobanova, A. V., Strelnikov, L. S., Dolya, V. G., & Strilets, O. P. (2016). Studying the properties of the microbial consortium "kombucha" and creation new products on their basis. *Tropical Issues Of New Drugs Development*, 1, 362–363.

Lončar, E. S., Petrović, S. E., Malbaša, R. V., & Verac, R. M. (2000). Biosynthesis of glucuronic acid by means of tea fungus. *Nahrung-Food*, 44(2), 138–139.

Lyu, S. Y., & Park, W. B. (2005). Production of cytokine and NO by RAW 264.7 macrophages and PBMC in vitro incubation with flavonoids. *Archives of Pharmacal Research*, 28(5), 573–581. https://doi.org/10.1007/BF02977761

Ma, X., Hua, J., & Li, Z. (2008). Probiotics improve high fat diet-induced hepatic steatosis and insulin resistance by increasing hepatic NKT cells. *Journal of Hepatology*, 49(5), 821–830. https://doi.org/10.1016/j.jhep.2008.05.025

Malbaša, R., Loncar, E., & Kolarov, L. J. (2002). L-lactic, L-ascorbic, total and volatile acids contents in dietetic Kombucha beverage. *Romanian Biotechnological Le Ers*, 7(5).

Malbaša, R., Lončar, E., Djurić, M., Klašnja, M., Kolarov, L. J., & Markov, S. (2006). Scale-Up of Black Tea Batch Fermentation by Kombucha. *Food and Bioproducts Processing*, *84*(3), 193–199. https://doi.org/10.1205/fbp.05061

Malbaša, R. V., Loncar, E. S., Vitas, J. S., Canadanovi, Č., & Brunet, J. M. (2011). *Influence of starter Cultures on the antioxidant Activity of kombucha beverage*.

Mao, R., Wu, L., Zhu, N., Xinran, L., & Yong, L. (2019). Naked oat (avena nuda L.) Oligopeptides: Immunomodulatory effects on innate and Adaptive immunity in mice via cytokine secretion, Antibody production, and Th cells stimulation. *Nutrients*, 11(4), 927. https://doi.org/10.3390/nu11040927

Marsh, A. J., O'Sullivan, O., Hill, C., Ross, R. P., & Cotter, P. D. (2014). Sequence-based analysis of the bacterial and fungal compositions of multiple kombucha (tea fungus) samples. *Food Microbiology*, 38, 171–178. https://doi.org/10.1016/j.fm.2013.09.003

Mau, J. L., Chang, C. N., Huang, S. J., & Chen, C. C. (2004). Antioxidant properties of methanolic extracts from Grifola frondosa, Morchella esculenta and Termitomyces albuminosus mycelia. *Food Chemistry*, *87*, 111–118.

Mau, J. L., Lin, H. C., & Song, S. F. (2002). Antioxidant properties of several specialty mushrooms. *Food Research International*, 35(6), 519–526. https://doi.org/10.1016/S0963-9969(01)00150-8

May, A., Narayanan, S., Alcock, J., Varsani, A., Maley, C., & Aktipis, A. (2019). Kombucha: A novel model system for cooperation and conflict in a complex multi-species microbial ecosystem. *PeerJ*, 7(9). https://doi.org/10.7717/peerj.7565. e7565.

Mayser, P., Fromme, S., Leitzmann, G., & Gründer, K. (1995). The yeast spectrum of the 'tea fungus Kombucha. *Mycoses*, 38(7–8), 289–295. https://doi.org/10.1111/j.1439-0507.1995.tb00410.x

Morales, D. (2020). Biological activities of kombucha beverages: The need of clinical evidence. *Trends in Food Science and Technology*, 105, 323–333. https://doi.org/10.1016/j.tifs.2020.09.025

Mosser, D. M. (2003). The many faces of macrophage activation. *Journal of Leukocyte Biology, 73*(2), 209–212. https://doi.org/10.1189/jlb.0602325

Mo, H., Zhu, Y., & Chen, Z. (2008). Microbial fermented tea - a potential source of natural food preservatives. *Trends in Food Science and Technology, 19*(3), 124–130. https://doi.org/10.1016/j.tifs.2007.10.001

Murphy, K., Travers, P., & Walport, M. (2007). *Janeway's immunobiology* (*7th* ed.). *Garland Science.*

Naidu, A. S., Cooper, L. W., Finney, B. P., Macdonald, R. W., Alexander, C., & Semiletov, I. P. (2000). Organic carbon isotope ratio (δ13C) of Arctic Amerasian Continental shelf sediments. *International Journal of Earth Sciences, 89*(3), 522–532. https://doi.org/10.1007/s005310000121

Neffe-Skocińska, K., Sionek, B., Ścibisz, I., & Kołożyn-Krajewska, D. (2017). Acid contents and the effect of fermentation condition of Kombucha tea beverages on physicochemical, microbiological and sensory properties. *CyTA-Journal of Food, 15*(4), 601–607. https://doi.org/10.1080/19476337.2017.1321588

Nguyen, N. K., Nguyen, P. B., Nguyen, H. T., & Le, P. H. (2015). Screening the optimal ratio of symbiosis between isolated yeast and acetic acid bacteria strain from traditional kombucha for high-level production of glucuronic acid. *LWT-Food Science and Technology, 64*(2), 1149–1155. https://doi.org/10.1016/j.lwt.2015.07.018

Nieman, D. C., Henson, D. A., Maxwell, K. R., Williams, A. S., Mcanulty, S. R., Jin, F., Shanely, R. A., & Lines, T. C. (2009). Effects of quercetin and EGCG on mitochondrial biogenesis and immunity. *Medicine and Science in Sports and Exercise, 41*(7), 1467–1475. https://doi.org/10.1249/MSS.0b013e318199491f

Nopitasari, D. (2006a). *Pengaruh pemberian ekstrak buah phaleria Papuana terhadap aktivitas fagositosis makrofag Mencit balb/c. Skripsi. Jurusan pendidikan dokter. Fakultas Kedokteran.*

Nopitasari, D. A. (2006b). *The Effect of Phaleria papuana fruit Extract on macrophage phagocytosis Activity of balb/c mice. Scientific Writing of the Faculty of Medicine.*

Ortuño-Sahagún, D., Zänker, K., Rawat, A. K. S., Kaveri, S. V., & Hegde, P. (2017). Natural immunomodulators. *Journal of Immunology Research, 2017.*

Östman, E., Granfeldt, Y., Persson, L., & Björck, I. (2005). Vinegar supplementation lowers glucose and insulin responses and increases satiety after a bread meal in healthy subjects. *European Journal of Clinical Nutrition, 59*(9), 983–988. https://doi.org/10.1038/sj.ejcn.1602197

Peluso, I., Miglio, C., Morabito, G., Ioannone, F., & Serafini, M. (2015). Flavonoids and immune function in human: A systematic review. *Critical Reviews in Food Science and Nutrition, 55*(3), 383–395. https://doi.org/10.1080/10408398.2012.656770

Pfeiffer, T., & Morley, A. (2014). An evolutionary perspective on the Crabtree effect. *Frontiers in Molecular Biosciences*, 1. https://doi.org/10.3389/fmolb.2014.00017

Prescott, S. C., & Dunn, C. G. (1959). *Industrial microbiology*. MC Grow Hill Book Company.

Qureshi, A. A., Guan, X. Q., Reis, J. C., Papasian, C. J., Jabre, S., Morrison, D. C., & Qureshi, N. (2012). Inhibition of nitric oxide and inflammatory cytokines in LPS-stimulated murine macrophages by resveratrol, a potent proteasome inhibitor. *Lipids in Health and Disease*, 11. https://doi.org/10.1186/1476-511X-11-76

Reiss, J. (1994). Influence of different sugars on the metabolism of the tea fungus. *Zeitschrift Für Lebensmittel-Untersuchung Und -Forschung, 198*(3), 258–261. https://doi.org/10.1007/BF01192606

Rice-Evans, C. A., Miller, N. J., & Paganga, G. (1996). Structure-antioxidant activity relationships of flavonoids and phenolic acids. *Free Radical Biology and Medicine, 20*(7), 933–956. https://doi.org/10.1016/0891-5849(95)02227-9

Rinihapsari, E. D., & Catur, A. R. (2008). Fermentasi kombucha sebagai minuman kesehatan. *Media Farmasi Indonesia, 3*(2), 241–247.

Roche. (1998). *The history and spread of Kombucha.*

Römling, U., & Galperin, M. Y. (2015). Bacterial cellulose biosynthesis: Diversity of operons, subunits, products, and functions. *Trends in Microbiology, 23*(9), 545–557. https://doi.org/10.1016/j.tim.2015.05.005

Sai Ram, M., Anju, B., Pauline, T., Prasad, D., Kain, A. K., Mongia, S. S., Sharma, S. K., Singh, B., Singh, R., Ilavazhagan, G., Kumar, D., & Selvamurthy, W. (2000). Effect of Kombucha tea on chromate(VI)-induced oxidative stress in albino rats. *Journal of Ethnopharmacology, 71*(1–2), 235–240. https://doi.org/10.1016/S0378-8741(00)00161-6

Saichana, N., Matsushita, K., Adachi, O., Frébort, I., & Frebortova, J. (2015). Acetic acid bacteria: A group of bacteria with versatile biotechnological applications. *Biotechnology Advances, 33*(6), 1260–1271. https://doi.org/10.1016/j.biotechadv.2014.12.001

Schechter, B. (2010). In *Lymphocyte stimulation. Dalam: Differential sensitivity to radiation biochemical and immunological process. New York (USA)*: Plenum Press. Castellani A.

Sentjurc, M., Marjana, N., Henry, D., & Veronika, A. (2003). *Antioxidant Activity of sempervivum Tectorum and its components*. American Chemical Society.

Sofiakmi, R. L. Q., Ulfah, M., & Sasmito, E. (2015). The immunomodulatory activity test of kombucha mushroom black tea fermentation against lymphocyte cell proliferation of BALB/C mice strains by in vitro. Yogjakarta. *Internal Journal. Universitas Wahid Hasyim.*

Spedding, G. (2015). So What is Kombucha? An Alcoholic or a Non-Alcoholic Beverage? A Brief Selected Literature Review and Personal Reflection. *Brewing and Distilling Analytical Services*, (2).

Sreeramulu, G., Zhu, Y., & Knol, W. (2000). Kombucha fermentation and it's antimicrobial Activity. *J Agriculture Food Chemistry, 886*, 65–73.

Srihari, T., Karthikesan, K., Ashokkumar, N., & Satyanarayana, U. (2013). Antihyperglycaemic efficacy of kombucha in streptozotocin-induced rats. *Journal of Functional Foods, 5*(4), 1794–1802. https://doi.org/10.1016/j.jff.2013.08.008

Subaryono, S., Perangiangin, R., Suhartono, M. T., & Zakaria, F. R. (2017). Imunomodulator activity of alginate oligosaccharides from alginate sargassum crassifolium. *Jurnal Pengolahan Hasil Perikanan Indonesia, 20*(1), 63. https://doi.org/10.17844/jphpi.v20i1.16434

Tchobanov, I., Gal, L., Guilloux-Benatier, M., Remize, F., Nardi, T., Guzzo, J., Serpaggi, V., & Alexandre, H. (2008). Partial vinylphenol reductase purification and characterization from Brettanomyces bruxellensis. *FEMS Microbiology Letters, 284*(2), 213–217. https://doi.org/10.1111/j.1574-6968.2008.01192.x

Turvey, S. E., & Broide, D. H. (2010). Innate immunity. *Journal of Allergy and Clinical Immunology,* 125(2).

Vázquez-Cabral, B. D., Larrosa-Pérez, M., Gallegos-Infante, J. A., Moreno-Jiménez, M. R., González-Laredo, R. F., Rutiaga-Quiñones, J. G., Gamboa-Gómez, C. I., & Rocha-Guzmán, N. E. (2017). Oak kombucha protects against oxidative stress and inflammatory processes. *Chemico-Biological Interactions, 272*, 1–9. https://doi.org/10.1016/j.cbi.2017.05.001

Vázquez-Cabral, B. D., Moreno-Jiménez, M. R., Rocha-Guzmán, N. E., Gallegos-Infante, J. A., González-Herrera, S. M., Gamboa-Gómez, C. I., & González-Laredo, R. F. (2016). Mexican oaks as a potential non-timber resource for Kombucha beverages. *Revista Chapingo Serie Ciencias Forestales Y Del Ambiente, 22*(1), 73–86. https://doi.org/10.5154/r.rchscfa.2015.04.014

Veličanski, A. S., Cvetković, D. D., Markov, S. L., Tumbas Šaponjac, V. T., & Vulić, J. J. (2014). Antioxidant and antibacterial activity of the beverage obtained by fermentation of sweetened lemon balm (Melissa offi cinalis L.) tea with symbiotic consortium of bacteria and yeasts. *Food Technology and Biotechnology,* 52(4), 420–429. https://doi.org/10.17113/ftb.52.04.14.3611

Vijayaraghavan, R., Singh, M., Rao, P. V. L., Bhattacharya, R., Kumar, P., Sugendran, K., Kumar, O. M., Pant, S. C., & Singh, R. (2000). Subacute (90 Days) oral toxicity studies of kombucha tea. *Biomedical and Environmental Sciences,* 13(4), 293–299.

Villarreal-solo, S., Sandra, B., & Jalloul, B. (2011). Understanding kombucha tea fermentation: A review. *Journal of Food Science,* 83(3), 580–588.

Villarreal-Soto, S. A., Beaufort, S., Bouajila, J., Souchard, J. P., & Taillandier, P. (2018). Understanding kombucha tea fermentation: A review. *Journal of Food Science,* 83(3), 580–588. https://doi.org/10.1111/1750-3841.14068

Watawana, M. I., Jayawardena, N., Gunawardhana, C. B., & Waisundara, V. Y. (2015). Health, wellness, and safety aspects of the consumption of kombucha. *Journal of Chemistry.* https://doi.org/10.1155/2015/591869, 2015.

Watawana, M. I., Jayawardena, N., Ranasinghe, S. J., & Waisundara, V. Y. (2017). Evaluation of the effect of different sweetening agents on the polyphenol contents and antioxidant and starch hydrolase inhibitory properties of kombucha. *Journal of Food Processing and Preservation,* 41(1). https://doi.org/10.1111/jfpp.12752

Wood, B. J. B. (1998). *Microbiology of fermented food. London: Blackie Academic and Professional.*

Yamada, Y., Yukphan, P., Vu, H. T. L., Muramatsu, Y., Ochaikul, D., Tanasupawat, S., & Nakagawa, Y. (2012). Description of Komagataeibacter gen. nov., with proposals of new combinations (Acetobacteraceae). *Journal of General and Applied Microbiology,* 58(5), 397–404. https://doi.org/10.2323/jgam.58.397

Yavari, N., Assadi, M. M., Larijani, K., & Moghadam, M. B. (2010). Response surface methodology for optimization of glucuronic acid production using kombucha layer on sour cherry juice. *Australian Journal of Basic and Applied Sciences,* 4(8), 3250–3256. http://www.insipub.com/ajbas/2010/3250-3256.pdf.

Zubaidah, E., Afgani, C. A., Kalsum, U., Srianta, I., & Blanc, P. J. (2019). Comparison of in vivo antidiabetes activity of snake fruit Kombucha, black tea Kombucha and metformin. *Biocatalysis and Agricultural Biotechnology,* 17, 465–469. https://doi.org/10.1016/j.bcab.2018.12.026

Zubaidah, E., Apriyadi, T. E., Kalsum, U., Widyastuti, E., Estiasih, T., Srianta, I., & Blanc, P. J. (2018a). In vivo evaluation of snake fruit Kombucha as hyperglycemia therapeutic agent. *International Food Research Journal,* 25(1), 453–457. http://www.ifrj.upm.edu.my/25%20(01)%202018/(60).pdf.

Zubaidah, E., Dewantari, F. J., Novitasari, F. R., Srianta, I., & Blanc, P. J. (2018b). Potential of snake fruit (Salacca zalacca (Gaerth.) Voss) for the development of a beverage through fermentation with the Kombucha consortium. *Biocatalysis and Agricultural Biotechnology,* 13, 198–203. https://doi.org/10.1016/j.bcab.2017.12.012

Zubaidah, E., Aulia, A. R., & Widyaningsih, T. D. (2020a). *Aulia immuno-modulatory activity Comparison of black tea kombucha (Camelia sinensis) and arabica coffee leaves tea kombucha (Coffee arabica) for Salmonella typhi-infected mice.*

Zubaidah, E., Valencia, V., Rifa'i, M., Srianta, I., & Tewfik, I. (2020b). Investigating chemical changes during snake fruit and black tea kombucha fermentation and the associated immunomodulatory activity in Salmonella typhi-infected mice. *Potravinarstvo Slovak Journal of Food Sciences,* 14, 995–1000. https://doi.org/10.5219/1416

Zubaidah, E., & Wistiani, D. (2015). Karakteristik kimiawi dan mikrobiologis kombucha Dari berbagai daun tinggi fenol selama fermentasi. *Jurnal Pangan Dan Agroindustri,* 3(4), 1446–1457.

Further reading

Antibodies. (n.d.). Fusion Proteins, Cytokines, and Immunoglobulins (Vol. 125, pp. 314–323).

Bhattacharya, S., Gachhui, R., & Sil, P. C. (2013). Effect of Kombucha, a fermented black tea in attenuating oxidative stress mediated tissue damage in alloxan induced diabetic rats. *Food and Chemical Toxicology,* 60, 328–340. https://doi.org/10.1016/j.fct.2013.07.051

Chen, C. Z., Li, L., Lodish, H. F., & Bartel, D. P. (2004). MicroRNAs modulate hematopoietic lineage differentiation. *Science,* 303(5654), 83–86. https://doi.org/10.1126/science.1091903

dan. (2002). *Cultivation of tea fungus on malt extract madium.*

Di Cagno, R., Coda, R., De Angelis, M., & Gobbetti, M. (2013). Exploitation of vegetables and fruits through lactic acid fermentation. *Food Microbiology,* 33(1), 1–10. https://doi.org/10.1016/j.fm.2012.09.003

Dutta, D., & Gachhui, R. (2007). Nitrogen-fixing and cellulose-producing Gluconacetobacter kombuchae sp. nov., isolated from Kombucha tea. *International Journal of Systematic and Evolutionary Microbiology,* 57(2), 353–357. https://doi.org/10.1099/ijs.0.64638-0

Goh, W. N., Rosma, A., Kaur, B., Fazilah, A., Karim, A. A., & Bhat, R. (2012). Fermentation of black tea broth (kombucha): I. Effects of sucrose concentration and fermentation time on the yield of microbial cellulose. *International Food Research Journal,* 19(1), 109−117. http://www.ifrj.upm.edu.my/19%20(01)%202011/(15)IFRJ-2011-105%20Rajeev.pdf.

Ivanišová, E., Meňhartová, K., Terentjeva, M., Harangozo, Ľ., Kántor, A., & Kačániová, M. (2020). The evaluation of chemical, antioxidant, antimicrobial and sensory properties of kombucha tea beverage. *Journal of Food Science and Technology,* 57(5), 1840−1846. https://doi.org/10.1007/s13197-019-04217-3

Kurtzman, C. P., Robnett, C. J., & Basehoar-Powers, E. (2001). Zygosaccharomyces kombuchaensis, a new ascosporogenous yeast from Kombucha tea. *FEMS Yeast Research, 1*(2), 133−138.

Malbaša, R. V., Lončar, E. S., Vitas, J. S., & Čanadanović-Brunet, J. M. (2011). Influence of starter cultures on the antioxidant activity of kombucha beverage. *Food Chemistry,* 127(4), 1727−1731. https://doi.org/10.1016/j.foodchem.2011.02.048

Mao, J., Zhang, S. Z., Du, P., Cheng, Z. B., Hu, H., & Wang, S. Y. (2020). Probiotics can boost the antitumor immunity of CD8+T cells in BALB/c mice and patients with colorectal carcinoma. *Journal of Immunology Research.* https://doi.org/10.1155/2020/4092472, 2020.

A Review. Journal of the American Society of brewing Chemists. (n.d.). 1−10.

Sievers, M., Lanini, C., Weber, A., Schuler-Schmid, U., & Teuber, M. (1995). Microbiology and fermentation balance in a kombucha beverage obtained from a tea fungus fermentation. *Systematic and Applied Microbiology,* 18(4), 590−594. https://doi.org/10.1016/S0723-2020(11)80420-0

Sinisa, M. (2006). Use of tea fungus isolate as starter culture for obtaining kombucha. *Annals of The Faculty of Engineering Hunedoara, 3*, 73−78.

So what is kombucha? An Alcoholic or A non-alcoholic beverage? A Brief selected literature Review and personal reflection.(2018). https://bdastesting.com/app/uploads/2020/07/KombuchaWhitePaperFeb2018Version.pdf.

Chapter 13

Fermented plant-based foods (e.g., tofu, sauerkraut, sourdough)

Sujit Das[1], Mital R. Kathiriya[2], Kunal Gawai[2] and Subrota Hati[2]

[1]*Department of Rural Development and Agricultural Production, North-Eastern Hill University, Tura Campus, Shillong, Meghalaya, India;* [2]*Dairy Microbiology Department, SMC College of Dairy Science, Anand Agricultural University, Anand, Gujarat, India*

13.1 Introduction

Plant-based foods include fruits, vegetables, nuts, seeds, oils, whole grains, legumes, and beans. Consuming plant-based foods on a regular basis provides the vital nutrients such as protein, fats, carbohydrates, vitamins, and essential minerals. It also contains good amount of fiber and phytonutrients with immense health beneficial potential. Due to shorter shelf life of these nature's food, tons of the produce are wasted and optimum utilization will not be obtained specially in developing countries where latest technologies for the processing are not available. There are several processing and preservation techniques reported in literature to increase the shelf life of plant-based foods such as drying, addition of sugar, freezing, irradiation, addition of preservatives, pasteurization, concentration, thermal processing, high-pressure processing, fermentation, etc.

Out of all the mentioned methods, fermentation is one of the best-known preservation techniques used since ancient times to increase the keeping quality of the food with minimal loss in the nutrient compared to other preservation techniques. Fermentation also provides people with a variety of health benefits as well as distinct tastes, formulations, and other essential nutrients in the most natural way possible. Since fermentation takes place by microorganism, these help in the breakdown of complex food compounds producing various bioactive peptides, leading to considerable increase in the available nutrients of the food. Furthermore, literature has revealed that fermented foods are known to possess hypocholesterolemic activity, anticancer effects, antihypertensive effect, and antidiabetic effect. It also boosts immunity and longevity. In addition to this, during fermentation antinutritional factors like phytate and tannin (present in raw plant-based produce) are degraded, which leads to release of antinutritional factor-bound nutrients (minerals, trace elements, polyphenols, etc.) in the food matrix and hence available free for the absorption in the consumers' digestive tract. Plant-based fermented foods have evolved in various parts of the world based on raw material supply, environmental factors, and people's preferences. However, some of the products like tofu (fermented soya-based product), sauerkraut (fermented cabbage), and sourdough (fermented cereal) are popular and prepared globally.

This chapter reviews these plant-based fermented foods available presently and recent research studies cited on the different processing methods utilized in preparation of such foods. The physicochemical and functional properties of plant-based fermented foods are also discussed in this chapter along with the bioavailable components of particular food matrix.

13.2 Plant-based fermented foods currently available

Plant-based fermentations and their microbes are an underutilized resource in novel biotechnological applications. These plant fermentations are majorly driven by a wide range of lactic acid bacteria (LAB), according to significant breakthroughs in DNA sequencing technology and sequencing data analysis. We assume here that, based on the latest insights into the functional and genetic flexibility of these LAB, they can be used safely in fermented foods, also, different novel probiotics, synbiotics, and a variety of other industrial applications can be generated (Wuyts et al., 2020). Life domains have been found to be essential to the fermented end product's taste, texture, and smell. *Saccharomyces cerevisiae* yeast

Engineering Plant-Based Food Systems. https://doi.org/10.1016/B978-0-323-89842-3.00006-3

strains, for starters, are needed for baking bread and brewing beer. LAB, on the other hand, are the predominant bacteria involved in the fermentation of cheeses, yoghurts, and fermented vegetables (Duar et al., 2017) (Table 13.1). Except for a few popular examples like sauerkraut and kimchi, fermentations utilizing plant materials as the principal substrate are understudied compared to dairy-based food fermentations (De Filippis et al., 2017). Consequently, they remain an interesting LAB source, which generally drives such fermentation processes. Fermentation of plant matter is an ancient process of preservation that may have originated in Asia. The traditional Korean product, kimchi, is now the most popular fermented vegetable food in several Asian nations (Kwon & Kim, 2007). In Europe and North America, sauerkraut, cucumbers, and olives are the most common fermented vegetable commodities. Prior to the advent of modern food preservation innovations such as refrigeration, controlled environment storage, and freezing, inhabitants in northern climates with just one annual growth season had uninterrupted access to essential vitamins such as vitamin C, minerals, and dietary fibers through the intake of fermented vegetable items (Settanni & Corsetti, 2008).

13.2.1 Tofu

Tofu is typically obtained by coagulating soymilk, also known as soy curd. For the preparation of tofu, coagulants like acid and $CaCl_2$, $CaSO_4$, $MgCl_2$, and glucono-D-lactone are primarily used in Japan (Murugkar, 2014; Oboh, 2006). During the processing of tofu, a coagulant generates a soy protein gel matrix, which aids in the capture of water, lipids, and other elements found in the matrix to form curd, which is then pressurized to form solids (Guan et al., 2021). Soybean curd (tofu) is made from soymilk or soy juice, as per the Japan External Trade Organization (JETRO, 2011). By heating, the soymilk for 2 min or by a related process that has a better sterilization effect, soy milk should be sterilized. The packed tofu can be sterilized for 45 min by heating it at 90°C or by an alternative method of sterilization. The stored tofu in the water bath must be kept for refrigeration or put in chilled drinking water, which must be altered regularly. Pasteurization and packing aid in prolonging the shelf-life of tofu for a year. There is a complex relationship between the coagulation of soy protein due to the interrelation amid the ingredients viz. soybean variety, soy milk, temperature for cooking, pH, coagulants used, amount and time of solid, and coagulant content. The tofu's texture and taste vary with the usage of various coagulants. Generalized tofu texture should, however, be smooth, firm, and coherent, although rough and rubbery surface are not preferred by the consumers (Zheng et al., 2020).

TABLE 13.1 Comparison of nutritional profiles of plant-based fermented foods and fermented dairy foods.

	Plant-based fermented foods			Fermented dairy foods		
Nutritional parameters (per 100 g)	**Sauerkraut**	**Tofu**	**Sourdough**	**Yogurt**	**Buttermilk**	**Sour cream**
Calories	19	76	289	59	40	193
Total fat	0.1 g	4.8 g	1.8 g	0.4 g	0.9 g	20 g
Saturated fat	0 g	0.7 g	0.5 g	0.1 g	0.5 g	12 g
Cholesterol	0 mg	0 mg	0 mg	5 mg	4 mg	52 mg
Sodium	661 mg	7 mg	513 mg	36 mg	105 mg	80 mg
Potassium	170 mg	121 mg	128 mg	141 mg	151 mg	141 mg
Total carbohydrate	4.3 g	1.9 g	56 g	3.6 g	4.8 g	2.9 g
Dietary fiber	2.9 g	0.3 g	2.4 g	0 g	0 g	0 g
Sugar	1.8 g	0 mg	2.6 g	3.2 g	4.8 g	2.9 g
Protein	0.9 g	8 g	12 g	10 g	3.3 g	2.1 g
Vitamin C	24%	0%	0%	0%	1%	1%
Iron	8%	30%	19%	0%	0%	1%
Vitamin B6	5%	0%	5%	5%	0%	5%
Magnesium	3%	7%	7%	2%	2%	2%
Calcium	3%	35%	4%	11%	11%	11%
Vitamin D	0%	0%	0%	0%	0%	3%
Cobalamin	0%	0%	0%	13%	3%	5%

13.2.1.1 Firm tofu

Soymilk is used to make firm tofu (also known as "cotton tofu"), which is then coagulated with a synthetic or natural coagulant and pushed through cheese cloth. To extract excess water, it is partly coagulated and pressed and the moisture in the tofu varies from 76 to 81% (DeMan et al., 1986).

13.2.1.2 Silken tofu/soft tofu

Bean curd for soy milk is usually coarse and uneven in texture, whereas silken (Kinugoshi) tofu is fine and smooth in texture, as the name implies. Silken tofu is a variant created by solidifying all the soy milk that gives silken tofu a soft and fine texture. 87−90% of the moisture content is present in soft/silken tofu (Beddows & Wong, 1987; Zhang et al., 2018).

13.2.1.3 Dried tofu

Dried tofu is the lightest type of tofu, with a water content of less than 76%. Since the soybean curd is vigorously disrupted before pushing to remove excess water, dry tofu has a low moisture content (Cai & Chang, 1997).

13.2.1.4 Sufu (fermented tofu)

Fermented tofu is also known as sufu. With different strains of microorganisms, viz., *Actinomucor elegans, Mucor racemosus, Mucor sufu, Mucor dispersus, Mucor racemosus, Mucor wutuongkiao, Aspergillus* sp., the tofu is fermented. The final consistency of the product is dependent on the characteristics of strain used for fermentation (Chou & Hwan, 1994).

13.2.2 Sauerkraut

Vegetable fermentation enhances their digestibility, increases nutritional value, expands seasonal availability and, in addition to prolonging their shelf life, overall acceptability. To establish the microbiological and biochemical changes occurring during fermentation, a large number of vegetable fermentations have been extensively examined (Breidt et al., 2012, pp. 841−855). Fermentation of cabbage, cucumber, carrot, cauliflower, and many other blends of vegetables are used in these studies. LAB such as *Leuconostoc mesenteroides, Lactobacillus plantarum,* and *Lactobacillus brevis* perform most of these fermentations (Swain et al., 2014).

Cabbage (*Brassica oleracea* var. *Capitata*) is cultivated in the world's tropical regions and is a valuable source of vitamins A, B, C, minerals, and carotene (Jahangir et al., 2009). Shredded cabbage is subjected to salt-controlled bacterial fermentation in many parts of the world, resulting in the production of acid cabbage, also known as sauerkraut (Johanningsmeier et al., 2007). Acid, which acts as a preservative in addition to the creation of the desired flavor, is produced during fermentation. In America and European countries, sauerkraut is very prominent (Breidt et al., 2012, pp. 841−855). To make them more appetizing, digestive, and increase the flavor of other foods, it is also eaten as an adjuvant with other foods. The isothiocyanates developed during sauerkraut fermentation have been reported to hinder the proliferation of cancer cells (Peñas et al., 2012).

Cabbage is cultivated in most parts of the world, but the supply of cabbage lasts for a limited period of time and a significant part of the overall harvest is lost due to lack of processing due to its perishable nature. Cabbage is being used either as a cooked vegetable or as a raw salad. Unless properly extracted, vitamin C, which is an essential component of cabbage, is destroyed during cooking (VanGarde & Woodburn, 1994). Vitamin C and other nutrients can thus be retained by fermentation and the supply of cabbage can be expanded across the year (Bousquet et al., 2021).

13.2.3 Sourdough

Spontaneous "sour" dough fermentation is among the earliest cereal fermentations. The fermentation procedure is to leaven the dough in order to produce a more gaseous piece of dough as well as a more aerated loaf of bread. Later, beer yeast has been used to leaven dough (Kulp & Lorenz, 2003). The use of sourdough to enhance bread flavor, density, and shelf life has been reported, especially in rye baking (Lorenz & Brummer, 2003). Since most consumers can only withstand mild acidity in wheat products, using sourdough in wheat baking is an alternative and much more demanding way of improving bread quality. As a result, the basis for enhanced flavor is a regulated acidity level in wheat sourdough and resulting bread. In today's world, during the mechanization of the bakery, the first fermentation stage, the bulk proof, is reduced or even ruled out (Thiele et al., 2002). Few bakeries have created advanced brew mechanisms in order to catch up with the lack of flavors, and a portion of the dough is fermented for several hours to create a pleasant fragrance known as sponge, brew, or polish. After that, the sponge is mixed into the final dough with the rest of the ingredients. With the next step in development, specialist companies are now supplying the bakery industry with dried sourdoughs to prevent or simplify this complicated process in the bakery. Sourdough fermentation is a traditional method for enhancing the consistency of bread and the production of various wheat and rye breads (Thiele et al., 2002).

Sourdough is presently utilized to make breads, cakes, and crackers. The typical functionality of sourdough is composed primarily of its microflora, which is essentially defined by LAB and yeasts. Because of its microbial life, such dough is metabolically active. When flour and water are combined, these microorganisms ensure that the acid is processed and leavened. The reactions in sourdough are intricate (Hammes & Gänzle, 1998). Various flour characteristics and process parameters combine to produce very precise impacts on the sourdough microflora's metabolic behavior. Various flour attributes and processing methods combine to produce very specific effects on the sourdough microflora's metabolic activity. Biochemical modifications in the carbohydrate and protein components of the flour occur during fermentation due to the activity of microbial and indigenous enzymes (Gobbetti et al., 2020).

Dough starts to ferment spontaneously when flour and water are mixed together without the addition of a starter culture or a portion of a former sourdough culture (mother dough). The microflora of such a dough is subjective in terms of the flour's shape, source, and storage factors as well as the technical specifications of the fermentation process used, based on the microflora of the raw materials being used and prevailing hygienic conditions. For example, whole grain cereals and 100% extraction rye flour may consist of 10^4-10^6 CFU of unspecified bacteria per gram; in which 10^2-10^3 CFU/g refers to genus *Lactobacillus* (Salovaara, 2004). In naturally fermented doughs, the dominant microbes are homofermentative lactobacilli and pediococci, present in both wheat and rye sourdoughs, at the range of $3x10^7 - 3x10^9$ CFU per g. Typical homofermentative LAB in spontaneous sourdoughs are *L. casei, L. delbrueckii, L. farciminis, L. plantarum, and Pc. pentosaceus.* Distinctive heterofermentative LAB are *L. brevis, L. buchneri,* and *b. fermentum* (Bartkiene et al., 2020). Diverse strains of yeast (*S. cerevisiae* and *Pichia satoi*) have also been previously identified and characterized from spontaneous fermentations (Lau et al., 2021).

13.3 Processing methods employed to manufacture the fermented foods

The global functional food market is growing at a very rapid pace and a budding area for probiotic products. Advanced research strategies are underway to create products that could integrate probiotic species into them. While probiotic fermentation has a well-established market segment in the dairy industry, parallel technology implementation in the field of plant products is currently understudied and offers enormous implications for designing functional food product models. A huge proportion of plant-based fermented foods are locally produced in different parts of the world. Optimized technological advancements are, however, required to produce commercially viable fermented food products successfully (Gupta & Abu-Ghannam, 2012) (Fig. 13.1).

Edible plants, especially flavonoids and phytoestrogens, are abundant in phenolic metabolites. These compounds contain antioxidant, anticarcinogenic, antibacterial activities and have been shown to have significant therapeutic abilities against cardiovascular disorders, osteoporosis, and hormone-related cancers. Oligosaccharides and dietary fibers are abundant in certain plant products (cereals and legumes). These oligosaccharides could be used as supplies of nondigestible carbohydrates that can actively trigger the proliferation of colon microflora and act as prebiotics, in addition to enhancing several beneficial physiological effects (Hui & Evranuz 2012).

There is a significant need for testing in the manufacture of synbiotic products that contain both probiotics and prebiotics, which would probably confer the favorable impacts of both ingredients. To fill the knowledge gap, research should be performed with the intention of finding any additional benefits or health effects from this mixture of ingredients (Gupta & Abu-Ghannam, 2012). More research is needed to optimize the different forms of probiotic bacteria that have the ability to grow in plant-based products. The optimization of fermentation processes for the growth and development of probiotic bacteria, as well as the discovery of new potential health benefits in plant-based products, is another area in which there is a deficit of in-depth knowledge. These fermentation strategies may also help in the production of innovative nutraceuticals that can be refined and utilized as additives. Few of the processing methods of some plant-based fermented foods are elaborated below.

13.3.1 Fermentation process: sauerkraut

tShredded cabbage is put in a vessel and salt is applied. For the expulsion of cabbage juice, which includes fermentable sugars and related nutrients necessary for LAB, mechanical pressure can be implemented. The initial fermentation is commenced by heterofermentative gas-producing cocci (*Leuconostoc mesenteroides*). The activity of this group of LAB is

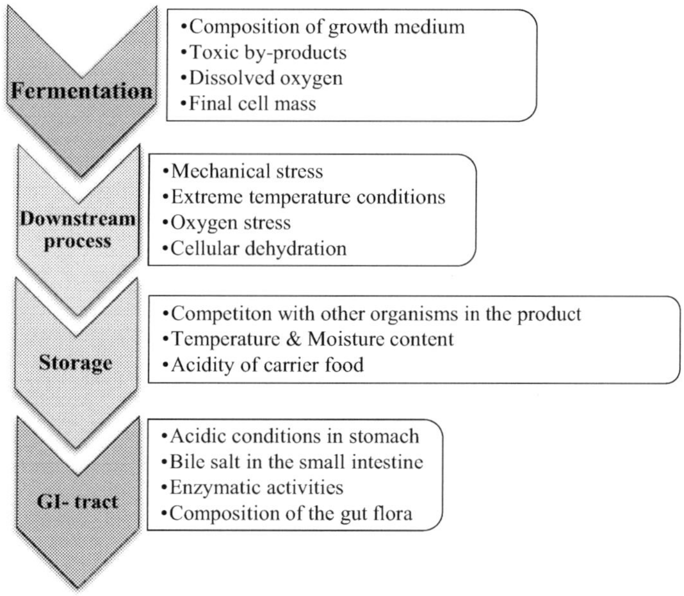

FIGURE 13.1 Factors affecting the viability of probiotics from production to the gastrointestinal tract.

lowered when the lactic acid exceeds 0.31%, while their enzymes continue to work. Subsequently, the lactobacilli (*L. plantarum, L. cucumeris,* and *L. brevis*) begin fermentation till the acidity ranges between 1.6% and 2.0%. To some degree, the high concentration of salt and low temperatures suppresses these bacteria (Zabat et al., 2018).

L. pentoaceticus eventually completes the fermentation step, taking the acidity to between 2% and 2.5% to complete the process (Daeschel Andersson & Fleming 1987). Sauerkraut fermentation produces lactic acid, as well as trace concentrations of acetic and propionic acids, a gas mixture (from which carbon dioxide is the predominant gas), significant volumes of alcohol, and a mix of aromatic esters (Satora et al., 2020). The acids, when mixed with alcohol, form esters, which add to the distinct sauerkraut taste. The acidity helps to control the growth of spoiled species, resulting in a longer shelf life for the commodity (Thakur et al., 2020) (Fig. 13.2).

13.3.2 Fermentation process: tofu

Several reviews are available on the tofu curd formation mechanism. The mechanism can be briefly summarized into steps as mentioned below:

- More than 50% of the proteins in soy milk comprise the total amount of large as well as medium-sized elements. Most of the particles are still intact until heating the raw soymilk. In the protein particles, mainly lipids are found (Tezuka & Ono 1995; Zheng et al., 2020).
- When soymilk is warmed to 90°C, the proteins denature and lipid droplets are distributed into the floating fraction. While nearly 75% of large particles are reduced to supernatant proteins, the quantities of medium-sized molecules rise resulting from the combination of the 7S subunit and the basic polypeptide of 11S from supernatant proteins, which primarily consist of 11 and 7S globulin (Guan et al., 2021).
- The main step is the addition of coagulant to the heated soymilk. Protein particles interact with lipid droplets at low coagulant(s) concentration(s), and the gel network is first developed to neutralize the negatively charged molecules by binding calcium to protein particles and to trigger protein aggregation due to a lowering in electrostatic repulsion. When approximately 50% of the proteins in soy milk are coagulated with coagulants, the lipids in the milk become trapped and inseparable due to interactions with particulate proteins. It contributes to a decrease in pH with the availability of more coagulants (Ono et al., 2010; Zhang et al., 2018) (Fig. 13.3).

13.3.3 Fermentation process: sourdough

Sourdough is a blend of flour (preferably wheat/rye flour) and water, and a blend of LAB and yeast ferments the mixture. The acidification is induced by a flour extraction rate of 0.54%−1.68% ash content (Siepmann et al., 2018). A decrease in

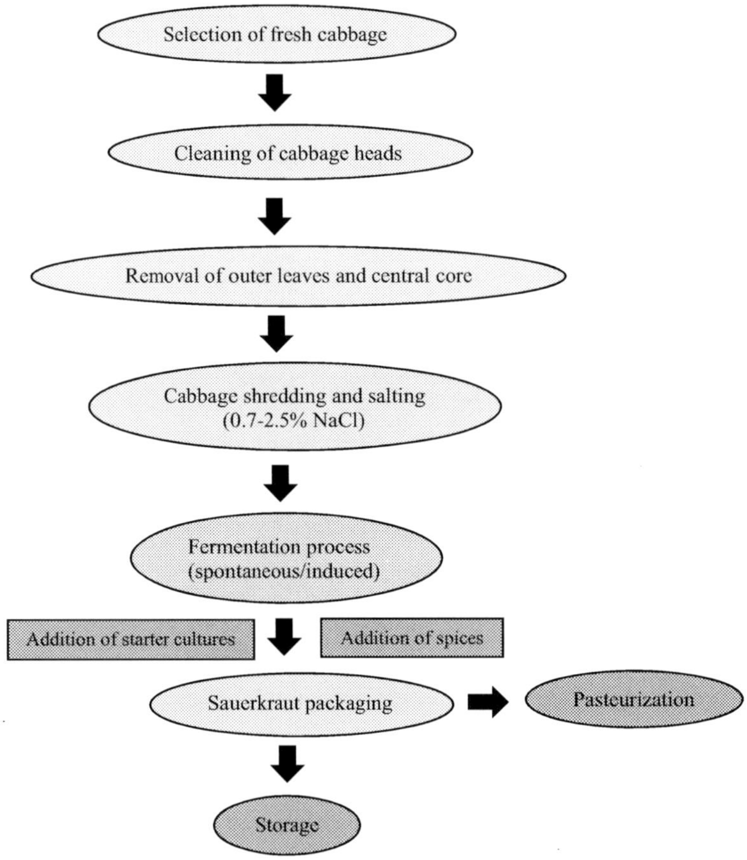

FIGURE 13.2 Scheme of sauerkraut production process.

the rate of extraction of flour can lead to a decrease in acid formation. The temperature of fermentation (25.5–35.5°C) could marginally affect acidification (Martínez-Anaya et al., 1996). Utilizing natural fermentation or starter fermentation, fermented bread can be produced. A mix of yeasts, primarily *Saccharomyces cerevisiae* and LAB, mainly the genera viz. *Lactobacillus, Leuconostoc, Pediococcus,* and *Weissella,* are generally found in indigenous flora in the natural fermentation of sourdough (Robert et al., 2009). However, with natural fermentation, the uniformity of the character traits of fermented bread, based on the microorganisms present in the system, is not easily controllable (Fig. 13.4a).

In industrial baking process, fermentation with a starter culture is favored because of being less cumbersome, has little or even less microbial composition variation, and has more uniform baked goods consistency (Minervini et al., 2010). There are high viable counts of microorganisms in a starter culture, which can be added to a substrate for creating specific features. It allows the fermentation process to be strictly controlled; the outcome is therefore predictable. *Lactococcus lactis*, introduced to industrial milk fermentation, is the first pure starter culture (Fernández-Peláez et al., 2020). A wide variety of starter cultures have now been produced for employing in industrial applications. *L. plantarum, L. casei, L. brevis, L. fructivorans,* and *L. fermentum* are comme-cially produced LAB starter cultures for making bread. Sourdough acidity, pH, and flavor variations are provided by different mixtures of those starter strains. Because of the rapid acidification and stable fermentation at various temperatures, *Lactobacillus plantarum* RTa12 and *Pediococcus pentosaceus* RTa11 were prescribed as starters for making gluten-free sourdough from amaranth flour (Sterr et al., 2009). In addition to the form of microorganisms, available nutrients are also essential for fermentation outcomes. Fermentable substrates may be generated by supplementing flour with some enzymes. Adding alpha-amylase, for instance, can aid in breaking down starch into maltose for yeast to develop in the dough. α-amylase, which can hydrolyze gluten proteins, is sometimes used along with proteases. Enzymes may originate from a range of sources and work at different temperatures (Gobbetti et al., 2020; Sahlström & Bråthen, 1997) (Fig. 13.4b).

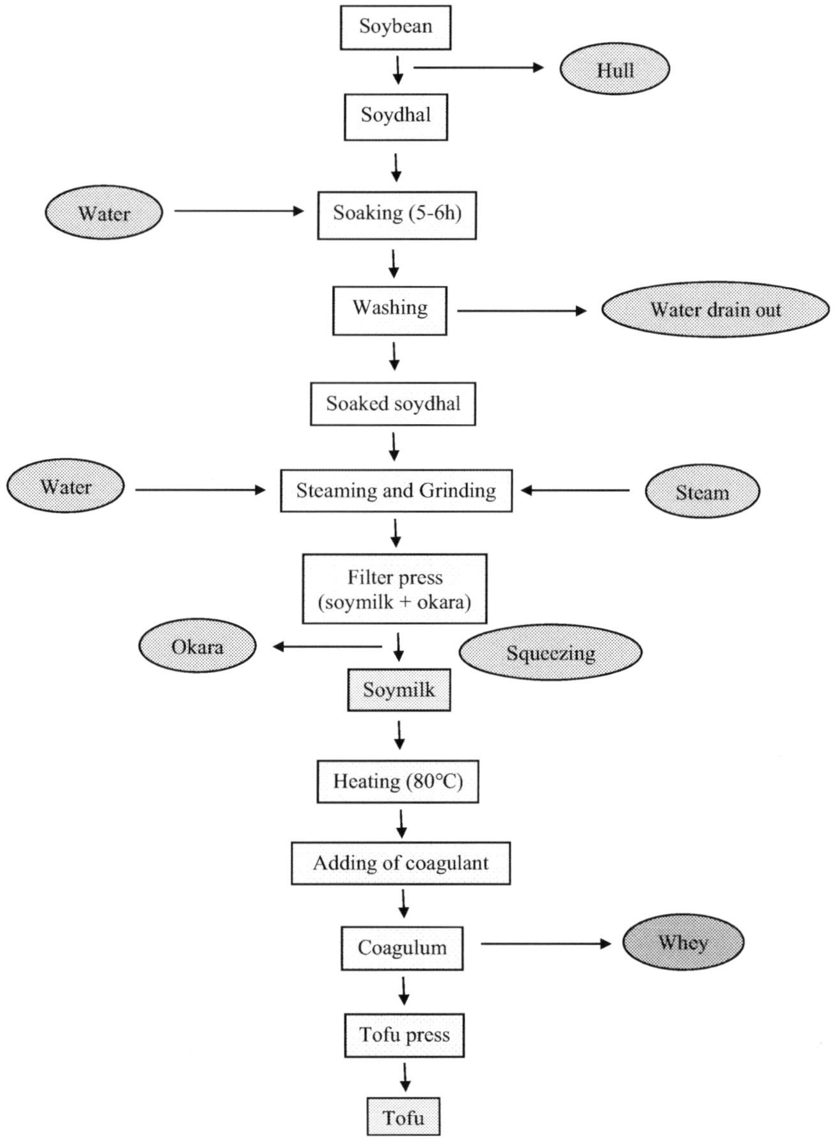

FIGURE 13.3 Process flow for the development of Tofu.

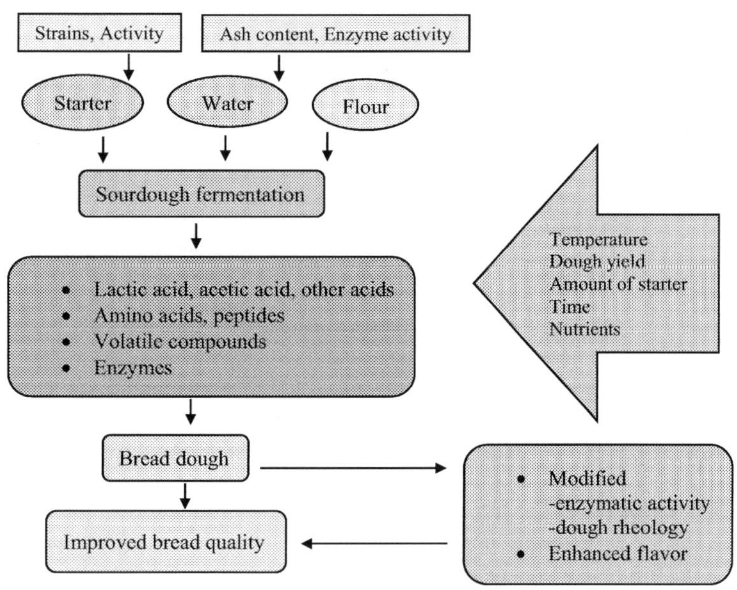

FIGURE 13.4A Fermentation of sourdough and its related influencing factors.

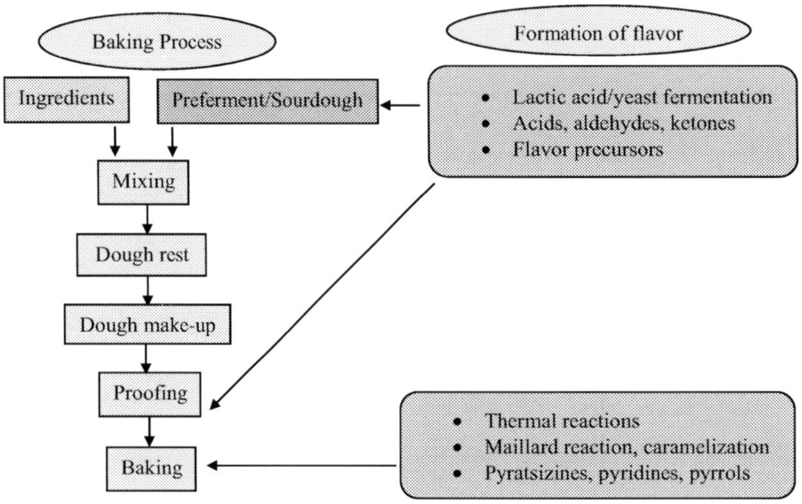

FIGURE 13.4B Bread flavor formation during the baking phase.

13.4 Physicochemical and sensory characteristics of plant-based fermented foods

Acceptable physicochemical and organoleptic quality of a food dominate compared to suitability or health promoting factors when it comes to consumer behavior (Heenan et al., 2004). In this regard, properties like viscosity, consistency/texture, color and appearance, taste, sweetness, and acidity should be considered seriously for successful development of plant-based fermented foods (Saint-Eve et al., 2019). Microorganisms influence the sensory characteristics of fermented foods in a variety of ways.

During fermentation, starter produces organic acid and reduces the pH of food matrix, this leads to sourness in the final product. Bacteria breakdown carbohydrate into acid, which results into reduced sweetness of the food and increased in the sour taste. The production of volatile flavor components tends to be the first mechanism considered for the development of flavor specific to a particular fermented food. The mechanism through which fermentation microorganisms affect flavor are: (1) Lowering the pH in lactic acid fermentations may reduce the activity or completely inactivate enzymes in the plant that generate either flavor components or flavor precursor compounds and (2) the fermentation microorganisms may directly metabolize precursor flavor compounds or flavor components themselves (McFeeters, 2004).

13.4.1 Tofu

Fermentation of soy disintegrates antinutritional factors by improving the bioavailability of isoflavones, promoting protein digestion, increasing the concentration of calcium in matrix, improving intestinal health, and boosting the immunity (Subrota et al., 2013). Besides its health benefits, fermentation decreases the intrinsic beany (unwanted) flavor of soymilk, its chalkiness, and content of flatulent sugars such as stachyose and raffinose (Jooyandeh, 2011). Tofu is prepared by coagulating soy with fermentation or different coagulant. The method utilized in preparation of soymilk and tofu affects significantly on tofu yield and quality. The heat-treated soya leads to give firmer texture to the tofu prepared out of it (Yoon & Kim, 2007). When tofu was prepared from the soybean that were initially frozen leads to give comparatively more acceptable body and texture than the tofu prepared from unfrozen soybean. It was observed that frozen soybean is comparatively more suitable for the tofu manufacturing (Noh et al., 2005). In China, people use fermented soy whey to make tofu as a coagulant. This is their traditional method of tofu preparation. Murugkar (2014) reported that improvement in composition and overall quality was observed when tofu was prepared from the sprouted soy milk.

Li et al. (2017) reported that *Lactobacillus plantarum* JMC-1 possess high acid producing activity compared to the other strain present in fermented soy whey. They have produced a tofu using strain JMC-1 into the sterile fermented soy whey stream as a coagulant for 16 h with pH 4.0 at 37°C. This tofu had a good flavor and texture compared to the tofu that made with the coagulants, magnesium chloride, and gypsum under the same tofu-making process. Li et al. found that the organic acids produced by *Lactobacillus plantarum* slowly released hydrogen ions, making the tofu have good flavor and texture (Li et al., 2017). Serrazanetti et al. (2013) found that LAB-fermented tofu can prolong the shelf life of tofu due the production of some antibacterial substances, such as acetic acid and limonene, during the fermentation process. Kobayashi et al. (2018) observed that fermenting soymilk increased the amount of isoflavone aglycones with potential antioxidant activity. Hou et al. (2016) suggested a modified method to prepare tofu having double network by utilizing enzymes like laccase and microbial transglutaminase. Obtained tofu received excellent sensory scores.

Tofu is actually tasteless and has sponge like qualities, so it can soak up all the delicious spices and flavor that one may like to add to it (Mitra et al., 2013). Sharma et al. (2015) suggested the addition of flavor ingredient like cinnamon, bay leaf, cardamom, sugar, and roasted cumin seed powder to soymilk to improve the taste of tofu. Mitra et al. (2013) found that flavor of tofu can be improved by replacing 30% soy milk by buffalo milk. In a study, the effect of *Lactobacillus casei* combined with salt coagulants ($MgCl_2$, $MgSO_4$, $CaCl_2$, $CaSO_4$) on the yield, texture, sensory factors, microstructure, and flavor were analyzed to evaluate the quality characteristics of fermented tofu. It was found that fermentation by *L. casei* showed significantly rise in the yield of tofu, i.e., 24.75%−31.26%. It was also observed that tofu prepared by fermentation had hard body and texture as well as uniformity in the structure. In L. *casei* + $MgSO_4$ tofu, 10 flavor compounds were detected, and the relative content (54.29%) of the four main flavor compounds was highest. Utilizing salt as coagulants contributes a lot to the texture and quality of fermented tofu. Therefore, fermentation of *L. casei* combined with salt coagulants significantly improved the quality characteristics of tofu (Wang et al., 2020).

13.4.2 Sauerkraut

Sauerkraut is lactic fermented and salted cabbage, also known as sour cabbage. The product is having unique sour flavor with increased shelf life due to action of LAB on cabbage sugar during fermentation. It is one of the popular cabbage preservation techniques in China, Central and Eastern Europe, the United States, and Asia (Wacher et al., 2010). Although sauerkraut is typically made with only salt and cabbage, other sauerkraut varieties may include other vegetables such as carrots, onions, garlic, and beets (Gardner et al., 2001).

The main constituents of white cabbage are carbohydrates (4.18−5.51 g/100 g), dietary fiber (1.9−2.9 g/100 g), followed by protein (1.27−1.37 g/100 g), minerals (0.3−0.7 g/100 g), fat (0.06−0.20 g/100 g), and vitamins, especially vitamin C (30−40 mg/100 g) (Souci et al., 2000). Cabbage is known for its high nutritional value and high levels of bioactive compounds. It also contains other phytochemicals, mainly phenolic compounds and glucosinolates (GLS). GLS, a group of nitrogen- and sulfur-containing plant metabolites, is responsible for the distinct flavor and odor of Brassica vegetables. GLS themselves are not biologically active but are hydrolyzed during the fermentation process to produce a wide range of biologically active compounds (Verkerk et al., 2009).

Sauerkraut should have approximately 1% acetic acid and 2% lactic acid, as this acidity ratio provides the expected sauerkraut flavor that consumers enjoy (Breidt et al., 2012, pp. 841−855). Adding L. *pentoaceticus* as a starter can result in an incomplete fermentation, which affects the flavor of the sauerkraut. When *Lactobacillus cucumeris* were used as a starter culture, it resulted in an incomplete fermented product, as well. In addition, the resulting product was bitter and had increased susceptibility to yeast spoilage. It was observed that undesirable flavors and softening of sauerkraut occurs due to insufficient salt addition to the cabbage. The obtained superior quality of sauerkraut typical salt concentrations added to cabbage should be between 2% and 10% (w/w) (Xiong et al., 2016). The unique flavors of sauerkraut are often derived from the salt concentration and the ratio of organic acids after a complete and adequate fermentation process.

13.4.3 Sourdough

Sourdough is a fermented flour mainly used in preparation of breads. It also helps to increase the shelf life, improve nutritional properties, and impart good sensory to the bread (De Vuyst et al., 2014). The sourdough is used widely in France, Germany, and Italy. Traditionally, breads were prepared by sourdough method. These breads were known to possess several health beneficial properties like reduction of risks associated with colorectal cancer, cardiovascular disorders, diabetes, and obesity.

Consumer satisfaction or acceptance for any novel product can be measured by the sensory analysis. The food having excellent sensory scores will have high chances for the acceptance in the market. Fermentation leads to production of many volatile compounds in sourdough, which ultimately gives flavor to the bread. It produces mainly acids, alcohols, aldehydes, esters, and ketones (Pico Bernal & Gómez 2015). The formation of acetic acid is the main factor that enhances pungent flavor and diminishes fresh flavor (Heiniö et al., 2003). On the other side, acidification plays a main role in proteolysis, the main factor enhancing roasted flavor, during sourdough fermentation (Thiele et al., 2002). Lactobacilli species contribute to flavor due to their ability to bring about acidification by the production of lactic and acetic acid. Additional flavor compounds can be generated due to conversion of amino acids like phenylalanine (sweet), isoleucine (acidic), asparagine, glycine, serine, and alanine (vinegar/sour) to aldehydes and ketones (Pétel et al., 2017).

Wheat flours obtained from two different varieties, that is, Mehregan as a hard and strong variety with high protein content and Sardari as a soft and weaker wheat with lower protein content, were used to prepare sourdough. It was observed that higher amounts of organic acids, i.e., lactic acid and acetic acid were produced in the sourdoughs obtained from Mehregan compared to those made from Sardari, which in turn led to their higher acidity and lower pH. Moreover, the breads made from Mehregan flour possessed higher moisture content, greater specific volume, and porosity and lower hardness than the samples produced from Sardari variety (Pazoki Ghouhe et al., 2019).

Bread prepared by sourdough method, using a starter containing *L. plantarum*, without dry yeast and fermented for 24 h, had high amounts of carboxylic and amino acids, which indicated that *L. plantarum* was an efficient microorganism in terms of carrying out both proteolysis and acidification in the medium during fermentation. This bread also received good sensory scores compared to other treatments (Limbad et al., 2020).

13.5 Bioavailable components from plant-based fermented foods

13.5.1 Tofu

Tofu is one of the best sources of plant protein, which is also rich in beneficial lipid, vitamins, and minerals, as well as other bioactive compounds, such as isoflavones, soyasaponin, and others. These components help in reducing the risk of many diseases like hypertension, hyperlipidemia, hypercholesterolemia, arteriosclerosis, coronary heart disease, breast cancer, and many others, when individual consume tofu in daily diet. Daily consumption of tofu is presently increasing by the people who value it for its characteristic taste and substantial nutritive value (Zhang et al., 2018). Activity of bioactive metabolites in soybeans has been widely reported in many research studies in their study. Epidemiological studies also clearly depict a direct relationship between the safeguard against breast cancer and soy intake (Wahyuni et al., 2016).

Tofu contains many naturally occurring phytochemicals, among them isoflavones are present in considerably large amounts. The enzyme β-glycosidase capable of catalyzing the bio-conversion of isoflavone glycoside to isoflavone aglycone contains a nonsugar component, aglycone. This enzyme can be used to improve the bioavailability of isoflavones during tofu preparation (Ryu et al., 2010). Isoflavones is a secondary metabolite product in soybeans and considered as a very important for humans. Isoflavones are subclass from large group of flavonoids that have estrogenic activity or phytoestrogens (Cederroth & Nef, 2009). There are four forms of the isoflavones found. These are glucosides (daidzin, genistin, and glycitin); acetylglucosides (acetyldaidzin, acetylgenistin, and acetylglycitin); malonylglucosides (malonyldadzin, malonylgenistin, and malonylglycitin); and structure unconjugated aglycone (daidzein, genistein, and glycitein) (Kao et al., 2007). Isoflavones usually found in soybean in the form of genistein, daidzein, and glycitein at concentrations in between 1.45 and 4.59 mg/g (Martino et al., 2011). In nonfermented tofu, the forms of isoflavones are reported are glucoside, genistin and daidzin (Wahyuni et al., 2016).

Isoflavones are primarily present as glucoside conjugates in soya-containing products, except products obtained from soya fermentation, like tempeh, tofu, or soya-based cheeses. When it consumed, it is not absorbed as such in GI tract. Its absorption requires hydrolysis of the sugar moiety by various gastric juices and by intestinal and bacterial β-glucosidases. The amount of isoflavones that are absorbed and conjugated is mainly into glucuronic acid and sulfuric acid. This increases its solubility and ability to circulate in the plasma. Intestinal bacteria metabolize small portion of the isoflavones in to equol 10 or o-desmethylangolensin (Vergne et al., 2007).

Soymilk contains phytate, the hexaphosphate form of myo-inositol, at the level of about 2 g per 100 g of protein and at 1–2 g per 100 g of protein in soy protein isolate. Phytate may interact and form proteinphytate complex with soy protein, thus lowering the protein digestibility and may create hurdle in essential mineral bioavailability. Along with this, phytate may affect the rheological and physical property like solubility of soy protein in a commercial food application (Tsumura et al., 2005). To make it available, isoflavones in tofu can bind to estrogen receptors in the body. It can reduce the pain caused by psychovasomotor especially bursts of heat in the chest usually experienced by women during menopause (Kridawati et al., 2019).

13.5.2 Sauerkraut

Sauerkraut a fermented food contains high levels of biologically active compounds like probiotics and antioxidants. It also possesses anticarcinogenic properties resulting in many proposed health benefits (Hayes et al., 2008; Tse & Eslick, 2014). Folate content increased due to fermentation with LAB. It is observed that commercial canned products of sauerkraut contain folate in the range of 5–21 μg/100g (Jägerstad et al., 2004).

Sauerkraut contains many by-products made from fermentation of sugars. These are organic acids, largely lactic and acetic acids (1%−2% (w/w)), along with propionic, malic, and succinic acids, ethanol, ethyl acetate, acetaldehyde, and carbon dioxide (Trail et al., 1996). Sauerkraut contains phenolic compounds such as gallic acid (0.44−1.06 mg gallic acid equivalents/100 g) (Ciska et al., 2005). Sauerkraut is considered a particularly good source of antioxidants. Among them, vitamin C content ranges from 14.7 to 75 mg/100 g fresh weight (fw), while phenolic compounds contribute as 0.44−1.06 mg gallic acid equivalents/100 g fw) (Peñas et al., 2017).

Fermentation of cabbage makes nutrients more bioavailable than its original form indicating most of the vitamins, mineral, and phytonutrients present in cabbage could become more bioavailable after fermentation process. Thus, sauerkraut is a super food though it is underestimated. Vitamin C in sauerkraut is 20 times higher than in the same serving as raw cabbage indicating more bioavailability of it to get absorbed by GI tract (Campbell-McBride, 2008). Additionally, with increased bioavailability, probiotic content in sauerkraut significantly enhances the value of it as a super food. It can be called sauerkraut as a vegetable with LAB (Douglas & Sanders, 2008). Starter culture is not needed as an inoculum for sauerkraut preparation because the cabbages are grown in soil, and thus it contains inherent micro flora to start fermentation. The lactic acid fermentation happening in cabbage thereby facilitates various chemical changes, which increase the bioavailability of vitamins and nutrients in sauerkraut. LAB contribution in releasing the nutrients present as such in cabbage by breakdown of sugar through process of fermentation, facilitating more ease way to absorb nutrient in the human GI tract (Orgeron et al., 2016).

Sauerkraut may possess methyl methanethiosulfinate, methyl methanethiosulfonate, acetaldehyde, ethanol, ethyl acetate, methanol, n-propanol, and 2-propanol are responsible for the characteristic aroma and flavor. Accumulation of these volatile and nonvolatile aroma or aroma-active compounds caused due to fermentation gives bitter, umami, sweet, sour, and salty tastes (Xiang et al., 2019). Various undissociated acid molecules presents in Sauerkraut penetrate through the bacterial plasma membrane as a function of their diffusion constant and thus confers the antimicrobial action. The acid in the cytoplasm dissociates to release protons and conjugate with bases in higher pH range disrupts the membrane protonmotive force, disabling the energy-yielding and transport process dependent upon it (Savard et al., 2002).

Lactic acid, which is present in significant amount, is responsible for the natural protection of the body against various infections, increases immunity, and gives better protection against digestion and liver diseases (Karovičová & Kohajdová, 2003). Lactic acid fermentation imparts stability and safety to the product and helpful in giving relief in disease infections such as diarrhea and salmonellosis (Nout & Ngoddy, 1997). Acetic acid can inhibit yeasts, molds, and bacteria than lactic acid (Caplice & Fitzerald, 1999). Both these lactic and acetic acid in sauerkraut production have antimicrobial effects though high amount of acetic acid are not desirable from organoleptic and quality aspects point of view. Along with this both acid and alcohol generate esters that are desirable and imparts typical flavors (Lee et al., 1999).

Breakdown of nondigestible material like cellulose, hemicellulose, and polygalacturonic and glucuronic acids in plant foods may lead to the improved bioavailability of mineral and trace elements. Fermented foods like yoghurt and sauerkraut may reduce the serum cholesterol concentration. It helps by reducing the intestinal absorption of dietary and endogenous cholesterol or inhibiting synthesis of it in liver (Kalantzopoulos, 1997).

Kalač et al. (2000) in their research reported that sauerkrauts prepared by using *Lactobacillus plantarum* and or mixed culture of *Lactobacillus plantarum*, *Bacillus casei*, *Enterococcus faecium* and *Pediococcus* spp. had significantly lower concentrations of biogenic amine like tyramine, putrescine and cadaverine.

13.5.3 Sour dough

Sour dough made up of wheat flour, a bran is broken down, releasing nutrients into the dough during the process of proofing. It is a time giving for making changes in dough. In this process, the phytic acid (phytin) needed around 90% neutralization for the minerals present in grains to be absorbed by the human body. Phytin can be neutralized by bacterial action naturally occurring and to some extent while baking process. All phytin present eliminated in naturally leavened bread, while in yeasted bread, about 90% remains (Kelly et al., 2014). In case of bread and backed cereals, taste and the flavor are the most important attributes conferring the quality attribute of the product. Sourdough fermentation has been recognized for improving flavor and structure of rye and wheat breads (Brummer & Lorenz, 2003). Sourdough fermentation improves nutritional value or promote healthfulness of cereal foods. Wholemeal flour is rich in fiber, minerals, vitamins, and many phytochemicals such as phenolic compounds, sterols, tocopherols, and tocotrienols, lignans, and phytic acid. Souring of flour in bread making, mouthfeel and palatability can be improved without degrading nutritional quality of important components (Salmenkallio-Marttila et al., 2001). In sour dough making process, fermentation of wholewheat flour for 2 days can almost completely degrade the phytate and increase the bioavailability of iron (WHO/FAO 2004, pp. 246−278).

During the sourdough bread, complex carbohydrates and proteins are broken down into more digestible simple sugars and into amino acids. Enzymes developed in the stage of proofing are not lost completely as the center of the loaf remains at a lower temperature than the crust. Enzymes partly from lactobacillus make good-quality bread by digestion of all complex carbohydrate foods including other grains, beans, and vegetables. Organic acids namely citric, lactic, acetic, butyric, propionic, and formic acids produced during fermentation of cereals (e.g., tobwa), vegetables (e.g., sauerkraut), and some soy sauces, it has the potential to form soluble ligands with trace minerals in the gastrointestinal tract. Thus, further helps in absorption of nonheme iron and zinc (Gibson, 2007).

The primary metabolic end products of LAB are L-lactic and acetic acids with small concentration of citric and malic acids. The ratio of lactic acid to acetic acid is important for imparting final aroma and flavor of bread (Linko et al., 1997). LAB produce lactic and acetic acids more slowly when incorporated as a mixed culture with yeasts when compared their role as a monoculture (Merseburger et al., 1995). Enzymatic oxidation or autoxidation of the lipid fraction of the wheat generates certain aldehydes by heterofermentative LAB (Frankel, 1982) and bacterial metabolism while some homofermentative LAB, except some strain like *Lb. delbrueckii subsp. Delbrueckii,* primarily contribute ethyl acetate and diacetyl (Hansen & Hansen, 1993).

Sourdough bread significantly reduced the serum glucose and insulin levels and increases satisfaction levels than the other bread (Bituh, 2011). Rye flour in greater extent has been used for sourdoughs preparation, and in certain cases it is produced by wheat flour (Kelly et al., 2014). Rye flour contains high amount of proteins and soluble fiber. This decelerates the breakdown of carbohydrates imparting the feeling of fullness after consumption of less bread. Similarly, it has high amounts of iron, calcium, zinc, and a large amount of B vitamins (Expert answer, 2010).

The LAB strains usually used in sourdough preparation vary in metabolism and aroma compounds. Culture used singly ferment dough for 15 h at 30°C, followed by mixing, and additional 10 h fermentation has shown rise in typical sourdough volatile compounds (Merotha et al., 2004). Stolz et al. (1993) investigated the sour dough making using both wheat doughs and simple substrates including maltose and glucose. They observed the production of volatile aroma compounds during fermentation with combinations of single strains of *S. cerevisiae* and *C. guilliermondii*, *Lb. plantarum*. They reported only with yeasts strains in wheat bread, seven volatiles were found in abundant amount. These are acetaldehyde, acetone, ethyl acetate, ethanol, hexanal, isobutyl alcohol, and propanol. They also reported that *S. cerevisiae* produced more volatiles than *C. guilliermondii*, but quantity of volatile flavor compounds can be improved by the addition of glucose and sucrose less by maltose. Addition of enzymes to sourdough sponges can also enhance bread volatile compounds (Martínez-Anaya et al., 1996). Moreover, addition of fructose and citrate to sourdough enhanced the production of volatiles by LAB including ethanol, which was eliminated in baking and 2-methyl-1-propanal with no effect on the production of lactic and acetic acids (Gobbetti et al., 1996). Frasse et al. (1993) reported that in a French yeast sourdough, more than 40 flavor components were identified. Among them 20 alcohols, 7 esters, 6 lactones, 6 aldehydes, 3 alkanes, and a single sulfur compound identified. LAB used during fermentation also produced EPS or gums are used as texturizing, antistaling, or prebiotic additives in bread production. These can be better replaced to hydrocolloids (Chavan & Chavan, 2011).

Sourdough bread has achieved higher sensory score due to a higher content of volatiles (Hansen et al., 1989; Hansen & Hansen, 1996). The contents of methylpropanol (isobutanol), 2- and 3-methylbutanol (iso-pentanols), ethyl acetate, and ethyl lactate in 3 bakery sourdoughs were clearly related to the amounts of these compounds in the sourdoughs (Hansen & Lund, 1987; Lund et al., 1987). Nonvolatile compounds produced during sourdough fermentation, including organic acids, produced by homo- and heterofermentative bacteria that acidify, decrease pH, and contribute aroma to the bread dough (Gobbetti et al., 1995). While volatile compounds include alcohols, aldehydes, ketones, esters, and sulfur. The generation of sufficient amounts of volatile compounds during fermentation requires about 12–24 h, while fermentation by baker's yeast alone can be completed within a few hours. The flavor profile of volatiles and nonvolatile components in sourdoughs is influenced by the activity of the LAB and the sourdough yeasts (Chavan & Chavan, 2011).

13.6 Applications associated with the plant-based fermented foods

13.6.1 Tofu

Isoflavone aglycone in higher amount in enzyme-treated tofu can benefit consumers when compared to the conventionally processed one. Ryu et al. (2010) suggested that the use of β-glycosidase during tofu processing is useful for the conversion of glycoside forms of isoflavones to readily bioavailable aglycone forms. Tsumura et al. (2005) recommended that a phytase treatment can be used to change the gel-forming ability of soy protein that can be applied to prepare new food products such as phytate-reduced tofu with high nutritional quality.

In a study on prostate cancer, the chemopreventive effect of the soybean mediated by mechanisms of modulations of gene expression in which isoflavones derived from soybean cake exhibited antiproliferative effects on prostate cancer cells. In that study was observed the decreasing levels of the cyclin B1 expression. Thus, daily consumption of soy in the form of tofu and other products may be useful strategy for chemopreventive benefit of bioactive compounds in soybeans (Wahyuni et al., 2016).

David et al. (1998) stated that the based on epidemiological data that one serving of soy daily (4 ounces of tofu containing approximately 36 mg of isoflavones) may lower the cancer risk. Higher consumption of isoflavones and tofu moderately lowering risk of developing CHD, and in women, the favorable association of tofu was more noticeable in young women or postmenopausal women without hormone use. Increased in consumption of rich in isoflavone-may useful in benefits for reducing risk of coronary heart disease. Along with this tofu is integrated as such healthy plant-based diets provide many excellent plant proteins and aid in the prevention of coronary heart disease. Tofu also contains several bioactive components such as polyunsaturated fats, fiber, and minerals that may act synergistically with isoflavones through various pathways to modulate CHD risk (Ma et al., 2020).

13.6.2 Sauerkraut

Sauerkraut has higher levels of vitamins C, E, and phenolic compounds, which act as potent free radical scavengers to prevent oxidation (Podsedek, 2007). Apart from this, the presence of vitamin C helps in reducing C-reactive protein (CRP) levels involved in inflammation. Both these phenolic compounds and vitamin C acts neutralizing superoxide and hydroxyl radicals by donating an electron for eight human enzymes (Ellulu et al., 2015).

Sauerkraut contains high levels of anticarcinogenic compounds. Sauerkraut at low-salt concentration levels improved ascorbigen content. Highest concentration of ascorbigen observed in lower sodium sauerkraut (0.5% NaCl) (Martinez-Villaluenga et al., 2009). Sauerkraut is effective against different types of cancer having the chemo-preventive effects. The antiestrogenic activity of sauerkraut extracts at low concentrations (5–25 ng/mL) in oestrogen-dependent human breast cancer (MCF-7) cells has been reported (WHO/FAO 2004, pp. 246–278). Few evidence are available on the effectiveness of sauerkraut on the IBS. One research indicated that both pasteurized and unpasteurized sauerkraut reduced IBS severity; this effect does not appear to be mediated by gastrointestinal microbiota, and it needs further research to elaborate the mechanisms of this effect on gastrointestinal disorders (Dimidi et al., 2019).

13.6.3 Sourdough

The sourdough fermentation has better acceptability in flavor profile when compared with chemically acidified breads (Lund, 1989). The synergistic metabolic activities among the microorganisms gave desired acidification or souring, which influence the texture, flavor profile, and also increased shelf-life (Katina, 2005, p. 569). The flavor of sourdough wheat bread is richer and more aromatic than in wheat bread due to long fermentation period gives the most desirable sensory characteristics at pH 4.0–5.5 and at 140°C (Przybylski & Kamin'ski, 1983). The loaves made with the inoculation of 5%–10% with *Lb. plantarum* and 5%–15% sourdoughs fermented with *Lb. sanfrancisco* are preferred in odor and taste (Hansen & Hansen, 1996). Sourdough of wheat bread crumb on sensory evaluation basis showed sourdough fermented with the heterofermentative *Lb. sanfranciscensis* had a pleasant, mild, sour odor, and taste, whereas sourdough bread fermented with the homofermentative *Lb. plantarum* had an unpleasant metallic sour taste. But, when sourdough was supplemented with sourdough yeast *S. cerevisiae*, the wheat bread received a more aromatic flavor (Chavan & Chavan, 2011).

A typical Western diet contains dietary fiber (DF) as less than 20 g/d, whereas the recommended daily intake of it is 25–30 g. Most people take too little or not sufficient fiber diet. These low levels of DF contribute in occurrence of many diseases, including dental problems, constipation to obesity, colorectal cancer, coronary heart disease, and type 2 diabetes. To overcome this problem sourdough can be modified with the bran fraction of the grain (rich in fiber) in such a way that maximum amounts of bran can be utilized in breads. The most common source of DF in bread making is cereal bran, especially wheat bran (Lopez et al., 2003). The mineral availability of sourdough-baked products can also be improved due to solubilization of the phytate complex by chemically or microbiological acidification of sourdough. EPS content is also improved by using such strain like *L. sanfranciscensis*, contributing improvement in nutritional properties of sourdough fermented products metabolized by bifidobacteria (Korakli et al., 2001). Sourdough LAB has shown hydrolyzing activities as increase in prolamin peptides showed difficulty in human cereal intolerance (Di Cagno et al., 2002).

The flavor of sourdough wheat bread is richer and more aromatic than nonfermented wheat bread (Brummer & Lorenz, 1991). The optimal use of condition making suitable for fermentation can improve the taste and flavor of the sourdough bread (Chavan & Chavan, 2011). The concentration of 2-phenylethanol as one of the most potent odorants of wheat breadcrumb also increases in sourdough breadcrumb (Gassenmeier & Schieberle, 1995). Ash content has a considerable effect on acidification (Clarke et al., 2002), and increase in acidity increases the solubility of the glutenin fraction extracted from wheat flour, which also affects gluten's swelling property (Axford et al., 1968). The application of LAB in sourdough preparation has a positive effect on bread staling. One such positive effect is an improvement in loaf-specific volume, which inversely associated with the rate of staling (Axford et al., 1968; Corsetti et al., 2001). Advantage also associated LAB in sourdough through generation of many antimicrobial substances, such as organic acids, CO_2, ethanol, hydrogen peroxide, diacetyl, fatty acids, phenyllactic acid, reuterin, and fungicins (Chavan & Chavan, 2011; Schnurer & Magnusson, 2005).

The metabolites of the LAB from fermented foods can have a broad array of applications in addition to their large enzymatic diversity. Biogenic amines, such as those generated by *Lactobacillus brevis* in fermented carrots or other bacteria in sauerkraut, are examples of these (Wuyts et al. 2018). These biogenic amines are undesirable in fermented foods because they can induce malodor and headaches, but they could be employed as therapeutic agents to boost cardiovascular health and wellness, but this is still debatable (Majcherczyk & Surówka 2019). The bacteria that break down the food produce various waste products as it ferments or decomposes. Alcohol is one of these by-products. A variety of fermented foods, like soy sauce, have a lot of alcohol in them. Although the amount of alcohol in fermented foods is usually relatively smaller amount but it has been reported that even small amounts of alcohol have an adverse consequence on hosts' body cells (Kharobe 2018). Furthermore, histamine generated from *Lactobacillus reuteri*, a biogenic amine, was identified to decrease pro-inflammatory TNF production, potentially leading to better anti-inflammatory therapies for chronic immune-mediated disorders (Nilsson & Persson 2019). Similarly, instead of live probiotics, the bacteriocins or isolated secondary antimicrobial metabolites could be employed as purified molecules, which could be a more appealing valorization method in terms of formulations and administration (Wuyts et al. 2020).

13.7 Summary and future perspectives

Probiotic products are a potential growth field in the international market for functional foods, which is increasing at a rapid pace. Probiotic species are being researched extensively in order to create products into which they can be integrated. While probiotic fermentation in the dairy industry has a very well market niche, the application of a similar techniques in the area of plant products is still stunted, offering considerable possibilities to establish new functional food systems. A significant percentage of plant-based fermented foods is produced locally in various parts of the world. However, in order to produce commercially viable goods, optimized technological advancements are necessary.

Plants are high in phenolic molecules, mainly flavonoids and phytoestrogens. These molecules possess antioxidant, anticarcinogenic, and antibacterial activities and have been shown to have potential anti-inflammatory, anti-cancer, and anti-osteoporosis properties. Oligosaccharides and dietary fibers are abundant in few plant products, viz. cereals and legumes. These oligosaccharides may be used as non-digestible carbohydrate sources that can have a variety of physiological benefits as well as specifically stimulate colon microflora development and function as prebiotics. As a result, there is a great need for research into the development of synbiotic products that combine probiotics and prebiotics and possibly confer the favorable impacts of both ingredients. To close the knowledge gap, more studies should be conducted to see if this mixture of ingredients has any beneficial effects or health impacts. More research is required to better understand the various types of probiotic bacteria that can thrive in plant-based products. The enhancement of fermentation strategies for the growth and development of probiotic bacteria, as well as the introduction of novel potential health benefits in plant-based products, is another area where there is a gap of in-depth expertise. These fermentation methods can also be used to create new nutraceuticals that can be processed and utilized as additives.

References

Axford, D. W. E., Kolwell, K. H., Cornfor, S. J., & Elton, G. A. H. (1968). Effect of loaf specific volume on rate and extent of staling in bread. *Journal of the Science of Food and Agriculture, 19*, 95–101.

Bartkiene, E., Lele, V., Ruzauskas, M., Domig, K. J., Starkute, V., Zavistanaviciute, P., Bartkevics, V., Pugajeva, I., Klupsaite, D., Juodeikiene, G., & Mickiene, R. (2020). Lactic acid bacteria isolation from spontaneous sourdough and their characterization including antimicrobial and antifungal properties evaluation. *Microorganisms, 8*(1), 64.

Beddows, C. G., & Wong, J. (1987). Optimization of yield and properties of silken tofu from soybeans. I. The water: Bean ratio. *International Journal of Food Science and Technology, 22*(1), 15–21.

Bituh, M., Novotni, D., Cukelj, N., Smerdel, B., & Dujmi, F. (2011). Glycemic index and firming kinetics of partially baked frozen gluten-free bread with sourdough. *Journal of Cereal Science, 55*(2), 120–125.

Bousquet, J., Anto, J. M., Czarlewski, W., Haahtela, T., Fonseca, S. C., Iaccarino, G., Blain, H., Vidal, A., Sheikh, A., Akdis, C. A., & Zuberbier, T. (2021). Cabbage and fermented vegetables: From death rate heterogeneity in countries to candidates for mitigation strategies of severe COVID-19. *Allergy, 76*(3), 735–750.

Breidt, F., McFeeters, R. F., Perez-Diaz, I., & Lee, C. H. (2012). *'Fermented vegetables', Food microbiology: fundamentals and frontiers* (pp. 841–855).

Brummer, J. M., & Lorenz, K. (1991). *European developments in wheat sourdoughs, 36* pp. 310–314). Cereal Food World.

Brummer, J. M., & Lorenz, K. (2003). In K. Kulp, & K. Lorenz (Eds.), *'Preferments and sourdoughs in German breads' in handbook of Dough Fermentations* (pp. 247–267). New York: Marcel Dekker Inc.

Cai, T. D., & Chang, K. C. (1997). Dry tofu characteristics affected by soymilk solid content and coagulation time. *Journal of food quality, 20*(5), 391–402.

Campbell-McBride, N. (2008). Gut and psychology syndrome. *Journal of Orthomolecular Medicine, 23*(2), 90–94.

Caplice, E., & Fitzerald, G. F. (1999). Food fermentations: Role of microorganisms in food production and preservation. *International Journal of Food Microbiology', 50*, 131–149.

Cederroth, C. R., & Nef, S. (2009). Soy, phytoestrogens and metabolism: A review. *Molecular and Cellular Endocrinology, 304*(1-2), 30–42.

Chavan, R. S., & Chavan, S. R. (2011). Sourdough technology - a traditional way for wholesome foods: A review. *Comprehensive Reviews in Food Science and Food Safety, 10*, 170–183.

Chou, C. C., & Hwan, C. H. (1994). Effect of ethanol on the hydrolysis of protein and lipid during the ageing of a Chinese fermented soya bean curd—sufu. *Journal of the Science of Food and Agriculture, 66*(3), 393–398.

Ciska, E., Karamac, M., & Kosinska, A. (2005). Antioxidant activity of extracts of white cabbage and sauerkraut. *Journal of Food Nutrition Science, 55*, 367–373.

Clarke, C. I., Schober, T. J., & Arendt, E. K. (2002). Effect of single strain and traditional mixed strain starter cultures on rheological properties of wheat dough and bread quality. *Cereal Chemistry, 79*, 640–647.

Corsetti, A., Lavermicocca, P., Morea, M., Baruzzi, F., Tosti, N., & Gobbetti, M. (2001). Phenotypic and molecular identification and clustering of lactic acid bacteria and yeasts from wheat (species Triticum durum and *Triticum aestivum*) sourdoughs of Southern Italy. *International Journal of Food Microbiology, 64*, 95–104.

Daeschel, M. A., Andersson, R. E., & Fleming, H. P. (1987). Microbial ecology of fermenting plant materials. *FEMS Microbiology Reviews, 3*(3), 357–367.

David, B., Fournier, J. W., Erdman, Jr, & Gordon, G. B. (1998). Soy, its components, and cancer prevention: A review of the in vitro, animal, and human data. *Cancer Epidemiology, Biomarkers & Prevention, 7*, 1055–1065.

De Filippis, F., Parente, E., & Ercolini, D. (2017). Metagenomics insights into food fermentations. *Microbial biotechnology, 10*(1), 91–102.

De Vuyst, L., Van Kerrebroeck, S., Harth, H., Huys, G., Daniel, H. M., & Weckx, S. (2014). Microbial ecology of sourdough fermentations: Diverse or uniform? *Food microbiology, 37*, 11–29.

DeMan, J. M., DeMan, L., & Gupta, S. (1986). Texture and microstructure of soybean curd (tofu) as affected by different coagulants. *Food Structure, 5*(1), 11.

Di Cagno, R., de Angelis, M., Lavermicocca, P., De Vincenzi, M., Giovannini, C., Faccia, M., & Gobbetti, M. (2002). Proteolysis by sourdough lactic acid bacteria: Effect on wheat flour protein fractions and gliadin peptides involved in human cereal intolerance. *Applied Environmental Microbiology, 68*, 623–633.

Dimidi, E., Cox, S. R., Rossi, M., & Whelan, K. (2019). Fermented foods: Definitions and characteristics, impact on the gut microbiota and effects on gastrointestinal health and disease. *Nutrients, 11*, 1806.

Douglas, L. C., & Sanders, M. E. (2008). Probiotics and prebiotics in dietetic practice. *Journal of the American Dietetic Association, 108*, 510–521.

Duar, R. M., Lin, X. B., Zheng, J., Martino, M. E., Grenier, T., Pérez-Muñoz, M. E., Leulier, F., Gänzle, M., & Walter, J. (2017). Lifestyles in transition: Evolution and natural history of the genus lactobacillus. *FEMS Microbiology Reviews, 41*(Supp_1), S27–S48.

Ellulu, M. S., Rahmat, A., Patima, I., Khaza, A. H., & Abed, Y. (2015). Effect of vitamin C on inflammation and metabolic markers in hypertensive and/or diabetic obese adults: A randomized controlled trial. *Drug Design, Development and Therapy, 9*, 3405–3412.

Expert answer. (2010). White rye bread healthier than whole wheat? *Science Daily*. Available from: http://www.sciencedaily.com/releases/2010/02/100212210212.htm [February, 2021].

Fernández-Peláez, J., Paesani, C., & Gómez, M. (2020). Sourdough technology as a tool for the development of healthier grain-based products: An update. *Agronomy, 10*(12), 1962.

Frankel, E. N. (1982). Volatile lipid oxidation products. *Progress in Lipid Research, 22*, 4–33.

Frasse, P., Lambert, S., Richard-Molard, D., & Chiron, H. (1993). The influence of fermentation on volatile compounds in French bread dough. *LWT-Food Science and Technology, 26*, 126–132.

Gardner, N. J., Savard, T., Obermeier, P., Caldwell, G., & Champagne, C. P. (2001). Selection and characterization of mixed starter cultures for lactic acid fermentation of carrot, cabbage, beet and onion vegetable mixtures. *International journal of food microbiology, 64*(3), 261–275.

Gassenmeier, K., & Schieberle, P. (1995). Potent aromatic compounds in the crumb of wheat bread (French-type). Influence of preferment and studies on the key odrants during dough processing. *Lebnsm Unters Forsch, 201*, 241–248.

Gibson, R. S. (2007). The role of diet—And host-related factors in nutrient bioavailability and thus in nutrient-based dietary requirement estimates. *Food and Nutrition Bulletin, 28*(1 (Suppl)), S77–S100. The United Nations University.

Gobbetti, M., Corsetti, A., & De Vincenzi, S. (1995). The sourdough microflora. Characterization of homofermentative lactic acid bacteria based on acidification kinetics and impedance tests. *Italian Journal of Food Science, 2*, 91–102.

Gobbetti, M., De Angelis, M., Di Cagno, R., Polo, A., & Rizzello, C. G. (2020). The sourdough fermentation is the powerful process to exploit the potential of legumes, pseudo-cereals and milling by-products in baking industry. *Critical Reviews in Food Science and Nutrition, 60*(13), 2158–2173.

Gobbetti, M., Smacchi, E., & Corsetti, A. (1996). The proteolytic system of Lactobacillus sanfrancisco CB1: Purification and characterization of a proteinase, a dipeptidase, and an aminopeptidase. *Applied and Environmental Microbiology, 62*(9), 3220–3226.

Guan, X., Zhong, X., Lu, Y., Du, X., Jia, R., Li, H., & Zhang, M. (2021). Changes of soybean protein during tofu processing. *Foods, 10*(7), 1594.

Gupta, S., & Abu-Ghannam, N. (2012). Probiotic fermentation of plant-based products: Possibilities and opportunities. *Critical reviews in food science and nutrition, 52*(2), 183–199.

Hammes, W. P., & Gänzle, M. G. (1998). Sourdough breads and related products. In BJB Wood (Ed.), *1. Microbiology of fermented foods* (pp. 199–216). London: Blackie Academic and Professional.

Hansen, A., & Hansen, B. (1993). Flavour compounds in wheat sourdoughs. In C. Benedito De Barber, C. Collar, M. A. Martinez-Anaya, & J. Morel (Eds.), *Progress in food fermentation, proceedings of European Food chemistry VII.* Valencia: IATA CSIC. Progress in food fermentation.

Hansen, A., & Hansen, B. (1996). Flavour of sourdough wheat bread crumb. *Zeitschrift für Lebensmittel-Untersuchung und -Forschung', 202*, 244–249.

Hansen, A., & Lund, B. (1987). Volatile compounds in rye sourdough. In M. Martens, G. A. Dalen, & H. Russwurm (Eds.), *Flavour Science and Technology* (pp. 43–49). New York: Wiley.

Hansen, A., Lund, B., & Lewis, M. J. (1989). *Flavor of sourdough rye bread crumb, 22* pp. 141–144). Lebensmittel-Wissenschaft & Technologie.

Hayes, J. D., Kelleher, M. O., & Eggleston, I. M. (2008). The cancer chemopreventive actions of phytochemicals derived from glucosinolates. *European Journal of Nutrition, 47*, 73–88.

Heenan, C. N., Adams, M. C., Hosken, R. W., & Fleet, G. H. (2004). Survival and sensory acceptability of probiotic microorganisms in a nonfermented frozen vegetarian dessert. *LWT-Food Science and Technology, 37*(4), 461–466.

Heiniö, R. L., Liukkonen, K. H., Katina, K., Myllymäki, O., & Poutanen, K. (2003). Milling fractionation of rye produces different sensory profiles of both flour and bread. *LWT-Food Science and Technology, 36*(6), 577–583.

Hou, J. J., Yang, X. Q., Fu, S. R., Wang, M. P., & Xiao, F. (2016). Preparation of double-network tofu with mechanical and sensory toughness. *International journal of food science & technology, 51*(4), 962–969.

Hui, Y. H., & Evranuz, E.Ö. (Eds.). (2012). *Handbook of animal-based fermented food and beverage technology, 1.* CRC press.

Jägerstad, M., Jastrebova, J., & Svensson, U. (2004). Folates in fermented vegetables - a pilot study. *LWT-Food Science and Technology, 37*(6), 603–611.

Jahangir, M., Kim, H. K., Choi, Y. H., & Verpoorte, R. (2009). Health-affecting compounds in Brassicaceae. *Comprehensive Reviews in Food Science and Food Safety, 8*(2), 31–43.

Japan External Trade Organization (JETRO). (2011). *Specifications and standards for food, food additives etc. Under the food sanitation Act 36 (abstract).* Available from: www.jetro.go.jp/ext_images/en/reports/regulations/pdf/foodext2010e.pdf [March 28 2016].

Johanningsmeier, S., McFeeters, R. F., Fleming, H. P., & Thompson, R. L. (2007). Effects of Leuconostoc mesenteroides starter culture on fermentation of cabbage with reduced salt concentrations. *Journal of food science, 72*(5), M166–M172.

Jooyandeh, H. (2011). Soy products as healthy and functional foods. *Middle-East Journal of Scientific Research, 7*(1), 71–80.

Kalač, P., Ščka, J., Křížek, M., & Pelikánová, T. (2000). The effects of lactic acid bacteria inoculants on biogenic amines formation in sauerkraut. *Food Chemistry, 70*, 355–359.

Kalantzopoulos, G. (1997). Fermented products with probiotic qualities. *Anaerobe, 3*, 185–190.

Kao, T. H., Wu, W. M., Hung, C. F., Wu, W. B., & Chen, B. H. (2007). Anti-inflammatory effects of iso-flavone powder produced from soybean cake. *Journal of Agricultural and Food Chemistry, 55*(26), 11068–11079.

Karovičová, J., & Kohajdová, Z. (2003). Lactic acid fermented vegetable juices. *Horticultural Science (Prague), 30*(4), 152–158.

Katina, K. (2005). *Sourdough: a tool for the improved flavour, texture and shelf-life of wheat bread* (p. 569). Academic dissertation, Department of Food Technology, University of Helsinki, ESPOO 2005 VTT Publications.

Kelly, C., Bourguignon, D. J. F., Smith, I. M. J., Morris, R. C. S., & Casserly, S. C. S. (2014). *Sourdough Bread Fermented Foods- Group 3.* Available from: http://fermentedfoods3.blogspot.co.nz/2012/04/sourdough-bread.html. (Accessed 23 February 2021).

Kharobe, V. B. (2018). Advantages and disadvantages of traditional fermentation of dairy products. *International Journal of Recent Trends in Science and Technology, 8*, 52–55.

Kobayashi, M., Shima, T., & Fukuda, M. (2018). Metabolite profile of lactic acid-fermented soymilk. *Food and Nutrition Sciences, 9*(11), 1327–1340.

Korakli, M., Rossmann, A., Ganzle, M. G., & Volgel, R. F. (2001). Sucrose metabolism and exopoly-sachrides production in wheat and rye sourdoughs by Lactobacillus sanfraciccensis. *The Journal of Agricultural and Food Chemistry, 49*, 5194–5200.

Kridawati, A., Rahardjo, T. B. W., Hardinsyah, Damanik, R., & Hogervorst, E. (2019). Comparing the effect of tempe flour and tofu flour consumption on estrogen serum in ovariectomized rats. *Heliyon, 5*(6), 01787.

Kulp, K., & Lorenz, K. (2003). *Handbook of dough fermentations.* New York: Marcel Dekker Inc.

Kwon, E. A., & Kim, M. H. (2007). Microbial evaluation of commercially packed kimchi products. *Food Science and Biotechnology, 16*(4), 615–620.

Lau, S. W., Chong, A. Q., Chin, N. L., Talib, R. A., & Basha, R. K. (2021). Sourdough microbiome comparison and benefits. *Microorganisms, 9*(7), 1355.

Lee, J. H., Lee, S. K., Park, K. H., Hwang, I. K., & Ji, G. E. (1999). Fermentation of rice using amylolytic Bifidobacterium. *International Journal of Food Microbiology, 50*(3), 155–161.

Limbad, M., Gutierrez Maddox, N., Hamid, N., & Kantono, K. (2020). Sensory and physicochemical characterization of sourdough bread prepared with a coconut water kefir starter. *Foods, 9*(9), 1165–1188.

Linko, Y., Javanainen, P., & Linko, S. (1997). Biotechnology of bread baking. *Trends in Food Science & Technology', 8*, 339–344.

Li, C., Rui, X., Zhang, Y., Cai, F., Chen, X., & Jiang, M. (2017). Production of tofu by lactic acid bacteria isolated from naturally fermented soy whey and evaluation of its quality. *LWT-Food Science and Technology, 82*, 227–234.

Lopez, H. W., Duclos, V., Coudray, C., Krespine, V., Feillet-Coudary, C., Messager, A., Demigne, C., & Remesy, C. (2003). Making bread with sourdough improves mineral bioavailability from reconstituted whole wheat flour in rats. *Nutrition, 19*, 524–530.

Lorenz, K., & Brummer, J. M. (2003). In K. Kulp, & K. Lorenz (Eds.), *'Preferments and sourdoughs for German breads' in Handbook of dough fermentations* (pp. 247–270). New York: Marcel Dekker.

Lund, B., Hansen, A., & Lewis, M. J. (1987). The influence of yeast and sourdough fermentation on volatile compounds in rye bread crumb. In L. Munck (Ed.), *Cereal science and technology* (pp. 185–195). Nordic Cereal Congress, The Danish Cereal Society. Copenhagen: proceedings from 23.

Lund, D. (1989). Food processing: from art to engineering. *Food Technology, 43*(1), 242–308.

Majcherczyk, J., & Surówka, K. (2019). Effects of onion or caraway on the formation of biogenic amines during sauerkraut fermentation and refrigerated storage. *Food Chemistry, 298*, 125083.

Ma, L., Liu, G., Ding, M., Zong, G., Hu, F. B., Willett, W. C., Rimm, E. B., Manson, J. E., & Sun, Q. (2020). Isoflavone intake and the risk of coronary heart disease in US men and women results from 3 prospective cohort studies. *Circulation, 141*, 1127–1137.

Martínez-Anaya, M. A., de Barber, C. B., & Esteve, C. C. (1996). Effect of processing conditions on acidification properties of wheat sour doughs. *International journal of food microbiology, 22*(4), 249–255.

Martinez-Villaluenga, C., Peñas, E., Frias, J., Ciska, E., Honke, J., Piskula, M. K., Kozlowska, H., & Vidal-Valverde, C. (2009). Influence of fermentation conditions on glucosinolates, ascorbigen, and ascorbic acid content in white cabbage (*Brassica oleracea* var. capitata cv. Taler) cultivated in different seasons. *Journal of Food Science, 74*(1), 62–67.

Martino, H. S. D., de Morais, C. L., Ribeiro, S. M. R., de Souza, D. M. I., Piovesan, N. D., & De Mejía, E. (2011). Nutritional and bioactive compounds of soybean: Benefits on human health. In *Soybean and Health* (pp. 465–488). London, UK: IntechOpen.

McFeeters, R. F. (2004). Fermentation microorganisms and flavor changes in fermented foods. *Journal of food science, 69*(1), FMS35–FMS37.

Merotha, C. B., Hammesa, W. P., & Hertela, C. (2004). Characterisation of the microbiota of rice sourdoughs and description of Lactobacillus spicheri sp. nov. *Systematic and Applied Microbiology, 27*, 151–159.

Merseburger, T., Ehret, A., Geiges, O., Baumann, B., & Schmidt Lorenz, W. (1995). Microbiology of dough preparation vii. production and use of preferments to produce wheat-bread. *Mitteilungen aus dem Gebiete der Lebensmitteluntersuchung und Hygiene, 86*, 304–324.

Minervini, F., De Angelis, M., Di Cagno, R., Pinto, D., Siragusa, S., Rizzello, C. G., & Gobbetti, M. (2010). Robustness of Lactobacillus plantarum starters during daily propagation of wheat flour sourdough type I. *Food Microbiology, 27*(7), 897–908.

Mitra, J., Jha, A., Alam, T., Singh, D. S., Ranjan, S. K., Pathak, S., & Naz, A. (2013). Sensory and textural properties of tofu manufactured by blending buffalo milk. *Asian Journal of Dairy & Food Research*, 135–138.

Murugkar, D. A. (2014). Effect of sprouting of soybean on the chemical composition and quality of soymilk and tofu. *Journal of food science and technology, 51*(5), 915–921.

Nilsson, B. O., & Persson, L. (2019). Beneficial effects of spermidine on cardiovascular health and longevity suggest a cell type-specific import of polyamines by cardiomyocytes. *Biochemical Society Transactions, 47*(1), 265–272.

Noh, E. J., Park, S. Y., Pak, J. I., Hong, S. T., & Yun, S. E. (2005). Coagulation of soymilk and quality of tofu as affected by freeze treatment of soybeans. *Food Chemistry, 91*(4), 715–721.

Nout, M. J. R., & Ngoddy, P. O. (1997). Technological aspects of preparing affordable fermented complementary foods. *Food Control, 8*, 279–287.

Oboh, G. (2006). Coagulants modulate the hypocholesterolemic effect of tofu (coagulated soymilk). *African Journal of biotechnology, 5*(3), 290–294.

Ono, T., Onodera, Y., Chen, Y., & Nakasto, K. (2010). Tofu structure is regulated by soymilk protein composition and coagulant concentration. In K. R. Cadwallader, & S. K. C. SKC Chang (Eds.), *Chemistry, texture and flavor of soy* (pp. 219–229). Washington, DC: American Chemical Society. ACS Symposium Series 1059.

Orgeron, R., Corbin, A., & Scott, B. (2016). Sauerkraut: A probiotic super food. *Functional Foods Health and Disease, 6*, 536–543.

Pazoki Ghouhe, F., Naghipour, F., & Faraji, A. (2019). Investigating on the effect of using sourdough made from hard and soft wheat flour on physicochemical and sensory characteristics of baguette bread. *' Food Science and Technology, 16*(94), 77–89.

Peñas, E., Martinez-Villaluenga, C., & Frias, J. (2017). In J. Frias, C. Martinez-Villaluenga, & E. Peñas (Eds.), *'Sauerkraut: production, composition, and health benefits' in Fermented Foods in Health and Disease prevention* (pp. 557–569). Academic Press, Elsevier.

Peñas, E., Martinez-Villaluenga, C., Frias, J., Sánchez-Martínez, M. J., Pérez-Corona, M. T., Madrid, Y., Cámara, C., & Vidal-Valverde, C. (2012). Se improves indole glucosinolate hydrolysis products content, Se-methylselenocysteine content, antioxidant capacity and potential anti-inflammatory properties of sauerkraut. *Food Chemistry, 132*(2), 907–914.

Pételt, C., Onno, B., & Prost, C. (2017). Sourdough volatile compounds and their contribution to bread: A review. *Trends in Food Science & Technology, 59*, 105–123.

Pico, J., Bernal, J., & Gómez, M. (2015). Wheat bread aroma compounds in crumb and crust: A review. *Food Research International, 75*, 200–215.

Podsedek, A. (2007). Natural antioxidants and antioxidant capacity of Brassica vegetables: A review. *LWT-Food Science and Technology, 40*, 1–11.

Przybylski, R., & Kamiński, E. (1983). Thermal degradation of precursors and formation of flavour compounds during heating of cereal products. Part II. The formation and changes of volatile flavour compounds in thermally treated malt extracts at different temperature and pH. *Die Nahrung, 27*, 487–496.

Robert, H., Gabriel, V., & Fontagné-Faucher, C. (2009). Biodiversity of lactic acid bacteria in French wheat sourdough as determined by molecular characterization using species-specific PCR. *International journal of food microbiology, 135*(1), 53–59.

Ryu, Y. G., Won, B., Park, H. R., Ghafoor, K., & Park, J. (2010). Effects of the β-glycosidase reaction on bio-conversion of isoflavones and quality during tofu processing. *Journal of the Science of Food and Agriculture, 90*(5), 843–849.

Sahlström, S., & Bråthen, E. (1997). Effects of enzyme preparations for baking, mixing time and resting time on bread quality and bread staling. *Food Chemistry, 58*(1-2), 75–80.

Saint-Eve, A., Granda, P., Legay, G., Cuvelier, G., & Delarue, J. (2019). Consumer acceptance and sensory drivers of liking for high plant protein snacks. *Journal of the Science of Food and Agriculture, 99*(8), 3983–3991.

Salmenkallio-Marttila, M., Katina, K., & Autio, K. (2001). Effect of bran fermentation on quality and microstructure of high-fibre wheat bread. *Cereal Chemistry, 78*, 429–435.

Salovaara, H. (2004). In S. Salminen, & A. von Wright (Eds.), *'Lactic acid bacteria in cereal-based products' in Lactic acid bacteria—Microbiological and functional aspects* (3nd edition, p. 431.451). New York: Marcel Dekker. A Ouwehand.

Satora, P., Skotniczny, M., Strnad, S., & Ženišová, K. (2020). Yeast microbiota during sauerkraut fermentation and its characteristics. *International Journal of Molecular Sciences, 21*(24), 9699.

Savard, T., Beaulieu, C., Gadren, N. J., & Champagne, C. P. (2002). Characterization of spoilage yeasts isolated from fermented vegetables and inhibition by lactic, acetic and propionic acids. *Food Microbiology, 19*, 363–373.

Schnurer, J., & Magnusson, J. (2005). Antifungal lactic acid bacteria as bio-preservatives. *Trends in Food Science and Technology, 16*, 70–78.

Serrazanetti, D. I., Ndagijimana, M., Miserocchi, C., Perillo, L., & Guerzoni, M. E. (2013). Fermented tofu: Enhancement of keeping quality and sensorial properties. *Food Control, 34*(2), 336–346.

Settanni, L., & Corsetti, A. (2008). Application of bacteriocins in vegetable food biopreservation. *International Journal of Food Microbiology, 121*(2), 123–138.

Sharma, D., Aditi, R., Aruna, K., Shivani-Siddharth, H., & Tanya, S. (2015). Preparation of flavored tofu. *The Delhi University Journal of Undergraduate Research and Innovation, 1*, 102–106.

Siepmann, F. B., Ripari, V., Waszczynskyj, N., & Spier, M. R. (2018). Overview of sourdough technology: From production to marketing. *Food and Bioprocess Technology, 11*(2), 242–270.

Souci, S. W., Fachmann, W., & Kraut, H. (2000). *Food composition and nutrition tables* (6th ed.). Medpharm GmbH Scientific Publishers.

Sterr, Y., Weiss, A., & Schmidt, H. (2009). Evaluation of lactic acid bacteria for sourdough fermentation of amaranth. *International Journal of Food Microbiology, 136*(1), 75–82.

Stolz, P., Böcker, G., Vogel, R. F., & Hammes, W. P. (1993). Utilization of maltose and glucose by lactobacilli isolated from sourdough. *FEMS Microbiology Letters, 109*, 237–242.

Subrota, H., Shilpa, V., Brij, S., Vandna, K., & Surajit, M. (2013). Antioxidative activity and polyphenol content in fermented soy milk supplemented with WPC-70 by probiotic Lactobacilli. *International Food Research Journal, 20*(5), 2125–2131.

Swain, M. R., Anandharaj, M., Ray, R. C., & Rani, R. P. (2014). Fermented fruits and vegetables of Asia: A potential source of probiotics. *Biotechnology Research International*, 1–19.

Tezuka, M., Ono, T., & Ito, T. (1995). Properties of soymilk prepared from soybeans of different varieties. *Nippon Shokuhin Kogyo Gakkai-Shi, 42*(8), 556–561.

Thakur, P. K., Panja, P., Kabir, J., & Dhua, R. S. (2020). Studies on shelf life of sauerkraut. *Journal of Crop and Weed, 16*(2), 204–209.

Thiele, C., Gänzle, M. G., & Vogel, R. F. (2002). Contribution of sourdough lactobacilli, yeast, and cereal enzymes to the generation of amino acids in dough relevant for bread flavor. *Cereal Chemistry, 79*, 45–51.

Trail, A. C., Fleming, H. P., Young, C. T., & McFeeters, R. F. (1996). Chemical and sensory characterization of commercial sauerkraut. *Journal of Food Quality, 19*, 15–30.

Tse, G., & Eslick, G. D. (2014). Cruciferous vegetables and risk of colorectal neoplasms: A systematic review and meta-analysis. *Nutrition and Cancer, 66*, 128–139.

Tsumura, K., Saito, T., Tsuge, K., Ashida, H., Kugimiya, W., & Inouye, K. (2005). Functional properties of soy protein hydrolysates obtained by selective proteolysis. *LWT-food Science and Technology, 38*(3), 255–261.

VanGarde, S. J., & Woodburn, M. J. (1994). *Food preservation and safety: Principles and practice*. Iowa State University.

Vergne, S., Bennetau-Pelissero, C., Lamothe, V., Chantre, P., Potier, M., Asselineau, J., Perez, P., Durand, M., Moore, N., & Sauvant, P. (2007). Higher bioavailability of isoflavones after a single ingestion of a soya-based supplement than a soya-based food in young healthy males. *British Journal of Nutrition, 99*, 333–344.

Verkerk, R., Schreiner, M., Krumbein, A., Ciska, E., Holst, B., Rowland, I., De Schrijver, R., Hansen, M., Gerhäuser, C., Mithen, R., & Dekker, M. (2009). Glucosinolates in Brassica vegetables: The influence of the food supply chain on intake, bioavailability and human health. *Molecular Nutrition & Food Research, 53*(S2). pp. S219-S219.

Wacher, C. G., Díaz-Ruiz, & Tamang, J. P. (2010). In J. P. Tamang, & K. Kailasapathi (Eds.), *'Fermented vegetable products' in fermented foods and beverages of the world*. Boca Raton, FL: CRC Press.

Wahyuni, S., Ratna, H. A., & Raharjo, R. A. (2016). Antioxidant activity of isoflavones from tofu pulp waste. In *6th int'l conference on agriculture, environment and biological sciences (ICAEBS'16), dec. 21-22, 2016, kuala Lumpur (Malaysia)* (pp. 108–111).

Wang, Y., Yang, X., & Li, L. (2020). A new style of fermented tofu by Lactobacillus casei combined with salt coagulant. *3 Biotech, 10*(2), 1–8.

World Health Organization, & Food and Agricultural Organization of the United Nations (WHO/FAO). (2004). *Vitamin and mineral requirements in human nutrition* (2nd edition, pp. 246–278). US: WHO Library Cataloguing-in-Publication Data.

Wuyts, S., Van Beeck, W., Allonsius, C. N., van den Broek, M. F., & Lebeer, S. (2020). Applications of plant-based fermented foods and their microbes. *Current Opinion in Biotechnology, 61*, 45–52.

Wuyts, S., Van Beeck, W., Oerlemans, E. F., Wittouck, S., Claes, I. J., De Boeck, I., Weckx, S., Lievens, B., De Vuyst, L., & Lebeer, S. (2018). Carrot juice fermentations as man-made microbial ecosystems dominated by lactic acid bacteria. *Applied and Environmental Microbiology, 84*(12), e00134–18.

Xiang, H., Sun-Waterhousea, D., Waterhouse, G. I. N., Cui, C., & Ruan, Z. (2019). Fermentation-enabled wellness foods: A fresh perspective. *Food Science and Human Wellness', 8*, 203–243.

Xiong, T., Li, J., Liang, F., Wang, Y., Guan, Q., & Xie, M. (2016). Effects of salt concentration on Chinese sauerkraut fermentation. *LWT-food Science and Technology, 69*, 169–174.

Yoon, H. H., & Kim, M. (2007). Physicochemical and sensory properties of tofu prepared with heat-treated soybeans. *Journal of Texture Studies, 38*(3), 393–403.

Zabat, M. A., Sano, W. H., Wurster, J. I., Cabral, D. J., & Belenky, P. (2018). Microbial community analysis of sauerkraut fermentation reveals a stable and rapidly established community. *Foods, 7*(5), 77.

Zhang, Q., Wang, C., Li, B., Li, L., Lin, D., Chen, H., Liu, Y., Li, S., Qin, W., Liu, J., Liu, W., & Yang, W. (2018). Research progress in tofu processing: From raw materials to processing conditions. *Critical Reviews in Food Science and Nutrition, 58*(9), 1448–1467.

Zheng, L., Regenstein, J. M., Teng, F., & Li, Y. (2020). Tofu products: A review of their raw materials, processing conditions, and packaging. *Comprehensive Reviews in Food Science and Food Safety, 19*(6), 3683–3714.

Section V

Plant-based functional components

Chapter 14

Polyphenols, phytosterols, aromatics, and essential oils

Nirali Dedhia and Narendra Shah
Centre for Technology Alternatives for Rural Areas, IIT Bombay, Mumbai, India

14.1 Introduction

Polyphenols, sterols, aromatics, and essential oils are groups of compounds that can be extracted from various plants. These groups of compounds are associated with the identifying factors of a plant source, i.e., its flavor, aroma, and color. They are also associated with numerous preventive and therapeutic benefits in human diseases (Lorenzo et al., 2019) with latest reports on potential against the coronaviruses as well Mani et al., 2020. Therefore, these are of prime importance from culinary as well as medicinal point of view. These compounds have varying structures, physico-chemical, and sensory properties. The common factors among these four groups are their extraction techniques, physico-chemical, and biological properties. Chemical structures of representative molecules of each group are presented in Fig. 14.1. A brief description of each of these groups is as follows:

- Polyphenols is the cluster of compounds that are found ubiquitously in various plant sources. They are naturally present in the glycoside form, i.e., bound to a sugar moiety. Hydrolysis of glycosidic bond leads to the release of sugar moiety and aglycoside polyphenol (Wiesner et al., 2017). Some polyphenols have been associated with medicinal properties and as second-generation symbiotics (Sharma and Padwad, 2020). These have been classified as flavonoids and phenolic acids based on their chemical structures. Flavonoids are further classified into flavones, flavonones, flavonols, flavanols, isoflavones, and phenolic acids into hydroxybenzoic and hydroxycinnamic acids (Abbas et al., 2017).

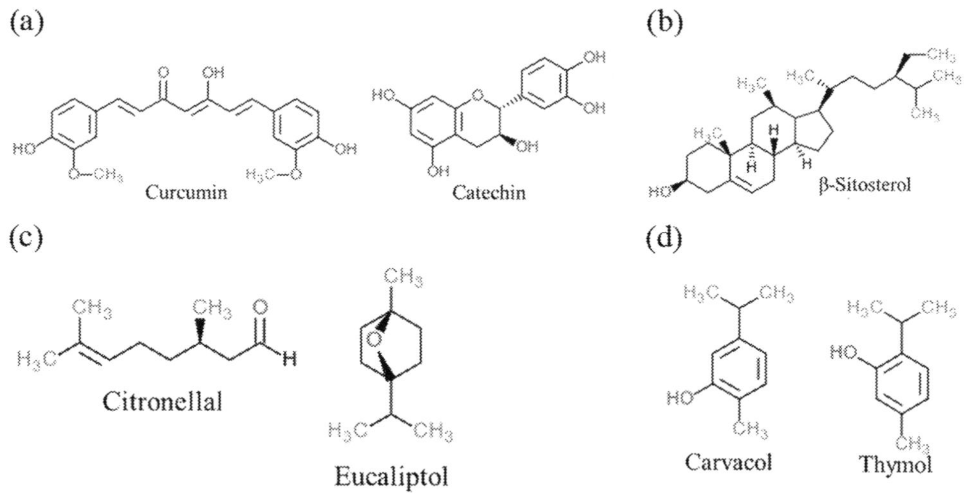

FIGURE 14.1 Chemical structures of illustrative compounds classified as, (a) polyphenols, (b) phytosterols, (c) aromatics, and (d) essential oils.

Engineering Plant-Based Food Systems. https://doi.org/10.1016/B978-0-323-89842-3.00016-6

- Phytosterols are sterols of plant origin. The typical steroid structure has a hydroxyl group at position 3, side chain at position 17, and contain one or more double bonds in the steroid skeleton (Bot, 2019). Structurally, they closely resemble cholesterol, but their side chain contains a methyl or ethyl group. They are associated with anti-inflammatory properties in various diseased conditions (Vilahur et al., 2019).
- Plant aromatics are a huge cluster of aromatic ring molecules, amino acids, and lignins that contain ketonic, aldehydic, hydroxyl, and amino groups. Aromatics can be precursors to polyphenolic compounds in the biochemical pathways of the plant (Rammohan et al., 2020). Aromatics help distinguish among varieties of a given plant species (Vivaldo et al., 2017).
- Essential oils are a cluster of aromatics, volatile, nonvolatile, and odorless substances extracted from plant sources through distillation process (Ríos, 2016). These are a complex mixture of secondary plant metabolites such as terpenes, phenolic compounds, and alcohols. Monoterpenes in essential oils can make up to 90% of the total content followed by the other compounds in smaller proportions (Falleh et al., 2020]).

14.2 Functional components and their health benefits

Some commonly found plant-derived components, *i.e.*, polyphenols, phytosterols, essential oils, and aromatics have demonstrated health benefits in humans and animals through various in vitro studies, animal studies, and clinical trials. Table 14.1 enlists some currently available functional components and their demonstrated health benefits along with their reported mechanism of action. Most of the plant-based functional components demonstrate antioxidant activities by affecting the intra- and intercellular signaling pathways. It has been suggested that the amounts in which foods containing polyphenols are consumed can help manifest the health benefits Fraga et al., 2019.

14.3 Methods employed to extract the functional components

The chemical nature and physical form of the materials to be extracted from botanicals, particle size of botanicals, pre-treatment given to botanicals, economic aspects, and environmental concerns influence the choice of a technique of extraction. The raw material cost, if the polyphenol, phytosterol extraction is a part of valorization of waste or is a niche material, also influences the choice of extraction technique to be used. Extraction processes are typical solid−liquid interaction systems. A generalized process flow diagram is presented in Fig. 14.2. For the purpose of extraction, the botanicals can be taken in their native form or preprocessed to dehydrated forms by heat or freeze-drying techniques. The dehydration step is convenient when the botanicals are to be transported for extraction or when the extractant is an organic solvent and presence of water may hamper the process. Another pretreatment step is cell lysis categorized under recent techniques such as homogenization, ultrasonication, and microwave treatment for improved efficiency. This is followed by solvent extraction. In case of phytosterols, an additional step of saponification is applied after solvent extraction for higher yield. This is followed by removal of solvent from the extract using various techniques such as lowering of vapor pressure, application of heat, centrifugation, or crystallization Fraga et al., 2019. The crude phytochemicals may be subjected to purification using column chromatography, or the crude mixture can be used as such to formulate into encapsulates, powders, topicals, etc. A comparison of the cost of production of extracts through different techniques is illustrated in Table 14.2. This is an important factor influencing the choice of technique.

14.3.1 Conventional techniques of extraction and process parameters

The conventional techniques of extraction of polyphenols, phytosterols, aromatics, and essential oils involve use of organic solvent(s) with or without water. Methanol, ethanol, acetone, acetonitrile, and iso-propyl alcohol may be used for this purpose. The interactions between botanicals and solvents can be improved with agitation or heating.

The physico-chemical properties of the extract decide the combination of solvents used for this purpose. A major concern with the solvent selection in a conventional method of extraction is the optimization of a balance between maximum yield and minimum solvent residues. The nature of solvent influences extraction of groups of compounds with varying polarity. The solvents with lower polarity are considered to be suitable for the extraction of lipophilic phenols, alternatively higher polar solvents such as ethanol can be used employing high pressure having cost implications. Ethanol is typically a solvent of choice due to its ability to dissolve a wide number of phenolic acids. The total phenolics and flavonoid extracted from pineapple, banana, and guava fruits varied with solvents used, i.e., methanol, ethanol, acetone, and their varying proportions with water Akhavan et al., 2018. Along with polarity, the acidity of extraction medium influences the yield of isoflavones from soybean Lin and Giusti, 2005. The temperature of extraction influences the yield as it reduces the viscosity of solvent, improving the liquid diffusion in solids and the subsequent extraction. Operations between 30 and 60°C are preferred. Above 60°C, degradation of polyphenols occurs due to redox reactions and polymerization Ozturk et al., 2018. Ethanol-based extraction systems are preferred when the final product is aimed for food or

TABLE 14.1 Illustrative bioactive constituents from plants that have proven health benefits.

Bioactives	Plant sources	Health benefits	Mechanisms of action	References
Curcuminoids	*Curcuma* species rhizome	Effective against inflammation, diabetes, allergies, arthritis and chronic illnesses	Regulation of various transcription factors, growth factors, inflammatory cytokines, protein kinases, and other enzymes	Aggarwal et al. (2007)
Gingerols	*Zingiber officinale* roscoe rhizome	Effective against cancer, inflammation, microbial infections, hyperlipidemia, neural disorders, angiogenesis	Regulation of cell apoptosis, DNA damage response, epigenetic modulation, cytoskeletal regulation and adhesion, cytokine response, immune responses, IGF-1 signaling, insulin signaling, angiogenesis signaling, proteasome signaling, atherogenesis, TGF-β signaling	Kiyama (2020)
Resveratrol	Grapes, berries, peanuts	Antioxidant, anticancer, cardiovascular protective, anti-inflammatory, antidiabetic, antiaging, neuroprotective properties	Interactions with cyclooxygenases, lipoxygenases, sirtuins, transcription factors, cytokines, DNA polymerase, adenylate cyclase and others	Luca et al. (2020)
Rutin	Buckwheat, rue, Japanese pagoda tree, crane flower	Antioxidant, anti-inflammatory, antidiabetic, antiadipogenic, neuroprotective, hormone therapy	Free radical scavenging effect, inhibition of production of nitric oxide, improving glucose homeostatis (increasing insulin level, and increasing glycogen con-tent in liver and muscle, but decreasing the glycogen content in kidney), phytoestrogen activity	Chua (2013)
Mangiferin	Mango tree, young plants of coffee, honeybush	Antioxidant, analgesic, antidiabetic, antiproliferative, chemopreventive, radioprotective, cardiotonic, immunomodulatory, diuretic, effective against brain ischemia, liver damage	Catechol of mangeferin complexes with iron for antioxidant effect, facilitate production of alpha tumor necrosis factor and nitric oxide, down-regulate phosphorylation of NF-κB and JAK1—STAT1/3 pathways, activating the nuclear factor E2 pathway and inhibiting NLRP3 inflammasome activation	Khare and Shanker (2016)
Pepper essential oil	Black pepper, white pepper	Antioxidant, anti-inflammatory, antibacterial, antifungal activity	Free radical scavenging activity, monoterpenes having antibacterial effects	Li et al. (2020)
Peppermint leaf essential oil	Peppermint leaves	Antibacterial, antifungal, antiviral, antioxidant, radiation protective, antidiabetic, hepatoprotective effect, neuroprotective properties	Protection from radiation-induced chromosomal damage, free radical scavenging activity, serum glucose and cholesterol lowering effect, decrease in alanine transaminase, aspartate transaminase, alkaline phosphotase, urea, creatinine, uric acid, bilirubin and malondialdehyde and increase in superoxide dismutase, glutathione peroxidase and gamma glutamyl transferase, in vitro AChE inhibitory, calcium regulatory, GABA receptor and nicotinic receptor binding	Mahendran and Rahman (2020)
β-sitosterol	Rice bran, wheat germ, peanuts, soybeans	Serum cholesterol lowering, anti-inflammatory effects, enhanced immune functions	Structurally similar to serum cholesterol and helps in reduction of serum levels, Evoke T-helper 1 cell, increase T-helper 2 cells, attenuates the phosphorylation of nuclear factor-kB p65, inhibits intracellular adhesion molecule 1 expression in tumor necrosis factor-α-stimulated human aortic epithelial cells	Brüll and Mensink (2009), Loizou et al. (2010)
Quercetin and its derivatives	Cocoa nibs, taro, cassava, onion, mango	Anticancer, antioxidant, antiviral, antidiabetic activity	Free radical scavenging, inhibition of human cathepsin B	Magar and Sohng (2020)

AChe: acetylcholinesterase; GABA: gamma-aminobutyric acid; IGF: insulin-like growth factor; TGF: transforming growth factor.

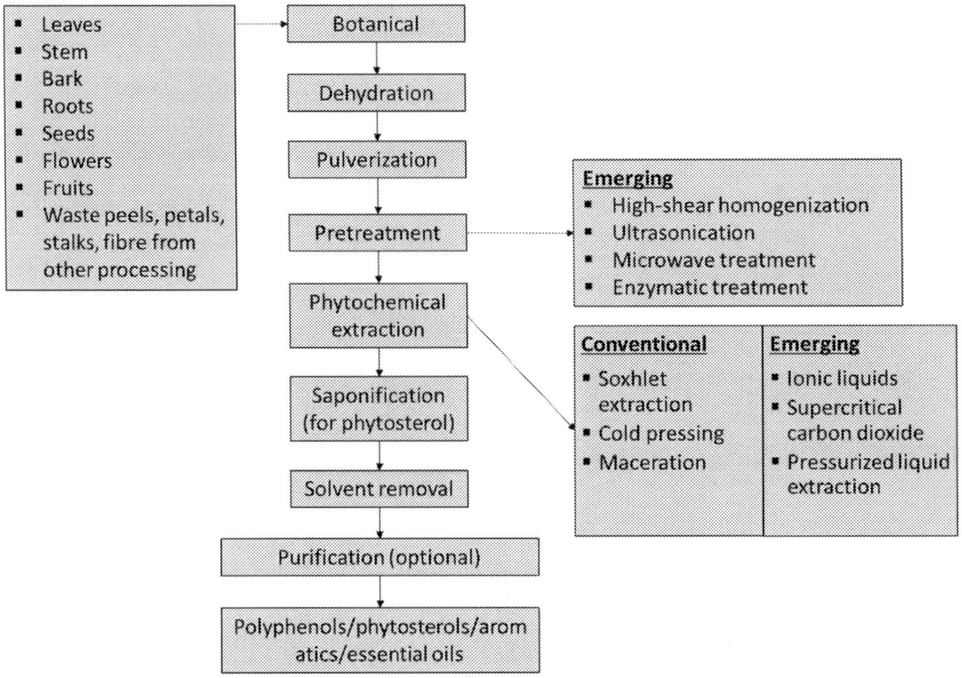

FIGURE 14.2 A general logic for the processes involved in extraction of polyphenols, phytosterols, aromatics, and essential oils from plant sources.

TABLE 14.2 Cost of production of various extracts from plant-based materials.

Substrate	Substance	Extraction technique	Extraction scale	Production cost	References
Oregano	Essential oils	Water distillation	8.14 kg/tonne material	Production cost-\$6.71/kg	Moncada et al. (2016)
Rosemary	Essential oils	Water distillation	8.43 kg/tonne material	Production cost-\$6.57/kg	
Aquilaria sinensis	Volatile oils	Supercritical CO_2	5289 kg/y capacity	Total capital investment-\$ 7.11 million	Gwee et al. (2020)
Turmeric rhizome	Curcuminoids	Pressurized liquid	500 L extractor	Cost of manufacturing-\$88/kg	Osorio-Tobón et al. (2014)
Purple yam	Anthocyanins	Ultrasonication	500 L extractor	Cost of manufacturing-\$124/kg	Ochoa et al. (2020)
		Low pressure solvent extraction	500 L extractor	Cost of manufacturing-\$263/kg	
		Soxhlet extraction	500 L extractor	Cost of manufacturing-\$765/kg	

pharmaceutical use. The selection of solvent is of importance from the point of process handling and environmental hazard Capello et al., 2007.

By virtue of their chemical structure, phytosterols and phytostanols are extracted in a slightly different manner compared to polyphenols. Edible oil deodorized distillates are subjected to saponification, esterification, or trans-esterification for separation of campesterol, stigmasterol, and β-sitosterol. Saponification provides a concentrated fraction of the sterol Moreau et al., 2002. This process can be performed either by heating water with botanicals at high temperatures (200–260°C) and high pressures (1.5–50 MPa) or by heating the extracted oil with an alkali at 90 to 120°C. The crude phytosterols may then be refined using solvent extraction, crystallization, or chromatographic techniques Dikshit et al., 2020. Thus, the solid–liquid extraction of oil is followed by liquid–liquid extraction of phytosterols.

Cold pressing of botanicals for extraction of essential oils is an age-old technique. It is accompanied by low yields and generation of some heat Reyes-Jurado et al., 2015. The water emulsion so obtained is centrifuged to remove water from the essential oil. Steeping of botanicals in boiling water or subjecting to steam to extract their oil followed by decantation for removal of water is another popular technique Mahato et al., 2019. Different techniques lead to essential oils of varying composition and aromatics. Solvents can be chosen based on the quality desired Sicari and Poiana, 2017. These methods can lead to thermal degradation, hydrolysis, and water solubilization of some fragrance constituents or solvent residues, affecting the quality of the final product Zhang et al., 2012.

14.3.2 Recent developments in greener extraction technology

Recent developments or emerging technologies employed for the extraction of phytochemicals focus on improving yield, quality of the extracts, reducing quantity of solvents used, and minimal impact on environment. Various emerging technologies either employ greener solvents with better recovery and recyclability or employ improved mechanical shearing for cell lysis to improve extraction efficiency Ameer et al., 2017. The methods can be classified based on their principle of operation and are described as below:

14.3.2.1 Supercritical carbon dioxide extraction

Supercritical carbon dioxide ($SC-CO_2$)-assisted extraction technique employs carbon dioxide in its supercritical state as a solvent for extraction. It is considered as a greener technique due to the ease of recyclability of carbon dioxide for subsequent extractions. The technique operates at a relatively low temperature and helps extract materials with minimal damage to bioactive. Higher extraction of thymol and better aroma profile have been reported from ajwain and thyme by $SC-CO_2$ vis-à-vis conventional hydrodistillation Morsy, 2020. Theoretical modeling is generally performed for extraction of phytochemical(s) of interest to estimate the process operating parameters before performing the actual extraction Jafarian Asl et al., 2020. The process of $SC-CO_2$ extraction starts with compression of CO_2 gas at the specific pressure, followed by its entry into the extraction vessel containing the botanicals, which have been preset at a specific temperature. The extraction gas passes through the vessel and then into the collection unit where the release is at a lowered pressure. The extract is collected, and the gas can be recycled back to the compressor after filtration. It is an adjustable technique wherein the concentration of various components in the extract can be tailored by modifying extraction time, solvent flowrate, process temperature and pressure, extractor diameter-to-length ratio, bed porosity, etc. Dąbrowski et al., 2019. Increase in pressure of $SC-CO_2$ (above 72 bar and up to 800 bar, also based on equipment design) increases density and hence solvation power of solvent. However, at certain higher pressures, the yield may decrease due to the channeling of $SC-CO_2$ through the sample. An increase in the temperature of extraction increases the vapor pressure and mass transfer coefficients of extractives from the botanical materials. However, a balance between yield and quality, i.e., minimal degradation of heat labile phytochemicals needs to be framed Nagavekar and Singhal, 2019. Temperature–pressure combination influences the time of extraction. This may be a process of diffusion of the phytochemical from the plant matrix into the $SC-CO_2$ (Nagavekar and Singhal, 2018).

The nonpolar solvent CO_2 majorly extracts nonpolar materials. Materials in a range of polarity can be extracted by mixing co-solvents or entrainers such as acetone, methanol, iso-propanol, and ethanol in the gas-line or in the sample before extraction. Entrainer concentrations are calculated on the basis of per unit weight of extraction sample or per unit weight of $SC-CO_2$ gas at the operating conditions or simply per unit time of extraction. These function by changing the polarity, density, and solvatochromic properties. The solvatochromic properties of the entrainers are: hydrogen bond donor ability, hydrogen bond acceptor ability, and their dissociation constants under supercritical conditions (de Castro et al., 1994). Entrainers such as ethanol, acetone, and methanol have been reported to increase the $SC-CO_2$ extraction efficiency of piperine oleoresin from *Piper nigrum* of 63%–95.9%, 77.6%, and 97.6%, respectively. Though acetone is more polar than ethanol, the yield is lower due to different solvatochromic properties (Nagavekar and Singhal, 2018). The choice of entrainer solvent is also based on their generally regarded as safe (GRAS) status for food or nonfood use Vijayan et al., 2019]. Vegetable oil as entrainer has also been explored. In such scenario, either the extraction material is wetted with vegetable oil or the meal is mixed with bioactive for co-extraction. Oil serves dual-purpose viz. facilitating extraction by increasing dissolution of bioactive in it and as diluent for the extracted oleoresin/essential oil Nagavekar and Singhal, 2020.

14.3.2.2 Ionic liquid–assisted extraction

Ionic liquids (ILs) are molten salts at temperatures below 100°C. These can have organic anion and inorganic or organic cation. These can be simple or binary salts. ILs have an advantage over conventional organic solvents due to their electrical conductivity, good thermal stability, negligible vapor pressure, and a wide liquidus range. They are also considered greener for their recyclability property (Earle and Seddon, 2002; Ventura et al., 2017). The solubility of cellulose in certain ILs such as 1-butyl-3-methylimidazoliumchloride, 1-allyl-3-methylimidazoliumchloride, and ethyl-3-methylimidazolium acetate has been widely reported Zhang et al., 2017. This solubility adds to the advantage of using ILs for extraction of materials inside the cellulosic membrane of plant cells. ILs, their aqueous mixtures, and alcoholic mixtures are used for solid—liquid extraction of bioactives Ventura et al., 2017). They can be a part of hydrodistillation process or added during maceration before hydrodistillation to enhance yield such as that of cumin essential oil Ascrizzi et al., 2017. Parameters of importance for improving yield include IL-botanical ratio, time, and temperature of extraction. The extraction involves dissolution of botanicals in the ILs followed by vacuum distillation of the essential oils or bioactive. In a laboratory set-up, organic solvent in minute quantities can be added to facilitate extraction. The distillation is followed by water-assisted precipitation of biopolymers followed by mechanical separation. This can help recycle the ILs back for the next batch (Bica et al., 2011; Usuki and Munakata, 2017). ILs have also been reported for deterpenation of simulated system of citrus essential oils Arce et al., 2006.

14.3.2.3 Pressurized liquid extraction

Pressurized liquid extraction (PLE) is a technique that involves use of one solvent or a mixture at higher pressure and temperature above their boiling point for extraction of various constituents from plant or animal matter. During PLE, the extraction solvent remains in liquid state throughout the operation, even above its boiling temperature. This allows for increased solubility and improved desorption kinetics of the bioactives from their matrices. These operation conditions enhance the extraction process as compared to Soxhlet extraction and hydrodistillation Bittencourt et al., 2019. After the pressurized solvent exposure, the system is purged with nitrogen to remove remainder solvents and oil. However, during this purging process, losses in essential oils have been reported (Dawidowicz et al., 2012; Wianowska, 2014). Selected extraction of constituents is also possible by tailoring the operating conditions Eikani et al., 2013. Organic solvents or their aqueous mixtures are generally employed. The polarity of the solvent, time of extraction, and solid—liquid ratio define the extraction yield and its composition Jacques et al., 2008). The time required for this process is a fraction of the time required for conventional Soxhlet extraction. For instance, the time and solvent taken by PLE using hexane vis-à-vis Soxhlet extraction by hexane to extract essential oils from turmeric leaves are 17 min, 33 mL and 16 h, 150 mL, respectively Zaibunnisa et al., 2009). The essential oils produced by PLE vary in aromatics, polyphenols, and eventually biopotential as compared to conventional techniques (Bendif et al., 2018; Corazza et al., 2018). Processes integrating SC-CO_2 and PLE for extraction of turmeric products have demonstrated decrease in costs of manufacture Osorio-Tobón et al., 2016.

PLE in a cyclic manner has also been explored for extraction of essential oils and polyphenols from botanicals. The process, also termed as rapid solid—liquid dynamic extraction, involves periodic pressurization of solvent, which is in contact with the botanical. This technique has been reported to involve the use of a Naviglio extractor at room temperature unlike PLE, which employs elevated temperatures. This technique has been developed towards sustainability and energy efficient technologies (Gallo et al., 2019; Naviglio et al., 2017). PLE using pressurized hot water or organic solvent and a membrane filter assembly for concentration of polyphenols have been reported. This technique leads to extracts with retained biological activity, environmentally clean, cost-effective, involve less energy inputs, and generate lesser waste. Nanofiltration and ultrafiltration of agro-wastes such as fruit pomace lead to polyphenol extracts with retained biological activity (Avram et al., 2017; Cassano et al., 2018).

14.3.2.4 Three phase partitioning

Three-phase partitioning (TPP), as the name indicates, is a technique of simultaneous separation of oleaginous material, water-soluble components, and proteins from a given material. A simple process involves mixing of the extracting material with water of desired pH to form an aqueous slurry. This is followed by addition of salt and then *t*-butanol and mixed well. Normally, *t*-butanol is miscible with water; however, in the presence of salt, it forms a separate layer. The components separate as: upper layer of *t*-butanol contains dissolved oil, middle layer of proteins precipitated with the salt, and the lower aqueous layer of water soluble molecules such as carbohydrates, fibers, gums, etc. Harde and Singhal, 2012. *t*-Butanol has a higher boiling point, thus reducing the addition of volatile organic compounds to the atmosphere even with the use of open systems. The separation in TPP is facilitated by chilling rather than heating, which further reduces the chances of contributions to atmospheric pollutants Kurmudle et al., 2011. The pH, temperature, and time of extraction along with the proportion and type of salt and solvent influence the efficiency of extraction. Oleoresins of plant materials such as turmeric and ginger have been extracted using TTP (Kurmudle et al., 2011; Varakumar et al., 2017).

14.3.2.5 Cell lysis techniques

Cell lysis techniques, as the name suggests, cause breakdown of cell structure by incorporation of energy. This leads to enhanced transfer of mass and hence yields of extraction. Techniques such as ultrasonication, high-pressure homogenization, pulsed electric field (PEF), microwave treatment, and enzymatic treatment are commonly reported to aid release of smaller molecules from the macromolecular matrix. Other cell lysis techniques include osmotic-shock, use of agents such as detergents and alkali, thermal lysis, freeze-thaw lysis etc. Gomes et al., 2020.

Microwave heating has been reported to loosen the cell wall and facilitate higher extraction yields that are governed by irradiation power, time, and type of material. The ratio of solvent-to-plant material contributes significantly to the energy intensification in microwave treatment Olalere et al., 2018. This technique works on the principle of ionic conduction and dipole interaction of electric and magnetic fields. The ionic conduction gives stiff resistance to the ionic movement generating a heating effect on the irradiation medium and subsequently effective extraction Olalere et al., 2017. Microwave-assisted extraction (MAE) can be performed stand-alone or in conjugation with hydro-distillation or ILs for extraction of essential oils (Boumahdi et al., 2021; Franco-Vega et al., 2021; Gonzalez-Rivera et al., 2021). It is considered as a safe, cheap, and rapid technique of extraction with minimal damage to heat-sensitive metabolites Olalere et al., 2018.

The ultrasound treatment is based on the principle of formation of high-frequency ultrasonic waves that cause cavitation effect due to the alternate expansion and contraction cycles. These cycles disrupt the cells of plant material subjected to ultrasound and aid extraction of desired components (Santos et al., 2015). Ultrasound-assisted extraction (UAE) leads to an increased swelling of plant materials in the solvent as well as cause damage to the middle lamellae for enhanced extraction yield. Most organic solvents do not undergo significant changes when subjected to ultrasound treatment (Toma et al., 2001). Thus, this treatment reduces extraction times and to some extent increases selectivity of extraction. It has been reported in conjugation with solvent extraction techniques such as traditional organic solvents, water, and super critical fluid—assisted extraction of plant materials for oleoresin and essential oils (Ananingsih et al., 2020; Menezes Maciel Bindes et al., 2019; Santos et al., 2015).

High shear homogenization is a mechanical technique of cell disruption in water-based systems. The homogenization pressures range between 20 and 120 MPa. The assembly involves a piston pump and a homogenizer valve. When the cell suspension is pressurized through a narrow gap, it leads to large fluid velocity that leads to cell-wall shearing. It is easily scalable to the industrial level. This technique is popularly used for bacterial, yeast, and algal cell disruption prior to extraction of biomolecules from them Gomes et al., 2020). Its use in plant cell disruption is focused mainly on protein and polysaccharide extraction Donsì and Velikov, 2020).

Enzyme-assisted extraction has emerged as a technique to degrade the macromolecules in the cell matrix, especially the cellulosic membrane to release the oleoresins and bioactives. Using this technique, an efficient extraction is achieved with increased yield and/or reduced time of extraction. Single enzymes such as cellulase, α-amylase, β-glucosidase, xylanase, β-glucanase, pectinase, or enzymatic systems such as pectinolytic or celluloytic complex have been reported to aid extraction (Marathe et al., 2019). The enzymatic pretreatment is followed by a conventional ethanolic extraction (Santamaría et al., 2000) or TTP (Varakumar et al., 2017) or SC-CO_2 extraction Dutta and Bhattacharjee, 2015).

Pulsed electric field treatment is a technique reported for preservation of foods through reduction of microbial load. Recently, it has been reported as a technique for extraction of polyphenols from grapes and turmeric rhizome. The technique involves exposure of the substrate to short pulses of high voltages that lead to the formation of pores in the plant cells to aid diffusion of constituents and hence enhance the extraction (Poompavai and Sree, 2020; Ricci et al., 2018).

14.3.3 Challenges in extraction of functional components

The major challenge associated with any extraction technique is the retention of the property of the functional component with minimum degradation. In the 21st century, the added challenge is the development or use of extraction technology, which has minimum impact on environment as well as consumer health (Chemat et al., 2019). The green technology strategies researched for extraction of functional components have been successful at the laboratory scale. The major challenge lies with the scale-up feasibility of the technologies proposed. A study Fomo et al., 2020 compared SC-CO_2, PLE, PEF, MAE, and UAE techniques for industrial processing. The study concluded SC-CO_2 to be more reliable in spite of some limitations and challenges in terms of extraction yield and solubility. Use of ILs as entrainer solvents instead of organic solvents may help improve the yields. Scale-up of ultrasonication from 50 mL to 5 L showed batch-type process recorded higher extraction yield of anthocyanins from Starkrimson fruit peel than continuous-mode (Belwal et al., 2019).

14.4 Enhancement of bioavailability of the extracted functional components

Essential oils, polyphenols, phytosterols, and aromatics extracted from plant sources are known to have various health benefits as discussed and are popularly termed as bioactives. These, however, by virtue of their chemical structure possess functional difficulties of poor water solubility, heat, light, and oxidation sensitivities. The poor water solubility is the major cause of poor absorption of these bioactives by the intestinal cells. Bioavailability is that fraction of the bioactive that is absorbed and available for exerting the desired physiological action. This term is used in reference to in vivo absorption. For in vitro studies, the term bioaccessibility is widely used. Solubility, stability in gastric digestion, and encapsulation efficiencies are indirect estimates deciding the absorption of bioactives in the small intestine. These parameters are hence used to predict the potential bioavailability of a bioactive formulated in a delivery system.

The major function for a bioactive delivery system is to protect the bioactive, improve its stability, and maximize its absorption while considering the safety, texture, taste, appearance, and shelf-life (Gonçalves et al., 2018). The release of the bioactive from the matrix can be either burst release, sustained release, targeted release, or triggered release (Gonçalves et al., 2018). The matrices of the delivery systems need to fulfill certain criteria viz. ease of handling, effective dispersion or emulsification, nonreactivity with bioactive, ability to seal the bioactive, proper swelling, solubility and dispersibility under certain conditions for controlled release, inexpensive, and food-grade status for their selection Mourtzinos and Biliaderis, 2017). These different delivery systems can be categorized on the basis of their nature and structure as lipid-emulsifier systems, biopolymer-based systems, and gel-based systems. Table 14.3 enlists the different techniques of bioactive delivery and their effect on bioaccessibility/bioavailability. Schematic representations of the various forms of delivery systems are presented in Fig. 14.3.

TABLE 14.3 Different formulations to improve the bioavailability of plant-based extracts.

Method	Bioactive	Carrier	Description	Improvement in bioavailability (direct/indirect)	Reference
Emulsions	Curcumin	Soy lecithin	Co-delivery of curcumin and essential oils	↑Solubility, thermal, and phytochemical stability, 10 times higher bioaccessibility of emulsified curcumin than free form	Yang et al. (2021)
Nanostructured lipid carriers	Quercetin	Medium chain triglyceride, glycerol monostearate, soy lecithin	Emulsification, evaporation, and then solidification	In topical delivery, ↑permeation and ↑retention in epidermis and dermis	Chen-yu et al. (2012)
Solid lipid nanocarriers	Curcumin	Palmitic acid, cholesterol, *N*-trimethyl chitosan	Emulsification of curcumin + palmitic acid + cholesterol with *N*-trimethyl chitosan solution followed by lyophilization	Stability in room and refrigerated conditions, controlled release in simulated intestinal fluid, ↑oral bioavailability and brain distribution	Ramalingam and Ko (2015)
Self-emulsifying drug delivery system	*Kaempferia parviflora* rhizome extracts	Diethylene glycol monoethyl, polyoxyl-35 castor oil, caprylic/capric glyceride	Dissolution and agitation	↑ Dissolution of bioactive, storage stability of 6 months and ↑ absorption in Wistrar rats	Chairuk et al. (2020)
Coacervation	Vitamin B9	Lactoferrin and β-lactoglobulin	Complex coacervation of bioactive + protein I mixing with protein II of opposite charge	Physical stability in food matrices, ~50% increase in plasma levels in rats Than only B9, ↑stability, ↑ solubility in GI tract	Chapeau et al. (2017)

TABLE 14.3 Different formulations to improve the bioavailability of plant-based extracts.—cont'd

Method	Bioactive	Carrier	Description	Improvement in bioavailability (direct/indirect)	Reference
Liposome	7,8-Dehydroflavone	Glycosylated or crosslinked lactoferrin	Effective H-bonding between bioactive and phospholipid	EE $\uparrow$ to 98%, $\uparrow$ stability, $\uparrow$ AO capacity, slowed in vivo release rate and $\uparrow$ bioaccessibility in GI digestion in Caco-2 cell monolayers	Chen et al. (2020)
Impregnation in bioaerogel	Mixture of β-sitosterol, stigmasterol, and campesterol	Wheat starch aerogel	Impregnation of phytosterols in starch aerogel by supercritical fluid assembly to form nanoparticles	Reduced crystallinity to increase solubility $\uparrow$ bioaccessibility to 35% from 3%, $\uparrow$ zeta potential	Ubeyitogullari et al. (2019), Ubeyitogullari and Ciftci (2017)
Yeast encapsulation	Resveratrol	*Saccharomyces cerevisiae*	Yeast cells devoid of cytoplasmic material are pressurized in the presence of bioactive for its encapsulation	$\uparrow$ In vitro stability, solubility, sustained release, $\uparrow$ in vitro AO capacity	Shi et al. (2008)
	Curcumin	*Saccharomyces cerevisiae*	Reticulated β-glucan network provides skeleton and maintains cell rigidity and external protein layer acts as a sieve to maintain porosity of cell wall	More protection than β-cyclodextrin and modified starch encapsulation, $\uparrow$ protection under heated and oxidative, light conditions, gradual release	Paramera et al. (2011)
Excipient emulsions	Quercetin	Sodium caseinate Whey protein isolate Soy proteins	Incorporation of bioactive in protein emulsions	Good solubility, stability, bioaccessibility order: caseinate > whey protein > Soy protein	Chen, McClements, Zhu, et al. (2018)
	Curcumin	Whey protein isolate Sodium caseinate	Incorporation of bioactive in protein emulsions	$\uparrow$ Bioaccessibility of curcumin in emulsified form, highest adsorption for caseinate emulsions	Zou et al. (2015)
Pickering emulsions	Curcumin	Magnetic cellulose nanocrystals	High-speed homogenization and ultrasonication for emulsification of curcumin + red palm olein	39% $\uparrow$ Curcumin release under applied magnetic field, 2-fold increase in cell toxicity up to about 82% compared	Low et al. (2019)
Hydrogel	Quercetin	Whey protein isolate, lotus root amylopectin	Whey protein hydrogel coated with lotus root amylopectin, encapsulating quercetin	High EE of 92.4%, stable in stomach and be effectively released in small intestine	Liu et al. (2020)

Continued

TABLE 14.3 Different formulations to improve the bioavailability of plant-based extracts.—cont'd

Method	Bioactive	Carrier	Description	Improvement in bioavailability (direct/indirect)	Reference
Emulgel	Epigallocatechin-3-gallate; Quercetin	Saccharose (for osmotic stress balance) gelatin (as a gelling agent), polyglycerol polyricinoleate (a lipophilic surfactant), gliadin nanoparticles (hydrophilic emulsifier)	Water-in-oil emulsion formation followed by water-in-oil-in water emulsion formation that is gelled	↑Stability and 2X bioaccessibility of Epigallocatechin-3-gallate, ↑ solubility and 4X bioaccessibility of quercetin in simulated GI system	Chen, McClements, Wang, et al. (2018)
Organogel	Curcumin		Curcumin organogel with span 20-saturated medium chain triglycerides, monostearin. This was emulsified in water with modified starch, WPI, or Tween 20	9-Fold increase in oral bioavailability in mice compared to unformulated curcumin	Yu and Huang (2012)
Liposome-hydrogel	Quercetin + linseed oil	Chitosan	Chitosan solution added to liposomes of bioactive with phosphatidyl choline followed by bead formation	↑ Stability in vitro GI tract, inhibition of rapid release of fatty acids, ↑ chemical stability	Huang et al. (2020)
Nanoparticles	Mixture of β-sitosterol, stigmasterol, campesterol, and brassicasterol	Sodium caseinate; Whey protein isolate; Soy proteins	High speed dispersion of hexane phase containing phytosterols into aqueous phase containing proteins, followed by microfluidization and hexane removal	Small size, ↑ EE, good water redispersion behavior and low crystallinity, ↑ bioaccessibility than free form, best in caseinate	Cao et al. (2016)

AO, antioxidant capacity; *EE*, encapsulation efficiency; *GI*, gastro-intestinal.

The different categories of bioactive delivery forms are elaborated as follows:

14.4.1 Lipid-based delivery systems

Lipids can be structured in 3-dimensional supramolecular assemblies to protect active lipids or others substances and enhance their bioavailability. The mechanism of interaction of the lipids and the core can be noncovalent interactions such as electrostatic interactions, hydrogen (H)-bonding, London and van der Waals forces, π-π stacking, host—guest, and hydrophobic interactions by the conditions of ionic strength, pH, temperature, etc. These structures can be in the form of micelles, vesicles, and liquid crystals (Magri et al., 2020). Nanocapsules with lipid formulations have been recommended over micro-sized capsules due to the increased surface area, improved solubility, high bioavailability, and sustained release of nutraceuticals Akhavan et al., 2018). Lipid-based systems have some advantage over polymer-based systems in nano-dimension production such as fewer organic solvent residues, lower toxicity, better solubility for lipophilic bioactive ingredients, and higher absorption of bioactives from the small intestine Akbarzadeh et al., 2013).

14.4.1.1 Emulsions

The lipid-emulsifier based delivery systems have been explored for improving solubility and bioavailability of bioactives. These emulsified systems may be thermodynamically and kinetically unstable, posing a challenge to the method of

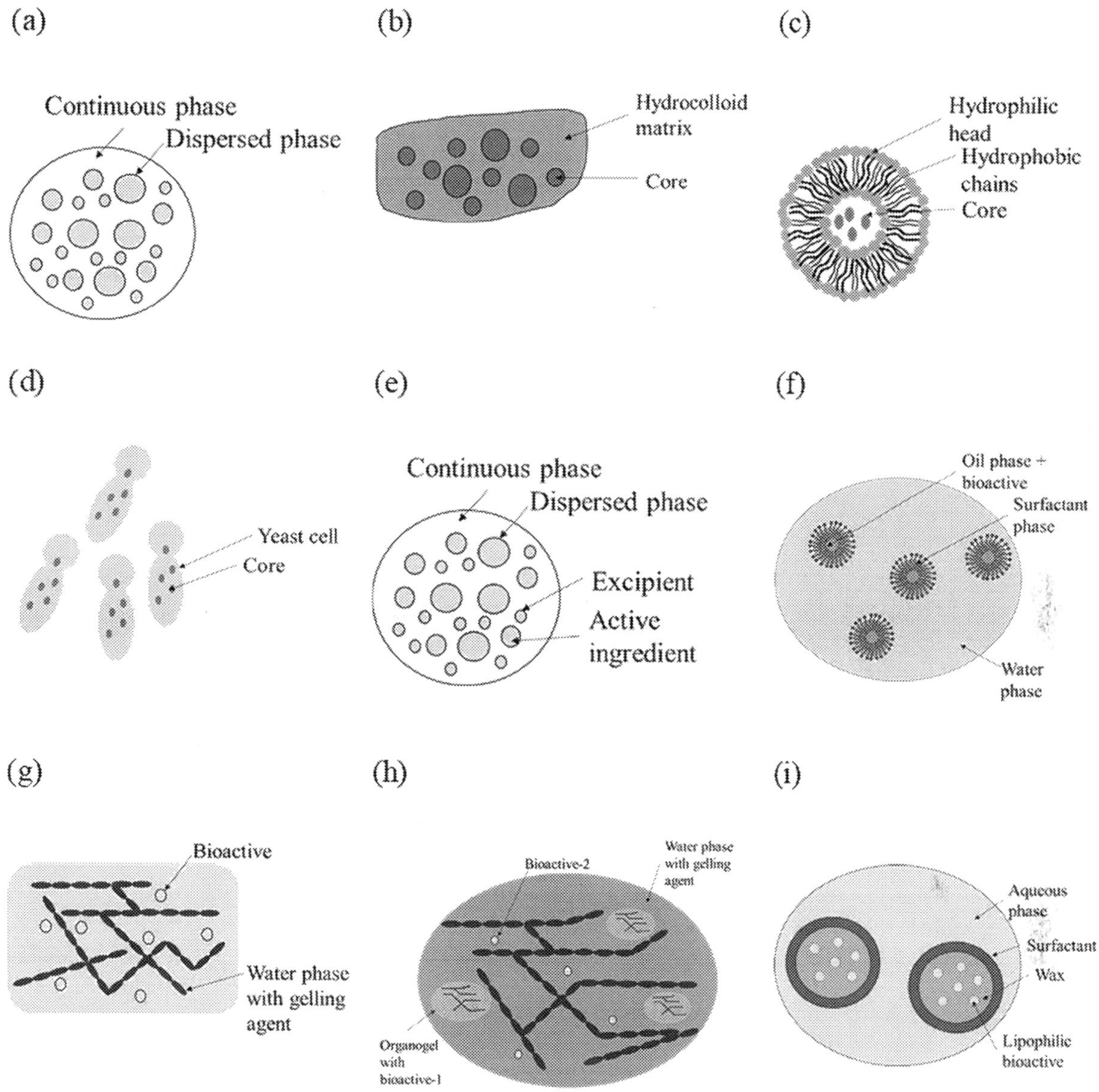

FIGURE 14.3 Schematic representation of forms of bioactive delivery to enhance their solubility, stability, alter particle size, bioaccessibility and/or bioavailability such as (a) emulsions, (b) coacervation, (c) liposomes, (d) yeast encapsulation, (e) excipient emulsions, (f) self-emulsifying delivery, (g) hydrogel, (h) bigel, and (i) nanostructured lipid carrier.

production. Nevertheless, the achieved bioavailabilities have always encouraged research in developing them. Emulsified delivery systems containing food-grade ingredients such as lecithin can be easily prepared using shearing techniques of homogenization He et al., 2020), ultrasonication, and microchanneling Manickam et al., 2020). Rapeseed oil bodies have been reported as novel carriers for the controlled release of lipophilic bioactives from emulsions during gastrointestinal digestion He et al., 2020). Micellar formulations of *trans*-resveratrol using surfactants have been reported for improved oral bioavailability in healthy humans Calvo-Castro et al., 2018). Newer developing systems involve the use of polymeric units to avoid precipitation, handling issues, and restricted lymphatic transport. Medium chain triglycerides and antioxidants are also added to formulations to avoid oxidation (Rajpoot et al., 2020).

14.4.1.2 Self-emulsifying drug delivery systems

Self-emulsifying drug delivery systems (SEDDs) have been researched and worked up on for their resultant improved solubility, dissolution, and oral absorption of poorly water-soluble drugs, which is also applied for bioactives. The process of preparation has one simple step of slight agitation of a blend of bioactive (with the lipid) with surfactants and co-surfactants in the aqueous phase to form an isotropic mixture (Rajpoot et al., 2020. In the human body, the dilution of SEDDs happens in the gastrointestinal tract where the agitation is provided by the digestive motion of the stomach and intestine. The lipids used in SEDDs preparation can be fatty acid esters or medium or long chain saturated, partially unsaturated hydrocarbons in liquid, semisolid, or solid states at room temperatures. Nonionic surfactants with hydrophilic—lipophilic balance on the higher side are preferred in these formulations (Gonçalves et al., 2018).

14.4.1.3 Liposomes

Liposomes are spherical lipid vesicles with an aqueous core enclosed by one or more lipid bilayers. Their classification is on the basis of dimensions (micro- and nano-), number of bilayers, and charge on phospholipid (Simão et al., 2015). Liposome technology is another lipid-based carrier system wherein the lipid encapsulates the bioactive. This technology is popular due to its industrial scalability and targetability. Liposomes can accommodate both water- and oil-soluble materials for protection and targeted delivery Mozafari and Khosravi-Darani, 2007. Their main constituents are natural or synthetic phospholipids whose amphiphilic properties facilitate the formation of liposomes in aqueous solutions by hydrophobic effect rising from the contact of nonpolar hydrocarbon chains with the surrounding aqueous environment Pattni et al., 2015. A simple process of liposome preparation involves mixing of phospholipid with water, followed by addition of the extract. This is followed by formation of vesicles by repeated extrusion through polycarbonate filters. Lecithin liposomes of phenolic extracts with improved stability and bioavailability have been reported (Rafiee et al., 2017). Other techniques of production of liposomes include thin-film hydration, ethanol injection, reverse-phase evaporation, high-pressure homogenization, and supercritical carbon dioxide (Situ et al., 2017).

14.4.1.4 Lipid nanocarriers-solid-lipid nanoparticles and nanostructured lipid carriers

Solid—lipid nanoparticles are a variant of oil-in-water emulsions, wherein the lipid phase is in solid state at physiological temperature. NLCs comprise lipids in both solid and liquid state in the emulsion. SLNs feature high stability, biocompatibility, targeted and sustained delivery, retention of bioactive molecules of bioactive ingredients with affordable, and straight scale-up techniques. These however also show loss of bioactive via expulsion, loss of polymorphic states, gelation, low free space, aggregation, and particle enlargement. NLCs are tailored in a manner to have better retention of bioactives, more crystals to provide for more space for bioactives (Akhavan et al., 2018; Lasoń et al., 2013; Milsmann et al., 2018; Tamjidi et al., 2013).

14.4.2 Biopolymer-based delivery systems

Proteins and polysaccharides have been explored as encapsulates for the delivery of bioactives with the intention of protection, appropriate solubilization, and targeted delivery. The formulations are biphasic in nature and are kept as such or dried to powders. Proteins demonstrate the said property by virtue of their amphiphilic nature and the associated surface activity at oil—water interface. Polysaccharides cause thickening of the aqueous phase as well as influence the fate of the formulation during the gastrointestinal digestion by: (i) forming a protective coat around the lipid core to prevent its digestion by lipases, (ii) altering the coalescence of oil droplets in the formulation, (iii) possible sequestration of digestive enzymes, bile salts, etc., and (iv) changing the mass transport of the digestive components by increasing the viscosity or hydrogel formation in the system (Anal et al., 2019; Taheri and Jafari, 2019). The various forms of biopolymer-based bioactive delivery systems are described as follows:

14.4.2.1 Emulsions

Emulsions are thermodynamically (microemulsions) or kinetically (nanoemulsions) stable colloidal form of delivery systems that can be prepared using biopolymers for delivery of food and pharmaceutical ingredients. These have different physicochemical properties based on their constituents that help deliver the desired effect. These can be prepared using simple processing techniques of mixing, shearing, and high-pressure homogenization. The increased solubility, targeted delivery, and reduced size (in case of nanoemulsions) improve the bioavailability of the bioactive delivery through emulsions (Raikos and Ranawana, 2017). Native, hydrolyzed, chemically or enzymatically modified proteins, and polysaccharides have been widely reported to function as emulsifiers to improve the delivery of polyphenols, phytosterols, and many lipophilic bioactives (Raikos and Ranawana, 2017).

14.4.2.2 *Conventional encapsulates*

Conventionally, encapsulation of food or pharmaceutical ingredients and nutraceuticals by proteins and/or polysaccharides has been performed using spray drying, spray chilling, or freeze-drying techniques (Fang and Bhandari, 2012). The final product is an active ingredient in the core encapsulated by the shell of protein, polysaccharides, and/or emulsifiers. The process involves emulsification of the active ingredient with the biopolymeric emulsifier(s) followed by removal of the solvent of the continuous phase by heating, cooling, or freeze drying, respectively, to obtain the powdered form. The biopolymers used and methods of emulsification and drying influence the encapsulation efficiency, particle size, agglomeration, solubility, controlled release, bioaccessibility, and bioavailability of the active core (Eun et al., 2020; Ribeiro et al., 2019).

14.4.2.3 *Nanoparticles*

Biopolymeric nanoparticles are an emerging mode of delivery of bioactives. These can be formulated by fragmentation of larger particles to nanosize (10—1000 nm) or antisolvent precipitation to nanoparticles. These are stable colloidal particles, which have demonstrated improved efficacy, solubility, and bioavailability due to their reduced size (Dima et al., 2020; McClements, 2015). Both soluble proteins (bovine serum albumin, human serum albumin, β-lactoglobulin) and insoluble proteins (zein, gliadin) have been reported to form bioactive nanoparticles with demonstrated changes in the properties of the bioactives delivered through them (Cao et al., 2016).

14.4.2.4 *Coacervates*

Coacervation is a microencapsulation process involving two liquid phases in a colloidal system. This technique utilizes phase separation of one or more incompatible polymers from the initial coating polymer solution under specific pH, temperature or composition of the solution. The incompatible polymer(s) is added to the coating polymer solution, and the dispersion is stirred. Changes in the physical parameters lead to the separation of incompatible polymer and deposition of dense coacervate phase surrounding the core material resulting in the formation of microspheres. Chemical or enzymatic cross-linking agents can be used for strengthening the walls of microspheres. The more important processing factors to be considered for coacervation technique are the volume of the dispersed phase, rate of addition of the incompatible polymer to the coating polymer solution, stirring rate of the dispersion, and core material to be encapsulated (Gouin, 2004; John et al., 2011; Oliveira et al., 2007).

14.4.3 Gelled systems

Gels for the delivery of bioactives and micronutrients are a three-dimensional network with a trapped liquid phase. The particles of interest that are to be delivered are embedded in this network. Based on the type of particles embedded and the liquid phase, the following are the classifications of various gelled delivery systems (Mao et al., 2020).

14.4.4 Hydrogels

Hydrogels are the three-dimensional network of biopolymers (proteins and polysaccharides) that trap abundant water molecules in them. Physical or chemical cross-linking is involved among the biopolymers to trap the molecules. The gelation of the biopolymers can be induced by temperature, pressure, acids, salts, enzymes, ethanol, etc. (McClements, 2017). The process of encapsulating a bioactive in a hydrogel involves dissolution of the biopolymer in aqueous phase followed by addition of bioactive dissolved in a suitable solvent in it. The process can be conducted under heated conditions followed by cooling to set the gel (Liu et al., 2020). Different types of bioactive molecules or particles can be trapped inside hydrogels to tailor their physicochemical and delivery properties (Zhang et al., 2015).

14.4.4.1 *Emulgels*

Emulgels or emulsion gels are a type of hydrogels with trapped emulsified oil droplets. The oil droplets contain the lipophilic bioactive to be delivered. Emulgels can be formed by faster droplet aggregation or by gelling the continuous phase of an emulsion. The former case may involve pickering emulsions or high oil content, and in the latter case, mostly proteins or polysaccharides in the continuous phase undergo gelation. The oil droplets also contribute to the gel strength and hence the critical gelation concentration of biopolymers in emulgels is lower than those used in preparing hydrogels (Ajazuddin Alexander et al., 2013; Mao et al., 2020). Bioactives can be delivered through emulgels. The gel matrix provides a barrier to physical disintegration and enzymatic action upon ingestion. Their nongreasy texture makes them suitable for topical delivery of drugs (Ajazuddin Alexander et al., 2013). These are recommended to overcome the limitations of gel-based formulations and mostly find applications in topicals and cosmetics (Sharma et al., 2018).

14.4.4.2 Organogels

Organogels have gained much popularity for their nutritional and sensory properties of replacing saturated fats with unsaturated oil systems of similar physico-functional but better nutritional properties (Cui et al., 2019). Organogels are three-dimensional gel structures of lipids making them a lipophilic system of delivery of bioactives or drugs. Both organic liquid solvent and lipophilic particles can be dispersed in the structure (Lupi et al., 2013). The gel network holds the oil via interactions of hydrogen bonding, van der Waals forces, p—p stacking, and capillary forces (Lupi et al., 2013). Lipophilic systems are preferred over hydrophilic systems due to their higher loading capacity of lipophilic bioactives. The choice of oil and gelators decide the properties of the organogels (Cui et al., 2019). A simple process of preparation can involve mixing of lipophilic bioactive with oil and gelator such as monostearin, span-20, followed by heating and mixing and subsequent cooling to set the organogel (Yu and Huang, 2012). Water-in oil emulsions have also been formulated in organogels. These are similar to emulgels only with a lipid gel structure in this case, which is aqueous in the case of emulgels. These are formed by emulsifying water phase in the melted organogel followed by cooling to set the gel (Ojeda-Serna et al., 2019).

14.4.4.3 Bigels

Bigels are emulsions wherein the aqueous and lipid phases exist in the gelled forms. The density and viscosity of the individual phases are matched in a manner to have computability and stability for a prolonged period. This system combines the benefits of both hydrogels and organogels. It is useful for the delivery of both lipophilic and hydrophilic ingredients. Most reports of bigel formulations are for pharmaceutical drug or probiotics delivery. This system can be explored for bioactive delivery as well (Behera et al., 2014; Shakeel et al., 2018).

14.4.5 Miscellaneous carriers

14.4.5.1 Cocrystallization

Cocrystallization is a physical entrapment technique, which involves noncovalent interactions among the active ingredient and the conformer. This method has been used for delivery of active pharmaceutical ingredients. It has now been reported for the enhanced bioavailability of bioactives such as curcumin and quercetin. The process involves mixing of stoichiometric amounts of the active and conformer in a solvent followed by prolonged mixing for cocrystalization. This process has been reported under the influence of supercritical solvent such as SC-CO_2 as well (Ribas et al., 2019; Smith et al., 2011).

14.4.5.2 Inclusion complexation

Inclusion complexes (ICs) involve complexation of a hydrophobic material in the core of cyclodextrins (CDs). CDs are cyclic oligosaccharides with a toroid-shaped molecular structure that are able to form noncovalent shell-core ICs with various molecules. The most common CDs have either 6, 7, or eight glucopyranose units in the ring and are referred to as α-, β-, and γ-CD, respectively. The formation and stability of ICs depends on size, shape match, the binding forces between the host and guest molecules, and the chemical environment. CDs are natural, nontoxic, and slowly biodegradable molecules, and, consequently, they are used in a variety of areas such as enhancement of drug solubility and stability, bioavailability, controlled drug delivery, reduction of drug toxicity, stabilization of volatile or unstable flavors, and masking or removal of unwanted tastes and odors, in membrane filters (Yamamoto et al., 2012). Essential oils or vitamins can be complexed with CDs followed by spray or freeze drying or rotary evaporation or SC-CO_2 drying to obtain a powdered product (Pires et al., 2018).

14.4.5.3 Yeast encapsulation

Yeast cells are spherical or ellipsoidal with a diameter of approximately 2—5 μm. The rigid cell wall and inner plasma membrane makes yeast an excellent encapsulating agent. The thin (>10 nm) plasma membrane mainly consists of phospholipids that are organized in lipid bilayers, in addition to sterols and globular proteins. A major role of the plasma membrane is selective permeability to what enters or leaves the cell. This permeability barrier for a permeating molecule is attractive for encapsulation (Paramera et al., 2011). Yeast encapsulation can be accomplished with no additives, with only the involvement of the yeast, water, and the active ingredient. The encapsulation process generally involves mixing of the yeast cells (alive, plasmolyzed, or nonplasmolyzed, wet, or dried) or the yeast walls with the active ingredient in a water or water/organic solvent solution, followed by stirring for several hours under controlled temperature (usually 20—60°C).

The stirrer speed is also controlled to obtain the desired droplet size for optimum diffusion. During the encapsulation process, the cells lose their viability within 2 h, while the cell wall and the membrane remain intact. The loss of viability of yeast cells does not influence their ability as encapsulant, since the encapsulation process is mostly driven by passive diffusion (Bishop et al., 1998).

14.5 Summary and future direction

The increasing consumer trend of plant-based diets, plant-based food ingredients, and plant-based nutraceuticals is fueling a great deal of research and development in this direction. The impact of the production line on ecology has led to the development of greener techniques at the laboratory and industrial scale toward the extraction of polyphenols, phytosterols, essential oils and aromatics from plant-based materials. These plant-based components, i.e., bioactives have demonstrated health benefits and are formulated in different delivery forms for human consumption. The current trend of bioactive extraction and formulation in efficient delivery forms is on a rise. This has helped valorize some indigenous plants and led to better economic returns to the producers. With this, studies need to focus on the effect of the natural plant matrix on the delivery of its constituent bioactives to reduce the ecological impact of extraction and subsequent formulations. The newer techniques of enhancing bioavailabilities can also be thoroughly gauged for any detrimental health effects on consumption. Future research can also be directed toward further scientific analysis of traditional medicinal knowledge of various global cultures.

References

Abbas, M., Saeed, F., Anjum, F. M., Afzaal, M., Tufail, T., Bashir, M. S., Ishtiaq, A., Hussain, S., & Suleria, H. A. R. (2017). Natural polyphenols: An overview. *International Journal of Food Properties,* 20(8), 1689–1699. https://doi.org/10.1080/10942912.2016.1220393

Aggarwal, B. B., Sundaram, C., Malani, N., & Ichikawa, H. (2007). Curcumin: The Indian solid gold. In B. B. Aggarwal, Y.-J. Surh, & S. Shishodia (Eds.), *The molecular targets and therapeutic uses of curcumin in health and disease* (pp. 1–75). Springer US. https://doi.org/10.1007/978-0-387-46401-5_1

Ajazuddin, Alexander, A., Khichariya, A., Gupta, S., Patel, R. J., Giri, T. K., & Tripathi, D. K. (2013). Recent expansions in an emergent novel drug delivery technology: Emulgel. *Journal of Controlled Release,* 171(2), 122–132. https://doi.org/10.1016/j.jconrel.2013.06.030

Akbarzadeh, A., Rezaei-Sadabady, R., Davaran, S., Joo, S. W., Zarghami, N., Hanifehpour, Y., Samiei, M., Kouhi, M., & Nejati-Koshki, K. (2013). Liposome: Classification, preparation, and applications. *Nanoscale Research Letters,* 8(1), 102. https://doi.org/10.1186/1556-276X-8-102

Akhavan, S., Assadpour, E., Katouzian, I., & Jafari, S. M. (2018). Lipid nano scale cargos for the protection and delivery of food bioactive ingredients and nutraceuticals. *Trends in Food Science & Technology,* 74, 132–146. https://doi.org/10.1016/j.tifs.2018.02.001

Alothman, M., Bhat, R., & Karim, A. A. (2009). Antioxidant capacity and phenolic content of selected tropical fruits from Malaysia, extracted with different solvents. *Food Chemistry,* 115(3), 785–788. https://doi.org/10.1016/j.foodchem.2008.12.005

Ameer, K., Shahbaz, H. M., & Kwon, J.-H. (2017). Green extraction methods for polyphenols from plant matrices and their byproducts: A review. *Comprehensive Reviews in Food Science and Food Safety,* 16(2), 295–315. https://doi.org/10.1111/1541-4337.12253

Anal, A. K., Shrestha, S., & Sadiq, M. B. (2019). Biopolymeric-based emulsions and their effects during processing, digestibility and bioaccessibility of bioactive compounds in food systems. *Food Hydrocolloids,* 87, 691–702. https://doi.org/10.1016/j.foodhyd.2018.09.008

Ananingsih, V. K., Soedarini, B., & Karina, E. (2020). Separation of oleoresin from nutmeg using ultrasound assisted extraction and hexane as solvent. *IOP Conference Series: Materials Science and Engineering,* 854, 012029. https://doi.org/10.1088/1757-899X/854/1/012029

Arce, A., Marchiaro, A., Rodríguez, O., & Soto, A. (2006). Essential oil terpenless by extraction using organic solvents or ionic liquids. *AIChE Journal,* 52(6), 2089–2097. https://doi.org/10.1002/aic.10844

Ascrizzi, R., González-Rivera, J., Pomelli, C. S., Chiappe, C., Margari, P., Costagli, F., Longo, I., Tiné, M. R., Flamini, G., & Duce, C. (2017). Ionic liquids, ultra-sounds and microwaves: An effective combination for a sustainable extraction with higher yields. The cumin essential oil case. *Reaction Chemistry & Engineering,* 2(4), 577–589. https://doi.org/10.1039/C7RE00075H

Avram, A. M., Morin, P., Brownmiller, C., Howard, L. R., Sengupta, A., & Wickramasinghe, S. R. (2017). Concentrations of polyphenols from blueberry pomace extract using nanofiltration. *Food and Bioproducts Processing,* 106, 91–101. https://doi.org/10.1016/j.fbp.2017.07.006

Behera, B., Sagiri, S. S., Singh, V. K., Pal, K., & Anis, A. (2014). Mechanical properties and delivery of drug/probiotics from starch and non-starch based novel bigels: A comparative study. *Starch - Stärke,* 66(9–10), 865–879. https://doi.org/10.1002/star.201400045

Belwal, T., Huang, H., Li, L., Duan, Z., Zhang, X., Aalim, H., & Luo, Z. (2019). Optimization model for ultrasonic-assisted and scale-up extraction of anthocyanins from Pyrus communis 'Starkrimson' fruit peel. *Food Chemistry,* 297, 124993. https://doi.org/10.1016/j.foodchem.2019.124993

Bendif, H., Adouni, K., Miara, M. D., Baranauskienė, R., Kraujalis, P., Venskutonis, P. R., Nabavi, S. M., & Maggi, F. (2018). Essential oils (EOs), pressurized liquid extracts (PLE) and carbon dioxide supercritical fluid extracts (SFE-CO2) from Algerian Thymus munbyanus as valuable sources of antioxidants to be used on an industrial level. *Food Chemistry,* 260, 289–298. https://doi.org/10.1016/j.foodchem.2018.03.108

Bica, K., Gaertner, P., & Rogers, R. D. (2011). Ionic liquids and fragrances – direct isolation of orange essential oil. *Green Chemistry,* 13(8), 1997–1999. https://doi.org/10.1039/C1GC15237H

Bishop, J. R., Nelson, G., & Lamb, J. (1998). Microencapsulation in yeast cells. *Journal of Microencapsulation,* 15(6), 761–773. https://doi.org/10.3109/02652049809008259

Bittencourt, G. M., Firmiano, D. M., Fachini, R. P., Lacaz-Ruiz, R., Fernandes, A. M., & Oliveira, A. L. (2019). Application of green technology for the acquisition of extracts of araçá (psidium grandifolium mart. Ex DC.) using supercritical CO2 and pressurized ethanol: Characterization and analysis of activity. *Journal of Food Science,* 84(6), 1297–1307. https://doi.org/10.1111/1750-3841.14584

Bot, A. (2019). Phytosterols. In L. Melton, F. Shahidi, & P. Varelis (Eds.), *Encyclopedia of food chemistry* (pp. 225–228). Academic Press. https://doi.org/10.1016/B978-0-08-100596-5.21626-0

Boumahdi, Y., Moghrani, H., Nasrallah, N., Ouarek, S., & Maachi, R. (2021). Microwave-assisted hydrodistillation of the essential oil from Algerian Pimpinella anisum seeds. *Flavour and Fragrance Journal,* 36(1), 34–46. https://doi.org/10.1002/ffj.3614

Brüll, F., & Mensink, R. (2009). Plant sterols: Functional lipids in immune function and inflammation? *Clinical Lipidology, 4*(3), 355–365. https://doi.org/10.2217/clp.09.26

Calvo-Castro, L. A., Schiborr, C., David, F., Ehrt, H., Voggel, J., Sus, N., Behnam, D., Bosy-Westphal, A., & Frank, J. (2018). The oral bioavailability of trans-resveratrol from a grapevine-shoot extract in healthy humans is significantly increased by micellar solubilization. *Molecular Nutrition & Food Research,* 62(9), e1701057. https://doi.org/10.1002/mnfr.201701057

Cao, W.-J., Ou, S.-Y., Lin, W.-F., & Tang, C.-H. (2016). Food protein-based phytosterol nanoparticles: Fabrication and characterization. *Food & Function,* 7(9), 3973–3980. https://doi.org/10.1039/C6FO00861E

Capello, C., Fischer, U., & Hungerbühler, K. (2007). What is a green solvent? A comprehensive framework for the environmental assessment of solvents. *Green Chemistry,* 9(9), 927–934. https://doi.org/10.1039/B617536H

Cassano, A., Conidi, C., Ruby-Figueroa, R., & Castro-Muñoz, R. (2018). Nanofiltration and tight ultrafiltration membranes for the recovery of polyphenols from agro-food by-products. *International Journal of Molecular Sciences,* 19(2), 351. https://doi.org/10.3390/ijms19020351

Chairuk, P., Tubtimsri, S., Jansakul, C., Sriamornsak, P., & Weerapol, Y. (2020). Enhancing oral absorption of poorly water-soluble herb (Kaempferia parviflora) extract using self-nanoemulsifying formulation. *Pharmaceutical Development and Technology,* 25(3), 340–350. https://doi.org/10.1080/10837450.2019.1703134

Chapeau, A.-L., Bertrand, N., Briard-Bion, V., Hamon, P., Poncelet, D., & Bouhallab, S. (2017). Coacervates of whey proteins to protect and improve the oral delivery of a bioactive molecule. *Journal of Functional Foods,* 38, 197–204. https://doi.org/10.1016/j.jff.2017.09.009

Chemat, F., Abert-Vian, M., Fabiano-Tixier, A. S., Strube, J., Uhlenbrock, L., Gunjevic, V., & Cravotto, G. (2019). Green extraction of natural products. Origins, current status, and future challenges. *TrAC Trends in Analytical Chemistry,* 118, 248–263. https://doi.org/10.1016/j.trac.2019.05.037

Chen, X., McClements, D. J., Wang, J., Zou, L., Deng, S., Liu, W., Yan, C., Zhu, Y., Cheng, C., & Liu, C. (2018a). Coencapsulation of (−)-Epigallocatechin-3-gallate and quercetin in particle-stabilized W/O/W emulsion gels: Controlled release and bioaccessibility. *Journal of Agricultural and Food Chemistry, 66*, 3691–3699. https://doi.org/10.1021/acs.jafc.7b05161

Chen, X., McClements, D. J., Zhu, Y., Chen, Y., Zou, L., Liu, W., Cheng, C., Fu, D., & Liu, C. (2018b). Enhancement of the solubility, stability and bioaccessibility of quercetin using protein-based excipient emulsions. *Food Research International,* 114, 30–37. https://doi.org/10.1016/j.foodres.2018.07.062

Chen, Y., Xia, G., Zhao, Z., Xue, F., Gu, Y., Chen, C., & Zhang, Y. (2020). 7,8-Dihydroxyflavone nano-liposomes decorated by crosslinked and glycosylated lactoferrin: Storage stability, antioxidant activity, in vitro release, gastrointestinal digestion and transport in Caco-2 cell monolayers. *Journal of Functional Foods,* 65. https://doi.org/10.1016/j.jff.2019.103742, 103742.

Chen-yu, G., Chun-fen, Y., Qi-lu, L., Qi, T., Yan-wei, X., Wei-na, L., & Guang-xi, Z. (2012). Development of a Quercetin-loaded nanostructured lipid carrier formulation for topical delivery. *International Journal of Pharmaceutics,* 430(1), 292–298. https://doi.org/10.1016/j.ijpharm.2012.03.042

Chua, L. S. (2013). A review on plant-based rutin extraction methods and its pharmacological activities. *Journal of Ethnopharmacology,* 150(3), 805–817. https://doi.org/10.1016/j.jep.2013.10.036

Corazza, G. O., Bilibio, D., Zanella, O., Nunes, A. L., Bender, J. P., Carniel, N., dos Santos, P. P., & Priamo, W. L. (2018). Pressurized liquid extraction of polyphenols from Goldenberry: Influence on antioxidant activity and chemical composition. *Food and Bioproducts Processing,* 112, 63–68. https://doi.org/10.1016/j.fbp.2018.09.001

Cui, M., Mao, L., Lu, Y., Yuan, F., & Gao, Y. (2019). Effect of monoglyceride content on the solubility and chemical stability of β-carotene in organogels. *LWT,* 106, 83–91. https://doi.org/10.1016/j.lwt.2019.02.042

Dawidowicz, A. L., Czapczyńska, N. B., & Wianowska, D. (2012). The loss of essential oil components induced by the Purge Time in the Pressurized Liquid Extraction (PLE) procedure of Cupressus sempervirens. *Talanta,* 94, 140–145. https://doi.org/10.1016/j.talanta.2012.03.008

de Castro, M. D. L., Valcárcel, M., & Tena, M. T. (1994). Theoretical and practical aspects of supercritical fluid extraction. In M. D. L. de Castro, M. Valcárcel, & M. T. Tena (Eds.), *Analytical supercritical fluid extraction* (pp. 79–168). Springer. https://doi.org/10.1007/978-3-642-78673-0_3

Dikshit, S., Bubna, S., Gupta, A., & Kumar, P. (2020). Advances in various techniques for isolation and purification of sterols. *Journal of Food Science and Technology,* 57(7), 2393–2403. https://doi.org/10.1007/s13197-019-04209-3

Dima, C., Assadpour, E., Dima, S., & Jafari, S. M. (2020). Bioactive-loaded nanocarriers for functional foods: From designing to bioavailability. *Current Opinion in Food Science,* 33, 21–29. https://doi.org/10.1016/j.cofs.2019.11.006

Donsì, F., & Velikov, K. (2020). Mechanical cell disruption of mustard bran suspensions for improved dispersion properties and protein release. *Food & Function,* 11(7), 6273–6284. https://doi.org/10.1039/D0FO00852D

Dąbrowski, G., Czaplicki, S., & Konopka, I. (2019). Fractionation of sterols, tocols and squalene in flaxseed oils under the impact of variable conditions of supercritical CO2 extraction. *Journal of Food Composition and Analysis,* 83, 103261. https://doi.org/10.1016/j.jfca.2019.103261

Dutta, S., & Bhattacharjee, P. (2015). Enzyme-assisted supercritical carbon dioxide extraction of black pepper oleoresin for enhanced yield of piperine-rich extract. *Journal of Bioscience and Bioengineering,* 120(1), 17–23. https://doi.org/10.1016/j.jbiosc.2014.12.004

Earle, M. J., & Seddon, K. R. (2002). Ionic liquids: Green solvents for the future. In , *Vol 819. Clean solvents* (pp. 10–25). American Chemical Society. https://doi.org/10.1021/bk-2002-0819.ch002

Eikani, M. H., Golmohammad, F., Amoli, H. S., & Sadr, Z. B. (2013). An experimental design approach for pressurized liquid extraction from cardamom seeds. *Separation Science and Technology,* 48(8), 1194–1200. https://doi.org/10.1080/01496395.2012.734365

Eun, J.-B., Maruf, A., Das, P. R., & Nam, S.-H. (2020). A review of encapsulation of carotenoids using spray drying and freeze drying. *Critical Reviews in Food Science and Nutrition,* 60(21), 3547–3572. https://doi.org/10.1080/10408398.2019.1698511

Falleh, H., Ben Jemaa, M., Saada, M., & Ksouri, R. (2020). Essential oils: A promising eco-friendly food preservative. *Food Chemistry,* 330, 127268. https://doi.org/10.1016/j.foodchem.2020.127268

Fang, Z., & Bhandari, B. (2012). 4—spray drying, freeze drying and related processes for food ingredient and nutraceutical encapsulation. In N. Garti, & D. J. McClements (Eds.), *Encapsulation technologies and delivery systems for food ingredients and nutraceuticals* (pp. 73–109). Woodhead Publishing. https://doi.org/10.1533/9780857095909.2.73

Fomo, G., Madzimbamuto, T. N., & Ojumu, T. V. (2020). Applications of nonconventional green extraction technologies in process industries: Challenges, limitations and perspectives. *Sustainability,* 12(13), 5244. https://doi.org/10.3390/su12135244

Fraga, C. G., Croft, K. D., & Kennedy, D. O. (2019). The effects of polyphenols and other bioactives on human health. *Food & Function,* 10(2), 514–528. https://doi.org/10.1039/C8FO01997E

Franco-Vega, A., López-Malo, A., Palou, E., & Ramírez-Corona, N. (2021). Effect of imidazolium ionic liquids as microwave absorption media for the intensification of microwave-assisted extraction of Citrus sinensis peel essential oils. *Chemical Engineering and Processing - Process Intensification,* 160, 108277. https://doi.org/10.1016/j.cep.2020.108277

Gallo, M., Formato, A., Giacco, R., Riccardi, G., Luongo, D., Formato, G., Amoresano, A., & Naviglio, D. (2019). Mathematical optimization of the green extraction of polyphenols from grape peels through a cyclic pressurization process. *Heliyon,* 5(4), e01526. https://doi.org/10.1016/j.heliyon.2019.e01526

Gomes, T. A., Zanette, C. M., & Spier, M. R. (2020). An overview of cell disruption methods for intracellular biomolecules recovery. *Preparative Biochemistry & Biotechnology,* 50(7), 635–654. https://doi.org/10.1080/10826068.2020.1728696

Gonçalves, R. F. S., Martins, J. T., Duarte, C. M. M., Vicente, A. A., & Pinheiro, A. C. (2018). Advances in nutraceutical delivery systems: From formulation design for bioavailability enhancement to efficacy and safety evaluation. *Trends in Food Science & Technology,* 78, 270–291. https://doi.org/10.1016/j.tifs.2018.06.011

Gonçalves, A., Nikmaram, N., Roohinejad, S., Estevinho, B. N., Rocha, F., Greiner, R., & McClements, D. J. (2018). Production, properties, and applications of solid self-emulsifying delivery systems (S-SEDS) in the food and pharmaceutical industries. *Colloids and Surfaces A: Physicochemical and Engineering Aspects,* 538, 108–126. https://doi.org/10.1016/j.colsurfa.2017.10.076

Gonzalez-Rivera, J., Duce, C., Campanella, B., Bernazzani, L., Ferrari, C., Tanzini, E., Onor, M., Longo, I., Ruiz, J. C., Tinè, M. R., & Bramanti, E. (2021). In situ microwave assisted extraction of clove buds to isolate essential oil, polyphenols, and lignocellulosic compounds. *Industrial Crops and Products,* 161, 113203. https://doi.org/10.1016/j.indcrop.2020.113203

Gouin, S. (2004). Microencapsulation: Industrial appraisal of existing technologies and trends. *Trends in Food Science & Technology,* 15(7), 330–347. https://doi.org/10.1016/j.tifs.2003.10.005

Gwee, Y. L., Yusup, S., Tan, R. R., & Yiin, C. L. (2020). Techno-economic and life-cycle assessment of volatile oil extracted from Aquilaria sinensis using supercritical carbon dioxide. *Journal of CO2 Utilization,* 38, 158–167. https://doi.org/10.1016/j.jcou.2020.01.002

Harde, S. M., & Singhal, R. S. (2012). Extraction of forskolin from Coleus forskohlii roots using three phase partitioning. *Separation and Purification Technology,* 96, 20–25. https://doi.org/10.1016/j.seppur.2012.05.017

He, S., Zhou, S., Guo, W., Wang, Y., Liu, C., Wang, R., & Xiao, F. (2020). Investigation of curcumin emulsion stability and gastrointestinal digestion prepared with rapeseed oil body. *Journal of Food Process Engineering,* 43(12), e13556. https://doi.org/10.1111/jfpe.13556

Huang, J., Wang, Q., Chu, L., & Xia, Q. (2020). Liposome-chitosan hydrogel bead delivery system for the encapsulation of linseed oil and quercetin: Preparation and in vitro characterization studies. *LWT,* 117, 108615. https://doi.org/10.1016/j.lwt.2019.108615

Jacques, R. A., Dariva, C., de Oliveira, J. V., & Caramão, E. B. (2008). Pressurized liquid extraction of mate tea leaves. *Analytica Chimica Acta,* 625(1), 70–76. https://doi.org/10.1016/j.aca.2008.07.002

Jafarian Asl, P., Niazmand, R., & Jahani, M. (2020). Theoretical and experimental assessment of supercritical CO2 in the extraction of phytosterols from rapeseed oil deodorizer distillates. *Journal of Food Engineering,* 269, 109748. https://doi.org/10.1016/j.jfoodeng.2019.109748

John, R. P., Tyagi, R. D., Brar, S. K., Surampalli, R. Y., & Prévost, D. (2011). Bio-encapsulation of microbial cells for targeted agricultural delivery. *Critical Reviews in Biotechnology,* 31(3), 211–226. https://doi.org/10.3109/07388551.2010.513327

Khare, P., & Shanker, K. (2016). Mangiferin: A review of sources and interventions for biological activities. *BioFactors (Oxford, England),* 42(5), 504–514. https://doi.org/10.1002/biof.1308

Kiyama, R. (2020). Nutritional implications of ginger: Chemistry, biological activities and signaling pathways. *The Journal of Nutritional Biochemistry,* 86, 108486. https://doi.org/10.1016/j.jnutbio.2020.108486

Kurmudle, N. N., Bankar, S. B., Bajaj, I. B., Bule, M. V., & Singhal, R. S. (2011). Enzyme-assisted three phase partitioning: A novel approach for extraction of turmeric oleoresin. *Process Biochemistry,* 46(1), 423–426. https://doi.org/10.1016/j.procbio.2010.09.010

Lasoń, E., Sikora, E., & Ogonowski, J. (2013). Influence of process parameters on properties of Nanostructured Lipid Carriers (NLC) formulation. *Acta Biochimica Polonica,* 60(4), 773–777.

Li, Y., Zhang, C., Pan, S., Chen, L., Liu, M., Yang, K., Zeng, X., & Tian, J. (2020). Analysis of chemical components and biological activities of essential oils from black and white pepper (Piper nigrum L.) in five provinces of southern China. *LWT,* 117, 108644. https://doi.org/10.1016/j.lwt.2019.108644

Lin, F., & Giusti, M. M. (2005). Effects of solvent polarity and acidity on the extraction efficiency of isoflavones from soybeans (Glycine max). *Journal of Agricultural and Food Chemistry,* 53(10), 3795–3800. https://doi.org/10.1021/jf050030r

Liu, K., Zha, X.-Q., Shen, W.-D., Li, Q.-M., Pan, L.-H., & Luo, J.-P. (2020). The hydrogel of whey protein isolate coated by lotus root amylopectin enhance the stability and bioavailability of quercetin. *Carbohydrate Polymers,* 236, 116009. https://doi.org/10.1016/j.carbpol.2020.116009

Loizou, S., Lekakis, I., Chrousos, G. P., & Moutsatsou, P. (2010). β-Sitosterol exhibits anti-inflammatory activity in human aortic endothelial cells. *Molecular Nutrition & Food Research,* 54(4), 551–558. https://doi.org/10.1002/mnfr.200900012

Lorenzo, J. M., Khaneghah, A. M., Gavahian, M., Marszałek, K., Eş, I., Munekata, P. E. S., Ferreira, I. C. F. R., & Barba, F. J. (2019). Understanding the potential benefits of thyme and its derived products for food industry and consumer health: From extraction of value-added compounds to the evaluation of bioaccessibility, bioavailability, anti-inflammatory, and antimicrobial activities. *Critical Reviews in Food Science and Nutrition,* 59(18), 2879–2895. https://doi.org/10.1080/10408398.2018.1477730

Low, L. E., Tan, L. T.-H., Goh, B.-H., Tey, B. T., Ong, B. H., & Tang, S. Y. (2019). Magnetic cellulose nanocrystal stabilized Pickering emulsions for enhanced bioactive release and human colon cancer therapy. *International Journal of Biological Macromolecules,* 127, 76–84. https://doi.org/10.1016/j.ijbiomac.2019.01.037

Luca, S. V., Macovei, I., Bujor, A., Miron, A., Skalicka-Woźniak, K., Aprotosoaie, A. C., & Trifan, A. (2020). Bioactivity of dietary polyphenols: The role of metabolites. *Critical Reviews in Food Science and Nutrition,* 60(4), 626–659. https://doi.org/10.1080/10408398.2018.1546669

Lupi, F. R., Gabriele, D., Baldino, N., Mijovic, P., Parisi, O. I., & Puoci, F. (2013). Olive oil/policosanol organogels for nutraceutical and drug delivery purposes. *Food & Function,* 4(10), 1512–1520. https://doi.org/10.1039/C3FO60259A

Magar, R. T., & Sohng, J. K. (2020). A review on structure, modifications and structure-activity relation of quercetin and its derivatives. *Journal of Microbiology and Biotechnology,* 30(1), 11–20. https://doi.org/10.4014/jmb.1907.07003

Magri, A., Petriccione, M., Cerqueira, M. A., & Gutiérrez, T. J. (2020). Self-assembled lipids for food applications: A review. *Advances in Colloid and Interface Science,* 285, 102279. https://doi.org/10.1016/j.cis.2020.102279

Mahato, N., Sharma, K., Koteswararao, R., Sinha, M., Baral, E., & Cho, M. H. (2019). Citrus essential oils: Extraction, authentication and application in food preservation. *Critical Reviews in Food Science and Nutrition,* 59(4), 611–625. https://doi.org/10.1080/10408398.2017.1384716

Mahendran, G., & Rahman, L.-U. (2020). Ethnomedicinal, phytochemical and pharmacological updates on Peppermint (Mentha × piperita L.)—a review. *Phytotherapy Research,* 34(9), 2088–2139. https://doi.org/10.1002/ptr.6664

Mani, J. S., Johnson, J. B., Steel, J. C., Broszczak, D. A., Neilsen, P. M., Walsh, K. B., & Naiker, M. (2020). Natural product-derived phytochemicals as potential agents against coronaviruses: A review. *Virus Research,* 284, 197989. https://doi.org/10.1016/j.virusres.2020.197989

Manickam, S., Sivakumar, K., & Pang, C. H. (2020). Investigations on the generation of oil-in-water (O/W) nanoemulsions through the combination of ultrasound and microchannel. *Ultrasonics Sonochemistry,* 69, 105258. https://doi.org/10.1016/j.ultsonch.2020.105258

Mao, L., Lu, Y., Cui, M., Miao, S., & Gao, Y. (2020). Design of gel structures in water and oil phases for improved delivery of bioactive food ingredients. *Critical Reviews in Food Science and Nutrition,* 60(10), 1651–1666. https://doi.org/10.1080/10408398.2019.1587737

Marathe, S. J., Jadhav, S. B., Bankar, S. B., Kumari Dubey, K., & Singhal, R. S. (2019). Improvements in the extraction of bioactive compounds by enzymes. *Current Opinion in Food Science,* 25, 62–72. https://doi.org/10.1016/j.cofs.2019.02.009

McClements, D. J. (2007). Critical review of techniques and methodologies for characterization of emulsion stability. *Critical Reviews in Food Science and Nutrition,* 47(7), 611–649. https://doi.org/10.1080/10408390701289292

McClements, D. J. (2015). Nanoscale nutrient delivery systems for food applications: Improving bioactive dispersibility, stability, and bioavailability. *Journal of Food Science,* 80(7), N1602–N1611. https://doi.org/10.1111/1750-3841.12919

McClements, D. J. (2017). Recent progress in hydrogel delivery systems for improving nutraceutical bioavailability. *Food Hydrocolloids,* 68, 238–245. https://doi.org/10.1016/j.foodhyd.2016.05.037

McClements, D. J., Bai, L., & Chung, C. (2017). Recent advances in the utilization of natural emulsifiers to form and stabilize emulsions. *Annual Review of Food Science and Technology,* 8, 205–236. https://doi.org/10.1146/annurev-food-030216-030154. Scopus.

Menezes Maciel Bindes, M., Hespanhol Miranda Reis, M., Luiz Cardoso, V., & Boffito, D. C. (2019). Ultrasound-assisted extraction of bioactive compounds from green tea leaves and clarification with natural coagulants (chitosan and Moringa oleífera seeds). *Ultr*asonics Sonochemistry, 51, 111–119. https://doi.org/10.1016/j.ultsonch.2018.10.014

Milsmann, J., Oehlke, K., Greiner, R., & Steffen-Heins, A. (2018). Fate of edible solid lipid nanoparticles (SLN) in surfactant stabilized o/w emulsions. Part 2: Release and partitioning behavior of lipophilic probes from SLN into different phases of o/w emulsions. *Colloids and Surfaces A: Physicochemical and Engineering Aspects,* 558, 623–631. https://doi.org/10.1016/j.colsurfa.2017.05.050

Moncada, J., Tamayo, J. A., & Cardona, C. A. (2016). Techno-economic and environmental assessment of essential oil extraction from Oregano (Origanum vulgare) and Rosemary (Rosmarinus officinalis) in Colombia. *Journal of Cleaner Production,* 112, 172–181. https://doi.org/10.1016/j.jclepro.2015.09.067

Moreau, R. A., Whitaker, B. D., & Hicks, K. B. (2002). Phytosterols, phytostanols, and their conjugates in foods: Structural diversity, quantitative analysis, and health-promoting uses. *Progress in Lipid Research,* 41(6), 457–500. https://doi.org/10.1016/s0163-7827(02)00006-1

Morsy, N. F. S. (2020). Production of thymol rich extracts from ajwain (Carum copticum L.) and thyme (Thymus vulgaris L.) using supercritical CO2. *Industrial Crops and Products,* 145, 112072. https://doi.org/10.1016/j.indcrop.2019.112072

Mourtzinos, I., & Biliaderis, C. G. (2017). Principles and applications of encapsulation technologies to food materials | thermal and nonthermal encapsulation methods | Taylor & Francis Group. In *Thermal and nonthermal encapsulation methods* (pp. 1–38). CRC Press. https://www.taylorfrancis.com/.

Mozafari, M. R., & Khosravi-Darani, K. (2007). An overview of liposome-derived nanocarrier technologies. In M. R. Mozafari (Ed.), *Nanomaterials and nanosystems for biomedical applications* (pp. 113–123). Springer Netherlands. https://doi.org/10.1007/978-1-4020-6289-6_7

MS, U., Ferdosh, S., Akanda, M. J. H., Ghafoor, K., A.H, R., Ali, M. E., Kamaruzzaman, B. Y., MB, F., Shaarani, S., & Sarker, M. Z. (2018). Techniques for the extraction of phytosterols and their benefits in human health: A review. *Separation Science and Technology,* 53(14), 2206–2223. https://doi.org/10.1080/01496395.2018.1454472

Nagavekar, N., & Singhal, R. S. (2018). Enhanced extraction of oleoresin from Piper nigrum by supercritical carbon dioxide using ethanol as a co-solvent and its bioactivity profile. *Journal of Food Process Engineering,* 41(1), e12670. https://doi.org/10.1111/jfpe.12670

Nagavekar, N., & Singhal, R. S. (2019). Supercritical fluid extraction of Curcuma longa and Curcuma amada oleoresin: Optimization of extraction conditions, extract profiling, and comparison of bioactivities. *Industrial Crops and Products,* 134, 134–145. https://doi.org/10.1016/j.indcrop.2019.03.061

Nagavekar, N., & Singhal, R. S. (2020). Simultaneous extraction of flaxseed spice blend using supercritical carbon dioxide: Process optimization, bioactivity profile, and application as a functional seasoning. *Separation and Purification Technology,* 248, 117030. https://doi.org/10.1016/j.seppur.2020.117030

Naviglio, D., Formato, A., Vitulano, M., Cozzolino, I., Ferrara, L., Zanoelo, E. F., & Gallo, M. (2017). Comparison between the kinetics of conventional maceration and A cyclic pressurization extraction process for the production of lemon liqueur using A numerical model. *Journal of Food Process Engineering,* 40(2), e12350. https://doi.org/10.1111/jfpe.12350

Ochoa, S., Durango-Zuleta, M. M., & Felipe Osorio-Tobón, J. (2020). Techno-economic evaluation of the extraction of anthocyanins from purple yam (Dioscorea alata) using ultrasound-assisted extraction and conventional extraction processes. *Food and Bioproducts Processing,* 122, 111–123. https://doi.org/10.1016/j.fbp.2020.04.007

Ojeda-Serna, I. E., Rocha-Guzmán, N. E., Gallegos-Infante, J. A., Cháirez-Ramírez, M. H., Rosas-Flores, W., Pérez-Martínez, J. D., Moreno-Jiménez, M. R., & González-Laredo, R. F. (2019). Water-in-oil organogel based emulsions as a tool for increasing bioaccessibility and cell permeability of poorly water-soluble nutraceuticals. *Food Research International,* 120, 415–424. https://doi.org/10.1016/j.foodres.2019.03.011

Olalere, O. A., Abdurahman, N. H., Alara, O. R., & Habeeb, O. A. (2017). Parametric optimization of microwave reflux extraction of spice oleoresin from white pepper (Piper nigrum). *Jour*nal of Analytical Science and Technology, 8(1), 8. https://doi.org/10.1186/s40543-017-0118-9

Olalere, O. A., Abdurahman, N. H., Yunus, R., bin, M., Alara, O. R., & Gan, C.-Y. (2018). Synergistic intermittent heating and energy intensification of scale-up parameters in an optimized microwave extraction process. *Chemical Engineering and Processing - Process Intensification,* 132, 160–168. https://doi.org/10.1016/j.cep.2018.08.011

Oliveira, A. C., Moretti, T. S., Boschini, C., Baliero, J. C. C., Freitas, L. A. P., Freitas, O., & Favaro-Trindade, C. S. (2007). Microencapsulation of B. Lactis (BI 01) and L. Acidophilus (LAC 4) by complex coacervation followed by spouted-bed drying. *Drying Technology,* 25(10), 1687–1693. https://doi.org/10.1080/07373930701590939

Osorio-Tobón, J. F., Carvalho, P. I. N., Rostagno, M. A., & Meireles, M. A. A. (2016). Process integration for turmeric products extraction using supercritical fluids and pressurized liquids: Economic evaluation. *Food and Bioproducts Processing,* 98, 227–235. https://doi.org/10.1016/j.fbp.2016.02.001

Osorio-Tobón, J. F., Carvalho, P. I. N., Rostagno, M. A., Petenate, A. J., & Meireles, M. A. A. (2014). Extraction of curcuminoids from deflavored turmeric (Curcuma longa L.) using pressurized liquids: Process integration and economic evaluation. *The Journal of Supercritical Fluids,* 95, 167–174. https://doi.org/10.1016/j.supflu.2014.08.012

Ozturk, B., Parkinson, C., & Gonzalez-Miquel, M. (2018). Extraction of polyphenolic antioxidants from orange peel waste using deep eutectic solvents. *Separation and Purification Technology,* 206, 1–13. https://doi.org/10.1016/j.seppur.2018.05.052

Paramera, E. I., Konteles, S. J., & Karathanos, V. T. (2011). Stability and release properties of curcumin encapsulated in *Saccharomyces cerevisiae*, β-cyclodextrin and modified starch. *Food Chemistry,* 125(3), 913–922. https://doi.org/10.1016/j.foodchem.2010.09.071

Pattni, B. S., Chupin, V. V., & Torchilin, V. P. (2015). New developments in liposomal drug delivery. *Chemical Reviews,* 115(19), 10938–10966. https://doi.org/10.1021/acs.chemrev.5b00046

Pires, F. Q., Pinho, L. A., Freire, D. O., Silva, I. C. R., Sa-Barreto, L. L., Cardozo-Filho, L., Gratieri, T., Gelfuso, G. M., & Cunha-Filho, M. (2018). Thermal analysis used to guide the production of thymol and Lippia origanoides essential oil inclusion complexes with cyclodextrin. *Journal of Thermal Analysis and Calorimetry,* 137(2), 543–553. https://doi.org/10.1007/s10973-018-7956-6

Poompavai, S., & Sree, V. G. (2020). Anti-proliferative efficiency of pulsed electric field treated curcuma longa (turmeric) extracts on breast cancer cell lines. *IETE Journal of Research,* 1–15. https://doi.org/10.1080/03772063.2020.1799873

Rafiee, Z., Barzegar, M., Sahari, M. A., & Maherani, B. (2017). Nanoliposomal carriers for improvement the bioavailability of high – valued phenolic compounds of pistachio green hull extract. *Food Chemistry,* 220, 115–122. https://doi.org/10.1016/j.foodchem.2016.09.207

Raikos, V., & Ranawana, V. (2017). Designing emulsion droplets of foods and beverages to enhance delivery of lipophilic bioactive components – a review of recent advances. *International Journal of Food Science & Technology,* 52(1), 68–80. https://doi.org/10.1111/ijfs.13272

Rajpoot, K., Tekade, M., Pandey, V., Nagaraja, S., Youngren-Ortiz, S. R., & Tekade, R. K. (2020). Chapter 9 - self-microemulsifying drug-delivery system: Ongoing challenges and future ahead. In R. K. Tekade (Ed.), *Drug delivery systems* (pp. 393–454). Academic Press. https://doi.org/10.1016/B978-0-12-814487-9.00009-0

Ramalingam, P., & Ko, Y. T. (2015). Enhanced oral delivery of curcumin from N-trimethyl chitosan surface-modified solid lipid nanoparticles: Pharmacokinetic and brain distribution evaluations. *Pharmaceutical Research,* 32(2), 389–402. https://doi.org/10.1007/s11095-014-1469-1

Rammohan, A., Reddy, J. S., Sravya, G., Rao, C. N., & Zyryanov, G. V. (2020). Chalcone synthesis, properties and medicinal applications: A review. *Environmental Chemistry Letters,* 18(2), 433–458. https://doi.org/10.1007/s10311-019-00959-w

Reyes-Jurado, F., Franco-Vega, A., Ramírez-Corona, N., Palou, E., & López-Malo, A. (2015). Essential oils: Antimicrobial activities, extraction methods, and their modeling. *Food Engineering Reviews,* 7(3), 275–297. https://doi.org/10.1007/s12393-014-9099-2

Ribas, M. M., Aguiar, G. P. S., Muller, L. G., Siebel, A. M., Lanza, M., & Oliveira, J. V. (2019). Curcumin-nicotinamide cocrystallization with supercritical solvent (CSS): Synthesis, characterization and in vivo antinociceptive and anti-inflammatory activities. *Industrial Crops and Products,* 139, 111537. https://doi.org/10.1016/j.indcrop.2019.111537

Ribeiro, A. M., Estevinho, B. N., & Rocha, F. (2019). Spray drying encapsulation of elderberry extract and evaluating the release and stability of phenolic compounds in encapsulated powders. *Food and Bioprocess Technology,* 12(8), 1381–1394. https://doi.org/10.1007/s11947-019-02304-z

Ricci, A., Parpinello, G. P., & Versari, A. (2018). Recent advances and applications of pulsed electric fields (PEF) to improve polyphenol extraction and color release during red winemaking. *Beverages,* 4(1), 18. https://doi.org/10.3390/beverages4010018

Ríos, J.-L. (2016). Chapter 1 - essential oils: What they are and how the terms are used and defined. In V. R. Preedy (Ed.), *Essential oils in food preservation, flavor and safety* (pp. 3–10). Academic Press. https://doi.org/10.1016/B978-0-12-416641-7.00001-8

Santamaría, R. I., Reyes-Duarte, M. D., Bárzana, E., Fernando, D., Gama, F. M., Mota, M., & López-Munguía, A. (2000). Selective enzyme-mediated extraction of capsaicinoids and carotenoids from chili guajillo puya (capsicum annuum L.) using ethanol as solvent. *Journal of Agricultural and Food Chemistry,* 48(7), 3063–3067. https://doi.org/10.1021/jf991242p

Santos, P., Aguiar, A. C., Barbero, G. F., Rezende, C. A., & Martínez, J. (2015). Supercritical carbon dioxide extraction of capsaicinoids from malagueta pepper (Capsicum frutescens L.) assisted by ultrasound. *Ultrasonics Sonochemistry,* 22, 78–88. https://doi.org/10.1016/j.ultsonch.2014.05.001

Shakeel, A., Lupi, F. R., Gabriele, D., Baldino, N., & Cindio, B. D. (2018). Bigels: A unique class of materials for drug delivery applications. *Soft Materials,* 16(2), 77–93. https://doi.org/10.1080/1539445X.2018.1424638

Sharma, V., Nayak, S. K., Paul, S. R., Choudhary, B., Ray, S. S., & Pal, K. (2018). 9—Emulgels. In K. Pal, & I. Banerjee (Eds.), *Polymeric gels* (pp. 251–264). Woodhead Publishing. https://doi.org/10.1016/B978-0-08-102179-8.00009-0

Sharma, R., & Padwad, Y. (2020). Plant-polyphenols based second-generation synbiotics: Emerging concepts, challenges, and opportunities. *Nutrition,* 77, 110785. https://doi.org/10.1016/j.nut.2020.110785

Shi, G., Rao, L., Yu, H., Xiang, H., Yang, H., & Ji, R. (2008). Stabilization and encapsulation of photosensitive resveratrol within yeast cell. *International Journal of Pharmaceutics,* 349(1), 83–93. https://doi.org/10.1016/j.ijpharm.2007.07.044.

Sicari, V., & Poiana, M. (2017). Comparison of the volatile component of the essential oil of kumquat (fortunella margarita swingle) extracted by supercritical carbon dioxide, hydrodistillation and conventional solvent extraction. *Journal of Essential Oil Bearing Plants,* 20(1), 87–94. https://doi.org/10.1080/0972060X.2017.1282841

Simão, A. M. S., Bolean, M., Cury, T. A. C., Stabeli, R. G., Itri, R., & Ciancaglini, P. (2015). Liposomal systems as carriers for bioactive compounds. *Biophysical Reviews,* 7(4), 391–397. https://doi.org/10.1007/s12551-015-0180-8

Situ, W., Song, X., Luo, S., & Liang, Y. (2017). A nano-delivery system for bioactive ingredients using supercritical carbon dioxide and its release behaviors. *Food Chemistry,* 228, 219–225. https://doi.org/10.1016/j.foodchem.2017.01.053

Smith, A. J., Kavuru, P., Wojtas, L., Zaworotko, M. J., & Shytle, R. D. (2011). Cocrystals of quercetin with improved solubility and oral bioavailability. *Molecular Pharmaceutics,* 8(5), 1867–1876. https://doi.org/10.1021/mp200209j

Taheri, A., & Jafari, S. M. (2019). Gum-based nanocarriers for the protection and delivery of food bioactive compounds. *Advances in Colloid and Interface Science,* 269, 277–295. https://doi.org/10.1016/j.cis.2019.04.009

Tamjidi, F., Shahedi, M., Varshosaz, J., & Nasirpour, A. (2013). Nanostructured lipid carriers (NLC): A potential delivery system for bioactive food molecules. *Innovative Food Science & Emerging Technologies,* 19, 29–43. https://doi.org/10.1016/j.ifset.2013.03.002

Toma, M., Vinatoru, M., Paniwnyk, L., & Mason, T. J. (2001). Investigation of the effects of ultrasound on vegetal tissues during solvent extraction. *Ultrasonics Sonochemistry,* 8(2), 137–142. https://doi.org/10.1016/S1350-4177(00)00033-X

Ubeyitogullari, A., & Ciftci, O. N. (2017). Generating phytosterol nanoparticles in nanoporous bioaerogels via supercritical carbon dioxide impregnation: Effect of impregnation conditions. *Journal of Food Engineering,* 207, 99–107. https://doi.org/10.1016/j.jfoodeng.2017.03.022

Ubeyitogullari, A., Moreau, R., Rose, D. J., & Ciftci, O. N. (2019). In vitro bioaccessibility of low-crystallinity phytosterol nanoparticles generated using nanoporous starch bioaerogels. *Journal of Food Science,* 84(7), 1812–1819. https://doi.org/10.1111/1750-3841.14673

Usuki, T., & Munakata, K. (2017). Extraction of essential oils from the flowers of osmanthus fragrans var. Aurantiacus using an ionic liquid. *Bulletin of the Chemical Society of Japan,* 90(10), 1105–1110. https://doi.org/10.1246/bcsj.20170202

Varakumar, S., Umesh, K. V., & Singhal, R. S. (2017). Enhanced extraction of oleoresin from ginger (Zingiber officinale) rhizome powder using enzyme-assisted three phase partitioning. *Food Chemistry,* 216, 27–36. https://doi.org/10.1016/j.foodchem.2016.07.180

Ventura, S. P. M., e Silva, F. A., Quental, M. V., Mondal, D., Freire, M. G., & Coutinho, J. A. P. (2017). Ionic-liquid-mediated extraction and separation processes for bioactive compounds: Past, present, and future trends. *Chemical Reviews,* 117(10), 6984–7052. https://doi.org/10.1021/acs.chemrev.6b00550

Vijayan, U. K., Varakumar, S., & Singhal, R. S. (2019). A comparative account of extraction of oleoresin from Curcuma aromatica Salisb by solvent and supercritical carbon dioxide: Characterization and bioactivities. *LWT,* 116, 108564. https://doi.org/10.1016/j.lwt.2019.108564

Vilahur, G., Ben-Aicha, S., Diaz-Riera, E., Badimon, L., & Padró, T. (2019). Phytosterols and inflammation. *Current Medicinal Chemistry,* 26(37), 6724–6734. https://doi.org/10.2174/0929867325666180622151438

Vivaldo, G., Masi, E., Taiti, C., Caldarelli, G., & Mancuso, S. (2017). The network of plants volatile organic compounds. *Scientific Reports, 7*(1), 11050. https://doi.org/10.1038/s41598-017-10975-x

Wianowska, D. (2014). The influence of purge times on the yields of essential oil components extracted from plants by pressurized liquid extraction. *Journal of AOAC International, 97*(5), 1310–1316. https://doi.org/10.5740/jaoacint.13-318

Wiesner, M., Hanschen, F. S., Maul, R., Neugart, S., Schreiner, M., & Baldermann, S. (2017). Nutritional quality of plants for food and fodder. In B. Thomas, B. G. Murray, & D. J. Murphy (Eds.), *Encyclopedia of applied plant sciences* (2nd ed., pp. 285–291). Academic Press. https://doi.org/10.1016/B978-0-12-394807-6.00128-3

Yamamoto, C., Neoh, T. L., Honbou, H., Furuta, T., Kimura, S., & Yoshii, H. (2012). formation of a polymer-coated inclusion complex of D-limonene and β-cyclodextrin by spray drying. *Drying Technology, 30*(15), 1714–1719. https://doi.org/10.1080/07373937.2012.699487

Yang, Q.-Q., Sui, Z., Lu, W., & Corke, H. (2021). Soybean lecithin-stabilized oil-in-water (O/W) emulsions increase the stability and in vitro bioaccessibility of bioactive nutrients. *Food Chemistry, 338*, 128071. https://doi.org/10.1016/j.foodchem.2020.128071

Yu, H., & Huang, Q. (2012). Improving the oral bioavailability of curcumin using novel organogel-based nanoemulsions. *Journal of Agricultural and Food Chemistry, 60*(21), 5373–5379. https://doi.org/10.1021/jf300609p

Zaibunnisa, A. H., Norashikin, S., Mamot, S., & Osman, H. (2009). An experimental design approach for the extraction of volatile compounds from turmeric leaves (Curcuma domestica) using pressurised liquid extraction (PLE). *LWT - Food Science and Technology, 42*(1), 233–238. https://doi.org/10.1016/j.lwt.2008.03.015

Zhang, X., Gao, H., Zhang, L., Liu, D., & Ye, X. (2012). Extraction of essential oil from discarded tobacco leaves by solvent extraction and steam distillation, and identification of its chemical composition. *Industrial Crops and Products, 39*, 162–169. https://doi.org/10.1016/j.indcrop.2012.02.029

Zhang, J., Wu, J., Yu, J., Zhang, X., He, J., & Zhang, J. (2017). Application of ionic liquids for dissolving cellulose and fabricating cellulose-based materials: State of the art and future trends. *Materials Chemistry Frontiers, 1*(7), 1273–1290. https://doi.org/10.1039/C6QM00348F

Zhang, Z., Zhang, R., Chen, L., Tong, Q., & McClements, D. J. (2015). Designing hydrogel particles for controlled or targeted release of lipophilic bioactive agents in the gastrointestinal tract. *European Polymer Journal, 72*, 698–716. https://doi.org/10.1016/j.eurpolymj.2015.01.013

Zou, J., Liu, W., Liu, C., Xiao, H., & Julian McClements, D. (2015). Designing excipient emulsions to increase nutraceutical bioavailability: Emulsifier type influences curcumin stability and bioaccessibility by altering gastrointestinal fate. *Food & Function, 6*(8), 2475–2486. https://doi.org/10.1039/C5FO00606F.

Chapter 15

Food processing interventions to improve the bioaccessibility and bioavailability of plant food nutrients

Gloria López-Gámez, Robert Soliva-Fortuny and Pedro Elez-Martínez

Department of Food Technology, Agrotecnio Center, University of Lleida, Lleida, Spain

15.1 Introduction

Plant-based foods can be considered as functional products due to their high content in bioactive substances (e.g., polyphenols and carotenoids) that exert positive effects on consumers health (Gong et al., 2020). It has been demonstrated that their consumption reduces the risk of suffering some types of cancer and cardiovascular diseases due to their antioxidant, anti-inflammatory, and antitumoral activities (Manach et al., 2004, 2004, 2004; Rao & Rao, 2007). Likewise, some minerals and vitamins cannot be synthetized by the organism in enough amount to achieve physiological requirements. Therefore, their consumption is essential for the correct functioning of the human body (Hashemi et al., 2019). However, in order to exert their positive effects in health, these compounds must be both bioaccessible and bioavailable.

The term bioavailability is defined as the part of ingested nutrient that reaches circulation and is available for utilization in physiological functions, which is evaluated by in vivo methods, using animals or human subjects. After the consumption of food containing the studied compound, the plasma is collected and analyzed (Carbonell-Capella et al., 2014). On the other hand, bioaccessibility is referred as the amount of a released compound from the food matrix during digestion and is available for absorption, which is usually determined throughout in vitro methods (e.g., Caco-2 cells) (Thakur et al., 2020). Upon ingestion, the bioaccessibility and bioavailability of carotenoids and polyphenols depend on several factors (Bohn, 2008, 2014). Carotenoids are lipophilic compounds that need to be released from the food matrix, solubilized, and incorporated into micelles for absorption (Sy et al., 2012). In addition, uptake into mucosal cells, packing into chylomicrons, and transport within the lymphatic system to reach target organs are necessary steps to be bioavailable. Cell walls and chromoplasts are the main barriers limiting carotenoids release from matrix (Schweiggert et al., 2012), whereas micellarization efficiency is affected by the type of carotenoid and the presence of lipids and fibers (Cervantes-Paz et al., 2016; Granado-Lorencio, Herrero-Barbudo, Blanco-Navarro et al., 2009). On the other hand, polyphenols are generally stored in vacuoles or associated with dietary fiber. Likewise, they must be released from the food matrix to be available for absorption, although incorporation to micelles is not required. Their bioaccessibility and bioavailability also depend on their polarity and interactions with cell wall polysaccharides, proteins, or starch (Ribas-Agustí et al., 2018). Furthermore, those with low molecular weight are easily absorbed (e.g., catechins, isoflavones) than those that need to be hydrolized by intestinal enzymes or microflora (e.g., glycosilated phenols) (Carbonell-Capella et al., 2014). Regarding water-soluble vitamins (vitamin C, B1, B6), these are sensitive to pH and temperature of the gastrointestinal tract. Some of them are bound to polysaccharides and need to be released before absorption (Khaneghah et al., 2019). Additionally, fat-soluble vitamins (vitamins A, E, K) absorption depends on their chemical form and physical state, as well as the interaction with other macromolecules, being the presence of food lipids beneficial to their absorption (Hashemi et al., 2019). Regarding minerals bioaccessibility, it can be negatively affected by the presence of antinutritional substances such as phytates, tannins, or fibers, which hinder their absorption (Raes et al., 2014).

Food processing may affect both the abundance of such compounds and the environment and microstructure of the food matrix, consequently leading to changes in their bioaccessibility (Barba et al., 2017). This chapter compiles updated information about the impact of different processing and preservation technologies on the bioaccessibility/bioavailability of

Engineering Plant-Based Food Systems. https://doi.org/10.1016/B978-0-323-89842-3.00014-2

bioactive compounds, vitamins, and minerals (Tables 15.1 and 15.2, Figs. 15.1 and 15.2). Delving into the relationship between food processing and bioaccessibility/bioavailability of such compounds is important to food scientists and food industry in order to develop strategies to obtain nutritious, attractive, and healthier plant-based products while meeting the consumers' demands.

TABLE 15.1 Effect of food processing on carotenoid and phenolic bioaccessibility of selected vegetal matrices.

Food matrix	Processing conditions	Bioaccessibility increase	Bioaccessibility decrease	References
Olive pomace	Micronization by ball milling 300 r/min for 5 h	Hydroxytyrosol, 3,4-DHPEA-EDA, oleuropein, luteolin, and apigenin	No decreases	(Sefrin Speroni et al., 2021)
Orange and orange juice	Blending, juicing and pasteurization (70°C at 80 L/h or 90°C at 60 L/h)	Flavonoids (juicing) and carotenoids (juicing and pasteurization)	No decreases	(Aschoff, Kaufmann, et al., 2015)
Carrot and tomato suspensions	Blending	Lycopene and β-carotene	No decreases	(Moelants et al., 2012)
Orange	Juicing	β-cryptoxanthin and total carotenoids	No decreases	(Aschoff, Rolke, et al., 2015)
Mango	Juicing	Chlorogenic acid and ferulic acid	No decreases	(Quirós-Sauceda et al., 2017)
Cashew apple fiber	Juicing	Total phenolic compounds	No decreases	(Silva de Lima et al., 2014)
Orange and mandarin	Juicing	No changes	No changes	(de Ancos et al., 2016)
Orange and clementine mandarin	Juicing	Orange juice: Phytoene, phytofluene, lycopene, β-carotene and β-cryptoxanthin	Mandarin juice: Phytoene, β-carotene and β-cryptoxanthin	(Rodrigo et al., 2015)
Carrot pulp and corn grit	Extrusion (80–160°C; 125, 175, 225 rpm)	β-carotene and lutein	No decreases	(Ortak et al., 2017)
Strawberry	Freezing (−20°C for a week)	Quercetin Per-3-Glu	*p*-coumaric and ferulic acids	(Balasooriya et al., 2020)
Orange juice	Ultrafreezing (−80°C no more than 3 days)	Phytoene, phytofluene, ζ-carotene, 9-cis-violaxanthin, zeaxanthin, 9-cis-anteraxanthin, β-carotene and total carotenoids	No decreases	(Mapelli-Brahm et al., 2018)
Pumpkin flour	Hot-air oven drying at 60°C for 24 h	Total phenolic compounds	No decreases	(Aydin & Gocmen, 2015)
Persimmon	Infrared drying (88 W), ultrasound-assisted vacuum drying (55°C), hot-air drying (55°C)	Total phenolic compounds	No decreases	(Kayacan et al., 2020)
Berry	Hot air drying (70°C for 12 h), microwave drying (136 W for 4 h), combined microwave-hot-air-drying (136 W for 2.5 h in microwave	Total phenolic compounds	Total phenolic compounds (combined microwave-hot-air-drying)	(Zhao et al., 2017)

TABLE 15.1 Effect of food processing on carotenoid and phenolic bioaccessibility of selected vegetal matrices.—cont'd

Food matrix	Processing conditions	Bioaccessibility increase	Bioaccessibility decrease	References
	and 70°C for 1 h in hot-air-drying)			
Tomato	Sun-dried (27–28°C for 4–7 days)	Lycopene	No decreases	(Karakaya & Yilmaz, 2007)
Carrot, sweet potato, yellow bell pepper and broccoli	Oven drying (60, 70, 80°C)	Carrot and yellow bell pepper: lutein, α-carotene and β-carotene	Sweet potato: α-carotene β-carotene and broccoli: lutein, β-carotene	(Zhang et al., 2018)
Carrot	PEF (5 pulses of 3.5 kV/cm)	Total carotenoids, α-carotene β-carotene, total phenolic compounds, caffeoylshikimic acid, caffeoylferuloylquinic acid, isoferulic acid, ferulic acid glucoside, ferulic acid caffeoyl glucoside and quercetin-3-O-galactoside	3-, 4-, 5-caffeoylquinic acid, caffeic acid arabinoside glucoside, caffeic acid Glu acetyl glucoside, ferulic acid and feruloylquinic acid derivative	(López-Gámez et al., 2021)
Tomato puree (5% olive oil)	PEF (0.02–2.31 kJ/kg) applied to whole tomato	Total carotenoids (0.38 kJ/kg), β-carotene (0.38 kJ/kg), lycopene (0.09–0.38 kJ/kg; 1.38–2.31 kJ/kg), lutein (0.09–0.38 kJ/kg) and γ-carotene (0.09–0.38; 0.83–2.31 kJ/kg)	Total carotenoids (0.02 and 0.5 kJ/kg), β-carotene (0.02 and 0.06 kJ/kg), lycopene (0.02 and 0.06 kJ/kg), lutein (0.06 and 0.5 kJ/kg), phytofluene (0.02, 0.06, 0.14, 0.5, 0.83, 1.38 and 2.31 kJ/kg), phytoene and δ-carotene (0.02–0.06 kJ/kg; 0.5–2.31 kJ/kg)	(González-Casado et al., 2018)
Mandarin juices	HPH (150 MPa reaching 68°C for 15 s)	Total and individual carotenoids	No decreases	(Sentandreu et al., 2020)
Tomato juice	HPH (200, 300, 400, and 500 bar) (2 cycles 15 min)	All-*trans*-lycopene and isomers (500 bar)	All-*trans*-lycopene and total lycopene (200 bar)	(Zhang et al., 2019)
Tomato juice	US (20 min; 50 μm; 25 Hz; 200 W, 400 W, 600, and 800 W)	All-*trans*-lycopene and isomers (800 W)	All carotenoids (400 W)	(Zhang et al., 2019)
Commercial pasteurised tomato pulp (with or without added sunflower oil)	US (30 min; 24 kHz; 100 μm; 71 W; 1462 J/cm^3)	Lycopene (in US-treated samples with 5% of oil)	No decreases in lycopene (in US-treated samples with 0%, 2.5% and 10% of oil)	(Anese et al., 2015)
Mango by-products (peel and paste)	US (30 min; 30% of amplitude; 9 W/mL)	Total carotenoids, β-cryptoxanthin, lutein and β-carotene (in peel and paste)	No decreases	(Mercado-Mercado et al., 2015)
Astringent persimmon	HHP (200 MPa at 25°C for 6 min)	Total and individual carotenoids	No decreases	(Cano et al., 2019)
Mango and papaya juice sweetened with *Stevia rebaudiana*	PEF (32 and 256 kJ/kg) and US (32 and 256 kJ/kg)	Total carotenoids (PEF (32 and 256 kJ/kg) and US (32 kJ/kg))	No decreases	(Buniowska et al., 2017)
Buriti juice	US (0, 0.9, 1.8, 2.7, and 3.6 kJ/cm^3)	β -carotene after all treatments	No decreases	(de Souza Carvalho et al., 2020)

Continued

TABLE 15.1 Effect of food processing on carotenoid and phenolic bioaccessibility of selected vegetal matrices.—cont'd

Food matrix	Processing conditions	Bioaccessibility increase	Bioaccessibility decrease	References
Tomato juice and kale-based juice	PEF (35 kV/cm for 1000 μs) and HPP (500 MPa for 3 min)	Lycopene (PÈF-treated tomato juice)	β-carotene (PÈF-treated tomato juice)	(Zhong et al., 2019)
Orange juice	HPP (20 or 150 MPa)	Lutein, zeaxanthin, zeinoxanthin, β-cryptoxanthin, α-carotene, β -carotene, phytoene, phytofluene, violaxanthin, antheraxanthin, luteoxanthin and mutatoxanthin	No decreases	(Stinco et al., 2020)
Calçots	US (40 kHz; 250 W at 0, 10, 25, 45 min)	No increases	Total phenolic compounds	(Zudaire et al., 2019)
Tomato, lettuce, green pepper, red pepper and zucchini	US (4°C, 40 kHz, and 250 W for 20 min)	Total phenolic compounds in green pepper and lettuce	Total phenolic compounds in tomato, red pepper and zucchini	(Lafarga et al., 2019)
Juçara based smoothie	US (10°C, 7 min at 220 W)	Total anthocyanins	No decreases	(de Oliveira Ribeiro et al., 2019)
Fruit juice-based beverage mixed with soymilk	Thermal treatment (90°C for 1 min), PEF (35 kV/cm electric field strength in bipolar mode, 4 μs pulse width, 200 Hz pulse frequency and 1800 μs total treatment time) and HPP (400 MPa for 5 min)	Thermal treatment: Total isoflavones, daidzin, genistin, daidzein and genistein; PEF: Total isoflavones, daidzin, genistin, daidzein and genistein; HPP: Total isoflavones, daidzin, genistin, daidzein, genistein and glycitein	No decreases	(Rodríguez-Roque et al., 2020)
Pomelo and kiwi juices	Thermal treatment (80°C for 30 min or 90°C for 30 s) and HPH (250 MPa)	Kiwi juice thermally-treated: Caffeoyl-glucoside (80°C)	Pomelo juice thermally-treated: Total phenolic compounds (80°C); Pomelo juice HPP-treated: Naringenin-rutinoside, isorhamnetin-rutinoside, feruloyl-glucoside and total phenolic compounds; Kiwi juice HPP-treated: Quinic acid	(Quan et al., 2020)
Apple	PEF and storage for 0 and 24 h: 0.01 kJ/kg	No increases	0 h: 4-caffeoylquinic acid	(Ribas-Agustí et al., 2019)
	1.8 kJ/kg	0 and 24 h: Phloretin xyloglucoside; 24 h: Total phenolic compounds, epicatechin, 5-caffeoylquinic acid and phloretin glycoside		
	7.3 kJ/kg	24 h: Total phenolic content		
Açai juice	US (0, 0.9, 1.8, 2.7 and 3.6 kJ/cm^3)	Total anthocyanin content	No decreases	(de Souza Carvalho et al., 2020)
Orange juice	HPP (20 or 150 MPa)	No increases	No decreases	(Stinco et al., 2020)
Mandarin juices	HPH (150 MPa reaching 68°C for 15 s)	Apigenin	No decreases	(Sentandreu et al., 2020)

HPH, High pressure homogenization; *HPP*, High pressure processing; *PEF*, Pulsed electric fields; *US*, Ultrasounds

TABLE 15.2 Effect of processing on the bioaccessibility of micronutrients in different vegetal matrices.

Food matrix	Processing conditions	Bioaccessibility increase	Bioaccessibility decrease	References
Orange and orange juice	Blending	No increases	Vitamin C	(Aschoff, Kaufmann, et al., 2015)
	Juicing	Vitamin C	No decreases	
	Pasteurization (70°C at 80 L/h or 90°C at 60 L/h)	Vitamin C	No decreases	
Cashew apple fiber	Juicing	Copper, iron and vitamin C	No decreases	(Silva de Lima et al., 2014)
Pomegranate	Juicing	No increases	Vitamin C	(Pérez-Vicente et al., 2002)
Strawberry	Juicing	No increases	Vitamin C	(Cassani et al., 2018)
Orange and mandarin	Juicing	No increases	Vitamin C	(de Ancos et al., 2016)
Strawberry	Freezing (−20°C for a week)	No increases	Vitamin C	(Balasooriya et al., 2020)
Peas and kidney beans	Extrusion (100 rpm; 150–155°C)	Calcium, magnesium, zinc, copper and phosphorous	No decreases	(Alonso et al., 2001)
Cowpea, mangalo and pigeon pea	Drum drying, microwave cooking (15 psi for 6 and 12 min) and oven cooking (15 psi for 20 and 40 min)	No increases	Zinc	(Carvalho Santos et al., 2017)
Kheer mix	Oven drying	Zinc and iron	No decreases	(Latha Panda et al., 2019)
Drumstick leaves	Oven drying	α-tocopherol	No decreases	(Sriwichai et al., 2017)
Chickpeas and cowpeas	Microwave heating (360 W for 30 min)	Iron	Zinc	(Hemalatha, Platel and Krishnapura, 2007)
Cowpea	Cooking (30 and 10 min at 100°C) and pressure cooking (4 and 13 min at room temperature)	Iron	No decreases	(Pereira et al., 2016)
Rice, spinach and green gram	Pressure cooking (15 psi for 10 min) and microwave cooking (360 W for 20 min)	No increases	Selenium	(Khanam & Platel, 2016)
Potato	Boiling (10–15 min)	No increases	Vitamin C	(Anese et al., 2015)
Apple	HPP (500 MPa for 2, 4, 8 and 10 min)	Iron (8 min) and zinc (8, 10 min)	Calcium (all treatments), iron (4 min) and zinc (2, 4 min)	(Briones-Labarca, Muñoz, & Maureira, 2011)
Brown rice	HPP (100, 300, 500 MPa for 10 min)	No increases	Iron	(Xia et al., 2016)

Continued

TABLE 15.2 Effect of processing on the bioaccessibility of micronutrients in different vegetal matrices.—cont'd

Food matrix	Processing conditions	Bioaccessibility increase	Bioaccessibility decrease	References
Fruit juice milk-based beverage	HPP (400 MPa for 5 min) and asteurization (90°C for 30 s)	Vitamin C (pasteurization)	α-tocopherol, γ-tocopherol, δ-tocopherol (HPP + soymilk) and itamin C (HPP + whole milk or soymilk)	(Cilla et al., 2012)
Gazpacho (vegetable soup)	PEF (35 kV/cm for 750 μs)	No increases	No decreases	(Sánchez-Moreno et al., 2005)
Orange juice	PEF (35 kV/cm for 750 μs)	No increases	No decreases	(Sánchez-Moreno et al., 2004)
Fruit juice-based beverage mixed with soymilk	PEF (35 kV/cm electric field strength in bipolar mode, 4 μs pulse width, 200 Hz pulse frequency and 1800 μs total treatment time), HPP (400 MPa for 5 min) and pasteurization (90°C for 60 s)	Vitamin C (HPP + milk)	Vitamin C (pasteurization)	(Rodríguez-Roque et al., 2015)
Cashew apple bagasse puree	US (500 W for 2, 6, 10 min)	Vitamin C	No decreases	(Fonteles et al., 2016)

HPP, High pressure processing; *PEF*, Pulsed electric fields; *US*, Ultrasounds.

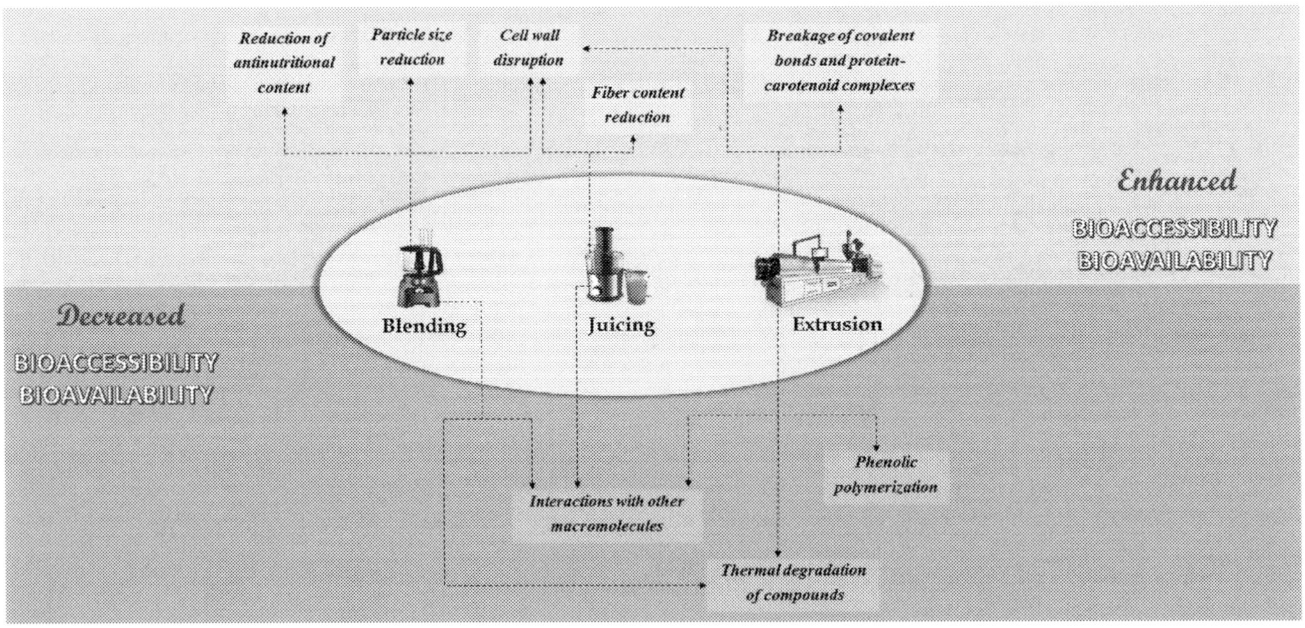

FIGURE 15.1 Compilation of the effects of food processing technologies on the bioavailability/bioavailability of bioactive compounds and micronutrients. Food processing (blending, juicing, and extrusion) lead to increase or decrease bioaccessibility/bioavailability of bioactive compounds and micronutrients throughout changes in their content and molecular structure or modifications in food matrix structure.

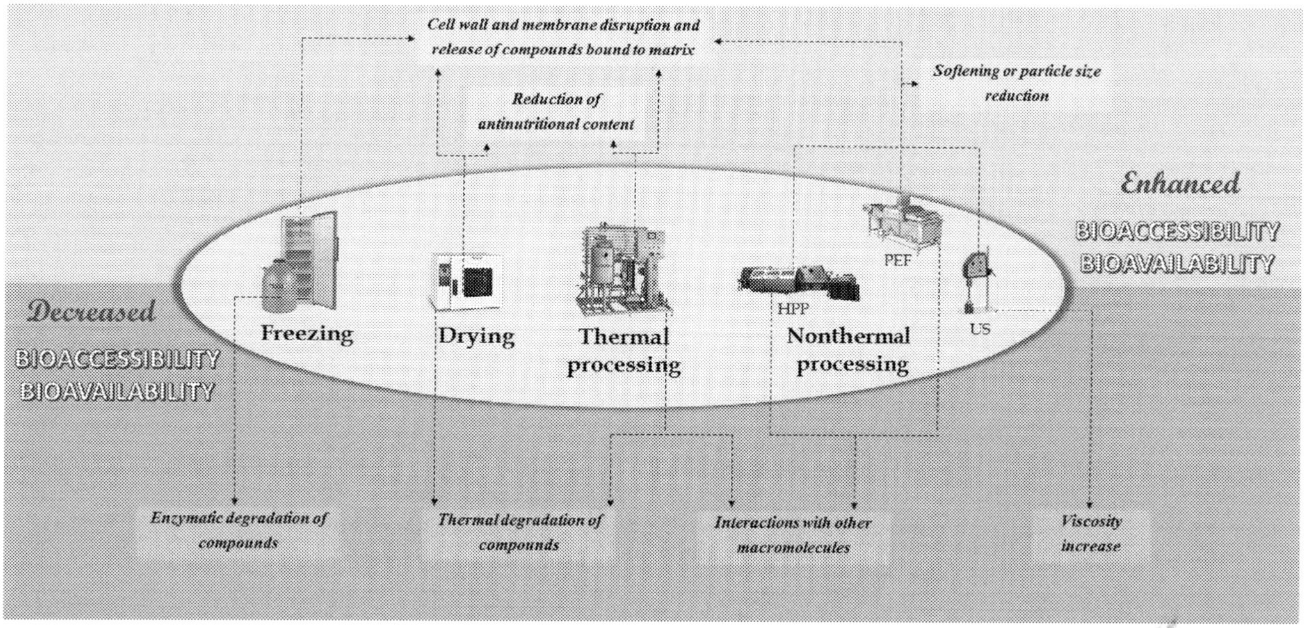

FIGURE 15.2 Compilation of the effects of food preservation technologies on the bioavailability/bioavailability of bioactive compounds and micronutrients. Food preservation technologies (freezing, drying, thermal processing, or nonthermal processing) lead to increase or decrease bioaccessibility/bioavailability of bioactive compounds and micronutrients throughout changes in their content and molecular structure or modifications in food matrix structure. *HPP*, high pressure processing; *PEF*, pulsed electric fields; *US*, ultrasounds.

15.2 Effects of food processing on bioaccessibility and bioavailability of bioactive compounds

15.2.1 Blending

Limited studies comparing bioaccessibility/bioavailability of bioactive compounds, minerals, and vitamins after different grades of mechanical processing have been done. Mechanical processing at high speed can lead to warm up the sample, causing bioactive compounds degradation. However, blending at low speed and other unit operations can also reduce particle size of solid plant-based foods (Ribas-Agustí et al., 2018), which facilitate the extractability of such compounds and probably favor their absorption.

15.2.1.1 Phenolic compounds

Recently, some studies performed with fruit pomaces have shown that particle size reduction increased the exposure and solubilization of phenolic compounds, which enhanced their bioaccessibility (Sefrin Speroni et al., 2021). However Aschoff, Kaufmann et al. (2015) reported that flavonoids bioaccessibility was similar between orange segments and homogenates. Obtained results indicated that homogenization was not enough to release those compounds tightly bound to membranes and albedo. Additionally, the presence of high amounts of dietary fibers can entrap polyphenols and limit their bioaccessibility/bioavailability (Carrillo et al., 2017; Quirós-Sauceda et al., 2014; Velderrain-Roddríguez et al., 2016); therefore, homogenization is not always appropriate, it depends on the matrix, processing parameters, and phenolic chemical structure.

15.2.1.2 Carotenoids

It has been extensively demonstrated that cell walls and chromoplasts are the main barriers for carotenoids absorption (Carrillo et al., 2017; Lemmens et al., 2010; Palmero et al., 2013). Generally, particle size reduction promotes their bioaccessibility since greater surface area will be available for digestive enzymes, leading to better digestion efficiency

(Lemmens et al., 2010). Nevertheless, the impact of mechanical processing on carotenoids absorption depends on the food matrix and chemical structure of studied carotenoids (Van Buggenhout et al., 2010). For instance Aschoff, Kaufmann, et al. (2015) reported that the bioaccessibility of carotenoids from orange was similar regardless mechanical processing degree (segments or blended). On the other hand, Moelants et al., (2012) found a correlation between a lower particle size and higher β-carotene and lycopene bioaccessibility in carrot- and tomato-derived suspensions, respectively. Furthermore, some authors have reported that high amounts of pectin can interfere with micelle formation by partitioning bile salts and fat in the pectin gel phase (Palafox-Carlos et al., 2011; Parada & Aguilera, 2007), which is necessary for their absorption.

Carotenoids must be released from food matrix during digestion and incorporated into bile salt mixed micelles in order to be absorbed. Therefore, the addition of milk to fruit beverages has been reported to increase bioaccessibility due to the solubilization of carotenoids into the fat content through the formation of lipid micelles (Granado-Lorencio, Herrero-Barbudo, Acién-Fernández et al., 2009; Rodríguez-Roque et al., 2015). Generally, oil improves micellarization, especially those with high unsaturation degree and long-chain fatty acids (e.g., olive oil). However, the added quantities must be optimized to such purpose since carotenoids may be entrapped in a lipid phase when lipid load is too high (Huo et al., 2007). Furthermore, it has been demonstrated that sugar addition can also improve bioaccessibility in both water- and milk-based beverages by affecting viscosity or micelle formation (da Costa & Mercadante, 2018).

15.2.1.3 Minerals

Antinutritional factors negatively affect bioaccessibility of mineral compounds. Therefore, reducing such factors by mechanical processing technologies will improve mineral bioaccessibility. Even though mineral content is not affected by this type of processing, cell wall degradation leads to changes in the location of minerals, antinutritional factors (e.g., phytates) and enzymes, which will affect their interactions, and probably will enhance mineral bioaccessibility (Raes et al., 2014). Other preparation techniques such as soaking can exert a positive effect on mineral bioaccessibility of pulses. Especially, higher bioaccessibility values have been reported in iron and zinc (Debon & Tester, 2001). This effect has been attributed to the activation of cell wall degrading enzymes, which degrade pectin and allow a better release of minerals from matrix (Martínez-Manrique et al., 2011). However, soaking impact on bioaccessibility would depend on temperature, time, and pH conditions. In this line, antinutritional compounds can also be leached out, bound to minerals, and reduce their bioaccessibility (Raes et al., 2014).

Furthermore, cereals addition to fruit juices can limit the absorption of compounds due to the formation of insoluble adducts with fiber (Jovani et al., 2000). On the other hand, lactose from milk-based beverages can enhance calcium bioavailability since it favors diffusion and passive transport throughout intestinal lumen. In addition, malic acid, organic acids, or fructose present in fruit juices favor the formation of soluble complexes with calcium, thus improving their bioaccessibility (Heaney et al., 1990; Perales et al., 2005).

15.2.1.4 Vitamins

Limited studies about the effects of mechanical processing on vitamins bioaccessibility/bioavailability are available. However, blending generally involves reductions in vitamin contents, especially vitamin C, due to oxygen and light exposure (Yaman et al., 2021). For instance, Aschoff, Kaufmann, et al. (2015) reported that vitamin C release during digestion of orange homogenates was significantly lower than in orange segments, which was attributed to oxygen exposure during blending.

Some vitamins need to be released from the food matrix before absorption (e.g., group B vitamins) since they are bound to polysaccharides or polypeptides (Yaman et al., 2021). Therefore, blending is likely to enhance their bioavailability. However, a reduction in particle size can negatively affect their absorption because more hydroxyl groups from fibers are exposed to interaction (Yaman et al., 2021). On the other hand, fat-soluble vitamins (e.g., vitamin D) need to be incorporated to micelles to be absorbed. The presence of phytates and tannins, which are highly present in pulses, inhibit pancreatic lipase activity and avoid micelles formation (Antoine et al., 2021). Mechanical processing could improve their absorption by decreasing the content of such antinutritional factors (Raes et al., 2014).

The addition of milk to fruit juices can decrease vitamin C bioaccessibility due to its interaction with proteins or minerals, which cause the formation of insoluble complexes (Claeys et al., 2013; Rodríguez-Roque et al., 2014). Conversely, the presence of phenolic compounds or other antioxidants could protect vitamin C from degradation during digestion (Rodríguez-Roque et al., 2015).

15.2.2 Juicing

Juicing of fruit and vegetables is used to obtain soluble compounds in a liquid medium separated from the solid fraction (Ribas-Agustí et al., 2018). Juices can be obtained by squeezing by hand in case of citric fruit or by different electric juicers devices when other fruit or vegetables are used. Centrifugal juicers reach high speed, which causes warm up and degradation of vitamins and bioactive compounds. On the other hand, masticating or triturating juicers apply slow speed, which allow to better preserve nutrients and bioactive compounds. Regarding compounds absorption, it can be differently affected by juicing, since some compounds would be better released from matrix, whereas others will remain entrapped in the insoluble fraction.

15.2.2.1 Phenolic compounds

In whole food matrices, phenolic compounds are usually bound to carbohydrates forming glycones. During digestion, phenolic compounds are detached from the sugar resulting in a more absorbable compound (Etcheverry et al., 2012). However, after juicing some compounds will remain in the insoluble fraction and therefore will not be present in the final product. Other compounds will be better extracted in their free form, which would affect their stability during digestion. Some phenolic compounds are sensitive to digestive pH conditions. Likewise, the absence of cell barriers makes them more available to be easily dialyzed or to interact with other minerals, dietary fiber or proteins, decreasing their bioaccessibility (Rodríguez-Roque et al., 2014). Phenolic compounds from fruit juices are expected to be more bioaccessible and bioavailable than those from fruit flesh because of their differences in fiber content (Quirós-Sauceda et al., 2014). Aschoff, Kaufmann, et al., (2015) found that flavanone bioaccessibility in orange juice increased by 450% compared to the orange fibrous matrix. Furthermore, phenolic bioaccessibility from cashew apple juice was higher (39%) than that obtained in cashew apple fiber (18.6%) (De Lima et al., 2014). Likewise, Quirós-Sauceda et al., (2017) determined that phenolic compounds from mango juices were more bioavailable than those from mango flesh. On the other hand, De Ancos et al. (2017) reported that phenolic compounds bioaccessibility was similar between pulps and juices from oranges and mandarins. Therefore, the impact of juicing on phenolic compounds bioaccessibility must be deeply investigated to understand the differences between matrices. Nevertheless, many polyphenols may not be dialysable, but they could be fermented in colon and exert their positive effects on health (Bohn et al., 2015).

15.2.2.2 Carotenoids

Pectin being one of the most important dietary fibers, not only plays an important role in maintaining cloudiness of fruit juices but also seems to impact the release of bioactive components. Recent studies draw functional relationships between total carotenoid bioaccessibility and water-soluble pectin characteristics in juice systems (Liu et al., 2019; Wellala et al., 2019). The amount of dietary fiber (e.g., pectin) can strongly affect micellarization of carotenoids, thus reducing their bioaccessibility (Rock & Swendseid, 1992). Fiber content in juices is usually lower than in whole fruit as has been reported by Aschoff, Kaufmann, et al. (2015), who determined that carotenoids bioaccessibility from orange segments and those homogenized was similar (11.9%), whereas their bioaccessibility in juices was highly superior (28.3%−39.5%). Likewise, β-cryptoxanthin was 1.8-fold more bioavailable from orange juice than from fresh oranges (Aschoff, Kaufmann, et al., 2015). However, the effect of mechanical or juicing processing in bioaccessibility will depend on the food matrix and carotenoids deposition. As an example, carotenoids of *Navel Cara* orange juice were more bioaccessible than those found in whole oranges, while the opposite was observed for *Clementine* mandarins (Rodrigo et al., 2015; Schweiggert et al., 2012).

15.2.2.3 Minerals

The impact of juicing on minerals bioaccessibility has been scarcely studied. However, it is known that the presence of oxalates, phytates, or tannins can negatively affect bioaccessibility and bioavailability of minerals (e.g., calcium) (Perales et al., 2005), although it depends on the type of matrix and mineral itself. For instance, De Lima et al., (2014) studied mineral bioaccessibility in cashew apple fiber and juice. The authors reported that despite cashew apple juice has a lower copper content than cashew apple fiber bagasse; the copper bioaccessibility in juice was almost four times higher than that observed in the bagasse. Furthermore, iron bioaccessibility was also higher in juice (11%) than in bagasse (1.2%). On the other hand, zinc bioaccessibility was similarly absorbed in both matrices. Lower bioaccessibility values in cashew apple fiber could be related to phytic or oxalic acid content found in seeds, peels, and insoluble fibers, which is capable of chelating minerals or due to formation of insoluble complexes with present phenolic acids that avoid absorption (Khouzam et al., 2011). Additionally, the presence of different minerals such as iron could affect the bioaccessibility of zinc.

15.2.2.4 Vitamins

The impact of juicing on the bioaccessibility of vitamins has not been extensively studied, although some publications about vitamin C bioaccessibility are available. Vitamin C is negatively influenced by oxygen, temperature, light, and pH, among other factors (Eitenmiller et al., 2008). Pérez-Vicente et al. (2002) and Cassani et al. (2018) reported a 65%–95% reduction in the absorbed content of vitamin C in pomegranate juice and strawberry juice, which was related to its instability under gastrointestinal conditions. In addition, De Ancos et al. (2017) reported that vitamin C was 22% more bioaccessible in orange pulp than in their corresponding juices, which could be related to their oxidation or interactions with metal ions during digestion (Rodríguez-Roque et al., 2013).

On the other hand, bioaccessibility of ascorbic acid in cashew apple juice was 26%, whereas in cashew apple fiber bagasse, it was not bioaccessible, which was attributed to the low level of tannins and phytates found in fruit juices, which hindered vitamins absorption (De Lima et al., 2014). Likewise, Aschoff, Kaufmann, et al., (2015) found that vitamin C bioaccessibility in a fresh orange juice was greater than that in an orange homogenate or in orange segments.

15.2.3 Extrusion

Extrusion combines high pressure with a moderately high temperature, to result in shearing forces for a short period of time. Its application can cause different effects on compounds absorption depending on matrix and compound, since their chemical alteration produced by thermal treatment could decrease their assimilation (Alonso et al., 2001).

15.2.3.1 Phenolic compounds

The application of high thermal and pressure conditions could lead to degrade phenolic compounds (Shahidi & Yeo, 2016), as has been demonstrated by Ortak et al., (2017). During extrusion, temperatures over 80°C can damage the structure of phenolic compounds or cause their polymerization, which leads to a decrease in their extractability (Brennan et al., 2011) and most likely in their bioaccessibility. However, it can also facilitate the conversion of bound phenolics into free forms throughout the breakage of covalent bonds (Brennan et al., 2011). Therefore, processing can favor their release from matrix during digestion and absorption, given that degradation of cell wall structure enhances the accessibility of digestive juices (Chen et al., 2019).

15.2.3.2 Carotenoids

Studies about the effect of extrusion on carotenoids bioaccessibility are limited. High pressure and temperature conditions can lead to degrade carotenoids, although this type of processing can also enhance their bioaccessibility. For instance, Ortak et al. (2017) reported that lutein and β-carotene bioaccessibilities were enhanced in an extruded carrot pulp and corn grit composite. Such results could be related to cell wall softening and particle size reduction, which favors the accessibility of digestive enzymes. Likewise, high temperatures could also denature proteins and facilitate the breakage of carotenoid–protein complexes (Faulks & Southon, 2005; Hedrén et al., 2002).

15.2.3.3 Minerals

Extrusion can affect mineral oxidation state and their interactions with other components such as phytic acid or dietary fibers, which influence their bioavailability (Gulati et al., 2020). Both phenolic compounds and phytic acid can bind minerals and reduce their bioavailability. However, extrusion causes dephosphorylation of phytic acid or degradation of polyphenols, which can improve minerals bioaccessibility (Alonso et al., 2001; Sandberg et al., 1987). For instance, it has been reported that calcium, magnesium, zinc, copper, and phosphorous were better absorbed in rats fed with extruded peas or kidney beans (Alonso et al., 2001). Authors attributed this enhancement to the reduction of tannins, phytates, and lectins or even the degradation of polyphenols caused by the application of heat.

15.2.3.4 Vitamins

Scarce information is available about the influence of extrusion on bioaccessibility of vitamins. Processing factors (e.g., temperature, residence time, pressure, among others), food matrix, vitamin chemical structure, or interactions affect stability of vitamins (Gulati et al., 2020). For instance, thiamine and pyridoxine are more sensitive to heat than riboflavin (Brennan et al., 2011). Vitamin C is also affected by temperature, but acidification can avoid its degradation (Sriburi & Hill, 2000). It can be speculated that bioaccessibility will be increased when the structural changes allow its release from the matrix and vitamins are not degraded or entrapped by other macromolecules.

15.3 Effects of food preservation on the bioaccessibility and bioavailability of bioactive compounds

15.3.1 Freezing

Freezing is a complex process that involves a reduction of the food product temperature and the consequent crystallization of water, the main component of plant tissues. This causes a significant reduction of the chemical reactions and microbial growth. In general, nutrients content is not affected, although limited information has been reported about its impact on their bioaccessibility/bioavailability.

15.3.1.1 Phenolic compounds

Generally, cryogenic freezing (e.g., with liquid nitrogen) allows to rapidly reaching low temperatures while maintaining quality and bioactive content, since ice crystals are smaller and more uniform than after slow freezing techniques (Stinco et al., 2013). However, the formation of ice crystals can also damage food matrix, rupturing cell walls and promoting the degradation of some bioactive compounds (Balasooriya et al., 2020; Davey et al., 2000), which can also affect their bioaccessibility. For instance, *p*-coumaric and ferulic acids decreased their bioaccessibility, whereas that of quercetin and Pel-3-Glu increased in frozen strawberries (Balasooriya et al., 2020). Therefore, such changes are highly dependent on individual compound and type of food matrix (Rickman et al., 2007).

15.3.1.2 Carotenoids

Freezing can cause structural rearrangements and disruption of cell membranes; therefore, it can also increase bioaccessibility through the release of membrane-bound carotenoid compounds (Lemmens et al., 2010; Lemmens, Van Buggenhout et al., 2009; Van Buggenhout et al., 2010). In this line, Leong & Oey (2012) reported higher release of lutein and β-cryptoxanthin in some frozen fruit and vegetables. However, the effect of freezing in juices would depend on the matrix, thawing conditions and freezing process, since different results have been obtained. On the one hand, freezing orange juices does not enhance carotenoids release nor their bioaccessibility, since cell structures have already been disrupted during juicing instead of during freezing as in raw fruit. Additionally, rheological properties can be altered by freezing, which would affect digestive enzymes activity (Stinco et al., 2013). On the other hand, Mapelli-Brahm et al. (2018) reported increases in carotenoids bioaccessibility after ultra-freezing orange juice, which was related to particle size decrease and the rupture of cell structures by ice crystals.

15.3.1.3 Minerals and vitamins

To the best of our knowledge, the impact of freezing on bioaccessibility of minerals and vitamins is under exploration. Vitamin losses during the freezing process occur mainly at the blanching stage due to thermal degradation or their release into the blanching medium (Davey et al., 2000). However, Balasooriya et al. (2020) reported a decrease in vitamin C content and bioaccessibility in frozen strawberries that had not been submitted to previous blanching. Therefore, its better contact with ascorbic acid oxidase or peroxidase may be responsible of such changes as well as vitamin C instability in the intestinal conditions (Cassani et al., 2018; Rickman et al., 2007).

15.3.2 Drying

Conventional hot air drying is a dehydration method in which water is removed by application of hot air. It is one of the most popular methods used in food preservation due to its low cost and easy application (Diamante et al., 2010). Drying influences food matrix, cellular structures, matrix interactions, and bioactive compounds stability, which subsequently affect their bioaccessibility (Betoret et al., 2015; Zhao et al., 2017). In this line, the effects on bioactive compounds depend on the type of drying process (e.g., hot air drying, infrared drying, sun drying), applied temperature, and food matrix (Zhao et al., 2017). In addition, freeze drying is another method for preserving vegetable food, which consists of removing water by sublimation under high vacuum and freezing temperatures. This avoids degradation of thermolabile compounds, but hot air drying is preferred by food industry because freeze drying is a more expensive process (García-Amezquita et al., 2018).

15.3.2.1 Phenolic compounds

Microstructural changes and cell disruption can promote the release of phenols and favor their contact with polyphenol oxidases (PPO) (Pico et al., 2019), which could lead to their degradation (Oghbaei & Prakash, 2013). However, the application of temperatures higher than 60°C can inactivate PPO (Mizobutsi et al., 2010); therefore, phenolic loss could be lower depending on the applied conditions. Besides the effects on phenolic content, drying process may alter the microstructure or promote the release of phenolic compounds bound to cell wall, thus increasing their bioaccessibility, as has been reported in hot air dried or microwave dried berries, in hot air dried, infrared dried, or freeze dried persimmon and hot air dried or freeze dried pumpkin flour (Aydin & Gocmen, 2015; Kayacan et al., 2020; Zhao et al., 2017).

15.3.2.2 Carotenoids

Drying can favor the release of carotenoids bound to chromoplasts into cytoplasm, hence enhancing their bioaccessibility, as reported for sun dried tomatoes by Karakaya and Yilmaz (2007). Nevertheless, it is known that carotenoids are extremely sensitive to oxygen, temperature, and light. Therefore, these will be more exposed to degradation during processing and digestion (Betoret et al., 2015) as observed in a mixture of hot air-dried green gram-amaranth leaves (Oghbaei & Prakash, 2013). Nevertheless, the impact of drying on carotenoids bioaccessibility is different depending on the food matrix. For instance, the bioaccessibility of carotenoids from hot air-dried carrots and yellow bell peppers was doubled, whereas that of sweet potato or broccoli was not significantly enhanced (Zhang et al., 2018). Structural characteristics and composition of each matrix are probably more prone to improve carotenoids bioaccessibility than drying. The presence of starch granules or a small cell size could limit the effect of drying on the microstructure and bioaccessibility of carotenoids (Zhang et al., 2018).

15.3.2.3 Minerals

Minerals bioaccessibility has not been extensively studied in food matrices submitted to drying processes. However, some studies have reported that this type of processing involves the application of high temperatures during a few seconds, reducing moisture content by 95% and softening the food matrix, releasing minerals from the protein complexes and reducing antinutritional factors (Lombardi-Boccia et al., 1995), which will affect their bioaccessibility. The impact will depend on the food matrix, mineral, and applied processing conditions. For instance, Santos et al. (2018) reported a decreased zinc bioaccessibility in dried pigeon pea, whereas no significant effect was observed in mangalo either cowpea. On the other hand, Panda et al., (2019) found that zinc bioaccessibility increased between 1.2 and 2.5-fold and iron doubled its bioaccessibility in drum dried kheer mixes (Indian traditional dessert). Additionally, a strong correlation was found between the dephytinising efficacy and iron and zinc bioaccessibilities ($R^2 > 0.9$), which confirms that phytate is the main inhibitor of their bioaccessibility (Panda et al., 2019).

15.3.2.4 Vitamins

Scarce information is available about the influence of drying on vitamins bioaccessibility. Heat application can degrade some thermolabile compounds (e.g., vitamin C) (Davey et al., 2000), although temperature/time conditions are generally not severe enough to induce losses. In this line, Sriwichai et al. (2017) reported that bioaccessibility of α-tocopherol was enhanced in sun dried drumstick leaves after drying at 38°C, which was not excessive to cause their degradation before digestion. In addition, drying causes microstructural changes that can lead to improve vitamins bioaccessibility because larger surface areas are exposed to digestive enzymes (Goh et al., 2009).

15.3.3 Thermal processing

Thermal processing is the most common technology applied in food preservation. During conventional heating of whole fruit and vegetables (e.g., blanching, cooking by steaming or boiling), heat is transferred by conduction resulting in thermal gradients, which can cause thermal damage to the product (Lemmens, Tibäck, et al., 2009). Likewise, pasteurization and sterilization are used for preserving liquid products; their application eliminates the presence of pathogens and inactivates enzymes. Furthermore, heating may influence the phytochemical content and their bioaccessibility by affecting the food matrix and microstructure (Barba et al., 2017).

15.3.3.1 Phenolic compounds

In general, phenolic compounds are thermolabile; thus, the content in thermally treated food matrices can be reduced. However, microstructural changes also occur, which favor the release of phenolics bound to polysaccharides (He et al., 2016). This could result in an enhancement of their bioaccessibility/bioavailability or their degradation in the gastrointestinal tract, since they are more available to be oxidized, transformed to other compounds or to form complexes with proteins or fiber (Rodríguez-Roque et al., 2014). In addition, changes in phenolics absorption depends on the food matrix and individual compound characteristics. For instance, Quan et al. (2020) reported decreases in phenolic bioaccessibility in thermally treated pomelo juice, whereas He et al. (2016) observed an increment in that of orange and grapes juices treated by heat.

Regarding isoflavones, their bioaccessibility can be enhanced after thermal treatments (microwave, ohmic heating, and among others) since those proteins to which isoflavones are bound, are better digested after such treatments (Kwok et al., 2002; Rodríguez-Roque et al., 2020).

According to Parada & Aguilera, (2007), bioaccessibility or bioavailability is determined by the balance between those compounds that have been degraded during processing and those that have been better released and absorbed due to structural changes.

15.3.3.2 Carotenoids

Generally, thermal processing increases carotenoids bioaccessibility since high temperatures weaken cell wall and membranes by β-elimination of pectin (Van Buggenhout et al., 2010), which allow an easy access of digestive enzymes to the food matrix, thus leading to carotenoids release. For instance, all carotenoids from pasteurized orange juice were more bioaccessible than those in fresh juices (Aschoff, Kaufmann, et al., 2015). Likewise, β-carotene bioaccessibility of carrots increased after cooking, sterilization, or pasteurization (Hedrén et al., 2002; Knockaert, Pulissery, et al., 2012) as well as lycopene bioaccessibility in pasteurized or sterilized purees (Colle et al., 2010). In addition, carotenoid solubilization in the micellar phase is essential for carotenoid bioaccessibility. Therefore, it can be negatively affected due to carotenoids isomerization or the formation of proteins aggregates that are not available during digestion, hindering their incorporation into micelles (Jayathunge et al., 2017; Palmero et al., 2014). Likewise, it has been demonstrated that the application of thermal treatments favors the transference of carotenoids to oil, although they are also more exposed to be degraded by temperature than inside plant cells (Mutsokoti et al., 2016).

15.3.3.3 Minerals

Most of the minerals are heat stable (Cilla et al., 2019); thus, their content is not usually affected by thermal processing. However, this type of processing can change the mobility and solubility of elements under the conditions of the gastrointestinal tract, which interfere on their bioaccessibility (Santos et al., 2018). The effects of heat treatments on bioaccessibility depend on the temperature-time applied conditions, food matrix, and type of mineral. On the one hand, if degrading enzymes (e.g., pectinases) are inactivated, this will avoid cell wall softening and mineral release during digestion. On the other hand, antinutritional factors could also be reduced, as has been reported in legumes and pulses by Rehman and Shah (2005) and Hemalatha et al. (2007), thus enhancing minerals bioaccessibility.

Most of the studies about the impact of thermal processing on mineral bioaccessibility has been performed on cereals and pulses. For instance, microwave heating decreased zinc bioaccessibility in chickpea and cowpea by 11% and 63%, respectively (Hemalatha et al., 2007). Zinc is usually attached to proteins and nucleic acids forming complexes, being phytic acid the main inhibitor to its availability (Drago, 2017). On the other hand, bioaccessibility of iron significantly increased (Hemalatha et al., 2007). Iron bioaccessibility in mangalo, pigeon pea, and cowpea was also favored after microwave, pressure, or conventional cooking (Pereira et al., 2016; Santos et al., 2018). Differences between bioaccessibilities of both compounds could be related to valence states of such metals (Lombardi-Boccia et al., 1995), their location inside grains or the type of linkages to proteins associated to each compound (Hemalatha et al., 2007). Additionally, ascorbic acid degradation could also be related to iron bioaccessibility reduction, since it is an enhancer of their absorption (Drago, 2017). Polyphenols can also react with divalent ions through carboxyl and hydroxyl groups forming insoluble complexes in intestine and reducing their bioavailability. Therefore, if these compounds are degraded during thermal processing, iron bioavailability can be enhanced (Pereira et al., 2016).

Regarding other minerals such as calcium, it has been reported that gelatinization of starch and protein denaturation caused by thermal treatments can favor the activity of digestive enzymes and improve calcium bioaccessibility in pulses (Santos et al., 2018). Nevertheless, different results have been obtained in fruit-based beverages with added milk (Cilla et al., 2011). The presence of caseinophosphopeptides, which are phosphorylated casein-derived peptides, can bind to divalent cations and maintain them soluble and bioavailable (Miquel et al., 2006). However, heat dephosphorylates casein, which reduces their binding to divalent cations (e.g., calcium) and their bioavailability (Seiquer et al., 2001).

Limited number of studies are available regarding selenium bioavailability in plant-food matrices. Khanam & Platel, (2016) investigated the effect of heat treatments (pressure and microwave cooking) on selenium bioaccessibility in green leafy vegetables, pulses, and cereals. It was decreased by half in rice, spinach, and green gram, which authors related to selenium volatile nature or changes in its chemical structure due to oxidation.

15.3.3.4 Vitamins

Some studies have demonstrated that thermal treatment can promote the release of vitamins from food matrices and enhance their bioaccessibility/bioavailability. However, vitamin C is a thermolabile compound (Hotz & Gibson, 2007), and it is also easily degraded by some oxidative enzymes (e.g., peroxidase) (Martínez-Hernández et al., 2013). Conflicting results have been reported. For instance, Anese et al., (2015) determined that the absorbed vitamin C content from boiled potatoes was reduced and Rodríguez-Roque et al. (2015) reported decreases in vitamin C bioaccessibility in thermally treated fruit-based beverages. This could be related to their exposure to intestinal pH conditions and interactions with metal ions (Pérez-Vicente et al., 2002). On the other hand, Aschoff, Kaufmann, et al., (2015) determined that vitamin C stability improved in pasteurized orange juice. Cilla et al., (2012) observed that pasteurization increased vitamin C bioaccessibility in milk- and soymilk fruit-based beverages. The authors attributed such results to the formation of milk protein aggregates by heat application, which can modify vitamin stability during digestion.

Little information is available on the stability of tocopherols. Granado-Lorencio, Herrero-Barbudo, Blanco-Navarro et al., (2009) observed that in vitro and in vivo results presented high variability and depended on the type of food matrix and digestive conditions. Cilla et al. (2012) reported a decrease in their bioaccessibility (α-tocopherol, δ-tocopherol, and ɣ-tocopherol) in milk fruit-based beverages, which was related to their degradation.

15.3.4 Nonthermal processing

Nonthermal processing such as pulsed electric fields (PEF), high-pressure processing (HPP), or ultrasounds (US) can be applied to solid and liquid food products, and they have an impact on compounds bioaccessibility. PEF consists of the application of short pulses of high voltage to a food product located between two electrodes, involving the electropermeabilization of cell membranes (Toepfl et al., 2005). US is another nonthermal processing technology that involves the generation of microbubbles that grow and collapse, causing cells disruption and formation of microscopic channels in tissues (Wiktor et al., 2018). Finally, HPP is able to inactivate microorganisms by subjecting foods to high hydrostatic pressures of up to 600 MPa for a short period, form a few seconds to a few minutes. HPP application leads to decrease molecular volume, which affects the integrity of cell walls and membranes, promoting the release of intracellular compounds (Barba et al., 2017).

15.3.4.1 Phenolic compounds

Information about the effect of nonthermal technologies on phenolic bioaccessibility is rather limited. Some authors have investigated phenolic bioaccessibility after applying nonthermal technologies to whole fruit and vegetables. Phenolic compounds must be released from the food matrix in order to be absorbed; therefore, PEF application can enhance their bioaccessibility. Thus, Ribas-Agustí et al. (2019) reported that phenolic bioaccessibility of PEF-treated apples submitted to 1.8 and 7.3 kJ/kg was higher than that of untreated apples. López-Gámez et al. (2021) also reported an increase of phenolic bioaccessibility after submitting carrots to PEF (5 pulses of 3.5 kV/cm). Such results were attributed to cell disruption and structural changes caused by electropermeabilization. An inverse correlation between the bioaccessibilities of some individual compounds and apple toughness was found (Ribas-Agustí et al., 2019).

The acoustic cavitation and cell wall disruption provoked by US application can significantly enhance bioaccessibility of phenolic compounds. Lafarga et al. (2019) observed that US (250 W for 20 min; 40 kHz; 4°C) improved phenolic bioaccessibility in lettuce and green pepper but not in other studied vegetables. On the other hand, phenols are more exposed to degradation during the gastrointestinal tract, as reported by Zudaire et al., (2019) in US-treated (40 kHz; 250 W at 10–45 min) calçots, a type of green onion. After 10 min of treatment, total phenolic content in digesta was lower than that of untreated calçots.

In liquid products, particle size reduction has been associated to enhanced phenolic bioaccessibility (Ribas-Agustí et al., 2018). Nevertheless, phenols would also be more exposed to be degraded or transformed than in solid foods (Bohn, 2014). New interactions between compounds and other macromolecules such as polysaccharides (starch, cellulose, and pectin), proteins and lipids (Jakobek, 2015), which negatively affect their bioaccessibility. Ribeiro et al., (2019) found that anthocyanins of US-treated juçara smoothies (220 W for 7 min) were more bioaccessible than those of untreated smoothies, although no differences were found in phenolics bioaccessibility. Despite phenolic release from matrix may be enhanced, high product viscosity most likely impaired their absorption. Likewise, an enhancement of anthocyanins bioaccessibility was reported in US-treated açai juices (3.6 kJ/cm^3) by de Souza Carvalho et al. (2020). In addition, fruit juices treated by PEF (256 kJ/kg) and US (32 and 256 kJ/kg) showed higher bioaccessibility than those untreated, which was attributed to their better release from food matrix (Buniowska et al., 2017). In addition, Rodríguez-Roque et al. (2015) also reported an enhanced phenolic bioaccessibility in PEF- or HPP-treated fruit beverages. Nevertheless, individual compounds were differently affected depending on their chemical structure. Authors proposed that processing caused changes in the phenol structure (hydroxylation, methylation, glycosylation, among others) that enable their absorption. Also, their degradation or conjugation may occur during digestion, causing a decrease in bioaccessibility.

15.3.4.2 Carotenoids

Thermal processing usually causes degradation or isomerization of carotenoids; therefore, non-thermal processing technologies (PEF, HPP, US) have been proposed as potential alternatives. It has been demonstrated that carotenoids bioaccessibility in fruit juices is positively affected by PEF and US application (Buniowska et al., 2017). Probably, cell walls disruption and the breakage of protein—carotenoid complexes can increase carotenoids release from matrix during digestion, which enhances their bioaccessibility (de Souza Carvalho et al., 2020). A relationship between a smaller particle size and higher carotenoid bioaccessibility has been found (Knockaert, Lemmens, et al., 2012; Lemmens et al., 2010; Moelants et al., 2012). For instance, Sentandreu et al. (2020) reported that HPP-treated mandarin juices (150 MPa reaching 68°C for 15 s) had 5-fold the total carotenoid bioaccessibility than those untreated due to particle size reduction.

In addition, these technologies can also cause isomerization, thus modifying bioaccessibility. For instance, isomerization of *trans* to *cis*-lycopene in tomato juice was favored by US application (800 W; 20 min; 25 kHz) (Zhang et al., 2019). Nevertheless, HPP avoided isomerization in tomato puree, decreasing carotenoids bioaccessibility (Knockaert, Pulissery, et al., 2012).

Limited information about the impact of PEF on carotenoids bioaccessibility in whole matrices is available. López-Gámez et al. (2021) reported an increase in carotenoids bioaccessibility after treating carrots with PEF (5 pulses of 3.5 kV/cm), which was attributed to microstructural changes in cell walls and carotenoids intracellular localization. Additionally, an enhanced bioaccessibility has been obtained in derived products prepared from PEF-treated whole matrices. Such results have been attributed to a better extractability of carotenoids as a consequence of electropermeabilization or to an enhanced initial carotenoid content (González-Casado et al., 2018). Nevertheless, such high content can also cause a competitive inhibition between carotenoids during micellarization and cause a decrease in bioaccessibility (Maiani et al., 2009).

Micelles formation can be negatively affected by dietary fiber (Palafox-Carlos et al., 2011), which binds to bile salts or interacts with lipase (Cervantes-Paz et al., 2016). The formation of fiber networks has been reported after applying HPP, PEF, and US, which reduces carotenoids bioaccessibility (Colle et al., 2010; Jayathunge et al., 2017). An inverse correlation has been reported between viscosity or consistency of HPP- or US-treated tomato products and lycopene bioaccessibility (Anese et al., 2013; Panozzo et al., 2013).

15.3.4.3 Minerals

Limited information has been reported about the effects of nonthermal processing technologies on minerals bioaccessibility. Available investigations are mainly focused on studying the influence of HPP on legumes or cereals. Different results have been obtained depending on the mineral, food matrix, time, and pressure conditions applied. On the one hand, HPP can cause the formation of noncovalent bonds, leading to microstructural changes that favor the accessibility of digestive enzymes (Xia et al., 2017). On the other hand, if HPP reduce the content of phytates and fibers (i.e., antinutritional factors), bioaccessibility of minerals could also be enhanced (Thakur et al., 2020). Some studies show that bioaccessibility of calcium and iron can be tripled or quintupled in case of zinc when carob beans are treated with HPP (500 MPa for 10 min), which could be related to microstructural modifications (Briones-Labarca, Muñoz, & Maureira, 2011). On the other hand, HPP reduces calcium bioaccessibility in HPP-treated apples (500 MPa for 2—10 min), whereas iron and zinc were differently affected depending on the intensity of the applied treatment. Their dialysability may be enhanced by cell wall disruption and the formation of low-molecular-weight complexes (Briones-Labarca, Venegas-Cubillos et al., 2011). In other study carried out by Xia et al., (2017), authors reported that copper bioaccessibility was increased by 2.87%—23.06% in HPP-treated brown rice (300 and 500 MPa for 10 min), but bioaccessibility of iron was decreased (100, 300, and 500 MPa for 10 min).

Bioaccessibility of minerals in liquid products has not been extensively studied, just Cilla et al. (2011) reported increases in bioaccessibility and bioavailability of phosphorous and calcium in HPP-treated (400 MPa for 5 min) milk-based fruit beverages compared to those thermally treated.

Therefore, no general recommendations can be given for improving mineral bioaccessibility under HPP processing. However, this technology has been suggested as an alternative to thermal treatments, since enhanced bioaccessibility and cell uptake have been reported.

15.3.4.4 Vitamins

Most of the studies about vitamins bioaccessibility/bioavailability after applying nonthermal technologies have been focused on vitamin C. For instance, Fonteles et al. (2016) reported an increased vitamin C bioaccessibility in apple bagasse treated by US, due to cell disruption and their better release during digestion. However, the influence of these technologies has been studied mostly in liquid products. Generally, the impact of PEF processing do not cause reductions in vitamin C bioavailability. Sánchez-Moreno et al. (2005) and Sánchez-Moreno et al. (2004) reported that consumption of both PEF-treated gazpacho (typical Spanish vegetable beverage) and PEF-treated orange juice led to same vitamin C concentration in plasma compared to the untreated products. Therefore, PEF is an alternative preservation technique that allows preserving nutritious value. Additionally, Rodríguez-Roque et al. (2015) also reported that neither HPP nor PEF processing caused changes in vitamin C bioaccessibility of fruit-based beverages with added milk. On the other hand, Cilla et al. (2012) observed that vitamin C bioaccessibility significantly decreased in HPP-treated fruit-based beverages with added whole milk or soymilk. This type of processing probably leads to the formation of aggregates with proteins from milk or other compounds leading to viscosity increases and making vitamin C less accessible during in vitro digestion. Additionally, Buniowska et al. (2017) reported that vitamin C from untreated or PEF- or US-treated fruit juices was not detectable after an in vitro digestion, likely due to its sensitivity to temperature. Nevertheless, despite the absence of vitamin C in the dialyzed fractions, in vivo absorption is expected to take place, as active transport phenomena and not just diffusion are involved (Davey et al., 2000).

15.4 Summary and future perspectives

In general, processing technologies may effectively enhance bioaccessibility and bioavailability of bioactive compounds and micronutrients of plant-based products. Nevertheless, type of food matrix and chemical characteristics of each compound are relevant factors to consider. Phenolic compounds bioaccessibility/bioavailability is usually enhanced after blending, drying, juicing, or nonthermal processing, since structural matrix changes (e.g., softening or particle size decrease) lead to a better release. However, it is also dependent on the specific compound and treatment intensity. Carotenoids bioaccessibility/bioavailability is generally improved when particle size is decreased or cell membranes are disrupted by blending, freezing, PEF, US, or HPP application. However, the formation of a fiber network could limit carotenoids incorporation to micelles. Regarding minerals, the reduction of antinutritional factors and matrix softening by blending, thermal processing, or HPP can favor their bioaccessibility/bioavailability in most studied matrices, although more studies are necessary to make a general recommendation. On the other hand, vitamin C bioaccessibility/bioavailability can be improved when structural changes induced by blending, heating, PEF, US, or HPP processing are induced and degradation by oxidative enzymes or entrapment within fiber networks are prevented. Further studies are needed about the effect of processing on other vitamins bioaccessibility/bioavailability different from vitamin C.

Therefore, it is necessary to select and optimize processing conditions in order to obtain products with enhanced health-promoting properties and delve into the changes that processing causes in their safety and quality attributes. Furthermore, studies to correlate structural properties of matrices after processing and compounds absorption should be carried out to easily optimize processing parameters and understand the structure-absorption relationship.

Acknowledgments

This research has been financed by the Ministerio de Economía y Competitividad and Ministerio de Ciencia, Innovación y Universidades (Spain) references AGL2013-44851-R and RTI2018-095560-BI00, respectively.

References

Alonso, R., Rubio, L. A., Muzquiz, M., & Marzo, F. (2001). The effect of extrusion cooking on mineral bioavailability in pea and kidney bean seed meals. *Animal Feed Science and Technology, 94*(1–2), 1–13. https://doi.org/10.1016/S0377-8401(01)00302-9

Anese, M., Mirolo, G., Beraldo, P., & Lippe, G. (2013). Effect of ultrasound treatments of tomato pulp on microstructure and lycopene in vitro bioaccessibility. *Food Chemistry, 136*(2), 458–463. https://doi.org/10.1016/j.foodchem.2012.08.013

Anese, M., Bot, F., Panozzo, A., Mirolo, G., & Lippe, G. (2015). Effect of ultrasound treatment, oil addition and storage time on lycopene stability and in vitro bioaccessibility of tomato pulp. *Food Chemistry, 172*, 685–691. https://doi.org/10.1016/j.foodchem.2014.09.140

Antoine, T., Icard-Vernière, C., Scorrano, G., Salhi, A., Halimi, C., Georgé, S., Carrière, F., Mouquet-Rivier, C., & Reboul, E. (2021). Evaluation of vitamin D bioaccessibility and mineral solubility from test meals containing meat and/or cereals and/or pulses using in vitro digestion. *Food Chemistry, 347*, 128621. https://doi.org/10.1016/j.foodchem.2020.128621

Aschoff, J. K., Kaufmann, S., Kalkan, O., Neidhart, S., Carle, R., & Schweiggert, R. M. (2015). In vitro bioaccessibility of carotenoids, flavonoids, and vitamin C from differently processed oranges and orange juices [*Citrus sinensis* (L.) osbeck]. *Journal of Agricultural and Food Chemistry, 63*(2), 578–587. https://doi.org/10.1021/jf505297t

Aschoff, Julian K., Rolke, C. L., Breusing, N., Bosy-Westphal, A., Högel, J., Carle, R., & Schweiggert, R. M. (2015). Bioavailability of β-cryptoxanthin is greater from pasteurized orange juice than from fresh oranges - a randomized cross-over study. *Molecular Nutrition & Food Research, 59*(10), 1896–1904. https://doi.org/10.1002/mnfr.201500327

Aydin, E., & Gocmen, D. (2015). The influences of drying method and metabisulfite pre-treatment on the color, functional properties and phenolic acids contents and bioaccessibility of pumpkin flour. *LWT - Food Science and Technology, 60*(1), 385–392. https://doi.org/10.1016/j.lwt.2014.08.025

Balasooriya, H., Dasanayake, K., & Ajlouni, S. (2020). Bioaccessibility of micronutrients in fresh and frozen strawberry fruits grown under elevated carbon dioxide and temperature. *Food Chemistry, 309*, 125662. https://doi.org/10.1016/j.foodchem.2019.125662

Barba, F. J., Mariutti, L. R. B., Bragagnolo, N., Mercadante, A. Z., Barbosa-Cánovas, G. V., & Orlien, V. (2017). Bioaccessibility of bioactive compounds from fruits and vegetables after thermal and nonthermal processing. *Trends in Food Science and Technology, 67*, 195–206. https://doi.org/10.1016/j.tifs.2017.07.006

Betoret, E., Betoret, N., Rocculi, P., & Dalla Rosa, M. (2015). Strategies to improve food functionality: Structure-property relationships on high pressures homogenization, vacuum impregnation and drying technologies. *Trends in Food Science and Technology, 46*(1), 1–12. https://doi.org/10.1016/j.tifs.2015.07.006

Bohn, T. (2008). Bioavailabilty of non-provitamin A carotenoids. *Current Nutrition and Food Science, 4*(4), 240–258. https://doi.org/10.2174/157340108786263685

Bohn, T. (2014). Dietary factors affecting polyphenol bioavailability. *Nutrition Reviews, 72*(7), 429–452. https://doi.org/10.1111/nure.12114

Bohn, Torsten, McDougall, G. J., Alegría, A., Alminger, M., Arrigoni, E., Aura, A., Brito, C., Cilla, A., El, S. N., Karakaya, S., Martínez-Cuesta, M. C., & Santos, C. N. (2015). Mind the gap—deficits in our knowledge of aspects impacting the bioavailability of phytochemicals and their metabolites—a position paper focusing on carotenoids and polyphenols. *Molecular Nutrition & Food Research, 59*(7), 1307–1323. https://doi.org/10.1002/mnfr.201400745

Brennan, C., Brennan, M., Derbyshire, E., & Tiwari, B. K. (2011). Effects of extrusion on the polyphenols, vitamins and antioxidant activity of foods. *Trends in Food Science and Technology, 22*(10), 570–575. https://doi.org/10.1016/j.tifs.2011.05.007

Briones-Labarca, V., Muñoz, C., & Maureira, H. (2011). Effect of high hydrostatic pressure on antioxidant capacity, mineral and starch bioaccessibility of a non conventional food: *Prosopis chilensis* seed. *Food Research International, 44*(4), 875–883. https://doi.org/10.1016/j.foodres.2011.01.013

Briones-Labarca, V., Venegas-Cubillos, G., Ortiz-Portilla, S., Chacana-Ojeda, M., & Maureira, H. (2011). Effects of high hydrostatic pressure (HHP) on bioaccessibility, as well as antioxidant activity, mineral and starch contents in Granny Smith apple. *Food Chemistry, 128*(2), 520–529. https://doi.org/10.1016/j.foodchem.2011.03.074

Buniowska, M., Carbonell-Capella, J. M., Frigola, A., & Esteve, M. J. (2017). Bioaccessibility of bioactive compounds after non-thermal processing of an exotic fruit juice blend sweetened with Stevia rebaudiana. *Food Chemistry, 221*, 1834–1842. https://doi.org/10.1016/j.foodchem.2016.10.093

Cano, M. P., Gómez-Maqueo, A., Fernández-López, R., Welti-Chanes, J., & García-Cayuela, T. (2019). Impact of high hydrostatic pressure and thermal treatment on the stability and bioaccessibility of carotenoid and carotenoid esters in astringent persimmon (*Diospyros kaki* Thunb, var. Rojo Brillante). *Food Research International, 123*(March), 538–549. https://doi.org/10.1016/j.foodres.2019.05.017

Carbonell-Capella, J. M., Buniowska, M., Barba, F. J., Esteve, M. J., & Frígola, A. (2014). Analytical methods for determining bioavailability and bioaccessibility of bioactive compounds from fruits and vegetables: A review. *Comprehensive Reviews in Food Science and Food Safety, 13*(2), 155–171. https://doi.org/10.1111/1541-4337.12049

Carrillo, C., Buvé, C., Panozzo, A., Grauwet, T., & Hendrickx, M. (2017). Role of structural barriers in the in vitro bioaccessibility of anthocyanins in comparison with carotenoids. *Food Chemistry, 227*, 271–279. https://doi.org/10.1016/j.foodchem.2017.01.062

Carvalho Santos, W. P., Moura Ribeiro, N., Muniz Batista Santos, D. C., Andrade Korn Graças, M., & Vieira Lopes, M. (2017). Bioaccessibility assessment of toxic and essential elements in produced pulses, Bahia, Brazil. *Food Chemistry, 240*, 112–122. https://doi.org/10.1016/j.foodchem.2017.07.051

Cassani, L., Gerbino, E., Moreira, M.d. R., & Gómez-Zavaglia, A. (2018). Influence of non-thermal processing and storage conditions on the release of health-related compounds after in vitro gastrointestinal digestion of fiber-enriched strawberry juices. *Journal of Functional Foods, 40*, 128–136. https://doi.org/10.1016/j.jff.2017.11.005

Cervantes-Paz, B., Ornelas-Paz, J.d. J., Pérez-Martínez, J. D., Reyes-Hernández, J., Zamudio-Flores, P. B., Rios-Velasco, C., Ibarra-Junquera, V., & Ruiz-Cruz, S. (2016). Effect of pectin concentration and properties on digestive events involved on micellarization of free and esterified carotenoids. *Food Hydrocolloids, 60*, 580–588. https://doi.org/10.1016/j.foodhyd.2016.04.038

Chen, Y., Ma, Y., Dong, L., Jia, X., Liu, L., Huang, F., Chi, J., Xiao, J., Zhang, M., & Zhang, R. (2019). Extrusion and fungal fermentation change the profile and antioxidant activity of free and bound phenolics in rice bran together with the phenolic bioaccessibility. *LWT, 115*, Article 108461. https://doi.org/10.1016/j.lwt.2019.108461

Cilla, A., Alegría, A., De Ancos, B., Sánchez-Moreno, C., Cano, M. P., Plaza, L., Clemente, G., Lagarda, M. J., & Barberá, R. (2012). Bioaccessibility of tocopherols, carotenoids, and ascorbic acid from milk- and soy-based fruit beverages: Influence of food matrix and processing. *Journal of Agricultural and Food Chemistry, 60*(29), 7282–7290. https://doi.org/10.1021/jf301165r

Cilla, A., Barberá, R., López-García, G., Blanco-Morales, V., Alegría, A., & Garcia-Llatas, G. (2019). Impact of processing on mineral bioaccessibility/bioavailability. In *Innovative Thermal and non-thermal processing, Bioaccessibility and Bioavailability of Nutrients and bioactive compounds* (pp. 209–239). Elsevier. https://doi.org/10.1016/B978-0-12-814174-8.00007-X

Cilla, A., Lagarda, M. J., Alegría, A., de Ancos, B., Cano, M. P., Sánchez-Moreno, C., Plaza, L., & Barberá, R. (2011). Effect of processing and food matrix on calcium and phosphorous bioavailability from milk-based fruit beverages in Caco-2 cells. *Food Research International, 44*(9), 3030–3038. https://doi.org/10.1016/j.foodres.2011.07.018

Claeys, W. L., Cardoen, S., Daube, G., De Block, J., Dewettinck, K., Dierick, K., De Zutter, L., Huyghebaert, A., Imberechts, H., Thiange, P., Vandenplas, Y., & Herman, L. (2013). Raw or heated cow milk consumption: Review of risks and benefits. *Food Control, 31*(1), 251–262. https://doi.org/10.1016/j.foodcont.2012.09.035

Colle, I., Van Buggenhout, S., Van Loey, A., & Hendrickx, M. (2010). High pressure homogenization followed by thermal processing of tomato pulp: Influence on microstructure and lycopene in vitro bioaccessibility. *Food Research International, 43*(8), 2193–2200. https://doi.org/10.1016/j.foodres.2010.07.029

da Costa, G. A., & Mercadante, A. Z. (2018). In vitro bioaccessibility of free and esterified carotenoids in cajá frozen pulp-based beverages. *Journal of Food Composition and Analysis, 68*, 53–59. https://doi.org/10.1016/j.jfca.2017.02.012

Davey, M. W., Van Montagu, M., Inzé, D., Sanmartin, M., Kanellis, A., Smirnoff, N., Benzie, I. F. F., Strain, J. J., Favell, D., & Fletcher, J. (2000). Plant L-ascorbic acid: Chemistry, function, metabolism, bioavailability and effects of processing. *Journal of the Science of Food and Agriculture, 80*(7), 825–860. https://doi.org/10.1002/(SICI)1097-0010(20000515)80:7<825::AID-JSFA598>3.0.CO;2-6

de Ancos, B., Cilla, A., Barberá, R., Sánchez-Moreno, C., & Cano, M. P. (2016). Influence of orange cultivar and mandarin postharvest storage on polyphenols, ascorbic acid and antioxidant activity during gastrointestinal digestion. *Food Chemistry, 225*, 114–124. https://doi.org/10.1016/j.foodchem.2016.12.098

De Ancos, B., Cilla, A., Barberá, R., Sánchez-Moreno, C., & Cano, M. P. (2017). Influence of orange cultivar and Mandarin postharvest storage on polyphenols, ascorbic acid and antioxidant activity during gastrointestinal digestion. *Food Chemistry, 225*, 114–124. https://doi.org/10.1016/j.foodchem.2016.12.098

De Lima, A. C. S., Soares, D. J., Da Silva, L. M. R., De Figueiredo, R. W., De Sousa, P. H. M., & De Abreu Menezes, E. (2014). In vitro bioaccessibility of copper, iron, zinc and antioxidant compounds of whole cashew apple juice and cashew apple fibre (*Anacardium occidentale* L.) following simulated gastro-intestinal digestion. *Food Chemistry, 161*, 142–147. https://doi.org/10.1016/j.foodchem.2014.03.123

Debon, S. J. J., & Tester, R. F. (2001). In vitro binding of calcium, iron and zinc by non-starch polysaccharides. *Food Chemistry, 73*(4), 401–410. https://doi.org/10.1016/S0308-8146(00)00312-5

Diamante, L. M., Ihns, R., Savage, G. P., & Vanhanen, L. (2010). A new mathematical model for thin layer drying of fruits. *International Journal of Food Science and Technology, 45*(9), 1956–1962. https://doi.org/10.1111/j.1365-2621.2010.02345.x

Drago, S. R. (2017). Minerals. In *Nutraceutical and functional food components: Effects of innovative processing techniques* (pp. 129–157). Elsevier Inc. https://doi.org/10.1016/B978-0-12-805257-0.00005-3

Eitenmiller, R. R., Ye, L., & Landen, J. W. (2008). *Vitamin analysis for the health and food sciences.*

Etcheverry, P., Grusak, M. A., & Fleige, L. E. (2012). Application of in vitro bioaccessibility and bioavailability methods for calcium, carotenoids, folate, iron, magnesium, polyphenols, zinc, and vitamins B 6, B 12, D, and E. *Frontiers in Physiology, 3*. https://doi.org/10.3389/fphys.2012.00317

Faulks, R. M., & Southon, S. (2005). Challenges to understanding and measuring carotenoid bioavailability. In *Biochimica et biophysica acta - molecular Basis of disease* (pp. 95–100). https://doi.org/10.1016/j.bbadis.2004.11.012. Vol. 1740, Issue 2.

Fonteles, T. V., Leite, A. K. F., Silva, A. R. A., Carneiro, A. P. G., Miguel, E. D. C., Cavada, B. S., Fernandes, F. A. N., & Rodrigues, S. (2016). Ultrasound processing to enhance drying of cashew apple bagasse puree: Influence on antioxidant properties and in vitro bioaccessibility of bioactive compounds. *Ultrasonics Sonochemistry, 31*, 237–249. https://doi.org/10.1016/j.ultsonch.2016.01.003

García-Amezquita, L. E., Tejada-Ortigoza, V., Campanella, O. H., & Welti-Chanes, J. (2018). Influence of drying method on the composition, physicochemical properties, and prebiotic potential of dietary fibre concentrates from fruit peels. *Journal of Food Quality*, 11. https://doi.org/10.1155/2018/9105237

Goh, L. P., Loh, S. P., Fatimah, M. Y., & Perumal, K. (2009). Bioaccessibility of carotenoids and tocopherols in marine microalgae, *Nannochloropsis* sp. and *Chaetoceros* sp. *Malaysian Journal of Nutrition, 15*(1), 77–86. http://nutriweb.org.my/publications/mjn0015_1/mjn15n1_art8.pdf.

Gong, X., Li, X., Xia, Y., Xu, J., Li, Q., Zhang, C., & Li, M. (2020). Effects of phytochemicals from plant-based functional foods on hyperlipidemia and their underpinning mechanisms. *Trends in Food Science and Technology, 103*, 304–320. https://doi.org/10.1016/j.tifs.2020.07.026

González-Casado, S., Martín-Belloso, O., Elez-Martínez, P., & Soliva-Fortuny, R. (2018). Application of pulsed electric fields to tomato fruit for enhancing the bioaccessibility of carotenoids in derived products. *Food and Function, 9*(4), 2282–2289. https://doi.org/10.1039/c7fo01857f

Granado-Lorencio, F., Herrero-Barbudo, C., Acién-Fernández, G., Molina-Grima, E., Fernández-Sevilla, J. M., Pérez-Sacristán, B., & Blanco-Navarro, I. (2009). In vitro bioaccesibility of lutein and zeaxanthin from the microalgae *Scenedesmus almeriensis*. *Food Chemistry, 114*(2), 747–752. https://doi.org/10.1016/j.foodchem.2008.10.058

Granado-Lorencio, F., Herrero-Barbudo, C., Blanco-Navarro, I., Pérez-Sacristán, B., & Olmedilla-Alonso, B. (2009). Bioavailability of carotenoids and α-tocopherol from fruit juices in the presence of absorption modifiers: In vitro and in vivo assessment. *British Journal of Nutrition, 101*(4), 576–582. https://doi.org/10.1017/S000711450803078X

Gulati, P., Brahma, S., & Rose, D. J. (2020). *Impacts of extrusion processing on nutritional components in cereals and legumes: Carbohydrates, proteins, lipids, vitamins, and minerals* (pp. 415–443). Elsevier BV. https://doi.org/10.1016/b978-0-12-815360-4.00013-4

Hashemi, S. M. B., Pourmohammadi, K., Gholamhosseinpour, A., Es, I., Ferreira, D. S., & Khaneghah, A. M. (2019). Fat-soluble vitamins. In *Innovative Thermal and non-thermal processing, Bioaccessibility and Bioavailability of Nutrients and bioactive compounds* (pp. 267–289). Elsevier. https://doi.org/10.1016/B978-0-12-814174-8.00009-3

Heaney, R. P., Recker, R. R., & Weaver, C. M. (1990). Absorbability of calcium sources: The limited role of solubility. *Calcified Tissue International, 46*(5), 300–304. https://doi.org/10.1007/BF02563819

Hedrén, E., Diaz, V., & Svanberg, U. (2002). Estimation of carotenoid accessibility from carrots determined by an in vitro digestion method. *European Journal of Clinical Nutrition, 56*(5), 425–430. https://doi.org/10.1038/sj.ejcn.1601329

Hemalatha, S., Platel, K., & Srinivasan, K. (2007). Influence of heat processing on the bioaccessibility of zinc and iron from cereals and pulses consumed in India. *Journal of Trace Elements in Medicine and Biology, 21*(1), 1–7. https://doi.org/10.1016/j.jtemb.2006.10.002

He, Z., Tao, Y., Zeng, M., Zhang, S., Tao, G., Qin, F., & Chen, J. (2016). High pressure homogenization processing, thermal treatment and milk matrix affect in vitro bioaccessibility of phenolics in apple, grape and orange juice to different extents. *Food Chemistry, 200*, 107–116. https://doi.org/10.1016/j.foodchem.2016.01.045

Hotz, C., & Gibson, R. S. (2007). Traditional food-processing and preparation practices to enhance the bioavailability of micronutrients in plant-based diets. In *Journal of nutrition* (pp. 1097–1100). American Institute of Nutrition. https://doi.org/10.1093/jn/137.4.1097. Vol. 137, Issue 4.

Huo, T., Ferruzzi, M. G., Schwartz, S. J., & Failla, M. L. (2007). Impact of fatty acyl composition and quantity of triglycerides on bioaccessibility of dietary carotenoids. *Journal of Agricultural and Food Chemistry, 55*(22), 8950–8957. https://doi.org/10.1021/jf071687a

Jakobek, L. (2015). Interactions of polyphenols with carbohydrates, lipids and proteins. *Food Chemistry, 175*, 556–567. https://doi.org/10.1016/j.foodchem.2014.12.013

Jayathunge, K. G. L. R., Stratakos, A. C., Cregenzán-Albertia, O., Grant, I. R., Lyng, J., & Koidis, A. (2017). Enhancing the lycopene in vitro bioaccessibility of tomato juice synergistically applying thermal and non-thermal processing technologies. *Food Chemistry, 221*, 698–705. https://doi.org/10.1016/j.foodchem.2016.11.117

Jovani, M., Alegría, A., Barberá, R., Farré, R., Lagarda, M. J., & Clemente, G. (2000). Effect of proteins, phytates, ascorbic acid and citric acid on dialysability of calcium, iron, zinc and copper in soy-based infant formulas. *Nahrung - Food, 44*(2), 114–117. https://doi.org/10.1002/(sici)1521-3803(20000301)44:2<114::aid-food114>3.0.co;2-n

Karakaya, S., & Yilmaz, N. (2007). Lycopene content and antioxidant activity of fresh and processed tomatoes and in vitro bioavailability of lycopene. *Journal of the Science of Food and Agriculture, 87*(12), 2342–2347. https://doi.org/10.1002/jsfa.2998

Kayacan, S., Karasu, S., Akman, P. K., Goktas, H., Doymaz, I., & Sagdic, O. (2020). Effect of different drying methods on total bioactive compounds, phenolic profile, in vitro bioaccessibility of phenolic and HMF formation of persimmon. *LWT, 118*, 108830. https://doi.org/10.1016/j.lwt.2019.108830

Khanam, A., & Platel, K. (2016). Bioaccessibility of selenium, selenomethionine and selenocysteine from foods and influence of heat processing on the same. *Food Chemistry, 194*, 1293–1299. https://doi.org/10.1016/j.foodchem.2015.09.005

Khaneghah, A. M., Hashemi, S. M. B., Es, I., Gholamhosseinpour, A., Loizzo, M. R., Giardinieri, A., Pacetti, D., Pourmohammadi, K., & Ferreira, D. S. (2019). Water-soluble vitamins. In *Innovative Thermal and non-thermal processing, Bioaccessibility and Bioavailability of Nutrients and bioactive compounds* (pp. 241–266). Elsevier. https://doi.org/10.1016/B978-0-12-814174-8.00008-1

Khouzam, R. B., Pohl, P., & Lobinski, R. (2011). Bioaccessibility of essential elements from white cheese, bread, fruit and vegetables. *Talanta, 86*(1), 425–428. https://doi.org/10.1016/j.talanta.2011.08.049

Knockaert, G., Lemmens, L., Van Buggenhout, S., Hendrickx, M., & Van Loey, A. (2012). Changes in β-carotene bioaccessibility and concentration during processing of carrot puree. *Food Chemistry, 133*(1), 60–67. https://doi.org/10.1016/j.foodchem.2011.12.066

Knockaert, G., Pulissery, S. K., Colle, I., Van Buggenhout, S., Hendrickx, M., & Loey, A. V. (2012). Lycopene degradation, isomerization and in vitro bioaccessibility in high pressure homogenized tomato puree containing oil: Effect of additional thermal and high pressure processing. *Food Chemistry, 135*(3), 1290–1297. https://doi.org/10.1016/j.foodchem.2012.05.065

Kwok, K. C., Liang, H. H., & Niranjan, K. (2002). Optimizing conditions for thermal processes of soy milk. *Journal of Agricultural and Food Chemistry, 50*(17), 4834–4838. https://doi.org/10.1021/jf020182b

Lafarga, T., Rodríguez-Roque, M. J., Bobo, G., Villaró, S., & Aguiló-Aguayo, I. (2019). Effect of ultrasound processing on the bioaccessibility of phenolic compounds and antioxidant capacity of selected vegetables. *Food Science and Biotechnology, 28*(6), 1713–1721. https://doi.org/10.1007/s10068-019-00618-4

Lemmens, L., Tibäck, E., Svelander, C., Smout, C., Ahrné, L., Langton, M., Alminger, M., Van Loey, A., & Hendrickx, M. (2009). Thermal pretreatments of carrot pieces using different heating techniques: Effect on quality related aspects. *Innovative Food Science and Emerging Technologies, 10*(4), 522–529. https://doi.org/10.1016/j.ifset.2009.05.004

Lemmens, L., Van Buggenhout, S., Oey, I., Van Loey, A., & Hendrickx, M. (2009). Towards a better understanding of the relationship between the β-carotene in vitro bio-accessibility and pectin structural changes: A case study on carrots. *Food Research International, 42*(9), 1323–1330. https://doi.org/10.1016/j.foodres.2009.04.006

Lemmens, L., Van Buggenhout, S., Van Loey, A. M., & Hendrickx, M. E. (2010). Particle size reduction leading to cell wall rupture is more important for the β-carotene bioaccessibility of raw compared to thermally processed carrots. *Journal of Agricultural and Food Chemistry, 58*(24), 12769–12776. https://doi.org/10.1021/jf102554h

Leong, S. Y., & Oey, I. (2012). Effects of processing on anthocyanins, carotenoids and vitamin C in summer fruits and vegetables. *Food Chemistry, 133*(4), 1577–1587. https://doi.org/10.1016/j.foodchem.2012.02.052

Liu, X., Liu, J., Bi, J., Yi, J., Peng, J., Ning, C., Wellala, C. K. D., & Zhang, B. (2019). Effects of high pressure homogenization on pectin structural characteristics and carotenoid bioaccessibility of carrot juice. *Carbohydrate Polymers, 203*, 176–184. https://doi.org/10.1016/j.carbpol.2018.09.055

Lombardi-Boccia, G., De Santis, N., Di Lullo, G., & Carnovale, E. (1995). Impact of processing on Fe dialysability from bean (*Phaseolus vulgaris* L.). *Food Chemistry, 53*(2), 191–195. https://doi.org/10.1016/0308-8146(95)90787-8

López-Gámez, G., Elez-Martínez, P., Quiles-Chuliá, A., Martín-Belloso, O., Hernando-Hernando, I., & Soliva-Fortuny, R. (2021). Effect of pulsed electric fields on carotenoid and phenolic bioaccessibility and their relationship with carrot structure. *Food and Function, 12*(6), 2772–2783. https://doi.org/10.1039/d0fo03035j

Maiani, G., Castón, M. J. P., Catasta, G., Toti, E., Cambrodón, I. G., Bysted, A., Granado-Lorencio, F., Olmedilla-Alonso, B., Knuthsen, P., Valoti, M., Böhm, V., Mayer-Miebach, E., Behsnilian, D., & Schlemmer, U. (2009). Carotenoids: Actual knowledge on food sources, intakes, stability and bioavailability and their protective role in humans. *Molecular Nutrition and Food Research, 53*(2), 194–218. https://doi.org/10.1002/mnfr.200800053

Manach, C., Scalbert, A., Morand, C., Rémésy, C., & Jiménez, L. (2004). Polyphenols: Food sources and bioavailability. *The American Journal of Clinical Nutrition, 79*(5), 727–747. https://doi.org/10.1093/ajcn/79.5.727

Mapelli-Brahm, P., Stinco, C. M., Rodrigo, M. J., Zacarías, L., & Meléndez-Martínez, A. J. (2018). Impact of thermal treatments on the bioaccessibility of phytoene and phytofluene in relation to changes in the microstructure and size of orange juice particles. *Journal of Functional Foods, 46*, 38–47. https://doi.org/10.1016/j.jff.2018.04.044

Martínez-Hernández, G. B., Artés-Hernández, F., Gómez, P. A., & Artés, F. (2013). Induced changes in bioactive compounds of kailan-hybrid broccoli after innovative processing and storage. *Journal of Functional Foods, 5*(1), 133–143. https://doi.org/10.1016/j.jff.2012.09.004

Martínez-Manrique, E., Jacinto-Hernández, C., Garza-García, R., Campos, A., Moreno, E., & Bernal-Lugo, I. (2011). Enzymatic changes in pectic polysaccharides related to the beneficial effect of soaking on bean cooking time. *Journal of the Science of Food and Agriculture, 91*(13), 2394–2398. https://doi.org/10.1002/jsfa.4474

Mercado-Mercado, G., Blancas-Benitez, F. J., Velderrain-Rodríguez, G. R., Montalvo-González, E., González-Aguilar, G. A., Alvarez-Parrilla, E., & Sáyago-Ayerdi, S. G. (2015). Bioaccessibility of polyphenols released and associated to dietary fibre in calyces and decoction residues of Roselle (*Hibiscus sabdariffa* L.). *Journal of Functional Foods, 18*, 171–181. https://doi.org/10.1016/j.jff.2015.07.001

Miquel, E., Alegría, A., Barberá, R., & Farré, R. (2006). Casein phosphopeptides released by simulated gastrointestinal digestion of infant formulas and their potential role in mineral binding. *International Dairy Journal, 16*(9), 992–1000. https://doi.org/10.1016/j.idairyj.2005.10.010

Mizobutsi, G. P., Finger, F. L., Ribeiro, R. A., Puschmann, R., Neves, L. L. de M., & Mota, W. F. da (2010). Effect of pH and temperature on peroxidase and polyphenoloxidase activities of litchi pericarp. *Scientia Agricola, 67*(2), 213–217. https://doi.org/10.1590/S0103-90162010000200013

Moelants, K. R. N., Lemmens, L., Vandebroeck, M., Van Buggenhout, S., Van Loey, A. M., & Hendrickx, M. E. (2012). Relation between particle size and carotenoid bioaccessibility in carrot- and tomato-derived suspensions. *Journal of Agricultural and Food Chemistry, 60*(48), 11995–12003. https://doi.org/10.1021/jf303502h

Mutsokoti, L., Panozzo, A., Van Loey, A., & Hendrickx, M. (2016). Carotenoid transfer to oil during thermal processing of low fat carrot and tomato particle based suspensions. *Food Research International, 86*, 64–73. https://doi.org/10.1016/j.foodres.2016.05.019

Oghbaei, M., & Prakash, J. (2013). Effects of processing and digestive enzymes on retention, bioaccessibility and antioxidant activity of bioactive components in food mixes based on legumes and green leaves. *Food Bioscience, 4*, 21–30. https://doi.org/10.1016/j.fbio.2013.08.001

Ortak, M., Caltinoglu, C., Sensoy, I., Karakaya, S., & Mert, B. (2017). Changes in functional properties and in vitro bioaccessibilities of β-carotene and lutein after extrusion processing. *Journal of Food Science and Technology, 54*(11), 3543–3551. https://doi.org/10.1007/s13197-017-2812-4

Palafox-Carlos, H., Ayala-Zavala, J. F., & González-Aguilar, G. A. (2011). The role of dietary fiber in the bioaccessibility and bioavailability of fruit and vegetable Antioxidants. *Journal of Food Science, 76*(1), R6–R15. https://doi.org/10.1111/j.1750-3841.2010.01957.x

Palmero, P., Lemmens, L., Hendrickx, M., & Van Loey, A. (2014). Role of carotenoid type on the effect of thermal processing on bioaccessibility. *Food Chemistry, 157*, 275–282. https://doi.org/10.1016/j.foodchem.2014.02.055

Palmero, P., Lemmens, L., Ribas-Agustí, A., Sosa, C., Met, K., De Dieu Umutoni, J., Hendrickx, M., & Van Loey, A. (2013). Novel targeted approach to better understand how natural structural barriers govern carotenoid in vitro bioaccessibility in vegetable-based systems. *Food Chemistry, 141*(3), 2036–2043. https://doi.org/10.1016/j.foodchem.2013.05.064

Panda, S. L., Jaiswal, A., & Lakshmi A, J. (2019). Compositional and processing effects in promoting the bioaccessibility of iron and zinc of ready to cook high protein kheer mix. *LWT, 109*, 186–193. https://doi.org/10.1016/j.lwt.2019.04.026

Panozzo, A., Lemmens, L., Van Loey, A., Manzocco, L., Nicoli, M. C., & Hendrickx, M. (2013). Microstructure and bioaccessibility of different carotenoid species as affected by high pressure homogenisation: A case study on differently coloured tomatoes. *Food Chemistry, 141*(4), 4094–4100. https://doi.org/10.1016/j.foodchem.2013.06.099

Parada, J., & Aguilera, J. M. (2007). Food microstructure affects the bioavailability of several nutrients. *Journal of Food Science, 72*(2), R21–R32. https://doi.org/10.1111/j.1750-3841.2007.00274.x

Perales, S., Barberá, R., Lagarda, M. J., & Farré, R. (2005). Bioavailability of calcium from milk-based formulas and fruit juices containing milk and cereals estimated by in vitro methods (solubility, dialyzability, and uptake and transport by caco-2 cells). *Journal of Agricultural and Food Chemistry, 53*(9), 3721–3726. https://doi.org/10.1021/jf047977y

Pereira, E. J., Carvalho, L. M. J., Dellamora-Ortiz, G. M., Cardoso, F. S. N., & Carvalho, J. L. V. (2016). Effect of different home-cooking methods on the bioaccessibility of zinc and iron in conventionally bred cowpea (*Vigna unguiculata* L. Walp) consumed in Brazil. *Food & Nutrition Research, 60*, 29082. https://doi.org/10.3402/fnr.v60.29082

Pérez-Vicente, A., Gil-Izquierdo, A., & García-Viguera, C. (2002). In vitro gastrointestinal digestion study of pomegranate juice phenolic compounds, anthocyanins, and vitamin C. *Journal of Agricultural and Food Chemistry, 50*(8), 2308–2312. https://doi.org/10.1021/jf0113833

Pico, J., Xu, K., Guo, M., Mohamedshah, Z., Ferruzzi, M. G., & Martinez, M. M. (2019). Manufacturing the ultimate green banana flour: Impact of drying and extrusion on phenolic profile and starch bioaccessibility. *Food Chemistry, 297*, 124990. https://doi.org/10.1016/j.foodchem.2019.124990

Quan, W., Tao, Y., Qie, X., Zeng, M., Qin, F., Chen, J., & He, Z. (2020). Effects of high-pressure homogenization, thermal processing, and milk matrix on the in vitro bioaccessibility of phenolic compounds in pomelo and kiwi juices. *Journal of Functional Foods, 64*, 103633. https://doi.org/10.1016/j.jff.2019.103633

Quirós-Sauceda, A. E., Ayala-Zavala, J. F., Sáyago-Ayerdi, S. G., Vélez-De La Rocha, R., Sañudo-Barajas, J. A., & González-Aguilar, G. A. (2014). Added dietary fiber affects antioxidant capacity and phenolic compounds content extracted from tropical fruit. *Journal of Applied Botany and Food Quality, 87*, 227–233. https://doi.org/10.5073/JABFQ.2014.087.032

Quirós-Sauceda, A. E., Oliver Chen, C. Y., Blumberg, J. B., Astiazaran-Garcia, H., Wall-Medrano, A., & González-Aguilar, G. A. (2017). Processing 'ataulfo' mango into juice preserves the bioavailability and antioxidant capacity of its phenolic compounds. *Nutrients, 9*(10). https://doi.org/10.3390/nu9101082

Raes, K., Knockaert, D., Struijs, K., & Van Camp, J. (2014). Role of processing on bioaccessibility of minerals: Influence of localization of minerals and anti-nutritional factors in the plant. *Trends in Food Science and Technology, 37*(1), 32–41. https://doi.org/10.1016/j.tifs.2014.02.002

Rao, A. V., & Rao, L. G. (2007). Carotenoids and human health. *Pharmacological Research, 55*(3), 207–216. https://doi.org/10.1016/j.phrs.2007.01.012

Rehman, Z. U., & Shah, W. H. (2005). Thermal heat processing effects on antinutrients, protein and starch digestibility of food legumes. *Food Chemistry, 91*(2), 327–331. https://doi.org/10.1016/j.foodchem.2004.06.019

Ribas-Agustí, A., Martín-Belloso, O., Soliva-Fortuny, R., & Elez-Martínez, P. (2018). Food processing strategies to enhance phenolic compounds bioaccessibility and bioavailability in plant-based foods. *Critical Reviews in Food Science and Nutrition, 58*(15), 2531–2548. https://doi.org/10.1080/10408398.2017.1331200

Ribas-Agustí, A., Martín-Belloso, O., Soliva-Fortuny, R., & Elez-Martínez, P. (2019). Influence of pulsed electric fields processing on the bioaccessible and non-bioaccessible fractions of apple phenolic compounds. *Journal of Functional Foods, 59*, 206–214. https://doi.org/10.1016/j.jff.2019.05.041

Ribeiro, L.d. O., Pinheiro, A. C. B., Brígida, A. I. S., Genisheva, Z. A., Vicente, A. A. M., Teixeira, J. A. C., de Matta, V. M., & Freitas, S. P. (2019). In vitro gastrointestinal evaluation of a juçara-based smoothie: Effect of processing on phenolic compounds bioaccessibility. *Journal of Food Science and Technology, 56*(11), 5017–5026. https://doi.org/10.1007/s13197-019-03974-5

Rickman, J. C., Barrett, D. M., & Bruhn, C. M. (2007). Nutritional comparison of fresh, frozen and canned fruits and vegetables. Part 1. Vitamins C and B and phenolic compounds. *Journal of the Science of Food and Agriculture, 87*(6), 930–944. https://doi.org/10.1002/jsfa.2825

Rock, C. L., & Swendseid, M. E. (1992). Plasma β-carotene response in humans after meals supplemented with dietary pectin. *American Journal of Clinical Nutrition, 55*(1), 96–99. https://doi.org/10.1093/ajcn/55.1.96

Rodrigo, M. J., Cilla, A., Barberá, R., & Zacarías, L. (2015). Carotenoid bioaccessibility in pulp and fresh juice from carotenoid-rich sweet oranges and mandarins. *Food and Function, 6*(6), 1950–1959. https://doi.org/10.1039/c5fo00258c

Rodríguez-Roque, M. J., de Ancos, B., Sánchez-Moreno, C., Cano, M. P., Elez-Martínez, P., & Martín-Belloso, O. (2015). Impact of food matrix and processing on the in vitro bioaccessibility of vitamin C, phenolic compounds, and hydrophilic antioxidant activity from fruit juice-based beverages. *Journal of Functional Foods, 14*, 33–43. https://doi.org/10.1016/j.jff.2015.01.020

Rodríguez-Roque, Janeth, María, De Ancos, B., Sánchez-Vega, R., Sánchez-Moreno, C., Elez-Martínez, P., & Martín-Belloso, O. (2020). In vitro bioaccessibility of isoflavones from a soymilk-based beverage as affected by thermal and non-thermal processing. *Innovative Food Science & Emerging Technologies, 66*, 102504. https://doi.org/10.1016/j.ifset.2020.102504

Rodríguez-Roque, M. J., Rojas-Graü, M. A., Elez-Martínez, P., & Martín-Belloso, O. (2013). Changes in vitamin C, phenolic, and carotenoid profiles throughout in vitro gastrointestinal digestion of a blended fruit juice. *Journal of Agricultural and Food Chemistry, 61*(8), 1859–1867. https://doi.org/10.1021/jf3044204

Rodríguez-Roque, M. J., Rojas-Graü, M. A., Elez-Martínez, P., & Martín-Belloso, O. (2014). In vitro bioaccessibility of health-related compounds from a blended fruit juice-soymilk beverage: Influence of the food matrix. *Journal of Functional Foods, 7*(1), 161–169. https://doi.org/10.1016/j.jff.2014.01.023

Sánchez-Moreno, C., Cano, M. P., De Ancos, B., Plaza, L., Olmedilla, B., Granado, F., Elez-Martínez, P., Martín-Belloso, O., & Martín, A. (2004). Pulsed electric fields-processed orange juice consumption increases plasma vitamin C and decreases F2-isoprostanes in healthy humans. *Journal of Nutritional Biochemistry, 15*(10), 601–607. https://doi.org/10.1016/j.jnutbio.2004.04.007

Sánchez-Moreno, C., Cano, M. P., De Ancos, B., Plaza, L., Olmedilla, B., Granado, F., Elez-Martínez, P., Martín-Belloso, O., & Martín, A. (2005). Intake of Mediterranean vegetable soup treated by pulsed electric fields plasma vitamin C and antioxidant biomarkers in humans. *International Journal of Food Sciences and Nutrition, 56*(2), 115–124. https://doi.org/10.1080/09637480500081811

Sandberg, A. S., Andersson, H., Carlsson, N. G., & Sandstrom, B. (1987). Degradation products of bran phytate formed during digestion in the human small intestine: Effect of extrusion cooking on digestibility. *Journal of Nutrition, 117*(12), 2061–2065. https://doi.org/10.1093/jn/117.12.2061

Santos, W. P. C., Ribeiro, N. M., Santos, D. C. M. B., Korn, M. G. A., & Lopes, M. V. (2018). Bioaccessibility assessment of toxic and essential elements in produced pulses, Bahia, Brazil. *Food Chemistry, 240*, 112–122. https://doi.org/10.1016/j.foodchem.2017.07.051

Schweiggert, R. M., Mezger, D., Schimpf, F., Steingass, C. B., & Carle, R. (2012). Influence of chromoplast morphology on carotenoid bioaccessibility of carrot, mango, papaya, and tomato. *Food Chemistry, 135*(4), 2736–2742. https://doi.org/10.1016/j.foodchem.2012.07.035

Sefrin Speroni, C., Rigo Guerra, D., Beutinger Bender, A. B., Stiebe, J., Ballus, C. A., Picolli da Silva, L., Lozano-Sánchez, J., & Emanuelli, T. (2021). Micronization increases the bioaccessibility of polyphenols from granulometrically separated olive pomace fractions. *Food Chemistry, 344*, 128689. https://doi.org/10.1016/j.foodchem.2020.128689

Seiquer, I., Aspe, T., Vaquero, P., & Navarro, P. (2001). Effects of heat treatment of casein in the presence of reducing sugars on calcium bioavailability: In vitro and in vivo assays. *Journal of Agricultural and Food Chemistry, 49*(2), 1049–1055. https://doi.org/10.1021/jf001008v

Sentandreu, E., Stinco, C. M., Vicario, I. M., Mapelli-Brahm, P., Navarro, J. L., & Meléndez-Martínez, A. J. (2020). High-pressure homogenization as compared to pasteurization as a sustainable approach to obtain Mandarin juices with improved bioaccessibility of carotenoids and flavonoids. *Journal of Cleaner Production*, 262. https://doi.org/10.1016/j.jclepro.2020.121325

Shahidi, F., & Yeo, J. D. (2016). Insoluble-bound phenolics in food. *Molecules, 21*(9). https://doi.org/10.3390/molecules21091216

de Souza Carvalho, L. M., Lemos, M. C. M., Sanches, E. A., da Silva, L. S., de Araújo Bezerra, J., Aguiar, J. P. L., das Chagas do Amaral Souza, F., Alves Filho, E. G., & Campelo, P. H. (2020). Improvement of the bioaccessibility of bioactive compounds from Amazon fruits treated using high energy ultrasound. *Ultrasonics Sonochemistry, 67*, 105148. https://doi.org/10.1016/j.ultsonch.2020.105148

Sriburi, P., & Hill, S. E. (2000). Extrusion of cassava starch with either variations in ascorbic acid concentration or pH. *International Journal of Food Science and Technology, 35*(2), 141−154. https://doi.org/10.1046/j.1365-2621.2000.00360.x

Sriwichai, W., Collin, M., Tranbarger, T. J., Berger, J., & Avallone, S. (2017). Improvement of the content in bioaccessible lipophilic micronutrients in raw and processed drumstick leaves (*Moringa oleifera* Lam.). *LWT - Food Science and Technology, 75*, 279−285. https://doi.org/10.1016/j.lwt.2016.09.001

Stinco, C. M., Fernández-Vázquez, R., Heredia, F. J., Meléndez-Martínez, A. J., & Vicario, I. M. (2013). Bioaccessibility, antioxidant activity and colour of carotenoids in ultrafrozen orange juices: Influence of thawing conditions. *LWT - Food Science and Technology, 53*(2), 458−463. https://doi.org/10.1016/j.lwt.2013.04.003

Stinco, C. M., Sentandreu, E., Mapelli-Brahm, P., Navarro, J. L., Vicario, I. M., & Meléndez-Martínez, A. J. (2020). Influence of high pressure homogenization and pasteurization on the *in vitro* bioaccessibility of carotenoids and flavonoids in orange juice. *Food Chemistry, 331*, 127259. https://doi.org/10.1016/j.foodchem.2020.127259

Sy, C., Gleize, B., Dangles, O., Landrier, J.-F., Veyrat, C. C., & Borel, P. (2012). Effects of physicochemical properties of carotenoids on their bioaccessibility, intestinal cell uptake, and blood and tissue concentrations. *Molecular Nutrition & Food Research*, 1385−1397. https://doi.org/10.1002/mnfr.201200041

Thakur, N., Raigond, P., Singh, Y., Mishra, T., Singh, B., Lal, M. K., & Dutt, S. (2020). Recent updates on bioaccessibility of phytonutrients. *Trends in Food Science and Technology, 97*, 366−380. https://doi.org/10.1016/j.tifs.2020.01.019

Toepfl, S., Heinz, V., & Knorr, D. (2005). Overview of pulsed electric field processing for food. In *Emerging Technologies for food processing* (pp. 69−97). Elsevier Ltd. https://doi.org/10.1016/B978-012676757-5/50006-2

Van Buggenhout, S., Alminger, M., Lemmens, L., Colle, I., Knockaert, G., Moelants, K., Van Loey, A., & Hendrickx, M. (2010). In vitro approaches to estimate the effect of food processing on carotenoid bioavailability need thorough understanding of process induced microstructural changes. *Trends in Food Science and Technology, 21*(12), 607−618. https://doi.org/10.1016/j.tifs.2010.09.010

Velderrain-Rodríguez, G., Quirós-Sauceda, A., Mercado-Mercado, G., Ayala-Zavala, J. F., Astiazarán-García, H., Robles-Sánchez, R. M., Wall-Medrano, A., Sayago-Ayerdi, S., & González-Aguilar, G. A. (2016). Effect of dietary fiber on the bioaccessibility of phenolic compounds of mango, papaya and pineapple fruits by an in vitro digestion model. *Food Science and Technology, 36*(2), 188−194. https://doi.org/10.1590/1678-457X.6729

Wellala, C. K. D., Bi, J., Liu, X., Liu, J., Lyu, J., & Wu, X. (2019). Juice related water-soluble pectin characteristics and bioaccessibility of bioactive compounds in oil and emulsion incorporated mixed juice processed by high pressure homogenization. *Food Hydrocolloids, 93*, 56−67. https://doi.org/10.1016/j.foodhyd.2019.02.011

Wiktor, A., Gondek, E., Jakubczyk, E., Dadan, M., Nowacka, M., Rybak, K., & Witrowa-Rajchert, D. (2018). Acoustic and mechanical properties of carrot tissue treated by pulsed electric field, ultrasound and combination of both. *Journal of Food Engineering, 238*, 12−21. https://doi.org/10.1016/j.jfoodeng.2018.06.001

Xia, Q., Wang, L., Xu, C., Mei, J., & Li, Y. (2016). Effects of germination and high hydrostatic pressure processing on mineral elements, amino acids and antioxidants in vitro bioaccessibility, as well as starch digestibility in brown rice (*Oryza sativa* L.). *Food Chemistry, 214*, 533−542. https://doi.org/10.1016/j.foodchem.2016.07.114

Xia, Q., Wang, L., Xu, C., Mei, J., & Li, Y. (2017). Effects of germination and high hydrostatic pressure processing on mineral elements, amino acids and antioxidants in vitro bioaccessibility, as well as starch digestibility in brown rice (*Oryza sativa* L.). *Food Chemistry, 214*, 533−542. https://doi.org/10.1016/j.foodchem.2016.07.114

Yaman, M., Çatak, J., Uğur, H., Gürbüz, M., Belli, İ., Tanyõldõz, S. N., Yõldõrõm, H., Cengiz, S., Yavuz, B. B., Kişmiroğlu, C., Özgür, B., & Yaldõz, M. C. (2021). The bioaccessibility of water-soluble vitamins: A review. *Trends in Food Science and Technology, 109*, 552−563. https://doi.org/10.1016/j.tifs.2021.01.056

Zhang, Z., Wei, Q., Nie, M., Jiang, N., Liu, C., Liu, C., Li, D., & Xu, L. (2018). Microstructure and bioaccessibility of different carotenoid species as affected by hot air drying: Study on carrot, sweet potato, yellow bell pepper and broccoli. *LWT, 96*, 357−363. https://doi.org/10.1016/j.lwt.2018.05.061

Zhang, W., Yu, Y., Xie, F., Gu, X., Wu, J., & Wang, Z. (2019). High pressure homogenization versus ultrasound treatment of tomato juice: Effects on stability and in vitro bioaccessibility of carotenoids. *LWT, 116*, 108597. https://doi.org/10.1016/j.lwt.2019.108597

Zhao, G., Zhang, R., Liu, L., Deng, Y., Wei, Z., Zhang, Y., Ma, Y., & Zhang, M. (2017). Different thermal drying methods affect the phenolic profiles, their bioaccessibility and antioxidant activity in *Rhodomyrtus tomentosa* (Ait.) Hassk berries. *LWT - Food Science and Technology, 79*, 260−266. https://doi.org/10.1016/j.lwt.2017.01.039

Zudaire, L., Lafarga, T., Viñas, I., Abadias, M., Brunton, N., & Aguiló-Aguayo, I. (2019). Effect of ultrasound pre-treatment on the physical, microbiological, and antioxidant properties of calçots. *Food and Bioprocess Technology, 12*(3), 387−394. https://doi.org/10.1007/s11947-018-2217-z

Section VI

Plant-based food — future outlook

Chapter 16

3D printing of plant-based foods

Fernanda C. Godoi[1], Sangeeta Prakash[2] and Bhesh R. Bhandari[2]

[1]*Tessenderlo Innovation Center, Tessenderlo Group, Tessenderlo, Belgium;* [2]*School of Agriculture and Food Sciences, The University of Queensland, Brisbane, QLD, Australia*

16.1 Introduction

Food and Agriculture Organization (FAO) of the United Nations reports that the livestock sector is responsible for 18% greenhouse gas emissions (measured in CO_2 equivalent) and accounts for over 8% of global human water consumption. Hence, there is an urgent need to develop sustainable frameworks that can reduce the environmental impact caused by livestock. Global production of meat is estimated to reach 465 million tonnes by 2050 and that of milk will be 1043 million tonnes (Steinfeld et al., 2006). The *2020 World Population Data Sheet* projects that the world population will increase from 7.8 billion in 2020 to 9.9 billion by 2050 (IISD, 2020). The global population growth drives the demand for trustworthy and sustainable protein sources. The rising understanding of the inefficiency in protein conversion through livestock meat production led to the development of alternative proteins solutions: plant-based foods (Choudhury et al., 2020). In 2019, the US plant-based milk ($2.0 billion sales) was the most established of all plant-based segments, followed by "plant-based dairy" ($1.4 billion) and then plant-based meat ($939 million). The least developed category, although the fastest-growing, is plant-based eggs ($10 million) (GFI, 2020).

The fast-evolving plant-based food market ignites the development of technologies that can increase the functionality of essential plant-based constituents. Companies like Novameat created 3D-printed meat through the extrusion of filaments made from peas, seaweed, and rice that imitates the taste and mouthfeel of meat (Montes & Moynihan, 2018). Legendary Vish created 3D-printed salmon fillets using plant-based ingredients like mushroom proteins and algae extracts (Holmyard, 2020). 3D printing technology brings functionality to plant-based ingredients in a sustainable way. Unlike subtractive manufacturing, 3D printing uses only the necessary material when the layers are added to the food construct. Besides, 3D printing enables reusing food waste converting it into new food products. In the Netherlands, a start-up company called uprinting uses residual foods to create 3D-printed foods (Upprinting, 2021). Recent publications reveal the potential of 3D printing to design texture (Pant et al., 2021), customize nutritional content (Srivastava, 2020), and encapsulate probiotics (Liu et al., 2020).

This chapter brings a good overview of the recent developments of 3D-printed plant-based foods. Section 16.2 describes the principle of extrusion-based 3D printing. Section 16.3 reports the feasibility of essential plant-based constituents (proteins and carbohydrates) for 3D printing. Section 16.4 highlights the potential of 3D printing technology to design texture and enhance the encapsulation of micronutrients and probiotics in plant-based formulations. Section 16.5 describes the advances in 4D printing.

16.2 Extrusion-based 3D printing

Extrusion processes are used in most 3D food printing approaches. In extrusion-based 3D printing, food formulations (known as edible inks) are extruded through a nozzle as a continuous viscoelastic filament. A 3D shape is formed driven by binding mechanisms within the layers deposited at regions predetermined by a 3D model (Godoi et al., 2016). The accommodation of layers may involve phase change or not. When the edible ink (melted state) is too liquid, solidification mechanisms are adopted to ensure self-sustaining properties of the 3D shape under construction. Solidification and/or binding mechanisms occur upon cooling by crystallization or glass transition (Bhandari et al., 1997), enzymatic cross-linking (Dong et al., 2020), or gelling mechanisms (Fan et al., 2020). The binding mechanisms within the extruded

Engineering Plant-Based Food Systems. https://doi.org/10.1016/B978-0-323-89842-3.00001-4

filaments for materials such as dough (Yang et al., 2018) and vegetal pastes do not involve phase change and are rather controlled by the material cohesive property. In most cases, printability is driven by a decay in viscosity under shear forces followed by partial (or nearly total) recovery after extrusion, characterizing a shear-thinning behavior.

16.2.1 3D printing stages

3D food printing is typically carried out in five steps: (S1) technology selection, (S2) modeling, (S3) slicing, (S4) G-code, and (S5) 3D printing (Fig. 16.1). The first step (S1) involves the selection of extruding methods (pneumatic or mechanical) based on the physical state of the edible ink. Pneumatic configurations can handle high-viscosity formulations with the extruding force limited only by the air pressure capabilities of the system. Mechanical methods are driven by a stepper motor and can be divided into two types: piston- and screw-driven systems. Compared to a pneumatic-driven system where there is a delay of the compression of the gas volume, the piston-driven configuration usually promotes more direct control over the printable formulation (Murphy & Atala, 2014). In screw-driven systems, the printable formulation can be continuously fed either in particulate state or paste-like consistency. The extrudability and homogeneity of the ink can be controlled by adjusting the speed of the shafts in the extruder (Mantihal et al., 2017).

The second step (S2) consists of modeling using a computer-aided design (CAD) software (e.g., TinkerCad or SketchUp). The resulting 3D model is converted into STL file and sliced (step S3) to generate a G-code (step S5). As the name suggests, the G-code is composed of geometric commands responsible for guiding the print-head location (X-Y-Z). Besides, the G-code contains information about the flow rate and temperature of the printer head and printer bed. A successful printing job (step S5) relies on the synergism within food ingredients, uniform flow throughout the nozzle opening, and instant recovery after extrusion (Godoi et al., 2019).

16.2.2 Extruding forces and controllable printing variables

Food inks are complex systems usually composed of multiple components and ingredients with time-dependent properties. This makes the printing process challenging because the printer variables are not easily changed during the printing job. Fig. 16.2 shows the three main extrusion forces used to 3D print food materials: piston-driven, screw-driven, and pneumatic systems extrusion. The flow rate is controllable by increasing the pressure or speed of the step motor, respectively, of pneumatic and piston-driven or screw-driven systems. Piston-driven and pneumatic systems are more indicated for printing small quantities per batch of the formulation. Both configurations provide direct control over the flow of paste-like materials through the nozzle. The advantage of the screw-driven system relies on the possibility of continuous feeding and printing of inks in powder format.

Printing speed, the diameter of the extruding nozzle, the temperature of the print bed, and the print head are examples of controllable printing parameters. All parameters together should be taken into consideration and aligned with the properties of the food materials to achieve a smooth printing job and high-quality 3D-printed construct. Most of the 3D food printers can operate from room temperature to around 150°C. These values enable 3D printing of a variety of food materials (with a broad spectrum of melting and/or gelation temperature) with viscosities ranging from 0.001 to 250 Pa s (Derossi et al., 2019). Higher viscosity requires lower printing speeds which, in turn, can be controlled by the nozzle diameter. Larger the nozzle diameter, faster printing moves are required to withstand flow increments. 3D printing of intricate constructs entails the use of smaller nozzle diameters (0.2 mm to a maximum 1 mm). High-viscosity materials require higher printing pressures. When the pressure needed to extrude material surpasses the pressure given by the step motor, even if the material can be extruded, its actual extrusion rate will be different from the set parameters (Guo et al., 2019).

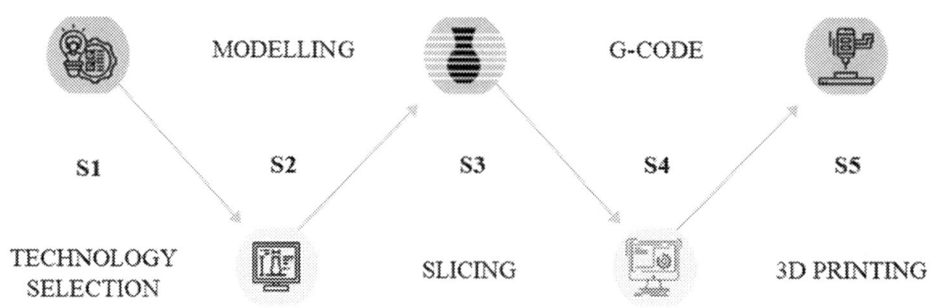

FIGURE 16.1 3D food printing steps.

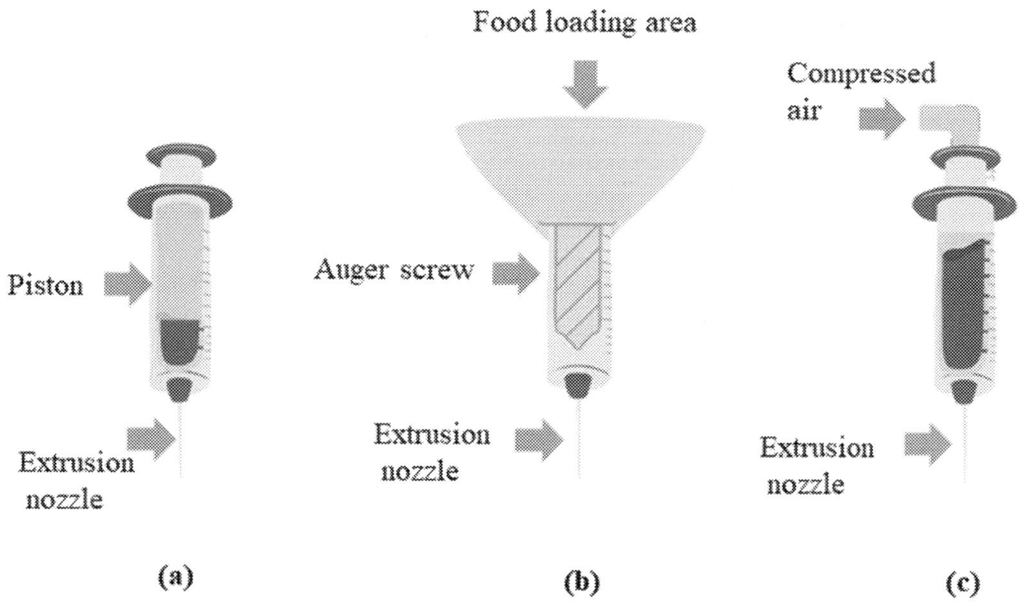

FIGURE 16.2 Main extrusion forces used to 3D print food materials: (A) piston-driven, (B) screw-driven, and (C) pneumatic systems extrusion.

16.2.3 Food ink quality and assessment of the 3D-printed construct

Rheological properties are key factors influencing the printing behavior of food inks. A printable food formulation exhibit both flow and shape retention properties. During extrusion, inks should flow with minimal internal resistance (shear-thinning behavior). After extrusion, the flow should be interrupted by the development of internal forces opposing deformation. The yield stress and storage modulus (G′) are parameters associated with the elastic shape retention. Increased yield stress usually enhances filament formation and stiffness of the 3D construct (Schwab et al., 2020).

Filament formation and shape fidelity can be determined by image analysis of pictures obtained with an optical microscope or high-resolution cameras. Ideal printing consistency is characterized by a uniform extruded filament in the three dimensions and smooth surface. At conditions of under gelation or low viscosity, the ink will show liquid-like behavior and formation of a droplet at the nozzle tip. When the concentration is too high (or at conditions of over-gelation), a stiff filament with fractures throughout its length is formed. Ouyang et al. (2016) showed that the printability can be associated with the level of circularity of the internal square area in a lattice structure. A circularity index of 1 means ideal printability. At conditions of under gelation and over gelation, circularity index is above and below 1, respectively (Ouyang et al., 2016).

16.3 Essential plant-based constituents and their feasibility for 3D printing

In a multicomponent system, differences in the fraction of proteins, carbohydrates, and fat will influence the melting behavior, glassy state, and plasticization of the food inks. This section focuses on the role of plant-based proteins and carbohydrates in the design of 3D-printed food.

16.3.1 Plant proteins

Plant-based proteinaceous powders (PPPs) are usually presented as flours (<65% protein, dry weight basis), concentrates (>65% protein, dry weight basis), or isolates (>90% protein, dry weight basis). PPPs can be used in food inks to customize nutritional value and printability. The nutritional value is enhanced by the content of essential amino acids and bioactive peptides. Printability, however, is tunned by functional properties such as water-holding capacity, fat binding, and gelation mechanisms of proteins.

Plant proteins can be divided into four groups (Fig. 16.3) according to their sources: pulses, cereal, oilseed, and biomass. The right combination of plant proteins in a single paste-like formulation or loaded in multiple nozzles can supply enough essential amino acids in the 3D-printed food. Group 1, pulses contain high amounts of protein (18%–32%) and are

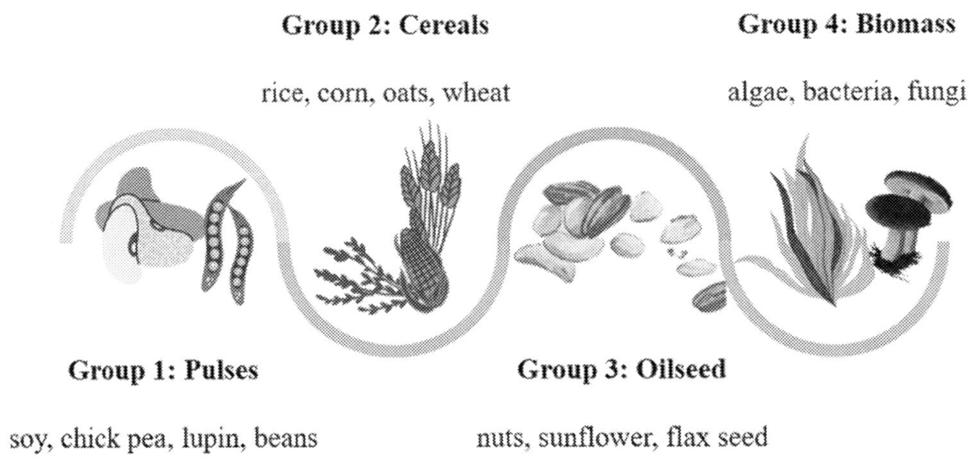

FIGURE 16.3 Plant-proteins classified into four groups according to their sources: pulses, cereal, oilseed, and biomass.

a valuable sources of energy, dietary fiber, minerals, and vitamins (Boye et al., 2010). Group 2 (cereals) is mainly composed of wheat, barley, rye, rice, maize, and sorghum. Of these, the dough-forming properties of wheat's storage proteins, "gluten" is remarkable (Bekes et al., 2004). Oil processing by-products (Group 3) are rich in proteins of excellent nutritive and bioactive opening opportunities to develop healthier products or new functional foods. Group 4, single-cell proteins (SCPs), also known as dried cells of microorganisms (algae, bacteria, actinomycetes, and fungi) are a healthy source of vitamins, carotenes, and carbohydrates (Upadhyaya et al., 2016).

PPPs are sensitive to the application of external stress (temperature or mechanical strength) or compounds (e.g., strong acid or base, crosslinking agents). Many plant proteins are globular, suggesting that controlled denaturation methods can expose buried reactive groups and enhance functional properties such as emulsifying capacity, foaming, and gelation. The formation of denatured aggregates and gelation properties are key factors governing the mechanical and binding properties of printed layers (Godoi et al., 2019). The least gelling concentration (LGC) measures the ability of PPPs to form a network between denaturated molecules aggregates containing large amounts of trapped water. A lower LGC indicates an enhanced gelling capacity (Raikos et al., 2014).

The gelling behavior of PPPs is dependent on the environment pH. At the isoelectric point (pI), the overall charge of proteinaceous polymers is zero. At pH values far from the pI, the protein surface is charged, leading to significant repulsive forces that can compromise the gelling capacity. Raikos et al. (2014) studied the effect of pH on the gelling behavior of pulse (green pea, fava bean, lupin, and hemp) and cereal (wheat and buckwheat) flours at different concentrations. Their study revealed that wheat and buckwheat (flours with the highest carbohydrate content) depicted the lowest LGC. The authors hypothesized that carbohydrates can positively influence the gelling behavior by decreasing the thermodynamic affinity of proteins for water molecules and favoring protein–protein interactions. Except for hemp flour, all PPPs are gelled within a concentration range of 100–140 g/L. The repulsive forces induced by pH changes played a more significant role for lupin flour with a decreased gelling capacity at pH 10. Hemp flour gave rise to a paste-like product even at the upper concentration range tested (200 g/L) (Raikos et al., 2014). These values are useful and provide an estimated initial concentration value for 3D printing processing involving pulse and cereal proteins. The gelation speed can be increased by augmenting the temperature of the printing head. The 3D printer user, however, should bear in mind that a fast gelation mechanism requires high translation speeds of the printing head.

The interplay between formulation and 3D printer parameters defines the printability of the food material. It is fundamental to understand the gelling behavior of the proteins in the system. Concentration, temperature, and pH should be monitored to avoid issues related to the under- or over-gelation. Under-gelation (or low-viscosity) results in poor shape-fidelity and self-structuring properties. Over-gelation, however, can be associated with high printing pressures and blockage of the printing nozzle. The ideal printability can be achieved by (1) blending several ingredients, which synergistically will provide the right rheology for printing or (2) pretreating the food-ink aiming to enhance gelling properties. Oyinloye and Yoon (2021) found optimum printing conditions by blending alginate with pea protein (80:20). They concluded that the filament thickness is responsible for the residual-stress component and total deformation across the printed construct (Oyinloye & Yoon, 2021). Fan et al. (2020) studied the effect of microwave pretreatment on the printability of the soybean protein isolate-strawberry ink system (Fig. 16.4). Their results demonstrated a combined

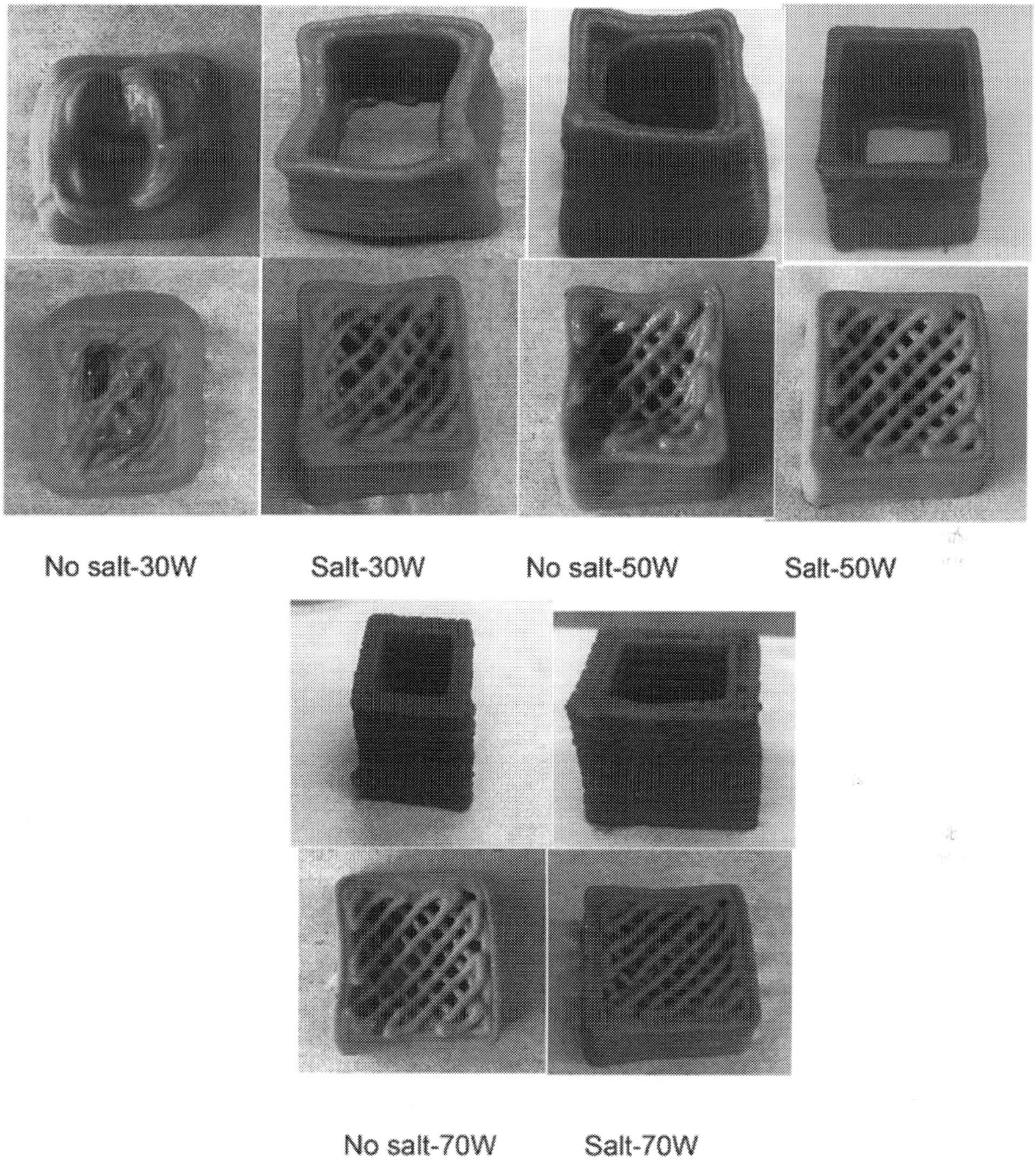

FIGURE 16.4 3D-printed soy protein and strawberry formulations: effect of microwave pretreatment at different power levels (W) and salt on the printability. *From Fan, H., Zhang, M., Liu, Z., & Ye, Y. (2020). Effect of microwave-salt synergetic pre-treatment on the 3D printing performance of SPI-strawberry ink system.* LWT, 122, *109004. https://doi.org/10.1016/j.lwt.2019.109004.*

positive effect between salt and microwave treatment on printing accuracy and structuring properties of the ink. As can be seen by Fig. 16.4 (left picture on the top), in the absence of salt and low microwave voltage (30 W), the 3D-printed layers were not self-sustaining and collapsed after deposition. Optimized printing conditions were achieved for a salted formulation treated with 70 W microwave (Fan et al., 2020).

16.3.2 Carbohydrates

The carbohydrate fraction in food formulations may vary from simple sugars (e.g., glucose) to polydisperse carbohydrate oligomers such as maltodextrins. 3D printing can be optimized by playing with the ratio between low- and high-molecular-weight carbohydrates (Godoi et al., 2016). The literature reports that sugar has a plasticizer effect in starch-rich systems by acting as a co-solute with starch and causing decay on the Tg. Pure starches show Tg ranging up to 243°C, while simple sugars such as fructose present Tg as low as 5°C (Bhandari et al., 1997). The 3D printing temperature of powder form of the ink is an important parameter that can be estimated based on the Tg of the formulated food ink. The mobility of the molecules is enhanced at conditions a few degrees above the Tg. The enhanced flow above the Tg enables a printing job without interruptions caused by nozzle clogging. Immediately after leaving the nozzle, the food filament reaches the print bed that should set at a temperature below the Tg of the mixture. The temperature gradient between the food and print bed allows shape fidelity by controlling in molecular level the mobility of the constituents in the powder-based extruded ink.

The printability of starch-based paste formulations has been reported in several works (Feng, Wu, et al., 2020; Feng, Zhang, et al., 2020; Liu et al., 2019). The increase in starch concentration (not higher than 25% w/w) has been associated with printability enhancements due to the formation of a gel structure with suitable rheological properties for printing. Maniglia and collaborators explored modification routes (Maniglia, Castanha, et al., 2020) to improve the printability of starch. Ozone treatment (Maniglia et al., 2019) and dry heating treatment (DHT) (Maniglia, Lima, et al., 2020) were found to enhance the printability of cassava starch and wheat starch, respectively.

Ozone processing promoted starch oxidation mainly by the depolymerization of cassava starch molecules through the cleavage of (1 → 4)-α-d and (1 → 6)-α-d glycosidic linkages and minor oxidation of hydroxyl groups to carbonyl and carboxyl groups. DHT, however, resulted in slight depolymerization of starch and reduction of the crystalline zones in the wheat starch granule. Optimized printing conditions were achieved with 10.7% wt cassava starch suspension ozonated for 30min pregelatinized at 65°C or 95°C for 5 min (Maniglia et al., 2019). In the case of wheat starch, optimized printability was achieved for formulation made of 5% modified wheat starch by DHT (DHT_2h or DHT_4h). For both cassava and wheat starches, the 3D printing job was conducted at 20°C using a nozzle with 0.8 mm diameter. The printing speed, however, was 20 mm/s for cassava starch and 5 mm/s for wheat starch. Before printing, the hydrogels were stored in the refrigerator for 7 days (cassava starch) or 1 day (wheat starch). Fig. 16.5 illustrates the enhanced printability depicted by the ozonated cassava starch.

The printability of food inks can also be enhanced by the addition of plant-based hydrocolloids that are commonly used to enhance melting, gelatinization, and retrogradation of starch in doughs. As an example, xanthan gum can induce dough strengthening by increasing water absorption and the ability of the dough to retain gas. Locust bean gum is known to provide higher baked product yields by improving the final texture and increasing the viscosity of the dough (Mandala et al., 2007). In extrusion-based techniques, hydrocolloids can be combined with food ingredients giving rise to different rheological behavior of the mixture. Pant et al. (2021) have used xantham gum and locust bean gum to produce 3D-printed vegetable pastes of soft texture for dysphagic diet. The vegetable pastes used by Pan et al. (2021) were made of garden pea, carrot, or bok choy with water content of 80%, 90%, and 96%, respectively. Garden peas were 3D-printed without thickener agents and showed minimal water leakage. Xanthan gum was used to improve the printing behavior of the carrot-based formulation, while combined xanthan gum with locust bean gum was added into bok choy for printing stable shapes (Pant et al., 2021).

16.4 Infill percentage and pattern: texture design and encapsulation of micronutrients

16.4.1 Texture design

One of the biggest advantages of 3D food printing over conventional food manufacturing methods is the ability to build internal structures. 3D printing food with different infill geometries and infill patterns not only saves food ink but enables the creation of new textures. The infill density and pattern influence the mechanical properties of the final 3D printed object. Mantihal et al. (2019) controlled the density and weight of the 3D-printed chocolate by changing the infill percentage. For a 60% infill, the authors reported stronger structural stability of chocolate constructs printed using the star- and honeycomb-infill patterns in comparison with the Herbert infill pattern (Mantihal et al., 2019).

Recently, Feng, Wu, et al. (2020) blended yam powder with potato processing by-product at different mass ratios (10:0, 9:1, 8:2, and 7:3) to 3D print snacks with infill levels ranging from 20% to 80%. After printing, the snacks were air-fried. The experimental matrix conducted by the authors has also included variations in the infill pattern: parallel-structures

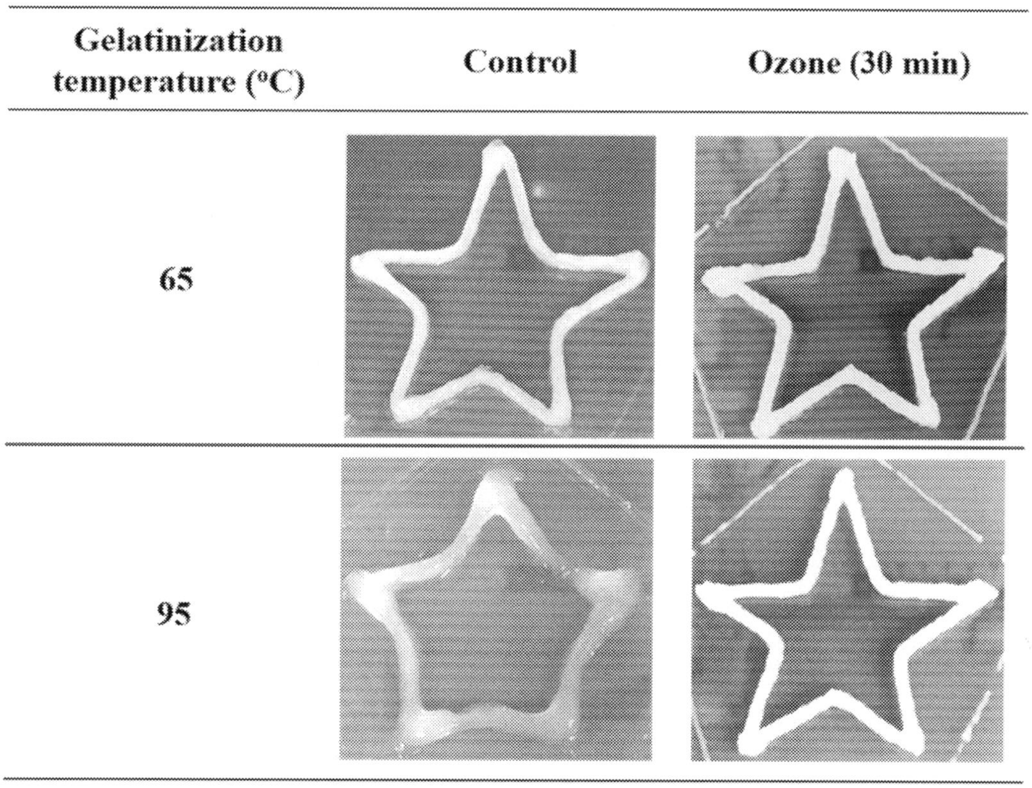

FIGURE 16.5 3D-printed cassava starch gels: control (native starch) and starch ozonated for 30min, pregelatinized at 65 and 95°C, and stored for 7 days at 5°C before printing. *From Maniglia, B. C., Lima, D. C., Matta Junior, M. D., Le-Bail, P., Le-Bail, A., & Augusto, P. E. D. (2019). Hydrogels based on ozonated cassava starch: Effect of ozone processing and gelatinization conditions on enhancing 3D-printing applications.* International Journal of Biological Macromolecules, 138, *1087–1097. https://doi.org/10.1016/j.ijbiomac.2019.07.124.*

(Rectilinear (0 degree) and Wiggle (0 degree)), cross-structures [Rectilinear (90 degrees) and Wiggle (90 degrees)], and complex structures (Triangular and Honeycomb). The authors observed that the infill percentage and pattern influenced the mechanical properties of the fried 3D printed snack. At low infill level (20%), parallel structures internal structures resulted in a harder fried snack (46 N), followed by cross-structures and complex structures (31 N). At medium and high infill level, triangular internal structures led to higher hardness values (148 N) than cross-structures, parallel-structures, and Honeycomb structures (99 N). Overall, the hardness of air-fried products decreased with the increased porosity induced by the infill design (Feng, Wu, et al., 2020). Fig. 16.6 shows 3D-printed blends of yam powder with potato processing by-product (at various infill patterns and percentages) before and after air-frying.

16.4.2 Encapsulation of micronutrients and probiotics

The encapsulation of micronutrients and probiotics into traditional foods has attracted the attention of consumers seeking products that provide more than basic nutrition. Particularly, baking goods can be enriched with active ingredients, offering a valuable nondairy alternative of supplementation vehicle. However, baking can compromise the functionality of these heat-sensitive active ingredients due to high temperatures involved during the process. Throughout baking, the heat transport in the dough is governed by the classic evaporation–condensation mechanism. Different temperature and moisture content in the inner part and the outer layer of the dough give rise to the macroscopic features in bread (soft and porous crumb and crispy and dense crust) (Zhang et al., 2019).

Zhang et al. (2018) studied the effect of baking conditions and storage on the viability of *Lactobacillus plantarum* supplemented bread. The researchers observed higher bacteria survival in the bread crust than in the crumb despite the higher temperature in the crust during baking under certain conditions. Thus, the thermo-stability of *Lactobacillus plantarum* in the crust was associated with reduced moisture content and glassy microstructure acting as a protective shell to the bacteria during baking (Zhang et al., 2018). These findings bring insights into the use of 3D printing to boost the heat resistance of bacteria during baking. Besides playing an important role in texture design (Section 16.4.2), the infill pattern

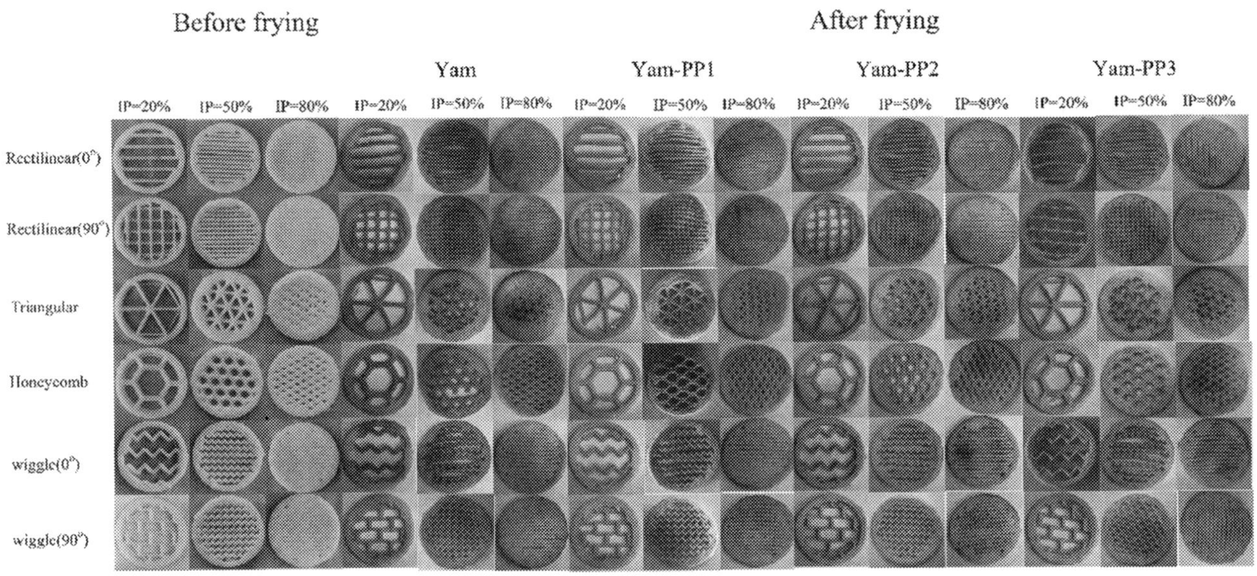

FIGURE 16.6 3D-printed blends of yam powder with potato processing by-product (mass ratios 10:0, 9:1, 8:2, and 7:3) before and after air-frying. Infill levels ranging from 20% to 80%. Infill pattern: parallel structures (Rectilinear (0 degree) and Wiggle (0 degree)), cross-structures (Rectilinear (90 degrees) and Wiggle (90 degrees)), and complex structures (Triangular and Honeycomb). *From Feng, C., Zhang, M., Bhandari, B., & Ye, Y. (2020). Use of potato processing by-product: Effects on the 3D printing characteristics of the yam and the texture of air-fried yam snacks.* LWT, 125, *109265. https://doi.org/10.1016/j.lwt.2020.109265.*

and percentage can be strategically used to find the right balance between water evaporation—condensation mechanism and bioactivity kinetics of probiotics. Zhang et al. (2018) encapsulated probiotics into dough formulations 3D printed with honeycomb and concentric internal designs. The authors observed the surface-to-volume ratio depicted by the honeycomb structure accelerated the baking process and favored the viable counts of probiotics at the endpoint of baking (defined by 6% moisture decay at 145°C) (Zhang et al., 2018).

Moreover, 3D printing can be used as an encapsulation method. Probiotics can be introduced at precise locations during layer-by-layer deposition. The probiotic encapsulation is facilitated when multiple monocentric nozzles or a single concentric nozzle is used (co-extrusion). When multiple nozzles are used (Fig. 16.7A), the first extruded layer is composed of an encapsulating hydrogel. The consecutive layer is extruded by a second cartridge loaded with a probiotic formulation. Thus, the encapsulation is completed with the deposition of a third layer by the cartridge containing the encapsulating hydrogel. The pattern and order of layering deposition may vary to enhance the encapsulation capacity of the technique. Fig. 16.7B illustrates the encapsulation of probiotics using a concentric nozzle. The process is called co-extrusion process and can be performed under the same conditions as the extrusion process shown by Fig. 16.7A. Probiotics encapsulated by co-extrusion are protected by a hydrogel wall resembling a core—shell structure. Silva et al. (2018) reported that, in comparison with the extrusion process, co-extrusion increased probiotic (*Lactobacillus acidophilus* LA3) survival by about 80% in gastrointestinal fluids. The 3D printing extrusion and co-extrusion techniques can be used for the encapsulation of not only probiotics but micronutrients in general (Silva et al., 2018).

Liu et al. (2021) studied the effect of printing parameters and storage time on the viability of probiotics (*Bifidobacterium animalis* subsp. *Lactis*) encapsulated into 3D-printed mashed potatoes. The authors observed that the tighter nozzle diameter (0.6 mm) used in the printing job affected the process productivity and the viability of probiotic microorganisms. Above 55°C, the probiotic viability of BB-12. The 3D-printed structures containing probiotics remained stable after storage for 12 days at 5°C. The probiotic content in the final construct was above 9.773 log CFU/g, which is higher than the recommended least dose in probiotic foods (6 log CFU/g) (Liu et al., 2020).

Oliveira et al. (2020) studied the effect of 3D printing on the bioactivity of baked functional foods. The researchers observed a synergistic effect between encapsulation, time, temperature, number of layers, and infill of the printed cookies on the moisture and antioxidant activity of cookies enriched with a grape skin extract. The experimental design shown in Fig. 16.8 demonstrated that enhancements in both antioxidant activity and phenolic content are accomplished with 4-layer cookies and 30% infill, baking for 10 min at 180°C. The bioactivity and the antioxidant content, respectively, increased by 115% and 173% when compared to the free extract cookies. The production of new antioxidant species was associated with the improved heat transfer caused by the designed porosity (infill percentage), which, in turn, accelerated water removal and the starting of Maillard reactions (Oliveira et al., 2020).

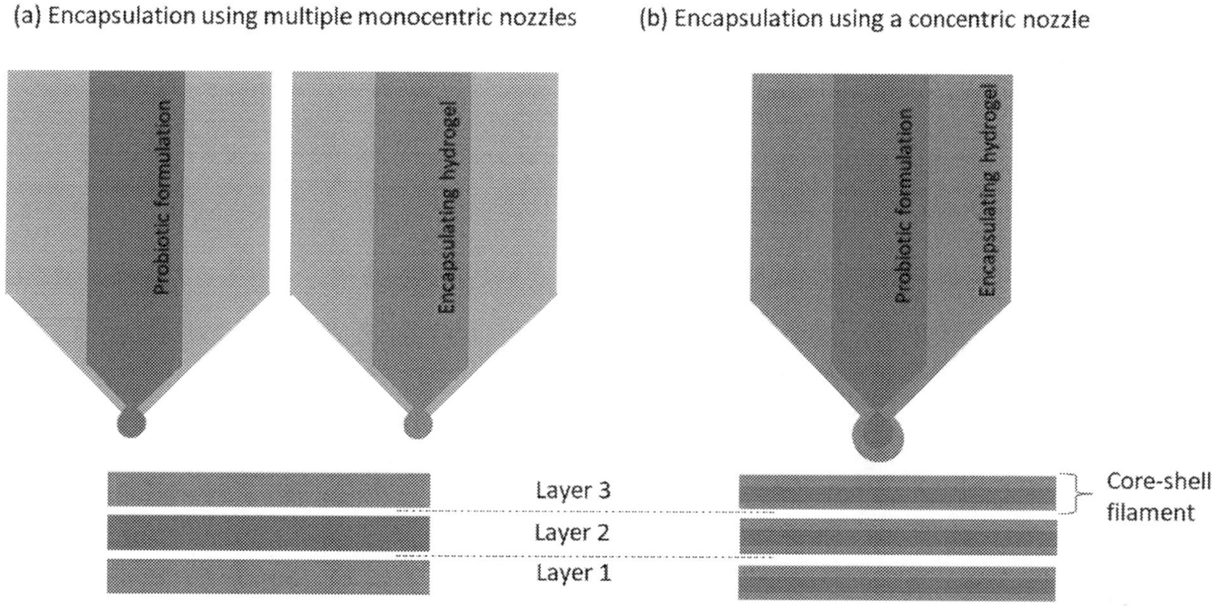

FIGURE 16.7 Schematic illustration of 3D printing: (A) extrusion and (B) co-extrusion techniques for probiotic encapsulation.

The literature using 3D printing technology for the encapsulation of probiotics and micronutrient is scarce (Liu et al., 2020; Oliveira et al., 2020; Vieira et al., 2020). 3D printing provides a new dimension in the development of functional foods. Opportunities can be visualized in the optimization of probiotics and/or micronutrients load. As well as, the upscaling of the process taking into consideration printing parameter and time-dependent properties of food materials.

16.5 4D printing

3D printing allows the creation of objects through the layer-by-layer deposition of formulations. 3D printing of materials possessing distinct expansion/shrinkage properties when exposed to an external stimulus (thermal, hygral or pH) can induce a structural shape shifting. The synergy between material science and 3D printing led to the 4D printing concept. 4D printing, however, is not only limited to shape-shifting (Fig. 16.9). 4D printing relates to a time-dependent change in the physical or chemical composition of 3D-printed shapes made of stimulus-responsive active materials (or smart materials) (Joshi et al., 2020).

4D printing finds a correlation with natural shape-morphing systems. As an example, the natural movement of plant parts caused by an external stimulus but unaffected in direction by it. Inspired by these vegetal systems, researchers around the world have printed plant-based architectures that yield complex four-dimensional shapes when exposed to an external stimulus (Chen, Zhang, Guo, & Chen, 2021; Sydney Gladman et al., 2016). These geometries are usually printed with plant-based composite hydrogels that are programmed with localized, anisotropic swelling pattern leading to fourth-dimensional printing pathways (Sydney Gladman et al., 2016).

Chen, Zhang, Guo, and Chen (2021) used pumpkin pure as smart material with shrinking due to the water loss during an air-drying process. Their work explored the potential of 3D printing to yield a complex 4D shape over time by the precise control of thickness and printing path (printed line angles) (Fig. 16.10). The authors observed self-bending and self-twisting depending on the print path adopted. Tubular deformation was observed when the lines were printed at 90 degrees angle. Distortion took place when the printing path was programmed within 0 degree $< \theta <$ 90 degrees or 90 degrees $< \theta <$ 180 degrees (Chen, Zhang, Guo, & Chen, 2021).

More recently, Liu et al. (2021) studied the shape shifting of 3D-printed starch-based structures induced by dielectric microwave, infrared, and hot air drying models (Fig. 16.11). The dielectric properties were also studied correlated with the 4D transformation of the 3D-printed constructs for the microwave stimulus. Dielectric constant (ε') was assigned to the ability of the material to store electromagnetic energy and dielectric loss factor (ε''), reflecting material's ability to convert electromagnetic energy into heat. The bending degree was positively affected by the shrinking ratio and negatively affected

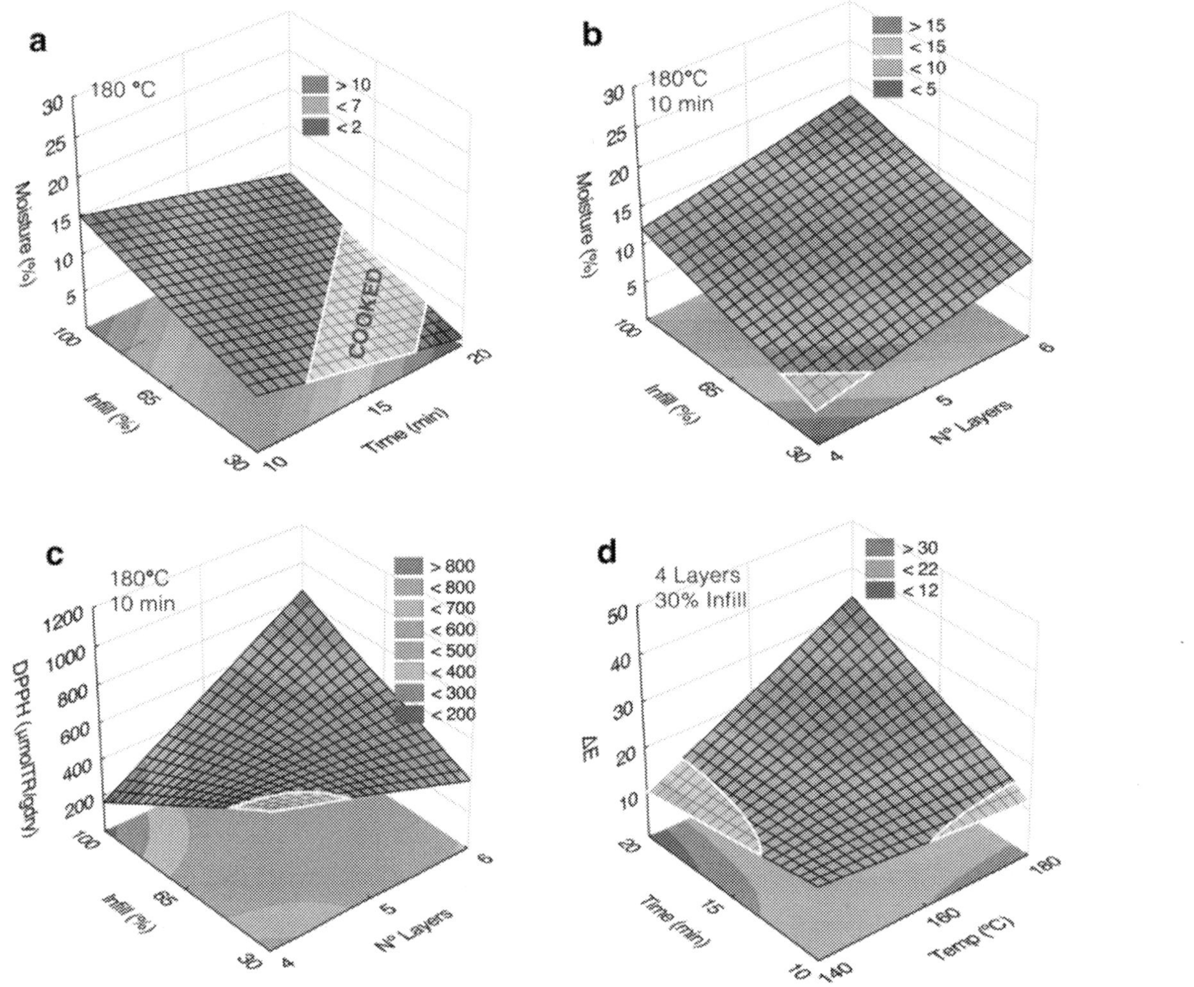

FIGURE 16.8 Surface curves of the combined effect of cooking and design variables on features of the 3D-printed cookies enriched with a grape skin extract. (A) Effect of the infill and time on the moisture when baking at 180°C. (B) Effect of the infill and N° Layers on the moisture when baking at 180°C for 10 min. (C) Effect of the infill and N° Layers on the antioxidant activity (DPPH) of cookies baked at 180 °C for 10 min. (D) Effect of the time and temp on the 4-layer and 30% infill cookies. *From Oliveira, S. M., Gruppi, A., Vieira, M. V., Matos, G. S., Spigno, G., Pastrana, L. M., Teixeira, A. C., & Fuci, P. (2020). How additive manufacturing can boost the bioactivity of baked functional foods. June. https://doi.org/10.1016/j.jfoodeng.2020.110394.*

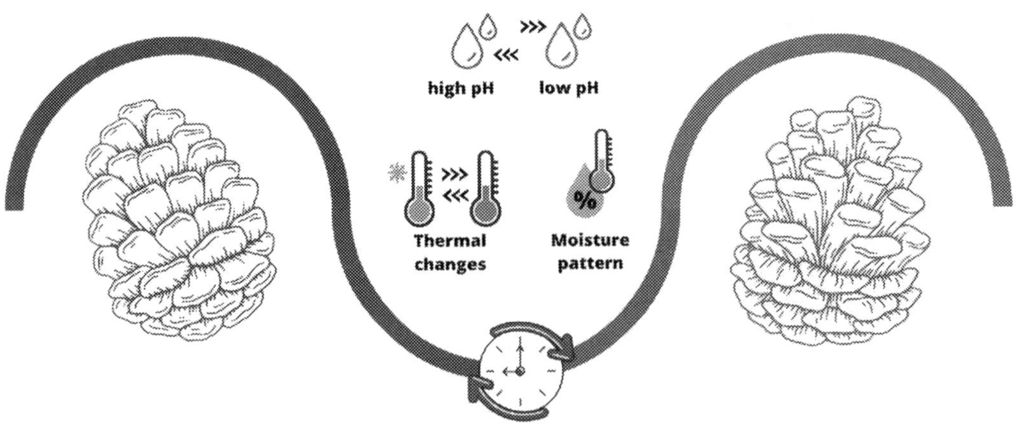

FIGURE 16.9 Schematic representation of the 4D printing concept.

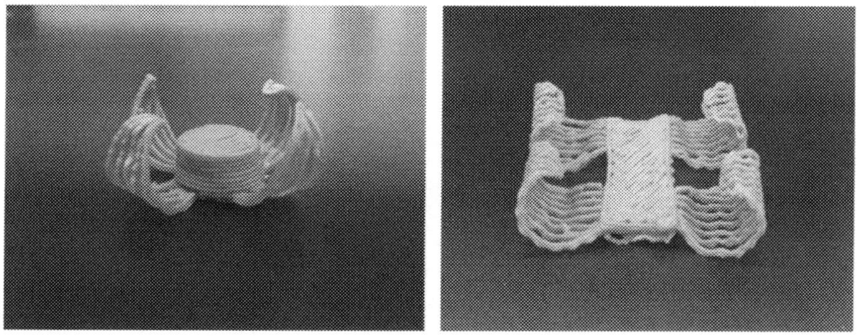

FIGURE 16.10 Complex deformation of the double-layer structure of pumpkin/paper. *From Chen, F., Zhang, M., Liu, Z., & Bhandari, B. (2021). 4D deformation based on double-layer structure of the pumpkin/paper.* Food Structure, 27, *100168. https://doi.org/10.1016/j.foostr.2020.100168.*

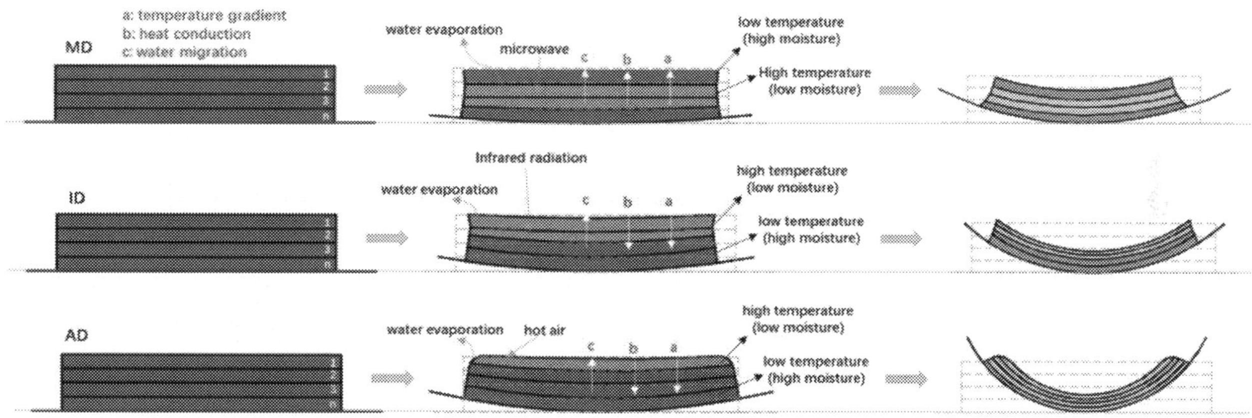

FIGURE 16.11 Shape change (forth-dimension bending angle) of 3D-printed starch-based formulations triggered by microwave (MD), infrared (IR), and hot air dehydration stimulus. *From Liu, Z., He, C., Guo, C., Chen, F., Bhandari, B., & Zhang, M. (2021). Dehydration-triggered shape transformation of 4D printed edible gel structure affected by material property and heating mechanism.* Food Hydrocolloids, 115, *106608. https://doi.org/10.1016/j.foodhyd.2021.106608.*

by increments in ε'' values. The microwave stimulus resulted in a rapid dehydration rate; however, the low shrinking ratio and internal heating model did not favor shape bending. Conversely, the hot air stimulus presented excellent shrinking ratio. For the hot air dying model, the dehydration took place from the surface to inward inducing a *bending accumulation effect* that enhanced the bending degree of the 3D-printed structure. Shorter dehydration time and smaller shrinkage ratio of infrared dried samples than microwave dried samples. As a result, the infrared stimulus promoted a smaller bending angle than the one induced by the microwave stimulus. Fig. 16.11 illustrates the effect of microwave, infrared and hot air dehydration stimulus on the bending angle of the 3D-printed shapes (Liu et al., 2021).

Plant-based foods have been used in 3D printing processes to induce not only shape-shifting but color/flavor changes under the exposure of an external stimulus (Ghazal et al., 2019). Ghazal et al. (2019) 3D-printed pH-responsive hydrogels composed of anthocyanin-potato starch blended with lemon juice gel. A time-dependent color change was observed from purple to red (fourth dimension) either due to the low pH promoted by the lemon juice gel or by spraying a solution with pH ranging from 2 to 2.5 on the 3D-printed construct. Chen, Zhang, Guo, and Chen (2021) studied the use of microwave stimulus to induce a color change in 3D printed lotus root hydrogel containing curcumin. The heat generated by microwave promoted an alkaline environment. As a result, the hydrogel color changed from yellow to red due to the deprotonation of curcumin (Chen, Zhang, Guo, & Chen, 2021). Phuhongsung et al. (2020) studied the flavor development (interpreted as a 4D effect) of 3D-printed plant-based structures under exposure to microwave heating. Printable formulations were prepared with soybean protein isolate, k-carrageenan, and vanilla flavor. An electronic nose, electronic tongue, and GC−MS demonstrated that the flavor of 3D-printed products were affected by the levels of microwave irradiation. Notably, four new generated flavor compounds (1-Octen-3-ol, maltol, ethyl maltol, and eugenal) were identified when the vanilla flavor was added (Phuhongsung et al., 2020).

In summary, a wide range of plant-based food materials, monolithic or combined with others, can be used as smart materials in such a way to respond when exposed to external stimuli such as temperature, humidity, and differences in pH. The 3D-printed structure can be stimulated for self-assembly, reconfiguration, and replication through eco-friendly energies. This opens new opportunities, such as a substantial volume reduction for storage, for transformations developed from a flat-pack 4D-printed structures (Momeni et al., 2017).

16.6 Summary and future directions

This chapter provided practical guidance about printing parameters and relevant materials properties of essential plant-based foods (proteins and carbohydrates). The structure and rheological properties of plant-based foods can be optimized to achieve printability. 3D printing technology offers many possibilities for the plant-based food market segment in at least four dimensions: (1) 3D-printed plant-based meat, (2) customized nutrition, (3) sustainability, and (4) decentralized production. The ability to design intricate internal structures enabled companies like Novameat and LegendaryVish to develop, respectively, 3D-printed steak and salmon with 100% plant-based ingredients. Customization of nutritional content by printing encapsulated vitamins into a gel reached commercialization in 2019 with the launch of Rem3dy Health, a UK company that stands behind the Nourished brand. Sustainability is claimed by Uprinting Food, a Dutch company using plant-based residual food to 3D print tasting and good-looking foods. Regarding decentralized production, 3D printing permits that everyone can be a developer and producer at once (prosumer trend). In a decentralized food manufacturing model, the consumer has control over the ingredients used to produce food.

To disrupt the four dimensions listed above in the food sector, 3D printing needs to overcome challenges related to scalability, printability of materials, and consumer perception. We believe, however, that a digital revolution and synergy between several areas of expertise will ignite the commercial realization of 3D printing using plant-based and all food-ingredient groups. 3D-printed structures may need post-processing (such as thermal treatment), which can trigger desirable or undesirable changes during this postprocessing step. 4D printing concept introduces to control the formulation, design, and printing parameters to achieve the desirable changes in shape and textural properties of 3D-printed products during postprocessing step. There are few works reported in this field; however, there is an ample opportunity to expand the research on 4D printing of food materials.

References

Bekes, F., Wrigley, C., & Wrigley, C. (2004). *CEREALS: protein chemistry* (pp. 254–262). Elsevier. https://doi.org/10.1016/B0-12-765490-9/00130-0

Bhandari, B. R., Datta, N., & Howes, T. (1997). Problems associated with spray drying of sugar-rich foods. *Drying Technology, 15*(2), 671–684. https://doi.org/10.1080/07373939708917253

Boye, J., Zare, F., & Pletch, A. (2010). Pulse proteins: Processing, characterization, functional properties and applications in food and feed. *Food Research International, 43*(2), 414–431. https://doi.org/10.1016/j.foodres.2009.09.003

Chen, C., Zhang, M., Guo, C., & Chen, H. (2021). 4D printing of lotus root powder gel: Color change induced by microwave. *Innovative Food Science & Emerging Technologies, 68*, 102605. https://doi.org/10.1016/j.ifset.2021.102605

Chen, F., Zhang, M., Liu, Z., & Bhandari, B. (2021). 4D deformation based on double-layer structure of the pumpkin/paper. *Food Structure, 27*, 100168. https://doi.org/10.1016/j.foostr.2020.100168

Choudhury, D., Singh, S., Seah, J. S. H., Yeo, D. C. L., & Tan, L. P. (2020). Commercialization of plant-based meat alternatives. *Trends in Plant Science, 25*(11), 1055–1058. https://doi.org/10.1016/j.tplants.2020.08.006

Derossi, A., Caporizzi, R., Ricci, I., Severini, C., Godoi, F. C., Bhandari, B. R., Prakash, S., & Zhang, M. (2019). *Chapter 3 - Critical variables in 3D food printing* (pp. 41–91). Academic Press. https://doi.org/10.1016/B978-0-12-814564-7.00003-1

Dong, X., Pan, Y., Zhao, W., Huang, Y., Qu, W., Pan, J., Qi, H., & Prakash, S. (2020). Impact of microbial transglutaminase on 3D printing quality of *Scomberomorus niphonius* surimi. *Lwt, 124*(September 2019). https://doi.org/10.1016/j.lwt.2020.109123, 109123–109123.

Fan, H., Zhang, M., Liu, Z., & Ye, Y. (2020). Effect of microwave-salt synergetic pre-treatment on the 3D printing performance of SPI-strawberry ink system. *LWT, 122*, 109004. https://doi.org/10.1016/j.lwt.2019.109004

Feng, C., Zhang, M., Bhandari, B., & Ye, Y. (2020). Use of potato processing by-product: Effects on the 3D printing characteristics of the yam and the texture of air-fried yam snacks. *LWT, 125*, 109265. https://doi.org/10.1016/j.lwt.2020.109265

Feng, L., Wu, J., Song, J., Li, D., Zhang, Z., Xu, Y., Yang, R., Liu, C., & Zhang, M. (2020). Effect of particle size distribution on the carotenoids release, physicochemical properties and 3D printing characteristics of carrot pulp. *LWT, 110576*. https://doi.org/10.1016/j.lwt.2020.110576

GFI. (2020). *Plant-based category dollar sales (summary). SPINSscan Natural and Specialty Gourmet (proprietary), SPINSscan Conventional Multi Outlet (powered by IRI), 52 weeks ending 2019-Dec-29*. https://www.gfi.org/marketresearch.

Ghazal, A. F., Zhang, M., & Liu, Z. (2019). Spontaneous color change of 3D printed healthy food product over time after printing as a novel application for 4D food printing. *Food and Bioprocess Technology, 12*(10), 1627–1645. https://doi.org/10.1007/s11947-019-02327-6

Godoi, F. C., Bhandari, B. R., Prakash, S., & Zhang, M. (2019). *Chapter 1 - An introduction to the principles of 3D food printing* (pp. 1–18). Academic Press. https://doi.org/10.1016/B978-0-12-814564-7.00001-8

Godoi, F. C., Prakash, S., & Bhandari, B. R. (2016). 3D printing technologies applied for food design: Status and prospects. *Journal of Food Engineering, 179*, 44–54. http://www.sciencedirect.com/science/article/pii/S0260877416300243.

Guo, C., Zhang, M., & Bhandari, B. (2019). Model building and slicing in food 3D printing processes: A review. *Comprehensive Reviews in Food Science and Food Safety*, 18. https://doi.org/10.1111/1541-4337.12443

Holmyard, N. (2020). *Legendary Vish creates 3D-printed salmon analogs made from mushroom proteins.* Algae extracts https://www.seafoodsource.com/news/foodservice-retail/legendary-vish-creates-3d-printed-salmon-analogs-made-from-mushroom-proteins-algae-extracts.

IISD. (2020). *World population to reach 9.9 billion by 2050.* https://sdg.iisd.org/news/world-population-to-reach-9-9-billion-by-2050/.

Joshi, S., Rawat, K., C, K., Rajamohan, V., Mathew, A. T., Koziol, K., Kumar Thakur, V., & A.S.S, B. (2020). 4D printing of materials for the future: Opportunities and challenges. *Applied Materials Today, 18*, 100490. https://doi.org/10.1016/j.apmt.2019.100490

Liu, Z., Bhandari, B., Prakash, S., Mantihal, S., & Zhang, M. (2019). Food Hydrocolloids Linking rheology and printability of a multicomponent gel system of carrageenan-xanthan-starch in extrusion based additive manufacturing. *Food Hydrocolloids, 87*(August 2018), 413–424. https://doi.org/10.1016/j.foodhyd.2018.08.026

Liu, Z., Bhandari, B., & Zhang, M. (2020). Incorporation of probiotics (*Bifidobacterium animalis* subsp. Lactis) into 3D printed mashed potatoes: Effects of variables on the viability. *Food Research International, 128*, 108795. https://doi.org/10.1016/j.foodres.2019.108795

Liu, Z., He, C., Guo, C., Chen, F., Bhandari, B., & Zhang, M. (2021). Dehydration-triggered shape transformation of 4D printed edible gel structure affected by material property and heating mechanism. *Food Hydrocolloids, 115*, 106608. https://doi.org/10.1016/j.foodhyd.2021.106608

Mandala, I., Karabela, D., & Kostaropoulos, A. (2007). Physical properties of breads containing hydrocolloids stored at low temperature. I. Effect of chilling. *Food Hydrocolloids, 21*(8), 1397–1406. https://doi.org/10.1016/j.foodhyd.2006.11.007

Maniglia, B. C., Castanha, N., Le-Bail, P., Le-Bail, A., & Augusto, P. E. D. (2020). Starch modification through environmentally friendly alternatives: A review. *Null*, 1–24. https://doi.org/10.1080/10408398.2020.1778633

Maniglia, B. C., Lima, D. C., da Matta Júnior, M., Oge, A., Le-Bail, P., Augusto, P. E. D., & Le-Bail, A. (2020). Dry heating treatment: A potential tool to improve the wheat starch properties for 3D food printing application. *Food Research International, 137*, 109731. https://doi.org/10.1016/j.foodres.2020.109731

Maniglia, B. C., Lima, D. C., Matta Junior, M. D., Le-Bail, P., Le-Bail, A., & Augusto, P. E. D. (2019). Hydrogels based on ozonated cassava starch: Effect of ozone processing and gelatinization conditions on enhancing 3D-printing applications. *International Journal of Biological Macromolecules, 138*, 1087–1097. https://doi.org/10.1016/j.ijbiomac.2019.07.124

Mantihal, S., Prakash, S., & Bhandari, B. (2019). Textural modification of 3D printed dark chocolate by varying internal infill structure. *Food Research International, 121*(September 2018), 648–657. https://doi.org/10.1016/j.foodres.2018.12.034

Mantihal, S., Prakash, S., Godoi, F. C., & Bhandari, B. (2017). Optimization of chocolate 3D printing by correlating thermal and flow properties with 3D structure modeling. *Innovative Food Science and Emerging Technologies.*

Momeni, F., M.Mehdi Hassani, N. S., Liu, X., & Ni, J. (2017). A review of 4D printing. *Materials & Design, 122*, 42–79. https://doi.org/10.1016/j.matdes.2017.02.068

Montes, L., & Moynihan, Q. (2018). This meat substitute is printed with a pea and rice protein paste and a 3D printer. In *Business Insider*. https://www.businessinsider.com/this-fake-meat-is-printed-in-a-lab-using-vegetables-and-a-3d-printer-2018-11?r=US&IR=T.

Murphy, S. V., & Atala, A. (2014). 3D bioprinting of tissues and organs. In *Nature Biotechnology*. https://doi.org/10.1038/nbt.2958

Oliveira, S. M., Gruppi, A., Vieira, M. V., Matos, G. S., Spigno, G., Pastrana, L. M., Teixeira, A. C., & Fuci, P. (2020). *How additive manufacturing can boost the bioactivity of baked functional foods.* https://doi.org/10.1016/j.jfoodeng.2020.110394. June.

Ouyang, L., Yao, R., Zhao, Y., & Sun, W. (2016). Effect of bioink properties on printability and cell viability for 3D bioplotting of embryonic stem cells. *Biofabrication.* https://doi.org/10.1088/1758-5090/8/3/035020

Oyinloye, T. M., & Yoon, W. B. (2021). Stability of 3D printing using a mixture of pea protein and alginate: Precision and application of additive layer manufacturing simulation approach for stress distribution. *Journal of Food Engineering, 288*, 110127. https://doi.org/10.1016/j.jfoodeng.2020.110127

Pant, A., Lee, A. Y., Karyappa, R., Lee, C. P., An, J., Hashimoto, M., Tan, U.-X., Wong, G., Chua, C. K., & Zhang, Y. (2021). 3D food printing of fresh vegetables using food hydrocolloids for dysphagic patients. *Food Hydrocolloids, 114*, Article 106546. https://doi.org/10.1016/j.foodhyd.2020.106546

Phuhongsung, P., Zhang, M., & Bhandari, B. (2020). 4D printing of products based on soy protein isolate via microwave heating for flavor development. *Food Research International, 137*. https://doi.org/10.1016/j.foodres.2020.109605, 109605–109605.

Raikos, V., Neacsu, M., Russell, W., & Duthie, G. (2014). Comparative study of the functional properties of lupin, green pea, fava bean, hemp, and buckwheat flours as affected by pH. *Food Science & Nutrition, 2*(6), 802–810. https://doi.org/10.1002/fsn3.143

Schwab, A., Levato, R., D'Este, M., Piluso, S., Eglin, D., & Malda, J. (2020). Printability and shape fidelity of bioinks in 3D bioprinting. *Chemical Reviews, 120*(19), 11028–11055. https://doi.org/10.1021/acs.chemrev.0c00084

Silva, M. P., Tulini, F. L., Martins, E., Penning, M., Fávaro-Trindade, C. S., & Poncelet, D. (2018). Comparison of extrusion and co-extrusion encapsulation techniques to protect Lactobacillus acidophilus LA3 in simulated gastrointestinal fluids. *LWT, 89*, 392–399. https://doi.org/10.1016/j.lwt.2017.11.008

Srivastava, A. (2020). *Nourished: World's first personalised 3D-printed vitamin is here and why you should try it? Silicon canals.* https://siliconcanals.com/news/startups/nourished-3d-printed-vitamin-uk/.

Steinfeld, H., Gerber, P., Wassenaar, T., Castel, V., Rosales, M., & Haan, C. (2006). Livestock's long shadow: Environmental issues and options. In , *5(1)*. *Communication Division*. FAO. https://doi.org/10.1890/1540-9295(2007)5[4:D]2.0.CO;2, 7–7.

Sydney Gladman, A., Matsumoto, E. A., Nuzzo, R. G., Mahadevan, L., & Lewis, J. A. (2016). Biomimetic 4D printing. *Nature Materials, 15*(4), 413–418. https://doi.org/10.1038/nmat4544

Upadhyaya, S., Tiwari, S., Arora, N., & Singh, D. (2016). *Microbial protein: A valuable Component for future food security.*

Upprinting.(2021). https://www.upprintingfood.com/press.

Vieira, M. V., Oliveira, S. M., Amado, I. R., Fasolin, L. H., Vicente, A. A., Pastrana, L. M., & Fuci, P. (2020). Food Hydrocolloids 3D printed functional cookies fortified with Arthrospira platensis. *Evaluation of Its Antioxidant Potential and Physical-Chemical Characterization*, (September 2019), 107. https://doi.org/10.1016/j.foodhyd.2020.105893

Yang, F., Zhang, M., Prakash, S., & Liu, Y. (2018). Physical properties of 3D printed baking dough as affected by different compositions. *Innovative Food Science & Emerging Technologies*. https://doi.org/10.1016/j.ifset.2018.01.001

Zhang, L., Boom, R. M., Chen, X. D., & Schutyser, M. A. I. (2019). *Recent developments in functional bakery products and the impact of baking on active ingredients*. https://doi.org/10.4995/ids2018.2018.7593. September.

Zhang, L., Taal, M. A., Boom, R. M., Chen, X. D., & Schutyser, M. A. I. (2018). Effect of baking conditions and storage on the viability of *Lactobacillus plantarum* supplemented to bread. *LWT, 87*, 318–325. https://doi.org/10.1016/j.lwt.2017.09.005

Chapter 17

Plant-based foods—future outlook

Christos Soukoulis
Environmental Research and Innovation (ERIN) Department, Luxembourg Institute of Science and Technology, Esch-sur-Alzette, Luxembourg

17.1 Introduction

According to UN's probabilistic population projections, the global population is forecasted to increase by 25% to almost 9.7 billion by 2050 (https://population.un.org/wpp/Download/Probabilistic/Population/), posing significant challenges regarding food security and malnutrition issues. In addition, the expanding urbanization and the shift in the consumers' diet patterns, as a result of socio-economical changes, create an alarming environment for ensuring the sustainable and resilient character of the food industry ecosystems as well as maintaining the public health and well-being indicators. Currently, public stakeholders, e.g., WHO-FAO, United Nations, European Commission etc., are working on the viable transformation of the food ecosystems to fulfill specific environmental, economic, and socio-cultural criteria. For example, EC has recently developed a policies roadmap in regard to the circular and sustainable management of food resources (Food 2030, 2018) that aims at the transformation of the conventional food ecosystems in a sustainable, ecologically friendly, and resilient manner without compromising in terms of food biodiversity and wholesomeness. This is outlined by the four primary goals of the Food 2030 initiative:

1. Nutrition for sustainable and healthy diets,
2. Food systems supporting a healthy planet,
3. Circularity and resource efficiency, and
4. Innovation and empowering communities.

As part of the Food 2030 initiatives frame, plant proteins (e.g., pulse, vegetable, cereal or pseudocereal isolated), edible freshwater unicellular (microalgae), and marine multicellular (seaweed) organism biomass are promising emerging food sources for tackling food security and creating alternative routes for personalized nutrition. In addition, optimization of the food production systems to minimize farm-to-fork food losses is another challenge to overcome in the near future. It is widely accepted that the adoption of a flexitarian diet regime, i.e., reduction of animal origin food in favor of plant-based food alternatives known as analogs, may contribute significantly toward tackling emerging challenges due to climate change and food security. Recommendations of the WHO European Region concerning shifting to predominantly plant-based diets have underpinned the need for:

1. Carrying out research on emerging diet regimes so to construct effective policies and guidelines plan for both food manufacturers and consumers,
2. Improving the labeling of plant-based foods,
3. Clearly defining, which plant-based foods can be enabled as the animal-origin food counterparts by considering processing, nutritional, and socio-cultural indicators, and
4. Developing or updating databases (e.g., foodDB) so as to create transparent mechanisms for inspecting the food supply and production chain (Wickramasinghe et al., 2021).

17.2 Clean-label issues in plant-based foods

To date, there is no established objective definition of the term "clean label"; instead, the clean label term is used to denote the consumers' familiarity with the food ingredients and production/processing method as well as the inferences that are

Engineering Plant-Based Food Systems. https://doi.org/10.1016/B978-0-323-89842-3.00002-6

drawn from the front of pack textual or visual claims and the back of backlist of ingredients or nutrition facts (Asioli et al., 2017). From an industrial stakeholder standpoint, the clean-label term may imply either the avoidance of chemically perceived ingredients such as E-number conveying food additives and preservatives or the use of organic and all-natural food ingredients that are minimally processed (Ingredion, 2014). Empirical consumer studies have demonstrated that the clean-label food trend is driven by six major factor categories including intrinsic and extrinsic product characteristics, biological, physiological, psychological, situational, and socio-cultural aspects (Asioli et al., 2017). Parameters such as the health conferring benefit claims, the absence of artificial ingredients or processing aids, and the sustainable character of the food are significant determinants of the consumers' preferences (Aschemann-Witzel, 2015). Exploring the consumers' categorization of plant-based reformulated foods (i.e., gelatine-free candy, dairy-free ice cream, vegan protein-rich beverages, and meat-free sausage), Aschemann-Witzel et al. (2019) unveiled that consumers tend to adopt a "line of reasoning" from unfamiliar to artificial and eventually, to unhealthy ingredients. In general, the association of a food ingredient with the clean label concept is driven by the classification of the ingredient in terms of its category type (e.g., sugar and syrups, flavoring agents or proteins) and techno-functionality as well as by the manner the ingredient is perceived by the consumers, i.e., harmless, unhealthy, basic, natural, etc (Aschemann-Witzel et al., 2019). A study exploring the association between the clean-label perception of several food ingredients (i.e., sugars, colorings, preservatives, stabilizers, and emulsifiers) and socio-demographic and consumer belief indicators revealed that stabilizers, emulsifiers, and colorings were considered as the least acceptable ingredients regardless of their natural source of origin, with the younger people or children/infant food buyers to be the most derogatory (Maruyama et al., 2021). On the other hand, highlighting the natural (e.g., plant or seaweed) origin of specific food ingredients such as sugars or stabilisers appeared to have a positive impact on their overall acceptability to the consumers.

Although the concept of protein fortification of conventional foods may be consorted with skepticism as regards their naturalness, healthiness, and safety, the exploitation of plant-sourced ingredients appears to partially alleviate the ambivalence of food fortification/enrichment. The ability of plant proteins to outperform animal protein sources (e.g., dairy or meat) stems from their direct link to the sustainable, ethical production, and health-promoting connotation (Banovic et al., 2018). The communication strategies adopted for promoting a plant-based diet lifestyle, i.e., sustainable, healthier, or with a focus on substitution of an animal-origin ingredient are narrowly associated with consumers' acceptability. Peschel et al. (2019) demonstrated that, depending on the adopted food manufacturer communication frame, differences in the size and structure of the consumers' attitude associative networks are observed. As a general trend, focus on food ingredient substitution enacts a strong communication of health and sustainability notions, and therefore, consumer acceptability vectors such as the sensorial quality, cost efficiency, environmental friendliness, and ethics need to be underpinned. On the other hand, a health-mediated communication plan needs to evoke better the connotations linked to environmental aspects, authenticity, geographical, and botanical origin (Peschel et al., 2019).

17.3 3D printed plant-based foods

Additive manufacturing (AM), commonly known as 3D printing, is a digitally controlled robotic manufacturing process employed in the construction of complex composite products using Computer-Aided Design (CAD) constructed three-dimensional models (Ngo et al., 2018). 3D printing technology has rapidly evolved over the last decade particularly in the fields of biomedicine, tissue and bone engineering, aerospace, automotive, biomechanics, and smart structure prototyping (Ngo et al., 2018).

In regard to the food industry sector, 3D printing has undergone a rapid expansion the last decade owing to the effectiveness of smart digital technologies coupled to 3D printers to construct complex macro-, meso-, or micro-structures, which in turn can confer specific technological, health, socio-cultural, and economic benefits Fig. 17.1 such as:

1. Processing and handling feasibility of intricate food shape or structures, which are impossible or uneconomical to be produced under conventional manufacturing,
2. Engaging the principles of Industry 4.0 principles into the conventional food production and retail chain,
3. Increasing the sustainability and eco-friendliness of mass food production, e.g., optimized management of water, energy and food resources, reduction of food waste losses, etc.,
4. Improving the nutritional fingerprint of processed food without compromising quality,
5. Introducing, in a cost and quality effective way, emerging or alternative food ingredients (e.g., insects, microalgae, food industry side stream recovered ingredients, etc.) into the food production chain,
6. Complementing or customizing dietary plans by considering individuals' lifestyle, genotype, phenotype, or metabiotypes, i.e., new foods for fast recovery or improved performance of athletes and sportsmen, space mission foods,

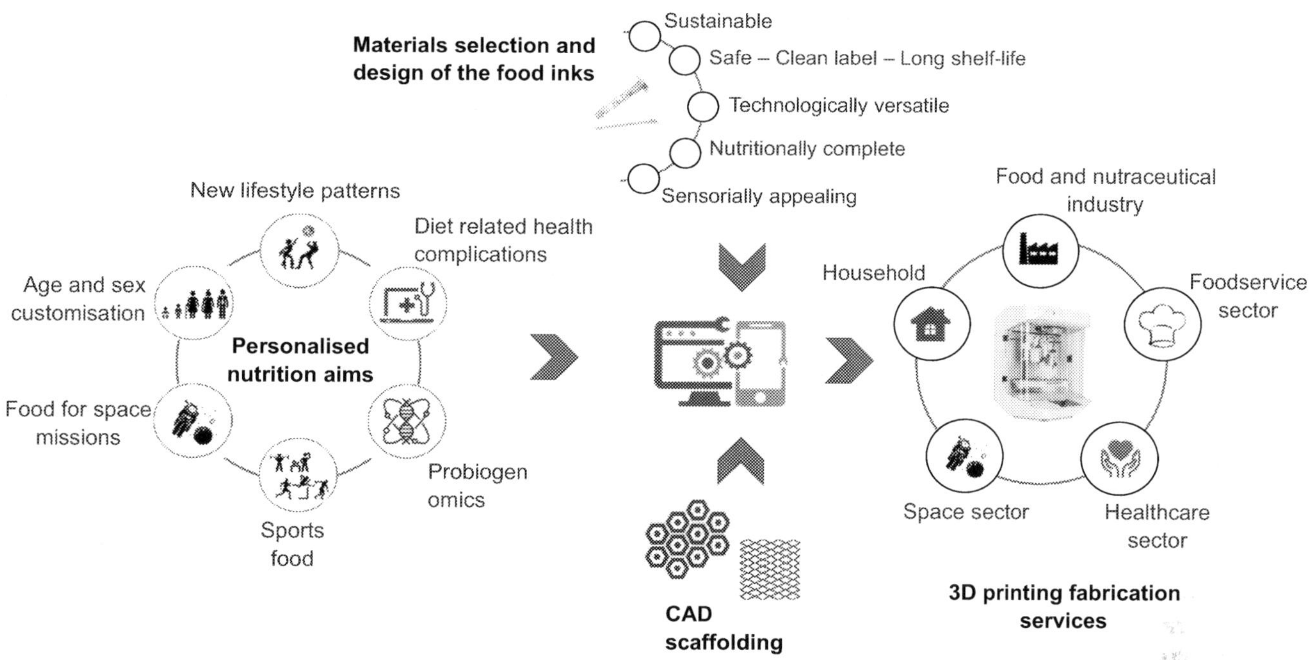

FIGURE 17.1 Schematic illustration of a 3D printing driven strategy for customised food innovation.

foods for tackling aging associated health complications such as dysphagia, sarcopenia, osteoporosis, cellular oxidative stress—associated chronic disease, etc.,

7. Alleviating food safety issues associated with the presence of allergens or microbial spoilage owing to the customisable inks' composition, the short-termed food production, and the build-in cleaning protocols of the new generation 3D printing devices,
8. Creating new multisensorial perception experiences for the consumers, i.e., gustatory, olfactory, mouthfeel, tactile, and visual, and
9. Promoting social attachment through food messaging (Dankar et al., 2018; Gramlich et al., 2014; Liu et al., 2017; Pérez et al., 2019; Portanguen et al., 2019).

17.4 3D printing technologies for plant-based foods

The steadily expanding market of novel products and materials produced via additive manufacturing has given rise to several 3D printing technical approaches, including binder jetting, directed energy deposition, material jetting, powder bed fusion, sheet lamination, and photopolymerization (Portanguen et al., 2019). As for food product applications, only four 3D printing technologies have been successfully employed: extrusion-based printing, inkjet printing, binding jetting, and selective sintering (Le-Bail et al., 2020). In addition, bioprinting, i.e., the layer-by-layer deposition of living cells, e.g., plant cell cultures via inkjet, microextrusion or laser-assisted printing is an additional promising technique for the development of plant-based food of bespoke bio-functionality (Godoi et al., 2016).

17.5 Extrusion-based printing

Extrusion-based printing or fused deposition modeling (FDM) has found the majority of applications in the food sector due to its high compatibility with most of the soft food materials and thermoplastic structuring biopolymers (e.g., starches, plant and animal proteins, thermo-reversible gelling polysaccharides, etc.) typically employed in food manufacturing. In general, the method is based on the forced ultrathin layer-by-layer nozzle deposition of soft materials (e.g., pastes, purees, soft gels, bakery, pasta dough, etc.) so as to form a 3D food structure (Godoi et al., 2016). Hot melts (e.g., chocolate, thermoreversible gelling agents) and hydrogels (ionotropically or electrostatically complexed) can also be printed via extrusion, though additional technological and technical prerequisites may have to be met. For example, extrusion printing of chocolate melt requires accurate control of the cocoa butter polymorphism in order to achieve 3D structures of acceptable textural and shelf-life characteristics. On the other hand, the structure conformational and the chemical structure of the hydrogel-forming biopolymers have been reported as being essential drivers of the structural and textural integrity of the 3D-printed food objects.

As illustrated in Fig. 17.2, the accuracy and precision of the extrusion-based printing are the resultant of the inherent physicochemical, rheological, thermophysical, and sensory properties of the printable food ingredients and the optimized technical settings of the 3D printing device (Liu et al., 2017). For instance, Liu, Bhandari, Prakash, et al. (2018) reported that the precision of the 3D printing process and the shape retention of mashed potato extruded objects were directly associated with the pseudoplasticity, the yield stress, and the elastic modulus of the food inks. When the minimum flow behavior and viscoelasticity characteristics cannot be satisfactorily fulfilled, additional structuring agents such as starch, starch derivatives, or gums (e.g., xanthan, gellan, carrageenan etc.) can be employed as co-structuring inclusions of the food inks.

17.6 Selective sintering-based printing

Selective sintering-based printing is classified into laser sintering (SLS) or hot air sintering and melting (SHASAM), depending on the source that is used for the generation of heat. In both cases, a powder bed is selectively scanned employing hot air or infrared laser jetting, inducing the binding of the fused powder microparticles. Once the cross-section scan of the powder bed is completed, the powder bed is dropped, and a new layer of powder is rolled over the top of the fused matrix. The layer-by-layer sintering powder deposition process is continued until the completion of the desired food prototype (Le-Bail et al., 2020; Liu et al., 2017). Selective sintering printing methods enable the fast construction of free-standing complex food structures without the need for postprinting curing (Le-Bail et al., 2020). On the other hand, many conditions should be satisfied in terms of the food powder properties (i.e., particle size, flowability, bulk density, wettability, compressibility, and thermoplasticity) and sintering processing factors. As for SLS printing, parameters such as the laser source, the laser power, the laser beam size, and wavelength, as well as the powder bed thickness, are important technical settings of the printing device so as to achieve the highest food printing fidelity (Awad et al., 2020). Selective sintering-based printing can be implemented on both crystalline and amorphous food particulates such as starch, sugar, lipids, etc., with the consolidation of the food matrix to be achieved at the melting (T_m) or glass transition (T_g) temperature, respectively.

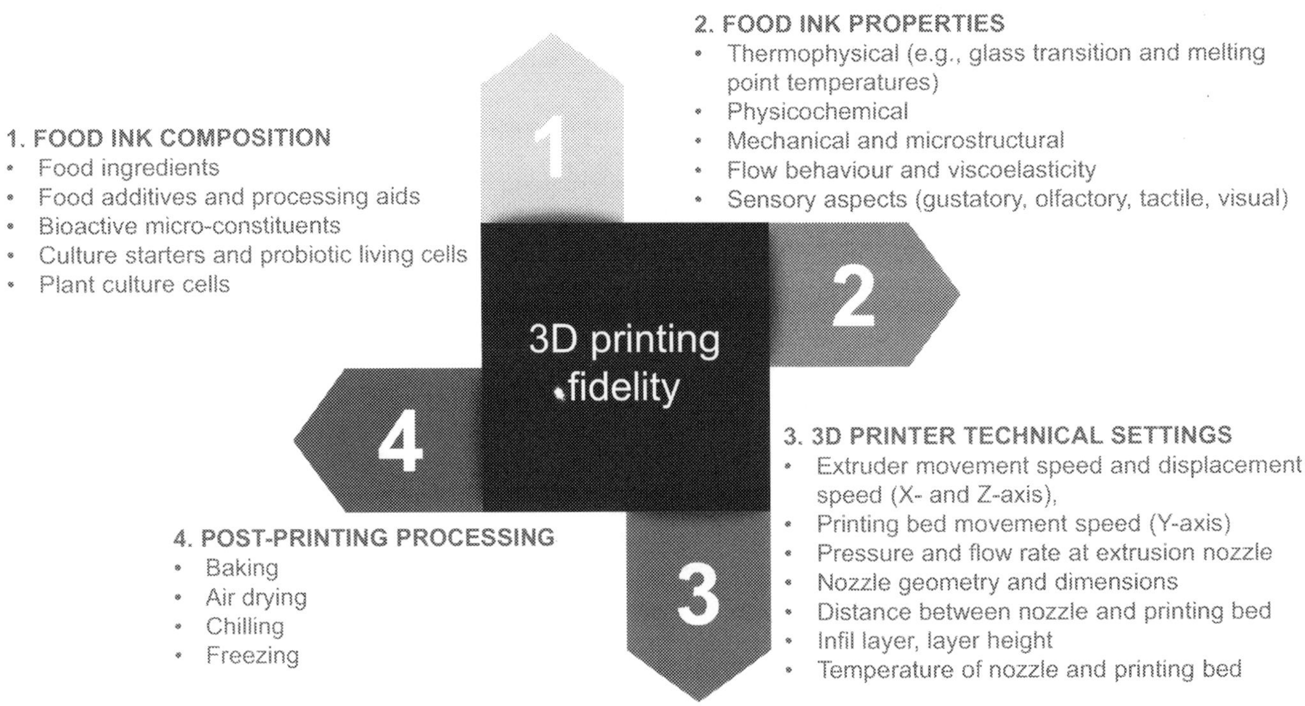

FIGURE 17.2 Schematic representation of the major technical and technological parameters that influence the fidelity of extrusion-based printed plant–based foods.

17.7 Binder jetting

Binder jetting is a layer-by-layer process that also implements a powder bed deposited on the fabrication platform. A liquid binding formulation is selectively ejected onto the powder bed employing an inkjet print head resulting in the binding of the powder particles. Then, a new powder bed layer using a counter-rotating roller or a vertically moving build chamber is created, and the entire process is repeatedly progressing until the food construct is completed (Godoi et al., 2016; Le-Bail et al., 2020). Due to the absence of any thermal process, binder jetting can be effectively applied for the printing of thermolabile ingredients such as food bioactives (Antic et al., 2021). On the downside, current binder jetting technology is considered very laborious as several intertwined parameters such as powder flow behavior, compressibility, wettability, and stability need to be carefully selected in order to achieve a surface roughness-free printing accuracy (Antic et al., 2021).

17.8 Bioprinting

Bioprinting aims at the manufacture of living functional tissues and organs for biomedical applications by embedding living cells and bioactive agents into interconnected 3D scaffolds (Yilmaz et al., 2021). In a technological context, fused deposition modeling, binder jetting, and powder bed fusion are the most common bioprinting technologies. As for food applications, bioprinting has been deployed in the fabrication of cultured edible meat (Fima, 2021; Forgacs et al., 2014; Kang et al., 2021) and cell plant-based food prototypes (Park et al., 2020; Vancauwenberghe et al., 2019).

17.9 Ingredients for plant-based food inks

Both conventional food ingredients (polysaccharides, proteins, lipids, sugars, etc.) and nonconventional materials such as microalgal biomass, insect-based powders or agri-food industry side stream recovered materials, e.g., proteins and dietary fibers of cereal, oilseed cake, fruit, and vegetable waste, can be used for 3D printing assisted food product development Table 17.1. The physical state, i.e., liquid, semi-solid or dry particulate, is essential for setting up the technical parameters of the 3D printing process and designing the food matrix composition and structure. As aforementioned, semi-solid and liquid materials are predominantly employed for extrusion assisted printing, whilst thermoplastic free-flowing dry particulates are required in the case of laser sintering printing. As aforementioned, the success of the 3D printing is directly associated with the physicochemical, rheological, thermophysical and mechanical properties of the food ingredients, whilst the modification of their sensory characteristics during the 3D printing process, e.g., heating induced formation of Maillard reaction flavor compounds and pigments, decoloration of natural pigments such as carotenoids and anthocyanins, etc. should also be considered thoroughly.

17.10 Starch and plant origin polysaccharides

Native or pregelatinized starch (rice, potato) as well as precooked starchy soft materials, e.g., mashed vegetable roots such as potato, yam, taro etc., are probably the most common food ingredients employed in 3D printing food fabrication (Feng et al., 2019). In most of the cases, starch pastes of varying botanical origin and genotype, amylose to amylopectin ratio, degree of crystallinity, etc., have been used as model systems for understanding and optimising the physicochemical, thermophysical, and technical parameters that define the pre-, and postprinting fidelity, including postprinting processing such as baking, frying, or steaming (Liu et al., 2019; Liu, Chen, et al., 2020; Liu, Tang, et al., 2020; Maniglia et al., 2020, 2021). Liu, Chen, et al. (2020) investigated the printability of potato starch suspensions (10%–30% wt) using a hot extrusion printing device. Although a certain potato starch solids amount was required to promote satisfactory extrudability, the use of excessive potato starch solids led to printing failures. Similarly, the nozzle temperature should be sufficiently high to permit complete starch gelatinization without hindering the inter- and supramolecular interactions between the starch chains. The addition of food additives, i.e., soluble dietary fibers exerting fair thermoplastic, thickening, and gelling properties, has been extensively studied as an alternative approach for enhancing the printability and postprinting shape retention of starch (Huang, Zhang, & Guo, 2020; Huang, Zhang, Bhandari, et al., 2020; Liu et al., 2019; Liu, Tang, et al., 2020). In this context, the inclusion of sodium alginate into pastes prepared from different rice varieties (*japonica, indica* and waxy) resulted in the improvement of their extrudability, printing accuracy and shape retention after steaming. In general, pastes from *japonica* and *indica* rice varieties exerted a better ability to maintain their height and length properties after steaming, whereas they were characterized by lower gumminess and chewiness compared to their waxy rice counterparts (Liu, Tang, et al., 2020). Liu et al. (2019) assessed ternary model systems comprising potato starch, xanthan gum, and κ-carrageenan for their ability to reproduce 1D (straight and multi-angled), 2D (lattice scaffold), and 3D

TABLE 17.1 Examples of plant-based ink composition and food applications implementing the extrusion deposition 3D printing method.

Food application	Food-ink composition including additives	3D printing performance and techno-functional characteristics	References
Starch and plant origin polysaccharides			
Rice food analogs	Rice paste sodium alginate	Rice pastes exerted a shear-thinning behavior allowing high printability and object shape retention after steaming. Sodium alginate addition enhanced the printing fidelity. Textural and shape retention properties of the 3D-printed objects after steaming was dependent on rice variety	Liu, Tang, et al. (2020), Liu, Chen, et al. (2020), Liu, Zhang, et al. (2020), Liu, Bhandari, et al. (2020)
Mashed potato construct	Gelatinized potato flakes	The inks allowed precise redevelopment of the dimensional properties of the targeting geometries 3D-printed mashed potato prototypes exhibited notably different microstructural and textural characteristics as compared to their casted counterparts	Liu, Bhandari, Prakash, et al. (2018), Liu, Zhang, Bhandari, et al. (2018), Liu, Zhang, and Yang (2018)
Potato starch prototypes (1D, lattice scaffold, hollow cylinder)	Gelatinized potato starch Xanthan gum κ-Carrageenan	The 3D printing process was divided into three major stages: extrusion (influenced by yield stress, viscosity and pseudoplasticity), shear and temperature recovery, and self-supporting stage (driven by elastic modulus) The presence of κ-carrageenan and xanthan gum improved the printing fidelity and self-supporting ability of the 3D printed prototypes	Liu et al. (2019)
Potato starch gels	Potato starch (native)	Potato starch suspensions exerted good printability (>10% wt). The nozzle temperature and the potato starch suspensions need to be thoroughly selected for achieving a good printing fidelity	Liu, Tang, et al. (2020), Liu, Chen, et al. (2020), Liu, Zhang, et al. (2020), Liu, Bhandari, et al. (2020)
Plant proteins			
Soy protein gel cubes	Soy protein isolate κ-Carrageenan vanilla flavor	Good printing fidelity was achieved for formulations containing >2% κ-carrageenan. 3D-printed objects were structurally collapsed following microwaving at > 80W	Phuhongsung, Zhang, and Bhandari (2020), Phuhongsung, Zhang, and Devahastin (2020)
Pea protein isolate fortified banana paste	Microwave dried banana powder Pea protein isolate ascorbic acid	The incorporation of pea protein isolate (15% wt) maximized the dimensional fidelity and shape retention of the 3D-printed objects Excessive pea protein isolate addition impaired the postprinting structure recovery resulting in the discontinuous printed lines	Kim et al. (2021)

Protein fortified food inks	Protein to sodium alginate to vegetable/fruit powder ratio 15:1:2 protein ingredients: Pea, peanut and whey protein isolate, wheat hydrolyzed protein, or casein Hawthorn, sweet potato or spinach freeze-dried powder	Peanut protein isolate exhibited the best printing performance without compromising in terms of textural, olfactory and gustatory modalities of the 3D-printed objects	Chen et al. (2021)
Vegetable and fruit pulps			
Vegetable-based gels	Xanthan gum freeze-dried spinach powder	All mixtures showed smooth extrudability regardless of particle size of spinach powder The printing resolution and structural integrity of the printed objects were reduced as the function of spinach powder particle size was increased	Lee et al. (2019)
Vegetable-based gels	Xanthan gum, hydroxypropyl methylcellulose (HPMC), locust bean gum or guar gum Freeze-dried vegetables (broccoli, spinach or carrot)	With the exception of xanthan gum, the rheological behavior and printing performance of the hydrocolloid-vegetable powder mixtures was influenced by the vegetable type. Overall, xanthan gum allowed a good printing fidelity and 3D object shape retention regardless of the amount of vegetable powder used	Kim et al. (2018)
Cereal ingredients			
Cereal snacks	Wheat flour dough	Baked snacks maintained their major structure features before and after baking. The dimensional and textural properties of the snacks were strongly influenced by the infill percentage and layer height	Severini et al. (2016)
Rectangular cereal snacks with cubic hollows	Wheat flour dough containing olive oil and sodium chloride	High fidelity in reproducing the 3D CAD model was achieved Shrinkage and increase in the porosity of the 3D snack models took place during baking	Derossi et al. (2020)
Living plant cells			
Lettuce tissue cubic-shaped prototypes	LM pectin calcium chloride lettuce leaf plant cells bovine Serum Albumin (BSA)	Good accuracy and reproducibility in constructing food composite structures structural and mechanical integrity is reduced by the presence of the plant cells. The viability of the plant cells was adversely related to the LM pectin content	Vancauwenberghe et al. (2019)
Carrot tissue mimicking lattice scaffolds	Sodium alginate calcium chloride carrot calli preparation (CC) MS medium	High printing resolution and structure conformity were achieved at 1:1 to 2:1 CC to sodium alginate ratios The plant cell proliferation was satisfactory dissociation of the plant cells from the embedding scaffold after certain cell density The intercellular adhesion was very limited to imitate natural plant tissues	Park et al. (2020)

Continued

TABLE 17.1 Examples of plant-based ink composition and food applications implementing the extrusion deposition 3D printing method.—cont'd

Food application	Food-ink composition including additives	3D printing performance and techno-functional characteristics	References
Microalgal biomass			
Cookies	Cookie batter enriched with different forms of *Arthrospira platensis* (raw biomass, extract or encapsulated form)	Microalgal biomass did not influence the printing fidelity cookies dimensional properties were well retained after baking no effects on hardness and water activity were observed	Vieira et al. (2020)
Baked snacks	Snack batter enriched with *Chlorella* or *Spirulina* biomass	The addition of microalgal biomass enhanced the snacks printing fidelity Baking led to the partial color loss of the snacks	Uribe-Wandurraga et al. (2020)

(hollow cylinder) structures. Three major stages in the 3D printing process were identified, i.e., extrusion, recovery, and self-supporting, driven by different rheological parameters: yield stress and shear-thinning behavior, shear and temperature recovery, and elastic modulus, respectively. The addition of xanthan and κ-carrageenan improved the shear-thinning behavior, viscosity, and gelation temperature of the food inks, parameters that were important for achieving an optimal extrudability and shape retention performance. In a recent study, Huang, Zhang, Bhandari, et al. (2020) investigated comparatively the effect of different hydrocolloids (xanthan gum, guar gum, sodium carboxymethylcellulose, and agar) on the rheological behavior, microstructure, and extrusion printing behavior of brown rice paste. In the absence of hydrocolloids, the 3D-printed objects (cylinders of 25% infill percentage) exhibited a partially collapsed structure with the deviations in height and diameter being as high as 14.1% and 12.7%, respectively. With the exception of xanthan gum, the addition of hydrocolloids enhanced the printing fidelity of the 3D structures, with the observed dimensional deviations to be less than 4% and 10% for diameter and height, respectively.

Modification of food ingredients through soft processing methods such as pulsed electric field (PEF), microwave, high pressure, or ultrasound can also provide alternative pathways in improving their printing fidelity. As such, Maniglia et al. (2021) investigated the impact of PEF on the printability of wheat and cassava starch pastes. The PEF pretreatment of the starch powders resulted in botanical origin associated effects on printing performance. Only in the case of PEF treated wheat starch, significant improvements in the overall printing accuracy and precision were identified, which was ascribed to the modification of the rheological and textural characteristics of the starch pastes. Dry heat treatment (DHT) of wheat starch prior to extrusion-mediated printing was found as having significant advantages as concerned to the extrudability and 3D object construction reproducibility (Maniglia et al., 2020). This was primarily associated with the ability of DHT to promote the rheological and pasting properties owing to the starch granules expansion and starch chain depolymerization.

17.11 Vegetable and fruit preparations

In principle, starchy soft materials constitute the best plant food commodities for 3D printing applications, with mashed potatoes being the most studied food matrix (Feng et al., 2020; Liu, Bhandari, Prakash, et al., 2018; Liu, Zhang, Bhandari, et al., 2018; Liu, Zhang, & Yang, 2018; Liu, Bhandari, et al., 2020; Liu, Zhang, et al., 2020). This is due to the easiness of potato starch granules to undergo complete gelatinization in the temperature range typically implemented in hot extrusion-based printing and the presence of phosphoric acid side groups in the potato amylopectin chains promoting its water retention properties. In addition, mashed potato is characterized by satisfactory extrudability due to its viscoelastic, shear softening behavior, as well as good binding and self-standing structuring properties as a result of the ability of the starch polymer chains to undergo intermolecularly (polymer—polymer) hydrogen bridging interactions (Feng et al., 2019). Parameters such as the yield stress (τ_0) and the elastic modulus (G') of mashed potato have been identified as critical for achieving a high printing accuracy and precision (Liu, Zhang, & Yang, 2018). The printing fidelity 3D objects composed of mashed potato blended with hydrocolloids (xanthan gum and κ-carrageenan) were successfully constructed, with their cross-sectional and longitudinal microstructure integrity to be associated with instrumental texture, i.e., hardness, gumminess, and Young's modulus (Liu, Bhandari, Prakash, et al., 2018). In a recent study, Feng et al. (2020) developed yam paste inks fortified with a potato by-product composed of cellulose, hemicellulose, pectin, and starch. The addition of the potato by-product reduced the extrusion swelling and increased the water absorption index and elastic modulus, ameliorating the printing accuracy and reproducibility of complex 3D structures with varying infill patterns. The addition of the potato by-product assisted in preventing the structure deformation of the 3D objects occurring during air frying, notwithstanding the substantial role of the directional printing path.

Food inks based on vegetables or fruit purees mixed with food additives and bioactive ingredients (e.g., prebiotics, soluble dietary fibers, bioactive microconstituents, low glycemic index sweeteners, beneficial living cells, etc.) have been conceptualized to impart bespoke cross-modal sensory (visual, olfactory, gustatory, oro-tactile) characteristics and at the same time to improve the biofunctional fingerprint (e.g., antioxidant, anti-inflammatory, immunomodulating, antilipidemic, antiglycemic, etc.) of the 3D-printed objects. 3D-printed mashed potato composites comprising fruit preparations based on the dual extrusion method have been successfully developed (Liu, Zhang, Bhandari, et al., 2018). In a recent study, Chen et al. (2021) investigated the sensory, textural, and olfactory characteristics of 3D-printed vegetable-based prototypes fortified with different types of plant or animal origin protein isolates (peanut, pea, hydrolyzed wheat, whey, and casein). Among the tested ingredients, the peanut protein isolate was the best in achieving high printing reproducibility without impairing the sensory, textural, and olfactory (electronic nose sensor response) characteristics of the 3D-printed objects. The impact of printing parameters (i.e., nozzle diameter and nozzle holding temperature) and postprinting storage time on the viability of probiotic cells (*Bifidobacterium animalis* subsp. *Lactis*) incorporated in mashed potato-based 3D objects were successfully studied (Liu, Bhandari, et al., 2020). Although the 3D printing was identified as a technologically viable strategy for creating novel probiotic foods, the implementation of low diameter nozzles, and long holding times at high temperature (55°C) resulted in significant mechanical and heat-induced injuries of the probiotic cells.

17.12 Plant proteins

Proteins are food macromolecules that possess essential structural and functional roles in the human body, while they fulfill interesting technological functionalities in food such as thickening, gelation, emulsification, foaming, water, and oil holding ability. Besides, proteins contribute to the development of the primary organoleptic features of food, such as color, flavor, and texture (Loveday, 2019). Hitherto, proteins isolated from cereals (wheat, corn, barley, oat, rice), pseudocereals (chia, amaranth), legumes (pea, soybean, fababean, lentil, lupin), and oilseeds (flax, rapeseed, sunflower, hempseed) exhibit the highest feasibility for food product applications. By analogy with the animal proteins, 3D extruded plant protein-based food relies on protein ingredients that can undergo heat-induced gelation when exceeding the least concentration of gelation (LCG). In this context, globular protein materials such as soy or pea protein isolates have been successfully implemented in extrusion deposition 3D printing (Chuanxing et al., 2018; Fan et al., 2020; Oyinloye & Yoon, 2020; Phuhongsung, Zhang, & Bhandari, 2020; Phuhongsung, Zhang, & Devahastin, 2020). This is due to their easiness to undergo a modification of their native globular structure conformation (dissociation and unfolding) enhancing the ability of the peptides to interact via interchain covalent or noncovalent interactions leading to the formation of soluble protein aggregates. Apart from LCG, parameters such as the ionic strength, the pH, the heat process severity, and the presence of other biopolymers are among the most important parameters that need to be optimized for ensuring satisfactory extrudability and 3D printing resolution. Phuhongsung, Zhang, and Devahastin (2020) demonstrated that the presence of sodium chloride in soy protein isolates based food inks enhanced their extrudability due to its ability to lower the viscosity and increase the transverse (spin-spin) relaxation time of the protein gels. Common thickening or gelling agents such as κ-carrageenan, xanthan gum, sodium alginate, starch etc., can also be employed as co-structurants in plant protein-based food inks. The inclusion of κ-carrageenan in soy protein isolate heat-set gels was associated with a tangible improvement of their extrudability and printing resolution performance, while it prevented the structural collapse and texture deterioration of the 3D objects upon microwave postprinting processing (Phuhongsung, Zhang, & Devahastin, 2020). Binary blends of sodium alginate with pea protein isolate exhibited a remarkably better capacity to reproduce the dimensional and structural features of 3D models compared to the individual biopolymer exemplars (Oyinloye & Yoon, 2020). The addition of 1% – 2% of pea protein isolate in potato starch inks promoted their thermoplastic properties allowing ease extrusion, good support of the filament deposition, and imparted acceptable textural properties, e.g., cohesiveness, hardness and adhesiveness to the end 3D-printed food objects (Chuanxing et al., 2018). Salt- or acid-set (e.g., δ-gluconolactone or lactic acid starter mediated acidification) protein hydrogels can also be employed in the creation of 3D extruded food objects. Finally, the use of transglutaminase (an enzyme that catalyses the acyl-transfer reaction between the ε-amino group of lysine and the γ-carboxyamide group of glutamine in proteins) is considered as an alternative route in promoting the 3D printing fidelity of protein gel food objects without food additives (Portanguen et al., 2019).

17.13 Living plant cells

Plant cell culture (PCC) technology is a recently established technology for producing food biomass and food relevant secondary plant metabolites such as phenolic compounds, alkaloids and isoprenoids, pigments, and flavorings (Nordlund et al., 2018). The PCC technology principle is based on the proliferation of cells obtained from plant organs tissues or transformed (hairy) roots in a bioreactor environment (Nosov, 2012). The steadily growing interest in PCC technology for the food industry sector has been associated with several advantages, including the following: (a) production of the plant cell biomass (and the secondary metabolites thereof) is conducted under highly controlled manufacturing conditions and shows no seasonality and (b) manufacture of PCC-based food ingredients, which allows sustainable management of natural resources and has a very low carbon footprint (Eibl et al., 2018).

Yet, in the predatory stage, the use of plant cells for revisiting conventional plant-based food, exploiting the technological capabilities of 3D printing, has been investigated. Unambiguously, the greatest challenge in the development of plant cell-based 3D bio-inks is the identification of biopolymeric structuring agents that exert high biocompatibility with the living plant cells without compromising the precision and accuracy of the printing technology as well as the structural, mechanical, and sensory aspects of the end-products (Chung et al., 2013). Vancauwenberghe et al. (2019) were the first who implemented the extrusion-based printing technology for incorporating living lettuce (*Valerianella locusta*) leaf cultured cells and air bubbles into ionotropically crosslinked pectin hydrogels. Although the presence of the plant cells did not impair the printing accuracy and reproducibility markers, shortcomings associated with the structural and the mechanical integrity of the 3D-printed objects were noticed. In addition, the viability of the plant cells was adversely associated with the concentration of the ionotropic gelling agent, which was ascribed to mechanical shear-induced cell injuries due to the increase in viscoelasticity. Recently, Park, Kim, and Park investigated the feasibility of extrusion-based printing for imitating cellular plant tissues (Park et al., 2020). Sodium alginate-based bio-inks comprising cell lines exemplified with carrot callus tissue were fabricated via ionotropic gelation into 3D lattice scaffolds. A 1:1 to 2:1 plant cell to sodium alginate ratio was required to construct well-defined macroporous scaffolds without impairing the tissue regeneration capacity. The cell immobilization capacity of the 3D-printed scaffolds was directly associated with the cell density. Hence, exceeding a certain cell density, the dissociation of the plant cells from the embedding hydrogel matrix was observed. Although these studies demonstrate the potential of 3D printing in revisiting plant-based food products, further research is required to achieve structures that can mimic the structure conformation and functional properties of plant cell tissues satisfactorily.

17.14 Microalgal biomass

Microalgae constitute a group of morphologically and physiologically diversified unicellular, oxygen-evolving, photosynthetic organisms found in aquatic environments (Grossmann et al., 2020). Valorization of microalgae via the biorefinery concept has been evolved rapidly in the last decade owing to their high density in primary (i.e., lipids, proteins, and polysaccharides) and secondary metabolites, e.g., fatty acids (EPA, DHA), carotenoids, chlorophyll, phycobiliproteins etc., offering great opportunities to meet the ever-growing consumers' demand for natural healthy food (Caporgno & Mathys, 2018; Grossmann et al., 2020; Rahman, 2020). This can be mainly ascribed to:

1. The high density of microalgal biomass in valuable, for human nutrition, macromolecules, as most of the energy is utilized for metabolic actions instead of the synthesis of structural support components,
2. The very high photosynthetic rate, rapid growth, and short harvesting time achieved compared to land plants,
3. No need for arable land,
4. The sustainable management of natural and energy resources, and
5. The high adaptability of microalgae to extreme environmental conditions, e.g., light, temperature, salinity, nutrient deprivation, etc.

Hitherto, Chlorella, Arthrospira, Nannochloropsis, Dunaliella, Tetraselmis, and Phaeodactylum are the most cultivated microalgae species for commercial food commodities purposes. Snack batter inks fortified with *Chlorella vulgaris* or *Arthrorspira platensis* lyophilized biomass have been tested for their extrusion-based printability and shape retention before and after the baking process (Uribe-Wandurraga et al., 2020). The highest degree of printing accuracy was achieved when 3%–4% wt of microalgal biomass was added to the snack batters. On the other hand, *Chlorella* was found to be more efficient in achieving higher printability and shape retention rates after baking.

17.15 Consumers attitude toward 3D plant-based food

Notwithstanding the vast possibilities that additive manufacturing can offer to mass food production, its disruptive innovation character renders 3D-printed food susceptible to consumers' mistrust and unacceptability. The edibility and food preferences do not imply a de facto connection to the digestibility or nutritional profile of food substances, but instead, the consumers tend to adopt food choices that are closely associated with their social, cultural, geographical, and historical background contexts (Lupton & Turner, 2017). When a new food production technology or food ingredient is introduced, the risk of violating the socio-cultural vectors associated with food acceptability such as "familiarity," "personal identity," "social relations," "traditionality," "cultural background," etc., arises. On many occasions, the use of uncommon food bioresources, e.g., insects, microalgae or genetically modified food, can portray novel foods as "unnatural," "disgusting," or "unsafe" (Lupton & Turner, 2018).

Hitherto, only a few consumer research studies have focused on the consumers' attitudes toward novel 3D-printed food products by amalgamating data concerning cross-modal sensory (e.g., visual, gustatory, textural, or olfactory) characteristics, socio-cultural contexts associated with food neophobia/neophilia, as well as sociotechnical imaginaries. Digital food activism, i.e., the digital information and communication actions to disseminate new food production and processing technologies as well as food materials to the public, has demonstrated that 3D printed food is inextricably associated with consumers' controversy and transgression (Lupton & Turner, 2017). Although 3D-printed food is recognized for its creative, fun, convenient, and efficient character, offering opportunities in regard to personalized nutrition, waste reduction, and improvement of the environmental markers of mass food production, food security and overcoming ethical, cultural or religious constraints, and efforts to alleviate consumers mistrust are required (Lupton, 2017; Lupton & Turner, 2017). In their study, Gayler et al. (2018) demonstrated that early adopters, i.e., consumers that are motivated by novel and rewarding experiences showing less concern with the risks of new technologies, perceived 3D-printed food as an exciting opportunity to engage in the sensory and health personalization of their meals. A survey on the attitude of Canadian consumers toward 3D-printed food (i.e., willingness to buy or eat the product, healthiness, and degree of food processing), unveiled that neophilic consumers express a strong belief that 3D printing can assist in reducing the cost of food and tackle challenges associated with food security (Manstan & McSweeney, 2020). On the other hand, the neophobic consumers believed that 3D-printed food is unacceptable and not safe to consume (Manstan & McSweeney, 2020). In line with this, Brunner et al. (2018) reported that although communication scenarios on 3D printing technology and 3D printable food ingredients can assuage some of the consumers' mistrust, food technology neophobia appeared to be quite persisting even in consumers who were sufficiently familiar with the edible additive manufacturing concept. Highlighting the positive roles of 3D printing food attributes such as health, fun, creativity, and natural content are known as being significant moderators of food neophobia and food technology neophobia. Lee et al. (2021) implementing a value-attitude-behaviour (VAB) model demonstrated that the effects of "natural content" and "creativity" on perceived values were related to food neophobia, while "natural content" and hedonic sense cross-modality were moderators of food technology neophobia.

Visual evaluation of a vegetable meal based on 3D-printed carrots aroused significant consumers' uncertainty regarding the naturalness and wholesomeness of the meal and the unnecessary implementation of processing or use of food additives (Lupton & Turner, 2018). On the other hand, Severini et al. (2018) demonstrated that 3D printing of fruit and vegetable blend smoothies enhanced their overall perceived quality with the appearance to be the most impacted sensory modality, while it did not confer any adverse effects on the shelf-life, nutritional aspects, and microbiological safety of the smoothies. In a recent study, Caulier and co-workers exploring the consumers' attitudes in a real-life military setting demonstrated that sequential exposure (for a 4-week period) to 3D printing technology increased the overall acceptability of the 3D printed savoury or sweet snacks filled with vegetable or fruit preparations customization. Besides, the consumer empowerment, desired degree of product customization, and printing fidelity (as compared to the conventional product exemplars) are significant components of the consumers' acceptability of 3D-printed food.

17.16 Targeting potential markets and consumers (children, adults, and elderly)

The market of 3D food printing innovation is one of the fastest-growing sectors of the food industry with a CAGR pace to range from ca. 20%–55% until 2025 and a forecasted global market value of USD 1015 million by 2027 (Emergen Research, 2021). Although the 3D food printing market is highly segmented, the majority of the innovation opportunities are identified in food product customization applications for the foodservice sector (e.g., creation of novel cross-modal sensory experiences in terms of texture, shape, flavor, taste, or appearance) and healthcare sector (e.g., construction of personalized dietary plans for patients or specific consumer targets such as athletes, children, astronauts, etc.).In addition, 3D food printing technology can be effectively used for tackling food security challenges due to the reduction of the food biomass utilization and loss and the prolonged shelf-life (to several years) of the food inks, when powdered (spray or freeze dried) ingredients are deployed as the major constituents of the food inks.

17.17 3D food printing for adults and elderly nutrition plan customization

Advances in nutrigenomics and food science have contributed significantly to the enhancement of the human health, well-being, and longevity. On the other hand, the increase in life expectancy is inextricably associated with biological, physiological, psychological, and socio-behavioral changes as a result of the aging process. On some occasions, aging-related changes may trigger degenerative chronic disease complications, imposing the need for customized dietary plans. Diabetes type II, cardiovascular disease, obesity, osteoporosis, Alzheimer's disease, and eating disorders are only a few examples of nutrition-related health complications due to aging (Di Renzo et al., 2019). 3D food printing has been recently recognized for its potential to provide customized dietary solutions in the form of nutritionally balanced food products (i.e., salt/sugar/fat-reduced, protein fortified) or functional foods (i.e., comprising soluble dietary fibers and prebiotics, living probiotic cells, essential microconstituents such as carotenoids, polyphenols, minerals, water- and liposoluble vitamins, prebiotics, and probiotics).

3D printing has been identified as a promising method for creating structurally engineered food matrices (e.g., mashed vegetables or cereal pastes) able to convey beneficial bacteria (e.g., Lactobacilli or Bifidobacteria) to the human gut (Liu, Bhandari, et al., 2020; Yoha et al., 2021; Zhang et al., 2018). Due to the relatively low temperatures applied during the extrusion deposition process, 3D printing appears to induce minimal heat-induced cellular sub-lethality compared to other encapsulation methods such as spray drying (Yoha et al., 2021). In addition, the inclusion of pre-encapsulated probiotic cells (for example, in spray or freeze-dried microcages) in 3D-printed food constructs, which were subjected to severe heating such as air drying or baking after the printing process, has shown a promising capacity to preserve the biological activity of the probiotic cells during storage and in-vitro simulated gastrointestinal conditions (Yoha et al., 2021).

Despite their rich character in several bioactive microconstituents (carotenoids, polyphenols, and vitamins, among others), raw fruit, and vegetable ingested micronutrients are characterized by very low bioavailability rates (Cilla et al., 2018). For example, the percentage of carotenoids solubilized into the mixed micelles can be as low as 7% of the total amount of carotenoid content present in the ingested raw fruit or vegetable (Tyssandier et al., 2003). On the other hand, food processing (e.g., size reduction, thermal treatment, high pressure, microwave, pulsed electric fields, or ultrasound processing) may improve the bioaccessibility and cellular uptake of several food bioactive compounds, e.g., carotenoids, tocopherols, or polyphenols (Cilla et al., 2018). Fruit or vegetable-based 3D-printed inks have recently been associated with advantages regarding the bioaccessibility of carotenoids such as lutein and β-carotene (Feng et al., 2021).

On an annual basis, almost 8% of the population is diagnosed with dysphagia, which is a physiological state associated with difficulties in chewing and swallowing due to neurological, muscular, or structural pathologies. Texture modified foods (e.g., food matrices absent of fibrous or hard-textured elements) and thickened liquid food are the most common interventions for dysphagic patients, as they allow ease food matrix disintegration and bolus formations as well as a controlled flow rate during the food bolus swallowing preventing the choking risk. 3D food printing has shown potential in reproducing textures established by the National Dysphagia Diet Task Force for thickened liquid foods (Diañez et al., 2021). Texture-modified food matrices based on fresh vegetables (garden pea, carrot, and bok choy) complying with the International Dysphagia Diet Standardisation Initiative (IDDSI) and food hydrocolloids such as xanthan, κ-carrageenan, and locust bean gum were successfully developed via extrusion deposition 3D printing. The 3D-printed food objects exerted minimal water seepage, orally processable texture and microstructure without compromising their aesthetically pleasing, palatable, flavor appealing, and nutritionally complete character (Pant et al., 2021).

17.18 Dietary concepts for children

It is well-established that children eating behavior (i.e., food intake, choice, preference, liking, acceptance, and neophobic response) is subject to environmental, socio-cultural, and past food-related experiences (DeCosta et al., 2017). In addition, the role of parental strategies, the family's education and socio-economical status, the media influences, the existence of obesogenic environmental conditions, etc., are also important parameters that define the eating behavior pattern of children (Scaglioni et al., 2018). Currently, public health stakeholders initiatives for tackling the rise of childhood obesity are centralized to increasing the vegetable and fruits intake and reducing the energy-dense macronutrients intake such as sugars and fat. Edible additive manufacturing is considered as a promising technological approach to promoting healthy customization of food for children without compromising the major preference traits such as sweetness and saltiness, lower bitterness, indulging shape, and visual appearance, appealing texture, etc. As an example, a 3D-printed snack designed to meet the recommended daily allowance for snacks (i.e., 5%−10% of RDA) in terms of energy, vitamin D, calcium, and iron contents, was created using a food ink comprising banana puree, canned white beans, skim-milk powder, dried mushrooms, ascorbic acid and pectin solution (Derossi et al., 2018).

17.19 Space mission food

The expansion of the frontiers of space exploration has posed several challenges associated with the need of ensuring that the members of space missions are appropriately fed to sustain their long-termed activities under space and gravity restriction conditions. Space mission purposed food should fulfill several pre-requisites such as complete nutritional profile (in terms of both macro- and micro-nutrients), extended shelf-life, ease preparation, realistic food appearance and acceptable texture, taste and flavour characteristics. Customization of space mission food via 3D printing appears to be a very promising technological approach to reconstruct multilayer experiences associated with "earth" eating behavior under the impaired human olfactory and gustatory response at low gravity conditions (Jiang et al., 2020). Among the advantages of 3D food printing implementation for space mission food applications is that food can be realistically remodeled using exclusively powdered food ingredients. The latter can exert extreme shelf-life stability (for several years), ease transportability, and storability. In addition, 3D bioprinters can allow the consumption of meat-based and meat-analog food items using embryonic stem cells or single-cell proteins.

17.20 Summary and future directions

3D food printing is a niche and innovative process that offers the opportunity for food manufacturers to explore the customization of mass-produced commodities. Edible additive manufacturing allows the creation of novel complex food concepts in a quick and inexpensive manner. A survey from Deloitte Insights (https://www2.deloitte.com/us/en/insights/focus/3d-opportunity/3d-printing-in-the-food-industry.html) has pointed out that edible additive manufacturing is a revolutionary process that can simplify product differentiation, product customization and direct-to-consumer relationships. That may offer opportunities to create unique food product concepts for personalized nutrition, customized medical/dietary supplements or simplified distribution into hard-to-reach locations. Although the existing know-how on transforming conventional or emerging food materials into organoleptically appealing 3D-printed food objects keeps developing, it is envisaged that edible additive manufacturing coupled with nutrigenomics will play a ground-breaking role in customizing food for specific consumer groups such as astronauts, athletes, youngsters, and elderly adults. Lastly, edible additive manufacturing shows great potential in tackling emerging challenges associated with food security, malnutrition, and food bioresources preservation by minimizing food waste and allowing the effective transformation of food waste recovered ingredients into tangible food solutions.

References

Antic, A., Zhang, J., Amini, N., Morton, D. A. V., & Hapgood, K. P. (2021). Screening pharmaceutical excipient powders for use in commercial 3D binder jetting printers. *Advanced Powder Technology, 32*(7), 2469–2483. https://doi.org/10.1016/j.apt.2021.05.014

Aschemann-Witzel, J. (2015). Consumer perception and trends about health and sustainability: Trade-offs and synergies of two pivotal issues. *Current Opinion in Food Science, 3*, 6–10. https://doi.org/10.1016/j.cofs.2014.08.002

Aschemann-Witzel, J., Varela, P., & Peschel, A. O. (2019). Consumers' categorization of food ingredients: Do consumers perceive them as 'clean label' producers expect? An exploration with projective mapping. *Food Quality and Preference, 71*, 117–128. https://doi.org/10.1016/j.foodqual.2018.06.003

Asioli, D., Aschemann-Witzel, J., Caputo, V., Vecchio, R., Annunziata, A., Næs, T., & Varela, P. (2017). Making sense of the "clean label" trends: A review of consumer food choice behavior and discussion of industry implications. *Food Research International, 99*, 58–71. https://doi.org/10.1016/j.foodres.2017.07.022

Awad, A., Fina, F., Goyanes, A., Gaisford, S., & Basit, A. W. (2020). 3D printing: Principles and pharmaceutical applications of selective laser sintering. *International Journal of Pharmaceutics, 586*, 119594. https://doi.org/10.1016/j.ijpharm.2020.119594

Banovic, M., Lähteenmäki, L., Arvola, A., Pennanen, K., Duta, D. E., Brückner-Gühmann, M., & Grunert, K. G. (2018). Foods with increased protein content: A qualitative study on European consumer preferences and perceptions. *Appetite, 125*, 233–243. https://doi.org/10.1016/j.appet.2018.01.034

Brunner, T. A., Delley, M., & Denkel, C. (2018). Consumers' attitudes and change of attitude toward 3D-printed food. *Food Quality and Preference, 68*, 389–396. https://doi.org/10.1016/j.foodqual.2017.12.010

Caporgno, M. P., & Mathys, A. (2018). Trends in microalgae incorporation into innovative food products with potential health benefits. *Frontiers in Nutrition*, 5. https://doi.org/10.3389/fnut.2018.00058

Chen, Y., Zhang, M., & Phuhongsung, P. (2021). 3D printing of protein-based composite fruit and vegetable gel system. *Lebensmittel-Wissenschaft und-Technologie*, 141. https://doi.org/10.1016/j.lwt.2021.110978

Chuanxing, F., Qi, W., Hui, L., Quancheng, Z., & Wang, M. (2018). Effects of pea protein on the properties of potato starch-based 3D printing materials. *International Journal of Food Engineering, 14*(3). https://doi.org/10.1515/ijfe-2017-0297

Chung, J. H. Y., Naficy, S., Yue, Z., Kapsa, R., Quigley, A., Moulton, S. E., & Wallace, G. G. (2013). Bio-ink properties and printability for extrusion printing living cells. *Biomaterials Science, 1*(7), 763–773. https://doi.org/10.1039/c3bm00012e

Cilla, A., Bosch, L., Barberá, R., & Alegría, A. (2018). Effect of processing on the bioaccessibility of bioactive compounds — a review focusing on carotenoids, minerals, ascorbic acid, tocopherols and polyphenols. *Journal of Food Composition and Analysis, 68*, 3–15. https://doi.org/10.1016/j.jfca.2017.01.009

Dankar, I., Haddarah, A., Omar, F. E. L., Sepulcre, F., & Pujolà, M. (2018). 3D printing technology: The new era for food customization and elaboration. *Trends in Food Science and Technology, 75*, 231–242. https://doi.org/10.1016/j.tifs.2018.03.018

DeCosta, P., Møller, P., Frøst, M. B., & Olsen, A. (2017). Changing children's eating behaviour - a review of experimental research. *Appetite, 113*, 327–357. https://doi.org/10.1016/j.appet.2017.03.004

Derossi, A., Caporizzi, R., Azzollini, D., & Severini, C. (2018). Application of 3D printing for customized food. A case on the development of a fruit-based snack for children. *Journal of Food Engineering, 220*, 65–75. https://doi.org/10.1016/j.jfoodeng.2017.05.015

Derossi, A., Caporizzi, R., Paolillo, M., & Severini, C. (2020). Programmable texture properties of cereal-based snack mediated by 3D printing technology. *Journal of Food Engineering, 289*. https://doi.org/10.1016/j.jfoodeng.2020.110160

Di Renzo, L., Gualtieri, P., Romano, L., Marrone, G., Noce, A., Pujia, A., Perrone, M. A., Aiello, V., Colica, C., & De Lorenzo, A. (2019). Role of personalized nutrition in chronic-degenerative diseases. *Nutrients, 11*(8). https://doi.org/10.3390/nu11081707

Diañez, I., Gallegos, C., Brito-de la Fuente, E., Martínez, I., Valencia, C., Sánchez, M. C., & Franco, J. M. (2021). Implementation of a novel continuous solid/liquid mixing accessory for 3D printing of dysphagia-oriented thickened fluids. *Food Hydrocolloids, 120*, 106900. https://doi.org/10.1016/j.foodhyd.2021.106900

Eibl, R., Meier, P., Stutz, I., Schildberger, D., Hühn, T., & Eibl, D. (2018). Plant cell culture technology in the cosmetics and food industries: Current state and future trends. *Applied Microbiology and Biotechnology, 102*(20), 8661−8675. https://doi.org/10.1007/s00253-018-9279-8

Emergen Research. (2021). https://www.emergenresearch.com/industry-report/3d-food-printing-market.

Fan, H., Zhang, M., Liu, Z., & Ye, Y. (2020). Effect of microwave-salt synergetic pre-treatment on the 3D printing performance of SPI-strawberry ink system. *Lebensmittel-Wissenschaft und-Technologie, 122*. https://doi.org/10.1016/j.lwt.2019.109004

Feng, L., Wu, J., Song, J., Li, D., Zhang, Z., Xu, Y., Yang, R., Liu, C., & Zhang, M. (2021). Effect of particle size distribution on the carotenoids release, physicochemical properties and 3D printing characteristics of carrot pulp. *Lebensmittel-Wissenschaft und-Technologie, 139*. https://doi.org/10.1016/j.lwt.2020.110576

Feng, C., Zhang, M., & Bhandari, B. (2019). Materials properties of printable edible inks and printing parameters optimization during 3D printing: A review. *Critical Reviews in Food Science and Nutrition, 59*(19), 3074−3081. https://doi.org/10.1080/10408398.2018.1481823

Feng, C., Zhang, M., Bhandari, B., & Ye, Y. (2020). Use of potato processing by-product: Effects on the 3D printing characteristics of the yam and the texture of air-fried yam snacks. *Lebensmittel-Wissenschaft und-Technologie, 125*. https://doi.org/10.1016/j.lwt.2020.109265

Fima, S. (2021). *Physical manipulation of cultured muscle tissue*.

Food 2030, https://research-and-innovation.ec.europa.eu/research-area/environment/bioeconomy/food-systems/food-2030_en.

Forgacs, G., Marga, F., & Jakab, K. R. (2014). *United States patent: 8703216 - engineered comestible meat*, 8703216.

Gayler, T. D., Sas, C., & Kalnikaitė, V. (2018). User perceptions of 3D food printing technologies. In , *Vol. 2018. Conference on human factors in computing systems - proceedings*. Association for Computing Machinery. https://doi.org/10.1145/3170427.3188529

Godoi, F. C., Prakash, S., & Bhandari, B. R. (2016). 3d printing technologies applied for food design: Status and prospects. *Journal of Food Engineering, 179*, 44−54. https://doi.org/10.1016/j.jfoodeng.2016.01.025

Gramlich, L., Lamarche, B., Ma, D., & Tremblay, A. (2014). Communication and food messaging: The consumer disconnect "from scientific findings to useful consumer information". *Applied Physiology, Nutrition and Metabolism, 39*(4), iii−iv. https://doi.org/10.1139/apnm-2014-0062

Grossmann, L., Hinrichs, J., & Weiss, J. (2020). Cultivation and downstream processing of microalgae and cyanobacteria to generate protein-based technofunctional food ingredients. *Critical Reviews in Food Science and Nutrition, 60*(17), 2961−2989. https://doi.org/10.1080/10408398.2019.1672137

Huang, M.s., Zhang, M., Bhandari, B., & Liu, Y. (2020). Improving the three-dimensional printability of taro paste by the addition of additives. *Journal of Food Process Engineering, 43*(5). https://doi.org/10.1111/jfpe.13090

Huang, M. S., Zhang, M., & Guo, C. F. (2020). 3D printability of brown rice gel modified by some food hydrocolloids. *Journal of Food Processing and Preservation, 44*(7). https://doi.org/10.1111/jfpp.14502

Ingredion.(2014). https://www.ingredion.es/content/dam/ingredion/pdf-downloads/emea/87%20-%20The%20Clean%20Label%20Guide%20to%20Europe%20from%20Ingredion.pdf.

Jiang, J., Zhang, M., Bhandari, B., & Cao, P. (2020). Current processing and packing technology for space foods: A review. *Critical Reviews in Food Science and Nutrition, 60*(21), 3573−3588. https://doi.org/10.1080/10408398.2019.1700348

Kang, D. H., Louis, F., Liu, H., Shimoda, H., Nishiyama, Y., Nozawa, H., Kakitani, M., Takagi, D., Kasa, D., Nagamori, E., Irie, S., Kitano, S., & Matsusaki, M. (2021). Engineered whole cut meat-like tissue by the assembly of cell fibers using tendon-gel integrated bioprinting. *Nature Communications, 12*(1). https://doi.org/10.1038/s41467-021-25236-9

Kim, Y., Kim, H. W., & Park, H. J. (2021). Effect of pea protein isolate incorporation on 3D printing performance and tailing effect of banana paste. *LWT, 150*, 111916. https://doi.org/10.1016/j.lwt.2021.111916

Kim, H. W., Lee, J. H., Park, S. M., Lee, M. H., Lee, I. W., Doh, H. S., & Park, H. J. (2018). Effect of hydrocolloids on rheological properties and printability of vegetable inks for 3D food printing. *Journal of Food Science, 83*, 2923−2932. https://doi.org/10.1111/1750-3841.14391

Le-Bail, A., Maniglia, B. C., & Le-Bail, P. (2020). Recent advances and future perspective in additive manufacturing of foods based on 3D printing. *Current Opinion in Food Science, 35*, 54−64. https://doi.org/10.1016/j.cofs.2020.01.009

Lee, J. H., Won, D. J., Kim, H. W., & Park, H. J. (2019). Effect of particle size on 3D printing performance of the food-ink system with cellular food materials. *Journal of Food Engineering, 256*, 1−8. https://doi.org/10.1016/j.jfoodeng.2019.03.014

Lee, K. H., Hwang, K. H., Kim, M., & Cho, M. (2021). 3D printed food attributes and their roles within the value-attitude-behavior model: Moderating effects of food neophobia and food technology neophobia. *Journal of Hospitality and Tourism Management, 48*, 46−54. https://doi.org/10.1016/j.jhtm.2021.05.013

Liu, Z., Bhandari, B., Prakash, S., Mantihal, S., & Zhang, M. (2019). Linking rheology and printability of a multicomponent gel system of carrageenan-xanthan-starch in extrusion based additive manufacturing. *Food Hydrocolloids, 87*, 413−424. https://doi.org/10.1016/j.foodhyd.2018.08.026

Liu, Z., Bhandari, B., Prakash, S., & Zhang, M. (2018). Creation of internal structure of mashed potato construct by 3D printing and its textural properties. *Food Research International, 111*, 534−543. https://doi.org/10.1016/j.foodres.2018.05.075

Liu, Z., Bhandari, B., & Zhang, M. (2020). Incorporation of probiotics (*Bifidobacterium animalis* subsp. Lactis) into 3D printed mashed potatoes: Effects of variables on the viability. *Food Research International, 128*, 108795. https://doi.org/10.1016/j.foodres.2019.108795

Liu, Z., Chen, H., Zheng, B., Xie, F., & Chen, L. (2020). Understanding the structure and rheological properties of potato starch induced by hot-extrusion 3D printing. *Food Hydrocolloids*, 105. https://doi.org/10.1016/j.foodhyd.2020.105812

Liu, Y., Tang, T., Duan, S., Qin, Z., Li, C., Zhang, Z., Liu, A., Wu, D., Chen, H., Han, G., Lin, B., He, J., & Wu, W. (2020). Effects of sodium alginate and rice variety on the physicochemical characteristics and 3D printing feasibility of rice paste. *Lebensmittel-Wissenschaft und-Technologie, 127*. https://doi.org/10.1016/j.lwt.2020.109360

Liu, Z., Zhang, M., Bhandari, B., & Wang, Y. (2017). 3D printing: Printing precision and application in food sector. *Trends in Food Science and Technology, 69*, 83–94. https://doi.org/10.1016/j.tifs.2017.08.018

Liu, Z., Zhang, M., Bhandari, B., & Yang, C. (2018). Impact of rheological properties of mashed potatoes on 3D printing. *Journal of Food Engineering, 220*, 76–82. https://doi.org/10.1016/j.jfoodeng.2017.04.017

Liu, Z., Zhang, M., & Yang, C.h. (2018). Dual extrusion 3D printing of mashed potatoes/strawberry juice gel. *Lebensmittel-Wissenschaft und-Technologie, 96*, 589–596. https://doi.org/10.1016/j.lwt.2018.06.014

Liu, Z., Zhang, M., & Ye, Y. (2020). Indirect prediction of 3D printability of mashed potatoes based on LF-NMR measurements. *Journal of Food Engineering*, 287. https://doi.org/10.1016/j.jfoodeng.2020.110137

Loveday, S. M. (2019). Food proteins: Technological, nutritional, and sustainability attributes of traditional and emerging proteins. *Annual Review of Food Science and Technology, 10*, 311–339. https://doi.org/10.1146/annurev-food-032818-121128

Lupton, D. (2017). 'Download to delicious': Promissory themes and sociotechnical imaginaries in coverage of 3D printed food in online news sources. *Futures, 93*, 44–53. https://doi.org/10.1016/j.futures.2017.08.001

Lupton, D., & Turner, B. (2017). 'Both fascinating and disturbing': Consumer responses to 3D food printing and implications for food activism. In *Digital food activism* (pp. 151–167). Taylor and Francis. https://doi.org/10.4324/9781315109930_8

Lupton, D., & Turner, B. (2018). "I can't get past the fact that it is printed": Consumer attitudes to 3D printed food. *Food, Culture and Society, 21*(3), 402–418. https://doi.org/10.1080/15528014.2018.1451044

Maniglia, B. C., Lima, D. C., Matta Junior, M. D., Le-Bail, P., Le-Bail, A., & Augusto, P. E. D. (2020). Preparation of cassava starch hydrogels for application in 3D printing using dry heating treatment (DHT): A prospective study on the effects of DHT and gelatinization conditions. *Food Research International*, 128. https://doi.org/10.1016/j.foodres.2019.108803

Maniglia, B. C., Pataro, G., Ferrari, G., Augusto, P. E. D., Le-Bail, P., & Le-Bail, A. (2021). Pulsed electric fields (PEF) treatment to enhance starch 3D printing application: Effect on structure, properties, and functionality of wheat and cassava starches. *Innovative Food Science and Emerging Technologies*, 68. https://doi.org/10.1016/j.ifset.2021.102602

Manstan, T., & McSweeney, M. B. (2020). Consumers' attitudes towards and acceptance of 3D printed foods in comparison with conventional food products. *International Journal of Food Science and Technology, 55*(1), 323–331. https://doi.org/10.1111/ijfs.14292

Maruyama, S., Streletskaya, N. A., & Lim, J. (2021). Clean label: Why this ingredient but not that one? *Food Quality and Preference*, 87. https://doi.org/10.1016/j.foodqual.2020.104062

Ngo, T. D., Kashani, A., Imbalzano, G., Nguyen, K. T. Q., & Hui, D. (2018). Additive manufacturing (3D printing): A review of materials, methods, applications and challenges. *Composites B: Engineering, 143*, 172–196. https://doi.org/10.1016/j.compositesb.2018.02.012

Nordlund, E., Lille, M., Silventoinen, P., Nygren, H., Seppänen-Laakso, T., Mikkelson, A., Aura, A. M., Heiniö, R. L., Nohynek, L., Puupponen-Pimiä, R., & Rischer, H. (2018). Plant cells as food – a concept taking shape. *Food Research International, 107*, 297–305. https://doi.org/10.1016/j.foodres.2018.02.045

Nosov, A. M. (2012). Application of cell technologies for production of plant-derived bioactive substances of plant origin. *Applied Biochemistry and Microbiology, 48*(7), 609–624. https://doi.org/10.1134/S000368381107009X

Oyinloye, T. M., & Yoon, W. B. (2020). Stability of 3D printing using a mixture of pea protein and alginate: Precision and application of additive layer manufacturing simulation approach for stress distribution. *Journal of Food Engineering*, 288. https://doi.org/10.1016/j.jfoodeng.2020.110127

Pant, A., Lee, A. Y., Karyappa, R., Lee, C. P., An, J., Hashimoto, M., Tan, U. X., Wong, G., Chua, C. K., & Zhang, Y. (2021). 3D food printing of fresh vegetables using food hydrocolloids for dysphagic patients. *Food Hydrocolloids*, 114. https://doi.org/10.1016/j.foodhyd.2020.106546

Park, S. M., Kim, H. W., & Park, H. J. (2020). Callus-based 3D printing for food exemplified with carrot tissues and its potential for innovative food production. *Journal of Food Engineering*, 271. https://doi.org/10.1016/j.jfoodeng.2019.109781

Pérez, B., Nykvist, H., Brøgger, A. F., Larsen, M. B., & Falkeborg, M. F. (2019). Impact of macronutrients printability and 3D-printer parameters on 3D-food printing: A review. *Food Chemistry, 287*, 249–257. https://doi.org/10.1016/j.foodchem.2019.02.090

Peschel, A. O., Kazemi, S., Liebichová, M., Sarraf, S. C. M., & Aschemann-Witzel, J. (2019). Consumers' associative networks of plant-based food product communications. *Food Quality and Preference, 75*, 145–156. https://doi.org/10.1016/j.foodqual.2019.02.015

Phuhongsung, P., Zhang, M., & Bhandari, B. (2020). 4D printing of products based on soy protein isolate via microwave heating for flavor development. *Food Research International, 137*. https://doi.org/10.1016/j.foodres.2020.109605

Phuhongsung, P., Zhang, M., & Devahastin, S. (2020). Investigation on 3D printing ability of soybean protein isolate gels and correlations with their rheological and textural properties via LF-NMR spectroscopic characteristics. *Lebensmittel-Wissenschaft und-Technologie, 122*, 109019. https://doi.org/10.1016/j.lwt.2020.109019

Portanguen, S., Tournayre, P., Sicard, J., Astruc, T., & Mirade, P. S. (2019). Toward the design of functional foods and biobased products by 3D printing: A review. *Trends in Food Science and Technology, 86*, 188–198. https://doi.org/10.1016/j.tifs.2019.02.023

Rahman, K. M. (2020). Food and high value products from microalgae: Market opportunities and challenges. In *Microalgae Biotechnology for food, Health and high value products* (pp. 3–27). Springer Singapore. https://doi.org/10.1007/978-981-15-0169-2_1

Scaglioni, S., De Cosmi, V., Ciappolino, V., Parazzini, F., Brambilla, P., & Agostoni, C. (2018). Factors influencing children's eating behaviours. *Nutrients, 10*(6), 706. https://doi.org/10.3390/nu10060706

Severini, C., Derossi, A., & Azzollini, D. (2016). Variables affecting the printability of foods: Preliminary tests on cereal-based products. *Innovative Food Science and Emerging Technologies, 38*, 281–291. https://doi.org/10.1016/j.ifset.2016.10.001

Severini, C., Derossi, A., Ricci, I., Caporizzi, R., & Fiore, A. (2018). Printing a blend of fruit and vegetables. New advances on critical variables and shelf life of 3D edible objects. *Journal of Food Engineering, 220*, 89–100. https://doi.org/10.1016/j.jfoodeng.2017.08.025

Tyssandier, V., Reboul, E., Dumas, J. F., Bouteloup-Demange, C., Armand, M., Marcand, J., Sallas, M., & Borel, P. (2003). Processing of vegetable-borne carotenoids in the human stomach and duodenum. *American Journal of Physiology - Gastrointestinal and Liver Physiology, 284*(6), G913–G923. https://doi.org/10.1152/ajpgi.00410.2002

Uribe-Wandurraga, Z. N., Zhang, L., Noort, M. W. J., Schutyser, M. A. I., García-Segovia, P., & Martínez-Monzó, J. (2020). Printability and physicochemical properties of microalgae-enriched 3D-printed snacks. *Food and Bioprocess Technology, 13*(11), 2029–2042. https://doi.org/10.1007/s11947-020-02544-4

Vancauwenberghe, V., Baiye Mfortaw Mbong, V., Vanstreels, E., Verboven, P., Lammertyn, J., & Nicolai, B. (2019). 3D printing of plant tissue for innovative food manufacturing: Encapsulation of alive plant cells into pectin based bio-ink. *Journal of Food Engineering, 263*, 454–464. https://doi.org/10.1016/j.jfoodeng.2017.12.003

Vieira, M. V., Oliveira, S. M., Amado, I. R., Fasolin, L. H., Vicente, A. A., Pastrana, L. M., & Fuciños, P. (2020). 3D printed functional cookies fortified with Arthrospira platensis: Evaluation of its antioxidant potential and physical-chemical characterization. *Food Hydrocolloids, 107*, 105893. https://doi.org/10.1016/j.foodhyd.2020.105893

Wickramasinghe, K., Breda, J., Berdzuli, N., Rippin, H., Farrand, C., & Halloran, A. (2021). The shift to plant-based diets: Are we missing the point? *Global Food Security, 29*, 100530. https://doi.org/10.1016/j.gfs.2021.100530

Yilmaz, B., Al Rashid, A., Mou, Y. A., Evis, Z., & Koç, M. (2021). Bioprinting: A review of processes, materials and applications. *Bioprinting, 23*, e00148. https://doi.org/10.1016/j.bprint.2021.e00148

Yoha, K. S., Anukiruthika, T., Anila, W., Moses, J. A., & Anandharamakrishnan, C. (2021). 3D printing of encapsulated probiotics: Effect of different post-processing methods on the stability of *Lactiplantibacillus plantarum* (NCIM 2083) under static in vitro digestion conditions and during storage. *Lebensmittel-Wissenschaft und-Technologie*, 146. https://doi.org/10.1016/j.lwt.2021.111461

Zhang, L., Lou, Y., & Schutyser, M. A. I. (2018). 3D printing of cereal-based food structures containing probiotics. *Food Structure, 18*, 14–22. https://doi.org/10.1016/j.foostr.2018.10.002

Index

Note: 'Page numbers followed by "*f*" indicate figures those followed by "*t*" indicate tables.'

F

G

H

I

J

K

Q

R

S

T

U

V

W

X

Y

LEGAL INTERVIEWING
ANALYTICS AND EXERCISES

VERSION 1—GUARDIANSHIP CLIENT

Cristina Carmody Tilley
Distinguished Scholar in Residence
Loyola University of Chicago
School of Law

NITA®
NATIONAL INSTITUTE FOR TRIAL ADVOCACY

Reprint Permission
National Institute for Trial Advocacy
361 Centennial Parkway, Suite 220
Louisville, CO 80027
Phone: (877) 648-2632
Fax: (720) 890-7069

ISBN: 978-1-60156-782-6
eISBN 978-1-60156-783-3
FBA: 1782

Library of Congress Cataloging-in-Publication Data

Names: Tilley, Cristina C., author.

Title: Legal interviewing analytics and exercises. Version 1. Guardianship client / Cristina Carmody Tilley, Distinguished Scholar in Residence, Loyola University of Chicago, School of Law.

Other titles: Guardianship client

Description: Boulder, Colorado : National Institute for Trial Advocacy, [2017]

Identifiers: LCCN 2017002880 | ISBN 9781601567826 (alk. paper)

Subjects: LCSH: Interviewing in law practice--United States. | Attorney and client--United States.

Classification: LCC KF311 .T55 2017 | DDC 346.7301/8--dc23 LC record available at https://lccn.loc.gov/2017002880

ASPEN PUBLISHING

Official co-publisher of NITA.
aspenpublishing.com/NITA

Contents

PREFACE

This book is designed to teach the practical skill of legal interviewing, primarily through simulation exercises. Interviewing cannot be learned passively, but must be learned by doing.

The book breaks legal interviews into seven distinct segments: overview, chronology, theory development, action plan, friendly witness, hostile witness, and functionary. The first part of the book devotes one chapter to each segment. Each chapter delivers a short, punchy lesson on the reason for and objectives of the particular interview segment, along with tips on how best to approach it. Where helpful, the chapter includes an example of a relevant interview segment, or the work product that an attorney would produce at the conclusion of the segment.

The meat of the instruction is delivered through the two case files that follow the text, one a guardianship case and one a criminal case. Each file contains background material and evidence for its case. The materials are designed so the instructor can assign half of the class to represent the guardianship client and the other half to represent the criminal client. Those who represent the guardianship client will know only what they are able to learn via interviewing the client and her witnesses. The guardianship file will be given to those who represent the criminal client, so that they can act as the guardianship client and her witnesses during the in-class simulations. Meanwhile, the guardianship attorneys will be given the criminal file so they can play those roles when it is the other group's turn to conduct interviews.

Attorney materials are limited to excerpts from laws relevant to the client's case. This means that—as in real life—what the attorney knows about the case is limited to what he or she can unearth via interviews and research. Witness materials include background briefs for clients and third-party witnesses, as well as timelines and relevant supporting documents. Many of the witness briefs include instructions to resist divulging specific facts until the attorney poses a precise or relevant question.

For the case in which you will be acting as client / witness, the background materials assigned to you should be considered off-limits to those students who will be acting as the interviewing attorney. So, for example, if you are the client in the guardianship case, do not share your background information with the students who will be acting as the attorney in that case.

At first, you may find that you, in your role as attorney, missed much of the information the client could have provided. As you continue through the text and exercises, your understanding of the interview process and skill at completing it will increase. When acting as an interviewee, pay attention to the instructions regarding what material to resist divulging unless asked specifically. This roleplaying provides an additional insight to the psychology of clients and will also increase your skills as an interviewer.

Working in conjunction with your fellow students, this interactive course provides a practical, hands-on approach to one of the foundational skills needed by attorneys.

Electronic, color copies of all attachments can be found at the following website:

http://bit.ly/1P20Jea
Password: Interviewing1

Part One

Interviewing Theory

Introduction

0.1 Practicing Law and Practicing Interviews

When it comes to the actual practice of law, law school teaches you in reverse. In a skills class like Legal Writing and Reasoning or Trial Advocacy, you might be given a sample complaint alleging violation of the Equal Employment Opportunity Act as a tool to teach the mechanics of drafting a motion or introducing a document into evidence. In doctrinal classes, you might read the final disposition of the same employment case from an appellate court to learn how judges interpret federal discrimination statutes. In both situations, you are given a prepackaged cause of action as a basis for learning the law. But in the real world, a client is unlikely to walk into your office and announce that he has a retaliatory discharge case he'd like you to file on his behalf. He is far more likely to sit down and fume inarticulately about how unfair it was that he lost his job before imploring, "Is there anything you can do?"

How do you get from that first messy exchange with the client to the appellate court ruling that the employer is obligated to provide back pay? The first step is the legal interview. In most cases, the majority of the legwork should take place before any legal document is ever filed. During that foundational period of work, you learn about the client's situation and elicit factual details. With those details, you can test those facts against various legal theories to determine the possible legal routes available to the client. Just as important, during that foundational period you work with the client to determine what's important to him and, consequently, which of those routes he wants you to pursue. The firing outlined above might have been a violation of federal discrimination laws. It might also have breached a contract, inflicted emotional distress—or it might have arisen from the employee's own wrongdoing. You cannot know what kind of case the client has—or what kind of remedy the client wants—without interviewing the client (and possibly other potential witnesses).

Interviews sit at the intersection of doctrinal instruction and skills instruction. Don't be fooled by a law school curriculum that too often treats skills and doctrine as mutually exclusive. If poor interviewing fails to elicit the fact that the client consulted a psychiatrist for insomnia after being fired, you may neglect to pursue an intentional infliction of emotional distress claim. And if you lack the doctrinal expertise to know that breaches of contract do not compensate for uncertain injury, you may misrepresent the likelihood of your client getting a sizeable verdict in a contract action for the loss of a commission job where the worker had an uneven earning record—a situation that can leave you with a disappointed, angry client. It is far better to turn away a potential client with a bad claim than to frustrate that client with a lengthy, expensive, but unwinnable suit. You don't want to be the defendant in his next lawsuit—for malpractice.

In short, legal interviewing is too important to proceed by mere goodwill and intuition, though many lawyers approach it with only those resources. You may be a great conversationalist and a friendly person, but that doesn't guarantee that you are a competent interviewer. The good news is that, like any skill, legal interviewing can be deconstructed into several component parts and each of those parts can be studied and practiced. This book will break down legal interviews into discrete segments and provide a set of tools that can be used systemically to achieve the best results in each segment.

You may be a great conversationalist and a friendly person,
but that doesn't guarantee that you are a competent interviewer.

Primarily, however, this book is designed to facilitate a hands-on course in legal interviewing. To that end, the book also provides case materials related to two potential client matters. Using these materials, you will practice

conducting the various segments of legal interviews. Interviewing skills simply cannot be acquired passively. There is a world of difference between reading about these skills and actually applying them. You will almost certainly not be polished in your first few efforts, which is why you are better off making those efforts now, in a classroom or seminar setting, than waiting until you have a paying client across the desk from you. These exercises draw on real statutes and cases, teaching practical skills that will deepen your appreciation of the symbiosis between legal doctrine and legal skills.

0.2 Structuring Legal Interviews

Lawyers regularly treat interviewing as an activity restricted to the initial client meeting. We use an "intake form," often prepared by an administrative assistant, to guide us through a summary of the problem, and then assume that the interview phase of our representation is completed. But the truth is you never stop interviewing your clients. Every conversation, from the first meeting to the completion of the appellate process or the close of a deal, is an opportunity to learn more about the client's situation, her wishes, her strengths, and her weaknesses. And once the client has decided on a course of action, it is highly likely that we will be talking to other parties on her behalf—to potential witnesses, to contract counterparts, to family members, and the like. Each of these encounters is another interview situation, another opportunity to learn about your client, her claim, and the circumstances surrounding that claim. Think of all law-related conversations in terms of the various phases of an interview.

The truth is you never stop interviewing your clients.

When you walk into an initial client meeting, you know the reason that client called—she wants a will drawn up, or she was injured in a car accident. The case may seem straightforward, but keep an open mind as that first meeting begins. Early in the representation, neither you nor the client know exactly what type of work product you will produce on her behalf. You may try to negotiate a deal or resolution informally; you may put a request or demand into writing to exert more formal pressure; or you may ultimately file a lawsuit. If you are a defense attorney, you may cut a deal with prosecutors or go to trial. If you are a probate attorney, you may draw up a will or design a revocable trust. You may negotiate a contract for a corporate client, or make a regulatory filing. Learning about the client's situation and desires is the "pre-work product" phase of the representation.

Successful pre-work product interviews involve three operational segments: overview, chronology, and theory development. These three segments may take place during a single interview, or over the course of several interviews, depending on the complexity of the situation and your need to conduct research into available legal theories.

At the conclusion of these first three segments, you and your client will discuss the theories and options you feel are available to address her situation; after advising her of the costs and benefits of each, you may counsel her as she weighs her course of action. Ultimately, she will decide how she wishes to proceed. This is the "action plan" phase of the representation.

Finally, once the client has decided what you will do on her behalf, you will execute the action plan. If you are filing a complaint, this is when you will obtain documents or witnesses to support your claims. If your client is subject to administrative regulations, you may speak with agency personnel. In a criminal case, you will interview witnesses, examine evidence, and discuss the charges with the prosecution. This is the "work product" phase of the representation.

Not surprisingly, you draw on different skills during different phases of the representation.

- In the pre-work product phase, the goal is to get as much information and input as possible from the client. During this phase, it is crucial to develop trust and rapport with the client and to absorb as much factual and emotional information as possible.

- In the action plan phase, the goal is to synthesize for your client the various routes available to her and to help her decide how she wishes to proceed. In this phase, you give the client your straightforward assessment of her legal position and outline the way you work. That way, you both have a clear picture of how you expect representation to unfold.
- In the work product phase, the goal is to talk with third parties who are necessary to carrying out the client's wishes, be they potential witnesses, the adversary, or representatives of institutions. In this phase, your job may require you to size up those you come across to determine whether passivity or assertiveness is more likely to work in your client's favor. Preparation and strategy will strengthen your hand during these interviews.

0.3 Using the Right Tools

The most important tools that we'll be using during the first phase of legal interviewing (and every phase thereafter) are questions. Interviewers have a variety of questions at their disposal, and each does a different job.

Open questions do not limit the information sought, but ask the interviewee to speak at length and without restriction on a broad topic.

"Tell me about your relationship with your husband."
"Please describe the layout of the grocery store."
"How did you get to your current position as COO?"

Closed questions seek a single, concrete piece of information, and can be used to add detail and clarity to information gleaned from the answer to an open question.

"How long did you and your husband date before marriage?"
"How close was the bakery aisle to the deli?"
"Where did you get your MBA?"

Yes-No questions ask the interviewee to confirm or deny a piece of information.

"Did you sign a prenuptual agreement?"
"Was it raining when you entered the Whole Foods?"
"Had you signed a noncompete agreement with your previous employer?"

Leading questions propose a scenario and suggest that the interviewee agree to that scenario.

"You waited until she put on her wedding dress to present her with the prenuptial agreement, correct?"
"This record indicates that the mops in the custodial closet hadn't been replaced in four years, doesn't it?"
"The noncompete agreement only applies to subsequent jobs in marketing, right?"

As you can see, each type of question is designed to elicit a different response. Your selection of question allows you to exert some control over the tenor and direction of an interview. Depending on your goals for a particular interview or interview segment, you will make strategic choices about what kinds of questions to pose and how to organize your questioning.

Other available tools include logistics, body language, and facial expression. You have a great deal of latitude as to when and where you conduct interviews. Clients generally come to you in your office, where you may meet them from behind your desk, across a table in a conference room, or in two chairs facing one another. Each of these setups sends different signals—conveying respect, professionalism, or warmth. Depending on what you know about the client and the matter at hand, you can set the stage for an effective interview before anyone has said a word by conducting it in the best possible setting. This may also include meeting the client outside of your office. For example, many public interest lawyers representing displaced people or youth hold office hours at shelters or seek out homeless clientele where they live.

Body language also sets the stage for productive conversations. Mirroring the interviewee's positioning elicits rapport, whereas crossing one's arms signals judgment.[1] Eye contact suggests interest and attention; if you intend to take notes that will prevent you from maintaining eye contact, explain this to the interviewee up front and confirm that he understands that your lack of eye contact isn't for lack of interest.

Facial expression is an effective way to show that you are following the interviewee's train of thought, particularly if you think that a verbal interjection might break the flow of conversation. An open, friendly expression encourages the interviewee to keep talking. Reacting with honest concern or sympathy prompts further disclosure. But if you frown, looking over the client's shoulder, he will probably interpret that as disapproval and stop short—even if you were just pondering your strategy for his case, his reaction to your expression can rob you of valuable information.

Think of curiosity as a tool. As you listen to your client's story, don't be so clinical that you tamp down your humanity. If you're overly occupied with taking notes and sorting the narrative into legal pigeonholes, you may miss some interesting and useful facts.

The following four chapters of the book concentrate specifically on the first two phases of representation: pre-work product and the action plan. In both of these phases, you are speaking exclusively with the client. These conversations take place in a very different context than legal interviews of non-clients. You have a unique relationship with, and a unique obligation to, the client. Your subliminal attitudes about that relationship emerge in your conduct of client interviews.

Think of curiosity as a tool.

For years, clients have been telling us frankly that we "have poor listening skills," and that we "don't genuinely care about them or their case," that we "think we know everything" and "nitpick every word they say."[2] Many of these complaints can be traced back to a longstanding attitude of paternalism and hierarchy that attorneys often brought to the client relationship. In recent decades, however, the profession has shifted toward a "client-centered" model of representation.[3] This model discourages attorneys from imposing their values and goals onto clients and instead encourages them to identify and empathize with clients in order to serve the client's wishes. Of course, if the client wishes to enter a contract with the neighbor's cat, the competent attorney is not obliged to indulge him (and is likely obliged to refuse the request). But the new model tells us that if a client, left out of his cherished uncle's will in favor of an insinuating nurse, is reluctant to challenge the will and cause a family rift, you have not represented him well if you press him to contest the will and his cousins stop speaking to him (even if you win the case).

Like any conversation, a legal interview is a bilateral activity. One person talks while another listens; one person gets attention while the other gives it. It is not necessary—or desirable—for you to keep a death grip on your conversations with the client. Over the course of the representation, the conversational lead will shift back and forth between the lawyer and the client.[4] Early on, clients do more of the talking and get more of the attention. As the representation proceeds, the attorney gradually takes the lead role in asking questions and proposing options.

---

1. See, e.g., Lee Ross & Donna Shestowsky, *Contemporary Psychology's Challenges to Legal Theory and Practice*, 97 Nw. U. L. Rev. 1081, 1089, n.39 (2003).
2. Scott Edward Walker, *Top 10 Reasons Why Entrepreneurs Hate Lawyers* (www.venturehacks.com/articles/hate-lawyers, last visited Aug. 15, 2016); Oscar Michelin, *Why People Hate Lawyers (But Shouldn't)* (www.courtroomstrategy.com/2011/08/why-people-hate-lawyers-but-shouldnt/, last visited Aug. 15, 2016).
3. *See, e.g.,* David A. Binder et al., Lawyers as Counselors 3–11 (West 2012).
4. *See, e.g.,* Thomas L. Shaffer & James R. Elkins, Legal Interviewing and Counseling 57–60 (Thomson West 2005) (suggesting that the attorney's need to maintain control can actually inhibit effective interviewing).

CHAPTER ONE

THE OVERVIEW INTERVIEW

Physicians who begin exams by asking, "What brings you here today?" typically interrupt their patients' answers after just eighteen seconds.[1] Hijacking the conversation in this way hurts the patient: only 15 percent of them fully comprehend what their doctors tell them, and half leave office visits unsure of what steps they are supposed to take to advance their health.[2] Just as important, the doctor can suffer if she overpowers the patient early in the relationship: physicians with a "dominant tone" during office visits are more likely to be sued for malpractice than doctors who convey warmth.[3]

The same principles apply to attorney-client conversations. You cannot represent your client if you don't know what she's going through or what she wants. And you cannot know those things if you do most of the talking during the early phases of the relationship. Many young attorneys are anxious to establish authority and to "lawyer." They hold forth with advice or inundate a client with probing questions at the first relevant fact revealed. More seasoned attorneys understand that *acting* like a good lawyer during initial meetings may actually prevent them from *being* a good lawyer.[4]

The goals during the pre-work product phase of legal interviewing are twofold: first, you want to learn as much as possible about the problem or opportunity the client is facing; second, you want to establish trust and rapport with the client. Of these, the second goal is the more important. You can't always expect to get a complete picture of the client's situation in the first interview, but if you develop trust, the client is likely to reveal more complicated or uncomfortable information in subsequent interviews. Remember, if you try to dig too deep right away, you may damage the relationship in a way that actually inhibits later revelations. And the fact that never emerges in your attorney-client conversations may be the one that comes back to haunt you during later litigation or negotiation.

If you develop trust, the client is likely to reveal more complicated or uncomfortable information in subsequent interviews.

Although you are trying to accomplish some fairly intangible goals during the pre-work product phase, that doesn't mean you can't conduct this phase of the representation systematically. You can, and you must. Systematizing the process means identifying the tools at your disposal. It also means conducting the interview in distinct phases to achieve distinct goals, deploying the right tools in each.

1.1 THE OVERVIEW

The overview is the first segment of the client interview and leads off the pre-work product phase of the representation. This segment occurs immediately when you meet the client to discuss possible representation. Ideally, you want to achieve four goals in this conversation:

1) Identify the client's problem or problems.

1. Meredith Levin, *Tell the Doctor All Your Problems, but Keep It to Less Than a Minute*, N.Y. TIMES June 1, 2004.
2. *Id.*
3. *Id.*
4. *See, e.g.*, THOMAS L. SHAFFER & JAMES R. ELKINS, LEGAL INTERVIEWING AND COUNSELING 57–60 (THOMSON WEST 2005), at 49–55.

2) Identify the client's desired outcome.
3) Get some insight into the client's preferred path for achieving the outcome.
4) Begin to build trust and rapport.

Your primary job during the overview segment is to *listen*. To an outside observer, it may not look like you are doing much, but if you are listening actively and effectively, you will achieve your four goals.

As the overview segment initiates the relationship with the client, it is a good time to consider confidentiality. When the client initially contacts you for the purpose of obtaining representation, your duty to keep his communications confidential has begun.[5] Of course, the client may ultimately retain a different lawyer or you may decline his case for any of a number of reasons, but even if an attorney-client relationship never formally arises from the initial conversation, the information shared must be kept confidential. You know this. Make sure the client knows it, too. If the client is uncomfortable or you sense that he is holding back, this information may nudge the interview along. Of course, there are some exceptions to the duty of confidentiality. With few exceptions, it is not necessary or helpful to dwell on those exceptions when you explain confidentiality to the client. It is enough to tell him that as a rule you are obliged to keep his information confidential, and that should one of the very narrow exceptions to that rule arise, you will let him know immediately. The focus should be on the fact that you will likely *not* divulge what he has said. If the client engages you, the particulars of attorney-client confidentiality should be reiterated in the formal engagement letter (discussed in Chapter Four).

You will lean most heavily on open questions during the overview interview. Ask the client why he's there—and then let him tell you. As he talks, assure him that you are following his story and that you care about it. You can do this with empathetic responses when he relates surprising, positive, or negative information: if he tells you that his house burned down, it is normal and empathetic to tell him you are sorry and to venture that this must be a difficult time; if he tells you that he's been offered a contract to play violin for the Cleveland Symphony Orchestra, it is appropriate to congratulate him and remark on the honor of such an invitation. Interject occasionally with a repetition or summary of what he's just said. If he has just related that his company's largest supplier has been acquired by a competitor who is unlikely to extend him credit, you could, at an appropriate pause, mirror his narrative: "So you're concerned about your relationship with this key supplier going forward." He will either appreciate that you understand his position, or he will clarify if you haven't quite followed: "Actually, I'm ready to write off Apex Supply. I've met a few times with Beta Supply and I think we're closing in on a deal."

You might worry that if you don't intervene with comments or advice, clients will ramble at length. They probably will. That's not necessarily a bad thing. If you're listening effectively, you will learn some relevant facts about their situation. You will also gain understanding about why they want your help. As important, you may start to get a sense of their values and what makes them tick. If you sense from their "rant" that they are unhappy about a breached contract not because it cost them a great deal of money but because they feel disrespected, this may help you counsel them on a satisfactory remedy. Your willingness to listen is an important relationship-building tool. It indicates that you take them seriously and that their problems are important to you—both of which contribute to a trusting and open relationship. Further, don't waste your efforts to carry on a neutral and dispassionate discussion if a client is intent on telling a story and venting some emotion; they will inevitably return to their preferred topic repeatedly until they're done.

Your willingness to listen is an important relationship-building tool.

This phase of the pre-work product interview process belongs primarily to the client. Don't rush him through it. That said, at some point you must help the client form a preliminary statement of what he wants

5. *See, e.g.*, Paul J. Zwier & Anthony J. Bocchino, Fact Investigation: A Practical Guide to Interviewing, Counseling, and Case Theory Development 37–38 (NITA 2014).

from you. Typically, initial client meetings are booked for an hour; if the client is closing in on the one-hour mark and is either repeating information or has not yet outlined a discernable legal problem, pose a few closed questions to clarify the legal issue or issues.

Once you sense that the interviewee has reached the end of her story, it is your job to help the client "land." Summarize what the client has told you about her factual situation and her initial preference for resolving it, and then ask if there's any additional information or consideration you should be aware of:

> "I understand you've come here because you are considering asking your husband for a divorce. You have told me there is a prenuptial agreement between you and you're not sure about your financial position under the agreement, so you are torn about ending the marriage. This must be a stressful situation. It seems that you'd like to see how airtight the prenup is and where that would leave you in order to make an informed decision. Have I covered all of the considerations here? Is there any other information or option we should be exploring?"

If the client adds no new wrinkles, you can consider the overview segment complete.

Complete Exercise One: Overview Interview

Chapter Two

The Chronology

Assuming that the client agrees with your recap of her story, it is time to begin the second segment of the pre-work product interviewing process: taking a chronology of events. Work through the relevant events in chronological order to impose an objective order on the messy web of subjective facts that the client initially presented. As you work, delve for facts that the client may have omitted from her first recitation. She may have done this because she did not appreciate their legal relevance, or because she thinks they work against her, or because they are unpleasant for some reason. Organizing the facts chronologically allows you to spot gaps; fill them in at this point unless you intuit that they are best handled at a later date—if so, make a note of the missing information while the question is fresh in your mind.

Think of the chronology as a "horizontal" picture of your client's case. Don't worry during this phase about getting all of the atmospheric details about any particular event on the timeline. The completed chronology will help you identify (either on the spot or after additional research) those events that have potential legal significance. Talk with your client about those events during the theory development segment of the interview process. At that point, ask her to paint a more thorough and nuanced picture of those events. During this phase of the interview, you will build several "vertical" pictures of the case.

For now, pivot from the overview to the chronology by explaining that you'd like the client to review the situation from start to finish so that you have a detailed understanding to help you assess her options. Throughout the chronology, you will likely shift from open questions—which empower the client to relate the events—to closed questions—which empower you to slow down the narrative and zoom in on crucial time periods.

If your client is a methodical and detailed storyteller, you may again take a fairly passive role during the chronology phase of the interview. Usually, though, you will shift to a slightly more collaborative approach. Why? Clients may fall into any of three traps, and it's up to you to get them out.

1) They follow a **sidetrack**. The client believes that a particular event in the chronology is legally significant (he is probably right!). So he sets the scene, telling you who was there, what they said, where they stood, how he felt, what the weather was like, and so on.

These details may be fascinating, and they may wind up as the crux of your case. But if you get caught up in them now, you may neglect other important events on the timeline. And if you settle on a legal theory early in the chronology, it may distort your view of subsequent events. So if you see client heading down a sidetrack, intervene. First, assure him that you are interested in these details and that they are probably important. Explain that you will return to this event once you've gotten through the whole story, and then prompt him to return to the chronology: "You were telling me that you went to Kansas City for your first interview on July 12. What time did you arrive at the office?"

2) They **clump** multiple discrete events into one point on the timeline.[1] The client may be interested in the culmination of a series of events and perceive them as one development when they are actually distinct events. She may say "in October my husband encouraged me to have a child."

1. *See, e.g.,* David A. Binder et al., Lawyers as Counselors 3–11, at 185 (West 2012).

There are two tipoffs to possible clumping here: the expansive time identifier and the ambiguous verb. It is unlikely that a single event took place over the course of the thirty-one days in October, and while "encouragement" can happen during the course of a single conversation, it is much more likely to have been offered through a series of conversations and actions.

If your client refers to units of time that seem lengthy in relation to the underlying event, or describes as a single event a transaction that generally unfolds over time, take a verbal pickaxe to the statement and break it down into constituent events: "After you came home from the doctor on October 27, what happened the following day? And what took place immediately after that? Tell me specifically what words or actions your husband used in relation to fertility treatments during the first week of October."

3) They leave **gaps** on the timeline. In retelling their stories, clients may jump from what they consider one significant or favorable event to the next. They may regard intervening events as unimportant, may have forgotten them, or may be reluctant to share them with you. Depending on the timeframe for your client's story, if days or weeks or months on the timeline are empty, explore the gap that needs to be filled.

Encourage your client to fill in gaps by "framing" them: "I see that you resigned from ExIm on April 15 and that Vandelay Enterprises contacted you on July 1. What occurred between your resignation and that first phone call from Vandelay? How did you spend that second half of April?"

If the client left events off the timeline because he thought them unimportant, this simple prodding and demonstration of interest should be enough to get him to fill in the gap. If he has forgotten, you must try to jog his memory. Select a particular relevant subject and ask him to trace any event relating to that subject during the timeline gap:

> "Let's track any efforts you might have made during this period to find a new job. Did you communicate socially with any of your former colleagues from ExIm between April and July? Did you communicate with any members of your alumni network during that period? Did you change your job status on LinkedIn? Did you prepare an updated resume during this period? Okay, now that I understand that you were actively looking for a new job during this period, let's talk about your relationship with ExIm after you resigned. Did you continue to receive consulting fees from ExIm during this period? Did they contact you to work on any projects during the spring or summer?"

Another possibility is that the client left items off the timeline because he thinks they are unfavorable or he finds them difficult to discuss. While it may be easy to discern skipped events the client thought insignificant or unimportant, it may be much harder to discover events that the client is intentionally withholding for some reason. If after probing you still find gaps on the timeline, it is time to draw on the trust and rapport you have been building during the early stages of the interview. Here are some suggestions for broaching these subjects:

- Remind the client that all of the information he shares with you is confidential.
- Remind the client that, depending on how you move forward with the case, you will likely meet the adversary in negotiations or in court; explain that it is better if you know of any unfavorable information before the opponent raises it.
- Reiterate to the client that you understand this must be a difficult process and that it may stir up some unpleasant memories.

If you make such comments in a neutral fashion, without explicitly suggesting that the client is intentionally withholding information, you are unlikely to place the client on the defensive. This simple reassurance may be sufficient to encourage the client to fill in gaps on the timeline. If the client is not responsive, be wary of pushing the subject. As he warms up to you in future interviews, you can return to the timeline gap. Further, as you gather documents and third-party input, you may get the information from another source, or you may figure out the reason for client's reluctance so that you can address it with him in a more direct way.

Suppose that your client is being sued for violating the noncompete agreement he signed when he was the chief brand officer at ExIm Corp., a company that wholesales rattan furniture from the East Asia market to big box retail

stores and select other establishments. He left ExIm in April 2014 and in September 2014 took a job as chief operations officer at Vandelay Enterprises. Vandelay wholesales restaurant furniture to casual dining chain restaurants. The contract bars him from taking corporate marketing jobs with any of ExIm's competitors. He was recruited to head up operations at Vandelay, which has a distribution network to many of the same suburban malls that ExIm serviced. His knowledge of this network was the key reason Vandelay hired him. A month into the job, Panda Express contacted Vandelay to supply Asian-style tables and chairs for a major redesign. Your client was instrumental in completing the deal to sign Panda Express. ExIm contends that his promotion of Vandelay as a supplier of Asian-style rattan seating amounted to marketing to a competitor in violation of the agreement.

In conducting the chronology phase of the interview, your goal is to produce a document that looks something like this:

January 5, 2010:	**Contacted by headhunter recruiting for chief marketing officer position at ExIm Enterprises.**
January 12, 2010:	**First interview with ExIm director of personnel.**
January 20, 2010:	**First interview with ExIm CEO.**
January 25, 2010:	**Second interview with ExIm CEO. Employment terms discussed, including salary and stock options, as well as required noncompete contract.**
February 2, 2010:	**Job offer received. Memo proffered, setting forth employment terms with draft non-compete contract attached.**
February 5, 2010:	**Job offer accepted. Non-compete contract signed.**
March 2010–December 2013:	**Employed uneventfully at ExIm.**
January 2014:	**ExIm receives poor booth assignment at National Furniture Expo show, attendance and promotion at which is major item in your marketing portfolio. ExIm hospitality suite at show is sparsely attended.**
February 28, 2014:	**Annual wholesale orders for deck and patio furniture, typically completed by February 28, come in 40 percent below 2013 figures.**
March 13, 2014:	**Pier 1 declines to renew annual contract, which has historically represented 20 percent of ExIm sales.**
April 1, 2014:	**CEO advises that board discussed recruiting candidates to replace you as chief brand officer and will take up subject at next meeting on May 1, 2014.**
April 15, 2014:	**You resign from ExIm rather than risk being terminated.**
May 1, 2014:	**You meet with a headhunter to discuss future employment opportunities.**
May 15, 2014:	**You update your alumni profiles for your undergraduate and MBA institutions.**

May 25, 2014:	**You attend a Midwestern Marketing Association luncheon, where you tell acquaintances in the business that you have left ExIm.**
June 4, 2015:	**You attend a National Restaurant Association seminar on "New Trends in Seating," where you tell acquaintances in the business that you have left ExIm.**
June 28, 2015:	**You attend an International Council of Shopping Centers luncheon, where you tell acquaintances in the business that you have left ExIm.**
July 6, 2014:	**Personnel manager from Vandelay Enterprises asks you to interview for position as director of operations.**
July 12, 2014:	**First interview with personnel manager.**
August 2, 2014:	**First interview with CEO.**
August 8, 2014:	**Second interview with CEO. During discussion of key responsibilities, you mention noncompete contract with ExIm.**
September 3, 2014:	**CEO offers you position as director of operations.**
September 5, 2014:	**You accept position. Along with required employment paperwork, you return to personnel director copy of your ExIm noncompete contract for your file.**
October 5, 2014:	**Director of marketing (DOM) tells you that Panda Express has inquired about price and specifications for a 12,000-unit order of rattan chairs for a chainwide redesign.**
October 6, 2014:	**DOM asks you to prepare a memo for Panda Express regarding distribution of Vandelay chairs to shopping centers and strip malls, with particular focus on price efficiency and timeliness.**
October 14, 2014:	**DOM asks you to join him at lunch with Panda Express supply team to promote Vandelay for deal.**
October 20, 2014:	**You receive phone call from Panda Express director of infrastructure with questions about durability and color options for Vandelay chairs.**
October 30, 2014:	**Panda Express director of infrastructure calls to tell you the chain has decided to give the contract to Vandelay.**

If your client has brought with him any documents relevant to the matter, peruse them after completing the chronology. If the documents are dated, you can note their existence on the timeline. Even if they are not dated, your client may be able to orient them chronologically once he compares them with the timeline you have drafted. If the timeline notes the existence of documents your client has not brought, ask him to bring those available to the next meeting. With the timeline and the documents, use this next meeting to explore available legal theories that the documents might support or weaken.

Complete Exercise Two: Taking a Chronology

Chapter Three

Theory Development

You now have a detailed understanding of the events that led your client to his current plight. Depending on the clarity of those events and your expertise, you may already have a clear legal theory. On the other hand, some facts don't fit into a ready category (such as last will and testament or slip and fall case), or the law in the area is complex. You may have to conclude the meeting and do some intervening research to develop a menu of possible legal measures your client may pursue before resuming the interview in a second meeting. In either event, the next segment of the client interview requires you to thoroughly develop the facts and test whether one or more of these legal measures works for your client. This phase is known as theory development.

The chronology phase of the interview proceeds horizontally; the theory development phase develops vertically. Identify a few junctures on the timeline about which you need to know more to determine whether your client is likely to secure a competitive contract, face a regulatory penalty, or succeed in a lawsuit. Prepare for the theory development phase of the interview by listing for yourself the various legal measures that you are exploring and the necessary elements or requirements for each of those measures. With those elements in mind, review the timeline for indications whether these elements have been met. Then ask your client to elaborate on those timeline points. For clarity, take the client through the complete timeline for a single theory. Repeat the process from the beginning for the next theory if the first discussion did not elicit information relevant to that theory. Do not become distracted by new facts that support a different theory; jot a quick note to remind yourself for the next time through the timeline, but stay with the current theory to avoid confusing both yourself and your client.

Begin the theory development phase with open questions, allowing the client to recall as much as possible about the logistical and personal details associated with a single point on the timeline. If you need specific information that is not forthcoming or if the client's free recall is unclear, follow up with closed questions. Continue to hone your information with yes-no questions if necessary.

Although the theory development phase of the interview takes place verbally, now is the time to begin thinking visually as well. In many instances, you will have a better grasp on the viability of your theory if you can picture the relevant events. If your client has come to you about a car accident, ask him to draw a map or diagram of the accident scene and to show you the relative positions of cars and people on that map as he discusses it. If your client is considering a breach of contract case in connection with residential construction, ask to see the architectural plans or plat of survey so you have a full appreciation of how the failure to perform affected the project. If the client has brought in the available documents you requested, they may shed light on the matter. It might also be worthwhile to visit the scene or look at a relevant machine or device yourself.

Let's return to our friend, the operations director at Vandelay. Based on the chronology you took from him, as well as a copy of the noncompete agreement and some simple internet research, you are considering two responses to the breach of contract claim. First, your client did not commit a material breach because restaurant seating is so different from retail furniture that Vandelay is not a competitor of ExIm. Second, Vandelay is liable to your client for tortious interference with contract because they hired him under the guise that he would be responsible for operations but then thrust him into a marketing role that he could not refuse without losing his job. In order to assess the strength of these theories, you need to know more about your client's work on the Panda Express contract.

Test theory one with questions such as:

- **Open:** I'd like to understand more about how Vandelay and Panda Express got into business together. Can you tell me more about that?
- **Focused Closed:** As far as you know now, when did Vandelay first propose supplying the chairs for the Panda Express redesign?
- **Focused Closed:** When were you first asked to join the Vandelay team handling the Panda Express deal?
- **Broad Closed:** Did Panda Express specify the rattan chairs it wanted before or after you came into the deal?
- **Open:** Compare the rattan chairs that Vandelay supplies to restaurants with those sold for home use by retail stores.
- **Broad Closed:** Based on your knowledge of Panda Express's specifications, how suitable would the chairs that ExIm carries have been for the redesign project?
- **Focused Closed / Yes–No:** The chronology we completed shows that you took part in five meetings with Panda Express. Did the subject of using retail chairs come up at any of those meetings?
- **Open:** How did that conversation proceed?

Test theory two with questions such as:

- **Open:** You had your first interview with Vandelay in July 2014. Tell me how they described to you the position they were filling.
- **Broad closed:** What did Vandelay tell you about why they were recruiting you for the job?
- **Focused closed:** What kinds of questions did Vandelay ask about your work for ExIm?
- **Broad Yes–No:** Did you get the impression that Vandelay was more interested in your marketing background than in your operations expertise?
- **Open:** Describe the discussions you had with Vandelay about your noncompete agreement.
- **Yes–No:** According to trade industry news coverage, Panda Express was reported to be contemplating a major redesign as early as June 2014. Did Vandelay mention the Panda Express pitch to you during your first interview?

The trust and rapport you have been establishing begins to pay off during the theory development interviews. So far, you have listened to the client's subjective version of events during the overview segment and have taken an objective version of events during the chronology segment. As you explore the viability of various theories for the client, you may have to push for answers that the client realizes are unfavorable or for information that makes him uncomfortable. You have been empathetic so far, and hopefully he's begun to see you as a partner. Explain that it's easier to deal with potentially unfavorable information at this early phase than if it emerges when you are in face-to-face contact with the adversary, who will be looking to exploit any weaknesses. Keep in mind that getting this information may be a long-term process and may require multiple attempts on your part.

Complete Exercise Three: Theory Development

CHAPTER FOUR

THE ACTION PLAN

After one or two sessions with your client, you have achieved a good overview of why she is seeking legal help, a working chronology of the events associated with her need for legal help, and a detailed understanding of whether the facts support one or more realistic legal theories. Now it's time to decide with her what form your legal help will take so you can begin work.

The goal of this representation phase is to counsel the client towards a course of action. The course of action may not be limited to just one legal response to her problem. She may choose to try inexpensive, low-stakes responses and hold the more expensive options in reserve if her first-choice route is unsuccessful. Or it may turn out that pursuing one option compromises the availability of an alternate option. Whereas during the pre-work product phases of interviewing the client did more talking and you did more listening, during this discussion you will do most of the talking. Beware—this doesn't mean that you are doing the deciding! You are explaining to your client all of the options open to her, and the costs and benefits associated with each.

If you've laid the groundwork for a productive and trusting relationship, she should be comfortable asking you many follow-up questions. Your job is to give her clear and straightforward answers, representing the applicable law as transparently as possible. Further, you can help the client evaluate the legal options from two distinct and important perspectives: her values and your experience. At this point, you have some understanding of what is important to your client and how she feels about the issue at hand. Even if her perspective is still unclear to you, you should have established enough rapport to ask her directly about her values and feelings and get a helpful answer. The action plan you craft should be the product of her general approach to the problem. For example, among business clients, lawyers have a reputation for "kill[ing] deals" that the client wanted. In general, lawyers are trained to spot and avoid risk, while those in business are more tolerant of risks because they can carry payoffs. Beware of imposing your own risk preferences on a client. At the same time, the client is seeking the assistance of an expert who has handled similar matters in the past and has insights that she lacks. You do the client a disservice if you allow her to dictate a course of action that, in your experience, carries significant costs without making her fully aware of those costs and making sure she appreciates them.

Devising an action plan clarifies both attorney and client expectations for the remainder of the relationship. Clients often report frustration that their lawyers do "not communicat[e] clearly," and do "not keep [them] informed."[1] These are not explicit complaints about our listening skills, but they reflect a fundamental failure to appreciate the client's point of view. One attorney malpractice insurer has estimated that 75 percent of all grievances filed against lawyers result from poor communication.[2] The law is not always predictable, and the client cannot receive a guarantee of her preferred outcome. But unfortunate developments should not come as a complete surprise. Now is the time to get those possibilities on the table.

At this point, put your action plan into writing to share with the client. This written document, sometimes called an engagement letter, accomplishes several goals. First, it forces you to reduce to concrete terms what you understand the client to be asking of you. If you have misunderstood, the client will speak up when she sees it in

1. Scott Edward Walker, *Top 10 Reasons Why Entrepreneurs Hate Lawyers* (www.venturehacks.com/articles/hate-lawyers, last visited April 17, 2013); Oscar Michelin, *Why People Hate Lawyers (But Shouldn't)* (www.courtroomstrategy.com/2011/08/why-people-hate-lawyers-but-shouldnt/, last visited April 19, 2013).
2. Lawyers Mutual, *Attorney-Client Agreements Toolkit* 1 (2009).

black and white, and you can correct course before you've wasted time and soured the relationship. Second, it is a blueprint you can refer back to from time to time to avoid getting so bogged down in the minutia of the representation—did we get duplicates of that crucial agency report? Is the brief over the court's page limit? Did the client's ex-husband release the lien on their vacation home?—that you lose sight of the client's wishes. Third, it is an opportunity to clarify the logistical terms of the attorney-client relationship going forward. Depending on the timeline involved, will you update the client regularly? At what interval? Or would you prefer that she check in with you? How will the matter be financed? On a contingency basis, via a retainer with monthly billing? At what rate? To the best of your knowledge at this early point, who will be staffing the matter? Are you handling it on your own, or will you use paralegals, associates, or others? What are their billing rates, and is the client amenable? Once you have memorialized all of this information in a memo or a letter, you may keep a copy on file and send one to the client for her reference. Finally, many firms also use engagement letters to keep track of the scope of representation for all clients in order to identify and address potential conflicts between clients.[3]

A sample action plan letter might read as follows:

> Dear Mrs. Atwood,
>
> I am pleased that you have asked me to serve as your counsel as you consider changes to your marital status. This letter is designed to confirm in writing the nature of your contemplated legal action and the terms of our representation. As always, please do not hesitate to contact me with any questions you may have regarding the material below. I look forward to serving you with regard to these matters.
>
> As we discussed, you intend to initiate divorce proceedings against your husband, Richard Atwood, in Nita County Circuit Court. Your top priority is to retain physical custody of your children and to secure an adequate monthly maintenance payment. You are not interested in retaining the family home. According to our calculations, monthly expenses for your children and yourself are about \$15,000. The terms of your prenuptial agreement with Richard Atwood purport to cap monthly alimony and child support at \$8,000. You have asked us to initiate informal negotiations with Mr. Atwood's attorney prior to filing in hopes of reaching agreement on a monthly fee in excess of the prenuptial agreement so that you may file for an agreed divorce jointly and avoid making the terms of the dissolution public. Should this negotiation prove unsuccessful, you have authorized us to file a contested petition for divorce and, relatedly, to challenge the validity of the prenuptial agreement; first, on the grounds that it was invalid because it was signed under duress; and second, on the grounds that if valid, your husband breached the contract by interfering with your career opportunities. A negotiated divorce typically takes twelve months to conclude and does not require extensive expert involvement or court attendance. A contested divorce typically takes from twelve to twenty-four months to conclude and may require hiring expert witnesses and your repeated testimony in depositions and in court. You have stressed that you would prefer to keep your children from testifying but you have said you understand that as this matter proceeds, that may be unavoidable.
>
> I will be the primary attorney on your case, supported by a junior associate and a paralegal. Once we have begun to work on your behalf, I will update you on our progress monthly or more often if the need arises. I hope these status reports will keep you abreast of the matter, but should any questions arise in the interim, do not hesitate to contact me. It is in your best interest that our communications are confidential, so that we avoid inadvertent disclosures of important information or loss of the attorney-client privilege. As a result, please do not communicate sensitive matters to us where the conversation might be overheard or share our communications with your family or friends. Please limit your email communications with us to a personal computer; employers are often entitled to access emails sent from work computers.

---

3. *Id.*

We are gratified that you have delegated certain levels of authority to act on your behalf; however, there will be times when we will not be able to proceed without your full and sometimes written consent, or without your personal participation. We will attempt to advise you well in advance of these situations.

Our fees will be based on the hourly rate for each attorney and paralegal who has worked on this matter. My hourly rate is $400; junior associates typically bill at $150 to $250 per hour; and paralegals typically bill at $50 to $100 per hour. Should these rates change, we will advise you accordingly. Further, aside from charging you for our legal work, we will also bill you for goods and services we engage on your behalf, such as expert consultations or travel. As we have discussed, it is impossible to predict with accuracy the total fee for this matter from start to finish. We estimate that if the matter is concluded via negotiation, the total cost will be approximately $60,000 to $70,000. If the matter proceeds to litigation, we estimate the total cost at $150,000 to $250,000. Our bills are sent monthly. Please see the attached Billing and Fee Policy for more details.

Our goal is to provide you with outstanding legal services for a reasonable cost. The attorney-client relationship is one of mutual trust and confidence. If you ever have questions or concerns about our work on your behalf or a monthly statement that you receive from us, we welcome your inquiries. Thank you once again for allowing us to represent you.

Complete Exercise Four: Client Action Plan

Chapter Five

Witness Interviewing

5.1 Introduction to Witness Interviewing

5.1.1 *The Purpose of Witness Interviews*

Once you have a good working knowledge of your client's version of events and her preliminary preferences for resolving them, consider interviewing third parties. Of course, if the matter is headed toward litigation, you will make document requests and take depositions during the discovery phase. But procedural reforms have limited the depth and extent of discovery, giving an advantage to the attorney who is adept at early, informal investigation.[1] Early access to information may be useful in negotiating settlements midway through litigation, before discovery has been completed.

Most of these interviews will be with people who have knowledge that can either strengthen or weaken your client's legal position. If your client is in a litigation posture, you may be interviewing these third parties not just to learn what they know but to assess their potential as witnesses. Even if your client is not in a litigation posture, you may find it helpful to get the perspective of an outsider or an expert on important issues. For example, if you are representing a client in the purchase of a building, you may work with inspectors during the contingency period to unearth any material defects in the property. Finally, as you move through the work product phase of the representation, you may have to deal with your counterpart on the other side of the deal, or with a court clerk or agency worker. Think of these interactions as legal interviews, subject to the same systematic, strategic approach as client and witness conversations.

5.1.2 *Types of Witnesses*

Most third parties—we will call them witnesses for ease of reference though they may never take the stand—can be classified as either "friendly" or "hostile." A friendly witness is one who wants your client to succeed, while a hostile wants to weaken your client or assist his adversary. Be careful: not all friendly witnesses are actually helpful, and some hostile witnesses may be crucial to your case. We classify witnesses as friendly or hostile because it helps us structure the interview and draw on the most appropriate tools to benefit from these conversations.

Conduct third-party interviews purposefully. That is, go into the interview with a very clear idea of the information that you want to walk away with. To accomplish this, begin your legal work before the interview. For example, if you are filing a lawsuit for your client, write a preliminary draft of the complaint before you contact third-party interviewees. This allows you to discern where you have good information about elements of the cause of action and where you have gaps. Your client may have told you that a particular witness was involved in the relevant event; if you know exactly what gaps in the complaint you must plug, you can design your interview to get that information. You may also suspect that the witness has additional but less crucial details. And there is always the possibility that he has important knowledge about matters of which you are not yet aware. Understanding what you need

1. *See,* Paul J. Zwier & Anthony J. Bocchino, Fact Investigation: A Practical Guide to Interviewing, Counseling, and Case Theory Development 49 (NITA 2014).

from the interview and what you'd like from the interview allows you to prioritize your areas of conversation and make strategic decisions about when to broach topics, depending on how the interview proceeds.

5.1.3 Conducting Witness Interviews: Ethics and Mechanics

Whether a witness is friendly or hostile to your client's interests, the fact remains that he is not your client. This raises some ethical and mechanical considerations. Think these matters through as part of your preparation for the interview with these witnesses.

Attorney ethics rules address in some detail the lawyer's duties when speaking with non-clients. This book is by no means a comprehensive treatment of professional responsibility, but you should be familiar with the relevant principles so you can look into them if you sense you are treading on ethically dubious ground. The most important of these rules for your purposes covers conversations with non-clients who are represented by an attorney in connection with the matter at hand.[2] If you know that a witness is represented by a lawyer, the rule bars you from communicating with him absent consent from his attorney. Realistically, in many cases you will not know whether the person is represented prior to approaching him, meaning you have latitude to at least begin a conversation without crossing ethical boundaries. But this stricture can pose a real obstacle when your client is litigating against an organization. If an organization employee has "managerial responsibility," or if his behavior can be imputed to the organization for liability purposes, or if his statement could be considered an admission by the organization, then that person is considered "represented" by the organization's counsel. As a result, you are on notice prior to interviewing any of the organization's employees or ex-employees that there is a good chance they are considered "represented," and thus off-limits without consent from the organization's lawyer.

Assuming that the rules permit you to contact a witness, your conversation is governed by some additional ethical requirements. For example, you cannot misrepresent the facts or the law in a conversation with a witness.[3] Relatedly, you cannot exploit a witness's misunderstanding of your role—you are not neutral, and you do not represent the witness; you are an advocate for your client.[4] And you may not "embarrass, delay, or burden" the witness,[5] or try to interfere with opposing counsel's access to the witness or his information.

Assuming you are permitted to approach a witness and conduct your conversation with due regard for ethical principles, how are you going to leverage the information obtained as the matter moves forward? That is, will you memorialize the conversation and, if so, how will you do it? It is generally prudent to memorialize the conversation, for a few reasons. The witness's memory may fade if the matter takes several years to proceed. You may not be able to elicit the same information in a subsequent conversation, deposition, or testimony. A memorialization of the initial conversation could be used later in the litigation to refresh the witness's recollection before or during any testimony at trial.[6] Or should this tactic fail, a memorialization of the conversation could be admissible as a recorded recollection.[7] If the witness's recollection changes in a subsequent deposition or testimony, the original statement could be used to impeach him. Finally, documenting information that is strong or conclusive on a crucial point may be useful in negotiations, whether in service of completing a transaction or in service of settling a lawsuit.

---

2. *See, e.g.*, Model Rule of Professional Conduct 4.2, Communications with Persons Represented by Counsel.

3. *See,* MRPC 4.1, Truthfulness in Statements to Others.

4. *See,* MRPC 4.3, Dealing with Unrepresented Persons.

5. *See,* MRPC 4.4, Respect for Rights of Third Persons.

6. *See* Fed. R. Evid. 612, Writing Used to Refresh a Witness. This rule contemplates the use of written statements to refresh the recollection of a witness the attorney has called. It also permits the adverse party in most cases to inspect the written statement, cross-examine the witness about it, and introduce into evidence any portion that relates to the testimony in question. If the statement is claimed to cover material outside the witness's testimony, the court may review the statement and exclude the irrelevant portions.

7. *See* Fed R. Evid. 803, Exceptions to the Rule Against Hearsay. Pursuant to this rule, the prohibition on hearsay does not bar introduction of a "recorded recollection," defined as a record that "(A) is on a matter the witness once knew about but now cannot recall well enough to testify fully and accurately; (B) was made or adopted by the witness when the matter was fresh in the witness's memory; and (C) accurately reflects the witness's knowledge."

Is there ever a scenario in which you would opt against memorializing the witness's statement in some way? A witness may pass along information that is very damaging to your client's interests, but it is still generally useful to memorialize the statement. The witness exists and is likely to be discovered by your adversary whether you record the statement or not; having a concrete version of the witness's statement allows you to look for inconsistencies, implausibilities, and bias as you pursue the matter. At the same time, every jurisdiction has a different approach to the privilege that attaches to witness statements. The general rule is that attorney work product need not be disclosed to an adversary during discovery. But states (and individual judges) may make different calls on whether a witness statement is a privileged product of attorney research and strategy or an objective fact outside the scope of the privilege. So if you fear the statement is subject to disclosure and if it contains information that is both highly damaging and highly suspect, you may not wish to reduce it to the form of a witness statement.[8]

The question remains: if you decide to memorialize the conversation, how will you do it? You have a number of options.[9] You can ask the witness to draft his own statement or fill out a prepared form. You can prepare a written statement based on your notes of the conversation and ask the witness to sign it. You can record the conversation, either over the phone or in person, with the witness's permission. Many attorneys like to bring along a "prover" to listen to the conversation for purposes of later recall in the event the witness does not sign a statement or agree to a recording. As you have probably surmised, choosing a method to memorialize the statement is a matter of interview strategy. Only the most enthusiastic witness is likely to go to the trouble of drafting his own statement, and these do-it-yourself documents may not be sufficiently comprehensive or sufficiently detailed for your purposes. A reluctant or nervous witness is unlikely to agree to being recorded; requesting to may simply intensify his discomfort and reduce his cooperation. In most cases, attorneys draft written statements for the witness's signature; this method gives the attorney maximum control. He can choose what information to include and exclude. And he can—without making misrepresentations—paint a picture. A signed statement in which the witness says that your client was "strolling" in the park is likely to be much more favorable than one that describes your client as "loitering" in the park. If this witness is not cooperative and is unlikely to sign the written statement, a prover may be preferable, so long as his presence does not interfere with the flow of the interview. These choices are likely to depend on whether you anticipate the witness will be friendly or hostile.

5.2 The Friendly Witness

A friendly witness wants to help your client, so he wants to help you. That doesn't necessarily mean he will be able to help you. He may want to help but lacks information; he may want to help but offers only damaging information; he may want to help but be so zealous that he is unreliable; or despite what he's told the client, he may not want to help. You can expect interviews with friendly witnesses to progress—at least initially—as a client interview might. Typically, you will receive cooperation and complete answers. But your goal in the friendly witness interview is different than your goal in a client interview. As a result, your strategy will be different.

Having talked extensively with the client, you have several theories of the case that you plan to pursue; these are premised on the client's version of events, which is likely to be the client's best recollection. A friendly witness has an autonomous recollection of some or all of these events. His narrative of the situation can support your theories, and they can also undermine your theories. Your top goal is to get any information that will support your theories. But you cannot afford to be blind to information that may weaken your client's case. Other witnesses may have it, and your adversary is looking for those witnesses. Forewarned is forearmed, and the friendly witness can forewarn you. So your second goal is to open the door to potentially damaging information the witness may be able to share with you.

A third goal is to test the interviewee's credibility. This is important for two reasons. First, it indicates how much faith you can put in the interviewee's information. Second, if you are considering relying on this interviewee at a later stage of the proceedings—as a trial witness or as an estate executor, say—now is the time to get a sense of how

8. *See, e.g.*, Roderick D. Blanchard & Robert S. Blanchard, Litigation and Trial Practice 255 (West 2001).
9. *See id.* at 255–57.

the interviewee would perform in those roles. Recall again that "friendliness" is not an unalloyed good for your client. The witness may be unreliable precisely because he is biased in your client's favor. As the objective expert in your client's case, it is your job to make this determination.

When you prepare to interview the friendly witness, begin by researching him. A simple internet search can confirm his job, his professional and educational background, and whether he's commented publicly on the matter at hand. This information can help you assess the likelihood that he is friendly. You can also make some preliminary inferences about his socioeconomic situation, his politics, his hobbies and interests. This background information may subtly influence how you approach the interview to achieve the best result for your client. Combined with the information about the witness your client provides, you can prioritize the information you expect to extract from this witness.

Once you have researched the witness, you must initiate the conversation. You can make thoughtful choices about how to do this depending on your prediction of his support for your client and his own self-interest. You could phone him; if he's truly friendly, it's unlikely he will screen or dodge your call. Should you phone him at home or at work? If your client is a former colleague from the same company, the witness may not feel as free to speak from the office and a call to his house might be more productive. If the witness's husband is a friend of the husband your client is divorcing, she may be reluctant to discuss the matter at home while her husband is listening, and you might not want the husband to overhear. A phone call to the office is a better choice. Emails and texts are also options; they may be convenient but appear unprofessional for a first contact. And don't forget that they create a paper trail—legally privileged but technologically vulnerable. You are likely to learn more in a face-to-face discussion, where you can observe body language and display your own empathy to good effect. Also, it is less socially acceptable to stand up and walk out of a personal meeting than it is to hang up the phone. But where will you propose to hold the meeting? The more convenient and comfortable the location, the more likely the witness is to cooperate. An office environment might underline the importance of the issue and cause the witness to take the conversation seriously; on the other hand, an informal environment such as a coffee shop or a home might relax the witness so that he lowers his guard. Neither choice is per se better, but each sets a different tone, and your early research might give you a hint as to which will serve your purposes. Make the same deliberate choices about how you track the witness's answers: will you use a computer, a notebook, a recording device, or nothing? When will you take it out, and how will you explain or ask the witness for permission?

If the witness is friendly, begin the interview with open questions that let the witness extemporize. Most people enjoy holding the floor; letting the witness talk gives him a sense of control and authority that will enhance his good will toward you and build conversational rapport. As in client interviews, listen actively and respond appropriately. During his initial answers, you can get a read on his agenda or orientation toward your client, which will shape your subsequent questioning.

Once the witness has exhausted his free response to your open questions, go back and zero in on blanks in his recitation or ambiguities in his narrative. This is where your advance prioritization of the information you need from the witness will pay off. If you know that this witness was present at a meeting the client did not attend, his recollection of that event may be the single most important thing you need to get from this interview. During his initial response, he may have neglected to list all those present at the meeting. Even if there were other ambiguities in his initial response, be sure that your first closed-question follow-ups focus on this meeting. He may have limited time, or may get tired or impatient with your questions—if you have spent your follow-ups on less important events, you may leave the interview without the one piece of information you needed.

Once you have exhausted questions designed to bolster your client's case, consider throwing out a couple of lines to fish for damaging information. Phrase these questions so that the witness can relate potentially bad information while remaining in your client's camp. For example, let's suppose you have reason to believe that your client's job performance prior to his resignation was unsatisfactory. If you ask the friendly witness, "Was John doing a good job at work?" she might be motivated by loyalty to report that your client was doing a bang-up job. If this was not the case, you've lost an opportunity to get valuable information that you'll have to address later when your opponent raises it. And you've caused the witness to throw herself behind a statement that is not credible, which may

mean that in a subsequent deposition she repeats the lie in order to save face. Rephrase the question to get accurate information without compromising the witness: "If I talked to other people on John's team at ExIm, are there any complaints they might have about his work?" or "It seems that John's team did a lot of client entertainment at bars and clubs. Can you describe how he and the others conducted themselves on those nights?" These questions allow the interviewee to share negative assessments about John but to put the words in other people's mouths, or to put his behavior in a somewhat forgiving context. You have learned what you're likely to be up against, but may also have netted some clues about how to rebut it.

Finally, assess the interviewee's credibility. Pay attention to the witness throughout the first two segments of the interview. Did he have difficulty hearing you, or did she squint when perusing the restaurant menu? If her sensory abilities are compromised, her value as an eyewitness may be reduced. She may seem unusually excitable or socially awkward, which might make her less effective on the stand. Or she may use especially pejorative language about some of your client's adversaries, suggesting a personal stake in the matter. Explore these matters with her, starting with the least potentially offensive and working your way towards the most difficult. As your questions grow more personal and challenging, even a friendly witness may grow hostile and uncooperative. Again, if you sense that she is shutting down and that you may have just one question left before she closes the interview, move the most important bias question to the front of the queue to try to preserve that information.

5.2.1 Example

Suppose that the *Nita Tribune* has reported that former Commerce Secretary and mayoral son / brother Bill Daley is planning to file for bankruptcy. Daley has sued the Tribune for defamation and you represent the paper. You are preparing to file a motion to dismiss the claim. The paper will only be liable if reporter Peggy Pringle knew that the statement was untrue or disregarded the possibility that it was untrue. Editors at the paper have told you that Pringle's source was Daley's former business partner, John Quinn. Pringle knows the partner through their children's school and the partner mentioned it at a school fundraising party where other parents were present. You are interviewing a parent who took part in the conversation. He seems to be a friendly witness; he is willing to meet with you and socializes with Pringle.

To assess whether the Tribune knew or should have known the story was false, what information do you need or want from this witness? You need to know whether Quinn explicitly said Daley was filing for bankruptcy, or whether he said something vaguer that the reporter extrapolated (suggesting that she disregarded the possibility that her interpretation was inaccurate). You need to know the circumstances in which Quinn made the statement, again to assess whether the statement was objectively likely to have been understood as an exaggeration or rhetorical flourish (suggesting that she disregarded the possibility that her interpretation was inaccurate). And you need to know whether this interviewee is reliable—can you believe what he is saying and would jurors believe him if you called him to testify? Questions might unfold as follows:

Open: Tell me a little bit about this fundraising event.

Open: I understand that Mr. Quinn was chatting with a group of parents about Bill Daley. Can you describe the conversation to me?

Closed: What specific details about Daley's finances did he give you?

Closed: Was he talking about Daley's personal finances or business finances?

Closed: Did he actually use the phrase "filed for bankruptcy"?

Open: Could you describe Mr. Quinn's tone of voice and his demeanor as he told this story?

Closed: Did he seem to have any animosity towards Daley?

Closed: You mentioned that the fundraiser took place at a country club dinner, with cocktails and wine. Did you notice whether Mr. Quinn had been drinking?

Closed: Did he seem sober when he related this story?

Closed: How many people took part in this conversation?

Closed: What were the names of the other parents who heard Mr. Quinn make this statement?

Open: If I asked other parents at the school about Mr. Quinn, what kinds of words might be used to describe him?

Open: Can you tell me how well you know Ms. Pringle?

Open: What is the relationship between Ms. Pringle and Mr. Quinn like?

Closed: And how long have you known Ms. Pringle?

Open: If I asked other parents at the school about Ms. Pringle, how might they describe her?

Open: How crowded was the room that night?

Closed: Was it very loud?

Closed: Did you have any difficulty hearing or following the conversation that night?

Closed: Had you been drinking that night?

Complete Exercise Five-A: Friendly Witness

5.3 The Hostile Witness

At this point in putting together your case, a hostile witness is simply someone who resists talking with you. She may, however, have information that will significantly strengthen your client's position. Or she may have very damaging information, early knowledge of which will allow you to shield your client or adjust your strategy effectively. So you must do the best you can to get this information, even though the interviewee is reluctant.

In order to craft a strategy for prying information from the hostile witness, try to figure out during your interview prep *why* she is hostile. As with the friendly witness, knowledge is power. Learn as much as you can about the interviewee. Your client may be a wealth of information, as are electronically available resources. Typically, witnesses are roadblocks for one of three reasons: they dislike your client personally; they fear their own interests will be compromised if they assist the client; or they are afraid or contemptuous of the legal process. Each of these reasons for obstruction suggests different responsive strategies.

Probably the easiest of these problems to deal with is an interviewee's fear of or contempt for the legal process or lawyers generally. If the interviewee fears the legal system, your own empathy and low-key approach can do a great deal to reassure her. Explain that cooperation at this early stage doesn't oblige her to participate in later, more formal, parts of the legal process—in fact, you may explain (if it's true) that if your client gets good information now it may avoid the need for a drawn-out legal dispute. Contempt for the legal system is a little trickier to counteract; if the interviewee has spent her life hating lawyers, it's unlikely that an hour in your charming presence will wipe out that deep-seated prejudice. Instead of fighting that uphill battle, shift the interviewee's focus from you, and from the law, to another interest that will be served by her cooperation; this may nudge her along. Perhaps you are litigating a case that will influence the way schools treat students or one that has the potential to stop a real estate development she opposes. Urging the interviewee to help achieve a goal that will outlive the lawyers involved in the case may thaw her enough to have a productive conversation.

A similar approach might work if the interviewee simply dislikes your client. Appeal to her interest in some higher, impersonal value that has nothing to do with your client, such as exposing the truth about some event to public view, or furthering some cause she believes in. This is one spot where your research may come in handy. If your client has given generously to the Friends of the Parks, for example, you might urge her to participate in your client's whistleblowing case because whistleblowers have a long history of exposing corporate malfeasance with regard to pollutants (this is true even if your client's whistleblowing had to do with tax evasion).

The toughest of these objections to overcome is the interviewee's fear that cooperating with you will harm her in some way. Here—within the bounds of truth and accuracy—show her that cooperation is not a zero-sum game; she can help you without hurting herself. For example, she might be able to suggest other witnesses or point you to documents that are sufficient to prove your case in court without having to testify herself. On the other hand, it is possible that the same information that exonerates your client implicates the witness—if that is the case, do not mislead the witness.

Another strategy to secure cooperation from the hostile witness is to use leverage of some kind. This can be difficult to do ethically and effectively. This strategy requires you to allude subtly to the fact that the witness has a stake in the outcome of your case, and to suggest that early participation might be helpful to her whereas lack of participation may be harmful.

Leverage might work in a situation where the witness had a peripheral role in the underlying event, giving your client the option of responding in a formal way or deciding to overlook. For example, suppose your client is attempting to renegotiate a corporate lease with the owner of an office building and wants the owner to fund a significant reconfiguration of the space. You are interviewing the tenant relations director for the building management company, an outfit hired by the owner to deal with daily operations. You suspect that the management company mixed recyclables and regular garbage when disposing of waste in order to save money, despite the owner's representations in the current lease that waste would be recycled. You also believe the tenant relations director has details about a generous build-out allowance the owner gave to a new tenant last year. In your conversation with the director, you might explain that your client is a little confused about how the building recycling protocol works and has listed it as an item to clarify in upcoming negotiations with the owner, but that she's much more interested in securing a competitive build-out allowance in those talks. Following up that explanation by asking the director if he knows the maximum build-out allowance any tenant has received in the past year allows him to infer that if you get your top-priority contract item, you might not ask the owner questions about recycling that are to the director's disadvantage. To repeat, this is a tricky strategy to land without crossing ethical lines. And if it doesn't work, you have succeeded in turning a hostile witness even more hostile. Put it in your back pocket and think carefully before taking it out.

In general, there's little hope that you'll have a leisurely conversation with a hostile witness where you can explore myriad topics at will. So, unlike client conversations or friendly witness conversations, there's little to be gained by setting a collaborative tone. Instead, maximize your control over the circumstances, without further aggravating the witness by being overtly controlling. How do you pull this off?

First, manipulate logistics to your advantage. Don't ask the witness's permission to interview her by phoning or emailing—this gives her control over whether and how to cooperate. It is much harder to put someone off in a face-to-face encounter than it is over the phone or computer. Further, giving the witness advance warning of your interest lets her get her footing to reject you or to give you canned answers. Consider approaching the witness in person somewhere she is unguarded and is not expecting you. This is another area where research will pay off. If you learn the witness is scheduled to attend an event such as a luncheon, you might approach her there. If you find out where the witness works, you might have your morning coffee in the building plaza and approach her there. The goal is to catch the witness off guard without appearing menacing, in hopes of shaking loose a little candor.

Don't ask permission to ask questions. Immediately start the conversation about the information you want to know—create the expectation that she'll talk to you and she might do it without really thinking. If she resists, you can always backtrack and "explain" why she should help you, or deal with his objections.

You can also exert control over the situation by making a surgical strike. Know exactly what information you're seeking. Have a list in mind of the information you want from the witness, from most crucial to least important—you may only have one shot and you want to be ready with your key question at the moment you sense she's most willing to cooperate. That doesn't mean that you open with an incendiary question. Rather, have ready one warm-up question that lets you "take her temperature." You can go from there depending on the response you get. If the

witness is impenetrable, take your shot with the million-dollar question. If she's more open than you expected, gradually escalate the questions from neutral to confrontational. She has opened the door to building rapport and you can take the invitation.

Closed questions are your friend in hostile witness conversations. Unlike a friendly witness, where you have the time and latitude to let her spill because you know she wants to help, with a hostile witness, you cannot expect her to let you take your time and ask unlimited questions. A closed question may elicit a quick answer, and once she's started talking to you, your job is much easier. Your warm-up question might seek to "confirm" a fact that you know she knows is wrong—she might be compelled to correct you, and now you've established her as an authority.[10] Her ego might lead her to continue setting you straight. You can always shift to open questions if it seems to be going well.

You may be able to disarm the witness by thinking of yourself more as an investigator than as an advocate as you articulate your queries. If your question has an accusation nested in it, "forgive" whatever you are accusing her of within the confines of the question. For example, suppose you need to know whether the interviewee filed false reports with the Food and Drug Administration on behalf of her former employer.[11] Here's a question that seeks the information without placing blame: "Isn't it true that AmFood left you no choice but to file the 421 report with the incomplete lab data?" Of course, you are working as an advocate because the information you unearth is going to be deployed in your client's best interest. But framing your questions to reflect your zeal for your client's position may simply force a hostile witness to dig in her heels.

Finally, keep your cool. Even if the witness is completely unhelpful and hostile, you've obtained information about her orientation and demeanor. This could be helpful in impeaching her if she winds up being called as a witness, and you haven't made a hostile witness even more hostile. You have also kept the door open for future communications if the witness reconsiders down the line.

Complete Exercise Five-B: Hostile Witness

---

10. Recall that ethical rules do not allow you to misrepresent facts. But this does not prevent you from posing a question that suggests a known fact is open to doubt, especially when the fact is not relevant to the legal issue at hand. For example, if your client is suing someone for negligence as the result of a car accident and you are interviewing a witness to the accident, you might ask, "Would you describe the car as red?" when you have reason to believe the car was black. You have not told the witness that the car was red, or that anyone else has stated the car was red, and in fact the driver's negligence has nothing to do with the color of the car he drove. You have given the witness a starting point for discussing the accident, one that invites her to view herself as the authority by giving you the correct information. Conversely, if you are defending someone charged with a hit-and-run, and the police report specifies that the car in question was black, and your client drives a black car, it would cross ethical boundaries to say to a witness, "According to the police report, the car involved in the hit and run was red. Is that right?" In that case, you are misrepresenting a fact that is probative of your client's guilt in an effort to manipulate the witness's recollection of the relevant events.

11. Of course, if you are litigating against AmFood and you have reason to believe that the former employee would be considered "represented" by the organization's counsel, you would not pose these questions without that counsel's permission.

CHAPTER SIX

FUNCTIONARY DISCUSSIONS

Aside from dealing with fact witnesses, in the pre-trial or pre-negotiation phase of a client matter, you may have to work with other third parties who can stand in the way of your client's best interest: administrators or opposing counsel. Particularly as a young attorney, you may be responsible for filing papers with the court, negotiating logistics with the junior lawyer on your opponent's team, or satisfying administrative requirements. In all of these roles, you will do better if you use some of the same skills you've been developing to interview clients and witnesses. First, research before you make your ask. Second, prioritize your goals and give some advance thought to what you *must* get from the conversation and what you'd *like* to get from the conversation.

When dealing with court and agency personnel, remember that they are paid less than you are and know the system far better than you do; they may not be familiar with the details of your case, but they know the rules and regulations. Some of them may admire lawyers but many of them are disdainful, often with good reason—far too many attorneys do not bother to learn the rules of court and make administrative mistakes over and over again. Court and agency personnel may be public servants, but they do not see themselves as serving your needs; they serve the needs of whoever leads the bureaucracy of which they are a member. With that in mind, here are some tips:

- Work with their schedule, their needs, their rules—find out closing time and work backwards by an hour—don't ask any government employee to do anything at the last minute!
- Don't necessarily take their first answer as the only answer. These workers tend to get the same question a dozen times a day and eleven lawyers will take the answer they're given, meaning the worker is in the habit of swatting away lawyers with a rote reply. If the rote answer doesn't work for you, politely ask again, or ask differently, or propose an alternate solution. Once you rouse people out of their automated way of thinking, they may find it novel to work with you on an alternative.
- You will always do better in person than you will over the phone. If you are on the scene and you remain at the desk or counter after they think they've rebuffed you, you're indicating that as far as you're concerned the conversation is still ongoing. If you don't go away, they may do what you are asking just to get rid of you. It is essential that you remain polite and patient while you're parked there, or the strategy will backfire.
- Know the relevant court or agency rules so that you can invoke them if they help you or have work-arounds thought through if they hurt you—have copies of the rules with you, but don't take them out right away.
- "Bank" their mistakes as currency to deploy for what you want.
- Find common enemies and join forces with them against those enemies.
- Ask colleagues for advice—your problem may be a common one that has a well-known solution.
- Be humble. Most of these people have many years on the job, and they are not impressed with cocky young lawyers who think their degree makes them special. Smile, recognize that you are asking a favor, and thank the person for their time and assistance, even if you don't get your way.

When dealing with your junior counterpart on the opponent's legal team, cultivate the sense that the two of you belong to the same team. You work the same long hours, you take the same abuse from your bosses, you do the same tedious tasks. You are allowed to have a certain amount of loyalty to each other. With that said, here are some tips:

- Know his or her boss's preferences, stakes, and relationship so you can play on that.
- Be civil and helpful whenever possible (keep a record) so you can ask the same of them when need be (unless you're instructed otherwise by your partner).
- LISTEN! You never know when you're getting information that will be helpful later.
- Find common enemies.

This advice also applies to paralegals and even legal secretaries on the opponent's team. Don't make the mistake of disregarding those personnel: treat them respectfully and they can be a great resource; treat them with contempt and you may never get another ounce of cooperation from the opponent's legal team again.

Complete Exercise Six: Functionary

Part Two

Practical Application

Contents

Deborah Dashwood Attorney Materials

Nita Estates Code

Title 3. Guardianship and Related Procedures

Subtitle D. Creation of Guardianship

Chapter 1101. General Procedure to Appoint Guardian

Subchapter A. Initiation of Proceeding for Appointment of Guardian

Sec. 1101.001. Application for Appointment of Guardian; Contents.

(a) Any person may commence a proceeding for the appointment of a guardian by filing a written application in a court having jurisdiction and venue.

(b) The application must be sworn to by the applicant and state:

(1) the proposed ward's name, sex, date of birth, and address;

(2) the name, relationship, and address of the person the applicant seeks to have appointed as guardian;

(3) whether guardianship of the person or estate, or both, is sought;

(4) the nature and degree of the alleged incapacity, the specific areas of protection and assistance requested, and the limitation or termination of rights requested to be included in the court's order of appointment, including a termination of:

(A) the right of a proposed ward who is eighteen years of age or older to vote in a public election; and

(B) the proposed ward's eligibility to hold or obtain a license to operate a motor vehicle under Chapter 521, Transportation Code;

(5) the facts requiring the appointment of a guardian;

(6) the interest of the applicant in the appointment of a guardian;

(7) the nature and description of any kind of guardianship existing for the proposed ward in any other state;

(8) the name and address of any person or institution having the care and custody of the proposed ward;

(9) the approximate value and description of the proposed ward's property, including any compensation, pension, insurance, or allowance to which the proposed ward may be entitled;

(10) the name and address of any person whom the applicant knows to hold a power of attorney signed by the proposed ward and a description of the type of power of attorney;

(11) for a proposed ward who is a minor, the following information if known by the applicant:

(A) the name of each of the proposed ward's parents and either the parent's address or that the parent is deceased;

(B) the name and age of each of the proposed ward's siblings, if any, and either the sibling's address or that the sibling is deceased; and

(C) if each of the proposed ward's parents and adult siblings are deceased, the names and addresses of the proposed ward's other living relatives who are related to the proposed ward within the third degree by consanguinity and who are adults;

(12) for a proposed ward who is a minor, whether the minor was the subject of a legal or conservatorship proceeding in the preceding two years and, if so:

(A) the court involved;

(B) the nature of the proceeding; and

(C) any final disposition of the proceeding;

(13) for a proposed ward who is an adult, the following information if known by the applicant:

(A) the name of the proposed ward's spouse, if any, and either the spouse's address or that the spouse is deceased;

(B) the name of each of the proposed ward's parents and either the parent's address or that the parent is deceased;

(C) the name and age of each of the proposed ward's siblings, if any, and either the sibling's address or that the sibling is deceased;

(D) the name and age of each of the proposed ward's children, if any, and either the child's address or that the child is deceased; and

(E) if there is no living spouse, parent, adult sibling, or adult child of the proposed ward, the names and addresses of the proposed ward's other living relatives who are related to the proposed ward within the third degree by consanguinity and who are adults;

(14) facts showing that the court has venue of the proceeding; and

(15) if applicable, that the person whom the applicant seeks to have appointed as a guardian is a private professional guardian who is certified under Subchapter C, Chapter 155, Government Code, and has complied with the requirements of Subchapter G, Chapter 1104.

***

Subchapter B. Hearing; Jury Trial

Sec. 1101.051. Hearing.

(a) At a hearing for the appointment of a guardian, the court shall:

***

(1) inquire into the ability of any allegedly incapacitated adult to:

(A) feed, clothe, and shelter himself or herself;

(B) care for his or her own physical health; and

(C) manage his or her property or financial affairs;

(2) ascertain the age of any proposed ward who is a minor;

(3) inquire into the governmental reports for any person who must have a guardian appointed to receive funds due the person from any governmental source; and

(4) inquire into the qualifications, abilities, and capabilities of the person seeking to be appointed guardian.

(b) A proposed ward must be present at the hearing unless the court, on the record or in the order, determines that a personal appearance is not necessary.

(c) The court may close the hearing at the request of the proposed ward or the proposed ward's counsel.

Subchapter C. Agreed Guardianship

Sec. 1101.052. Agreed Guardianship, Application.

(a) If the proposed ward agrees that he or she meets the criteria outlined in 1101.051(1), the proposed ward may complete a voluntary application for guardianship.

(b) The voluntary application for guardianship will contain all of the information required by Sec. 1101.001(b) and will specify the scope of the guardianship agreed to pursuant to Sec. 1101.151 of this Act.

(c) In the event a proposed ward files a voluntary application for guardianship that meets the requirements of Sec. 1101.052(b), the court will dispense with a hearing and will issue an automatic order appointing the guardian.

Sec. 1101.053. Provision of Records Required; Use of Records.

(a) Before a hearing may be held for the appointment of a guardian, current and relevant medical, psychological, and intellectual testing records of the proposed ward must be provided to the attorney ad litem appointed to represent the proposed ward unless:

(1) the proposed ward is a minor or a person who must have a guardian appointed to receive funds due the person from any governmental source; or

(2) the court makes a finding on the record that:

(A) current or relevant records do not exist; and

(B) examining the proposed ward for the purpose of creating the records is impractical.

(b) Current medical, psychological, and intellectual testing records are a sufficient basis for a determination of guardianship.

(c) The findings and recommendations contained in the medical, psychological, and intellectual testing records are not binding on the court.

Subchapter C. Determination of Necessity of Guardianship; Findings and Proof

Sec. 1101.101. Findings and Proof Required.

(a) Before appointing a guardian for a proposed ward, the court must:

(1) find by clear and convincing evidence that:

(A) the proposed ward is an incapacitated person;

(B) it is in the proposed ward's best interest to have the court appoint a person as the proposed ward's guardian; and

(C) the proposed ward's rights or property will be protected by the appointment of a guardian; and

(2) find by a preponderance of the evidence that:

(A) the court has venue of the case;

(B) the person to be appointed guardian is eligible to act as guardian and is entitled to appointment, or, if no eligible person entitled to appointment applies, the person appointed is a proper person to act as guardian;

(C) if a guardian is appointed for a minor, the guardianship is not created for the primary purpose of enabling the minor to establish residency for enrollment in a school or school district for which the minor is not otherwise eligible for enrollment; and

(D) the proposed ward:

(i) is totally without capacity as provided by this title to care for himself or herself and to manage his or her property; or

(ii) lacks the capacity to do some, but not all, of the tasks necessary to care for himself or herself or to manage his or her property.

(b) The court may not grant an application to create a guardianship unless the applicant proves each element required by this title.

Sec. 1101.102. Determination of Incapacity of Certain Adults: Recurring Acts or Occurrences.

A determination of incapacity of an adult proposed ward, other than a person who must have a guardian appointed to receive funds due the person from any governmental source, must be evidenced by recurring acts or occurrences in the preceding six months and not by isolated instances of negligence or bad judgment.

Sec. 1101.103. Determination of Incapacity of Certain Adults: Physician Examination.

(a) Except as provided by Section 1101.104, the court may not grant an application to create a guardianship for an incapacitated person, other than a minor or person for whom it is necessary to have a guardian appointed only to receive funds from a governmental source, unless the applicant presents to the court a written letter or certificate from a physician licensed in this state that is:

(1) dated not earlier than the 120th day before the date the application is filed; and

(2) based on an examination the physician performed not earlier than the 120th day before the date the application is filed.

(b) The letter or certificate must:

(1) describe the nature, degree, and severity of the proposed ward's incapacity, including any functional deficits regarding the proposed ward's ability to:

(A) handle business and managerial matters;

(B) manage financial matters;

(C) operate a motor vehicle;

(D) make personal decisions regarding residence, voting, and marriage; and

(E) consent to medical, dental, psychological, or psychiatric treatment;

(2) in providing a description under Subdivision (1) regarding the proposed ward's ability to operate a motor vehicle and make personal decisions regarding voting, state whether in the physician's opinion the proposed ward:

(A) has the mental capacity to vote in a public election; and

(B) has the ability to safely operate a motor vehicle;

(3) provide an evaluation of the proposed ward's physical condition and mental function and summarize the proposed ward's medical history if reasonably available;

(4) state how or in what manner the proposed ward's ability to make or communicate responsible decisions concerning himself or herself is affected by the proposed ward's physical or mental health, including the proposed ward's ability to:

(A) understand or communicate;

(B) recognize familiar objects and individuals;

(C) perform simple calculations;

(D) reason logically; and

(E) administer to daily life activities;

(5) state whether any current medication affects the proposed ward's demeanor or the proposed ward's ability to participate fully in a court proceeding;

(6) describe the precise physical and mental conditions underlying a diagnosis of a mental disability, and state whether the proposed ward would benefit from supports and services that would allow the individual to live in the least restrictive setting; and

(7) include any other information required by the court.

(c) If the court determines it is necessary, the court may appoint the necessary physicians to examine the proposed ward. The court must make its determination with respect to the necessity for a physician's examination of the proposed ward at a hearing held for that purpose. Not later than the fourth day

before the date of the hearing, the applicant shall give to the proposed ward and the proposed ward's attorney ad litem written notice specifying the purpose and the date and time of the hearing.

(d) A physician who examines the proposed ward, other than a physician or psychologist who examines the proposed ward under Section 1101.104(2), shall make available for inspection by the attorney ad litem appointed to represent the proposed ward a written letter or certificate from the physician that complies with the requirements of Subsections (a) and (b).

SUBCHAPTER D. COURT ACTION

Sec. 1101.151. Order Appointing Guardian with Full Authority.

(a) If it is found that the proposed ward is totally without capacity to care for himself or herself, manage his or her property, operate a motor vehicle, and vote in a public election, the court may appoint a guardian of the proposed ward's person or estate, or both, with full authority over the incapacitated person except as provided by law.

(b) An order appointing a guardian under this section must contain findings of fact and specify:

(1) the information required by Section 1101.153(a);

(2) that the guardian has full authority over the incapacitated person;

(3) if necessary, the amount of funds from the corpus of the person's estate the court will allow the guardian to spend for the education and maintenance of the person under Subchapter A, Chapter 1156;

(4) whether the person is totally incapacitated because of a mental condition;

(5) that the person does not have the capacity to operate a motor vehicle and to vote in a public election; and

(6) if it is a guardianship of the person of the ward or of both the person and the estate of the ward, the rights of the guardian with respect to the person as specified in Section 1151.051(c)(1).

(c) An order appointing a guardian under this section that includes the rights of the guardian with respect to the person as specified in Section 1151.051(c)(1) must also contain the following prominently displayed statement in boldfaced type, in capital letters, or underlined:

"ANY PERSON WHO KNOWINGLY PRESENTS FOR PURPOSES OF ESTABLISHING PHYSICAL CUSTODY OR PROPERTY OWNERSHIP AN ORDER THAT IS INVALID OR NO LONGER IN EFFECT COMMITS AN OFFENSE THAT MAY BE PUNISHABLE BY CONFINEMENT IN JAIL FOR AS LONG AS TWO YEARS AND A FINE OF AS MUCH AS $10,000. ANY TRANSFERS OF TITLE BASED ON INVALID CLAIMS OF OWNERSHIP WILL BE RESCINDED BY THE COURT UPON APPLICATION."

***

Sec. 1101.156 Invalid Guardianship Orders.

Guardianship orders are to be considered invalid in the following circumstances:

(a) The holder of the order has engaged in forgery or undertaken other means to create a document that purports to be the product of the application process provided for herein without actually taking part in that process.

(b) the order was procured using any of the following measures:

(1) Making misrepresentations in an application for guardianship

(2) Conspiring with physicians or third-party experts to file inaccurate reports as addenda to an application for guardianship

(3) Use of duress or deceit to procure an agreed application for guardianship from a proposed ward.

Estates Code

Title 3. Guardianship and Related Procedures

Subtitle F. Evaluation, Modification, or Termination of Guardianship

Chapter 1202. Modification or Termination of Guardianship

Subchapter A. Termination and Settlement of Guardianship

Sec. 1202.001. Term of Guardian or Guardianship.

(a) Unless otherwise discharged as provided by law, a guardian remains in office until the estate is closed.

(b) A guardianship shall be settled and closed when the ward:

(1) dies and, if the ward was married, the ward's spouse qualifies as survivor in community;

(2) is found by the court to have full capacity to care for himself or herself and to manage the ward's property;

(3) is no longer a minor; or

(4) no longer must have a guardian appointed to receive funds due the ward from any governmental source.

(c) An order appointing a guardian or a successor guardian may specify a period of not more than one year during which a petition for adjudication that the ward no longer requires the guardianship may not be filed without special leave.

(d) A request for an order under this section may be made by informal letter to the court. A person who knowingly interferes with the transmission of the request to the court may be adjudged guilty of contempt of court.

(e) If a nonresident guardian of a nonresident ward qualifies as guardian under this title, any resident guardian's guardianship may be terminated.

Subchapter B. Application for Complete Restoration of Ward's Capacity or Modification of Guardianship

Sec. 1202.051. Application Authorized.

A ward or any person interested in the ward's welfare may file a written application with the court for an order:

(1) finding that the ward is no longer an incapacitated person and ordering the settlement and closing of the guardianship;

(2) finding that the ward lacks the capacity to do some or all of the tasks necessary to provide food, clothing, or shelter for himself or herself, to care for the ward's own physical health, or to manage the ward's own financial affairs and granting additional powers or duties to the guardian; or

(3) finding that the ward has the capacity to do some, but not all, of the tasks necessary to provide food, clothing, or shelter for himself or herself, to care for the ward's own physical health, or to manage the ward's own financial affairs and:

(A) limiting the guardian's powers or duties; and

(B) permitting the ward to care for himself or herself or to manage the ward's own financial affairs commensurate with the ward's ability.

Sec. 1202.052. Contents of Application.

An application filed under Section 1202.051 must be sworn to by the applicant and must state:

(1) the ward's name, sex, date of birth, and address;

(2) the name and address of any person serving as guardian of the person of the ward on the date the application is filed;

(3) the name and address of any person serving as guardian of the estate of the ward on the date the application is filed;

(4) the nature and description of the ward's guardianship;

(5) the specific areas of protection and assistance and any limitation of rights that exist;

(6) whether the relief being sought is:

(A) a restoration of the ward's capacity because the ward is no longer an incapacitated person;

(B) the granting of additional powers or duties to the guardian; or

(C) the limitation of powers granted to or duties performed by the guardian;

(7) if the relief being sought under the application is described by Subdivision (6)(B) or (C):

(A) the nature and degree of the ward's incapacity;

(B) the specific areas of protection and assistance to be provided to the ward and requested to be included in the court's order; and

(C) any limitation of the ward's rights requested to be included in the court's order;

(8) the approximate value and description of the ward's property, including any compensation, pension, insurance, or allowance to which the ward is or may be entitled; and

(9) if the ward is sixty years of age or older, the names and addresses, to the best of the applicant's knowledge, of the ward's spouse, siblings, and children or, if there is no known spouse, sibling, or child, the names and addresses of the ward's next of kin.

Sec. 1202.053. Citation Required.

When an application is filed under Section 1202.051, citation shall be served on:

(1) the ward's guardian; and

(2) the ward if the ward is not the applicant.

Sec. 1202.054. Informal Request for Order by Ward; Investigation and Report.

(a) A ward may request an order under Section 1202.051 by informal letter to the court. A person who knowingly interferes with the transmission of the request to the court may be adjudged guilty of contempt of court.

(b) On receipt of an informal letter under Subsection (a), the court shall appoint the court investigator or a guardian ad litem to investigate the ward's circumstances, including any circumstances alleged in the letter, to determine whether:

(1) the ward is no longer an incapacitated person; or

(2) a modification of the guardianship is necessary.

(c) The court investigator or guardian ad litem shall file with the court a report of the investigation's findings and conclusions. If the court investigator or guardian ad litem determines that it is in the best interest of the ward to terminate or modify the guardianship, the court investigator or guardian ad litem shall file an application under Section 1202.051 on the ward's behalf.

(d) A guardian ad litem appointed under this section may also be appointed by the court to serve as attorney ad litem under Section 1202.101.

Sec. 1202.055. Restriction on Subsequent Application Regarding Capacity or Modification.

A person may not reapply for complete restoration of a ward's capacity or modification of a ward's guardianship before the first anniversary of the date of the hearing on the last preceding application, except as otherwise provided by the court on good cause shown by the applicant.

***

Subchapter D. Hearing, Evidence, and Orders in Proceeding for Complete Restoration of Ward's Capacity or Modification of Guardianship

Sec. 1202.151. Evidence and Burden of Proof at Hearing.

(a) Except as provided by Section 1202.201, at a hearing on an application filed under Section 1202.051, the court shall consider only evidence regarding the ward's mental or physical capacity at the time of the hearing that is relevant to the complete restoration of the ward's capacity or modification of the ward's guardianship.

(b) The party who filed the application has the burden of proof at the hearing.

Sec. 1202.152. Physician's Letter or Certificate Required.

(a) The court may not grant an order completely restoring a ward's capacity or modifying a ward's guardianship under an application filed under Section 1202.051 unless the applicant presents to the court a written letter or certificate from a physician licensed in this state that is dated:

(1) not earlier than the 120th day before the date the application was filed; or

(2) after the date the application was filed but before the date of the hearing.

(b) A letter or certificate presented under Subsection (a) must:

(1) describe the nature and degree of incapacity, including the medical history if reasonably available, or state that, in the physician's opinion, the ward has the capacity to:

(A) provide food, clothing, and shelter for himself or herself;

(B) care for the ward's own physical health; and

(C) manage the ward's financial affairs;

(2) provide a medical prognosis specifying the estimated severity of any incapacity;

(3) state how or in what manner the ward's ability to make or communicate responsible decisions concerning himself or herself is affected by the ward's physical or mental health;

(4) state whether any current medication affects the ward's demeanor or the ward's ability to participate fully in a court proceeding;

(5) describe the precise physical and mental conditions underlying a diagnosis of senility, if applicable; and

(6) include any other information required by the court.

(c) If the court determines it is necessary, the court may appoint the necessary physicians to examine the ward in the same manner and to the same extent as a ward is examined by a physician under Section 1101.103 or 1101.104.

Sec. 1202.153. Findings Required.

(a) Before ordering the settlement and closing of a guardianship under an application filed under Section 1202.051, the court must find by a preponderance of the evidence that the ward is no longer partially or fully incapacitated.

(b) Before granting additional powers to the guardian or requiring the guardian to perform additional duties under an application filed under Section 1202.051, the court must find by a preponderance of the evidence that the current nature and degree of the ward's incapacity warrants a modification of the guardianship and that some or all of the ward's rights need to be further restricted.

(c) Before limiting the powers granted to or duties required to be performed by the guardian under an application filed under Section 1202.051, the court must find by a preponderance of the evidence that the current nature and degree of the ward's incapacity warrants a modification of the guardianship and that some of the ward's rights need to be restored.

Sec. 1202.153(b) Voluntary Application for Termination or Modification of Guardianship.

(a) If the court-appointed guardian is the party seeking a declaration that the order of guardianship for the ward be terminated or modified, the requirements of Sec. 1202.151-1202.153 are waived.

Sec. 1202.154. General Requirements for Order.

(a) A court order entered with respect to an application filed under Section 1202.051 to completely restore a ward's capacity or modify a ward's guardianship must state:

(1) the guardian's name;

(2) the ward's name; and

(3) whether the type of guardianship being addressed at the proceeding is a:

(A) guardianship of the person;

(B) guardianship of the estate; or

(C) guardianship of both the person and the estate.

(b) In an order described by this section, the court may not grant a power to a guardian or require the guardian to perform a duty that is a power granted to or a duty required to be performed by another guardian.

Sec. 1202.155. Additional Requirements for Order Restoring Ward's Capacity.

If the court finds that a ward is no longer an incapacitated person, the order completely restoring the ward's capacity must contain findings of fact and specify, in addition to the information required by Section 1202.154:

(1) that the ward is no longer an incapacitated person;

(2) that there is no further need for a guardianship of the person or estate of the ward;

(3) if the ward's incapacity resulted from a mental condition, that the ward's mental capacity is completely restored;

(4) that the guardian is required to:

(A) immediately settle the guardianship in accordance with this title; and

(B) deliver all of the remaining guardianship estate to the ward; and

(5) that the clerk shall revoke letters of guardianship when the guardianship is finally settled and closed.

ORDINANCE NO. __________

AN ORDINANCE REPEALING AND REPLACING CHAPTER 4–6 OF THE CITY CODE REGARDING PAWNSHOPS AND DEALERS IN SECONDHAND GOODS; CREATING AN OFFENSE AND PROVIDING A PENALTY.

BE IT ORDAINED BY THE CITY COUNCIL OF NITA CITY:

PART 1. The City Code is amended by repealing Chapter 4–6 and replacing it with a new Chapter 4–6 to read as follows:

CHAPTER 4–6. PAWNSHOPS AND SECONDHAND DEALERS. ARTICLE 1. PAWNSHOPS

§ 4–6–01 APPLICABILITY.

This article applies to a pawnshop licensed under Chapter 371 *(Pawnshop s)* of the Nita Finance Code.

§ 4–6–02 HOLD PERIOD SHORTENED FOR CERTAIN PAWNSHOPS.

The hold period for property purchased from the general public required by 7 Nita Administrative Code § 85.420 *(Purchase Transactions)* is shortened to seven days for a pawnshop that daily enters pawn and purchase ticket information in an electronic system prescribed by the chief of police.

§ 4–6–03 ADMINISTRATION.

The chief of police may adopt rules and forms to administer and enforce this article.

ARTICLE 2. SECONDHAND DEALERS.

§ 4–6–21 DEFINITIONS.

In this article:

(1) **APPLICANT** means a person who applies for a license under this article. If the applicant is not a natural person, the term includes:

(a) in the case of a partnership, the names of individual members of the partnership; and

(b) in the case of a corporation or other business association, the owners, directors and officers of the corporation or business association.

(2) **AUTOMOBILE ACCESSORY** means a hubcap, wheel cover, radar detector, recorded music player, removable automobile top, or an item that is crafted or designed for use in or on an automobile as an accessory item.

(3) **BUSINESS DAY** means a day that a secondhand dealer transactsbusiness with a member of the public, or is open to the public for the transaction of business.

(4) **CRAFTED PRECIOUS METAL** means jewelry, silverware, coins, bars, or any object crafted, in whole or in part, from gold, silver, platinum, palladium, iridium, rhodium, osmium, ruthenium, or their alloys.

(5) **JEWELRY** means a gem, a jewel, or an object made of a precious metal that is worn for adornment, including, but not limited to, a gem stone removed from a mounting.

(6) **MOTORCYCLE ACCESSORY** means an item that is crafted or designed for use in or on a motorcycle as an accessory item.

(7) **POWER TOOL** means a tool powered by electrical current, a battery, solar energy, or liquid or gaseous fuel, including, but not limited to, pneumatic and welding equipment.

(8) **REGULATED MERCHANDISE** means any of the following items of used merchandise:

(c) an item that bears or as originally manufactured bore a manufacturer's serial number;

(d) a video recording, including but not limited to a motion picture or a video game;

(e) an automobile accessory;

(f) a business machine;

(g) a crafted precious metal;

(h) an electronic device, including, but not limited to, a radio, a television, a telephone or mobile telephone, an audio or video recorder or player, a computer, a video camera, a mobile telephone with computing ability and connectivity, or a device that is used in playing a video or computer game;

(i) a household appliance;

(j) jewelry;

(k) a motorcycle accessory;

(l) a musical instrument or an accessory used with a musical instrument, including, but not limited to, sound amplification equipment;

(m) optical equipment, such as binoculars, a telescope, or a microscope;

(n) photographic equipment;

(o) a power tool;

(p) a sporting good, or;

(q) a weapon.

(9) **SECONDHAND DEALER** means a person buying, trading, selling,or accepting for sale on consignment regulated merchandise.

(10) **USED MERCHANDISE** means personal property that has been previously sold or transferred to a person for his own use or the use of another.

§ 4622 APPLICABILITY.

(A) This article does not apply to:

(1) a pawnshop licensed under Chapter 371 (Pawnshops) of the Nita Finance Code;

(2) sales involving the transfer of a motor vehicle title;

(3) a garage sale allowed by Section 25–2–902 (Garage Sales);

(4) a nonprofit business that sells only donated items;

(5) sales by, or on behalf of, a governmental entity;

(6) except as provided by Subsection (B), sales by a person whose only business transactions with regulated merchandise are a display space, booth, or table maintained for displaying or selling merchandise at a trade show, convention, festival, fair, circus, market, flea market, swap meet, or similar event for no more than a total of fourteen days in a calendar year;

(7) a person who occasionally sells a personal item to another person and does not engage in buying regulated merchandise for resale as a business; or

(8) a transaction that does not involve a person buying, trading, selling, or accepting for sale on consignment regulated merchandise.

(B) The chief of police may adopt rules consistent with this article to regulate a business described by Subsection (A)(6). The rules may provide for licensing the business, reporting regulated merchandise, and holding regulated merchandise before the merchandise may be sold or disposed of, as the chief finds appropriate. The rules may provide for a reduced or prorated licensing fee.

(C) The chief of police may by rule narrow or further define the term "regulated merchandise" in order to exclude from coverage by this ordinance items or kinds of property if the chief finds that reporting and holding the items or kinds of property does not enhance the ability of the police department to reduce property crimes and recover stolen goods. The chief may not expand the definition of regulated merchandise; however the chief may reinstate any item or kind of property into the definition of regulated merchandise that the chief has previously excluded.

(D) This article is cumulative of other law.

§ 4–6–23 LICENSE REQUIRED.

(A) A person may not do business as a secondhand dealer without a license.

(1) A secondhand dealer who conducts business at more than one location in the City must obtain a separate license for each business location.

(2) A secondhand dealer that does not have a permanent business location in Nita City must comply with Section 4–6–27(E).

(B) To obtain a license as a secondhand dealer a person must make a written application to the chief of police. The application must provide:

(1) the applicant's name and the name the of the business;

(2) if the applicant is a partnership, corporation, or other business association, the names of individual members of the partnership, or owners, directors and officers of the corporation or business association;

(3) a registration certificate for the business from the Secretary of State, if registration is required by law, and a copy of the business's assumed name certificate, if any;

(4) the physical address or location of the business and, if different, the mailing address of the business;

(5) the regular days and hours of operation of the business;

(6) a statement that the applicant, and any employee of the applicant whose duties include the purchase of regulated merchandise, has not been convicted of a violation of this chapter in the immediately preceding two calendar years;

(7) a statement that the applicant, and any employee of the applicant whose duties include the purchase of regulated merchandise, has not been convicted of an offense involving theft, burglary, robbery, fraud, or tampering with or falsification of evidence or official records;

(8) a notarized statement that all information provided by the applicant on the applicant's application is correct.

(C) The chief of police may require additional information or identification reasonably necessary to determine an applicant's eligibility for a license under this article, As a condition of licensing, the applicant must provide or obtain any information or authorization necessary for the chief of police to conduct a criminal background check on the applicant and on any employee of the applicant whose duties include the purchase of regulated merchandise.

(D) Each application must be accompanied by a nonrefundable fee. The amount of a fee authorized by this ordinance shall be set by ordinance.

(E) A license issued under this article expires on the second anniversary of the date the license is issued. A license that is revoked and reissued expires on the license's original expiration date. An applicant whose application expires must apply for a new license and pay a new fee. The chief of police may issue a temporary license while the application for a new license is being processed.

(F) The chief of police may deny a license, and may revoke a license previously issued under this article, if the chief finds that:

(1) the applicant's place of business is not in compliance with applicable laws;

(2) the applicant, or an employee of the applicant whose duties include the purchase of regulated merchandise, has been convicted of a violation of this chapter in the immediately preceding two calendar years;

(3) the applicant has supplied incorrect information on the applicant's application;

(4) the applicant, or an employee of the applicant whose duties include the purchase of regulated merchandise, has been convicted of a felony offense involving theft, burglary, robbery, fraud, or tampering with or falsification of evidence or official records, unless:

(a) the applicant or employee has been pardoned;

(b) two years have elapsed since the applicant or employee has been fully discharged the from the sentence imposed for the offense, including any term of incarceration, parole, supervision , or probation ordered by any court; or

(c) the criminal proceedings were deferred without a final adjudication of guilt.

(G) An applicant may appeal a denial or revocation of a license. An appeal must be:

(1) in writing;

(2) submitted to the chief of police not later than the fifteenth working day after the date of the denial or revocation of the license; and

(3) must clearly state why the chief's decision to deny or revoke the license was incorrect.

(H) On receipt of a written appeal, the chief of police shall schedule a hearing to be held no later than the thirtieth day after the date the chief receives the written appeal. The hearing shall be before a hearing examiner selected by the chief. The hearing examiner shall issue a written opinion not later than the fifteenth working day after the date the hearing concludes recommending that the denial or revocation of the license be upheld, or recommending that the chief issue or reinstate the license. The hearing examiner's opinion must include findings of fact and conclusions of law. The chief may accept or reject the hearing examiner's recommendation. Not later than the fifteenth working day after the date that chief receives the hearing examiner's written opinion, the chief

will either reaffirm the decision to deny or revoke the applicant's license or issue or reinstate the applicant's license. The chief's decision is final.

(I) A secondhand dealer must post the dealer's license in a place clearly visible to the public in each place of business.

§ 4–6–24 CHANGE IN INFORMATION.

(A) A secondhand dealer shall report a change in the information required to be submitted with a license application to the chief of police not later than the fifteenth working day after the date that the change occurs.

(B) If the change is either to the name of the business or to the physical location of the business, the secondhand dealer must obtain a new license, and pay a new license fee. The chief of police may issue a temporary license while the application for a new license is being processed.

§ 4–6–25 IDENTIFICATION REQUIRED: SELLER MUST ATTEST TO OWNERSHIP.

(A) At the time a secondhand dealer purchases or receives regulated merchandise in the course of business, the secondhand dealer shall require the person selling or transferring the regulated merchandise to the secondhand dealer to provide one of the following, which must be current and include the person's photograph:

(1) a driver's license issued by a state, territory , commonwealth, possession, or district of the United States;

(2) a personal identification certificate issued by the Nita Department of Public Safety, or a similar certificate issued by another state, territory, commonwealth, possession, or district of the United States;

(3) a passport issued by the United States or issued by another country and recognized by the United States;

(4) a United States military identification;

(5) a matricula consular issued by the Mexican Consulate.

(B) At the time a secondhand dealer purchases or receives regulated merchandise in the course of business, the second dealer shall require from the person selling, or transferring the regulated merchandise to the secondhand dealer a written statement attesting to the person's ownership of property at the time of the transaction.

(C) A secondhand dealer may not purchase or receive regulated merchandise except from a person who provides the identification and statement required by this section.

§ 4–6–26 INVENTORY.

(A) A secondhand dealer shall make each item of regulated merchandise purchased by or received by a secondhand dealer in the course of business open to inspection by a police officer during normal business hours.

(B) A secondhand dealer may not purchase, receive, or possess an item of regulated merchandise from which the manufacturer's identification or serial number or brand has been altered, removed, or obliterated by means other than by wear caused by ordinary use of the item by a consumer.

(C) If a police officer finds an item of regulated merchandise described by Subsection (B) in the possession of a secondhand dealer, the police officer may seize the item and hold it at the police department for sixty days. At the time of the seizure of the property, the police officer shall issue to the secondhand dealer a receipt for the property notifying the secondhand dealer that the property may be redeemed on or after the sixty-first day after the date on which the item was seized unless it is determined by the chief of police to be evidence in an investigation.

(D) A police officer may order a secondhand dealer to stop the sale or removal from the secondhand dealer's place of business any item of regulated merchandise for a period of sixty days or until the property is released by the officer. After receiving the order a secondhand dealer may not sell the item or permit the item to be removed from the secondhand dealer's place of business until the sixty-first day after the date the order was received by the secondhand dealer or until the item is released by the officer issuing the order.

(E) Except as provided by Subsection (F), a secondhand dealer may not sell an item of regulated merchandise purchased by or received by the dealer in the course of business, or permit the item to be removed from the secondhand dealer's place of business, before the eleventh day after the date that the secondhand dealer reports the item under Section 4–6–27(*Recordkeeping Required*). Until a secondhand dealer is permitted to sell an item under this section, the secondhand dealer must retain the item in its original condition.

(F) This subsection applies only to an item of crafted precious metal that has been photographed by a secondhand dealer and the item, including the photograph, has been reported electronically by the dealer as provided by Section 4–6–27 (*Recordkeeping Required*). A secondhand dealer may not sell an item of

crafted precious metal to which this subsection applies, or permit the item to be removed from the secondhand dealer's place of business, before the eighth business day after the date that the secondhand dealer reports the item. Until a secondhand dealer is permitted to sell an item under this section, the secondhand dealer must retain the item in its original condition. The photograph must be of sufficient clarity to reasonably allow a person to identify the item photographed. The photograph must conform to any specification adopted by the chief of police.

(G) A report under this section is not required if:

(1) a secondhand dealer purchases or receives in the course of business an item from another secondhand dealer or from a pawn shop;

(2) the secondhand dealer from whom the item is purchased or received has already reported the item as required by this section or by other law; and

(3) the secondhand dealer from whom the item is purchased or received has held the item for the period required by this section or by other law.

§ 4–6–27 RECORDKEEPING REQUIRED.

(B) A secondhand dealer shall maintain records required by this article in hard copy in a format prescribed by the chief of police. A secondhand dealer shall maintain each record required by this article for a period of three years from the date the record is created.

(C) In addition to the hard copy records required by Subsection (A), a secondhand dealer shall report each record required by this article by entel ing the record in an electronic system prescribed by the chief of police. The report must include a statement that that the secondhand dealer has had no unreported transactions since the secondhand dealer's last report. A secondhand dealer shall enter into the electronic system a record required to be created for a transaction no later than the end of the seventh business day after:

(1) the day of the transaction; or

(2) in the case of a transaction that occurred outside of Travis County, the day that an item purchased in the transaction is brought into the City.

(D) A record kept or reported under this article is a government record. By entering a record electronically a person attests to the truth of the1nformation stated in the record. Notwithstanding any other provision of the City Code, a record kept or reported under this article is considered to be under oath by the person required to keep and report the record regardless of the absence of verification, including a signature.

(E) Except as provided by Subsection (E), a secondhand dealer must keep the following records with respect to each transaction regarding an item of regulated merchandise:

(1) an accurate description of each item purchased or received, including the item's serial number;

(2) information about the person from whom the item is purchased or received, including:

(a) the person's name, date of birth, race, and sex;

(b) the person's current address;

(c) the kind and number of the person's identification required by Section 4–6–25(A);

(3) the date and time that the item is purchased or received by the secondhand dealer;

(4) if the item is purchased, the amount of consideration paid to seller;

(5) the statement required by Section 4–6–25(B).

(F) In lieu of the information required by Subsection (D)(2), if a secondhand dealer purchased an item of regulated merchandise from a vendor at a garage sale, flea market, or a similar event that is not covered by Subsection (G) of Section 4–6–26 (*Inventory*), the secondhand dealer must record the address of the place where the purchase was made. If a secondhand dealer purchases an item of regulated merchandise from a business with a permanent business location that is not covered by Subsection (G) of Section 4–6–26 (*Inventory*), the secondhand dealer must record the name of the business from which the item was purchased and the address of the place of business where the purchase was made. Section 4–6–25 (*Identification Required: Seller Must Attest To Ownership*) does not apply to a purchase under this subsection.

(G) In addition to complying with the other requirements of this article, a secondhand dealer that does not have a permanent business location in Nita City must notify the Chief of Police no later than the thirty-first day before the first day of any period of time in which the secondhand dealer will be open for business in the City. The notice must disclose:

(1) the physical location where the secondhand dealer will be conducting business; and

(2) how many days the secondhand dealer expects to be open for business.

§ 4–6–28 ADMINISTRATION.

The chief of police may adopt rules and forms to administer and enforce this article.

§ 4–6–29 PENALTY.

(A) A person commits an offense if the person violates this article or a rule adopted under this article. An offense under this article is a Class C misdemeanor, punishable by a fine not to exceed $500.00, as provided in Section 1–1–99 (*Offenses; General Penalty*).

(B) A secondhand dealer licensed under this article whose employee or agent violates a provision of this article is equally guilty with the offender and subject to the same penalty.

(C) A culpable mental state is not required, and need not be proved, for an offense under this section.

(D) Each day that a violation occurs or continues is a separate offense.

PART 2. The Chief of Police may adopt rules in anticipation of the effective date of this ordinance. This Part takes effect on June 18, 2012.

PART 3. Except as provided by Part 2, this ordinance takes effect on October 1, 2012.

PASSED AND APPROVED

§

§

______________________, 2012 § ______________________

Lee Leffingwell
Mayor

APPROVED:______________________
Karen M. Kennard
City Attorney

ATTEST:______________________
Shirley A. Gentry
City Clerk

Attorney Assignments

Exercise One: Overview Interview

Attorney instructions

Your aunt, Katharine White, phoned you last week to see if you could meet with a friend of hers. Mrs. White explained that the friend had recently moved in with her daughter in Oakdale, Nita, and was having some conflicts with her daughter. She would like to talk to you about the situation and get your advice on how she might resolve their problems. You make an appointment, and your aunt brings Deborah Dashwood to your office for a consultation.

Conduct the overview.

Exercise Two: Taking a Chronology

Attorney instructions

After speaking with Mrs. Dashwood for about thirty minutes, you have a general sense that her daughter has secured a guardianship without Mrs. Dashwood's cooperation. Mrs. Dashwood's primary concern seems to be that her daughter is being evasive about what she is doing with her mother's things; Mrs. Dashwood seems very attached to some of her possessions for sentimental reasons and her narrative has focused on this aspect of her situation.

Take a detailed timeline from Mrs. Dashwood to gain a clearer and more objective understanding of how the relevant events unfolded.

Exercise Three: Theory Development

Attorney instructions

You have now spoken extensively with Mrs. Dashwood about the fact that her daughter arranged to become her guardian. You have learned that Mrs. Dashwood objects to the situation primarily because she feels she has lost control over the disposition of property that is important to her, and also because she does not view herself as legally incompetent. Based on the event timeline you have taken, it seems that Marcy rushed to take this serious step. From what Mrs. Dashwood is telling you, you suspect that Marcy may have used deceit to get her mother to sign a voluntary guardianship application, which would render it invalid under Nita law. Further, even if the guardianship were not obtained via deceit, Deborah could challenge it on the merits by arguing that she is not incapacitated as defined by Nita law.

Conduct a theory development interview to obtain more information about the possibility that Marcy manipulated the guardianship process, and to obtain more information about whether Deborah meets the legal requirements to be found "incapacitated."

Exercise Four: Client Action Plan

Attorney instructions

You represent Mrs. Dashwood. You have already conducted a client intake interview and recorded a detailed timeline of events. Further, you followed up with a theory development interview that has allowed you to consider various approaches to her situation and that has led you to research Nita law on a number of matters. Review the Nita Estates Code regarding Guardianship, and the Nita City, Nita ordinance regulating pawnshops and second-hand dealers of goods to evaluate the legal options available to Mrs. Dashwood.

- **Prepare a short memo outlining Mrs. Dashwood's options for termination of the guardianship and recovery of her books and real property.**
- **Discuss with Mrs. Dashwood your current understanding of her situation, her goals for addressing it, and the legal consequences attached to various steps she might wish to take in pursuit of those goals. Work with her to develop an action plan that you can begin to execute to try to achieve her goals.**
- **Based on your in-class discussion with Mrs. Dashwood, draft an engagement letter outlining the terms of your engagement and your understanding of the work Mrs. Dashwood wishes you to perform on her behalf.**

Exercise Five-A: Friendly Witness

Attorney Instructions

Mrs. Dashwood wants her daughter's guardianship lifted. You could do this either by getting Marcy to voluntarily withdraw or by filing a request for termination on Mrs. Dashwood's behalf. Before you broach the subject of voluntary withdrawal with Marcy, you want a preliminary sense of the strength of your termination case—the stronger the basis for Mrs. Dashwood's request, the more likely it is that Marcy would prefer to stay out of court. A request for termination entails a physician assessment of Mrs. Dashwood, and you are reluctant to seek that assessment without more knowledge of her ability to care for herself.

Katharine White is Mrs. Dashwood's oldest and closest friend. You have asked her to talk with you about this matter.

Conduct the interview to get an initial read on how capable Mrs. Dashwood would be found under the Nita criteria. Keep in mind that Mrs. White could be called as a witness if the termination proceedings go to a hearing, and assess her strengths and weaknesses as a witness. Finally, see if Mrs. White might have information relevant to any other remedies Mrs. Dashwood is seeking.

Exercise Five-B: Hostile Witness

Attorney instructions

Mrs. Dashwood's primary goal is to have her books returned to her, her secondary goal is to get her home taken off the real estate market, and her final goal is get her daughter's guardianship formally revoked. She would prefer to have her daughter voluntarily resign as guardian but is willing to file a formal challenge if her daughter will not resign.

You have not yet spoken with Mrs. Dashwood's daughter, as you have decided that conversation might be more productive if you have additional information that you can get from other people. Today you are going to talk to Judy Smith, the Realtor Mrs. Dashwood's daughter has retained to list the home for sale. **The Multiple Listing Service listing for the house is attached.**

1) Real estate agents often assist in "staging" a house for sale, a process that can involve storing or selling furniture in the house that will be distracting to possible buyers. In short, Ms. Smith may have been involved in or have knowledge of how Mrs. Dashwood's daughter disposed of the books.
2) Real estate agents typically ask home sellers to sign a six-month listing contract. If the property is unsold after six months, the seller has the option of relisting with the agent or taking the listing elsewhere. The agent's fee is typically 7 percent of the sales price. One of the terms of the typical listing contract that the home seller signs is a warranty that the seller owns the property or is authorized to list it for sale.

Interview Ms. Smith. The top three pieces of information you seek in order to achieve Mrs. Dashwood's goals are:

3) What happened to the books and furniture in the house before it was listed for sale?
4) Under what circumstances would Ms. Smith consider taking the home off the market? That is, does she need the permission of her real estate company to unlist the property? Would her real estate company consider the contract void if it were suggested that the daughter got guardianship via fraud? If she is free to void the contract without approval from her company, under what circumstances would she do so?
5) Your theory of the case regarding the daughter's fraudulent guardianship is that the daughter is struggling financially and believes that managing her mother's finances may give her access to additional funds. What, if anything, can Ms. Smith tell you to support this theory? To undermine this theory?

realtor.nita®

2209 Sunnyslope Dr.

Nita City, NT 78703

$650,000

2 Bed 1 Full Bath 806 Sq Ft 6,273 Sq Ft Lot

Open House

Sun. 2 p.m.–4 p.m.

Tons of original charm! Single story cottage with original hardwood floors, crown molding and tons of natural light. Come in and make this your own space! Roomy fenced backyard with tall trees. Excellent location in the heart of Nita City

- easy access to dining, shopping, entertainment, Lady Bird Lake, Downtown and so much more!

General Information

Beds	2 Bed	Baths	1 Full Bath
House Size	806 Sq Ft	Lot Size	6,273 Sq Ft Lot
Price	$650,000	Price/sqft	$806
Property Type	Single Family Home	Year Built	YR-90
Neighborhood	Old West Nita City		

Stories 1

Bedrooms

- Bedrooms: *2*
- Num Main Level Beds: 2
- Master Main: Yes
- Master Description: See Agent

Bathrooms

- 1 Full Bathroom
- Baths Total: 1

Kitchen and Dining

- DiningDescription: Breakfast Area, Formal Dining ROom
- Num Dining: 2

Other rooms

- Rooms: Bedroom/Office, Living/Den
- Laundry Location: Main Level
- Num Living: 1

Interior Features

- Crown Molding
- Indoor Utilities

Building and Construction

- Levels or Stories: 1
- Construction: Frame
- Flooring: Tile – Hard Wood
- Foundation Details: Pier & Beam
- Roof: Composition Shingle

Exterior and Lot Features

- View: No View
- Patio-Uncovered
- Porch-Open
- Fence: Chain Link, Wood
- Trees: Large (Over 40 Ft)

Garage and Parking

- Parking Features: Parking: Outside
- Num Parking Spaces: 2

Heating and Cooling

- AC: Central Air
- Heating: Central Heat

Utilities

- Sewer: City At Street
- Utilities: Electricity On Property
- Water Source: City

Appliances

- Range-Free Standing

School Information

- Elementary School: Mathews
- School District: Nita City ISO
- Senior High School: Nita City
- Middle Intermediate School: O Henry

Other Property Info

- Nita County
- Westridge Subdivision
- Directions: Take Mopac to 5th St. Left on West Lynn St. Left on West 9 112 St.
- City: Nita City
- State: NI
- Steps: Exterior Steps
- Area: 18
- APN Number: 01110303120000

Multi-Unit Info

- Laundry Facilities: Stackable W/D Connections

Listing Provider

Listing agent	This listing is represented by
•	
Listed by	
Broker Location	Nita City, NT
Data Source	Nita City
Source's Properly ID	1746900
Data Source Copyright	© 2015, Nita City Board of REALTORS®, Inc., All Rights Reserved.

Public Records Property Information from local public records.

Beds		Baths	
House Size	806 Sq Ft	Lot Size	6,260 Sq Ft Lot
Year Built	1926	Price	
Properly Type	Single Family Home	Stories	1
Style	Not Available	Garage	
Units		Cooling	Central
Pool		Construction	Wood Frame
Heating	Central	Year Renovated	-
Rooms		Roofing	Composition Shingle
Fireplace			

Exercise Six: Functionary

Attorney instructions

You have learned from Judy Smith that Furniture Brokers Westlake consigned all of Mrs. Dashwood's possessions back in August. You visited Furniture Brokers today and found all of Mrs. Dashwood's furniture at the store. However, her hundreds of books—the items she values most highly—were nowhere to be found. You talked with proprietor Pamela Pickle, who told you that a new employee had sold Mrs. Dashwood's book and coin collections in a single lot the day before. The Nita City, Nita, ordinance governing secondhand dealers requires that all sales of rare coins be recorded and that dealers file sales records with the city police department. Mrs. Pickle looked in her records and found that the employee had inadvertently given to the police both the originals of the reports and the carbon copies that were meant to be retained in Furniture Brokers's internal files. The Nita ordinance is attached.

Do your best to get information from the Nita City Police Department that will help you retrieve Mrs. Dashwood's books.

PEOPLE V. MOORE INTERVIEWEE MATERIALS

Kelly Moore—Background Brief

You are Kelly Moore, a twenty-four-year-old African-American woman living and working in Nita. You are seeking legal help because you were arrested last week at a local drugstore and charged with shoplifting. The chain of events that led to the arrest is as follows:

You work at the Lakeview branch of Hertz Rent-A-Car, along with five other people. The branch is managed by Phyllis Kaminsky; she is about thirty years old and was recently named branch manager. Phyllis seems very content at Hertz and her career plan for the moment seems to be to work her way incrementally up the chain at the company. Meredith Boyle is about twenty-eight; she has worked at Hertz for the past seven years. Meredith and Phyllis are very good friends, and Meredith does not seem to have any plans to leave the Lakeview branch or to explore other career paths. Scott Michael is about twenty-five. He has been taking evening classes at the Nita Graduate School of Management for the past year and plans to seek a management consulting job upon graduation. You and Scott have a lot in common, and you have actually been dating for more than a month. Wayne Hill is about twenty-three and has just graduated from college. He is new to the branch and you know little about him. Kara Gabinski is about twenty-five; she has worked at the branch for about a year and is getting her teaching certificate at night. She is a hard worker but very quiet.

On Fridays, you and your coworkers often go out for a drink when the branch closes at 6 p.m. Last Friday, you went out, as usual, to Sheffield's. While most of your coworkers live in nearby apartments and take public transportation to work, you live at your parents' house in Edison Park to save money, so you drive to and from work. As a result, although you always join your friends for happy hour, you do not drink because you know you have a commute awaiting you at the end of the night.

Last Friday, Meredith became very intoxicated at happy hour. Concerned, you offered to drive her to her apartment in Andersonville before heading home. On the way home, she felt ill and pestered you to get her a Coke because she thought it would calm her nausea. You agreed and pulled over at the CVS at Broadway and Foster. You both went into the store. While you headed to the cooler to get a Coke, Meredith stopped in the novelty aisle and began playing with some of the toys and children's Easter costumes. In the cooler aisle, you quickly found Meredith's Coke, but spent a few minutes looking for a grapefruit-flavored LaCroix—your favorite drink, but one that some stores don't carry. While there, you could hear Meredith imitating cartoon characters and laughing uproariously. A moment later you saw her crouching behind a display at the end of the aisle, from which she pelted you with plastic Easter eggs. You quickly grabbed yourself a 7-Up, and then unsuccessfully tried to get her to calm down by sitting her on a bench near the registers. You got in line to pay for the drinks. A few minutes later, she joined you at the register, jostling and poking you and loudly calling your name. Just then, your phone rang. You put down the drinks and your bag and pulled out your phone—it was Scott, calling to see whether you had gotten Meredith home safely. While you chatted, Meredith was still singing and laughing so loud that Scott told you he could hear her. "She's a trainwreck," he said. "And you're a saint." Just then, the customer in front of you finished his transaction and the cashier gestured for you to approach. You gave the cashier money and were waiting for change. As the cashier was handing you your change, a security guard approached and asked you to join him in the store manager's office. You told Scott that someone was talking to you and you had to go. You agreed to walk to the office with the security guard. Meredith followed, but the guard looked her up and down and said, "Not you. You're free to go." You followed him through a door and down a hallway that led away from the sales floor and to a private room. There, he said that he suspected you of shoplifting and asked to see the tote bag that you were carrying. You gave it to him. He looked inside and pulled out four L'Oréal makeup items, all in the original packaging. You were surprised and told him that you had not taken the items. He said, "Tell it to the cops."

The manager called the Nita Police Department to report the incident and ask an officer to come to the store. He told you to stay seated, and asked for your name for his records. He took out a binder and entered your name on a log—from where you were sitting next to the desk, you could see that about half the names on the log appeared to be Hispanic. The security guard remained standing at the door until the police officer arrived. Twenty minutes later, an officer arrived, arrested you, and took you to the local station in his car. He booked you and allowed you

to make a phone call. You telephoned your parents, who arrived an hour later. They paid your bail, which is set by statute at $1,500 for the Class A misdemeanor of retail theft, and you were released around 11 p.m. They drove you to retrieve your car. You found Meredith passed out in the front seat, and you drove her home. She said very little.

Upon your release, the police presented you with a General Offense Report, which serves as the complaint against you. In Nita, courts need not hold a probable cause hearing for misdemeanor offenses such as retail theft. According to the report, a preliminary hearing in the case will be held May 23 at 10 a.m. In Nita, the preliminary hearing and arraignment may be combined, meaning that the state's attorney will be present and that you will be given the opportunity to enter a plea.

You are very upset about this situation, for a number of reasons. First and foremost, you consider yourself an honest and trustworthy person and you are mortified to have been accused of this crime. Second, you are very ambitious and you do not want this incident to derail your career plans. You were recently accepted to Georgetown Law School and you plan to begin your studies next year. This incident could be problematic for a couple of reasons. In the short-term, you need to keep this job. You graduated from the University of Nita in May YR-3 and you have been saving your Hertz salary for the past three years to pay for law school. You need the next four months' salary to reach the savings goal necessary to cover your education. It is inevitable that your coworkers will learn about this matter, and you do not want any fallout to jeopardize your job. In the long-term, you are dimly aware that when you apply for bar membership you will have to disclose any criminal convictions. If you are found guilty or plead guilty, it is likely that you will be given a fine and you could be forced to serve up to a year in prison. You are afraid that if you had to serve time, it might jeopardize your eventual enrollment at Georgetown. Even if you don't have to serve time, you are afraid that such a blot on your record will thwart your law career before it has begun, because your bar application could be denied. Finally, you suspect that the security guard was racially profiling you, a possibility you find unacceptable.

Note to interviewee: *When you describe what happened in the CVS, say that you "went in to buy the Coke" and "went to the register" without being specific about what Meredith was doing.* ***Do not*** *volunteer that Meredith was wandering around the store while you bought the Coke until the interviewer poses a question that specifically seeks details about where you were in the store or where Meredith was in the store.* ***Do not*** *be explicit that you were taking your change from the cashier when the security guard approached unless you are asked a question that specifically elicits this information—say "after I bought the Coke, the guard approached. . . ."* ***Do not*** *mention your plans to attend law school in the fall.*

Kelly Moore—Arrest Report

Nita City Police Department
General Offense Report & Complaint

Arrest Number 2932-1
Arrest Date 4/23/

Last Name	First Name	Middle	Alias
Moore	Kelly	Ann	—

Sex	Age	DOB	Place of Birth
F	24	11/13/	Tinley Pk IL

Height	Weight	Eye Color	Hair Color
5'7"	140	BR.	BL.

Social Security Number	Drivers License Number	State
555-90-2124	M520-3928-4772	IL

Address	City	State	Zip	Phone
2109 NORFOLK	CHI.	IL	60631	773-555-9842

Occupation	Employer
rental agent	Hertz

Scars, Marks & Tattoos
N/A

Time of Arrest	Place of Arrest	Arresting Officer
9:03 p.m.	CVS - Broadway & Foster	Sgt. Ted Weber

Booking Charge	Offense Type	Offense Code
retail theft	gen 1	0860

Complaint Charge	Elements
retail theft	knowing carriage of merch. from retail estb. w/ intent to deprive w/out pay

Bond Amount	Bond Disposition	Court Date
$1500	10% - cash pd.	5/23/

Signed by:

Ted Weber

KELLY MOORE—STORE DIAGRAM

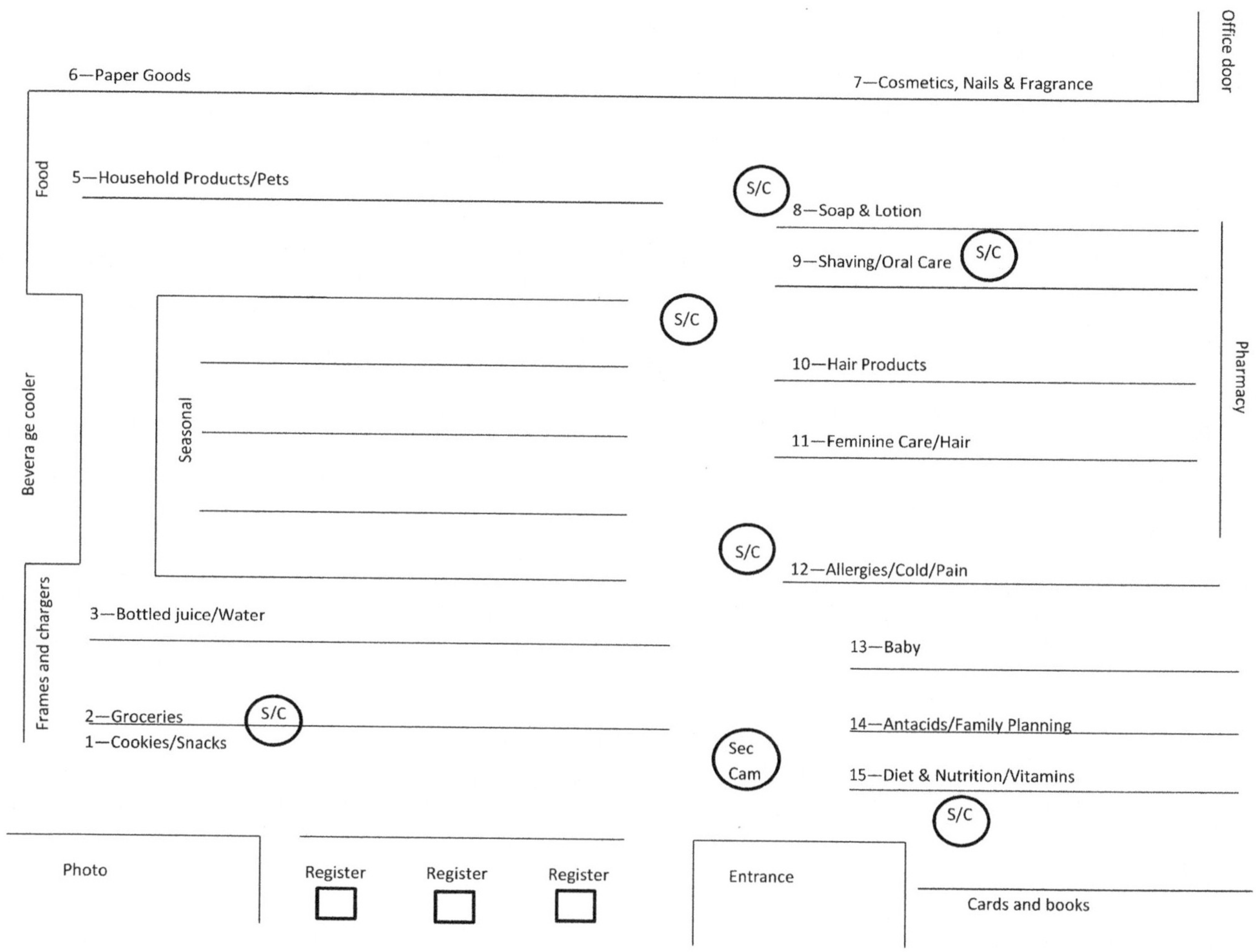

Kelly Moore—Timeline

April 23, 2014	
6 p.m.	**Kelly and coworkers depart their place of business, the Lakeview branch of Hertz Rent-A-Car, and proceed on foot to a local bar.**
6:10 p.m.	**Kelly and coworkers arrive at Sheffield's for happy hour. They remain until 8 p.m.**
8:10 p.m.	**Kelly retrieves her car from the Hertz parking garage. By previous arrangement with her coworkers, she drives back to Sheffield's to pick up intoxicated coworker Meredith Boyle, whom she plans to drive to her apartment in Andersonville. With the help of coworkers, Kelly gets Meredith into the passenger seat of the car.**
8:30 p.m.	**Kelly pulls into the parking lot at the CVS at Foster and Broadway. She and Meredith exit the car and go into the store to buy a Coke.**
8:31 p.m.	**Kelly goes to the store cooler to get the drink for Meredith. Meredith heads to the novelty aisle. In addition to Meredith's Coke, Kelly looks for a grapefruit-flavored LaCroix for herself.**
8:34 p.m.	**Meredith finds Kelly scanning the cooler and starts to pelt her with plastic Easter eggs. Kelly gives up on the LaCroix and grabs a 7-UP. She leads Meredith to a bench near the exit and tells her to sit there while she goes to the register to buy the drinks.**
8:36 p.m.	**While Kelly waits in line, Meredith stands up and heads towards the cosmetics department. There are several customers waiting to check out, so the line is rather lengthy.**
8:38 p.m.	**Still in line, Kelly hears her phone ring. She puts the two drinks and her bag on the ground and takes her phone out of her pocket. Her coworker and boyfriend Scott Michael has called to see if she has gotten Meredith home safely. She replies that they have pulled over at the CVS and that she's waiting in line to buy sodas but will probably get to Meredith's house in about twenty minutes. She promises to text him when she leaves Meredith's house. Scott tells Kelly he can hear Meredith speaking loudly in the background.**
8:40 p.m.	**The customer in front of Kelly completes her transaction and the cashier nods for Kelly to approach the register. Kelly picks up her bag and the drinks and goes to the register to pay.**
8:41 p.m.	**As Kelly is accepting her change from the cashier, a CVS security guard approaches Kelly and tells her he suspects her of shoplifting. Kelly says goodbye to Scott. The guard directs her to join him in the manager's office in the back of the store. He tells Meredith she is "free to go."**
8:41 p.m.	**As Kelly is walking to the office with the security guard, Meredith leaves CVS and goes to Kelly's car to wait for her.**

8:45 p.m.	**In the office, the security guard says to the manager, "I think we've got another one here." The guard asks for permission to go through Kelly's bag and she consents. He finds four items of L'Oréal makeup in their original packaging. CVS manager calls Nita City Police Department to report Kelly for shoplifting. He asks for her name and notes it on a form in one of his binders.**
9 p.m.	**Officer Ted Waber arrives at CVS and formally places Kelly under arrest. He puts her in his squad car.**
9:20 p.m.	**Officer Waber parks his car at the Greenway Park station and escorts Kelly inside. He takes a mugshot and fingerprints her. He puts her in the station holding pen while he processes paperwork.**
10:00 p.m.	**Officer Waber unlocks the holding pen. He gives Kelly a copy of the arrest report he has filed. He allows her to make a phone call. She calls her parents to explain the situation.**
11 p.m.	**Kelly's parents arrive at station house and pay Kelly's bail. She is released pending trial.**
11:15 p.m.	**Kelly's parents drive her to the CVS parking lot where she left her car. They find Meredith passed out in the passenger seat. Kelly's mother gets in the back seat, and together they take Meredith to her apartment at 5423 N. Ravenswood.**

Kelly Moore—Action Plan Brief

You have spoken extensively with your lawyer about the events leading to your arrest for shoplifting on April 23, YR-1. She has asked you to come to her office so that you can discuss your plan for addressing the charges. You do not know what options she might present to you, but your objectives, in order of priority are:

1) Avoid being convicted of or pleading guilty to a crime. This would seriously impair your ambition of becoming an attorney. Further, it is your position that you did nothing wrong and you are adamant that you have no association with this crime. This is your top priority and you are extremely risk-averse, so if achieving this requires you to forfeit some other objective, you will do so.
2) Retain your job at Hertz for the next four months. You badly need the money from this job to fund your Georgetown tuition. You know that Phyllis could figure out a pretext for firing you, or make your work life miserable if she chose to do so. If at all possible, you want to clear your name without destroying your livelihood. However, if the only way to clear your name would be to take a step that imperils your job, you would do it.
3) Get some recompense for the indignity, humiliation, and expense associated with this event. You are not a vengeful person, and you're not sure how far you would be willing to go to get recourse. But you suspect you were racially profiled. And if it turns out that this incident gets in the way of law school, or even if it just causes you to lose your job, then you'd like to see if there's some way to be compensated for that.

Scott Michael—Background Brief

You are Scott Michael. You work at the Lakeview branch of Hertz Rent-A-Car with Kelly Moore and Meredith Boyle. You were with them at Sheffield's the Friday of the events that led to Kelly's arrest. Kelly's attorney has contacted you and asked to meet to discuss what you know about the situation.

You were one of six coworkers who went to Sheffield's for happy hour that evening. You all walked in together at 6:15 and got a table near the window. Your custom as a group is to take turns buying pitchers of beer for the group; anyone who wants a different drink generally goes to the bar on his or her own. You recall that on that Friday Kelly went to the bar a few times; because she lives in a different neighborhood and drives home, she typically sticks to non-alcoholic drinks on these evenings, and you assumed that she was ordering Cokes.

Meredith bought a pitcher that night, and the group as a whole finished three pitchers. Meredith drank quite a bit, though you do not recall exactly how many beers she had. Meredith is generally a fairly heavy drinker, and this is not the first time that someone in the group has felt compelled to escort her home. You called Kelly later that night to make sure she had gotten Meredith home, but it turned out they had stopped at CVS. While you chatted with Kelly you heard Meredith in the background, and heard Kelly talking with the cashier. She then abruptly told you she had to go, and hung up.

You have spoken to both Kelly and Meredith about the events that took place after the group left the bar. Kelly phoned you on April 24 to ask if you could meet over the weekend to discuss something important. At dinner on Saturday, she told you all about her arrest. She said that the only thing she took off the shelves at CVS were the drinks that she wanted to purchase. She also said that after her parents paid her bail, they drove her to retrieve her car from its spot in the CVS parking lot. They found Meredith passed out in the passenger seat. They woke her up and dropped her at home. Kelly reported that Meredith seemed very confused and didn't say much on the ride home. Kelly said she wanted you to know what had happened but she was intent on keeping office gossip about the incident to a minimum. She asked you to meet outside of work because she was hoping not to discuss the matter in the office. She said she would not raise the subject at work, but thought it likely that Meredith would mention it.

You and Kelly began to date about six weeks ago, and you have found her to be generally honest and fair. For example, several months ago she found a pair of Gucci sunglasses in one of the rental cars while cleaning it out. She immediately put the glasses into the lost and found, though one of your coworkers observed that the owner could never trace them and that "finders keepers" was one of the perks of the job. You would be surprised if Kelly had shoplifted from the CVS.

You have known Meredith for about two years, through your work in the car rental office. She is also a friendly person and you have no reason to doubt her honesty (though she was the one who implied that Kelly should keep the sunglasses). You have noticed that she likes to go out frequently, and often gets intoxicated when she does. You have never known her to do anything dishonest, but she does get boisterous and like to pull pranks when she is out having drinks.

Meredith approached you on the Monday after the arrest. She asked whether Kelly was "mad" at her. You told Meredith that Kelly had not expressed any anger towards her. "But she blames me, right?" said Meredith. "I would if I were in her shoes," she said. She began to cry. "I shouldn't even be talking to you. Phyllis told me to let this blow over and that she'd take care of things if there was a problem. I'm sorry." You didn't say anything in response, but busied yourself with a task that required you to go another room in the office.

Meredith is especially close to Phyllis, who is the office manager. The two have worked together for three years and plan to stay at Hertz indefinitely.

Note to interviewee: *Answer any questions that Kelly's lawyer asks.* ***Do not*** *volunteer that you and Kelly are dating.* ***Do not*** *volunteer that Meredith approached you to discuss the situation.* ***Do not*** *volunteer that Meredith is friends with the office manager.* ***Do not*** *volunteer that Meredith was the one who suggested Kelly keep the sunglasses. Give the lawyer this information only if it is directly relevant to the precise question asked.*

Phyllis Kaminski—Background Brief

You are Phyllis Kaminski. You are thirty years old and have worked at Hertz Rent-A-Car for eight years. You started there shortly after college and did not take the job seriously for the first couple of years. Meredith Boyle started at about the same time as you and most of the other workers at the branch were recent college grads. No one was terrible at the job, but most of you were more interested in socializing than working. You often went out together after work, and Friday night happy hour was observed religiously. The branch manager, Bob Krumholz, was about a decade older than the rest of the employees, and you initially wrote him off as humorless and power-hungry.

About five years ago, you found yourself tiring of the constant partying after work; you realized that many of your coworkers considered the job a temporary gig until they could go back to graduate school or get a more challenging post at a different company. You slowly began to take the work more seriously and started volunteering to take on extra tasks. Two years ago, Bob asked whether you would be interested in attending the summer "Hertz University" program for those with management ambitions. You were flattered and found the three-week seminar very helpful. You received stellar annual reviews for your good work, and noticed the younger employees in the office looking up to you. Last year, Bob left Hertz for a marketing job at rival Avis. You applied for the branch manager position and got the job.

You now run a staff of five employees. Four of them are relatively new and much younger than you. They are all considerably more responsible and ambitious than you were when you started, and that makes your job relatively easy. They do not socialize nearly as much as you did in the early days, but you've managed to keep up the tradition of Friday night happy hour—it seems to build camaraderie and you get to keep tabs on office gossip, which can come in handy. There is one problem. You and Meredith became fast friends when you both started at Hertz and you are very loyal to her. During your first year on the job, the two of you went out socially several times a week and helped each other out when you made careless mistakes. But now you've got your act together and Meredith does not. She is constantly late for work and neglects company protocol. She still drinks to excess at the office happy hour and on her own time. Finally, she has developed a very antagonistic relationship with the new group of employees, particularly Kelly Moore. You suspect that she is frustrated to see these ambitious people passing her by, and you agree that they're a little entitled and snobbish. But they do good work, whereas Meredith does not. Still, you feel compelled to protect her.

On Monday, April 26, Meredith barged into your office first thing and closed the door behind her. "I'm freaking out," she said. "Kelly took me home last Friday and when we pulled over at CVS to get a Coke she was arrested for shoplifting." You are shocked—Kelly is a straight arrow and seems very honest. "The thing is, I think it's my fault," Meredith said. "I was wasted, so I don't remember things very clearly. But when I looked through my texts the next morning, I found one from my mother at about 8:30 reminding me that we were having a family photo taken the next day, and asking me to cover the tattoo on my neck. Then I remembered taking some concealer off the shelf at CVS, and maybe a few other things. And when I found Kelly in the checkout line, I was going to ask her to spot me for the makeup, but she was on the phone. I don't remember much more than that, but I didn't have the makeup when I got home."

You asked Meredith whether she had told Kelly of her possible involvement, but she explained that she was on probation for a public intoxication charge and the charge won't be lifted from her record if she is involved in another criminal matter. You advised her to keep her distance from Kelly, and said you'd do your best to quiet things down.

Corporate has recently suggested that if the branch's sales do not improve, you might have to let one of your workers go. As the most recent hire, Kelly is the obvious candidate for discharge. On Tuesday, you called a staff meeting and explained that corporate was contemplating a personnel reduction in the next several weeks. You had Meredith sit next to you, to make it clear that as a senior employee her job was safe. You told the workers you simply wanted to keep them up to date but you were looking at Kelly the whole time, hoping she would get the message that you were looking out for Meredith and that you had the authority to fire her. She is smart, and you

thought she would understand that she should keep Meredith out of her legal troubles. Frankly, you sympathize with Kelly and have lost patience with Meredith. But if you can protect Meredith by making a few broad hints, you'll do it for old time's sake.

Note to interviewee: Do not *volunteer that Meredith confided in you that she had a dim recollection of taking makeup off the shelf at CVS.* ***Do not*** *volunteer that Meredith is on probation. If the attorney asks you any specific questions for which these would be relevant answers, you may give them. If after a first round of questions, the attorney persuades you that it is in your interest to share what you know about the shoplifting, you may provide the information.*

MARTY WILTON—BACKGROUND BRIEF

You are Marty Wilton, the manager of the Foster & Broadway CVS store. You have worked for CVS for six years. You began as a cashier in the Oakdale store, and after two years successfully applied for the company's management trainee program. The program called for you to relocate to corporate headquarters in Rhode Island for one year, during which time you rotated on a monthly basis through various departments. You were then assigned to manage a Knoxville, Tennessee, branch for two years. At that branch, you were very successful in improving grocery sales via promotional placements. You also received a national award for your low rate of shrinkage—an industry term for preventing shoplifting. Based on this success, CVS transferred to your current location, which had been plagued by theft for several years. Your regional supervisor has hinted that if you can drive down theft at this store, you will be in line for the next assistant regional manager post that opens up.

Your store is located on a busy intersection in a congested part of the city. On weekend evenings, it is typically very crowded. Reviewing statistics for the past several years, you have noticed that shoplifting tends to spike on Friday and Saturday nights. As a result, in January you hired a weekend security guard. He wears a uniform and is stationed near the exit. You instructed him in January to speak to customers on their way in to remind them that they were being watched, but not to detain anyone. When you checked statistics in March, it appeared that his presence was deterring theft—shrinkage had dropped by 10 percent. However, your personal goal was to drop shrinkage by 25 percent in the first six months of YR-0. In order to accomplish this, you instructed the guard to detain customers whose shopping behavior was suspicious. You learned at your corporate training that if a store developed a reputation for shoplifting arrests, repeat offenders would avoid the store, and you explained this to the guard. "If we had arrests every weekend this spring, we'd have the problem fixed before summer," you laughed. The guard detained about four people every weekend from then on, and called the Nita City Police to arrest customers on five of the eight weekends in March and April. In fact, you visited the local police station right before St. Patricks's Day to speak with the community policing officer. You explained that your store had a high shoplifting rate, that you were cracking down, and that you would be grateful for his support. He agreed to alert the weekend desk sergeants so they would respond quickly if they received a call from your security guard about suspected shoplifters.

You are very happy with the results of your initiative, and the regional manager congratulated you on the steady decline in theft reflected in the May numbers. At the same time, in the monthly Managers' Directive bulletin circulated in May, corporate warned against loss prevention strategies that could damage the CVS public image or expose it to civil liability. It explained that a group of former CVS security guards in New York had sued the company, contending that they had been directed to racially profile shoppers. The practice has come under scrutiny in recent years after New York shoppers at Macy's and Barney's sued those stores for racial profiling. Corporate asked managers to make sure their store policies could not be misconstrued as profiling. You became a little alarmed when you received this bulletin because you have noticed that the majority of people the guard has detained at your store are African-American and Hispanic. Since your initiative has been very successful and you have received the accolades you sought, you told the guard last week that you were discontinuing the program.

In addition to the guard you recently hired, your store has a set of security cameras, something CVS has installed in every branch store. The footage is kept on servers in each store, and CVS corporate gives individual managers discretion to handle public requests to see the footage. You have not looked at the footage from the night Kelly Moore was arrested, nor have the police or state's attorney's office asked for it. At this point, it appears that the state assumes the security guard's eyewitness account implicating Kelly is sufficient to get a conviction.

Note to interviewee:* *Do not *volunteer that you recently terminated the security guard program.* ***Do not*** *volunteer that CVS was recently sued for racially profiling shoppers.* ***Do not*** *volunteer that you have discretion to share the security footage.* ***If*** *the attorney makes a persuasive case that the footage will help him get a confession that clears his client but preserves your shoplifting numbers, you may* ***agree*** *to share the footage. You are free to set whatever terms you like for doing so—you can give the attorney a copy of the tape, or require that he watch it on your server.*

About the Author

Cristina Carmody Tilley is a former litigator at Mayer Brown in Chicago, where she specialized in product liability and securities class action defense work at both the trial and appellate levels. She is a Distinguished Scholar in Residence at Loyola University Chicago Law School, where she teaches Remedies, Contracts, and Media Law. She earned her law degree from Northwestern University School of Law, where she served as Editor in Chief of the *Northwestern University Law Review*. Before studying law, she was a news reporter for seven years at United Press International, *The National Law Journal*, and other publications.

Marrying the interviewing techniques she developed as a journalist with her litigation experience and doctrinal skills, she has spent the past several years developing Loyola's Interviewing Skills curriculum. This book draws on insights gained over years of interviewing subjects from all walks of life and shaping the resulting information into clear and persuasive narrative form.

Made in the USA
Monee, IL
13 September 2026

62976961R10063

Printed in the United States
by Baker & Taylor Publisher Services